Biomimicry for Optimization, Control, and Automation

Springer
London
Berlin
Heidelberg
New York
Hong Kong
Milan
Paris
Tokyo

Kevin M. Passino

Biomimicry for Optimization, Control, and Automation

With 365 Figures

Springer

Kevin M. Passino

Department of Electrical and Computer Engineering, 416 Dreese Laboratories,
The Ohio State University, 2015 Neil Ave., Columbus, OH 43210, USA
http://www.ece.osu.edu/~passino

British Library Cataloguing in Publication Data
Passino, Kevin M.
Biomimicry for optimization, control, and automation
1. Control systems 2. Adaptive control systems 3. Intelligent control
systems
I. Title
003.5
ISBN 1852338040

Library of Congress Cataloging-in-Publication Data
Passino, Kevin M.
Biomimicry for optimization, control, and automation / Kevin M. Passino.
p. cm.
Includes bibliographical references and index.
ISBN 1-85233-804-0 (alk. paper)
1. Control systems. 2. Mathematical optimization. I. Title.
TS156.8.P245 2004
629.8—dc22 2004041694

ISBN 1-85233-804-0 Springer-Verlag London Berlin Heidelberg
Springer-Verlag is a part of Springer Science+Business Media
springeronline.com

Typesetting: Camera-ready by author
Printed and bound in the United States of America
69/3830-543210 Printed on acid-free paper SPIN 10967713

To Annie, my best friend, soul-mate, wife, *y cielito*
and to our sweet children who fill our hearts with joy and
whom we love so much,
Carina, Juliana, Jacob, and Zacarias

Prefacc

Biomimicry uses our scientific understanding of biological systems to exploit ideas from nature in order to construct some technology. In this book, we focus on how to use biomimicry of the functional operation of the "hardware and software" of biological systems for the development of optimization algorithms and feedback control systems that extend our capabilities to implement sophisticated levels of automation. The primary focus is not on the modeling, emulation, or analysis of some biological system. The focus is on using "bio-inspiration" to inject new ideas, techniques, and perspective into the engineering of complex automation systems.

There are many biological processes that, at some level of abstraction, can be represented as optimization processes, many of which have as a basic purpose automatic control, decision making, or automation. For instance, at the level of everyday experience, we can view the actions of a human operator of some process (e.g., the driver of a car) as being a series of the best choices he or she makes in trying to achieve some goal (staying on the road); emulation of this decision-making process amounts to modeling a type of biological optimization and decision-making process, and implementation of the resulting algorithm results in "human mimicry" for automation. There are clearer examples of biological optimization processes that are used for control and automation when you consider nonhuman biological or behavioral processes, or the (internal) biology of the human and not the resulting external behavioral characteristics (like driving a car). For instance, there are homeostasis processes where, for instance, temperature is regulated in the human body. Another example is the neural network for "motor control" that helps keep us standing (balancing). In the cognitive process of planning in the brain, there is the evaluation of multiple options (e.g., sequences of actions), and then the selection of the *best* one. The behavior of attentional systems can be seen as trying to dynamically focus on the *most important* entity in a changing environment. Learning can be seen as gathering the *most useful* information from a complex noisy environment, or as a process of constructing the *best* possible representation of some aspect of the environment for use in decision making. Evolution can be viewed as a stochastic process that designs optimal and robust organisms according to Darwin's principle of "survival of the fittest" (i.e., the best-suited organisms for the environment survive to reproduce). Both learning and evolution can be viewed as optimization processes that lead to adaptations, over short and long time scales, respectively. Foraging can be modeled as a sequential optimization process of making the best choices about where to go to find nutrients so as to maximize energy intake per time spent foraging, and how to avoid threats (e.g., getting eaten) at the same time. In cooperative ("social") foraging, animals work together to help the group find resources. Sometimes such social animals operate in cohesive "swarms" to forage and avoid threats. In competitive foraging, the forager must make the best decisions in the presence of its adversaries in order to survive.

In this book, we will explain how to model such biological processes, and how to use them to develop or implement methods for optimization, control, and automation. We will be quite concerned with showing that our methods are verifiably correct (e.g., so that, if we use them in an engineering application, we know they will operate correctly and not necessarily have the same, possibly high, error rate as their biological counterparts). This drives the decision to include a significant amount of material on engineering methodology, simulation-based evaluations, and modeling and mathematical verification of properties of the systems we study (e.g., stability analysis and an emphasis on robustness). The overall goal is to expand the horizons for optimization, control and automation, but at the same time to be pragmatic and keep a firm foundation in traditional engineering methods that have been consistently successful. Generally, the focus is on achieving high levels of "autonomy" for systems, not on the resulting "intelligence" of the system. It is hoped that this book will show you that the synthesis of the biomimicry viewpoint with traditional physics and mathematics-based engineering, offers a very broad and practically useful perspective, especially for very complex automation problems.

For all these reasons, this book is likely to be primarily useful to persons interested in the areas of "intelligent systems," "intelligent automation," or what has been called "intelligent control" (essentially, the viewpoint here is that "biomimicry for optimization, control, and automation," the title of this book, is the definition of the field of "intelligent control"). While this book will likely be of most interest to engineers and computer scientists, it may also be interesting to some in the biological sciences and mathematical biology.

A Quick Glance at Key Concepts and Topics

If you want to get a better sense of what this book is about, first scan the table of contents, then go to the first few pages of each *part* and read the "Sequence of Essential Concepts" (their concatenation tells the basic "story" of this book, and it may be useful to reread that story as you progress through the book). This gives a high-level view of what you can learn by reading this book, and help some readers decide on which parts to focus on. For an even more detailed sequence of key concepts, scan the "side notes" that are in the margins throughout the entire book. These notes state in a concise way the important concepts.

Overview of the Book

Part I serves as the introduction to the book and establishes the philosophy of the general methodologies that are used. First, we provide a detailed overview of the control engineering methodology for traditional feedback systems and complex automation systems. Next, scientific foundations for biomimicry for "intelligent control" are established. We overview some ideas from biology, neuroscience, psychology, behavioral/sensory ecology, foraging theory, and evolution that have been particularly useful in providing biologically inspired control methods. Also, we explain how to exploit human expertise on how to control systems ("human-mimicry"), and use this to achieve automation.

In Part II, we study methods to automate those biological control functions and human expertise that do not involve learning. First, we introduce the basics of neural networks and explain how they can serve as the "hard-wiring" for implementing control functions in animals. Next, we introduce fuzzy and expert control, and provide a design example to clarify how a heuristic rule-based controller synthesis methodology works. We discuss how to perform Lyapunov stability analysis of neural and fuzzy control systems. We show how autonomous robots can perform path planning for obstacle avoidance, and how planning concepts can be used in the closed-loop via model predictive control. Then we introduce attentional systems, where animals seek to manage the complexity of sensory information via focusing and filtering. We introduce a model of an organism in a predator/prey environment that wants to "pay attention," so that it can keep track of predator/prey locations. We introduce several attentional strategies (resource allocation methods), simulate their behavior, discuss attentional strategy design, and perform stability analysis.

In Part III, we introduce learning. We overview the psychology and neuroscience of learning. We focus on incorporating aspects of learning that arise from function approximation to improve performance in control systems while they are in operation. We define several heuristic adaptive control (learning control) strategies that are based on the underlying neuroscience and psychology of learning (e.g., reinforcement learning). We cover least squares, steepest descent, Newton, conjugate gradient, Levenberg-Marquardt, and clustering methods for training approximators. We study basic issues in learning related to generalization, overtraining/overfitting, online and offline learning, and model validation. Next, we show how the learning methods can be used to adapt the parameters of the controller (or estimator) structures, defined in Part II, to create adaptive decision-making systems (i.e., ones that can learn to accommodate problems that arise in the environment). We explain how least squares and gradient optimization procedures can be used to tune approximators to achieve adaptive control (i.e., automatic tuning of the controller in response to plant uncertainties) for nonlinear discrete-time systems. Finally, we show how to develop stable, continuous-time adaptive control systems that use fuzzy systems or neural networks as approximators.

In Part IV, we explain how the genetic algorithm can be used to simulate evolution and solve optimization problems. Next, we discuss general issues in stochastic optimization for design of control and automation systems. For this, we first overview the relevant biology of learning and evolutionary theories, including the concept of "highly optimized tolerance" and robustness trade-offs for complex systems, and synergies between evolution and learning (e.g., evolution of learning, the Baldwin effect, and evolved instinct-learning balance). We show how the "response surface methodology" for nongradient optimization and design can be used to understand robustness trade-offs in control and learning system design. We show how nongradient and "set-based" stochastic optimization methods can be used for robust design, and give an example of evolution of instinct-learning balance. Moreover, we discuss the use of evolutionary and stochastic optimization methods for "Darwinian design of physical control systems." Next, we show how online "set-based" stochastic optimization algorithms can achieve a type of online evolution of controllers to achieve real-time evolutionary adaptive control.

In Part V, after explaining the basics of foraging theory and foraging search strategies, we show how "taxes" (motion) of populations of foraging *E. coli* bacteria can achieve optimization, either as individuals or as groups (swarms). We show how to simulate social foraging bacteria, and how they can work together cooperatively to solve an optimization problem. We discuss the basics of the modeling and stability analysis of foraging swarms, and study an application to cooperative control for multiple autonomous vehicles (robots). Next, we discuss animal fighting behaviors and game theory models of cooperative and competitive foraging. In the final section, we discuss intelligent foraging. This section is designed to provide an integrated view of the methods studied in the book, point to future research directions, and to provide several challenging design problems, where the student is asked to integrate the methods of the book to develop and evaluate a group of social foraging vehicles or two competing intelligent teams.

Topics Not Covered

It is impossible to cover all the relevant biomimicry topics in one book, even with the relatively narrow focus of optimization, control, and automation. Here, various choices have been made about what to include, choices which depend on my energy level, my own expertise (or lack thereof), my experience with applications, the need to limit book length, the availability of other good books, and level of topic maturity in the literature at the time of the writing of this book. This led to little or no attention given to the following topics: (i) combinatorial optimization and dynamic or linear programming, and its use in intelligent systems (e.g., in path planning, learning, and foraging); (ii) Bayesian belief networks (e.g., their use in decision making); (iii) temporal difference learning and "neuro-dynamic programming;" (iv) sensor management and multisensor integration; (v) immune systems and networks (e.g., in learning and connections with evolution); (vi) construction or evolution of the *structure* of neural and fuzzy systems (i.e., automated approximator structure construction); (vii) learning automata; (viii) evolutionary game theory and evolutionary dynamics; and (ix) study of other control processes in biological systems (e.g., in morphogenesis, genetic networks, inside cells, motor control, and homeostasis). Some of these topics are relatively easy to understand, once you understand this book, while others would require a much more significant time investment. Considering (ix), this book is generally stronger for biomimicry of the "higher level" functionalities of biological systems (e.g., we may study the motile behavior of a bacterium during foraging, but ignore the control processes inside the cell that are used to achieve the motile behavior). Regardless, for several of the above topics, there are exercises or design problems that help introduce them to the reader and show how they are relevant to the topics in this book. Also, references are provided to the interested reader who wants to study the above topics further.

Bibliographic References

To avoid distractions and produce a smooth flow of the text, citations and explanations of, for example, what was done when and how each research contribution built on and related to others, are generally not placed within the text. The referencing style adopted here is more like the one used for a typical textbook, rather than a research monograph. At the same time, however, each part ends with a "For Further Study" section that is an annotated bibliography. For these sections, note the following:

- The main sources used for each chapter, and some related ones, are included. Sources that most significantly affected the content and approach of this book are highlighted.
- There are recommendations on which books or papers to use to get introductions to some topics, and more detailed treatments of others. In particular, there are key references given in control theory and engineering, and in several of the foundational "bio" topics.
- Sources for applications of the methods are highlighted, as many of the techniques in this book have been used successfully in a wide array of problems, too many to reference here.
- There is recommended reading for topics that were not covered here (see above list).
- There are many lists of references that would help support the pursuit of research into the topics studied in this book.
- Connections among a variety of topics are highlighted in the context of referencing research areas.
- A number of times the fuller lists of references for topics are found in the references that are cited, not here. This helped to keep the bibliography size here more manageable.

A large number of researchers have contributed significantly, and over many years, to the topics that are studied here. While it is impossible to succeed in referencing every paper, the annotated bibliographies are meant to help and encourage the reader to search for other sources and investigate in a scholarly way the literature. Clearly, a danger of producing a book with a broad topical scope such as this, is that it becomes impossible to give proper treatment to the literature. The annotated bibliographies, however, try to remedy this problem by highlighting connections to fields and providing key "gateway" references that will lead the reader deeper into the literature. Finally, we must all accept that the topics here are currently still being researched; hence, it is imperative that the researcher use current modern searching and bibliographic aids to come fully up to date on any topic discussed here. Putting this book on paper implies it will age. As this book gets older, the quality of the bibliographic referencing will inevitably degrade.

Relationships to Other Fields

This is not a book on biology, neuroscience, psychology, cognitive neuroscience, behavioral/sensory ecology, or evolution. We simply borrow ideas from these fields to solve optimization, control, and automation problems (and sometimes we will take the freedom of significantly distorting basic biologically motivated functionalities when they are employed to provide solutions to engineering problems). Moreover, we use the biological metaphor to provide a cohesive framework to think about a wide variety of control and automation problems. How much biology, neuroscience, psychology, etc., are covered here? The attempt is to provide just enough to make a strong metaphor and form a scientific foundation for existing methods. The interested reader is encouraged to study the "bio" part of biomimicry for control and automation at greater depth in the "For Further Study" sections at the end of each part.

This is not a book on computer science or "artificial intelligence" in the traditional (classical) sense of what those fields have focused on. We do not even bother to add the word "artificial" in front of the terms "neural network," "planning system," or "attentional system," as it is always clear from the context whether we are talking about a biological system or a model of such a system that we will implement on a computer. The primary goal here is generally *not* to produce computer programs that perfectly emulate biological systems, as it often is in the field of artificial intelligence and some other fields (e.g., mathematical biology), where simulation of an organism is used to help understand the underlying science. The goal is to produce control and decision-making algorithms that operate reliably with high performance for high technology systems. The background of the author is control science, theory, engineering, and practice (and hence, physics, differential equations, system theory, stability theory, optimization theory, stochastic processes, etc.), and not, for example, object-oriented programming and databases. The content of the book will naturally reflect this.

To the Systems and Control Engineer or Scientist

This is a book on engineering. It is not, however, a book on engineering technology (e.g., we spend no time discussing the best available computer, sensing, and actuation technologies to implement the methods). This is not a book on the mathematics that form a foundation for control engineering and automation. We do, however, provide an introduction to Lyapunov stability theory and discuss its use in neural, fuzzy, expert, and planning systems for control. We show how to model and analyze stability for attentional systems. We study stable adaptive fuzzy/neural control, but note that the interested reader should see the "For Further Study" section of Part III for more references in this area, since here we only introduce some basic ideas. We study stability analysis of swarms. The more mathematically inclined reader should view this book as pointing to many opportunities to formalize and prove that various control methods operate in a reliable fashion, so that they are stable and robust (it is recognized that many methods in this book are essentially in their infancy, and significant additional work, some of which should be along the lines of mathematical studies, should be performed).

This is not a "handbook" for a practicing control engineer, where immediate solutions to practical problems are provided. An explicit decision was made not to provide extensive practical case studies; instead the focus is on providing an introduction to basic concepts and principles that will be useful in guiding the design of practical systems and providing insights into solving challenging control and automation problems. We do, however, take a somewhat pragmatic approach in that code is provided and discussed for a wide variety of algorithms (e.g., neural networks, fuzzy systems, planning systems, attentional systems, learning, evolution, foraging, swarms, games, etc.); this code may certainly be useful in speeding a practicing engineer toward a solution to an automation problem.

The "intelligent control" community, both in academia and industry, should find that this book will provide an up-to-date view of the field, show the major approaches, provide good references for further study, and provide a nice outlook for thinking about future research directions. It is hoped that this book will provide a unified and balanced treatment of the topic of intelligent control (and I apologize a priori for the inevitable appearances of my own biases that come from my own research and experience with implementations and applications). The goal is to make you aware of the many approaches, but to do so in a common framework that helps you approach very complex automation problems.

For the conventional control engineering community, it is hoped that this book will provide a different, and useful, perspective on the enterprise of control engineering. The utility of the entire biomimicry approach is, however, open to debate and ultimately will only show its value if it stands the test of time, or, at least in the short term, provides effective solutions to current challenging real-world problems.

Organization, Prerequisites, and Usage

Each part includes an overview that summarizes the essential concepts to be learned, and has a "For Further Study" section (see discussion above). Many chapters include a set of both exercises and design problems. Exercises or design problems that are particularly challenging (e.g., considering how far along you are in the text), long (since they may require significant reading in the literature), or that require you to help define part of the problem are designated with a star ("⋆") after the title of the problem. Some of these topics could serve as thesis research ideas; however, you would certainly want to evaluate the current literature before pursuing any one of them. In addition to helping to solidify the concepts discussed in the chapters, the problems at the ends of the chapters are sometimes used to introduce new topics (e.g., immune networks, approximator structure construction, stability of planning systems, and other topics in foraging). We require the use of computer-aided design (CAD) in many of the design problems at the ends of the chapters (e.g., via the use of Matlab or some high-level language). The reader will naturally find that it will be most convenient to use Matlab, as there is a significant amount of Matlab code provided at the Web site for the book. This code can be quite useful, since it is likely easier to modify that code to suit your needs, rather than to completely rewrite the code.

The necessary background for the book includes courses on differential equations and classical control. Courses on nonlinear stability theory, adaptive control, and game theory would be helpful but are not necessary. Hence, much of the material here can be covered in an undergraduate course. For instance, one could easily cover most of Part II in an undergraduate course, as it requires very little background besides a basic understanding of signals and systems (including Laplace and z-transform theory). Also, most of Part III can be covered once a student has taken a first course in control (a course in nonlinear or adaptive control would be helpful but is not necessary). One could cover the basics of intelligent control by adding topics from this book to the end of a standard undergraduate or graduate course on control. Basically, however, the book is appropriate for a first-level graduate course in biomimicry for optimization, control, and automation ("intelligent control").

This book has been used in a 10-week (quarter system) graduate-level course on intelligent control and for undergraduate independent studies and design projects. In addition, portions of the text have been used for short courses and workshops on intelligent control at conferences, universities, and companies (where the focus is particularly pragmatic).

World Wide Web Site: Matlab Code Available

You can access code used to develop most of the examples in this book and partial solutions to a number of homework problems via the Web site:

`http://www.ece.osu.edu/~passino/ICBook/ic_index.html`

There, you will find Matlab code for neural networks, fuzzy systems, planning systems, attentional systems, least squares and gradient methods (including data sets to work with), several types of direct and indirect adaptive controllers, genetic algorithms, pattern search and stochastic nongradient optimization methods, genetic adaptive controllers, foraging methods for optimization, swarm simulation, games and game-theoretic modeling of foraging, and more.

It was not the objective of the author to provide code that is efficient, in the sense that it minimizes the use of memory or computation time. In some cases, simple changes will result in significant computational speed-up or savings of memory. The objective is to provide code to help educate the reader on the topics, and to help move students quickly toward doing useful simulations. So the code is primarily designed to give the student a type of "hands-on" experience with the methods. My experience in teaching these topics is that if the students are required to code "from scratch," they will not get as much done or learn as much; hence, I could not cover as many topics. Of course, the danger is that the student will simply use the code and not understand it. The homework problems are, however, designed to try to make sure that this problem is avoided (e.g., by asking for changes to the code that would require the student to clearly understand the existing code). Regardless, the instructor must be alerted to this problem and be diligent in assessing students' true understanding of the methods and code.

Feedback on the Book

You are encouraged to provide a description of any errors you may find. For this, please send e-mail to passino@ee.eng.ohio-state.edu or regular mail to: Kevin M. Passino, Dept. of Electrical and Computer Engineering, The Ohio State University, 2015 Neil Ave., Columbus, OH 43210. The author will keep an errata for the book posted on the Web site.

Acknowledgments

Many conversations, presentations/seminars, papers, and books have influenced the writing of this book, and unfortunately, it is clear to me that I can only roughly outline these. People who have directly influenced my own thinking on control systems are Profs. P. Antsaklis, A. Michel, C. Rohrs, and M. Sain at the University of Notre Dame, and the control systems group here at Ohio State University (Profs. J. Cruz, H. Hemami, H. Özbay, Ü. Özgüner, A. Serrani, V. Utkin, and S. Yurkovich). Prof. P. Antsaklis has had a broad impact on intelligent and hybrid systems and hence, on the field and this book. Several years ago, Prof. Yurkovich and I collaborated on a variety of industry-funded projects that focused on the application of intelligent control methods to real-world problems. I have benefited from the work and perspectives of Prof. Özgüner in the area of autonomous vehicles, hierarchies, and large-scale systems. Profs. Cruz and Özbay provided some helpful comments on Chapter 1, and Prof. Cruz provided me help with game theory. Prof. D. Orin at OSU provided feedback on the book, by showing me a robotics perspective on intelligent control, and I benefited from early collaborations with Prof. G. Rizzoni at OSU on automotive applications. Also, I would like to acknowledge the impact of many pleasant discussions over many years about intelligent control with: S. Adibhatla, J. Albus, P. Antsaklis, M. Betancur, D. Driankov, J. Farrell, M. Kokar, K. Lee, F. Lewis, A. Meystel, D. Orin, Ü. Özgüner, T. Parisini, M. Polycarpou, F. Proctor, T. Samad, E. Sanchez, R. Sanz, J. Spall, L.X. Wang, G. Yen, S. Yurkovich, and many others (too many to mention here, but listed in the bibliography at the end of the book). Dr. Jim Spall provided help with the "simultaneous perturbation stochastic approximation" (SPSA) section and also provided part of a homework problem for that. Prof. M. Polycarpou, Prof. D. Jacques, Prof. M. Pachter, and I, worked together on some topics in distributed intelligent control systems and this book benefited from that collaboration. Dr. J. Albus has been a collaborator over many years via involvement in funding projects here at OSU; his broad perspective and contributions have significantly affected the field and hence, this book. I would especially like to thank Profs. Panos J. Antsaklis, Kwang Y. Lee, and Xuping Xu for class-testing this book, and their students who provided specific feedback.

I would like to thank all the OSU graduate students who took a course from this book (e.g., those in the Spring quarters of 1999, 2001, and 2003), especially the ones who provided corrections and suggestions for improvement, and the persons who attended tutorial workshops on this topic that were based on this book (e.g., at the IEEE Conf. on Decision and Control, the American Control Conference, the IEEE Conf. on Control Applications, and several times in Colombia). I have learned much from my own graduate students, who have done research with me in this area throughout the years and it is a pleasure to acknowledge them here: Burt Andrews*, Anthony Angsana, Michael Baum*, Scott C. Brown, Kevin L. Burgess, G. Andrew Bushong, Yixin Diao*, Jorge Finke*, Sriram Ganapathy, Veysel Gazi*, Alvaro Gil*, Mustafa K. Guven, David L. Jenkins, Waihon Andrew Kwong, Eric G. Laukonen, Jeffrey R. Layne, Yanfei Liu*, Yang Liu*, Alfonsus D. Lunardhi, William K. Lennon, Amit T. Magan, Manfredi Maggiore*, Brandon J. Moore, Mathew L. Moore, Sashonda R. Morris, Vivek G. Moudgal, Hazem Nounou, Raúl Ordóñez, LaMoyne L. Porter, Nicanor Quijano, Jeffrey T. Spooner, Doug Torzewski, and Jon Zumberge. Several of my graduate students critiqued various parts of the book while it was being written and provided valuable feedback. These individuals have stars next to their names in the list above. The collaborations with my past graduate students Jeffrey R. Layne [299, 302, 301, 303, 300], Jeffrey T. Spooner [486, 485], Raúl Ordóñez, and Manfredi Maggiore [484] have had an especially significant impact on the writing of the parts on adaptive control (indeed, portions of those listed references served as the first drafts of my writing on adaptive control methods using fuzzy systems and neural networks). More than any other parts of the book, the section on discrete-time adaptive control and the chapter on stable adaptive control were heavily influenced by papers written with Jeff Spooner; the author gratefully acknowledges his kind collaboration on these. Yixin Diao helped me with the use of the Matlab neural networks toolbox. Yanfei Liu performed the simulations for the cooperative robot swarm and provided those figures. Scott C. Brown, Raúl Ordóñez, and Veysel Gazi assisted in the development and running of a laboratory course on intelligent control at OSU that accompanied the course that is taught from this book. More recently, Nicanor Quijano, Alvaro Gil, Jorge Finke, Sriram Ganapathy, and Yanfei Liu assisted in developing a controls laboratory and doing research for NIST, where several experiments for networked and decentralized control were studied. Also, thanks to Kike, Nicanor, and Alvaro for fun help with the Spanish in the dedication.

I would like to thank Carla Spoon, Jacqueline Edwards, Allan Abrams, Francesca Warren, and Oliver Jackson for all their help in the editing and production of this book. A number of companies granted us permission to use published figures and this is indicated in the relevant figure captions throughout the book. We thank those publishers for their permissions.

The author would like to acknowledge the financial support from a variety of sponsors during the course of the development of this textbook, some of whom provided the opportunity to apply some of the methods in this text to challenging real-world applications, and others where a course on the topics covered in this book was given. These sponsors include AFOSR, AFRL, Air Products and Chemicals Inc., Battelle Memorial Institute, DARPA, Dayton Area Graduate Studies Institute, Delphi Chassis Division of General Motors, Ford Motor Company, General Electric Aircraft Engines, NIST (Intelligent Systems Division), NSF grants (e.g., IRI-9210332, EEC-9315257, and IIS-0208664), NASA Glenn Research Center, The Center for Automotive Research and Intelligent Transportation (CAR-IT) at The Ohio State University, and The Ohio Aerospace Institute (in a teamed arrangement with Rockwell International Science Center and AFRL).

Finally, I would most like to thank my very special wife, Annie, for her friendship, support, and love. She and our wonderful children, Carina, Juliana, Jake, and Zac, gave me the extremely valuable gift of constant but gentle reminders that there are many other fun and *much* more important things to do than write a book! ☺

Kevin Passino
Columbus, Ohio
April, 2004

Contents Overview

Contents

Part I

INTRODUCTION

Part Contents

Sequence of Essential Concepts

- Conventional control systems engineering uses an iterative design methodology involving mathematical models, simulation, analysis, and experimentation to construct controllers. The challenge is to produce control systems that achieve verifiable high performance operation and are robust in the sense that they can cope with uncertainty (e.g., disturbances, noise, and inaccuracies in the mathematical models used to derive the control laws that are implemented). Such challenges exist for single isolated control loops and for hierarchical and distributed feedback controls that exist in complex automation problems.

- Intelligent control is based on two types of biomimicry. First, the areas of biology, cognitive neuroscience, psychology, foraging, group behavior of organisms, and evolution provide concepts that can be used to establish the functionality of sophisticated decision-making systems for high technology automation. Second, there are often situations where humans, or groups of humans, have significant expertise on how to solve an automation problem and this knowledge can be automated in computer algorithms to replace them (i.e., there is often an opportunity to mimic actions of humans who solve control problems by distilling their knowledge into controllers). Biomimicry represents a somewhat different route to control system development, and hence to automation. However, in terms of end functionality, the resulting "intelligent controllers" sometimes have certain characteristics that are similar to controllers obtained with conventional control engineering methods. Ultimately, the engineering goal is achieving verifiable high levels of autonomy no matter what design path is taken to get there.

Chapter 1

Challenges in Computer Control and Automation

Chapter Contents

Automation has had, and will continue to have, a significant impact on society. Starting with the industrial revolution, fueled by the computer revolution, and continuing today in many roles in the commercial and industrial sectors, automation has played a key role in business, industry, and national economies. It has provided many benefits and associated problems (e.g., adverse environmental impacts). In this book we focus on the use of computers to achieve automation, with special emphasis on the development of computer programs (algorithms) called "controllers" that take a central role in coordinating and specifying the behavior and ordering of activities in automated systems. Controllers currently play a key role in many automated systems such as modern manufacturing systems (e.g., in fine positioning of robot arms and in sequencing operations on assembly lines), automotive systems (e.g., engine, transmission, and brake system controls), aircraft (e.g., engine and flight controls), autonomous vehicles (e.g., land-based or underwater), and chemical processes (e.g., to control temperature of a batch of reacting chemicals), to mention a few.

"Control" is in fact a very generic term that is used in many ways in computer automation, and in everyday life. For example, there are "micro-controllers" that are multifunction computer chips used in a wide variety of industrial and commercial applications (for us, these provide a way to implement the algorithms we focus on developing). Sometimes, in computer systems, a "controller" is part of the software or hardware that influences or guides the behavior of other subsystems. For instance, there are "controllers" in various components of operating systems or databases. Moreover, we encounter the operation of controllers quite frequently in our daily lives. There are controllers for dispensing soft drinks when enough money has been put in a machine, and there are traffic controllers that decide how to switch traffic lights given information about traffic loading conditions. Notice that in each case the controller is needed to make the system function properly, it plays a central role in the implementation of the system, and it provides some useful type of automation. We do not omit from consideration all these types of controllers; however, you will see that we will pay special attention to controllers for certain types of *dynamical* systems, and ones that use information "fed back" from the system in order to make decisions.

1.1 The Role of Traditional Feedback Control Systems in Automation

Control plays a central role in automation. Here, we outline traditional feedback control systems, the "work horse" central to many automated processes. Following that, we outline how controllers are increasingly becoming central to the automation of complex industrial processes. As technology has progressed, higher levels of automation have been achieved, and control is playing even more complex roles in such systems.

In this book we shall be primarily concerned with "feedback control systems," where some variable of the process is sensed and a controller tries to manipulate some input to the process (often called the "plant" by analogy with the plant in chemical process control) so that the sensed variable is regulated to some value. Some examples will serve to solidify the idea.

First, consider temperature control in a home as shown in Figure 1.1(a). For this, there is a thermostat where you can set the desired temperature (sometimes called the "set point" or "reference input"), and that normally holds a temperature sensor that measures the actual temperature. Then, via furnace or air conditioner controls (the "actuators," i.e., the devices that carry out the action specified by the control input), the control algorithm turns on the heat (turns off the cooling) when the temperature gets below the specified temperature set-point, and turns it off (in the case of cooling, turns it on) when the temperature goes above the set point. It tries to maintain the sensed temperature as close to the desired temperature as possible, even if there are significant "disturbances" such as changes in the outside (ambient) temperature, the presence of people or other heat sources such as appliances, or the opening and closing of doors and windows. In older implementations of temperature controls, a mechanical or other electrical system is used; however, in some modern systems the controller is implemented in a computer algorithm. The computer algorithm that gathers the sensed temperature information and sends signals to switch the furnace (or air conditioner) on or off is a "controller," or what is sometimes called a "control algorithm." The algorithm is the central element in the temperature control system for your home. If designed properly, it will ensure comfort and efficient operation (to keep your utility bills low and conserve energy). The use of such temperature control systems in homes has relieved humans of the task of continually regulating the temperature (e.g., by putting more logs on the fire or opening a window) and provided for comfortable living spaces.

Feedback controllers are decision-making systems that gather information from the environment to decide how to change it to achieve some goal.

Second, consider the cruise control system on an automobile as shown in Figure 1.1(b). For this, there is a way that the driver can input the desired speed, typically, via a mechanism on the steering column. There is a speedometer that can measure the speed of the vehicle. Then there is often a computer that automates the function of regulating the vehicle speed to the desired speed (early implementations were entirely mechanical, or done with circuits, not computers). This computer takes the desired speed and sensed speed and decides what throttle input to produce to ensure that the sensed speed "tracks" (follows) the desired speed. A good cruise control system will succeed in regulating the speed even if there are significant disturbances such as road grade variations, wind, or "plant parameter variations" such as changes in the weight of the vehicle (via the addition of cargo or passengers). A cruise control system provides a useful automation service that relieves the driver of a tedious speed regulation task.

It is an interesting and important fact that while at first glance temperature and speed control may appear to be very different problems, the two fit nicely into the same sort of "functional block diagram" as shown in Figure 1.1, and at a conceptual level, operate in a similar fashion.

The basic problem studied in this book is how to design the controller to achieve certain "design objectives" for a given plant. While this may appear

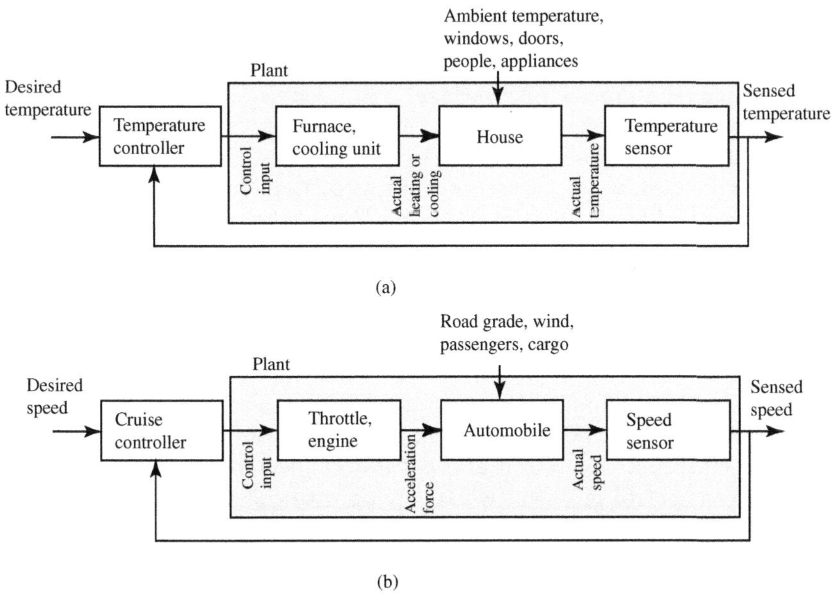

Figure 1.1: Example control systems: (a) temperature control in a home, and (b) cruise control for an automobile.

to be a simple task in some cases, in practical applications there are many complicating factors. In the next section we explain how the process to be controlled can be quite complex, thereby dictating certain complexities in the controller. In Section 1.2 we highlight the wide variety of design objectives that arise for real-world applications. From these you will begin to see how complicated the controller design task can be; challenging design objectives typically create a challenging design problem. Next, in Section 1.3, we outline the standard methodology that is used to design control systems. This will further clarify the challenging nature of the design problem by pointing out the issues involved in making sure that a control system will perform according to the specifications while in operation, where it may be subjected to adverse conditions and other unforeseen situations.

1.2 Design Objectives for Control Systems

Assuming you use feedback control, the design objectives (also called "closed-loop specifications" or "performance objectives"), i.e., what you want the control system to achieve, can involve the following factors, each explained via the temperature or automobile cruise control examples of the previous section.

1.2.1 Tracking

Often, one of the primary reasons for implementing a feedback control system is to automate the process of "tracking" the reference input signal. Tracking refers to the ability of the controller to manipulate the plant input so that the plant output stays as close as possible to the reference input. For the examples of temperature and cruise control in the last section, the controllers typically try to achieve tracking of constant signals that are changed infrequently. For instance, you will often set the desired temperature in your home and leave it at that value for weeks or months. In your automobile you may drive to a highway and then set your desired speed and leave it there for an hour or more. However, this is still considered to be a tracking problem. There are typically environmental effects that move the output from the set point, and the control system must then manipulate the control input to move the output back to the set point. For instance, in temperature control there can be doors and windows that are opened and closed and also natural ambient temperature influences. For the cruise control problem, while driving with a constant speed set point you may encounter a hill so that the controller will have to increase the throttle to keep the speed constant going up the hill, and then decrease it to maintain constant speed going down the hill.

There are many other types of tracking problems, some that require continuous tracking of a dynamically changing reference signal. For instance, you are engaged in a type of tracking problem right now! You have an objective of reading these lines and to do so you have generated a type of reference input to your eyes and the muscles there move your eyes to the desired point. Humans perform many types of tracking activities (e.g., in baseball when the hitter tries to "keep his eye on the ball"). In technological applications there are many instances where we want to achieve tracking. For instance, in aircraft flight control there may be a flight path that minimizes fuel consumption or travel time, and the control system flies the plane to track this trajectory, even if there are wind disturbances and air density changes. Similar examples exist for autonomous vehicles and other applications.

Tracking, the ability to produce inputs that guide the plant to produce desired outputs, is often a central objective.

Finally, it is important to note that without feedback control, it is impossible to achieve tracking, or at least it can be very difficult if there are certain adverse conditions as we discuss below. When a feedback signal is not used, the controller is called a "feedforward controller," the resulting control system is called an open-loop control system, and one such system for the cruise control problem is shown in Figure 1.2. Here, we remove the feedback path, and hence we can save the expense of implementing a speed sensor (yes, feedback control generally costs money since you need to implement a sensor to provide the feedback information). There are times when such "open-loop" control systems are useful, particularly when it is possible to specify sequences of inputs independent of what happens in the plant. However, for the cruise control application, if you start going up a hill and the desired speed is held constant, then without a speed sensor (or other sensor) and feedback path, the controller cannot know that the speed of the vehicle has decreased so it will not increase the throttle to counteract the road grade change and keep the velocity at the desired set point (i.e., the desired speed cannot be tracked). With feedback control, when we start going up a hill, the controller will slightly increase the throttle to try to keep the vehicle at the desired speed. Also, if we start going down a hill and the desired speed is held constant and the speed increases above the set point, then the throttle input can be decreased to reduce the speed to the desired value. Clearly, however, control system tracking abilities in this case will be limited by the steepness of the hills the car encounters.

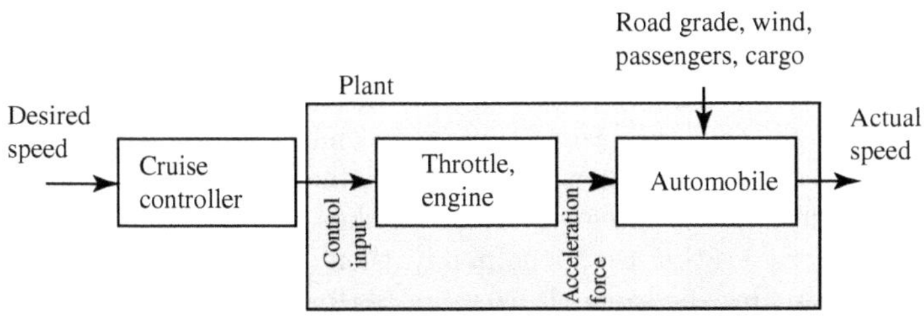

Figure 1.2: Feedforward cruise controller for an automobile.

1.2.2 Reducing Effects of Adverse Conditions

One of the key features of a feedback control system is its ability to reduce the effects of certain adverse conditions and uncertainty that can arise in the plant and its environment. If a control system has such capabilities it is said to be "robust" and a controller that acts to achieve these features is said to be a "robust controller." Consider the following properties that we typically want a control system to achieve:

The ability to achieve good performance, even in the presence of adverse plant conditions, is often a central objective.

- *Disturbance rejection properties:* For the cruise control problem, the ability to reject disturbances could mean that the control system will be able to damp the effects of winds or road grade variations (which are "uncertainties" since we normally do not measure these and use them in making control decisions). Basically, the need for disturbance rejection demands the use of feedback control. To achieve disturbance rejection, it is necessary to take the appropriate steps during controller design to ensure that this property is met.

- *Insensitivity to plant parameter variations:* For the cruise control problem, insensitivity to plant parameter variations means, for instance, that the control system will be able to compensate for changes in the total mass of the vehicle that may result from varying the number of passengers or the amount of cargo. Again, note that since we do not normally measure the number of passengers and the weight of the cargo, we view them as providing influences on the performance of the control system that are "uncertain" in the sense that we do not know the timing or extent of their influence a priori. Typically, it is impossible to achieve an insensitivity to plant parameter variations without using feedback information from the plant.

To expand on our example, suppose that a person without cargo takes a ride in a vehicle on a long stretch of flat land (e.g., western Kansas) and holds the throttle input constant so that she is going exactly 75 mph, which is also the desired speed set point. Next, suppose that she repeats the experiment with everything the same (in particular, the throttle setting), but adds three other people and their heavy luggage. Clearly, with the same throttle input, the speed will be lower since the engine has to work harder to keep the same speed with a heavier load (e.g., the tires will press more firmly on the road, making them more difficult to turn). Moreover, this scenario can be used to emphasize the ability of a feedback control system to reduce the effect of disturbances. Notice that if she repeats the experiment but has a strong constant head wind (common in Kansas) rather than a calm day, then clearly the speed will be lower if the throttle remains the same. If we use a feedback controller in each of these cases, the speed sensor would provide an indication that the speed decreased, and then the controller would provide an increased throttle input to maintain the desired speed.

1.2.3 Behavior in Terms of Time Responses

While the abilities to achieve tracking and reduce the effects of adverse conditions are some of the key features of feedback control systems, the behavior of the system, in terms of shapes of time responses of system signals, is usually used to specify how we want a closed-loop system to behave. Some of the most important ways to characterize the behavior of systems are given in the following, using the cruise control example discussed earlier:

Quality of system behaviors is often represented via characteristics of its dynamical operation.

- *Stability:* In the cruise control problem, having a stability property could guarantee that on a level road the actual speed will converge to the desired set point. There are, however, many ways to characterize stability, so that in other contexts "stability" may simply mean that all the system variables remain bounded by a fixed constant. Or, an oscillating signal could be considered "unstable." In the cruise control example, suppose that the desired speed is 75 mph and consider Figure 1.3. In each of the four plots, the solid thick line represents the desired speed we would like to track. Then, in Figure 1.3(a), we show three types of responses that could be considered to be "unstable." In one, the actual speed does not converge to the desired speed but to a different constant value. In another, the speed oscillates between zero and about 90 mph, and in yet another, the speed appears to quickly grow unbounded (obviously, in the actual system there would be limits to how big it could get). We would normally want to design our control system so that the closed-loop system does not exhibit unstable behavior.

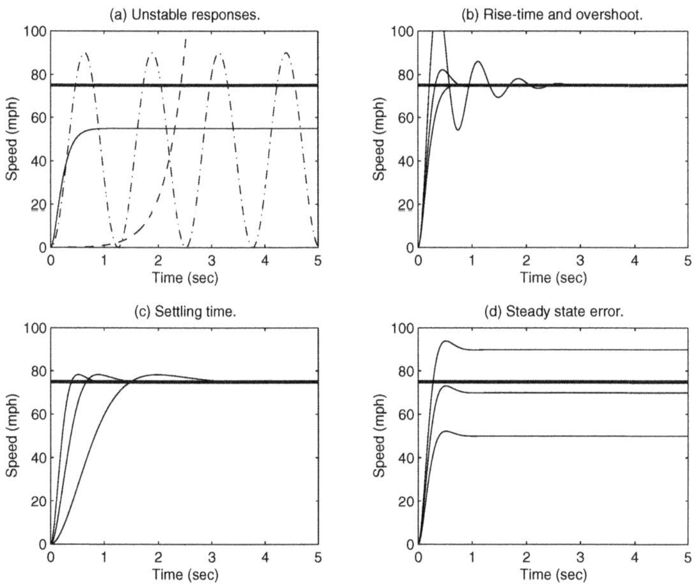

Figure 1.3: Desired speed (thick solid line at 75 mph) and possible closed-loop time responses resulting from different controller designs. The plots are of the vehicle speed while the cruise control system is in operation.

- *Rise-time:* For example, in the cruise control problem, rise-time is a measure of how long it takes for the actual speed to get close to the desired speed when there is a step change in the set point speed (e.g., the time to go from 10% of the final value of 75 mph to 90% of the final value). In Figure 1.3(b), we show three different responses with three different rise-times, some longer than others. Clearly, a short rise-time would normally be considered to be the best; however, physical characteristics of the vehicle (e.g., the size of the engine) will limit how short the rise-time can be.

- *Overshoot:* For example, in the cruise control problem, when there is a step change in the set point, overshoot is a quantification of how much the speed will increase above the set point. In Figure 1.3(b), we show three different responses with three different amounts of overshoot, some larger than others, and one with no overshoot. Clearly, too much overshoot is not good, as you could get a speeding ticket if it is too large. Note that when we get a better (smaller) rise-time, we get a worse (larger) overshoot. There is a "trade-off" between these two design objectives, and such trade-offs are often encountered in control system design. Creating a good control design often amounts to achieving an appropriate balance between competing objectives.

- *Settling time:* For example, in the cruise control problem, settling time is the amount of time it takes for the speed to become within 1% of the set point. In Figure 1.3(c), we show three different responses with three different settling times, some longer than others. Clearly, we would typically like to have a short settling time; however, as shown in Figure 1.3(b), if you get a short rise-time it can sometimes be more difficult to get a short settling time.

- *Steady-state error:* For example, in the cruise control problem, if you have a level road, can the error between the set point and actual speed actually go to zero? Or, if there is a long positive road grade, can the controller eventually achieve the set point? In Figure 1.3(d), we show three different responses with three different steady state errors. Clearly, we would typically like to have no steady state error, or at least a very small one, independent of the desired speed. Special attention must be paid to this performance objective in design to ensure that this property holds.

Generally, when we begin the design process we may have in mind the many possible time responses that we might get for different controller choices (e.g., all the ones shown in Figure 1.3). In the design problem, we try to pick a controller so that a response is obtained that meets the performance objectives; we design the controller so that it provides a response in Figure 1.3 that represents that the cruise control system will behave properly. It is important to note that it is not sufficient to simply achieve a single adequate response for one set of conditions. The actual objective is to achieve this response, or something close to it, even if there are disturbances, plant parameter variations, or sensor noise. In other words, we would like the performance, in terms of time responses, to be "robust" to many different conditions, including adverse conditions. We also want to have stability be robust under these same conditions. The problem of achieving such performance and stability robustness goals significantly complicates the control design problem; indeed, it is often the central issue to focus on in design.

Finally, note that in practical control systems there are many other design objectives; we will discuss some of these for complex hierarchical control systems in Section 1.4. Moreover, in the next subsection we provide some general engineering goals.

1.2.4 Engineering Goals for Control Systems

While the above factors are used to characterize the conditions that are typically used to indicate whether or not a control system is performing properly, there are other issues that must be considered that are often of equal or greater importance when you develop control systems. These include the following:

There are many broader engineering goals that are critical to recognize in practice.

- *Cost:* How much money will it take to implement the controller, or how much time will it take to develop the controller? Costs of sensors, actuators, the controller, and other interfaces must be taken into consideration.

Also, the number of person-years needed for development, implementation, commissioning, and maintenance must be considered.

- *Computational complexity:* How much processor power and memory will it take to implement the controller? In some practical applications, controllers can quickly burden the available processors (e.g., in current automotive applications, relatively simple inexpensive processors are used to achieve a variety of control functions in the engine, transmission, etc.). Clearly, in such situations, the concerns of computational complexity can significantly impact how you design your controller.

- *Manufacturability:* Does your controller have any extraordinary requirements with regard to manufacturing the hardware that is to implement it? It should be designed so that it is easy to manufacture.

- *Reliability:* Will the system always perform properly? What is its "mean time between failures"? What causes these failures? Sensors, actuators, communication links, or controller? How can you design the system so that the number of failures is minimized? This can be particularly important in safety-critical applications such as aircraft control where redundant hardware is often used. Is your controller simply "too aggressive"? Does it try to achieve the best possible time responses, without giving enough attention to the need to be conservative to ensure that adverse conditions will be adequately dealt with, even ones that you cannot envision at this time? Sometimes experienced control engineers express such concerns when a new controller is developed and there are extraordinary performance claims.

- *Maintainability:* Will it be easy to perform maintenance and routine adjustments to the controller? As we discussed earlier, extra code is often added to the controller for a maintenance interface.

- *Adaptability:* Can the same design be adapted to similar applications so that the cost of later designs can be reduced? For instance, for the cruise control problem, will it be easy to modify the controller to fit on different vehicles so that the development can be done just once (this is sometimes called the "calibration problem")? The control engineer then views her task as that of designing one controller that fits all the plants in a certain class (e.g., all mid-sized vehicles that they produce). In practice, the design is sometimes provided with a set of instructions on how to tailor the design to each particular application.

- *Expandability:* How much redesign work will we have to do to be able to add new hardware or functionality to the control system? Will we have to start over? Or, are the existing controller's functions and code established in a manner that makes it possible to easily add new functionality? Is the control system easy to interface to other systems (is it open?)?

- *Understandability:* Will the right people be able to understand the approach to control? For example, will the people who implement, maintain, or test it be able to fully understand it? The importance of this characteristic cannot be overstated for some applications. You may have the best controller imaginable, but if it cannot be clearly explained to certain key personnel, it will not be chosen for implementation.
- *Politics:* Is your boss biased against your approach? Can you sell your approach to your colleagues? Is your approach too novel and does it thereby depart too much from standard company practice? Is your approach too risky?

Not only must a particular approach to control satisfy the basic performance objectives, but the above issues must also be taken into consideration; these can often force the control engineer to make some very practical decisions that can significantly affect how, for example, the ultimate cruise controller is designed. It is important, then, that the engineer has these issues in mind early in the design process to ensure that the best possible controller is delivered.

1.3 Control System Design Methodology

When confronted with a control problem, a control engineer generally follows a relatively systematic design procedure. In this section we outline this procedure.

1.3.1 Understand Plant and Specify Design Objectives

For our simple example of automobile cruise control, the control engineer first seeks to gain an intuitive understanding of the plant's dynamics as shown in Figure 1.4. This often occurs by having the control engineer involved in the design of the plant, or at least she should talk to the designers or study the plant specifications. Moreover, she may enhance her understanding by performing experiments with the plant, or by observing it while in operation, possibly with an existing control system she is trying to improve the performance of. After gaining a clear understanding of the plant's operation, capabilities, and limitations, the control engineer should establish the design objectives (and sometimes there are iterations needed to come to a clear enough understanding of the plant in order to make a reasonable statement of the control objectives). Sometimes, this task is left entirely to the control engineer, and other times it involves the customer and possibly management. The design objectives could include many aspects, and many possibilities were discussed in the last section. In some projects, the engineer may be able to request that improvements be made to the plant (e.g., via improved or additional sensors or actuators) to make sure that the design objectives can be met. Other times, the control engineer may have to "soften" some design objectives, even late in the design process, if it appears to be impossible to meet the objectives.

Modeling forms the foundation for control design methodology. No model is perfect; however, even uncertainties can be represented so that they may be considered in the design process.

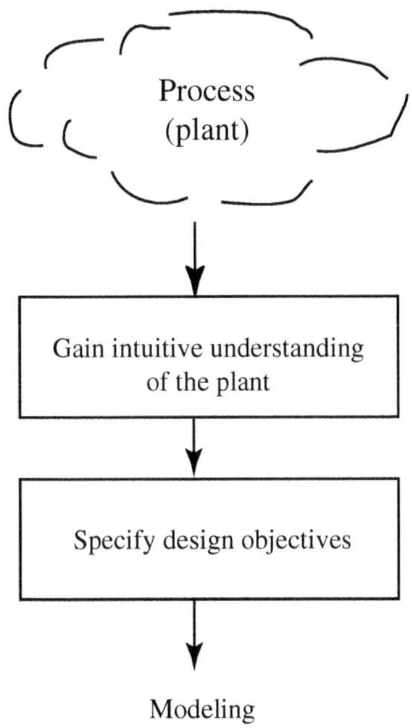

Figure 1.4: Early steps in the control system design procedure.

1.3.2 Construct Models and Uncertainty Representations

The modeling steps that are used are depicted in Figure 1.5. We will discuss these steps in the context of the automobile cruise control problem presented earlier. Typically, the modeling process begins with the development of a "truth model" of the automobile dynamics (which may model vehicle and power train dynamics, the effect of road grade variations, wind effects, weight changes in the automobile, etc.) that you generally try to make as accurate as possible (so that it represents the truth). Generally, physical modeling principles, "system identification" (see references in the "For Further Study" section at the end of Part I) using data from the system, and approximations are used to specify the truth model. The final form of the truth model may be a computer program that simulates the system on a digital computer, an analog computer simulation, a scaled-down hardware model, or a combination of these. Currently the most common choice is the use of a simulation program on a digital computer.

It must be emphasized that the truth model cannot be made to be perfectly accurate. It is a model, and even if actual plant data is used in its specification, there will *always* be some characteristic of the plant that is being ignored in the specification of the truth model. It is important to recognize that the richness of possible behaviors of the plant can only typically be captured by setting various parameters in the truth model and executing the model many times (e.g., changing the wind effects on the automobile and considering all

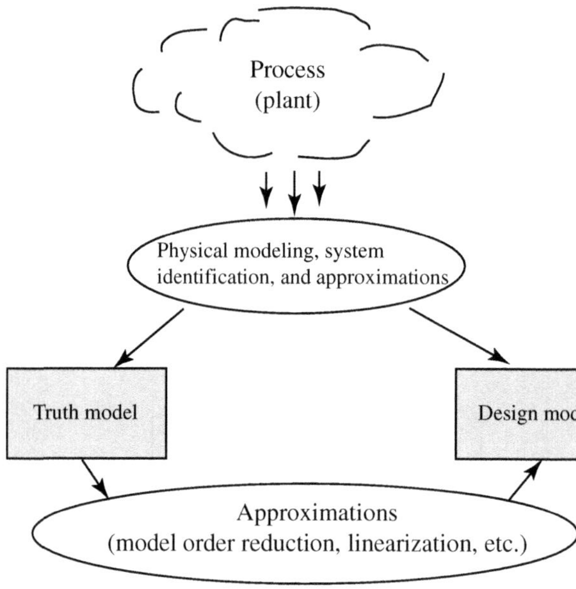

Figure 1.5: The modeling procedure.

the different effects that different wind speeds and directions can have). A good truth model will model "adverse" conditions, disturbances, noise, plant parameter variations, etc. We think of this as a model of the uncertainty in the physical process. Finally, we note that there are certain applications where a truth model is not produced, and this will be discussed more below. Basically, you only produce the truth model if you will need it somewhere in the design process.

One place where a truth model is sometimes used is in the process of producing a "design model." A design model is normally a simplified version of the truth model and it is used in the controller synthesis procedure. It is simplified by using approximation methods such as model order reduction or linearization. Normally, it is simpler than the truth model since controller synthesis procedures typically require certain forms of mathematical models (e.g., linear or affine nonlinear models). Sometimes, the design model is produced directly without the development of a truth model. Often, the design model may include some type of representation of uncertainty, but not as detailed as one for the truth model. This "uncertainty model" may be a noise model, it may incorporate certain unknown but bounded additive or multiplicative terms, or other features. The uncertainty model that is used as a component of the design model should generally be simpler than the one in the truth model (e.g., it may be linear), but still capture the truth model and plant uncertainties (but of course this may not be possible since the plant is nonlinear, so uncertainties often enter in a nonlinear fashion and can only be modeled accurately that way). Regardless, by incorporating representations of uncertainty into the truth and design models, and thereby later taking the uncertainty into account in the controller design process, it is hoped that a "more robust" control system is realized. (Here by "more robust," we mean that we can increase the size of the disturbance or perturbation, or the amount of uncertainty, and it can be tolerated in the sense that performance or stability will not significantly suffer.)

1.3.3 Analyze Model Accuracy and System Properties

The steps are illustrated in Figure 1.6. First, evaluation of model accuracy is a very important step in the design process since if you design based on the model(s), and the model(s) is (are) not sufficiently accurate, then the designed controller is not likely to succeed. The typical approach to study model accuracy is to use analytical methods and also perform repeated simulations and comparisons to actual plant data, especially under a variety of conditions to study the accuracy of the uncertainty model. The problem is, however, that the model *cannot* be perfect, and it is difficult to know how accurate it must be to develop a good controller. For the truth model, we generally aim to create the most accurate model possible. For the design model, we make explicit choices to ignore certain aspects of the system dynamics in order to make the control design procedure feasible. Ultimately, however, our inability to know how accurate our model must be is one of the key reasons why iteration is needed in the overall design process.

A model should be simple, but accurate enough to succeed in control design.

Often, before starting the controller construction process, the control engineer will analyze system properties to gain insights into how the plant behaves. This analysis may be conducted on the truth model or the design model, and in fact can lead to adjustments of the design model if it does not exhibit the same essential properties as the truth model. Typical system properties studied for the design model include the following:

Models help provide intuition on how the plant behaves.

- *Stability:* Is the plant stable? If you provide it a bounded input, will it produce a bounded output? If you start its operation near an "equilibrium" (i.e., values of the system variables where the system will not move from when placed there), will it stay close to the equilibrium (stability in the sense of Lyapunov) and will it converge to the equilibrium (asymptotic stability)? Unstable plants can present challenges for certain applications.

- *Controllability:* Are the "states" of the plant (certain system variables) able to be steered by the control inputs to any location you would like them? If you cannot steer the system state to certain places, you may not be able to meet certain control objectives.

- *Observability:* Can we see where the states are, or have been, by observing the plant inputs and outputs? If you cannot determine where a state is, then it may be impossible to know how to steer it to where you want it, so certain objectives may be impossible to achieve. Moreover, if some state (system) variable can become unbounded without proper control and that

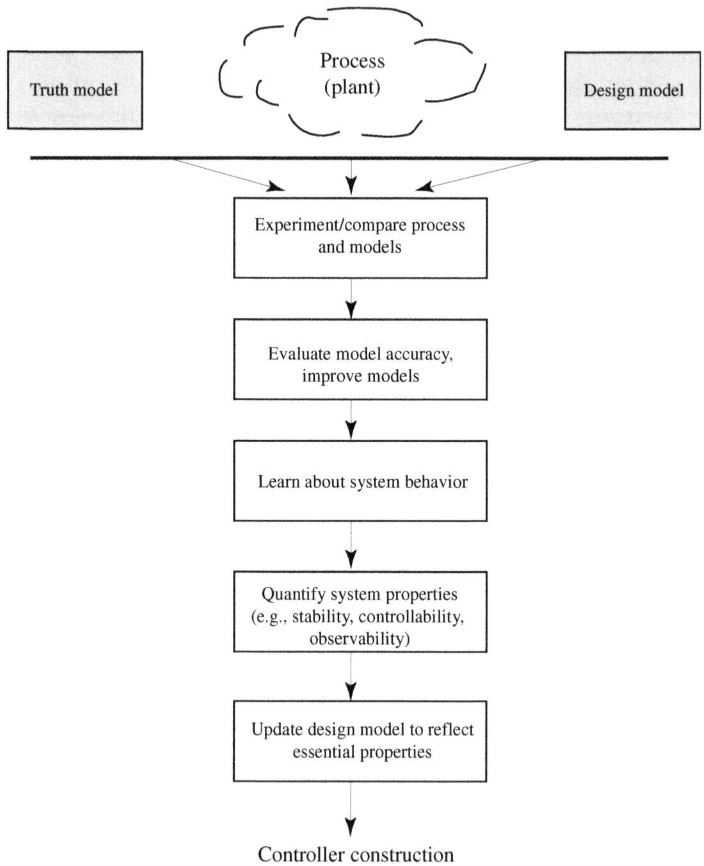

Figure 1.6: Model evaluation and adjustment process.

variable is not observable, then it is clear that the control method cannot be successful.

- *Rate of operation:* How fast can the system react to the inputs? What types of transient behavior are expected?

This list is certainly not exhaustive (e.g., minimum phase characteristics, stochastic effects, etc., are also important for some applications).

For most control methods to be applicable, some of the above properties must hold (e.g., you may need a controllable system to be able to apply the synthesis procedure). Understanding the above properties helps the designer know which approach to use and what issues to pay special attention to. Moreover, it alerts the engineer to the times when a plant redesign may be needed to be able to achieve the control objectives. For instance, it may indicate that another sensor must be purchased and implemented to make sure that some key variable is observable. Or, it may mean that we need to add an actuator to ensure that we can properly steer the state. Other times, such analysis or simulation with the truth model may indicate that we need a more accurate sensor or a faster actuator. Sometimes such analysis can even indicate that the plant itself needs to be redesigned to make it possible for the controller to achieve its objectives.

1.3.4 Construct and Evaluate the Control System

Next, the designer constructs a controller and evaluates it using mathematical analysis, simulations, and experimentation. The overall procedure is outlined in the flowchart in Figure 1.7, and is discussed next.

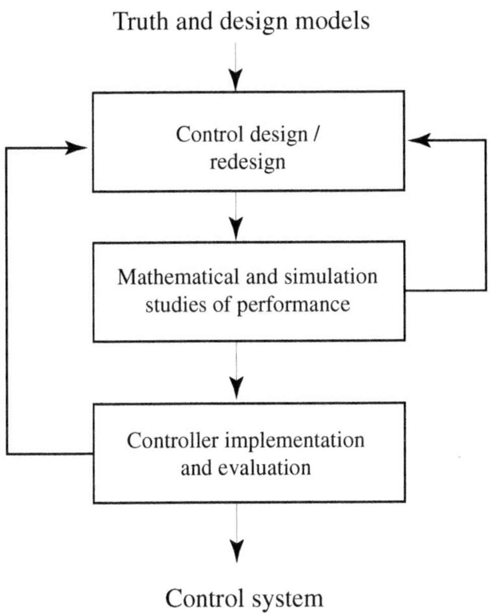

Figure 1.7: Controller construction and evaluation process (for convenience, it ignores the possibility of iterative improvement of models).

Controller Synthesis

At this step, we use the design model to construct the controller. Conventional control has provided numerous successful methods for constructing controllers for dynamic systems from models. Some of these are listed below, and we provide a list of references at the end of this part for the reader who is interested in learning more about any of these topics. We highlight these methods here so that the reader is aware of the rich body of concepts and methods that have been used in the past and are currently being used to design control systems. The

There are many successful conventional control design methodologies, and some control problems are candidates for more than one method.

focus of this book is *not* on these methods, even though many of these methods form a foundation for the intelligent control methods studied here. The popular conventional control methods are the following:

- *Proportional-integral-derivative (PID) control:* A high percentage of the controllers in operation today are PID controllers (or at least some form of PID controller like a P or PI controller). This approach is often viewed as simple, reliable, and easy to understand. Often, heuristics are used quite effectively to tune PID controllers (e.g., the Zeigler-Nichols tuning rules).
- *Classical control:* Lead-lag compensation, Bode and Nyquist methods, root-locus design, and so on.
- *State-space methods:* State feedback, observers, and so on.
- *Optimal control:* Linear quadratic regulator, use of Pontryagin's minimum principle or dynamic programming, and so on.
- *Robust control:* H_2 or H_∞ methods, μ-synthesis, quantitative feedback theory, loop shaping, and so on.
- *Nonlinear methods:* Feedback linearization, Lyapunov redesign, sliding mode control, backstepping, and so on.
- *Adaptive control:* Model reference adaptive control, self-tuning regulators, nonlinear adaptive control, and so on.
- *Stochastic control:* Minimum variance control, linear quadratic Gaussian (LQG) control, stochastic adaptive control, and so on.
- *Discrete event systems:* Petri nets, automata, supervisory control, infinitesimal perturbation analysis, and so on.
- *Hybrid systems:* Control of systems that can most conveniently be represented by a combination of continuous time differential equations and discrete event system models.

For some classes of design models, the construction procedure is entirely systematic (mechanical) and many computer-aided control system design software packages are available, but most of the time the controller construction process is iterative. The iterations begin by constructing a controller, then performing a preliminary evaluation of whether some of the performance objectives are met. Typically, even if the evaluation is only made with respect to the design model, there is a need to tune the design by adjusting some design parameters, at least to shape the transient response. Use of the truth model or experimentation may indicate the need for additional tuning or redesign, as we discuss next.

Analysis of Closed-Loop Control System Performance

When a controller that seems to be a viable candidate is constructed, it is normally evaluated via mathematical and simulation-based analysis before implementation. Normally, mathematical studies are conducted using the design model, while simulation-based evaluations are done using the truth model.

Analysis entails the use of mathematics, simulations, and experimentation; each has its own advantages and disadvantages.

Sometimes, the design procedure is guaranteed via mathematical proofs to result in a stable control system. Other times, stability of the closed-loop system must be studied after the controller is constructed. Many types of mathematical nonlinear analysis can be used to verify the performance of the closed-loop system, but its form is typically dictated by the form of the mathematical design model used. For instance, if a linear design model is used, then there is a rich set of tools for the analysis of system properties. When an affine nonlinear design model that satisfies certain assumptions is used, there are generally fewer tools, but several sorts of analysis are still possible (including stability analysis). Some mathematical approaches can actually be used to show that the performance of the system will be maintained even in the presence of uncertainty (using a model of the uncertainty).

It is important to emphasize, however, that any conclusions you reach via mathematical analysis are only for the mathematical model (e.g., stability is a property of the model) and not necessarily for the physical system. If the design model (perhaps including a model of uncertainty) you use in the mathematical analysis is a very accurate representation of the physical system, then you can generally transfer your conclusions to the physical system (assuming, of course, that a proper implementation of the controller is achieved). On the other hand, as is typical in practical applications, if the design model has significant inaccuracies, then we cannot automatically expect that the properties we find to hold for the design model will hold for the physical implementation of the control system. This is why even though fewer analysis tools are available when a nonlinear design model is used, it is generally a more accurate model, so such analysis can certainly be valuable. Indeed, one main goal of the research community is to develop modeling, analysis, and design tools for as general of a class of nonlinear (and uncertain) models as possible (the research is considered mostly complete for the linear plant case).

There is a similar problem in transferring conclusions of analysis via the simulation-based study of performance with the truth model. The accuracy of the truth model will dictate how valid it is to transfer our analysis conclusions to the physical control system. If we have a very accurate truth model, then the conclusions of the analysis will tend to transfer to the physical system; however, if the model has inaccuracies, the conclusions may not be valid for the physical system. Moreover, there is an additional complication. Simulations are only conducted for a finite amount of computer time, and only a finite number of simulations can be conducted. Hence, even if your truth model is extremely accurate, it is impossible to completely evaluate the effects of disturbances, noise, and adverse conditions on the performance of the closed-loop system. Hence, while a complicated model of system uncertainties can be used in the truth model, it is typically impossible to fully exercise this model to completely evaluate the robustness of the closed-loop system. Generally, we use the simulation-based analysis method simply to improve our confidence that the closed-loop system will perform properly.

Experimental Evaluation of System Performance

Next, the control system is implemented and evaluated, and we try to design experiments to test whether the closed-loop system meets the required objectives, and whether it seems to be robust in achieving this performance. Given a finite amount of resources and deadlines that the control engineer must meet, clearly it is impossible to test every aspect of the control system, such as the influence of every possible plant parameter variation or disturbance. This is where the mathematical and simulation-based analysis done earlier can help to raise our confidence that the physical control system will behave as expected.

Typically, at every stage of the process, and particularly at the implementation stage, problems may be uncovered that require us to return to some earlier design stage and repeat design steps. There is an attempt to make the control design process completely systematic so that it will succeed the first time, but due to model inaccuracies and unexpected adverse conditions, it is virtually always necessary to tune the implemented controller using clever heuristics and insights gained earlier in the design process. We cannot overemphasize the importance of understanding the plant very well, how various controllers have operated on the plant in the past, and the conditions that can arise and result in degraded performance.

Basically, the design procedure is concluded when the engineer has demonstrated that the control objectives have been met, and the controller (the "product") is approved for manufacturing and distribution. Next, we summarize the iterative design process.

1.3.5 Summary: The Iterative Design Procedure When Using Mathematical Models

To summarize the controller design process, we provide the flowchart in Figure 1.8. This flowchart basically outlines the general steps used to construct control systems. We start with a process (plant) for which we want to achieve some type of automation. We gain an intuitive understanding of the plant and establish the performance objectives. We model the plant, often producing both truth and design models, but sometimes just one of these. We test the validity of the model(s) and study properties of the system. Next, we use the design model to construct a controller, but of course the truth model can be used in simulation for controller tuning. We evaluate the control system using mathematical and simulation-based approaches, and if it is not possible to produce an acceptable design, we may refine the control objectives (usually in consultation with the customer and only after significant design efforts, and when we gain some confidence that the current design objectives are impossible to meet), change the model(s) (e.g., by adding the representation of an additional plant characteristic that the controller must accommodate for), or redesign the controller (possibly with a different controller synthesis technique). Next, when we have an acceptable design, we implement it and evaluate whether it works properly for the actual physical system. If it does not, we may have to change the design objectives or model(s), or redesign the controller. The final product is the control system that provides the originally desired automation function.

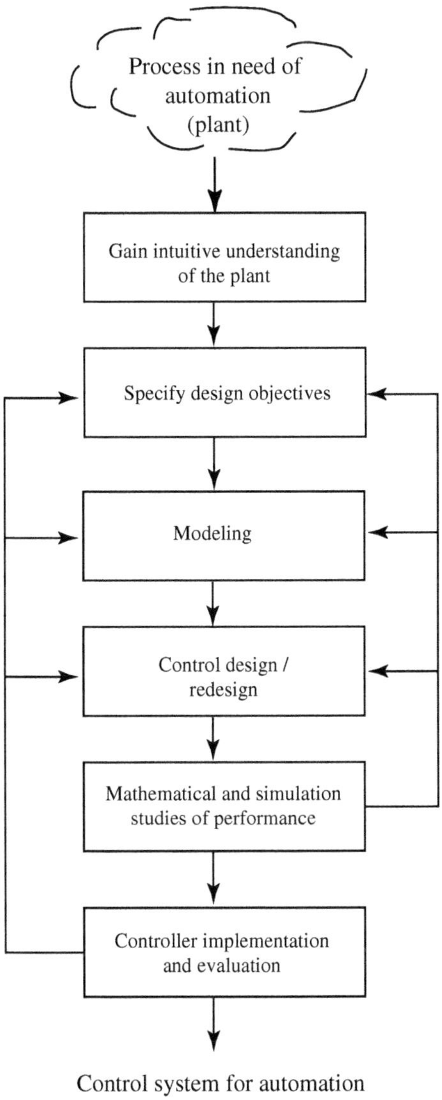

Figure 1.8: Flowchart of control system design steps.

1.3.6 Methodology Without Mathematical Models: The Use of Heuristics

In many practical control problems it is difficult or impossible to obtain a mathematical model and in this case, we rely on the use of heuristics for control design (e.g., in PID control).

Suppose that it is difficult, impossible, or undesirable to develop a mathematical model of the process to be controlled. Difficulties in creating a model may arise simply due to the time and expense required to develop a good model. Other times, the process to be controlled is extremely complex and the physics of various components of the process may not be well-understood, and this can make it difficult to produce an accurate enough mathematical model (in control engineering practice, this is a well-recognized problem). Still other times, the performance requirements are not very demanding, and hence it is possible to avoid the modeling process and heuristically construct a controller that will perform "good enough." In such a case, if a model is not needed to construct a controller, there is simply no need to produce it. The plant may simply demand some cleverly constructed rules on how to control it, and we may not need a mathematical model to assist in the construction of such rules, or the later evaluation of the resulting control system (for more discussion on this, see [2]).

Heuristic Control Design Methodology

How does the control design process change if you do not have (or use) a mathematical model? The design process for this case is shown in Figure 1.9. Typically, we need to gain an intuitive understanding of the plant before we can write down the design objectives, but sometimes there is iteration between these two stages. The step where we gain an intuitive understanding of the plant is quite important, as it provides the information that allows us to construct a controller in one of the two following ways:

- Sometimes there is a human operator who has been manually performing the control task and who is considered an "expert" at conducting this task. If there is such a person, then the control engineer can interview him and code his knowledge into the controller to automate his control task.

- Other times there is no such expert human operator, just the control engineer who has insights into how to properly control the process. In this case, the engineer simply writes down her intuitive ideas about how to control the process, and directly implements these.

We evaluate the performance of the closed-loop system via implementation studies, and iterate on the design process as needed. As for the conventional control system design methodology where mathematical models are used, we note that while the diagram indicates that you can change the design objectives on every iteration, normally you would not do this until you have exhausted many design possibilities and you gain some confidence that the stated design objectives cannot be met (or can be modified to demand *higher* performance). Typically, the iteration that is used the most is the one where we implement

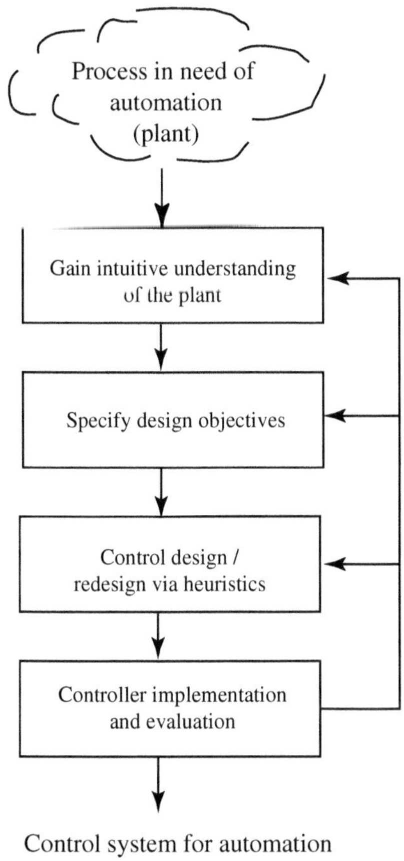

Figure 1.9: Heuristic control design methodology.

the controller and then redesign using heuristics. The reason for this is that in practice, most of the time we learn something from implementing and testing some version of a controller, and this knowledge can then later be coded into the controller to improve performance. We either learn via our own experimental analyses, or via the assistance of a human operator who may help evaluate how the controller emulates what she does when she controls it manually.

Now, notice that with no models, we clearly cannot conduct mathematical or simulation-based analysis. The only way we can evaluate the performance of the system is by implementing it on the actual system. Since we cannot use mathematical or simulation-based analysis, we may not be as confident of the performance of the resulting system; however, we must remember that such *analysis is only as valid as the model that was used to conduct it.* Moreover, in many practical applications, an implementation-based evaluation is sufficient to convince us that we have a good design. Indeed, this is how PID controllers are typically designed in practice, and PID controllers account for a high percentage of all controllers currently in operation! This methodology is certainly sound, at least for some practical applications, even though it may not be acceptable for some safety-critical applications.

Close Relationships With Traditional Design Methodology

The heuristic design methodology is quite similar to one that relies on the use of models. After all, models are the product of the use of heuristics ("art") and science.

It must be emphasized that the heuristic design methodology does not represent a significant departure from standard control engineering methodology, where mathematical models are used. What is different, is the explicit inclusion of heuristics *throughout* the design process, not only at the end when you try to implement and tune the controller on the physical process. (Of course, you could view the use of models as equivalent to the use of heuristics because a model can be viewed as a type of heuristic.) Even though we can use a heuristic method for nonlinear controller construction, this does not in any way mean that we are willing to ignore all the conventional control approaches discussed above. While a heuristic method often provides a "first cut" at how to control a system, and performs adequately for the first generation of the system, when more is learned from how the controller operates on the actual physical plant, and more time is dedicated improving the quality of the solution, it may be possible to make incremental improvements, or perhaps a conventional approach may emerge (e.g., importance of the problem may justify the expenditure needed to develop a good model and a conventional method). Sometimes, however, the opposite can occur: a crude mathematical model is used, a controller is constructed, significant tuning/modifications are required to get a reasonable level of performance, but ultimately the design is not satisfactory. Much is learned in this process, and more heuristic ideas about how to achieve good control are gathered, and then such ideas are used directly in a controller (e.g., in the form of rules).

Using a Truth Model in the Heuristic Control Design Methodology

Finally, we must highlight the following two important issues related to the above design methodology, both of which highlight the fact that truth models are sometimes used in the heuristic design methodology:

1. In practice, the above methodology is sometimes modified by producing a truth model that can be used for simulation-based analysis and tuning of the controller. Indeed, there are some applications for which significant effort has been put into the development of truth models, and these models are very accurate, but are sometimes not very useful in the conventional control design process since determination of a useful design model is difficult or impossible. Or, if it is possible, it only leads to controllers that can provide a somewhat inferior level of performance (e.g., in this case, it may be that design models that are compatible with controller synthesis methods do not represent the essential characteristics of the system that will lead to a successful controller design). Exercise 1.4 directs you to

produce a flowchart describing the heuristic design methodology for the case where a truth model is used.

2. In textbooks, it is convenient to present the design methodology when we have a truth model, since then it is easy to assign homework problems, and change plant conditions to illustrate various points about design that would be difficult or impossible to illustrate for an actual experiment since readers are not likely to have the same experiment at their disposal. Of course, you could view our use of truth models as if we were working with an actual experiment; however, it is certainly better to interpret them for what they are, models, not the physical experiment, and remember that upon implementation we can expect that we may uncover problems that require us to iterate on the controller design.

1.4 Complex Hierarchical Control Systems for Automation

In some practical applications, the plant is quite complex and hence, many additional functionalities are needed for automation beyond those used in traditional single or even multiple loop or decentralized feedback control systems.

Experience has shown that it is often convenient to view the controller for complex automation problems as being hierarchical as shown in Figure 1.10. Each block in the figure represents a distinct function performed by the controller and the arrows represent communication links, possibly between functions implemented on computers in physically separated parts of the plant. For illustration, we show three "levels" to the hierarchy, with higher levels generally supervising the operation of the lower levels. The human interface here can be quite complex. It could allow someone to monitor all aspects of operation of the system via a graphical interface, provide the individual with information about the overall health of the system, and work with the individual to specify reasonable and achievable goals for automation. These goals are typically broken down into subgoals, and the responsibility to meet these is then given to the middle level. Then, the actions needed to meet the subgoals can often be broken down and traditional control systems at the lowest level can be used to achieve them, possibly with some coordination between the low level controls provided by the middle level. Throughout the hierarchy there may be a need for a whole variety of functions. For instance, there may be failure detection and identification algorithms, estimators, pattern recognizers, optimization algorithms, etc.

Computer decision-making can be very complex, hierarchical, and distributed, and be designed to solve complex automation problems, even when the "plant" contains intelligent human adversaries.

Some examples of systems whose controllers can be viewed as hierarchical are the following:

- *Robots:* For a robot that is modeled after a human, we have low level algorithms at the robot extremities for force feedback functions in grasping objects, higher level path planning functions for motions of the robot arms,

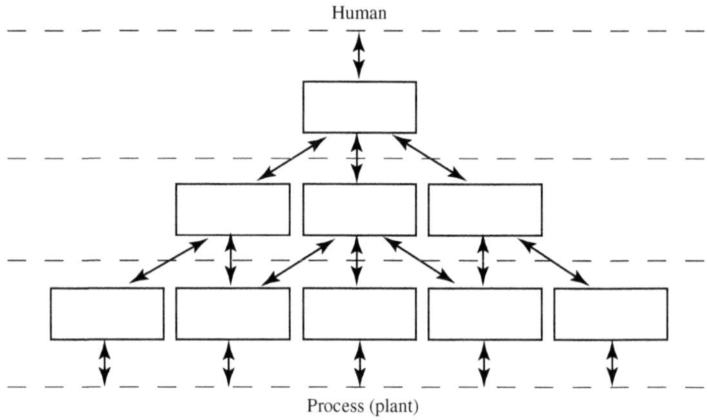

Figure 1.10: Hierarchical control system.

and even higher level mission planning functions for setting overall goals, coping with failures, and minimizing the use of resources.

- *Autonomous vehicles (ground, air, or underwater):* Here, there are the low level control systems for vehicle guidance to keep the vehicle on the path that is specified by a higher level route planner. There are subsystems for failure handling, route optimization, multiple vehicle coordination, and mission planning.

- *Manufacturing and process control:* Here, there are typically physically distributed components of the overall process that require control systems and signal processing. There is often a computer network that allows for the coordination of plant-wide activities to meet overall system objectives. Moreover, there is typically a well-developed user interface for the plant operator.

- *Networks of intelligent agents:* The process that the hierarchical controller interfaces to may not be just some piece of hardware; it could be composed of hardware and some other intelligent (biological) agents. For instance, there may be an adversary or adversaries that is part of the "process" in Figure 1.10. In some cases, you may model them as being a type of disturbance, while in other contexts, it is useful to represent them for what they are: intelligent adversaries whose intent you may want to try to model and counteract. In such formulations, "game theory" has proven to be useful.

Next, we discuss architectures for hierarchical control systems and design methodology.

1.4.1 Functional Architectures

This section serves as a general introduction on how to structure hierarchical intelligent controllers. In several ways it is modeled on our treatment of traditional feedback control systems where we first explained the basics of control systems, and how to draw the functional block diagram for those.

What Functions Are Needed for Control and Automation?

It turns out that in practical applications, it is often the case that components of the overall automated process that are "traditional" feedback control systems (e.g., the ones shown in Figure 1.1 on page 11) often comprise less than one tenth of the overall code needed to implement the system. The remainder of the code is used for a variety of functions, including the following:

- To interface to humans (e.g., for inputs to guide the operation of the system, a graphical user interface, and perhaps a special mode of operation for the maintenance or repairs person).
- To interface to other systems (and hence, there is a need to implement communications interfaces). Some examples include the need to interface to diagnostics equipment for routine maintenance, or interfaces to databases for data logging, process monitoring, and data analysis.
- To control parts of the system by sequencing operations and guiding the overall behavior of the system. For instance, many systems require discrete actions to be taken to ensure proper sequencing of events in the plant (e.g., to sequence the ordering of assembly of a product).
- To handle exceptions to normal operation of the control system (e.g., start-up and shutdown of the process).
- To cope with special operating conditions. For example, it must be able to cope with situations when the control should operate differently in order to meet some user demand. Or, many systems operate in several different "modes" of operation, and there may be the need to switch in different controllers to cope with the special needs of these different modes.
- To provide a "safety net" in safety-critical applications to monitor the process and take action to ensure that disasters can be avoided. When there are system failures or plant changes, this may entail detecting and diagnosing those problems, and subsequently trying to accommodate for the problems, or at least switch to a controller that can maintain some acceptable, although degraded, level of performance until the problem can be fixed.

For instance, in aircraft jet engine control, the control loops needed for regulation of fan speeds (that are roughly proportional to thrust) via fuel flow comprise only a small part of the code needed for the overall jet engine control problem. There is a significant amount of code needed for controlling the special situations of start-up and shutdown, and for switching or tuning controllers for different modes of operation (e.g., for takeoff, climb, cruise, or energy-saving cruise). There are also many rules that are implemented that disallow certain control actions (e.g., use of too much or too little fuel at various operating conditions). Moreover, there is a significant amount of code used to monitor that the engine is within normal operating parameters and if it is not, this code provides failure indications for the maintenance crews or pilot.

The jet engine control problem is often "centralized" in that there is one controller and all the sensed information about the engine is available to this controller (aside from, perhaps, low-level actuator controls). The fact that there is a diversity of components to a real automation system is most clearly demonstrated by studying hierarchical and "decentralized" (sometimes called "distributed") control systems, as we do next. Such systems are distributed in the sense that there may be different controllers implemented in physically separated parts of the plant so that even if one sensed variable is available to one portion of the system, it may not be available to another part of the system, and decisions are made locally, not by one centralized controller that has all the information available about the system.

Why Do Hierarchies and Distribution Arise?

There are a wide variety of hierarchical control systems that are implemented in industrial applications (e.g., in process control and manufacturing systems), robotics, and automated vehicles, to mention a few. Hierarchies naturally arise in complex control systems due to the following issues:

- There is a need in the design of complex systems to split large tasks into smaller, more manageable tasks (the principle of "divide and conquer").
- Goals and priorities often split the performance of tasks into different parts, where each "subtask" may have a different precedence, and performing some tasks may necessitate that others are not completed. This can set up a type of "behavioral hierarchy" where to succeed, the system ranks and orders tasks according to whether they will help achieve the ultimate goal.
- There are often components of the system that are physically distributed, and there is a need for a local controller for high-rate tightly coupled feedback control. Moreover, there is a need to coordinate the actions of these local controllers to achieve overall objectives, and this naturally gives rise to a hierarchy of controls, and the need for communications to the remote controllers.
- Complexity of implementing the automation system often dictates the need to spread the responsibilities for achieving various tasks across more than one computer system and this then can give rise to a hierarchy and distribution of functions.

- Typically, the interface to a human is much different from the traditional control system interface to components of the plant, so there tends to be a natural split of functions. Moreover, since the desires of the human operator are often considered paramount, the interface to the human tends to supervise the traditional control system operation to meet those needs. This creates a natural hierarchy, perhaps analogous to what one often finds in a business management hierarchy with bosses and employees.
- Experts at control can typically "abstract" their reasoning dynamically and reconsider and modify the approach they are taking to control. Hence, even at the "software level," there are natural hierarchies that seem to arise.

In addition to all these reasons, sometimes there are cultural influences that make us think of distributing tasks hierarchically. Generally, however, from an engineering perspective, the formation of hierarchies seems to be a good idea for managing complexity and organizing the controller's development and operation.

1.4.2 Organizing the Controller Functional Architecture

It is often convenient to organize a controller for a complex automation problem into a hierarchical structure, such as the one shown in Figure 1.11. This is a "functional architecture" for the controller since a different function is performed in each box, and the arrows indicate the directions of communications between the various subsystems. Even though it could be, it is not necessarily a "software architecture" that indicates how the software for the controller will be separated, or a "hardware architecture" that indicates how the computer hardware that implements the system will be grouped and interconnected. What we specifically mean by functions and communications should be clear from the examples that we will give below.

At the top of the functional architecture in Figure 1.11 is the interface to other systems and the user (e.g., human operator). At the bottom is the interface to the plant to be controlled. In between, we split the functionalities into three "levels" (sometimes called "layers") simply for the purpose of illustration. In some applications you may need to split the functions into a different number of levels. The highest level is often called the "management level" (or sometimes the "organization level") since it interfaces to humans and other systems, and dictates the overall operation of the lower levels of the controller (think of the analogy with business management). The next level down is called the "coordination level," and it is used to interface to the management level and coordinate the actions of the lower level. The "execution level" contains the interfaces to the plant, and low level control algorithms (e.g., traditional feedback control systems such as the ones we discussed in the last section). Moreover, it may also contain other signal processing and system monitoring functions.

The boxes at different levels may have different types of processing, of varying complexities. Note that at each level, we show a different configuration of

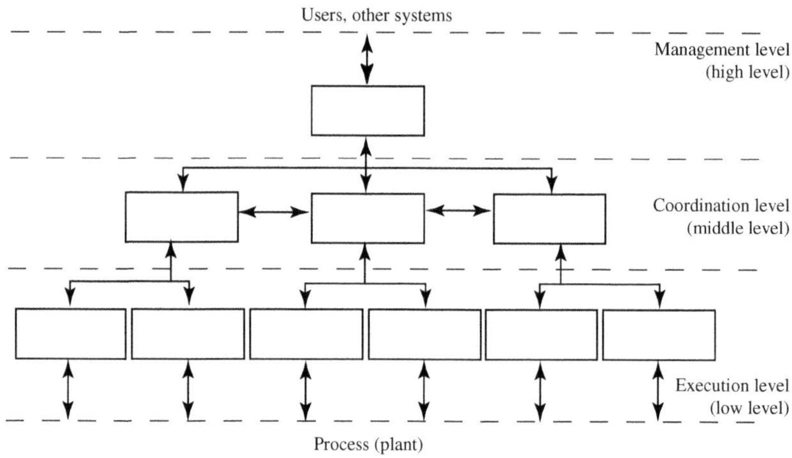

Figure 1.11: A typical hierarchical distributed control system.

blocks and communications. At the management level we have all the functions in one centralized system that interfaces to the user, and to the coordination level. At the coordination level we have three blocks that communicate with each other, and the management and execution levels. At the execution level we have multiple subsystems that are supervised by the coordination level. Ultimately, the method for structuring the functional hierarchy arises from the particular application domain, and constraints dictated by software and hardware for implementation of the controller.

1.4.3 Fundamental Operational Characteristics

There are several fundamental characteristics that have been identified for complex hierarchical distributed control systems. It is important to recognize these as they help us understand the dynamical operation of these complex systems, and sometimes offer ideas on how to design them. Some of the basic features are the following:

- There is generally a successive delegation of duties from the higher to lower levels, and the number of distinct tasks typically increases as we go down the hierarchy.

- Higher levels are often concerned with slower aspects of the system's behavior and with its larger portions, or broader aspects.

- There is then a smaller contextual horizon at lower levels—that is, the control decisions are made by considering less information.

- Higher levels are typically concerned with longer time horizons than lower levels.

- At the higher levels there is typically a decrease in time-scale density, a decrease in bandwidth or system rate, and a decrease in the decision (control action) rate.

- There is typically a decrease in the granularity of models used—or, equivalently, an increase in model abstractness at the higher levels. Such models are used in system development, or in decision-making and estimation.

Next, we give two examples to help clarify how hierarchies arise, how to create a functional architecture, and how the above fundamental characteristics manifest themselves in actual applications. We will in particular expand the temperature control problem discussed earlier (Figure 1.1(a)) by showing how temperature control problems "scale up" for a large complex building. Moreover, we will discuss the area of "intelligent transportation systems," a much more significant level of automation for automobiles than the cruise control discussed earlier via Figure 1.1(b). This is a system that is currently under study, and one that will incrementally be implemented over many years. We will, however, show that a functional architecture can provide a way to show how a fully operational system could be constructed.

Before discussing the examples, however, we would like to alert the reader to an interesting and important fact: even though building temperature control and intelligent transportation systems are quite different, there are many similarities in how we structure the controllers for these problems, and how we think about their operation. Try to understand the fundamental relationships between the two applications as you read about them. Also in the following two sections it will become clear that at the conceptual level, there are many similarities between the design objectives and design methodologies for these systems.

1.4.4 Example: Building Temperature Control

As an example, note that temperature controls can be much more complex in practice than the one described in Figure 1.1(a) since the complexity of the plant will tend to dictate the need for more sophisticated controls. For instance, consider the need to implement temperature controls for a large complex building. For this, there may be a need to control both the heating and air conditioning, in addition to humidity. Clearly, the effects of surrounding rooms and the outside temperatures can affect different parts of the building in complex ways. Moreover, there may be rooms where very tight temperature regulation is required (e.g., a computer room) and we may be able to incorporate certain schedules of use of the building or ambient weather conditions into how the controls should operate. Good controls are critical in such large applications, as significant savings can be realized if the system is properly designed (especially if the same temperature control technology is used in many different buildings).

A candidate control system for a large building is shown in Figure 1.12. We have a "local" temperature control system for each room, similar to the one described in Figure 1.1(a), and these make up the execution level. Next, we have temperature controllers for "zones," which are comprised of sets of rooms (e.g., adjacent rooms). These zone controllers, at the coordination level, are used to localize temperature control decisions to regions of the building (e.g., if an event is held in one part, or if there is a need for tighter regulation of temperature elsewhere). At the management level, the master temperature controller coordinates the control of temperature in the entire building. This component serves as a supervisor to the lower level zone controllers by, for example, fixing temperature set points. It accepts weather and event scheduling information and massages it for use in the control system. For example, if a large event is to be held in one part of the building when the ambient temperature is hot, and many people will attend, it changes set points to cool the building ahead of time so that the temperature is more comfortable during the event. Or, if it knows that doors between several rooms will be open, one of these rooms opens to the outside, and it is winter, it may choose to have the room with the door to the outside at a higher temperature to ensure that the other rooms are isolated from temperature fluctuations due to periodic opening of the door. Clearly, the possibility of a diverse set of events, each requiring different rooms, could demand that we dynamically change what is considered a "zone" so the hierarchy itself could change over time (or at least the one we consider here could be considered a special case of one that included all the possibilities).

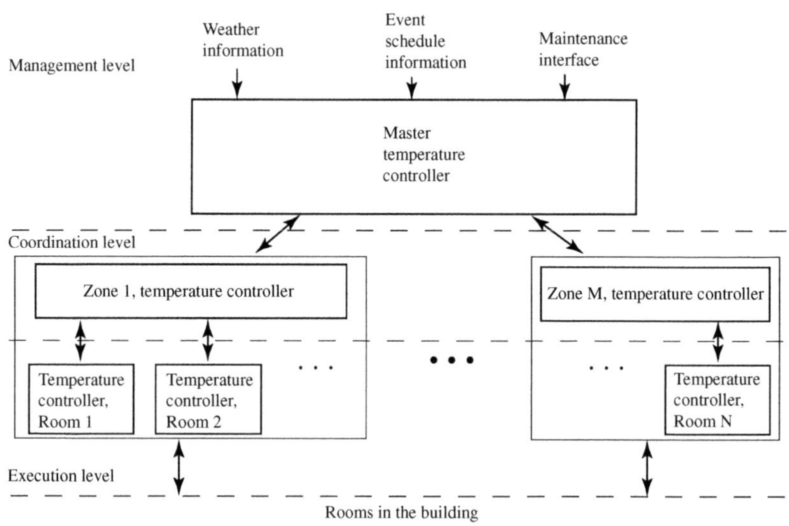

Figure 1.12: Temperature control system for a building.

Notice that the master temperature controller can also provide a "maintenance interface" for diagnosing problems with the building temperature control system. Such a task is included in the control functions since the temperature sensors, and behavior of the various controllers, gives an indication of possible problems with the mechanical parts of the system. (For example, if a temperature controller repeatedly turns on the heating in a room, and there are no unusual circumstances, the temperature in the room should increase and if it does not, there must be some failure in a certain part of the system.)

In summary, it is important to recognize that feedback control typically lies at the center of the flow of information in complex automation systems. Feedback control can provide additional information about process operation since it tries to make changes to the system and it observes the responses of the system to these changes, and this information can be useful for many other functions (e.g., diagnosing the health of the automated process).

While a controller is central to the automation of temperature control in a building, the algorithms needed in the "high level" master temperature controller are quite different from the "low level" (numeric) control algorithms for single-room temperature control. The decision-making systems needed for the master and perhaps zone controllers are typically discrete in nature and must be able to incorporate abstract (non-numeric) information and objectives. In Part II we provide two approaches to design and implement such components: the rule-based system and planning system. Ultimately, however, the problem is a general one of how to program computers to achieve automation; the methods in this book are only intended to show a few methodologies to develop such decision-making systems for automation. These should then serve to clarify how other computer algorithms (in the latest most popular programming language, or the one that the boss wants you to use) could solve the problems equally well.

1.4.5 Example: Intelligent Transportation Systems

Next, let us examine an intelligent transportation system problem of automating a highway system (contrast this with traditional feedback control for speed regulation in Figure 1.1(b)). One possible general functional architecture for automated highway systems is shown in Figure 1.13. Here, suppose that we have many vehicles operating on a large roadway system in the metropolitan area of a large city. Each vehicle is equipped with a (1) *vehicle control system* that can control the brakes, throttle, and steering to automate the driving task (for normal operation or collision avoidance). In addition, suppose that there is a (2) *vehicle information system* in each vehicle that provides information to the driver (e.g., platoon lead vehicle information; vehicle health status; information on traffic congestion, road construction, accidents, weather, road conditions, lodging, and food; etc.) and information to the overall system about the vehicle (e.g., if the vehicle has had an accident or if the vehicle's brakes have failed). For the roadway there are (3) the *traffic signal controllers* (e.g., for intersections and ramp metering) and (4) the *roadway information systems* that provide information to the driver and other subsystems (e.g., automatic signing systems that provide rerouting information in case of congestion, road condition warning systems, accident information, etc.). These four components form the "execution level" in the hierarchical controller, and clearly these components will be physically distributed across many vehicles, roadways, and areas of the metropolitan area.

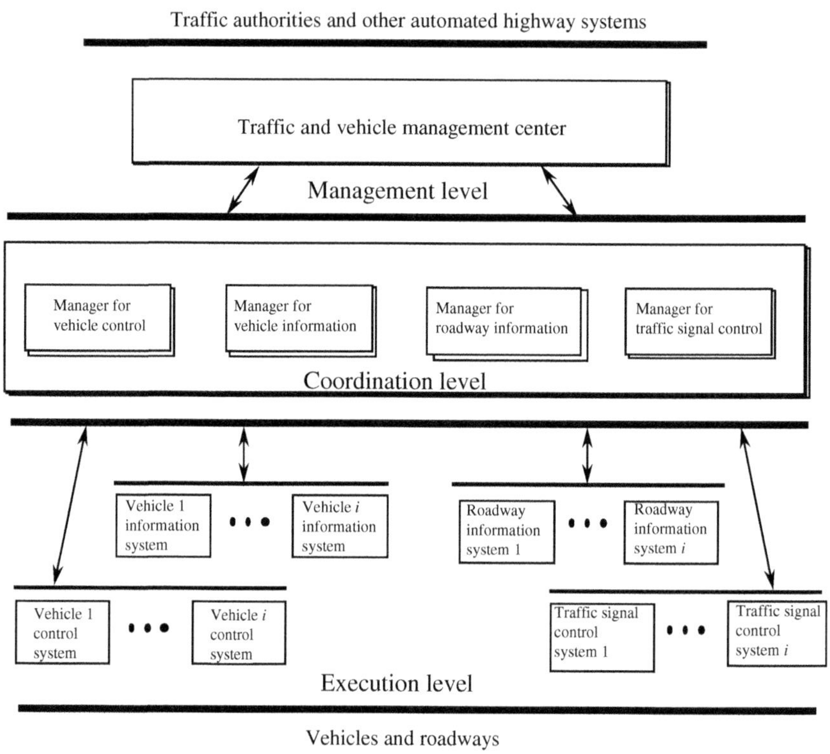

Figure 1.13: Hierarchical distributed controller for an intelligent transportation system (figure taken from [406], © IEEE, used with permission).

In the coordination level shown in Figure 1.13, there is a manager for vehicle control that (1) may coordinate the control of vehicles that are in close proximity to form "platoons," maneuver platoons, and avoid collisions; and (2) provide information about such control activities to the rest of the system. In addition, there is a manager for vehicle information that (1) makes sure that appropriate vehicles get the correct information about road, travel, and traffic conditions; and (2) manages and distributes the information that comes in from vehicles on accidents and vehicle failures (e.g., so that the control manager can navigate platoons to avoid collisions). The manager for traffic signal control could (1) utilize information from the roadway information system (e.g., on accidents or congestion) to adaptively change the traffic light sequences at several connected intersections to reduce congestion; and (2) provide information to the other subsystems about signal control changes (e.g., to the vehicle information systems). The manager for roadway information (1) provides information on road conditions, accidents, and congestion to the other subsystems; and (2) provides information from the other subsystems to the roadway for changeable message signs (e.g., rerouting information from the traffic signal control manager). As indicated in Figure 1.13, there are multiple copies of each of the managers and the entire coordination level as needed for different areas in the metropolitan region.

There is a management level which provides for high-level management of traffic flow. It provides an interface to other automated highway systems (perhaps in rural areas or other nearby metropolitan areas) and to traffic authorities (e.g., to provide information to police and emergency services on accidents and to input information on construction, weather predictions, and other major events that affect traffic flow). It can interact with traffic authorities to advise them on the best way to avoid congestion given current weather conditions, construction, and expected traffic loads. It can monitor the performance of all the lower-level subsystems in the coordination and execution levels, and suggest corrective actions if there are problems.

Notice that in terms of the fundamental characteristics, we find a successive delegation of duties as we go down the hierarchy of the controller in Figure 1.13. For example, high-level tasks at the management level may involve reconfiguring traffic signaling due to construction and weather. The coordination-level manager for roadway information and traffic signal control may develop a new signaling strategy. This strategy would be implemented in the execution level on the changeable message signs (to inform drivers) and the traffic signal control strategy. The higher levels of the hierarchy are often concerned with slower and broader aspects of the system behavior (of course, in an accident situation, the traffic and vehicle management center would react as quickly as possible to alert emergency vehicles). The lower levels of the system have a smaller "contextual horizon" since they consider much less information in making decisions. Also, the decision rate tends to be higher at the lower levels (e.g., the rate at which control corrections are made as a vehicle automatically steers around a curve may be on the order of milliseconds, while the decision rate at the management level may be on the order of minutes or hours).

Clearly, there is the need for a significant amount of interdisciplinary activity to implement such a complex control system that involves a wide range of technologies and falls beyond the traditional scope of control problems. There is probably no single control technique that can be used to solve the diversity of problems found in a complex automated highway system problem.

1.5 Design Objectives for Automation

This section and the next mirror those for traditional feedback control systems where we defined design objectives for traditional control systems (e.g., stability, rise-time, overshoot, disturbance rejection) and then defined the control system design methodology. Here, we focus on design objectives for complex hierarchical control systems, and in the next section we discuss design methodologies for complex automation systems.

The quality of the response of a single loop feedback control system is relatively easy to specify in terms of measures defined earlier. However, sometimes the performance objectives are not so easily quantified for complex control systems, such as the building temperature control system in Figure 1.12, or the intelligent transportation system in Figure 1.13. For such complex control systems, a multifaceted quantification of performance objectives is needed.

Design objectives for general automation systems typically include a mixture of requirements on dynamics, leading to a quantification of "autonomy."

The performance objectives for the higher levels of the control system may include the following:

- *Dynamically changing composition of the specifications of the traditional control systems:* One part of the specifications may indicate, using measures like those discussed above, how each low level control system should behave. In some applications, however, this is not a simple conjunction of the objectives of each control system, but may dictate that if one set of low level control systems is perfectly meeting its objectives, another set may not have to. Generally, higher level objectives may modify, even dynamically over time, low level control objectives.

- *Proper sequencing of events:* Typically, correct operation of the higher level depends on performing tasks in a specific order, and if it is not performed in that order, performance of the overall system can degrade.

- *Rate of operation:* Often, we want to guarantee that decision making and other activities at the higher level will occur at a rate that is fast enough so that it will not keep the lower levels waiting. If it does delay the lower levels, then it is possible that there will be many different negative consequences (e.g., low level controllers that do not meet their local control goals that are stated in terms of time responses).

- *Proper parsing and use of information:* The high level must properly use the information that it gathers, either from the controlled system or from the external interfaces, to ensure the highest possible performance levels. It must parse the information and decide what information can be ignored (a problem of what to pay attention to), and it must apply the proper information at the proper points in the system, at the proper times.

- *Orderly and efficient operation:* The higher level system must conduct its activities in an orderly and efficient manner. In actual systems, there is often more than one way to perform a task, and generally the operation should be more similar to that of an excellent symphony orchestra, rather than a chaotic flurry of hastily conducted activities (yes, there is an art to creating a complex control system, but this art must be conducted in the context of the existing science of control, and it is constrained by available technology).

- *The ultimate goal—autonomy:* For many systems our ultimate goal is to achieve as much as possible with our automation system. Sometimes this means automating as many tasks as possible so that the human is fully relieved of the task (e.g., in automated highway systems, the full automation of the driving task). Other times the goal is simply to get the system to perform its functions without any outside intervention by another system or human, under conditions as broad as possible, and as for as long as possible. Typically, for different systems there are different "degrees of autonomy" that have been achieved.

While these are high level performance objectives, the general engineering design goals for control systems of Section 1.2.4 apply here also.

1.6 Software Engineering for Complex Control Systems

While the controller design procedure outlined for single-loop control systems was largely developed for traditional low level feedback control systems, it is stated in a general enough manner to be useful for relatively complex control systems. In large computer automation projects, however, project management and design methodology issues must be given special attention since the process involves the construction of a large software program by teams of engineers and scientists. In this section we briefly outline some basic ideas from software engineering, and relate these to the enterprise of control engineering for complex automation systems.

1.6.1 Software Engineering Methods

Probably the most popular software engineering methodologies are the "waterfall" and "spiral" techniques (others include the "build and fix method," the "rapid prototype method," and the "incremental method"). Software engineering methodologies are called "software processes" and hence the waterfall and spiral methods are sometimes called "process models;" we will not use this terminology since it conflicts with existing control systems terminology.

Waterfall Method

The waterfall method has been used for a variety of software design projects, and some project managers even require certain documentation and reports at the end of each stage. A version of the waterfall method is depicted in Figure 1.14. Apparently, it is called the "waterfall" method simply based on the shape of the functional block diagram where the arrows flow down and to the right (note that in early versions of the diagram there were no dashed arrows). The methodology starts with the software design team coming to an understanding of the customer's requirements for the software. Next, these are converted into specifications for the software and the software project phases are planned. Following this, you perform software design. Then, portions of the code are implemented and tested. The various portions of the code are integrated and then further tests are done. Then, we reach the useful life of the software where it is put into operation.

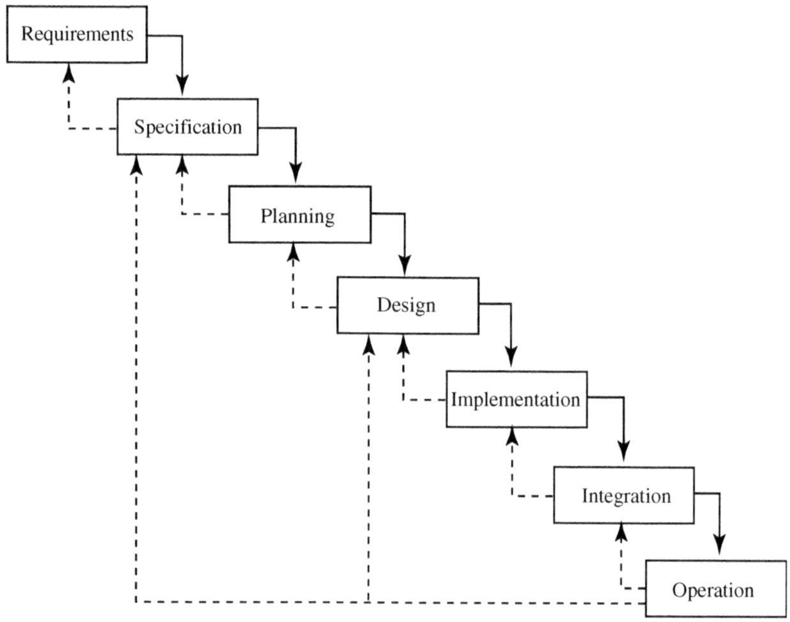

Figure 1.14: Waterfall software engineering methodology.

Notice that we have also added dashed arrows to the diagram. These represent iterations that can occur. For instance, there is often an iteration between two adjacent stages (e.g., we iterate between design and implementation to improve on the design as problems are encountered in implementation). Moreover, in the operations mode we learn more about whether the software is working properly, and we may get new requests for changes to the requirements, and hence, specifications from the customer. In this case, we repeat several parts of the process. Of course, we will also return to redesign the software if normal operations uncover a problem not previously found.

Clearly, there are relationships between this software engineering procedure and that of the design of traditional feedback control systems. Indeed, the waterfall method may be perfectly appropriate for the development of traditional single-loop control systems (not surprising, since from one perspective control engineering *is* software engineering, at least for digital controllers). More complex automation systems that involve, for instance, hierarchical distributed control for industrial processes, however, demand a more sophisticated method.

Spiral Method

The waterfall method is sometimes criticized for being too simple to represent the methodology that is needed for the design of complex software systems. This, and the rise of other methodologies like rapid prototyping, gave rise to the "spiral model," which we will call the spiral method (again, named by the form of the diagram used to summarize the methodology). In many ways, other methods, such as the waterfall or rapid prototyping methods, are a subset of the spiral method.

The spiral method is depicted in Figure 1.15. First, note that the figure is broken into four quadrants, labeled I, II, III, and IV. As you move out radially, costs of software development (time and money) increase. The spiral provides the path along which software development proceeds, and we will explain the methodology by following it, starting at the black dot on the axis between the second and third quadrants.

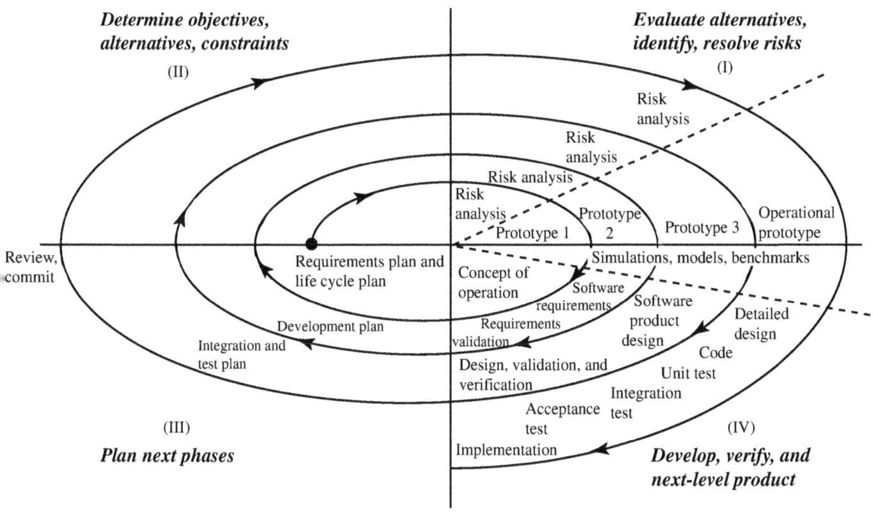

Figure 1.15: Spiral software engineering methodology ([72], © IEEE, used with permission).

We start in the quadrant II to determine objectives, alternatives, and constraints each time we pass through that quadrant. When we start the project, we work with the customer to establish project objectives. In the process of setting up the objectives we may discuss several alternate solutions (and this early in the process, we may not commit to a single alternative), and all the constraints must be identified (e.g., the customer may want the project completed according to a certain schedule, the software designers may only be able to commit certain resources to the project, or certain real-time constraints may need to be met).

Next, moving into the quadrant I, we evaluate the alternatives, and identify and resolve risks (evaluation of constraints and risks is central to the spiral method and so some call it a "risk-based" method). Risks can come in many different forms in each phase of the project. Early in the process it may be that the software development team feels that one alternative is "risky" since it appears that it may require many resources of the company (e.g., employee time, and possibly the purchase of additional support software). Risk may also quantify concerns about whether members of the team have the expertise to complete the project according to one of the alternative methods, or if a certain software architecture, the given hardware, or computer language will measure up to the task. If the risks for development of the project are seen as too large, the project may be terminated at any point. Often, however, evaluation of risks leads to the development of a prototype, which is roughly thought of as a simple version of the software to be developed that has only a subset or simplified version of the full desired functionality (it is a rapid prototype). This prototype often helps to clarify the project objectives, uncover more constraints, and identify other risks.

Moving into the quadrant IV, we evaluate the prototype using simulations, models, and benchmarks. We try to assess the performance of the prototype to clarify what will be needed later in development, and to solidify the overall concept of how the system will operate. Next, moving into quadrant III, we specify the requirements plan and life-cycle plan (i.e., what will be required at different phases of the project, and what the overall plan for use of the software, including maintenance, will be). Before passing into quadrant II for the "second phase" of development (i.e., another loop around the spiral) we review the project, and decide whether or not to commit resources to the development project.

Starting the second phase, we determine objectives, alternatives, and constraints, but this time with the benefit of the use of the results of the last phase, and in particular our previous analysis of constraints, risk, and the evaluation of the prototype. We may refine the overall project objectives, sometimes by working closely with the customer. We may decide that certain alternatives are not viable, and we may develop more alternatives to solving new problems that were uncovered in the last phase. We reevaluate our constraints, to make sure that they are being met, and to create any new ones that are needed. We move to quadrant II and perform risk analysis and develop a more detailed prototype so that we move in the direction of trying to meet the overall project objectives. We then evaluate the prototype, and try to establish software requirements and validate that the requirements that we set are appropriate. Next, we solidify a development plan since the overall objectives and constraints of the project should be clear at this point.

We then start the third phase after a review. We again clarify objectives, alternatives, and constraints, assess risks and build another more detailed prototype. We evaluate this more sophisticated prototype and establish the product design, and verify and validate this design. Next, we produce a plan of how we will integrate the various software components (often different parts of the software are developed separately, possibly by different team members, or simply to divide and conquer large tasks), and how we will test the overall system. After a review and consideration of near end-of-project objectives, alternatives, and constraints, we perform a risk analysis and develop an operational prototype. We evaluate the prototype using simulations, models, and benchmarks to try to uncover any problems before we move toward implementation. In the final steps we get into the details of low level design and coding. Next, we test the various units (components) of the software, and integrate the software components and test them. Next, we conduct tests that help us to decide if we should accept the software as complete (this may involve the customer), and then we implement the code. Following this, there will be a need for maintenance and further improvements based on what we learn while the code is in operation.

1.6.2 Software Vs. Control Engineering Methodologies

It should be clear that there are many relations between the software engineering process, viewed either as the waterfall or spiral methods, and the control design process. The two are not, however, entirely the same. Notice that the software engineering process *sometimes* ignores the use and differences between truth and design models, dynamics, disturbances, noise, and even robustness. That these issues are sometimes ignored is not surprising. The spiral method was developed for use in general software development and these issues could be considered to be details specific to a particular type of software development.

It is interesting to notice that the control engineering process *sometimes* ignores practical issues in how to manage a large software engineering project, broad issues such as use of financial and employee resources, certain issues in how to structure the software so that it is easier to construct, maintain, and extend, and issues related to integrating different pieces of software that were developed separately and how these must be tested (i.e., integration and test). Generally, we must recognize that for complex automation applications, software development issues become very important and are not just an "implementation detail."

1.6.3 Complex Control System Design Methodology

Clearly, aspects of the conventional control design methodology and the software engineering methodology are needed for the development of complex control systems for modern automation systems. The precise methodology to follow is, however, ultimately tied to the application that you are focusing on. Below, we provide a sample methodology for one class of problems, those that need a hierarchy of distributed controls, such as the building temperature regulation problem in Figure 1.12. This is provided simply to show one possible methodology. Do not attach too much significance to this particular methodology. What is important is that you understand the key elements of the control and software engineering methodologies, the importance of having and following an agreed-upon methodology, and the fact that the various elements can be rearranged to provide a design methodology for virtually any automation project that you encounter.

Control and software engineering methods can be integrated for automation system development.

There are several issues that need special attention in the development of hierarchical distributed control systems, and some of these are the following:

1. The need to establish, early in the design process, a hierarchy for the control system operation: The key factors influencing how this hierarchy is structured are the physical layout of the plant, the functionalities of various parts of the plant and how they are inter-related, the available computing resources (hardware and software) and how they are distributed, the available communication links, and the subsystem and overall performance objectives.

2. The need to develop low level controls first: Sometimes, you would start by using the standard control engineering design procedure to develop "local" temperature controls for individual rooms. Other times, such development may come later.

3. The need for communications: The issue of what information is available and how it is transferred between components of the system and to the human user must be confronted.

4. The need for coordination: Given information from a wide variety of sources, and the wide variety of control objectives that exist, both for local controllers and the overall automation objectives, you must establish a method to coordinate the operation of the low level parts of the system to achieve overall goals.

5. The need for an (possibly complex) interface to the user: This will enable the user to specify broad system objectives, obtain health information about the system, and monitor the operation of the system.

How do we incorporate these features into the overall design process? We could modify the flowchart for the design process for traditional feedback control systems shown in Figure 1.8, modify the spiral diagram, or create an entirely new diagram to represent the methodology. Here, we will simply modify the spiral method to accommodate the aspects of control engineering for complex systems (in Exercise 1.6, you are requested to provide a flowchart that incorporates the spiral method and control design procedure steps). We will call the method the "spiral method for control engineering." It is pictured in Figure 1.16.

Here, in the first step where we determine objectives, alternatives, and constraints, we also define a vision of how the control functional hierarchy would ultimately be implemented. Note that the wedge that is outlined with a dotted line is significantly different from the corresponding portion of the spiral method in Figure 1.15. Basically, after risk analysis, we conduct low level control designs according to the design steps outlined in Figure 1.8 (hence, the conventional control design steps are incorporated into the spiral method to

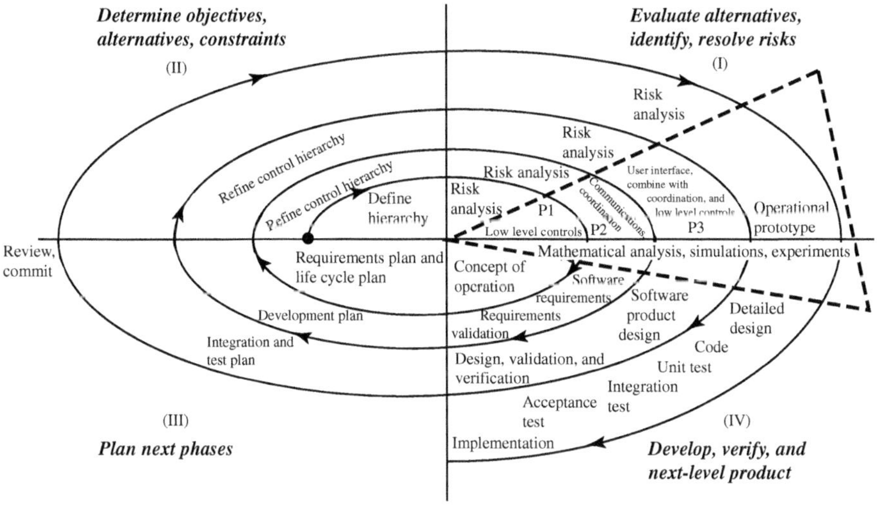

Figure 1.16: Spiral method for control engineering (edited version of [72], © IEEE, used with permission).

develop prototype 1, denoted in the figure by "P1"). Depending on the application, we may do this for every low level control system. Note that we indicate in quadrant IV the use of mathematical analysis, simulations, and experimentation to evaluate (and perhaps iterate) on the design, and indeed, we consider this to be an integral part of the corresponding part in quadrant I where the prototypes are developed (i.e., there can be iteration between quadrants I and IV before moving further around the spiral).

Next, combining our view of how the low level controls perform, and the control functional hierarchy, we develop a concept of how the system is to operate. We develop the requirements and life cycle plan and review the project to decide whether to commit to another design phase. Next, we reevaluate objectives, alternatives, and constraints and based on what we learned in the first phase, we refine the control hierarchy, especially with a view towards incorporating communications and achieving coordination, which are implemented next, after a risk analysis. The coordination level is evaluated both independently, and in conjunction with the operation of the execution level. Then, after establishing the overall software requirements, and validating the requirements, we determine a development plan for the remainder of the project.

After review of the project we make further refinements to the hierarchy, this time focusing on incorporation of the "third level," which we will refer to as the management level. After a risk analysis, we develop the user interface and combine it with the coordination and execution levels. We evaluate the overall design using mathematical analysis, simulations, and experimentation. The remainder of the process is similar to the standard spiral method.

To summarize, in Figure 1.16 we add the need to define and refine the control hierarchy. We use prototypes ("P1," "P2," and "P3") of increasing complexity and propose that the first prototype only include the low level controls, the second one include the communications and coordination level, and the third one include the user interface and the entire management level. The portion of the diagram contained in the wedge outlined with dotted lines incorporates the control design steps in the first phase of the design. In the second and third phases, it also incorporates those design steps, but now it may require different analysis methods (e.g., ones for "discrete event systems"(DES) since the higher levels may have such components).

Overall, we see that this provides a way to embed the control engineering methodology into the software engineering methodology so that all the key issues are addressed. Clearly, there are many other ways to structure the diagram that represents the automation software development.

1.7 Implementing Complex Control Systems

There are many application-specific challenges you will encounter in trying to implement a control system (e.g., interfacing issues with the plant and operator), and these can be particularly numerous and difficult for hierarchical distributed control systems such as the ones discussed in this part.

Some of the challenges that are typically encountered include:

- The various components of the control system may be implemented on different computer platforms (e.g., a workstation or personal computer) that use different operating systems.
- There is often the need to establish communications between the different platforms that are efficient and flexible.
- There are system timing issues where different portions of the overall system may be operating at different sampling times and in synchronous or asynchronous modes.
- It must be easy to establish hierarchies in the controller.
- It must be easy to develop the complex control systems so that coding and debugging can be implemented by *teams* of programmers (e.g., you may need special monitoring, design, and code generation tools).
- Sometimes, for large projects, it is useful to use software engineering methodologies (e.g., the spiral method).

Methodology and tools are available for decentralized networked implementations of control systems.

While these are some of the generic problems encountered, there are certainly application-specific ones also.

The National Institute of Standards and Technology in the U.S. developed a software package, called the real-time control system (RCS) library, that can be quite useful in the development and implementation of hierarchical and distributed control systems. While we only briefly overview it here, and recommend that interested readers see the "For Further Study" section for a reference with all the details, it is interesting to note that it is a free software package that can be used to solve many of the problems listed above, and is well-developed enough to use in practical industrial control and automation problems.

The RCS software tool allows for controller implementation on a wide variety of platforms, and it uses a "neutral message language" to ensure consistent and flexible cross-platform communications between "modules" that implement the various components of the system. Timing issues are easy to deal with via the given timer functions. Hierarchies are simple to establish, especially with the "design tool," which is a graphical environment for creating hierarchical controllers. The controller operation can be monitored and directed by a human operator at all its levels (and modules) via a graphical "diagnostics tool." Automatic code generation is provided, and tools are available to facilitate team-based development of a complex controller.

1.8 Hybrid System Theory and Analysis

A "hybrid system" is one that has components that are easiest to model with a mix of different models. Recently, this has often come to mean a mix of conventional differential equation models with discrete event system (DES) models. For instance, a hybrid model would be one that is composed of an ordinary differential equation and an automata model. There are many processes that are actually hybrid in nature and, hence, dictate the need for hybrid models. For example, the automated highway system has low-level steering, braking, and throttle controls (all typically using differential equation models in design), and high-level controls for platoon maneuvering, check-in, and check-out procedures, and so on (which often require automata-type models). Moreover, there are many times when it may be desirable to implement a hybrid controller for a system that may be easily described by conventional differential equations. In this case, the closed-loop system becomes hybrid, due to the presence of a hybrid controller. We see that a system could be hybrid, due to the plant, the controller, or both.

Hybrid models arise in biomimicry from multiscale (spatial and temporal) models in biology.

A hybrid control system is shown in Figure 1.17. Notice that there is a continuous time system and DES component to the plant, with an "interface." This interface has a function ϕ, which maps conditions in the continuous time system to events in the DES. It also has a function ψ that maps conditions in the DES to changes to the continuous time system. The controller has two components. There is C_d, the controller for the discrete event component of the plant, and C_s, the controller for the continuous time component of the plant. (This is only one way to set up the controller structure.) The controller has a similar interface to the plant model and a similarly structured model could be used for the closed-loop system.

Modeling and analysis of hybrid systems could offer valuable verification approaches for automation systems.

As for the other applications in this book, there is often the need to conduct stability analysis of hybrid systems. It is an important research direction to develop modeling, analysis, and design methodologies that will be guaranteed

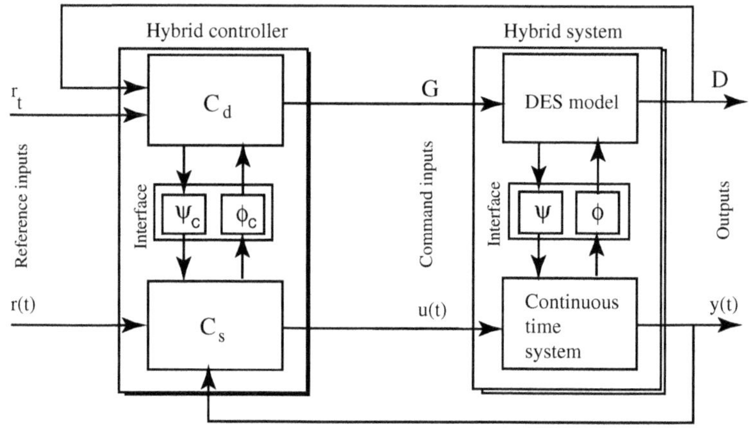

Figure 1.17: A hybrid controller for a hybrid plant.

to produce stable and high performance hybrid systems. If this problem can be solved, then we will also know much more about how to conduct stability analysis for complex hierarchical intelligent control systems. In the mean time, it is interesting to point out that the techniques studied in this book provide methods to heuristically construct controllers for hybrid systems and hence, intelligent control is generally a viable approach for hybrid systems design (e.g., some of the methods outlined above for hybrid hierarchical intelligent control system design could apply).

1.9 Exercises

Exercise 1.1 (Control System Components and Performance Objectives):

(a) Define a functional block diagram for a control system for a system that will automatically steer an automobile along a road. Define your sensors, actuators, and plant. Define the reference input and controller.

(b) For (a), define what could be meant by "this control system performs well." Include the following in your quantification of "good performance": tracking objectives, how the system should respond to adverse conditions (that you will define), and what the time responses should look like.

Exercise 1.2 (Modeling Issues):

(a) Explain the difference between a truth model and a design model.

(b) Is a truth or design model ever a perfect representation of a physical system? Why?

(c) Explain the process needed to evaluate the accuracy of a model of a physical system.

(d) Explain why (under what conditions) it is necessary to modify the model that is used in the design of a control system.

Exercise 1.3 (Analysis Issues):

(a) Why are the conclusions that are reached via mathematical analysis of system properties not necessarily valid for the physical control system?

(b) Why are the conclusions that are reached via simulation-based analysis of system properties not necessarily valid for the physical control system?

(c) Why are the conclusions that are reached via experimental analysis of system properties not necessarily valid for the physical control system? Hint: Take into consideration robustness issues.

Considering (a)-(c), we would conclude that there is no way to be absolutely certain that the control system that you design will perform properly when it is physically implemented. The analysis methods simply increase our confidence that it will perform properly. At some point, however, we can become certain enough that we can achieve successful implementation of control systems. Indeed, there are many successful control systems in operation today!

Exercise 1.4 (Heuristic Design Methodology): In this chapter, specifically in Figure 1.9, we defined a heuristic design method for control systems where no models were used. Draw and explain a flowchart for a heuristic design methodology that uses a truth model, but not a design model. Develop your flowchart by modifying the earlier ones (e.g., Figure 1.8).

Exercise 1.5 (Hierarchical Intelligent Controller Functional Architectures): In this problem, design a functional hierarchy for an intelligent autonomous controller for two different applications.

(a) Draw the block diagram for the functional architecture for a multilayer hierarchical controller for solving a robot control problem (one where planning, learning, and low-level control is used). Provide a description of each block in the functional architecture and clearly explain the relationships between the various components of the controller. Also, clearly explain how commands and information flow in the controller, and between the controller and its environment (including the plant, other systems, and human operators). To do this, you will need to specify the type of robot you are considering and the types of tasks and performance it is to achieve.

(b) Repeat (a) but for an autonomous land or underwater vehicle problem (choose one).

Exercise 1.6 (Design Methodology for Complex Control Systems): In this chapter, we defined a spiral method for control engineering to define the methodology needed for designing complex control systems. Define, using a flowchart diagram, a methodology for complex control systems design. Be sure to include in your flowchart all the steps for conventional control system design, and the essential steps used in the spiral method for software engineering.

Exercise 1.7 (Alternative Architectures for Hierarchical Intelligent Control)*: The meaning of the "$\star$" is explained in the preface. Over the years there have been many different architectures proposed for complex intelligent control systems. Read about the ones in [21, 234, 520, 536, 20, 10, 489] and summarize the approaches in [520], [10], the one by Meystel in [21], and [20]. What are the common features of the approaches? What are the differences? Does each seem general enough to be suitable for most applications, or do some of them (which ones?) have certain limitations that will limit their applicability?

Exercise 1.8 (Design of Functional Architectures)*: In this problem, design your own general functional hierarchy for an intelligent autonomous controller and apply it to one problem.

(a) Complete Exercise 1.7 and then, using ideas from the common functional architectures, create your own.

(b) Use the architecture that you design in (a) and apply it to the specification of an architecture for a robotic, manufacturing, or vehicular system (choose one).

Exercise 1.9 (Mathematical Definition of Autonomy)*: There are several reasons why a mathematical definition of autonomy can be useful. First, since ultimately autonomy can be viewed as a type of performance specification, a precise definition can help guide the specification of performance requirements. Second, it can be useful to researchers trying to compare the merits of different control methodologies, architecture designs, and plant designs. Third, it may suggest methodologies for design and analysis.

(a) Based on your reading of this book, and particularly this chapter, provide a mathematical definition of autonomy. Include in this definition the ideas that there are "degrees of autonomy" that depend on the size of the regions of operation of the system, the robustness of the system in that region, and the amount of communications/interactions needed between an "autonomous subsystem" and other systems (and perhaps a human operator).

(b) Use the method to compare the degrees of autonomy in the hierarchical intelligent controllers for the robot and autonomous ground vehicle problems in Exercise 1.5.

Chapter 2

Scientific Foundations for Biomimicry

Chapter Contents

"Intelligent control" is the study of how to achieve control automation via the emulation of biological intelligent systems. Intelligent control comprises a set of methods to exploit biomimicry for automation. The "biologically inspired" methods can be classified according to natural cell, tissue, organ, organism, and population hierarchies in biology. For instance, we may borrow ideas from cells or organs (e.g., neurons and human brain functions), organisms (e.g., humans who are employed to solve a control problem at a chemical process control plant), or from the coordinated group behavior of organisms (e.g., modeling autonomous vehicle controls after those in microorganisms that dynamically seek nourishment and avoid harmful environments). Moreover, there are evolutionary adaptations that occur in all biological systems and these can be useful in improving control system performance via the long-term interaction with its environment. With such biological foundations and motivations, in intelligent control we will sometimes say that we are using "biomimicry" to construct our controller or automation system.

Before discussing biologically inspired methods, it is important to recognize that we are stepping outside the usual scientific foundations for control. Rather than solely relying on the foundations of mathematics and physics (and perhaps chemistry) as is done in traditional feedback control systems, in intelligent control we expand our view to include other foundations of control science—biology, physiology, psychology, neuroscience, and their related fields, as shown in Figure 2.1.

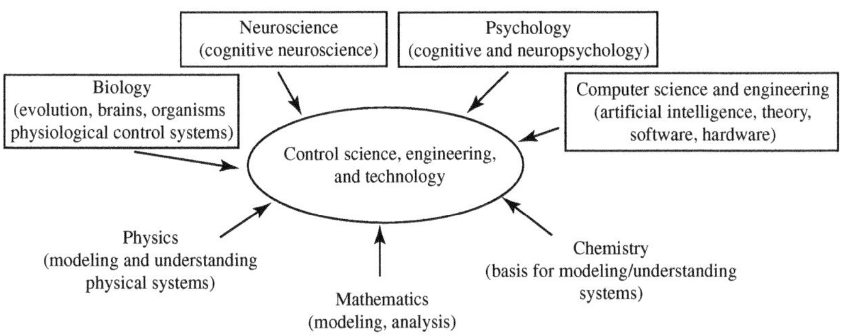

Figure 2.1: Mathematical, physical, and social sciences that affect control science, engineering, and technology.

It is also interesting to note that so far, as shown in the figure, control *technology* has been driven by computer technology and science (and, e.g., advances in electromechanical systems); however, biology may also provide some fundamental concepts and methods for improving computing technology (e.g., DNA computing or neural network-based integrated circuits) and hence, control technology.

2.1 Control Systems in Biology

To outline the sources for biological inspiration, it is best to first note that biologists often view their field as being hierarchical as mentioned above (i.e., hierarchical with the lowest-to-highest level sequence being the cell, tissue, organ, organism, and population levels). Indeed, the population level is sometimes viewed as being hierarchical with the usually species-centric view that humans are at the top.

In the remainder of this chapter we will generally organize our discussions along the lines of a biological hierarchy. Some examples will serve to clarify sources of biomimicry for control and automation. At the cellular and organ levels, there are the following control functions:

There is a wealth of complex control systems in biological systems; some of their aspects seem familiar to a control engineer while others reach significantly beyond current control engineering practice, in complexity and robustness.

- In the (single-cell) *E. coli* (*Escherichia coli*) bacterium, there is sensing and locomotion involved in seeking nourishment and avoiding harmful chemicals. There is sensing via the recognition of chemicals (essentially a type of shape recognition), internal decision-making, and actuation via locomotion (swimming). This example is discussed in more detail in Section 2.4.2 later in this chapter.
- In many plants there is tracking of light and nutrient sources (e.g., many plants turn toward light and roots grow towards nutrient sources) and reactions to gravity where a type of control system is involved.
- In animals there is homeostasis, for instance, thermoregulation, where the animal seeks to regulate its internal temperature to an optimal set point where it will function best physiologically. There is also thyroid and hormone regulation.
- Many animals have an "immune system" that has extraordinary abilities to recognize foreign substances and take actions to help the animal survive by controlling the density of antigens.
- In many animals, the pancreas is involved in the regulation of blood sugar levels (e.g., a diabetic is a person whose control system for blood sugar levels is not functioning properly).
- There are motor functions in two-legged animals that provide for balancing while standing.
- In the human brain there is the supervision of motor control for voluntary movement, supervision of the attentional system, and others.

You are asked to explore some of these characteristics further in Exercise 2.1.

At the organism and population levels of biology consider, for example, the following:

- Perhaps the most relevant and important organism to our studies is the human, since humans have been required to interface to the very technological control problems that we wish to solve (e.g., process operators manually perform control functions). We can also think of automating the actions of groups of humans who are performing some control task, or in some cases, emulating the actions of groups of microorganisms.

- At the population level we think of evolution as acting to shape every biological system; hence, it can have a profound impact on the design of control functions.

To date, the human nervous system (brain and motor functions), the use of human control expertise (whether that comes from a single human or a group), the emergent behaviors of groups of organisms, foraging, and evolution are the areas that have provided the most inspiration for the development of intelligent control methods, so each of these are discussed in more detail below.

2.2 Nervous Systems

The nervous system is the central command and control system of many organisms. Many of its key functions lie in the "brain," and we will be particularly interested in how it operates so as to produce intelligent behavior. The areas of psychology and cognitive science have studied human behavior for many years. In addition, at the physiological level there has been significant progress in neuroscience in determining the structure and functions of the brain of various animals and humans. Recently, these fields have been merging at several points to form the fields of "neuropsychology" and "cognitive neuroscience," where clinicians and researchers are trying to take a wholistic view, and exploit methods from many relevant areas, to shed light on how our brains operate and determine our behavior.

The computer is a useful metaphor for the brain. Long-term memory is a hard disk, and short-term memory is RAM. The computer processes inputs and generates outputs.

2.2.1 Sensory, Motor, and Brain Processes

Many sophisticated forms of biological intelligent systems have evolved a brain that has an ability to store, retrieve, and process information. For a human, the information comes in from the senses (sight, hearing, touch, smell, taste) and, depending on what we pay attention to, some information is stored for later use, while other information is discarded or soon forgotten. Broadly speaking, we store information in memory (e.g., short- and long-term memory) and use it to recognize patterns (e.g., visual or auditory), perform motor tasks, solve problems, plan ahead, focus attention, and perform creative tasks, to mention a few. Often, short-term memory is called "working memory" since it is where our current mental operations take place (you may think of it as a kind of scratch pad—the word "scratch" that you just read is there now). Working memory seems to be distributed to several locations in the brain, and each seems to serve a specific purpose. There is also "sensory memory" where we store, for instance, visual and auditory information for a short amount of time (e.g., less than one second) so that it can be preprocessed before being used (e.g., to help select what to attend to). Rehearsal of the rote repetitive type, or where a

Memory is essential for learning. Attention helps to manage complexity. It "filters" to obtain essential information to meet our goals, and to avoid overloading memory with useless information. It enables learning of essential information.

memory is recalled and "elaborated" on so that you think more deeply about it, results in transfer of information from short- to long-term memory. Our ability to organize information via "chunking" and hierarchies, in addition to our ability to visualize information, affects our ability to remember it. We have an ability to perform deduction (e.g., deriving information from information we have stored) or induction (e.g., learning general principles from examples we have encountered). All these processes are colored by our emotions and are affected by our motivations and goals, and taken as a whole, all these functions (and others not mentioned here) operate together in sophisticated ways so that "intelligent" behavior emerges.

Understanding brain physiology and function is one of the great frontiers of science. To date, there is reasonably good agreement on how to split the brain into its various functions (at least at an abstract level), and in fact, there are general "maps" that indicate which type of processing occurs in each part of the brain, and a fair amount is understood about lateralization (i.e., left-right brain theories) and specialization of functions. For instance, in the left cerebral cortex (the outer layer of the brain) there are four "lobes," each with its own specialized functions. The occipital lobe has components that are responsible for visual information processing. Other areas are responsible for smelling, hearing, taste, speech, reading, and auditory processing. There are several association areas that are involved in processing, integrating, and interpreting sensory inputs.

The somatosensory cortex area is for touch, and the primary motor cortex is for motor control (e.g., voluntary movement of the limbs and face, such as when you look over at your coffee cup and reach out to get a drink), and these two are split by the central sulcus. These areas are nicely visualized by the use of a "homunculus" of the body surface that can be obtained by systematic electrical stimulation to the motor and somatosensory cortices, and then by plotting the movements and sensations. This provides regions of the brain associated with senses and motor outputs. Then, it is interesting to note that in the motor cortex, there seems to be more cortex "real estate" dedicated to fine motor control (e.g., the fingers) and highly sensitive touch (e.g., the lips).

Accurate sensing and fine motor skills often require a greater quantity of neurons than do low resolution sensing and gross motor skills.

The frontal lobes are for inhibitory control of behavior and higher intelligence (e.g., at least some aspects of planning are implemented via the frontal lobes). Some think of the brain as being hierarchical with low level sensory processing, motor control, or instinctual components, intermediate processing (e.g., object recognition), and "high-level" processing in the cerebral cortex, especially in the frontal lobes. It is not, however, fully understood how the individual components operate, let alone how they dynamically function together to achieve complex cognitive reasoning.

2.2.2 Functional Operation of the Brain

For the purposes of this book, we will view the brain as an information processing system that takes inputs, processes them, and generates outputs. This does not imply that we literally think of the brain as a piece of silicon with electrical activity implementing software. We are interested in the connections to the neuroscience and what it has to offer. To explain this view in additional detail, consider Figure 2.2, where a functional block diagram of certain brain functions is shown. You should view this diagram in two ways. First, it shows basic functions of the bioelectrochemical processes. Second, it shows a type of connection to psychology and commonly observed human behaviors.

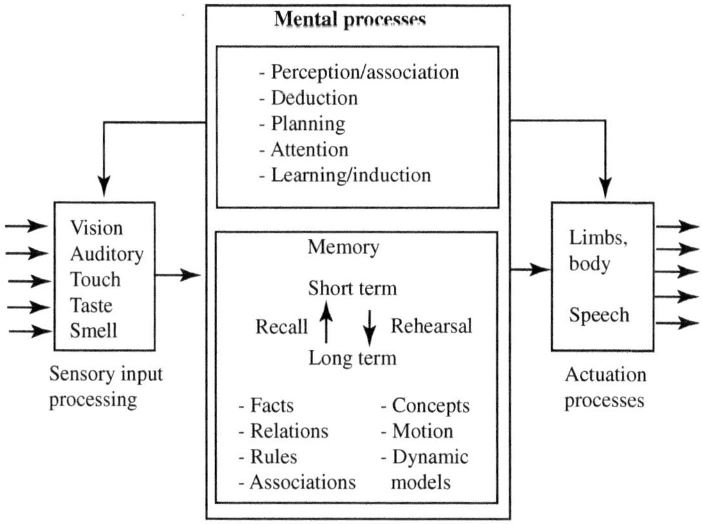

Figure 2.2: Functional block diagram of *some* brain functions.

The sensory input processing block brings information into the brain, and the actuation processes take actions on the environment. Many types of mental processes occur in between. For instance, if some visual information is gathered, certain perceptual processes are activated for association to classify or identify objects. Then, a deductive process may be invoked that considers what has been observed, and decides what to do about it. For instance, it may decide to say "apple" since that is what it has identified the object as, or it may decide to pick up the apple via the right hand. This is just one example of the type of information processing that can occur. In more complex situations, many different types of sensory information may be integrated, and the human may plan what to do about the current situation. To do this, a model stored in long-term memory may be used to predict (simulate into the future) how different strategies will help the human achieve her goal. Then, it may select the best approach, and take the subsequent actions. Such mental processes are significantly affected by what we are currently paying attention to, because our current focus actually helps to select what is sensed (one reason for the arrow from the "mental processes" box to the "sensory input processing" box), will make sure that only the most relevant information from memory is used (we all seem to have innate and learned prioritizations of information), and, if planning is used, may "prune" the tree of possibilities that we may need to consider in predicting

From a control perspective, perception, deduction, planning, attention, and learning are key candidates for physiological biomimicry.

the future. Attention affects many cognitive processes, as does learning; hence, they are central to how we function as humans. While engaged in many mental activities, we seem to have an ability to abstract and observe those processes, how effective they are, and decide if a different strategy could be better. This gives us the opportunity not only to store information in memory, but also to store and update reasoning strategies (and even learning strategies). This general abstraction capability is one of humans' greatest intellectual distinguishing features.

Clearly, we are ignoring many mental activities that we are engaged in every day (e.g., emotions); however, Figure 2.2 is useful in pointing out how to view the brain as a computer, a practice that winds its way through much of the literature, even though it is vigorously opposed by some philosophers. Engineers, however, often feel comfortable with the metaphor of the brain as a computer, and here we will use it, if for no other reason than to help teach the the various approaches to intelligent control.

2.2.3 The Neurophysiological Level

A network of neurons provides for information processing, and generating responses for specific patterns of stimuli.

At the cellular level, the neurons are the cells responsible for information processing in the brain. There are several different types of neurons in the brain, and many "glial" cells whose role seems to be to support the processing of the neurons. There are other cells responsible for waste removal and to do "housekeeping," for instance, by removing dead neurons. There are many neurons, and many interconnections between these neurons, and they communicate chemically and via electrical impulses. An example of a neuron is shown in Figure 2.3(a), along with a scanning electron micrograph of a neuron shown in Figure 2.3(b). The neuron is composed of "dendrites" that allow for connections to the "cell body" and an "axon" (when mature it is normally covered with Schwann cells, and it can be relatively long—in some organisms up to one meter) that allows for connections to other neurons via the terminal branches. The connection points are called the "synapses." Signals from other neurons are gathered on the dendrites, and under proper conditions, electrical impulses travel along the axon towards the terminal branches.

Motor control is a type of neural hierarchical distributed learning control system.

For some neurons it seems that the functions of the cell change with repeated use, thereby providing the ability to store information (i.e., memory via what is sometimes call "Hebbian learning"). The neurons interface via special cells to the sensory inputs (e.g., the rods and cones in your eyes) and are also connected to the motor system so that we can move our arms and legs via the neuronal control of muscle contraction and expansion. In fact, it seems that some of our motor functions are hierarchical and distributed: (i) in structure involving portions of the brain (premotor area, motor cortex, cerebellum, basal ganglia, brain stem, spinal cord, etc.); (ii) in generating action sequences (a sequence of actions will be implemented by a string of subsequences of actions, each possibly implemented in a different part of the body); and (iii) in motor learning (it seems that learning takes place at multiple levels of a hierarchy). Moreover, it seems that for the motor system, experts at a motor task adjust responses during an

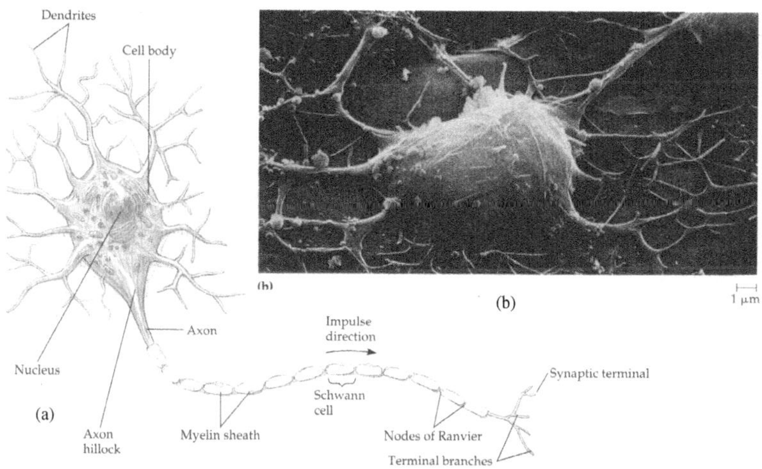

Figure 2.3: The neuron (nerve cell). In (a), a vertebrate motor neuron is shown, and in (b), a scanning electron micrograph of a neuron is shown (figure taken from [91], © 1987, 1990, 1993, 1996, and 1999, Benjamin/Cummings Pub. Co. Inc., and reprinted by permission of Addison Wesley Longman Publishers Inc.).

action, while novices do not, and experts tend to learn some type of model so that they do not rely on feedback as much.

There are many complex neurophysiological issues that are not very well understood in the brain. For example, evidence suggests that language is an instinct (i.e., we are somewhat "hard-wired" for language), but how is this "wiring" done, how much of it is there, and how can it be modified by education and experience? How many and to what extent are other functions hard-wired? It seems that our abilities to learn and plan are instinctual, at least to some extent (we do not need to learn how to learn or plan, but we may learn over our lifetime how to do these tasks better). However, is our sense of morality hard-wired? In life-span development of the brain (in the womb, as an infant, child, adolescent, and young adult), many interconnections between neurons are formed and, for example, if the connections are not used, then they may later disconnect (and hence, neurons are essentially "pruned"). There is evidence that the brain has a type of "functional plasticity" where if you lose a finger, the neurons for sensing adjust to dedicate themselves to the adjacent intact finger. This provides heightened sensitivity in the adjacent finger to help compensate for the lost finger; the very structure of the brain seems to be able to adapt. Neural connections in the brain are actually influenced by our environment, and even up to the age of 20, the structure of the neural connections (i.e., how neurons are interconnected) is adaptive. The interconnectivity is influenced by education, and what activities we are exposed to. Clearly, our nervous system is a type of very complex adaptive control system; properties of existing interconnections

Learning takes place via modifications to individual neurons, and changes to interconnectivity of a neural network.

can change to implement memory, and the very interconnections can change depending on our environment.

2.2.4 Hierarchical Neural Organization

While it is not always the case, at times it is useful (and accurate) to view a biological neural network as being arranged in a hierarchical fashion. For instance, in traditional studies of the brain many viewed parts as being "old" (e.g., motor functions) and other parts as being "new" (e.g., the cortex where higher functions such as planning take place). Old and new refer to evolutionary time, and it has been hypothesized that the older "worker functions" must have been placed in our ancestors well before higher level executive functions evolved. Here, we will not seek to explain the various hierarchical views of the human brain. Instead, we simply briefly discuss one part of a nervous system that is clearly hierarchical: the human motor system.

Neural network structures are sometimes organized in a hierarchical fashion.

The human motor control system is a hierarchical and distributed control system. It has local control functions for movement and higher level controllers that control gross motion. As shown in Figure 2.4 it is connected to many other parts of the brain, so that we can plan, learn, and execute motions. Motor learning, is in fact, quite interesting as it involves a type of adaptive model building in attaining skilled coordination of motion (e.g., in Olympic-caliber athletes).

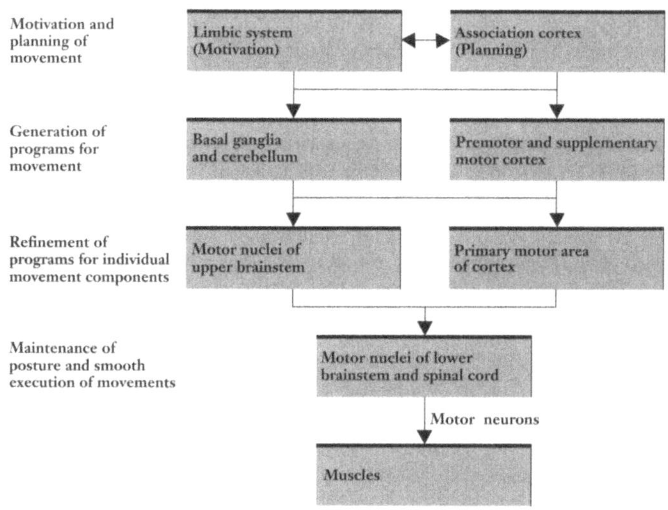

Figure 2.4: Hierarchy of motor control (figure taken from [223], © 1991, 1994, and 1999 by Worth Publishers Inc., and used with permission).

2.2.5 Brain Science: An Expanding Frontier

Although much progress has been made, the neuron itself is not yet completely understood. Moreover, the complex pattern of electrical pulses and chemical reactions that occur to achieve thought is not well understood. This is partly due to the fact that while we have an ability to probe the electrical activities of a single neuron or a few neurons, and via sophisticated scanning methods we can "see" how entire regions of the brain are used at various times (e.g., via metabolism), we cannot yet directly monitor many important and complex multiregion brain activities so that they can be studied in detail. The main barrier is the immense complexity of the brain (clearly, however, obtaining human subjects to study is problematic). Moreover, there are issues such as how evolution has shaped our brain, and what is the nature of consciousness, that are not likely to be well understood for a long time.

The main point is that we do not yet understand everything about how the brain operates. This, coupled with the fact that only crude models of cognitive and neural processing are used in the field of intelligent control, should make the reader hesitate to attribute a very strong connection between the systems that we will study in this book and cognitive neuroscience. You may, however, be comfortable with referring to the methods as "biologically inspired" or perhaps loosely based on some type of biomimicry. Regardless of the strength of the connections to biology, it seems likely that they will become more numerous and significant as research in intelligent control advances. In the meantime, it is interesting to learn about connections between systems and control theory and cognitive neuroscience, since it provides a different perspective on automation—after all, it could be argued that a human is the most complex automated system and it may pay to follow the human example to lead us to better solutions to automation problems (e.g., see Exercise 2.2, where you are asked to further explore the connections between cognitive neuroscience and control). For now, let us ask the following question: What basic ideas does cognitive neuroscience offer to the development of control systems?

2.2.6 Biomimicry for Automation: Cognition

In this book we will use simple models of neurons and show how they can be interconnected and adjusted to perform adaptive estimation and control. Moreover, in addition to the neurophysiological components (the "hardware level," if you subscribe to the computational theory of mind philosophy), we use ideas from the cognitive level ("software level") of the human brain as studied by cognitive neuroscience and psychology, particularly in rule-based inference, planning, attention, and learning.

The types of questions we study are as follows:

- *Neural networks:* Can neural networks that have no learning abilities still be profitably employed in control? What are some examples of neural networks that can be used in this way, and how would you specify (design) such a neural network? (In nature, evolution is employed for this design.)

Can such neural networks achieve orderly behavior (e.g., provably stable operation under a variety of conditions), such as is achieved by many neural (or biochemical) stabilization mechanisms in nature?

- *Deduction:* What deductive strategies are useful in control? In the relatively simple deductive strategy employed by rule-based control (e.g., via fuzzy or expert systems), how do we design the information used by the controller (the "rule-base") and the "inference mechanism" in order to achieve good closed-loop performance? How do we guarantee that a heuristic nonlinear controller synthesis approach will result in a stable closed-loop system?

- *Planning:* What types of models can be used in planning? How accurate must such models be? How do model errors affect prediction quality? How far ahead in time should predication take place? What is the best way to choose a plan? What criteria should be used for plan selection? Can we show that a planning strategy can successfully operate in a dynamic and uncertain environment? Can it successfully achieve a whole sequence of goals? How much uncertainty can it cope with?

- *Attention:* What is important to attend to? Which information can be safely ignored? What should the attentional strategy be for dynamic refocusing? What are the inputs to such a strategy (controller)? How do aspects of the environment (e.g., characteristics of predators/prey that the organism wants to attend to) affect how attention should be dynamically focused? What is the best way to design an attentional strategy so that an organism is maximally aware of its environment? What constraints exist on how much information can be attended to by an organism with fixed physiology? How does the filtering, selection, and refocusing in attentional systems help an organism cope with "information overload" and when does an organism with a fixed attentional strategy experience information overload? Can we prove that an attentional strategy will be able to maintain a certain accuracy in keeping track of stimuli in its environment? Does planning have a useful role in attentional strategies (e.g., via planning what to attend to)? On the other hand, do attentional strategies have useful roles for deductive and planning strategies? For example, how can attentional characteristics be integrated into deduction for focusing on what to plan for?

- *Learning:* How can information (e.g., data) gathered a priori or "online" be stored and later recalled in order to enhance performance? How do neural networks learn from examples? What types of "examples" are best to help enhance the quality of learning? Can we control which examples are presented to the learning system? What are the trade-offs between learning substrate flexibility, complexity, and tunability? For example, what is the minimal size neural network needed to learn some relationship to a certain degree of accuracy? What types of information can be learned by neural networks? What is the role of forgetting? How do we measure how much is learned and the accuracy of the learned information? How accurate must the learning be in order to achieve good performance? Can the information used by a deductive strategy be learned and profitably employed to adapt the strategy to achieve or maintain performance? Can our deductive strategies be adapted over short periods of time to cope with problems presented by the environment? Can we prove that online learning strategies lead to control decisions that result in stable closed-loop behavior? Can the learning of models during operation of a planning strategy enhance performance? Can it learn new planning strategies? Can a model be learned and profitably employed in an attentional strategy (e.g., via learning a model that is used to plan how to pay attention)? Can the very structure of the storage and learning strategy also be learned? For example, can we learn how to learn?

It is these and other such questions that will be discussed in this book. In the next section we provide a different perspective on these questions: If we consider a human who is controlling some process, can we profitably emulate the deductive, planning, attentive, and learning strategies he might use in order to automate control of that process? Hence, we consider biomimicry in more detail, but where the "bio" is typically a human who employs a range of cognitive skills to solve a control problem.

2.3 Organisms

Another major approach to intelligent control is to use a computer and other hardware to automate control tasks that have traditionally been solved by animals or humans. We think of observing how an organism solves a control problem, then use this as the inspiration for construction of a computer algorithm for control.

2.3.1 Organisms Vs. Computers for Control

We begin with what may seem to a control engineer to be odd questions. Why even use a computer and other hardware to try to achieve automation? Why not just use a human to perform a task? Or, why not employ some other biological organism to perform the task? These are not ridiculous questions. Biological organisms are introduced in some ecosystems to try to remove hazardous waste (i.e., to control the level of concentration of some noxious chemical). How far can the idea of employing biological systems for control go? Humans are currently employed in performing many control tasks (e.g., driving a truck, controlling nuclear power plants, etc.).

Consider the following reasons to replace humans (or other animals) with computers and other hardware for automation:

- There is a desire to remove the need for humans to perform simple, often repetitive, boring tasks so that they can pursue more enjoyable employment. (At least, that is the objective, and it is not to simply raise the unemployment rate.)

- Properly designed monitoring and control systems may be more reliable and less expensive. Machines typically perform more consistently than humans do. The energy supply can typically be electricity, which is often relatively easy to supply to the automation system. Animals and humans require nutrient sources that are likely to be more difficult to deliver, especially if the control task is to be performed over a long period of time. In addition, humans need rest and sleep, while machines do not (but they may need maintenance and replacement).

- Computer and other hardware is often not subject to some of the same constraints as humans are in performing control tasks. For example, humans cannot process many types of events at a rate that a computer can. Mechanical systems can be designed to provide greater forces and to be faster than humans.

- When it comes to verifying that the overall system will operate properly, with the human "in-the-loop," it then becomes necessary to analyze human responses in complex dynamical situations, and this is very difficult (human behavior is not very well understood and is certainly unpredictable to a certain extent).

- It may be a significant safety risk for a human to perform some control task; hence, this may necessitate computer automation (e.g., in some robotic applications, for the cleanup of hazardous waste or the clearing of land mines).

There are, however, reasons why it could be better to use a human to perform a control task. Humans have extraordinary visual and pattern matching capabilities for moving objects, can often be creative in solving problems, and can often successfully reason about control strategies in an abstract manner. For example, automated collision avoidance for automatically driven vehicles is currently quite a difficult problem, yet humans perform this task quite well.

To solidify the above advantages and disadvantages in employing a living organism for control, we next consider one possible application.

2.3.2 Example: Skinner's Pigeons for Missile Guidance

The famous behaviorist psychologist B.F. Skinner was given funding by the U.S. government in 1943 to try to determine the feasibility of using trained pigeons to guide a missile that was to be dropped from an aircraft to hit targets on the ocean (e.g., enemy ships). Figure 2.5 shows a pigeon being placed in the nose cone of the Pelican missile that Skinner was to test the pigeons on. The system in fact employed three pigeons and a majority voting scheme to combine their actions and guide the missile.

Figure 2.5: Pigeon being placed in nose cone of the Pelican missile test bed (figure taken from [106], © IEEE, used with permission).

First, the pigeons were trained to respond only to a particular type of target. This was done by projecting aerial photographs of targets on a screen, crossing beams of light at the centers of the targets, and then when a pigeon struck the center of the two beams, a signal was generated that rewarded the pigeons by swinging a tray of grain in front of them so they could eat. Several pigeons were trained in this fashion. Next, the pigeons would be deployed to provide real-time guidance of the missile as it descended towards its target. To do this, the pigeons were to peck at a movable screen (on gimbal bearings) that was connected to the missile guidance system, so that pecking at specific target locations on the screen would redirect the missile toward the target.

It turns out that when the project was evaluated for further funding, even though the pigeons performed perfectly, the project was discontinued. Some believe the reason had more to do with electrical and mechanical engineers (who were working on what we would now think of as more conventional designs) having difficulties changing their viewpoint. There must have been, however, serious concerns about the reliability of the pigeons (even though Skinner found them to work faster, in total darkness, or with loud sounds nearby when they were fed marijuana seeds), the expenses of training, and the problems with managing the animals en route (e.g., feeding, reactions to relatively high altitudes).

It is not difficult to speculate, however, that at least part of the reason that Skinner was funded in the first place must have been that at that time we did not have good sensors and pattern matching algorithms, and the pigeons' visual and cognitive systems were the key to their success. Hence, perhaps the objective was to try to exploit some of the things that pigeons could do well, that the current technology could not. Now, however, with the rise of radar and "automatic target recognition" technology, we can design computer-based electromechanical systems for missile guidance. Perhaps there is a fundamental principle here. We use humans or animals in places where our current technology is not very successful, but once the technology is improved to the point where its success and other advantages outweigh those of a human (or some animal) we replace the human with the technology. (Clearly, however, there are human factors and cultural aspects we are not considering.)

Assume that we decide to automate some control task that has typically been performed by a human (we ignore the case where an animal is used since it is much less common). How do we go about this? We could take the conventional approach to control design as explained in Section 1.3. This approach, however, has a critical dependence on the existence of a mathematical model for the plant. Most often, however, a human does not use an explicit mathematical model of the process to decide what control inputs to apply to the process; hence, when the control task has traditionally been performed by a human, it is often the case that a mathematical model for the plant does not exist (typically all that exists is a description of how the human goes about solving the control problem). It is then important to discuss in more detail why we may not have a model and what to do in order to design a control system in this case.

2.3.3 Human Control Expertise: "Human-Mimicry"

If we wish to emulate the control expertise of humans, it is useful to categorize the types of expertise that they may have, both to recognize the range of possibilities and to provide indications on how to implement the behavior. In broad terms, the general types of control expertise are the following:

Perception, deduction, planning, attention, and learning drive the behavior of the human control expertise we seek to emulate.

- *Rules:* Often human operators have "rules of thumb" about how to control a process. These are often in the form of, "*If* certain events and conditions occur, *then* take these corrective actions." For example, human operators often use this type of approach in making corrections when the process is in steady-state operation.

- *Deduction over more complex representations:* Humans often represent information in sophisticated "semantic networks," where similar information is stored nearby and where there seem to be links (some hierarchical) between related concepts and objects. In controlling some processes, humans may reason over such a representation to decide how to act, and not just use simple rules as described above. Moreover, expert human operators tend to have a better ability to move between more and less abstract levels of reasoning.

- *Planning using mental models:* In certain control problems, a human will build a mental model of how the process will react to actions, and will use this model to plan which actions to take. (This normally involves predicting into the future with the mental model how the process will respond to various actions, and then picking the input that will be best.) An expert human operator is also capable of anticipating future goals, and as discussed below, is able to learn better mental models of the process during operation of the system.

- *Attention to salient behaviors of the process:* Humans typically focus on the most important aspects of the behavior of the plant and make decisions based on these. This helps to manage the complexity of the vast amount of information that is dynamically gathered to indicate how a process is operating. We think of this as the human having "selective perception."
- *Learning how to control:* It is important to recognize that for each of the above types of expertise, there is the possibility that the expertise can be improved via learning. Humans initially learn how to control a process and if the process changes, they can learn how to control it so as to maintain performance. For instance, humans can learn new control rules or other mental representations of plant behavior that could be used in planning, or they could learn how to pay attention to the most important features of the plant's behavior. Moreover, expert human operators are said to resort to open-loop control more often than are novices who rely on closed-loop operation more frequently. Why? Experts do this since they have learned a very good mental model of the process and taking control actions *in advance of certain anticipated events* is often the best strategy in processes that have delays.

Clearly, there are strong connections between our understanding of cognitive neuroscience and what aspects of human operator behavior may be relevant to generating control actions. Next, we discuss the fact that while human operators may be quite creative in solving a control problem, there may also be situations where they may not perform very well. This provides a cautionary note about the value of *perfect* emulation of human behavior.

2.3.4 Human Operator Control Expertise: Quantity and Quality

While the use of human operator knowledge is sometimes a good source for the heuristic construction of a controller, there are two issues that need special attention. First, operator expertise varies, especially with the process that is controlled, and for some processes their control expertise may not be very useful. For instance, the human operator is typically balancing several different objectives, including productivity and company profits vs. safety. If given too high of a work load, the operator can fail to achieve a correct balance. Second, we certainly only seek to model human operators when they are doing their best, not when they make mistakes or are performing poorly (e.g., since they are "having a bad day").

Do Humans Always Have the Needed Control Expertise?

No, not always. There are control tasks that humans can currently do much better than a computer (e.g., obstacle avoidance while driving) and this may be due to, for instance, our unique visual processing capabilities or complex cognitive processing. For such applications it makes sense to first try to emulate and then automate what a human would do. This approach has been successful in several applications, especially for "first generation" automation. Often, however, in further refinements of the controller, improvements can be made that result in higher performance operation, and these improvements do not necessarily represent more accurate representations of what a human would do in performing the control task. In essence, over time, the system evolves independent of the original motivation to replace the human by emulating and automating his actions.

Computers may be able to perform better than humans in some cases, but in others humans may do better.

On the other hand, there are other processes where a computer can perform much better than a human (e.g., balancing a double inverted pendulum, or in damping the effects of wind currents on aircraft turbulence). These controllers perform better since detailed information about the system to be controlled (e.g., a mathematical model that has been shown to be reasonably accurate in predicting how the system will behave) is used to construct the controller, and since a computer (in concert with sensors and actuators) can act much faster than a human for many processes. It is important to realize that there are physiological delays in sensing, cognitive processing, and actuation that make it very difficult for a human to control some processes. Evolution has shaped us to be good at only certain survival tasks, some of which include control tasks. It is possible, however, for an engineer to come to a good understanding of the physical process (e.g., via physical modeling and system identification) and to then imagine how to control the process *as if they could act very quickly.* In this case we can still think of automating human control expertise (some of which may have come from modeling), and such an approach has proven to be useful in several applications.

It must be acknowledged, however, that there are some nonlinear processes that are very difficult to control, whose complexities, time delays, nonminimum phase behavior, and stochastic characteristics tax humans beyond their capabilities (e.g., the "working memory" in humans is only a finite size so that even with the most intelligent manipulations, human performance is constrained by this limitation). Indeed, such processes have perhaps benefited most from the methodology of producing a mathematical model and using this to specify a controller. In this book we firmly acknowledge the existence of such systems and emphasize that the conventional approach of using mathematical models to represent the plant and construct controllers has significant value for this type of system. However, we also highlight the fact that there are times when a mathematical model is very difficult, impossible, or undesirable to produce, and hence there is a need or desire to rely solely on human control expertise in constructing controllers.

Do You Want to Emulate What a Human Would Do?

No, not always. As mentioned above, evolution has not shaped us to solve all the possible control problems that we can encounter in a modern technological world. Moreover, humans often make mistakes in solving control problems, or are unreliable, or inconsistent in their performance.

In this book we have no interest in developing controllers that emulate humans who do not do a good job in solving a control problem. Our intent is not to model human behavior, but to solve the control problem. Ultimately, we often want an automatic system that replaces the human and performs better than any human ever could. Computers have a significant advantage over humans because they can be quite consistent in how they perform a task. Hence, we emphasize that we do not always want to emulate what a human would do in solving a control problem. We simply want to do good control engineering to get the best possible system.

The goal is not emulation of substandard human behavior; it is to design the best control system possible.

2.4 Groups of Organisms

We can think of two broad approaches to derive biological inspiration from group behavior of organisms. First, we will consider the case where there are simple control systems that are used on organisms, and how when they are together as a group and may communicate with each other and perhaps coordinate their actions, certain types of intelligent behavior seem to "emerge" (e.g., swarm patterns that emerge when groups of animals forage and avoid predators). Some call this the study of self-organization in distributed autonomous multiagent systems, and think of it as illustrating how intelligence manifests itself in a "bottom-up" fashion from many simple interacting systems. Often, the system is "flat" or "horizontal" in the sense that there is no special agent that supervises the actions of others. Second, we will consider the case where we emulate a group of organisms organized in a hierarchical fashion that performs a control task via the coordination of their actions. There are layers of bosses and workers, and intelligence is "top-down" and arranged in a vertical fashion with higher levels associated with more abstract thinking, and hence higher levels of intelligence. In either case, we think of the intelligence as emerging from the group's interactions; some say that "the whole is more than the sum of its parts."

2.4.1 Social Foraging and Emergent Swarm Behavior

Some animals search for and obtain nutrients in a way that maximizes

$$\frac{E}{T}$$

where E is energy obtained, and T is time spent foraging (or, they maximize long-term average rate of energy intake). Evolution optimizes foraging strategies since animals that have relatively poor foraging performance do not survive.

Foraging is an optimal decision-making process that has been fine-tuned via evolution.

Foraging could involve finding a "patch" of food (e.g., group of bushes with berries), deciding whether to enter it and search for food (do you expect a better one?), and when to leave the patch. Alternatively, foraging could involve selecting which prey to include in a diet, taking into consideration the relative abundance of the prey types and the expected time it takes to search for them. There are predators and risks, energy required for travel, and physiological constraints (sensing, memory, cognitive capabilities). Foraging scenarios can be modeled and optimal policies can be found using, for instance, dynamic programming. Search and optimal decision-making of animals for foraging can be broken into three basic types: cruise (e.g., tunafish, hawks), saltatory (e.g., birds, fish, lizards, and insects), and ambush (e.g., snakes, lions). In cruise search the animal searches the perimeter of a region, and in ambush it sits and waits. In saltatory search, an animal typically moves in some direction, stops (or slows down), looks around, and then changes direction. It searches throughout a whole region.

Some animals forage as individuals and others forage as groups. While to perform "social foraging" an animal needs communication capabilities, it can gain advantages in that it can essentially exploit the sensing capabilities of the group, the group can "gang up" on large prey, individuals can obtain protection from predators while in a group, and in a certain sense, the group can forage with a type of collective intelligence (it is sometimes said that the group "self-organizes" and intelligence seems to emerge). For instance, some of the following animals exhibit this characteristic:

- *Birds, Bees, Fish:* Flocks of birds, swarms of bees, and schools of fish exhibit complex dynamical spatial patterns of behavior that emerge when each animal in the group follows a simple set of rules (e.g., if each organism tries to move to the center of the group, tries to avoid collision with its neighbors, and tries to continue to move in the same general direction of the group, simulations have shown that emergent swarm behaviors arise that are reminiscent of how this occurs in nature). Also, other "herds" of animals (e.g., horses) have similar group behavior. Basically, via inter-organism communications, in social foraging animals cooperatively forage and protect each other from predators and risks and thereby increase energy intake per unit time spent foraging, or perhaps they reduce variance in intake.

- *Ants:* Groups of ants seem to have purposeful behavior when in fact it has been shown that there is not a management hierarchy as in human organizations; ants follow simple rules in locally coordinating their actions, and colony-level goal-seeking behaviors emerge. In particular, their social foraging behavior leads to more efficient acquisition of food.

Simple organisms in colonies that obey simple rules can sometimes achieve a type of collective intelligent behavior.

Note that there is a type of "cognitive spectrum" where some foragers have little cognitive capability, and other higher life forms have significant capabilities (e.g., compare the capabilities of a single ant with those of a human). Generally, endowing each forager with more capabilities can help them succeed in foraging, both as an individual and as a group.

Why is this relevant to control? First, it is often the case that control functions are being performed by each member of a group, and the "additive effect" of the local control and coordination is what results in an emergent group behavioral pattern. Second, there is an inherent distributed optimization process that results in the emergent behavior, and ideas from this optimization process can be used for control systems.

To clarify the types of behavior that can emerge when many "agents" with resident control systems act together, we will not now consider birds, bees, fish, herds, or ants; instead, in the next subsection we will consider a much simpler microorganism as an example: *E. coli* bacteria.

2.4.2 Example: Bacterial Chemotaxis

Here, we consider the foraging behavior of *E. coli*, which is a common type of bacteria (it lives in your gut) with a diameter of $1\mu m$ and a length of about $2\mu m$, and which under appropriate conditions can reproduce (split) in 20 minutes. Its ability to move comes from a set of up to six rigid 100-200 rps spinning flagella, each driven by a biological "motor." An *E. coli* bacterium alternates between running (at 10-20 μmeters/sec, but they cannot swim straight) and tumbling (changing direction). When the flagella rotate clockwise (counterclockwise), they unbundle (bundle into a "propeller") and hence tumble (run).

Chemotactic actions:

1. If in a neutral medium, alternate tumbles and runs $\Rightarrow$ Search
2. If swimming up a nutrient gradient (or out of noxious substances), swim longer (climb up nutrient gradient or down noxious gradient) $\Rightarrow$ Seek increasingly favorable environments
3. If swimming down a nutrient gradient (or up noxious substance gradient), then search $\Rightarrow$ Avoid unfavorable environments

In this way it can climb up nutrient "hills" and at the same time avoid noxious substances. The sensors it uses are receptor proteins which are very sensitive, and overall there is a "high gain" (i.e., a small change in concentration of nutrients can cause a significant change in behavior). The sensor averages sensed concentrations and computes a derivative. This is probably the best understood sensory and decision-making system in biology (it is understood and simulated at the molecular level).

What is the resulting emergent pattern of behavior for a whole group of *E. coli* bacteria? Generally, as a group they will try to find food and avoid harmful phenomena, and when viewed under a microscope, you will get a sense that a type of intelligent behavior has emerged since they will seem to intentionally move as a group.

Bacteria are often killed and dispersed and this can be viewed as part of their motility. Mutations in *E. coli* affect reproductive efficiency at different temperatures, and occur at a rate of about 10^{-7} per gene, per generation. *E. coli* occasionally engage in a type of "sex" called "conjugation" that affects characteristics of a population of bacteria. There are many other types of taxes that are used by other bacteria. For instance, some bacteria are attracted to oxygen (aerotaxis), light (phototaxis), temperature (thermotaxis), or magnetic lines of flux (magnetotaxis). Some bacteria can change their shape and number of flagella based on the medium to reconfigure to ensure efficient foraging in a variety of media.

E. coli and *S. typhimurium* can form intricate stable spatiotemporal patterns in certain semisolid nutrient media. They can radially eat their way through a medium if placed together initially at its center. Moreover, under certain conditions they will secrete cell-to-cell attractant signals so that they will group and protect each other. These bacteria can "swarm."

2.4.3 Emulation of Coordinated Behavior of Humans

Groups of humans may be able to coordinate their actions to manage and control some enterprise; it can be useful to mimic their collective behavior for automation.

In an analogous way to how we sought to emulate human control expertise so that we could automate the control functions that they are employed to perform, we can take a similar approach for *groups* of humans (e.g., as in business management shown in Figure 2.6). For instance, there may be groups of human operators ("workers") employed to coordinate the control of a chemical process control plant. In such a situation we could interview the group of process operators to gather their control expertise, and pay special attention to how they coordinate their actions (e.g., how they communicate with other operators, and how that information influences the decisions that they make). In such situations there is often a hierarchical nature to the management and execution of tasks. For instance, there may be humans with very little autonomy (workers), who have skills in certain areas, and who repetitively apply these. Their supervisors (bosses) try to coordinate their actions so that they are effective in achieving broader goals, and so that, for instance, if one worker learns how to perform a task better, this information can be shared with others. They must plan some of the worker's tasks, and try to make sure that these plans achieve broader objectives. It is the responsibility of the supervisors to make sure that the workers are efficient, and this can be viewed as the supervisor doing the right things to make sure that the workers can do well, or perhaps they might be more authoritarian and simply dictate to the workers what the workers must achieve. The supervisors could be viewed as "middle management" and this group will also have bosses, sometimes called "upper management." Upper management, which is sometimes just one person, is responsible for high-level oversight, trying to optimize the performance of the overall process, setting high-level goals, planning future actions of the lower levels of management and workers, and interfacing to the customer and other systems.

Proper design of groups of humans to achieve hierarchical, distributed, coordinated control involves many issues that we will not discuss in this book. Clearly, management science, organizational/industrial psychology, economics, business practices and other areas can play especially important roles.

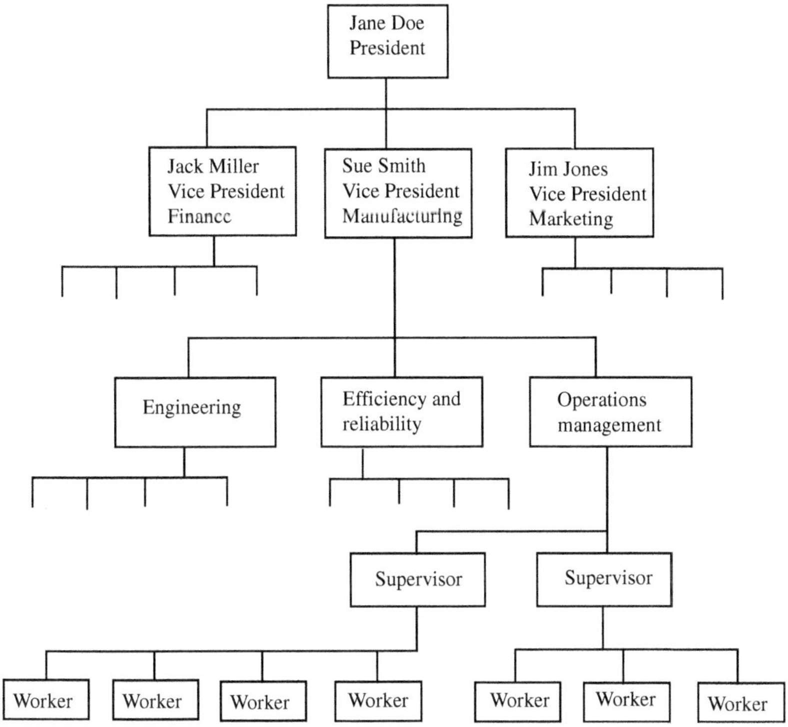

Figure 2.6: Hypothetical organization chart for business management.

2.4.4 Biomimicry for Automation

In this book we will use simple models of social foraging groups of organisms for distributed optimization and to provide ideas for how to achieve cooperative control. We can study hierarchies in each of the approaches. The types of questions we study are as follows:

- *Hierarchical biological/cognitive structures and organizations:* Can fixed neural networks that are arranged in a hierarchical fashion be used in control? Can a hierarchical deductive strategy, where there is reasoning over how to supervise a lower-level deductive strategy, be used for control? Can we conduct high-level planning that generates plans that are then implemented by multiple low-level planners that seek to achieve each task or goal in the plan specified by the higher-level planning system? Can attentional strategies that employ abstraction and focusing of a *set* of attentional strategies be useful? Can a hierarchical structure (e.g., hierarchical neural network) learn from examples? If we think of each human in an organization as an intelligent decision-maker endowed with deductive, planning, attentional, and learning capabilities, how do we design the hierarchical structure and individual strategies to ensure success by the whole enterprise (possibly at the expense of some individuals in the organization who may be more successful outside the organization)?

- *Intelligent social foraging:* Can foraging teach us how to design guidance strategies for vehicles (e.g., if we think of a goal position as providing nutrients, and obstacles in its environment as noxious substances that are to be avoided)? Can foraging strategies be used to choose inputs to a plant (e.g., if we define nutrients to be good performance, and noxious substances to be bad performance)? If we consider "good information" to be food for the foraging organism, can we design foraging strategies that can learn (e.g., learn models that are then used in adaptive deduction, planning, or attention)? Can social foraging strategies be useful for designing cooperative control strategies for groups of autonomous vehicles? What type of command and control hierarchy should be used by such a group (e.g., "flat," where every individual is the same as every other one, or hierarchical with a "leader" who tells each vehicle what to do)? What types of communications should be used? What quality of communications is needed in order for the group to be successful? For example, can success be obtained with noisy, band limited, and range limited strategies? What level of intelligence is needed by each individual to ensure group success? Is simple rule-based behavior sufficient, or do they need planning, attention, and learning capabilities? How do the designs change if there are two adversarial teams of foragers operating in the same environment?

2.5 Evolution

Evolution provides a unifying theme for, and affects virtually every aspect of biology, from *E. coli* bacteria to human physiology and cognition. D. Dennett appropriately points out that "biology is engineering" since evolution essentially incrementally designs organisms [139]. Clearly, there should be a strong role for evolution in intelligent control since it has shaped both our ability to manually control a process and the very biological systems that we mimic.

2.5.1 The Evolutionary Process

Evolution is a type of incremental adaptive process that persistently redesigns biological systems to be best-suited to reproduce in their environment.

Darwin's theories have been strengthened and extended over time (resulting in "neo-Darwinism" or the "modern synthesis") but are still basically intact. Darwin observed that species have great potential fertility, population sizes are largely fixed, but do change with climate changes, and that resources are limited. This leads to a struggle for life so that only a relatively few offspring survive. Moreover, Darwin observed that individuals in a population vary extensively and that the variation is heritable. This led him to conclude that individuals who possess characteristics that allow them to survive to reproduce leave more offspring than less fit individuals. Also, he concluded that this will then lead to a gradual change in a population, and that the favorable characteristics will tend to accumulate in the population over many generations.

As an example of the results of evolution, consider the polar bear. Polar bears have white fur. Why? Evolution has designed them this way. Polar bears live in cold climates, yet need to maintain a body temperature and white fur is not good for that. A dark fur would tend to keep them warmer. On the other hand, a polar bear is more fit for its environment with white fur since its prey cannot detect it as easily against a white (snow) background. Apparently, as they evolved from their ancestors with dark fur, parents that bore children that had slightly lighter colored fur tended to survive to reproduce (they had more food, so they were able to survive to produce more offspring). After many generations the fur turned white. Also, in this process the hairs that make up the fur coat evolved to become hollow! This helps with insulating the polar bear and in this way, evolution created an "optimal design" for its environment. This, of course, assumes that the evolution has converged on a design, but clearly it is very difficult to determine this.

Evolution is what glues biology together since it affects all living organisms. It is then important to have an abstract understanding of the evolutionary process. The general evolutionary process is abstractly depicted in Figure 2.7. Here, suppose that climate and population size are relatively fixed and that resources are limited. At the top of the figure, there is an oval that represents the population at generation k. In this population there are two dots that are representative of the frequency of occurrence of a certain type of trait of the organisms in the population at time k. The lines coming out of the bottom of the oval represent the offspring of the population. Here, there are different line types representing different inherited characteristics (note the heavier and lighter line thicknesses, solid, dash, and dash-dot types, ones with arrowheads, and ones with arrowheads and shaded circles on them; view each of these as a different offspring in the same species). Due to the struggle with limited resources, suppose that only the lines with black arrowheads represent the offspring that survive. Here, there is one arrow representing an individual who does not hold the "dot characteristic" but who survived anyway. The lines without arrows at the end represent organisms that had inherited characteristics that did not help them survive, so they died before reproducing. Clearly, having the dot characteristic was helpful for survival, so that at generation $k+1$, the population has a higher frequency of the dot characteristic. The process repeats in the same general way at generation $k+1$. Clearly, individuals with the dot characteristic will, over many generations, tend to be more numerous in the population; in this way we think of the whole population as adapting to survive in its environment. Finally, it should be highlighted that there seems to be a persistent temptation to assume that every organism in existence is "highly evolved" or "optimized" for its environment (i.e., that it is at an evolutionary equilibrium). Evolution is an ongoing process so there is little reason to think that any particular organism's design is at an equilibrium. Evolution does not occur instantaneously, and its associated "inertia" makes it difficult to know if an equilibrium is achieved. In addition, it is difficult to know what "noise" in the environment is driving the evolution, so the notion of a single equilibrium is likely inaccurate. Experimental verification that an organism is optimized is certainly difficult. Moreover, viewing evolution as being modeled by a stochastic nonlinear optimization process, there is no reason to think that long-term behavior is not oscillatory. Hence, you should be careful in drawing conclusions about the ultimate results of biological evolution.

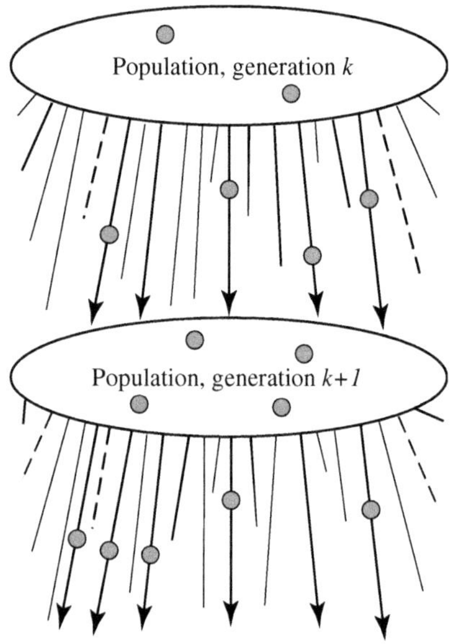

Figure 2.7: Graphical depiction of the biological evolutionary process.

2.5.2 Evolution of Behavior

Evolution also determines components of the behavior of animals, but ultimately animal behavior is certainly a combination of genetic and learned responses. Focus for a moment on what seem to be clear genetic influences on behavior. For instance, there seem to be species-typical behaviors where different members of the same species produce identical responses to the same environmental stimuli (these are called "fixed action patterns" in response to "sign stimuli"). For example, certain aggressive behaviors in animals, such as guarding their nests, follow such a pattern. Also, it seems that human emotional expressions are species-specific (e.g., given a picture of anyone in the world who is happy, almost anyone else in the world would identify the subject as such). Evolved behaviors seem to have a component of biological preparedness, but there is also a component where learning during a lifetime can change the natural tendency to respond in some way.

In addition, various characteristics of animals and their behavior are thought to undergo "coevolution," where their evolution is linked to that of another organism because they are in competition with that organism. The field of evolutionary game theory has provided models and explanations of how behavioral survival strategies evolve in animals. The behaviors that are most successful propagate; actually, the genes that propagate are the ones that lead to the types of behavior that result in survival and reproduction. It is interesting to note that the game-theoretic approach to the study of evolution is closely related to control systems since we can view organisms as a type of controller that is interacting with its environment, trying to generate inputs (behavior) that allows it to survive. The organism with the best strategy (control algorithm) wins and propagates its DNA through time.

2.5.3 Evolution of Intelligence

Evolution has shaped biology at all levels, from cells to organisms, behavior, and intelligence.

The evidence for evolution is as conclusive as any scientific theory, and has been shown in many forms, from laboratory experiments with bacteria and fruit flies, to fish and birds in their natural habitats. We can think of evolution as a type of adaptation over long periods of time. Higher level cognitive functions evolved since they gave us a differential advantage in survival. For instance, language capabilities evolved so we can pass information to our descendants and other individuals (e.g., via spoken language, books, and the library). Planning has evolved to give us an ability to build models in our minds of portions of our environment and to use these to predict and pick the best actions to ensure our survival (what portion of this environmental model is instinctual?). Attention evolved to provide us with ways to cope with the complexity of sensed information about our environment and to allow us to concentrate on a sequence of activities without distraction. Memory is adapted so that it stores and does not forget information that is likely to be useful for reproductive survival in the environment the human lives in. Learning allows us to be flexible (it gives us "plasticity," as does planning when it involves model building) in how to best cope with our environment over the span of our lives. Learning can give us a differential advantage for survival, so it is clear why this capability has evolved in humans. Overall, in terms of the relationships between learning and evolution, it seems that evolution has produced a type of trade-off in what is passed through the genes (and hence what becomes instinctual), and what must be learned via the organism interacting with its environment. An appropriate amount of learning has evolved to help ensure reproductive success.

Evolution designs and tunes learning processes. Learning can accelerate evolution; there are "synergistic" effects between these two adaptive processes.

Moreover, there is an additional effect due to our ability to learn called the "Baldwin effect" that hypothesizes that because learning capabilities are heritable, organisms able to learn can survive certain effects of their environment for a long enough period of time so that they can actually (via a mutation or incremental improvements) gain a genetic coding of the ability to survive these effects (it is better to have the instinct or natural resistance since it is on average more reliable than having to learn to avoid something that adversely affects our ability to survive). Learning then is not needed so the capability could be lost, or perhaps it could be adapted for use in other survival tasks. Overall, the Baldwin effect characterizes how learning can tend to accelerate evolution since over a single life span it raises the reproductive fitness that evolution operates on.

2.5.4 Example: Selective Breeding for Intelligence?

"Selective breeding," a practice used for thousands of years by plant and animal breeders, involves repeatedly (over several generations) mating animals that have a desired characteristic. After several generations, the effects of this strategy tend to accumulate and hence can affect behavior or intelligence, independent of environment and learning. For instance, even in one generation, for some animals, researchers have shown that a *single* gene controls certain behavioral aspects (e.g., it seems that a single heritable gene controls fearfulness in some dogs). Generally, however, for other behavioral characteristics it seems that many genes affect behavior and intelligence, and hence these are sometimes referred to as "polygenic characteristics."

While for plants, breeders select for bigger grain, or ears and kernels of corn, in animals they may select for docility and milk production in cows, or speed or work capabilities in horses. In controlled laboratory conditions scientists have bred fruit flies so that they will move toward or away from a light source, mice to be more or less inclined to fight, and rats to prefer alcohol or water. Clearly, such selective breeding could be used for many types of behavior and intelligence.

Perhaps the most interesting goal, however, is to try to breed for some aspect of intelligence such as an ability to learn or an ability to plan. A classic study of selective breeding for intelligence was conducted beginning in the 1920s by Robert Tryon, where he conducted an experiment to evolve rats to be better at learning how to navigate a maze. To do this, he started with a genetically diverse group of rats and tested them for their ability to learn a particular maze (the same maze was always used). For his test he scored each rat via the number of errors that they made in trying to find their way through the maze. Ones with many errors were called "dull" and ones with fewer errors were called "bright." Next, he mated the males and females that scored the best with each other and those who scored the worst with each other. In this way he sought to selectively breed to obtain a bright strain of rats and a dull strain. Some of his results are shown in Figure 2.8, where you can see that after only the seventh generation, there is a clear separation between the two strains of rats. He evolved a bright strain of rats! To control for the possibility that the offspring were learning how to be bright or dull from their mothers, he "cross-fostered" the rats by having some rats raised by a foster mother from another strain. Then he discovered that how the rats were raised did not affect their performance; bright rats raised by dull mothers still performed well, showing that the environment and lifetime learning did not corrupt the conclusions.

Some have raised concerns about this study since it is not clear what was selected for since the maze learning task also depends on sensory, motor, and motivational processes. The "bright" ones could have simply had better vision! Or, the "dull" ones may not have liked the taste of the food at the end of the maze. In fact, others have found that if you simply change the learning task, the dull ones may do just as well or better. Clearly, it is quite difficult to definitively show that you have evolved a more intelligent strain since there is such a wide

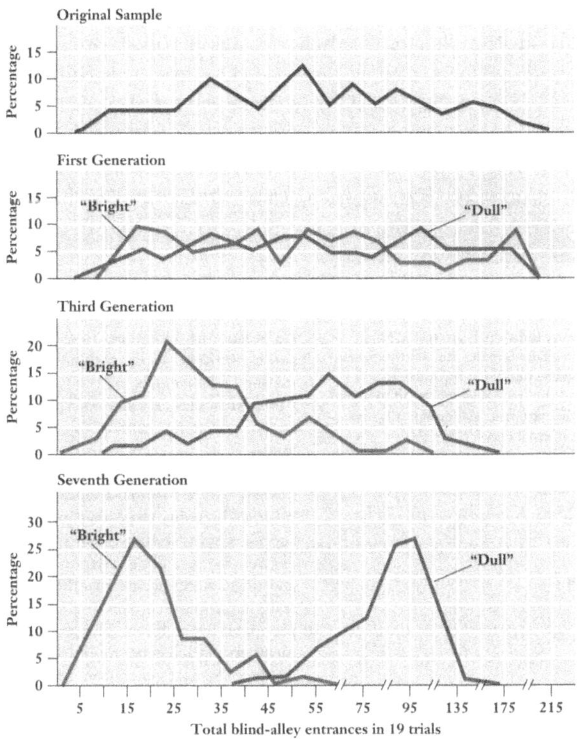

Figure 2.8: Results of selective breeding in rats for learning capability (figure taken from [223], © 1991, 1994, and 1999 by Worth Publishers Inc., and used with permission).

diversity of factors that characterize intelligence. Tryon's study is, however, quite interesting since no matter what characteristics actually evolved, some of them were characteristics that contribute to improved learning, which is a key aspect of intelligence.

2.5.5 Biomimicry for Automation: Evolution

Evolution shapes all biological organisms and hence can be used in many bio-inspired methods. Above, from the basic principles of evolution it should be clear that there are adaptive and optimization processes in evolution that may offer ideas to the design of components of control and automation systems. The notion that evolution provides for a biological version of a type of (stochastic) optimization process is especially fundamental. The types of questions we study are as follows:

- *"Darwinian" design:* Can an evolutionary strategy be used in computer-aided control system design to construct a controller (or estimator) to meet some specified performance objectives that are quantified with a fitness function? Could evolutionary principles be used in tuning parameters of a control strategy (e.g., tuning a planning strategy)? Could such an evolutionary design approach be used in a massively parallel experiment where each controller is thought of as an individual and each plant as its environment to obtain a novel approach to robust control system design?
- *Evolution and learning:* Can we evolve appropriate learning capabilities for a controller that fit what is needed for the aspects of the plant that are not known a priori and hence need to be learned online? Can evolution create a balance in a controller between what "instincts" are needed and what learning is needed? What are the effects of learning on evolution?
- *Evolution for online adaptation:* Can evolutionary strategies be used in online optimization approaches that result in evolution of good models or controllers for a plant (e.g., ones used in a planning strategy)? Can evolutionary and learning principles be used in tandem to achieve good performance?
- *Evolution of hierarchies and foragers:* Can we evolve the best command, communication, and control structure for a hierarchy of intelligent agents? Can we evolve "cognitive" and "physiological" characteristics (e.g., planning and learning algorithms, and vehicular hardware) of a social foraging swarm of vehicles?

2.5.6 Evolution of Control Technology: Global Perspective

Most people think of evolution as operating at the level of the individual organism, but others have proposed that it operates at the population level (i.e., adaptation occurs so that the whole population can survive, possibly via altruistic behavior that results in an individual's sacrifice), species level, or ecosystem level. Stretching these ideas even further, the control engineer who reads the literature on evolution will perhaps be tempted to view the development of the entire field of control systems engineering that has occurred over the past 50 years as a type of evolutionary process (from this view you may conclude that if you draw an analogy with evolution of life, we are perhaps at an early Cambrian period in the evolution of control systems). You may develop this view by using Dawkin's "memes" that are ideas that can survive and propagate (i.e., they are hypothesized to be analogous to genes in biological systems). For example, the most successful controllers or approaches to develop controllers "survive" in the sense that they are used more, and then are propagated since they are successful. Or, you may apply this to analysis methods for control systems (e.g., Lyapunov theory and related methods have grown and improved incrementally, and have survived the test of time due to their high fitness).

However, one must be careful with this sort of thinking since evolution in such a domain can certainly be quite different from that in living systems such as bacteria. There are many other influences on the field of control, including textbooks, education, computing technology, economics, politics, "hype" of new technology, and so on. While a full explanation of how the field of control has evolved is useful, we will leave this to historians and philosophers (or to the interested reader in Exercise 2.3).

2.6 A Control Engineering Viewpoint

While the connections to biology are interesting, and sometimes provide useful ideas for controller development, the focus here is firmly on control engineering for practical systems. In this section we seek to refocus our discussion back to control and automation engineering since this is the central focus of this book.

2.6.1 Why All the Biology? Do I Need It?

Some control engineers would complain that much of what is done in control engineering and automation and even the field of intelligent control is quite independent of the fields of neuroscience, psychology, foraging, evolution, or any other of the social or life sciences. Are these other areas just a distraction to the central focus of engineering control and automation systems? The answer is "no" for the following reasons:

Exploit biomimicry for functionalities and a cohesive view, get your hands dirty with simulations and implementations, and use disciplined mathematical approaches.

- Biomimicry provides a cohesive framework to think about the development of (complex) control and automation systems.
- Biological systems provide another viewpoint on the dynamics, functionality, design, and operation of high technology control systems for automation. Alternative perspectives are generally valuable in engineering to gain new insights on how to improve approaches and to deepen our understanding, especially for complex automation systems.
- Some of the concepts from biology seem more familiar to us, and hence help to teach the engineering methods.
- Intelligent control methods were originally inspired by biological systems, and while they may have departed over time from their original forms (or may never have been accurate models of their biological counterparts), researchers and practitioners can often return to the original biological system and use analogies (or more accurate models) to get ideas for how to improve the methods.
- By tracking the advancements in the foundational sciences, we may gain more ideas about how to engineer control and automation systems. Nature has things to teach us. The most complex and robust control and automation systems in existence are those in living organisms.

- The foundational sciences can sometimes provide a tool by which you can explain what you are doing to nonengineers (explanation via analogy is used very effectively by some people).
- The analogies are found to be *fun* by many students and control engineers! It is best to take this type of perspective as you learn the methods in this book. We all learn in different ways; some learn much more quickly and gain much deeper insights via the analogies with the foundational sciences. You must be cautioned on this, however. The treatment of and connections to biology are rather shallow at this point for several of the topics. This is true both in this book and in most of the field of intelligent control, so the analogies sometimes tend to break down if they are taken too far. However, by seeing how they break down you may get other research ideas.
- Strictly speaking, however, you do not need to know much about the foundational sciences to be able to work in this field, or apply the methods. (Try not to let the biology hinder your understanding rather than enrich it.) There are many ways to view the methods as being quite conventional; these will be pointed out below and throughout the book.

2.6.2 The Focus Is Engineering and Not the Foundational Sciences

As we emphasized above, we do not care if the neural, fuzzy, expert, planning, attentive, learning, or genetic systems model their biological counterparts—we are simply trying to get ideas from how they work to solve engineering problems. We do not care if the biofunctions are for control in the biological system (although they often tend to be); we will massage them into forms for control. We will not strictly adhere to the biological system analogy in the development of our control systems. In other words, we seek inspiration from biological systems, but when it is convenient we will not follow the functionalities or ideas that are suggested by biology. To give some firm examples to illustrate this point, we will use least squares methods to train neural networks or fuzzy systems, even though it is not likely that we learn from examples precisely in this manner. We may design a control system that uses a genetic algorithm with parameters set to values that are likely to be far different from how nature would have them. Basically, our focus is not on contributing to the sciences that are foundational to control. Our focus is on control engineering. We recognize the potential value of what we do to the underlying sciences (via analogies, methods, etc.), but we leave it to others to exploit this.

Finally, we note that our focus is quite different from many parts of the field of artificial intelligence (AI) and cognitive psychology. Often, AI researchers are actually concerned with modeling how humans think and act, along with the peculiarities and mistakes that they make (there is good reason to do this as it can be useful in understanding human behavior and can have applications in, for example, human-machine interface design). Here, we take a pragmatic engineering approach where the focus is always on trying to construct the best controller possible, and in particular we will focus on how to prove that it will deliver good performance under many different conditions so that it is robust. (Put another way, we use inspirations from biology, but we do not want "bugs" in our algorithms that implement our controllers—pun intended.) We take a conventional control engineering approach in doing this.

2.6.3 Close Relationships To Conventional Control

Since human cognition is used in conventional control system design, intelligence is embedded in conventional controllers and many of the methods developed prior to the introduction of the field of intelligent control actually emulate the way that humans would solve control problems if they were to do it themselves (e.g., consider the analogy between "model predictive control" and human planning functions in the frontal lobes). Due to this, there are actually close relationships between many of the methods in this book and others in the conventional control literature (e.g., there are close relationships between the methods of learning and adaptive control in this book and those in conventional adaptive control theory).

Figure 2.9 shows many relationships between conventional and intelligent control. For each part of the book, after the introduction, the biological motivation is provided, then the intelligent control methods, then some related conventional optimization and control methods. This figure shows the basic layout of the book.

It is useful to understand the basic science, and the corresponding field in conventional control, to understand the methods and ideas in intelligent control. This can help to show where the intelligent control methods offer advantages, and where a conventional approach may work better. Moreover, for researchers it is essential to know the corresponding methods in conventional control, otherwise there is the danger of producing solutions to problems that have already been adequately solved with conventional control methods. For example, if PID control works quite well for a certain class of problems, you probably should not waste the time to try to come up with some (possibly sophisticated) intelligent control (or advanced conventional) method to solve that type of problem.

2.6.4 Themes: Optimization, Adaptation, and Decision-Making

There are three central themes in this book: optimization, adaptation, and decision-making. There are several connections between these:

Optimization is a fundamental unifying concept in decision-making, adaptation, and design.

- Automation is achieved via control, often feedback control.
- From the biomimicry perspective, control can be thought of as a decision-making process.

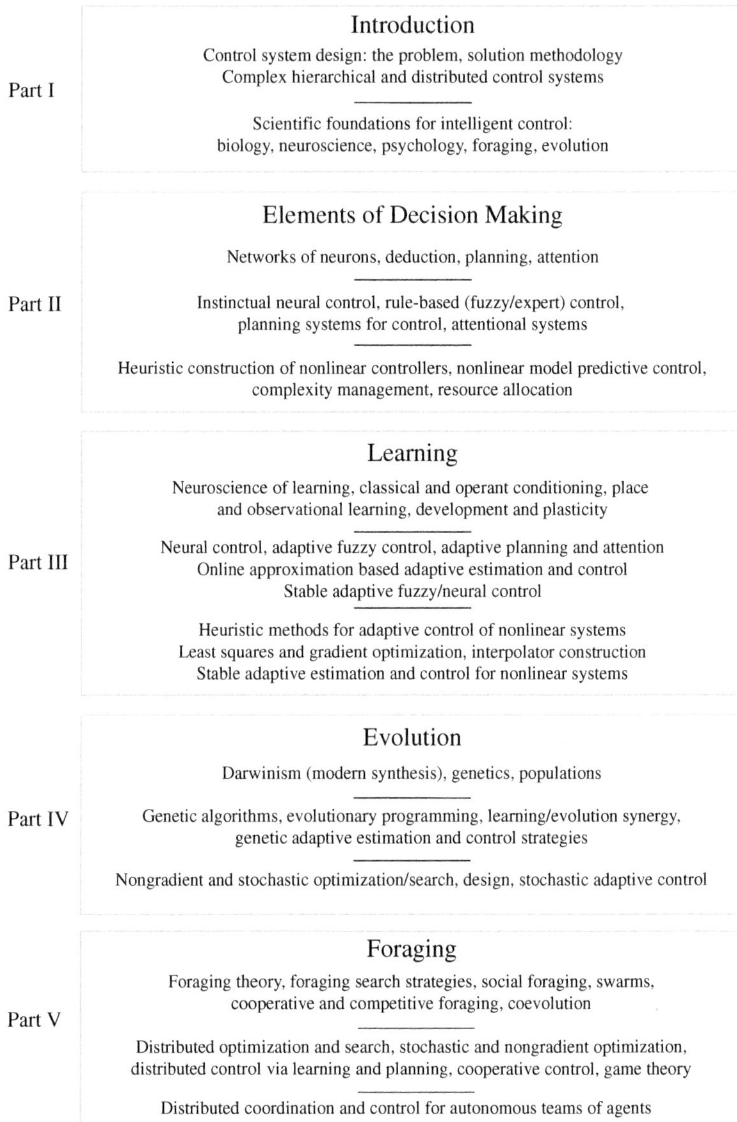

Figure 2.9: Relations among biological systems, intelligent control, and conventional control for each part of the book.

- Optimization is a fundamental process that operates to achieve learning (e.g., we try to find the best information) and hence adaptation.
- Optimization and adaptation are key features of many decision-making strategies (e.g., optimization can be used for learning a model that is used

in a planning or control strategy).

- Foraging is an optimization process where food is sought and predators are avoided. Optimization, learning, and decision-making are all used by an intelligent forager.
- Evolution is a design strategy, an optimization process, and an adaptation method. We can evolve optimization methods (e.g., a foraging strategy), adaptation methods (e.g., a learning strategy for control), and decision-making processes.

It is for these reasons that optimization is an especially important topic for this book. We will see that there are many types of biological optimization processes. There are gradient-type methods used in organisms for learning, methods for selecting the best plan, strategies for picking which object to pay attention to, foraging strategies, and the mother of all these: evolution.

The significant focus on optimization and adaptation, along with the concern for stability and adherence to conventional design methodology, will provide the control engineer with a firm connection to conventional control engineering methodology.

2.6.5 Intelligent Vs. Conventional Control? No Actual Conflict!

Note that while historically there have been times when the methodologies of conventional and intelligent controller synthesis were thought to be diametrically opposed to each other, the discussion here should convince you that they are not. The approaches complement one another at times, overlap at times and are very similar, or, other times, are most appropriate for very different application domains. Generally, too much time has been wasted on focusing on conflict, rather than how to exploit the best of both areas.

Regardless, when used for the same application, it is often very difficult to determine which is the best approach to control a system (whether comparing conventional methods with each other or to intelligent control methods) since such a determination is highly application dependent and involves many factors such as cost, understandability of the method, reliability, which offers the quickest solution that performs reasonably well, and others—not just the traditional measures of closed-loop system performance such as stability, robustness, rise-time, and overshoot. This fact has been ignored too many times.

2.6.6 The Goal is Not "Intelligence," It Is Autonomy

Finally, we briefly discuss the sometimes troublesome use of the word "intelligent" in intelligent control. We justify its use due to the fact that we automate human control intelligence or mimic how an intelligent biological system operates in order to solve a control problem. However, we must also highlight the fact that what we mean by "intelligence" is always in a state of flux, not just due to the fact that there is no precise definition of intelligence that is universally accepted (indeed, this is a research problem that is still receiving attention in psychology), but due to the continual influences of advances in automation technology and perceptions of the public.

As an example, while it may be difficult to imagine, years ago an automatic temperature regulation system for the home was considered "smart" since it performed a task that was usually performed by humans. For the first automobiles we had to regulate the speed manually, but as new technologies and innovations were introduced, the cruise controller was introduced and in its early days it certainly seemed "intelligent," at least to a certain degree. More recently, algorithms and sensors for what is called "intelligent cruise control" have been introduced and these will also slow the vehicle when it approaches another vehicle. Notice that even today designers are willing to add the word "intelligent" simply because it is performing what is considered a more advanced function, one that in the past a human had to perform. The marketing and sales department may even be enthusiastic about adding this word!

While a noble goal, we are not concerned with whether we achieve "intelligence." We only seek verifiably correct highly automated systems.

In many products, one can find this willingness by engineers and the general public to call something "intelligent" that was once performed by a human and is now performed by a computer control algorithm. Generally, we seem more willing to use the word "intelligent" if the algorithms and tasks seem relatively complex in relation to current automation capabilities. For instance, today people are hesitant to call their thermostat intelligent but are willing to refer to a system that automatically drives a vehicle from San Francisco to New York as intelligent, especially if it traveled east to west. It seems that this trend is not likely to end. We call the latest technologies intelligent, and old technologies seem boring and algorithmic.

Moreover, there is a tendency to view some systems as having a high degree of intelligence and others as having a low degree of intelligence. Perhaps this comes from how we view various biological organisms, but generally the more functions that a system has automated that were recently performed by humans, and the more "autonomous" the system is thus made, the more we are willing to attribute a high degree of intelligence to it. Clearly, the system we may be willing to call "highly intelligent" today, might not be considered as such ten years from now. It all depends on how much technology has advanced, and in particular, what is considered to be the state-of-the-art in automation at that time. Hence, our views of degrees of intelligence of systems changes over time.

Due to this apparent time-varying nature of "intelligence," the essential goal in this book is *not* to produce an "intelligent" controller. The goal is simply to do a good job in control design and the construction of reliable and highly efficient fully automated systems.

2.7 Exercises

Exercise 2.1 (Control Systems in Biology)*: Using [91] (or another appropriate reference) in an analogous manner to Exercise 1.1, draw a block diagram representing the control systems for the following components of biological systems. Be sure to properly label the block diagram.

(a) Homeostasis in the human body, specifically, thermoregulation.

(b) How plants can track light sources.

(c) How two-legged animals can balance while standing.

(d) The human immune system.

Exercise 2.2 (Systems and Control Theoretic Views on Cognitive Neuroscience)*: For background on stability theory, see Section 4.6 on page 141.

(a) Read the chapters in [206, 91, 268] on the human motor system. Draw a functional block diagram of the human motor system as a hierarchical control system. Clearly identify low level controls, communication links, and high level controls. Explain the role of learning in the system and discuss effects of evolution on the structuring of the system.

(b) Using [206, 404] as references specify a functional block diagram that views the "supervisory attentional system" in the human brain as a hierarchical control system.

(c) Using [206, 404] as a reference, and possibly other sources that you find in the library, explain how what psychologists call "attention deficit disorder" (ADD) could perhaps be explained as a type of instability that results from the improper functioning of the attentional system of the human brain. In your discussion consider the definitions of bounded-input bounded-output stability, stability in the sense of Lyapunov, and asymptotic stability. Typical drugs that are used to treat this condition are actually stimulants. Why would a *stimulant* help reduce the effects of ADD (e.g., improve concentration)?

Exercise 2.3 (Systems and Control Theoretic Views on Evolution)*:

(a) Read [139] and explain the "Baldwin effect," which focuses on some interactions between learning and evolution. For more details on this effect you may consider the discussion in [363] or some of the references there. Explain how this effect could be useful in control systems.

(b) Read [273, 272], focusing on the parts where stability is discussed ([272] is the "light" version of [273] and is very readable, so even though you may not finish [273], you should read all of [272]). Explain what the author believes the role of stability is in evolution. What role does he imply stability has in morphology? A continuation of this problem is given in Exercise 16.8 on page 752.

(c) Read [126]. Consider the global enterprise of control systems development and implementation. It began over 50 years ago and has "evolved" over the years to its present form today. Define a "meme" for and mechanisms of evolution that drive the development of (evolution of) control systems. For this, do not focus on the development of a specific control system or even a class of control systems. Consider the development of control systems by the worldwide engineering community. Discuss the effects of education, textbooks, computing technology, sciences, conferences, and journals. Can you use this type of evolutionary thinking to explain why PID controllers are in such widespread use? What are the essential factors that have led to this situation?

(d) This is a continuation of part (c). Read [220]. Now, using Gould's ideas about the success of simple organisms, expand your discussion on the success of PID controllers. Also, what are the "walls" that will ultimately limit the attainment of excellence in control engineering? Have we hit any of these walls?

Chapter 3

For Further Study

To deepen your knowledge of hierarchical intelligent control, its connections to cognitive neuroscience, and complex applications (e.g., vehicular or robotic) where the methods of this part are used, see [11, 78]. For more discussion on autonomy, see [22] and for a philosophical discussion on the advantages and disadvantages of biomimicry, physics, and mathematics in control and automation system development see [407]. To gain a better understanding of cognitive neuroscience, a good place to start is to read [421] and then to deepen your knowledge, read [206, 268]. There are many good books on general psychology (see, e.g., [223]) and you may want to pursue several of the topics you find there in more detail (e.g., in learning theory). To deepen your understanding of evolution, you could start with [537] and then read [185].

Conventional Control: The more that you understand about conventional control, the more you will be able to appreciate some of the finer details of the operation of intelligent control systems. We realize that all readers may not be familiar with all areas of control, so next we provide a list of books from which the major topics can be learned. There are many good texts on classical control [183, 292, 184, 153, 131, 29]. State-space methods and optimal and multivariable control can be studied in several of these texts and also in [187, 100, 15, 32, 341]. Robust control is treated in [398, 154, 557]. Control of infinite dimensional systems is covered in [181, 111]. Nonlinear control is covered in [277, 526, 33, 470, 518, 519, 257, 289]; stability analysis in [360, 357]; and adaptive control in [254, 289, 448, 30, 219, 376]. System identification is treated in [331] (and in the adaptive control texts), and optimal estimation and stochastic control are covered in [291, 315, 314, 228]. A relatively complete treatment of the field of control is in [311] and there are many further references there.

Hierarchical Intelligent Control: For a general introduction to the issues in hierarchical intelligent and autonomous control, see the books [21, 234, 520,

536, 517, 11] or articles [22, 20, 10, 489, 77, 78]. The general areas in conventional control of "large scale systems" (or "interconnected" or "decentralized" systems) [357] and game theory [47] are particularly relevant to the study of hierarchical intelligent control. There has probably been more applications-directed research activity in the use of hierarchical distributed controllers, and intelligent autonomous control, in the area of robotic systems (and "swarm robotics") [76, 77, 78, 44, 258, 23, 520, 144, 156, 73, 11, 483, 135] than any other application area. Connected to this research are studies on intelligent manufacturing systems [294, 11]. For an introduction to control system problems in intelligent vehicle and highway systems, see [177].

Planning systems are also discussed in [444, 387, 136, 11] (a nice discussion on hierarchical and adaptive planning is in [136]) and in [137]. Agents are discussed in [444, 535] and elsewhere. For some examples of distributed multiagent systems, see [409] where the FMS and computer network load balancing problems discussed in the section on multiagent systems are discussed in more detail. Another example of a distributed multiagent system is given in [17], where the authors study distributed adaptive fuzzy control for an FMS.

For more details on design and software implementation of complex hierarchical and distributed control systems, see [197] or the introductory article [365] (where other software packages are also discussed). A relevant study of software for real-time control (the "open control platform") is in [446].

Computer Science and Engineering: There are many subfields in electrical engineering and computer science that contribute to aspects of computer software and hardware development, concepts, and methodology that are relevant to the development of control systems. For instance, the area of computer languages, structured programming, and processor technologies has a significant impact. The emerging area of software architectures [465] can have an impact on how we structure large complex control system software. The important area of software engineering has a significant impact on the methodologies we use to develop and maintain computer software that is employed in large automation systems [475, 449] (the discussion here is based on [449, 72]). Moreover, of course, ideas from theoretical computer science (e.g., some work in automata theory) and AI [444, 387, 136] are sometimes used in the field of control systems development, and the relationship to AI was discussed in this chapter.

Discrete Event and Hybrid System Analysis: For an introduction to stability analysis of discrete event systems, see [409]. For more information on hybrid systems, see the special issue of the journal [18], the book [19], or [416, 415, 551, 75, 150, 149, 340, 151, 405]. The importance of providing a mathematical definition of autonomy (see Exercise 1.9), and some initial ideas on such a characterization, was first presented by the author in a panel discussion on autonomy at the 1998 IEEE Int. Symp. on Intelligent Control. Note that this definition would be useful in a hybrid systems analysis context, but significant future research is needed in autonomous control systems analysis via hybrid system methods to determine exactly how (for fun, in Exercise 1.9 you are asked to mathematically define autonomy).

Biological and Robot Intelligence: Questions about whether we will be able to implement high levels of human intelligence with computers have been discussed in many contexts and bodies of literature, including psychology, artificial intelligence, and robotics. Some of these discussions get rather philosophical. Others, such as the ones in [368, 11], have an engineering flavor where the focus is on raw computing capabilities, technological progress in automation, and how this is envisioned to rival intelligent human behavior in the future based on current progress for specific highly automated systems. The important conclusion that many reach from these types of studies is that levels of intelligence will be achieved in the near future that compete with humans along many dimensions so "intelligence" can be thought of as a goal, rather than some mystical unachievable functionality that is specific only to human behavior.

Cognitive Neuroscience and Evolution: A good introduction to biology is contained in [91] and introductory ideas in human physiology are given in [346]. Other areas that use biomimicry are discussed in [59]. There, the part on the use of DNA for computing is particularly interesting; however, for even more details on molecular computing, see [13, 195].

Under the heading of "cognitive neuroscience," for convenience, we will group the areas of cognitive psychology, neuroscience, neurobiology, neurophysiology, neuropsychology [318], the psychology of learning, cognitive science, cognitive neuroscience, and others. There is a huge literature in each of these areas. One place that has a nice synthesis of the ideas in these areas that are relevant to the area of intelligent control is in [206, 268]. A detailed treatment of the neuron is given in [312, 268]. There is also a popular literature on how the brain operates, and along these lines the reader may want to consider [421, 122]. To learn more about neuroscience, learning, and evolution of language, see [420].

Skinner's pigeon example was taken from [106]. More information on the capabilities of humans in performing control tasks, including some discussion on the advantages and disadvantages of using humans or computers and hardware to achieve automation, is given in [538].

Group behavior of organisms is discussed in the areas of swarm intelligence and artificial life [73, 6, 434, 313]. The authors in [73] give a particularly intriguing view of how ant colonies provide the inspiration for the solution to engineering problems. The description of the biology of the *E. coli* bacteria was taken from [342, 384, 8, 87].

Evolution is a field in biology that has a long and rich history. Scientific introductions to evolutionary biology are given in [185, 436] and a brief tutorial introduction is given in [91] and part of the writing in this part was based on that and [348]. For more details on evolutionary game theory, which has close ties to control theory, see [472, 473, 245, 534] (see also the "For Further Study" section at the end of Part V). For an introduction to ideas on how the human brain evolved, see [206, 162]. Evolutionary impacts of human memory are discussed in [16, 450]. For a readable introduction to the area of evolution, see [139], and the Baldwin effect is discussed there and also in [243, 53, 363].

There is, in fact, a significant amount of popular literature on the topic and much of this is referenced in [139]. Some interesting books include [126, 220] (which are discussed in Exercise 2.3), [127, 128, 129], and other books by Gould mentioned in [220]. A recent interesting perspective, one that is likely to resonate with a system/stability theorist, is provided in [273, 272]. If you are interested in the debate between creationism and evolution, [359] provides a recent study.

This book has its roots in the cybernetic tradition. Early references in cybernetics by Weiner and others are given in [421]. In cybernetics, in addition to the work of Weiner, the book [26] about the brain and the origins of adaptive behavior (e.g., learning) will likely be quite interesting to a system/stability theorist.

Finally, it is interesting to note that the fields of mathematical psychology and mathematical biology sometimes offer interesting perspectives and useful research to bridge the gap between biological systems and engineering (the mathematical models and analysis help build the bridge).

Part II

ELEMENTS OF DECISION MAKING

Part Contents

Sequence of Essential Concepts

- Networks of neurons for motor control used in many animals (e.g., for swimming or walking) are "hard-wired" (i.e., they do not support learning) and hence can be viewed as "instinctual neural controllers." Computer models of such neural networks can provide sophisticated stimulus-response characteristics that allow them to serve as general-purpose controllers. In biology, the "design methodology" for such hard-wired neural controllers is provided by evolution, while in engineering, we are confronted with the often complicated task of choosing the types and interconnections of the neurons so that they provide appropriate controller functions; this will later motivate the need for automatically learning or evolving neural network parameter values.

- Humans are employed to solve a wide array of feedback control tasks. The fuzzy and expert control methodologies provide two "rule-based control" methods to distill human control expertise into computers (e.g., rules about what actions to take in different situations); they provide methods for "human mimicry." From a control-theoretic perspective, they provide a heuristic construction procedure for nonlinear controllers. From a biological perspective, they provide a way to overcome the design difficulties for instinctual neural controllers via a convenient vehicle for the exploitation of domain-specific heuristics. They emulate the "software level" of deduction (or "behavioral level" of control tasks) while the neural networks we study emulate the physiological level.

- Humans who perform control tasks often use "mental models" of the environment (problem domain) to plan ahead and to select actions that appear to best lead to achieving their current goals. Controllers that use such planning rely on the use of a model of the plant (e.g., a design model) to predict how the plant will react to different inputs. Then, optimization methods are used to pick the sequence of inputs that best leads to achievement of goals. Finally, the first input (action) from that sequence is input to the plant and the process repeats. Model inaccuracies lead to poor predictions and hence, inputs that may not lead to achievement of goals; however, properly designed controllers use feedback to compensate for the model inaccuracies. Planning strategies provide for very general and widely applicable control and automation methods.

- Attentional systems allow an organism to focus on important information, allocate cognitive resources, and manage information complexity. There are elements of planning (learning) in attention and vice versa: an attentional system may plan (learn) what to pay attention to, and an attentional system can be used to decide what to plan (learn, respectively). Attentional mechanisms are a foundational component of intelligent systems and they can be employed in neural, fuzzy, expert, planning, and learning systems for control.

Chapter 4

Neural Network Substrates for Control Instincts

Chapter Contents

We first provide examples of how biological neural networks help to implement instinctual control functionalities in some simple organisms. Then, departing somewhat from biology, we introduce two types of neural network models that ignore many details of real neurons and their interconnections (e.g., voltage spikes and dynamics) to produce "firing rate models" with specific forms for "tuning curves." In this chapter we also ignore learning and evolution and thereby only model a special type of control instinct with limited functional capabilities (e.g., the functions and parameters of our models will not change over time). Also, we only model functionality of the neural network, and not the many other sensory and actuation functions needed for an organism to achieve control.

Next, we explain how our neuron models can be viewed as building blocks (sets of tuning curves) for creating a neural network that can implement a complex input-output mapping. To do this, we show how to take the mappings implemented by neurons and build by hand ("design") a controller for a specific engineering application. In biological systems, over long time periods, this "design" is a task of evolution; over short time periods such as a life time, it is the task of learning. Aside from ignoring learning and evolution, this design approach will at the same time represent an even more significant departure from biology as we ignore whether it has a biological basis (e.g., in development) and whether the constructed neural networks bear any similarity to those in any biological system. We only concern ourselves with constructing input-output mappings that will lead to the performance objectives being met. Biology provides building blocks with basic functionalities, and we pay no respect to whether we use these building blocks as biology would.

For specific engineering applications, we show how to simulate the neural network controlling the plant and evaluate whether the closed-loop system meets performance objectives. This will provide insights into neural network stimulus-response characteristics and their effect on closed-loop behavior. Moreover, it will serve as an introduction on how to evaluate control systems in simulation. Ultimately, however, certain difficulties in the design process, and the need for controllers with more sophisticated functionalities that will be highlighted in the simulations, will motivate the need to study learning in Part III and evolutionary methods in Part IV that automate the construction of neural networks.

4.1 Biological Neural Networks and Their Role in Control

First, we briefly outline some basics of how biological neural networks operate, specifically in controlling functions of a few simple organisms.

4.1.1 Neurons and Neural Networks

An invertebrate motor neuron was shown in Figure 2.3. There are, however, many different types of neurons, with, for example, the cell body at different

There are massively interconnected networks of neurons of many forms in organisms; engineering models often use the invertebrate motor neuron connected in special topologies.

locations. The "multipolar" configuration shown in Figure 2.3(a) is the one that is often modeled and then used in control engineering (and indeed, it underlies many other studies of neural networks in other engineering areas). Each neuron is composed of "dendrites" which allow for connections to the cell body, an "axon" which allows for connections to other neurons, and the connection points are called the "synapses." Generally, the cell body performs a type of summing and thresholding operation on signals obtained via the dendrites and provides an electrical signal (actually it is typically a sequence of pulses that are often called "spikes") that travels along the axon to other neurons. When such a signal is transmitted on the axon, the neuron is said to "fire." The inputs to the neuron on the dendrites can be "excitatory" (having a tendency to cause the neuron to fire) or "inhibitory" (having the tendency to restrict the firing of the neuron).

It is via the massive neural interconnectivity that complex reasoning and intelligence emerges.

The human brain is composed of a large, massively interconnected biological network of about 10^{11} neurons, each of which may have as many as 10^3 or 10^4 connections to other neurons (for a total of up to $100,000$ miles of neuron connections). These neurons dynamically interact with each other, change their properties over time (e.g., via learning), and even grow new connections to each other (e.g., during fetal development), to act as a sophisticated bio-electrical "computer" of sorts. While the interconnection of neurons in humans is extremely sophisticated (see Figure 4.1), here we only consider simple interconnections. For instance, the three neurons in Figure 4.2 are connected in a "feedforward" fashion (i.e., without connecting an axon of one neuron back to another neuron that has a path to that neuron), since this is a common interconnection strategy used in engineering applications.

Processing characteristics of individual neurons and network interconnections, and hence topology of the network, change via learning.

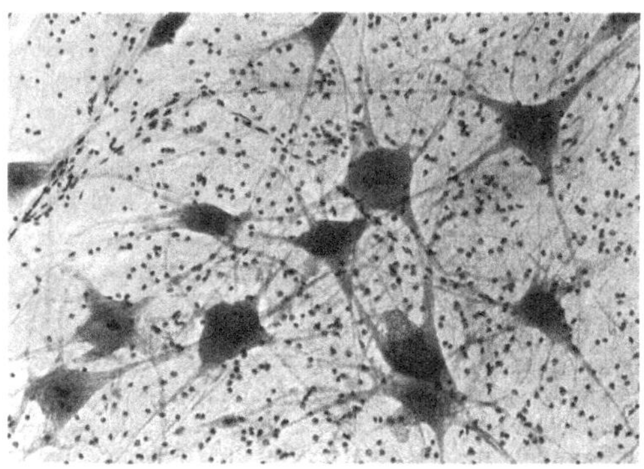

Figure 4.1: Network of motor neurons in the spinal cord, photograph taken through a microscope (figure taken from [223], © 1991, 1994, and 1999 by Worth Publishers Inc., and used with permission).

There are a number of neurons in our body and in other organisms that are "hard-wired" in the sense that their properties are fixed in a specific manner

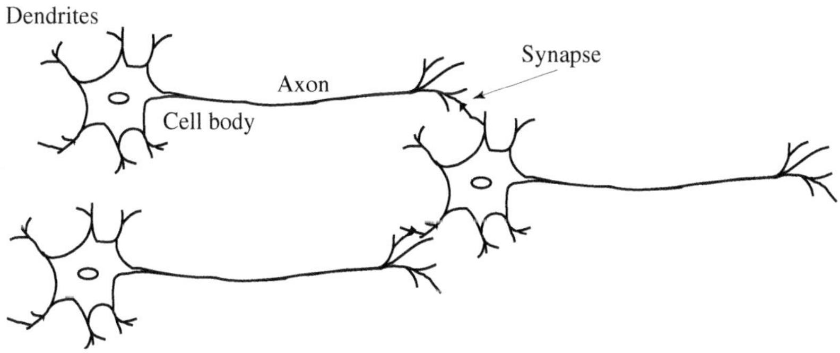

Figure 4.2: Three connected neurons, a simple biological neural network.

such that they have no ability to learn. For example, certain motor reflexes are implemented this way, and the functions that they implement are then sometimes said to be "instinctual" (i.e., they do not need to be learned). While there are a variety of functions in a human (or other organisms) that are instinctual or automatic, a perhaps more interesting case is when a neural network develops and neurons have an ability to learn. For instance, when humans are born they have very low neural network connectivity in their brain. As they develop and learn, the brain forms interconnections between different neurons and the resulting network defines the functional properties of the brain (yes, the environment does affect the actual connection structure in our brain). Moreover, to memorize information it is said that the interconnections between the neurons are modified and then fixed so that the information can later be recalled. This learning capability will be more fully investigated in Part III; here, the focus is on "instinctual" control functions, examples of which are given next.

4.1.2 Example: Instinctual Neural Control Functions in Simple Organisms

"Command systems" of neurons are used in biological systems for a variety of tasks, such as control of motion, locomotion, digestion, etc. Many animal behaviors, such as walking or swimming, result from a network of neurons called a "central pattern generator" that produces a pattern of signals that results in a rhythmic contraction and relaxation of muscles. More generally, instinctive responses are sometimes called "fixed action patterns" that are evoked by a "sign stimulus." Such behaviors can be quite complex, but are thought of as being rigidly evoked by a certain type of stimulus (i.e., the animal does not learn these responses, or forget them).

In this section we will show how in one organism, neurons can control movement and in another organism, how the neural network can respond to stimuli to produce movement. The goal here is simply to show neurons and neural networks "at work" in acting as controllers in biological systems.

Neurons That Control Swimming in a *Clione*

A simple neural "circuit" (interconnection of neurons) that can produce alternating contraction and relaxation in two different muscles that move "wing-like" structures called *parapodia* in the *Clione* (a small marine mollusc), is shown in Figure 4.3. There are two neurons for moving each wing, one for "upswing" and the other for "downswing." Each neuron has an inhibitory effect on the other so that when one is active, the other is not (i.e., it is inhibited by the other). The signals on the right show the basic electrical pattern between the neurons where when the voltage in the upswing neuron spikes (and actuates a muscle for moving the wing up), it inhibits the other neuron. However, after the upswing spike has decreased, the downswing neuron voltage increases and then spikes, which signals the muscle to move the wing down. The firing of one neuron to actuate the upswing inhibits the firing of the downswing neuron, and vice versa. This is called "reciprocal inhibition" and it is the key feature that allows the command system of neurons to generate rhythmic movement.

Neural networks with only a few neurons can control movements in simple organisms that are very useful for survival (e.g., locomotion for foraging or predator avoidance).

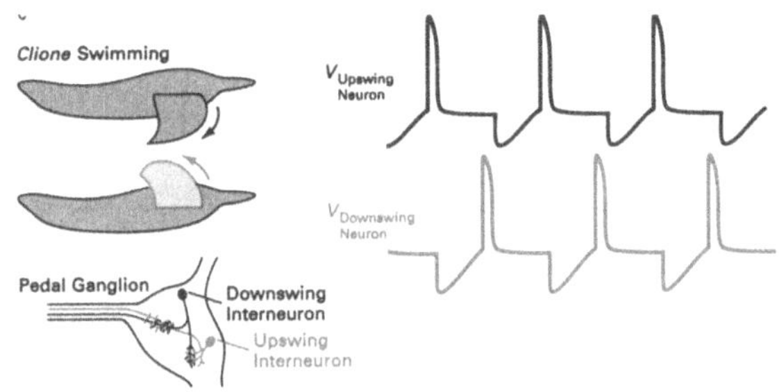

Figure 4.3: Command system of neurons (neural controller) for swimming in a *Clione* (figure taken from [312], © Oxford University Press, reproduced by permission).

Neuron Stimulus-Response Actions to Achieve Control in a Swimming Leech

The pulse-type voltage patterns ("spikes") in Figures 4.3 and 4.4 are typical for neurons; however, most engineering models do not represent neuron behavior to this level of detail.

The medicinal leech *Hirudo medicinalis* swims by making undulating motions with its body, somewhat like a snake or some fish (see Figure 4.4). The movements result from alternating contraction and relaxation of muscles that are located in the body wall of the leech. When it swims, there are rhythmic bursts from a central pattern generator that produce a "wave" of contraction that travels from the front to the rear of the leech. Reciprocal inhibition is used to produce the rhythmic motion in the leech, just as it was in the *clione* discussed in the last subsection.

The leech will start swimming if there is a brief strong mechanical stimulus applied to the body of the animal as shown in Figure 4.4. There, a sensory neuron detects the stimulus and starts firing (in the figure the "firing" is characterized by the spikes in the signal voltage of the sensory neuron). This sets off

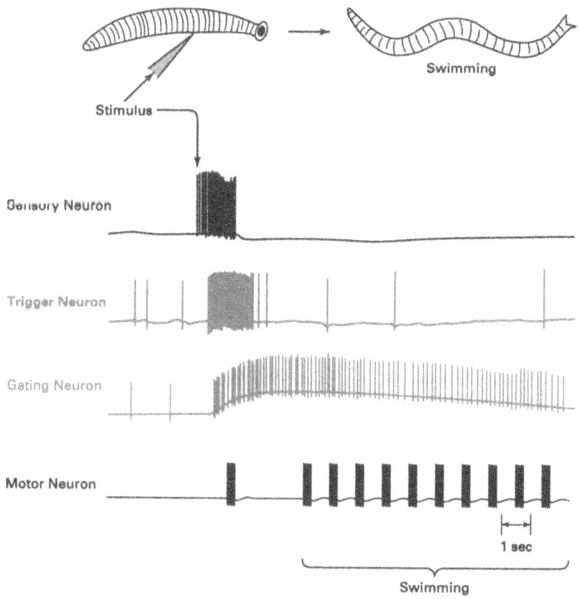

Figure 4.4: Neuron signaling connecting stimulus to swimming response in a medicinal leech *Hirudo medicinalis* (figure taken from [312], © Oxford University Press, reproduced by permission).

a "trigger neuron" that in turn starts a "gating neuron" to fire, and the firing of that gating neuron persists even when the stimulus is removed. When the gating neuron is active (i.e., when it fires), it sustains the rhythmic activity of the central pattern generator. The gating neuron makes a motor neuron active at regular one second intervals and these signal the muscle for swimming.

4.2 Multilayer Perceptrons

Next, we will explain how we model the physiological system of the neural network. It must be emphasized that the models here are not meant to be precise models of parts of a biological brain or neurons in any other organism. Essentially, they are "firing rate models" since they do not model voltage spikes, but have outputs that are thought of as being proportional to the frequency of the spikes. Moreover, they are "static" since they do not include, for example, dynamic systems to represent that currents or voltages in a neuron cannot change instantaneously. First, we consider a multilayer perceptron which is a feedforward neural network (e.g., it does not use past values of its outputs to compute its current output). It is composed of an interconnection of basic neuron processing units.

4.2.1 The Neuron

For a single neuron, suppose that we use x_i, $i = 1, 2, \ldots, n$, to denote its inputs and suppose that it has a single output y. Figure 4.5 shows the neuron. Such a neuron first forms a weighted sum of the inputs

$$\bar{x} = \left(\sum_{i=1}^{n} w_i x_i \right) + b$$

where w_i are the interconnection "weights" and b is the "bias" for the neuron. The signal x_i is the input to the ith dendrite and $w_i > 0$ represents an excitatory connection with larger w_i values representing dendrites that "amplify" their input signals more. Conversely, $w_i < 0$ represents an inhibitory input. The signal $\bar{x}$ represents a signal in the biological neuron that represents the combined effects of all the inputs from the dendrites.

The stimulus-response characteristics of a neuron can be changed by adjusting the weights w_i, bias b, or by using different types of activation functions f.

The processing that the neuron performs on this $\bar{x}$ signal is represented with an "activation function." This activation function is represented with a function f, and the output that it computes is

$$y = f(\bar{x}) = f\left(\left(\sum_{i=1}^{n} w_i x_i \right) + b \right) \tag{4.1}$$

Basically, the neuron model represents the biological neuron that "fires" (turns on and passes an electrical signal down the axon so that it can go to other neurons as shown in Figure 4.2) when its inputs are significantly excited (i.e., $\bar{x}$ is big enough). Normally, Equation 4.1 is represented as shown in Figure 4.5.

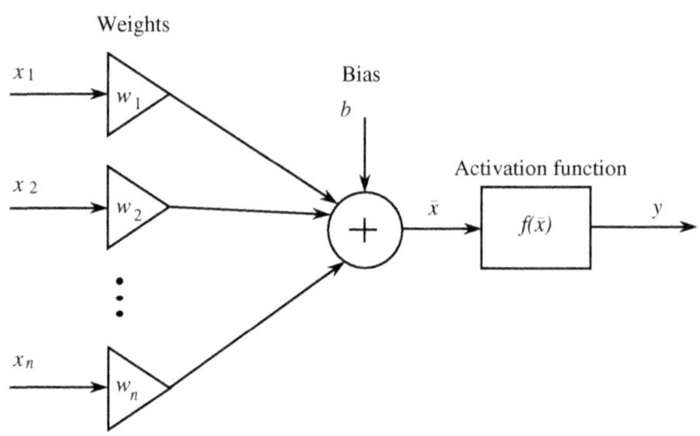

Figure 4.5: Single neuron model.

The manner in which the neuron fires is defined by the activation function f. There are many ways to define the activation function:

- *Threshold function:* For this type of activation function, we have

$$f(\bar{x}) = \begin{cases} 1 & \text{if } \bar{x} \geq 0 \\ 0 & \text{if } \bar{x} < 0 \end{cases}$$

so that once the input signal $\bar{x}$ is above zero the neuron turns on (see Figure 4.6).

- *Linear function:* For this type of activation function, we simply have

$$f(\bar{x}) = \bar{x}$$

and we think of the neuron being on when $f(\bar{x}) > 0$ and off when $f(x) \leq 0$ (see Figure 4.6).

- *Logistic function:* For this type of activation function, which is a type of "sigmoid function," we have

$$f(\bar{x}) = \frac{1}{1 + \exp(-\bar{x})} \tag{4.2}$$

so that the input signal $\bar{x}$ continuously turns on the neuron an increasing amount as the input increases as shown in Figure 4.6 (note that for the logistic function $f(0) = 0.5 \neq 0$ but $f(0) - 0.5 = 0$ where 0.5 can be modeled by another bias so that you can think of the logistic function as "turning on" in a similar way to how the other functions in Figure 4.6 turn on).

- *Hyperbolic tangent function:* There are many functions that take on a shape that is sigmoidal. For instance, one that is often used in neural networks is the hyperbolic tangent function

$$f(\bar{x}) = \tanh(\bar{x}) = \frac{1 - \exp(-2\bar{x})}{1 + \exp(-2\bar{x})}$$

which is shown in Figure 4.6.

Equation (4.1), with one of the above activation functions, represents the "computations" made by one neuron in a neural network. Notice that the input-output characteristics of a neuron in a multilayer perceptron are quite different from the biological neurons discussed in the last section. Along with the assumption that the weights, a bias, and a summing operation represent part of what happens in the dendrites and cell body, the activation function output (and hence firing of the neuron) is not represented as a voltage spike that travels down the axon, or a sequence of such spikes (such as in Figure 4.4) that might be frequency modulated by the overall activation level of the neuron (e.g., have higher frequency spikes for greater activation levels as is sometimes found in a biological neuron). Essentially, a larger input to the activation function here (for a sigmoid function) simply turns the neuron on to a greater extent; the neuron here is a very simple (abstract) model of the behavior of *some* biological neurons called a "firing rate model." It is interesting to note, however, that the specific shapes for the above activation functions (and some others) have been experimentally demonstrated for firing rate models of real neurons, so there is some biological justification for the form or the model we use here. The specific

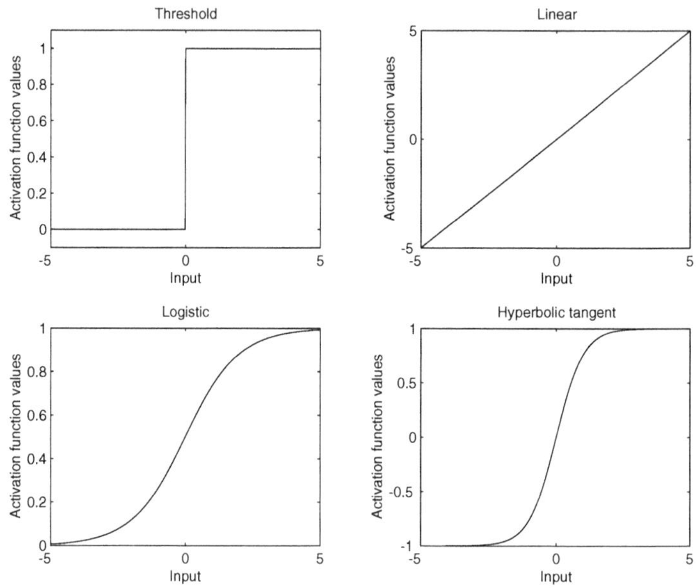

Figure 4.6: Activation functions for neurons.

shapes of the mappings produced by individual neurons are sometimes called "tuning curves" or "tuning functions" (they show how the neuron is "tuned" to a stimulus by showing for a whole range of stimulus inputs what the firing rate output of the neuron will be). Some neurons have tuning curves in the shape of sigmoids so that for some stimuli they are not on but as the stimulus changes, the neuron starts firing at a high rate above some threshold, if the slope of the sigmoid near the threshold is steep. Other neurons to be modeled in Section 4.4 have tuning curves in the shape of Gaussian functions so that they are "on" the most for a specific range of stimuli. See the "For Further Study" section at the end of this part.

Next, we define how we interconnect neurons to form a neural network—in particular, the feedforward multilayer perceptron.

4.2.2 Feedforward Network of Neurons

The basic structure for the multilayer perceptron is shown in Figure 4.7. There, the circles represent the neurons (weights, bias, and activation function) and the lines represent the connections between the inputs and neurons, and between the neurons in one layer and those in the next layer. This is a three-layer perceptron since there are three stages of neural processing between the inputs and outputs. The layer connected to the output is called the "output layer," and all the other ones are called "hidden" layers since they do not connect to the output.

The multilayer perceptron has inputs x_i, $i = 1, 2, \ldots, n$, and outputs y_j, $j = 1, 2, \ldots, m$. The number of neurons in the first hidden layer (see Figure 4.7) is n_1. In the second hidden layer there are n_2 neurons, and in the output layer there are m neurons. Hence, in an N layer perceptron there are n_i neurons in the ith hidden layer, $i = 1, 2, \ldots, N-1$.

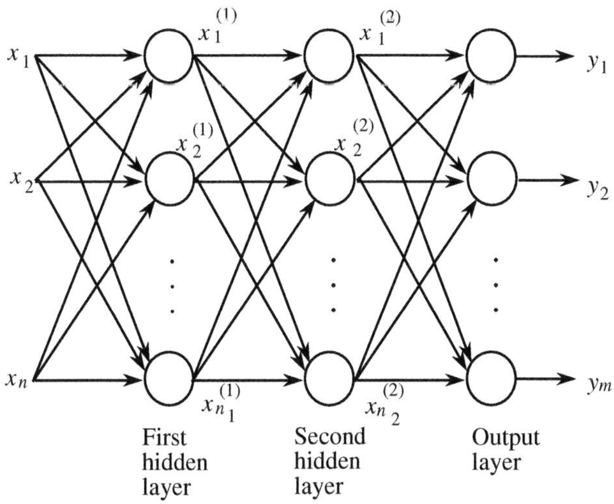

Figure 4.7: Multilayer perceptron model.

The neurons in the first layer of the multilayer perceptron perform computations, and the outputs of these neurons are given by

$$x_j^{(1)} = f_j^{(1)}\left(\left(\sum_{i=1}^{n} w_{ij}^{(1)} x_i\right) + b_j^{(1)}\right)$$

with $j = 1, 2, \ldots, n_1$. The neurons in the second layer of the multilayer perceptron perform computations, and the outputs of these neurons are given by

$$x_j^{(2)} = f_j^{(2)}\left(\left(\sum_{i=1}^{n_1} w_{ij}^{(2)} x_i^{(1)}\right) + b_j^{(2)}\right)$$

with $j = 1, 2, \ldots, n_2$. The neurons in the third layer of the multilayer perceptron perform computations, and the outputs of these neurons are given by

$$y_j = f_j\left(\left(\sum_{i=1}^{n_2} w_{ij} x_i^{(2)}\right) + b_j\right)$$

with $j = 1, 2, \ldots, m$.

The parameters (scalar real numbers) $w_{ij}^{(1)}$ are called the weights of the first hidden layer. The $w_{ij}^{(2)}$ are called the weights of the second hidden layer. The

w_{ij} are called the weights of the output layer. The parameters $b_j^{(1)}$ are called the biases of the first hidden layer. The parameters $b_j^{(2)}$ are called the biases of the second hidden layer, and the b_j are the biases of the output layer. The functions f_j (for the output layer), $f_j^{(2)}$ (for the second hidden layer), and $f_j^{(1)}$ (for the first hidden layer) represent the activation functions. The activation functions can be different for each neuron in the multilayer perceptron (e.g., the first layer could have one type of sigmoid, while the next two layers could have different sigmoid functions or threshold functions).

The simulus-response characteristics of a neural network can be changed via neuron parameters or their interconnections.

For convenience, we sometimes use

$$y = F_{mlp}(x, \theta)$$

to denote the multilayer perceptron where θ is a parameter vector that holds all the tunable weights and biases of the multilayer perceptron.

This completes the definition of the multilayer perceptron. Next, we will show how a multilayer perceptron's stimulus-response characteristics can be designed so that it can be used to regulate the heading of a ship.

4.3 Design Example: Multilayer Perceptron for Tanker Ship Steering

Here, we show how a neural network can be used to steer a ship that is traveling on the ocean. To do this, we first define the ship model and heading regulation problem. Next, we define the neural network, design its stimulus-response characteristics, and then evaluate how it performs in its ship steering task.

4.3.1 Tanker Ship Model and Heading Regulation

Our tanker ship heading regulation problem is shown in Figure 4.8. Here, the ship is moving forward in the indicated x direction at a nominal speed u, ψ denotes the heading angle (in radians), and δ is the rudder input (in radians). We will use ψ_r to denote the desired ship heading that is specified, for instance, by the captain (or route planner). The goal is to develop a control system that will ensure that ψ tracks ψ_r.

It is very important to achieve good heading regulation for ships since this reduces consumption of fuel. Steering performance can be affected by a variety of variables. It is known that the ship can travel at different speeds and this affects how the ship is steered (the rudder becomes less effective at very low speeds), that in general the ship weighs different on different trips (and heavy ships turn slower), that wind can be encountered on some trips and when wind hits the side of the tanker this can affect heading regulation, that water currents can affect steering, and that the sensor for ship steering provides a noisy measurement. Also, the rudder will only move between ± 80 degrees and this affects our ability to steer the ship.

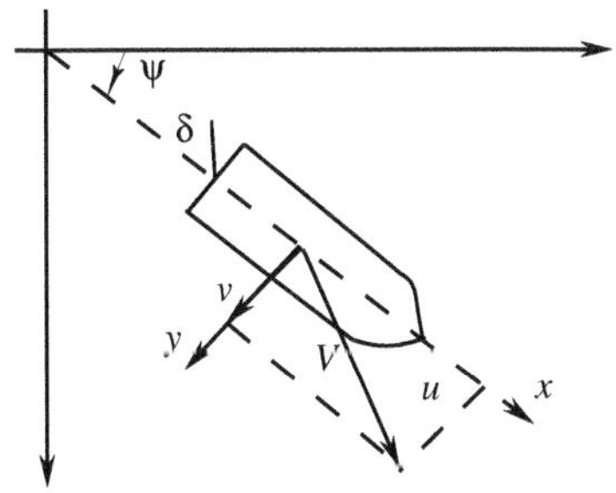

Figure 4.8: Tanker ship steering problem.

Ship Model

In order to study the behavior of the system, we will simulate it on a digital computer. To do this, we need to develop a computer program that is based on a nonlinear model of the ship; we will develop this model next. Often, ship dynamics are obtained by applying Newton's laws of motion to the ship. For very large ships, the motion in the vertical plane may be neglected since the "bobbing" or "bouncing" effects of the ship are small for large vessels. The motion of the ship is generally described by a coordinate system that is fixed to the ship [30] as shown in Figure 4.8.

A simple model of the ship's motion is given by

$$\dddot{\psi}(t) + \left(\frac{1}{\tau_1} + \frac{1}{\tau_2}\right)\ddot{\psi}(t) + \left(\frac{1}{\tau_1\tau_2}\right)\dot{\psi}(t) = \frac{K}{\tau_1\tau_2}\left(\tau_3\dot{\delta}(t) + \delta(t)\right) \tag{4.3}$$

where ψ is the heading of the ship and δ is the rudder angle. Assuming zero initial conditions, we can write Equation (4.3) as

$$\frac{\psi(s)}{\delta(s)} = \frac{K(s\tau_3 + 1)}{s(s\tau_1 + 1)(s\tau_2 + 1)} \tag{4.4}$$

where K, τ_1, τ_2, and τ_3 are parameters that are a function of the ship's constant forward velocity u and its length l. In particular,

$$\begin{aligned} K &= K_0\left(\frac{u}{l}\right) \\ \tau_i &= \tau_{i0}\left(\frac{l}{u}\right) \qquad i = 1, 2, 3 \end{aligned}$$

where we assume that for a tanker ship under "ballast" conditions (a very heavy ship), $K_0 = 5.88$, $\tau_{10} = -16.91$, $\tau_{20} = 0.45$, $\tau_{30} = 1.43$, and $l = 350$ meters [30]. For "full" conditions (a lighter ship), $K_0 = 0.83$, $\tau_{10} = -2.88$, $\tau_{20} = 0.38$, $\tau_{30} = 1.07$. If we do not say otherwise, we will simulate the tanker ship under ballast conditions. Also, we will assume that nominally the tanker ship is traveling in the x direction at a velocity of $u = 5$ m/s.

In normal steering, a ship often makes only small deviations from a straight-line path. Therefore, the model in Equation (4.3) was obtained by linearizing the equations of motion around the zero rudder angle ($\delta = 0$). As a result, the rudder angle should not exceed approximately 5 degrees, otherwise the model will be inaccurate. For our purposes, we need a model suited for rudder angles that are larger than 5 degrees; hence, we use the model proposed in [51]. This extended model is given by

$$\dddot{\psi}(t) + \left(\frac{1}{\tau_1} + \frac{1}{\tau_2}\right)\ddot{\psi}(t) + \left(\frac{1}{\tau_1\,\tau_2}\right)H(\dot{\psi}(t)) = \frac{K}{\tau_1\,\tau_2}\left(\tau_3\,\dot{\delta}(t) + \delta(t)\right) \quad (4.5)$$

where $H(\dot{\psi})$ is a nonlinear function of $\dot{\psi}(t)$. The function $H(\dot{\psi})$ can be found from the relationship between δ and $\dot{\psi}$ in steady state such that $\dddot{\psi} = \ddot{\psi} = \dot{\delta} = 0$. An experiment known as the "spiral test" has shown that $H(\dot{\psi})$ can be approximated by

$$H(\dot{\psi}) = \bar{a}\dot{\psi}^3 + \bar{b}\dot{\psi}$$

where $\bar{a}$ and $\bar{b}$ are real-valued constants and $\bar{a}$ is always positive. We choose the values $\bar{a} = \bar{b} = 1$. Also, we assume that the maximum deviation of the rudder angle is ± 80 degrees (or $80\pi/180$ radians).

Simulation of Nonlinear Systems

The ship model is nonlinear; hence, in order to simulate its behavior on a digital computer we need to discuss how to simulate nonlinear systems. In this subsection we give a brief overview of how to simulate general nonlinear systems. In the next subsection, we will return to the ship example and show how to develop a simulation for its behavior.

Suppose that the system to be simulated can be represented by the ordinary differential equation

$$\begin{aligned} \dot{x}(t) &= f(x(t), r(t), t) \\ y &= g(x(t), r(t), t) \end{aligned} \quad (4.6)$$

where $x = [x_1, x_2, \ldots, x_n]^\top$ is a state vector, $f = [f_1, f_2, \ldots, f_n]^\top$ is a vector of nonlinear functions, g is a nonlinear function that maps the states and reference input to the output of the system, and $x(0)$ is the initial state. Note that f and g are, in general, time-varying functions due to the explicit dependence on the time variable t. To simulate a nonlinear system, we will assume that the nonlinear ordinary differential equations are put into the form in Equation (4.6).

Euler's Method: Now, to simulate Equation (4.6), we could simply use Euler's method to approximate the derivative $\dot{x}$ in Equation (4.6) as

$$\begin{aligned} \frac{x(kh+h) - x(kh)}{h} &= f(x(kh), r(kh), kh) \\ y &= g(x(kh), r(kh), kh) \end{aligned} \quad (4.7)$$

Here, h is a parameter that is referred to as the "integration step size." Notice that any element of the vector

$$\frac{x(kh+h)-x(kh)}{h}$$

is simply an approximation of the slope of the corresponding element in the time varying vector $x(t)$ at $t = kh$ (i.e., an approximation of the derivative). For small values of h, the approximation will be accurate provided that all the functions and variables are continuous. Equation (4.7) can be rewritten as

$$\begin{aligned} x(kh+h) &= x(kh) + hf(x(kh), r(kh), kh) \\ y &= g(x(kh), r(kh), kh) \end{aligned}$$

for $k = 0, 1, 2, \ldots$. The value of the vector $x(0)$ is the initial condition and is assumed to be given. Simulation of the nonlinear system proceeds recursively by computing $x(h)$, $x(2h)$, $x(3h)$, and so on, to generate the response of the system for the reference input $r(kh)$.

Note that by choosing h small, we are trying to simulate the *continuous-time* nonlinear system. If we want to simulate the way that a digital control system would be implemented on a computer in the laboratory, we can simulate a controller that only samples its inputs every T seconds (T is not the same as h; it is the "sampling interval" for the computer in the laboratory) and only updates its control outputs every T seconds (and it would hold them constant in between). Normally, you would choose $T = \alpha h$ where $\alpha > 0$ is some positive integer. In this way, we simulate the plant as a continuous-time system that interacts with a controller that is implemented on a digital computer.

The Runge-Kutta Method: While Euler's method is easy to understand and implement in code, sometimes to get good accuracy the value of h must be chosen to be very small. Most often, to get good simulation accuracy, more sophisticated methods are used, such as the Runge-Kutta method with adaptive step size or predictor-corrector methods. In the fourth-order Runge-Kutta method, we begin with Equation (4.6) and a given $x(0)$ and let

$$x(kh+h) = x(kh) + \frac{1}{6}(k_1 + 2k_2 + 2k_3 + k_4) \tag{4.8}$$

where the four vectors

$$\begin{aligned} k_1 &= hf(x(kh), r(kh), kh) \\ k_2 &= hf\left(x(kh) + \frac{k_1}{2}, r\left(kh + \frac{h}{2}\right), kh + \frac{h}{2}\right) \\ k_3 &= hf\left(x(kh) + \frac{k_2}{2}, r\left(kh + \frac{h}{2}\right), kh + \frac{h}{2}\right) \\ k_4 &= hf\left(x(kh) + k_3, r(kh+h), kh+h\right) \end{aligned}$$

These extra calculations are used to achieve a better accuracy than the Euler method. We see that the Runge-Kutta method is very easy to use; it simply involves computing the four vectors k_1 to k_4, and plugging them into Equation (4.8). Suppose that you write a computer subroutine to compute the output of a fuzzy controller given its inputs (in some cases these inputs could include a state of the closed-loop system). In this case, to calculate the four vectors, k_1 to k_4, you will need to use the subroutine four times, once for the calculation of each of the vectors, and this can increase the computational complexity of the simulation. The complexity is reduced, however, if you can simulate the fuzzy controller as if it were implemented on a digital computer in the laboratory with a sampling interval of $T = \alpha h$ (see the discussion above). Also, sometimes $r(kh)$ is used in place of $r(kh + h/2)$ and $r(kh + h)$ in the above equations; this can be a reasonable approximation if r is constant most of the time and f and g are not time-varying functions (i.e., they do not have t as one of their arguments).

Generally, if the Runge-Kutta method has a small enough value of h, it is sufficiently accurate for the simulation of most control systems (and if an adaptive step size method is used, then even more accuracy can be obtained if it is needed). For more details on numerical simulation of nonlinear differential equations, see [332, 217, 508].

Simulating the Ship and a Digital Controller

Next, we need to convert the nth-order nonlinear ordinary differential equations representing the ship to n first-order ordinary differential equations; for convenience, let

$$a = \left(\frac{1}{\tau_1} + \frac{1}{\tau_2}\right)$$

$$b = \left(\frac{1}{\tau_1 \tau_2}\right)$$

$$c = \frac{K\tau_3}{\tau_1 \tau_2}$$

and

$$d = \frac{K}{\tau_1 \tau_2}$$

We would like the model in the form

$$\begin{aligned} \dot{x}(t) &= f(x(t), \delta(t)) \\ y(t) &= g(x(t), \delta(t)) \end{aligned}$$

where $x(t) = [x_1(t), x_2(t), x_3(t)]^\top$ and $f = [f_1, f_2, f_3]^\top$ for use in a nonlinear simulation program. We need to choose $\dot{x}_i$ so that f_i depends only on x_i and δ for $i = 1, 2, 3$. We have

$$\dddot{\psi}(t) = -a\ddot{\psi}(t) - bH(\dot{\psi}(t)) + c\dot{\delta}(t) + d\delta(t) \tag{4.9}$$

Choose

$$\dot{x}_3(t) = \dddot{\psi}(t) - c\dot{\delta}(t)$$

so that f_3 will not depend on $c\dot{\delta}(t)$ and

$$x_3(t) = \ddot{\psi}(t) - c\delta(t)$$

Choose $\dot{x}_2(t) = \ddot{\psi}(t)$ so that $x_2(t) = \dot{\psi}(t)$. Finally, choose $x_1(t) = \psi$. This gives us

$$\begin{aligned}
\dot{x}_1(t) &= x_2(t) = f_1(x(t), \delta(t)) \\
\dot{x}_2(t) &= x_3(t) + c\delta(t) = f_2(x(t), \delta(t)) \\
\dot{x}_3(t) &= -a\ddot{\psi}(t) - bH(\dot{\psi}(t)) + d\delta(t)
\end{aligned}$$

But, $\ddot{\psi}(t) = x_3(t) + c\delta(t)$, $\dot{\psi}(t) = x_2(t)$, and $H(x_2) = x_2^3(t) + x_2(t)$ so

$$\dot{x}_3(t) = -a\left(x_3(t) + c\delta(t)\right) - b\left(x_2^3(t) + x_2(t)\right) + d\delta(t) = f_3(x(t), \delta(t))$$

Also, we have $\psi = g(x, \psi_r) = x_1$. This provides the proper equations for the simulation. Next, suppose that the initial conditions are $\psi(0) = \dot{\psi}(0) = \ddot{\psi}(0) = 0$. This implies that $x_1(0) = x_2(0) = 0$ and $x_3(0) = \ddot{\psi}(0) - c\delta(0)$ or $x_3(0) = -c\delta(0)$.

For the ship steering problem, we let the integration step size be $h = 1$ sec. and $\alpha = 10$ so that $T = \alpha h = 10$ sec. (i.e., the controller is implemented on a digital computer with a sampling period of $T = 10$ sec. so that a new plant input is calculated every 10 sec. and applied to the rudder). We will use this same approach for all the simulations for the tanker ship in this book.

4.3.2 Construction of a Multilayer Perceptron for Ship Steering

Here, we construct a simple multilayer perceptron for steering the ship. To do this, we must first choose the controller inputs and outputs. We will assume that the only input for steering is the rudder angle δ (we will not consider the speed u to be an input; it will be fixed). Hence, δ is the only output of the multilayer perceptron. The choice of inputs to the multilayer perceptron depends on what variables of the tanker can be sensed. It is assumed that the reference input ψ_r is given (e.g., if you use a neurophysiological view you may think of it as being provided by the frontal lobes in the brain that perform route planning functions). To keep things simple we will assume that we can only sense the ship heading ψ (if you took a neurophysiological view, this would then have to be provided by visual processing or some other sensory input that is then connected to the neurons that make the steering decisions). With these choices, the control system block diagram for the ship steering problem is shown in Figure 4.9.

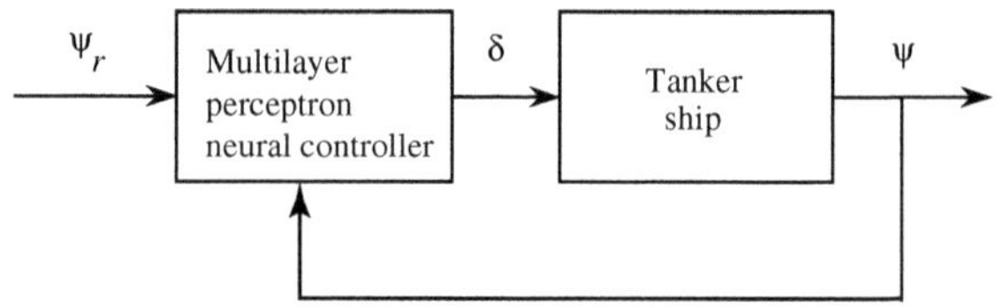

Figure 4.9: Control system for using a multilayer perceptron for tanker ship steering.

Structure Choice and The First Hidden Layer

From Figure 4.9, we see that the multilayer perceptron is a mapping from ψ_r and ψ to δ, the input to the ship. Suppose we denote the multilayer perceptron as $\delta = F_{mlp}(\psi_r, \psi)$ (note that for convenience we omit the arguments indicating the dependence on the parameters of the network). Our objective is to specify the mapping $F_{mlp}(\psi_r, \psi)$ by picking the number of layers of neurons, the number of neurons in each layer, and the specific weights, biases, and activation functions for all the neurons (from a neurophysiological view, it is evolution that specifies this mapping and we will simply construct one that we hypothesize evolution could have constructed). To do this, we will simply show one possible choice for the multilayer perceptron and explain some of the reasoning behind its construction. Consider Figure 4.10. There, we use a four layer perceptron with both linear and logistic sigmoidal activation functions. First, consider the first hidden layer. For this we choose $w_{11}^{(1)} = 1$, $w_{21}^{(1)} = -1$, and $b_1^{(1)} = 0$. Hence, we see that the output of the first layer is the heading error

A neural network can be designed to compare signals for use in "decision-making" (e.g., comparing a desired value to a sensed one).

$$e = \psi_r - \psi$$

To a control engineer this may seem to be an odd approach to implement a simple summing junction to provide the heading error; however, it is interesting that a neuron can provide a method to compare two signals, something that is certainly of fundamental importance in making control decisions for tracking and regulation.

Choosing Weights and Biases: Building Nonlinearities with Smooth Step Functions

Next, we explain how to pick the weights and biases for the remaining layers. To do this, view the perceptron in Figure 4.10 as having two "paths" of processing from the signal e that is the output of the first hidden layer to the output δ. Imagine that you remove the path on the bottom and first focus on constructing the path on the top. We will think of the top path as being used to regulate the ship heading when

$$e = \psi_r - \psi \geq 0$$

In this case, we want to have a *negative* rudder input. To see this, consider Figure 4.8 where you can see that a positive rudder input results in a *decrease*

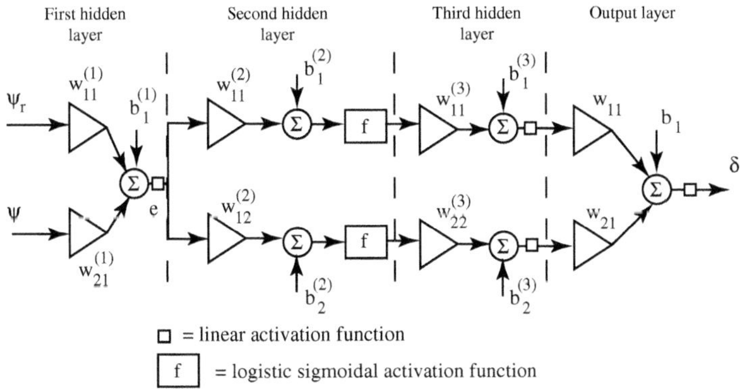

Figure 4.10: A multilayer perceptron for tanker ship steering.

in the heading ψ and a negative rudder input results in an *increase* in the heading ψ. Hence, if $\psi_r - \psi \geq 0$, we have $\psi_r \geq \psi$ so that we want to increase the size of ψ, so we use a negative rudder input δ. Next, note that for larger values of $|e| = |\psi_r - \psi|$, we will generally want larger values of δ since larger heading errors generally require larger rudder inputs to reduce them quickly. How do we choose the weights and biases in the top path to implement this type of control action?

Neural network construction can be viewed as "building" stimulus-response characteristics from basic neuron building blocks that are deformable via their parameters (here, we build functions from "smooth steps").

To specify values for the weights and biases, we think of each neuron as providing a type of "smooth switching" (a smooth step function as shown in Figure 4.6, e.g., for the logistic function) and tune the weights and biases to adjust these and build nonlinear control mappings. Note that when only the top path is considered, we have

$$\delta = w_{11} \left(\frac{w_{11}^{(3)}}{1 + \exp(-\bar{x})} + b_1^{(3)} \right) + b_1$$

where

$$\bar{x} = b_1^{(2)} + w_{11}^{(2)} e$$

The parameters in these equations affect the shape of the nonlinearity from e to δ in the following manner:

- b_1, $b_1^{(3)}$: Shift the mapping up and down.
- w_{11}, $w_{11}^{(3)}$: Scale the vertical axis.
- $b_1^{(2)}$: Shifts the smooth step (logistic function) horizontally, with $b_1^{(2)} > 0$ shifting it to the *left*.

- $w_{11}^{(2)}$: Scale the horizontal axis (you may think of this as a type of gain for the function, at least locally).

Using these ideas, choose

$$\begin{aligned}
b_1 &= b_1^{(3)} = 0 \\
w_{11} &= 1 \\
w_{11}^{(3)} &= -\frac{80\pi}{180} \\
b_1^{(2)} &= -\frac{200\pi}{180} \\
w_{11}^{(2)} &= 10
\end{aligned}$$

With these choices we get the nonlinear mapping shown in the top plot of Figure 4.11. The general shape of the function is appropriate to use as a controller for $e > 0$ since it provides negative rudder input values for positive values of error, and it provides a type of proportionality between the size of e and the size of δ. Notice that the choice of $w_{11}^{(3)}$ results in the perceptron providing a maximum negative rudder deflection of -80 degrees. The choice of $b_1^{(2)}$ simply shifts the function to the right, so that the value of the function near $e = 0$ provides $\delta \approx 0$. The value of $w_{11}^{(2)}$ affects the slope of the function as $e > 0$ increases in size; if $w_{11}^{(2)}$ were chosen to be larger, then it would reach the maximum negative value of -80 degrees quicker as the size of e increases. This completes the construction of the perceptron for the top path, which is dedicated to control for the case where $e \geq 0$.

Next, consider the bottom path of Figure 4.10 (imagine disconnecting the top path) which is dedicated to the case $e < 0$. Using the same ideas for the choice of the parameters above, select

$$\begin{aligned}
b_2^{(3)} &= \frac{80\pi}{180} \\
w_{21} &= 1 \\
w_{22}^{(3)} &= -\frac{80\pi}{180} \\
b_2^{(2)} &= \frac{200\pi}{180} \\
w_{12}^{(2)} &= 10
\end{aligned}$$

The resulting nonlinearity implemented by the bottom path is shown in the middle plot of Figure 4.11. First, note that its general shape is appropriate to use as a controller for the case where $e < 0$; as the size of e increases in the negative direction, increasingly positive values of a rudder input δ are provided to try to decrease the value of ψ to the given ψ_r. The values of $w_{22}^{(3)}$, $b_2^{(2)}$, and $w_{12}^{(2)}$ were chosen in a similar way as the corresponding values for the top path were chosen. The value of $b_2^{(3)}$ was chosen to shift the nonlinearity up

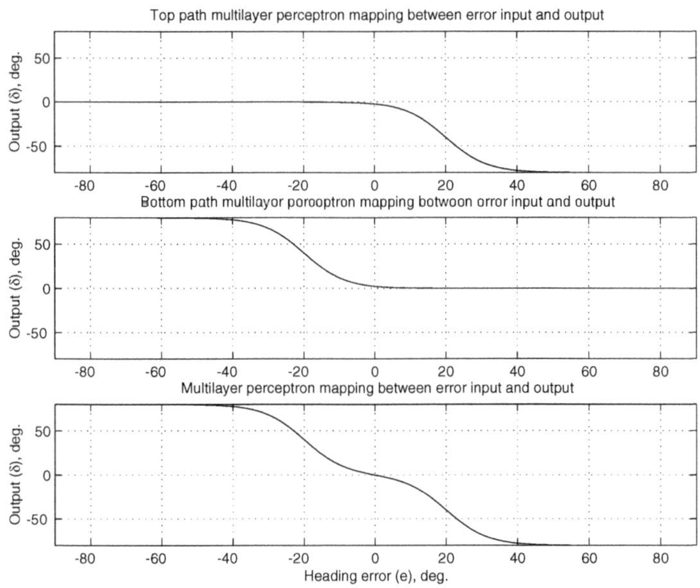

Figure 4.11: Multilayer perceptron mappings, top plot is for the top path of the perceptron from e to δ, middle plot is for the bottom path of the perceptron from e to δ, bottom plot is for the entire perceptron from e to δ.

by 80 degrees. The choice for w_{21} completes the specification of the output layer, which simply sums the functions generated by the top and bottom paths, and results in the overall mapping from e to δ shown in the bottom plot of Figure 4.11 (i.e., when both the top and the bottom paths in Figure 4.10 are used). Notice that due to the symmetry in our choices, $e = 0$ implies that $\delta = 0$ so that if the ship is going in the right direction, the rudder does not try to correct for the heading direction. This completes the construction of the multilayer perceptron for regulating the ship heading.

4.3.3 Multilayer Perceptron Stimulus-Response Characteristics

The multilayer perceptron construction procedure in the last subsection showed how to construct the controller

$$\delta = F_{mlp}(\psi_r, \psi)$$

Given a stimulus represented by particular values of ψ_r and ψ, the perceptron will react and provide a response δ according to how the function $F_{mlp}(\psi_r, \psi)$ is shaped. In this way the $F_{mlp}(\psi_r, \psi)$ nonlinearity implements a "control surface," which in this case has the shape shown in Figure 4.12.

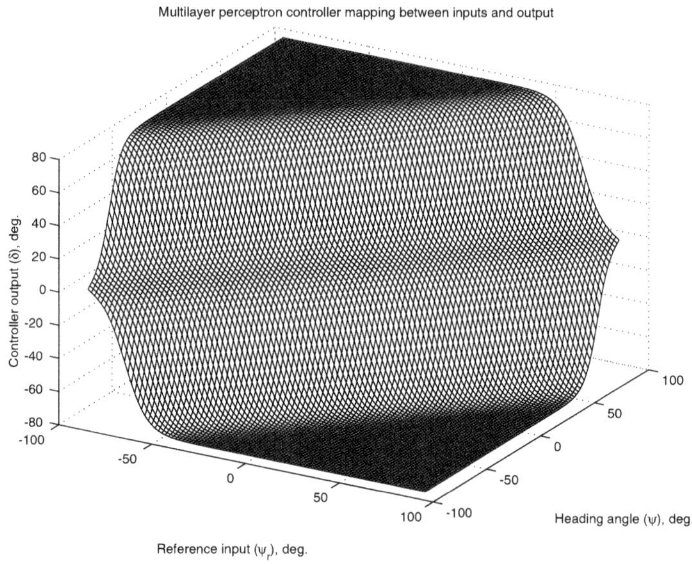

Figure 4.12: Control surface implemented by the multilayer perceptron for tanker ship steering.

The neural network stimulus-response mapping is generally nonlinear so it implements a nonlinear controller. To understand how this nonlinear controller might affect closed-loop behavior, it is important to have insights into the shape of the nonlinearity.

Figure 4.12 summarizes the input-output behavior of the multilayer perceptron. Consider some examples of how it behaves. Notice that if $e = \psi_r - \psi = 0$, the ship is heading in the correct direction, and the mapping in Figure 4.12 shows that the perceptron chooses $\delta = 0$ (i.e., it does not make any course corrections); this is due to the symmetry in our parameter choices for the top and bottom paths. If, on the other hand,

$$\psi_r = 50$$

degrees and

$$\psi = -50$$

degrees, then δ is at its maximum *negative* deflection so that it is trying to increase the heading angle ψ to get it pointed in the direction specified by ψ_r. Other combinations of values for ψ_r and ψ can be viewed in an analogous manner.

4.3.4 Behavior of the Ship Controlled by the Multilayer Perceptron

To evaluate how a neural network can regulate the ship heading, we will use simulation studies for a variety of operating conditions for the ship.

Closed-Loop Response, Nominal Conditions

If we use "nominal conditions," where we have "ballast" conditions, no wind, no sensor noise, and a speed of 5 meters/sec., we get a closed-loop response shown in Figure 4.13. For this, we use the multilayer perceptron in Figure 4.10 in the control system in Figure 4.9. Notice that the reference input ψ_r is set to zero for 100 sec. and then 45 degrees until $t = 2000$ sec. when it returns to a zero value. The actual ship heading responds quickly, but there is some overshoot past the desired value, some oscillations, and then the heading settles to the desired value. Note that the result of using a digital controller that only updates the control input every 10 sec. manifests itself as the "staircase" signal in the bottom plot of Figure 4.13. The middle plot also has a staircase form since we only stored and plotted one of every 10 values. The top plot appears smooth since we plot values each second.

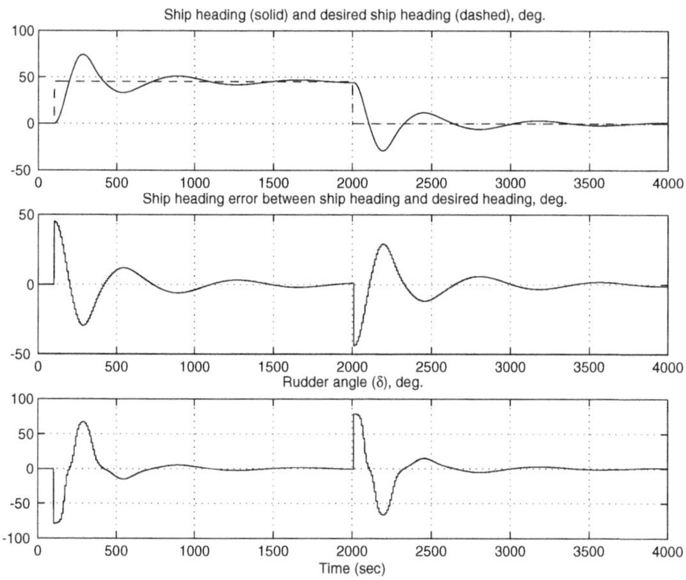

Figure 4.13: Closed-loop response resulting from using the multilayer perceptron for tanker ship steering.

This response may not be considered to be very good; however, for a first design it is reasonable. How do you improve the response? You tune the shape of the nonlinearity pictured in Figure 4.12 by tuning the weights and biases of the network, and possibly its structure (e.g., the number of layers, neurons, and types of activation functions). Another option would be to use more inputs to the controller, but we will consider this option when we study the use of a radial basis function neural network for this same application in Section 4.5.

For now, we will assume that this is a reasonably good design, at least for illustration purposes, and test its performance for other conditions.

Effects of Wind on Heading Regulation

Simulation-based evaluations of control systems should consider effects of a variety of adverse influences.

Next, consider the effects of a wind disturbance on the ship. Suppose that the wind is gusting. It hits the side of the ship and moves the ship a bit, which then pushes the rudder against the water which induces a torque to move the rudder. To model this, we add a disturbance onto the rudder angle input by adding

$$0.5\left(\frac{\pi}{180}\right)\sin\left(2\pi(0.001)t\right)$$

to what the multilayer perceptron controller commands as an input to the tanker ship (this is an additive sinusoid disturbance with an amplitude of 0.5 degrees and a period of 1000 sec.). In this case, we get the response in Figure 4.14. We see that the wind affects our ability to achieve very good regulation of the ship heading. In particular, it adversely affects the "steady-state behavior" of the control system (i.e., when the value of ψ_r is held constant for a long period of time, such as the times leading up to $t = 2000$ sec.) since the heading ψ does not properly converge to the desired heading ψ_r.

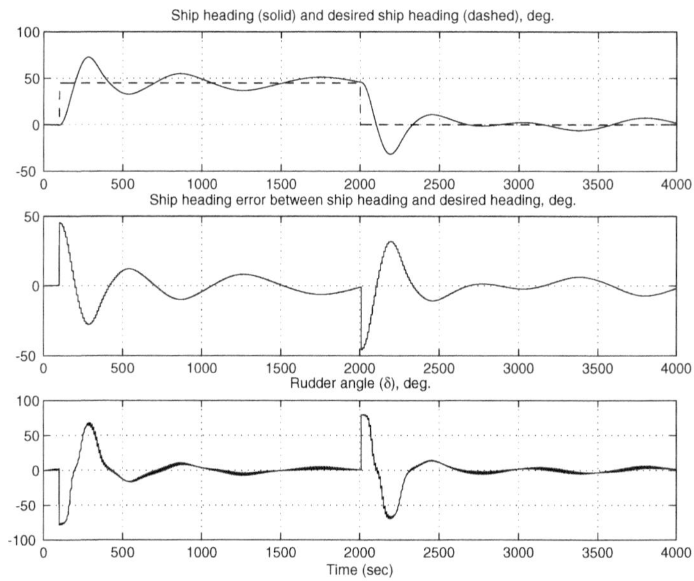

Figure 4.14: Closed-loop response resulting from using the multilayer perceptron for tanker ship steering, with wind.

Effects of Speed Changes on Heading Regulation

Next, consider the effect of a speed change on our ability to steer the ship. Generally, if you speed up the ship it is easier to steer, while if you slow it down, it becomes more difficult to steer because the rudder becomes less effective. If we use a speed of $u = 3$ meters/sec. (i.e., a decreased speed compared to the previous simulations), then we get the response in Figure 4.15. We see that the speed decrease causes a general slowing of the response since the rudder is not as effective in influencing the ship heading.

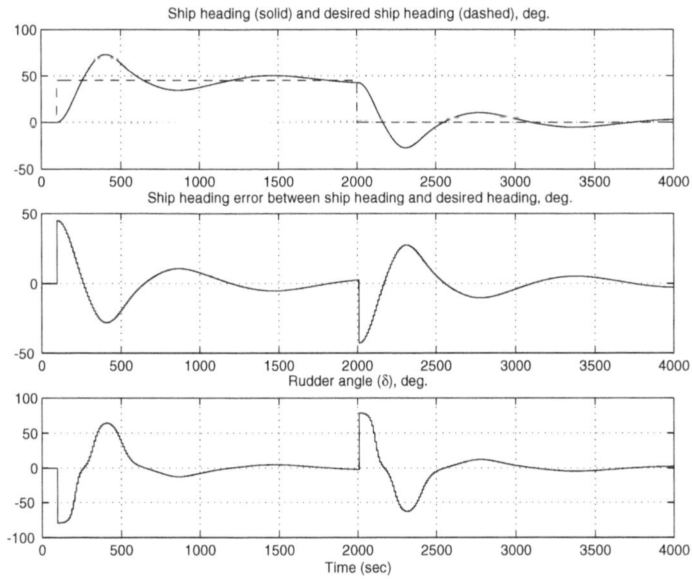

Figure 4.15: Closed-loop response resulting from using the multilayer perceptron for tanker ship steering with speed of 3 meters/sec.

Effects of Sensor Noise and Weight Changes on Heading Regulation

If you use an additive sensor noise uniformly distributed on $[-0.01, 0.01]$, there is little effect on the response so we do not show the plot. Of course, if you use a sensor with worse performance characteristics, then you will expect tracking errors to arise in an analogous manner to results for the wind.

Note that on different journeys, ships will weigh different amounts and the amount a ship weighs affects your ability to steer it. For the simulations up till now we have studied the case for "ballast" conditions. Next, we will consider the case of how the ship steers when it is under "full" conditions. In this case, when we use the multilayer perceptron that we tuned for ballast conditions on the full ship, we get the response in Figure 4.16. This shows how the multilayer perceptron controller, which was tuned for ballast conditions, performs quite poorly for full conditions. Why does it fail? It responds with too large of inputs for errors in the heading. In the beginning of the simulation when ψ_r first switches to 45 degrees at $t = 100$ sec. there is suddenly a positive error of $e = 45$ degrees, and the multilayer perceptron quickly reacts by putting in a maximum

Careful evaluations may uncover conditions for which the control system performs poorly.

negative value for the rudder to try to get it moving in the right direction. After a time it succeeds, but in doing this it has the ship heading moving too fast and it overshoots in the opposite direction. The multilayer perceptron then responds by putting a maximum positive value into the plant, which after an even longer period of time than in the case where the ψ value swung too far positive, it manages to move the ship heading in the opposite direction. This process repeats with the controller inducing a growing heading oscillation that results in the heading growing excessively large (which we intuitively think of as going "unstable," but of course strictly speaking, a simulation cannot generally prove instability since simulations run for a finite length of time while stability is an asymptotic property).

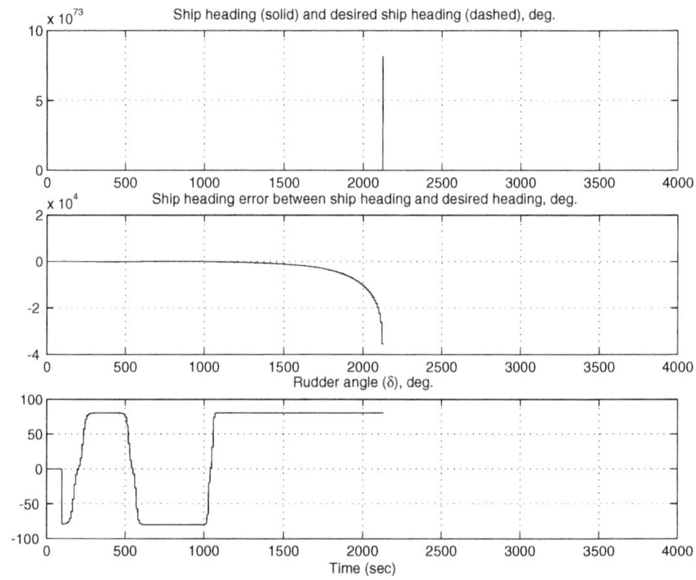

Figure 4.16: Closed-loop response resulting from using the multilayer perceptron for tanker ship steering, full rather than ballast conditions.

Clearly, the multilayer perceptron is not equipped for this condition. From a neurophysiological view, we could say that evolution has not encountered this situation frequently enough to result in a good design for the stimulus-response characteristics of the perceptron. From a control engineering perspective, we see that we need to reshape the nonlinearity $F_{mlp}(\psi_r, \psi)$ in Figure 4.12 so that the closed-loop response is adequate for all the possible conditions. For now, we will not consider performing such a design iteration. Instead, we will show how to use a different type of neural controller to regulate the ship heading using a different strategy.

4.4 Radial Basis Function Neural Networks

A locally tuned overlapping receptive field is found in parts of the cerebral cortex, in the visual cortex, and in other parts of the brain. The radial basis function neural network model is based on these biological systems (but once again, the model is not necessarily accurate, just inspired by its biological counterpart).

A radial basis function neural network is shown in Figure 4.17. There, the inputs are x_i, $i = 1, 2, \ldots, n$, and the output is $y = F_{rbf}(x)$ where F_{rbf} represents the processing by the entire radial basis function neural network. Let $x = [x_1, x_2, \ldots, x_n]^\top$. The input to the ith receptive field unit (sometimes called a radial basis function) is x, and its output is denoted with $R_i(x)$. The receptive field unit has what is called a "strength" which we denote by b_i. Assume that there are n_R receptive field units. Hence, from Figure 4.17,

$$y = F_{rbf}(x, \theta) = \sum_{i=1}^{n_R} b_i R_i(x) \tag{4.10}$$

is the output of the radial basis function neural network, and θ holds the b_i parameters and possibly the parameters of the receptive field units.

Stimulus-response characteristics of the radial basis function neural network are tuned by changing the b_i, R_i parameters, or the structure (e.g., R_i definitions and how they are combined).

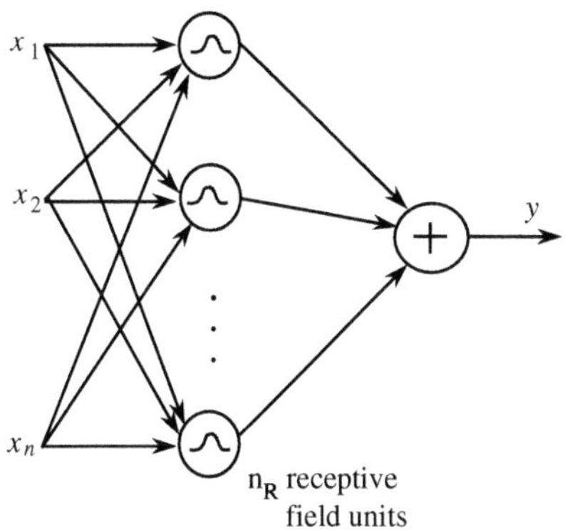

Figure 4.17: Radial basis function neural network model.

There are several possible choices for the "receptive field units" $R_i(x)$:

1. We could choose

$$R_i(x) = \exp\left(-\frac{|x - c^i|^2}{(\sigma^i)^2}\right) \tag{4.11}$$

where $c^i = [c_1^i, c_2^i, \ldots, c_n^i]^\top$, σ^i is a scalar, and if z is a vector then $|z| = \sqrt{z^\top z}$. For the case where $n = 1$, $c^1 = [c_1^1] = [2]$, and $\sigma^1 = 0.1$, $R_1(x)$ is shown in Figure 4.18(a). As x moves away from c_1^1, $R_1(x)$ decreases with the rate of decrease dictated by the size of σ^1 (a smaller value of σ^1 results in a steeper slope on the function, and so its value will decrease quicker as x moves away from c_1^1).

2. We could choose

$$R_i(x) = \frac{1}{1 + \exp\left(-\frac{|x-c^i|^2}{(\sigma^i)^2}\right)}$$

where c^i and σ^i are defined in choice 1. For the case where $n = 1$, $c^1 = [c_1^1] = [2]$, and $\sigma^1 = 0.1$, $R_1(x)$ is shown in Figure 4.18(b). Here, we see that the receptive field unit values are small where the values for choice 1 above are large, and vice versa.

3. In each of the above cases you can choose to make the σ^i also depend on the input dimension (which makes sense if the input dimensions are scaled differently). In this case for 1 above, for example, we would have $\sigma^i = [\sigma_1^i, \sigma_2^i, \ldots, \sigma_n^i]^\top$ and

$$R_i(x) = \exp\left(-\sum_{j=1}^{n} \frac{(x_j - c_j^i)^2}{(\sigma_j^i)^2}\right)$$

where σ_j^i is the spread for the jth input for the ith receptive field unit. This is the approach that we will use in the example in the next section.

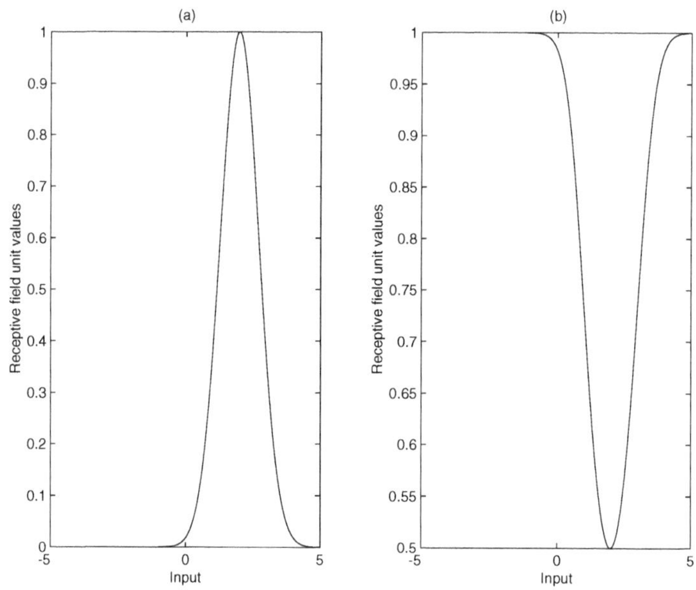

Figure 4.18: Example receptive field units.

There are also alternatives to how to compute the output of the radial basis function neural network. For instance, rather than computing the simple sum as in Equation (4.10), you could compute a weighted average

$$y = F_{rbf}(x, \theta) = \frac{\sum_{i=1}^{n_R} b_i R_i(x)}{\sum_{i=1}^{n_R} R_i(x)} \tag{4.12}$$

It is also possible to define multilayer radial basis function neural networks.

Finally, note that our radial basis function neural network model is developed in an analogous way to what our multilayer perceptron was, relative to biological neurons. It is a "firing rate model" that has had the receptive field unit function shapes experimentally validated by finding the "tuning curve" for an individual neuron (e.g., in the visual cortex). See the "For Further Study" section at the end of this part for more details.

4.5 Design Example: Radial Basis Function Neural Network for Ship Steering

This section parallels Section 4.3, but we will design a radial basis function neural network for ship heading regulation.

4.5.1 A Radial Basis Function Neural Network for Ship Steering

We will design the radial basis function neural network, study its stimulus response characteristics, and then show via simulations how it regulates the tanker ship heading.

Controller Input Choice and Control System Structure

Note that for the multilayer perceptron, we used ψ_r and ψ as inputs to the controller and then in the first layer, we formed the error e that served as an input to the second layer. Here, taking a more standard control engineering approach, we will use the error e as one input to the radial basis function neural network, and we will also use the derivative of that error. Hence, our inputs to the radial basis function neural network will be

$$e = \psi_r - \psi$$

and

$$\dot{e} = \dot{\psi}_r - \dot{\psi}$$

We will, however, use a backward difference approximation to the derivative which we will denote by $c(kT)$,

$$\dot{e} \approx \frac{e(kT) - e(kT - T)}{T} = c(kT)$$

where $T = 10$ sec. and k is an index for the time step (this is an Euler approximation of the derivative and T is the sampling period of the digital controller on which we will implement the controller). As is standard in discrete-time systems, for convenience we will often use "k" rather than "kT" as the argument for the signals. With this, we can denote the radial basis function neural network for the ship by

$$\delta(k) = F_{rbf}(e(k), c(k))$$

(we omit parameter vector argument for convenience) and use it in the control system shown in Figure 4.19.

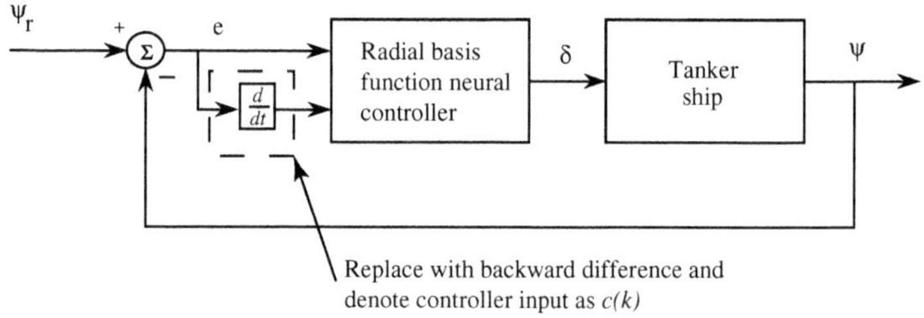

Figure 4.19: Radial basis function neural network used as a controller for ship heading.

Design of a Radial Basis Function Neural Network for Steering

Next, we construct a radial basis function neural network of Equation (4.10) with $n = 2$ inputs, and $n_R = 121$ so we will have to pick 121 strengths b_i, $i = 1, 2, \ldots, 121$. For the $R_i(e(k), c(k))$ we use Equation (4.11) and create a uniform grid for the c^i centers, $i = 1, 2, \ldots, 121$. To pick the grid points, assume that $e(k)$ lies in the range

$$e(k) \in [-\frac{\pi}{2}, \frac{\pi}{2}]$$

(which will hold if we do good regulation and do not get fast changes in ψ_r). Via simulations of the ship, the angular rate of movement is often such that

$$c(k) \in [-0.01, 0.01]$$

so we will make that assumption to guide our design choices. For convenience, we simply create a uniform grid with its four outer corners at $(-\frac{\pi}{2}, -0.01)$, $(-\frac{\pi}{2}, 0.01)$, $(\frac{\pi}{2}, 0.01)$, and $(\frac{\pi}{2}, -0.01)$ with $n_R = 121$ centers uniformly placed at the grid points (i.e., with 11 points along each input dimension). We show the centers of the receptive field units in Figure 4.20.

The parameters of receptive field units are often chosen via "gridding" the input space; this ensures that the controller will have a response for each input.

For the receptive field units we use spreads σ_j^i (i.e., so that the size of the spread depends on which input dimension is used) with

$$\sigma_1^i = 0.7 \frac{\pi}{\sqrt{n_R}}$$

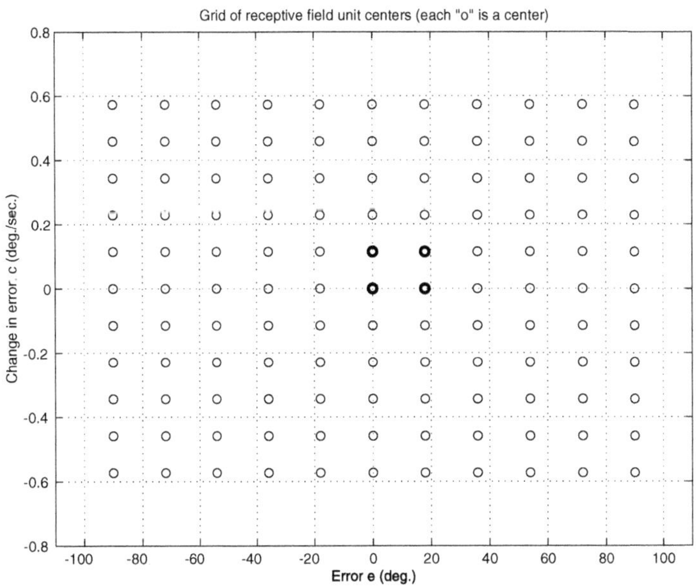

Figure 4.20: Receptive field unit centers.

and

$$\sigma_2^i = 0.7 \frac{0.02}{\sqrt{n_R}}$$

for $i = 1, 2, \ldots, 121$. For σ_1^i, the $\frac{\pi}{11}$ factor makes the spread size depend on the number of grid points along the e input dimension (similarly for σ_2^i), and the 0.7 factor was chosen to get a smooth interpolation between adjacent receptive field units (see more discussion on this point below). With these choices, as an example, consider the shape of receptive field unit $R_{73}(e, c)$ shown in Figure 4.21 (note that the receptive field unit index is found by starting in the lower left-hand corner of Figure 4.20 with 1, and counting up for the point directly above it, then when you reach the top of the first column, you go to the bottom of the next column). Notice that it simply has the shape of a Gaussian function. The center of this particular receptive field unit is the upper right-hand darkly shaded circle in Figure 4.20. (When comparing Figures 4.20 and 4.21, be careful to mentally rotate Figure 4.20 so that the plane appropriately aligns with the $R_{73}(0, 0) = 0$ plane in Figure 4.21.)

We can view construction of a radial basis function neural network as building a stimulus-response characteristic from tunable "spatially local" functions (e.g., Gaussian functions).

Next, we will consider how the input-output mapping of the radial basis function neural network is shaped by the choice of the scaling parameters b_i. For instance, note that b_{73} would simply scale the height of the receptive field unit in Figure 4.21. Consider the scaling and summation of the receptive field units with centers at the four darkly shaded circles in Figure 4.20 (the indices for these are 61, 62, 72, and 73). In particular, we compute

$$2R_{61}(e, c) + R_{62}(e, c) + 2R_{72}(e, c) + R_{73}(e, c)$$

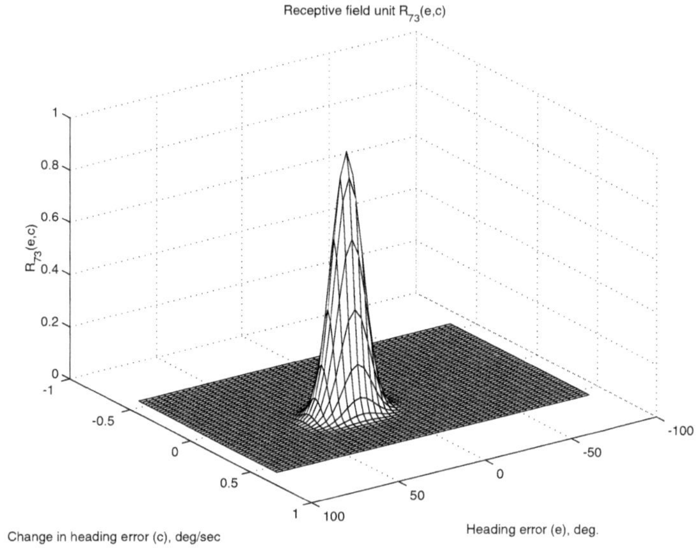

Figure 4.21: Mapping implemented by receptive field unit $R_{73}(e,c)$.

and plot it vs. e and c in Figure 4.22. Note that we scaled two of the receptive field units by 2 and in this way, we obtain a small region (near the center of the four darkly shaded circles in the (e,c) plane) that has a slope fixed by the parameters that we have chosen. In essence, we have designed a neural controller for this small region.

To design a radial basis function neural network for the ship steering problem, we simply need to choose the b_i, $i = 1, 2, \ldots, 121$, parameters to shape the mapping in the appropriate way. Suppose that we view the parameters as being loaded in a matrix

$$\begin{bmatrix} b_1 & b_{12} & b_{23} & b_{34} & b_{45} & b_{56} & b_{67} & b_{78} & b_{89} & b_{100} & b_{111} \\ b_2 & & & & & \cdots & & & & & b_{112} \\ \vdots & & & & & \cdots & & & & & \vdots \\ b_{11} & b_{22} & b_{33} & b_{44} & b_{55} & b_{66} & b_{77} & b_{88} & b_{99} & b_{110} & b_{121} \end{bmatrix}$$

and then choose this matrix to be

```
Columns 1 through 7

  1.3963    1.3963    1.3963    1.3963    1.3963    1.3963    1.3963
  1.3963    1.3963    1.3963    1.3963    1.3963    1.3963    1.0472
  1.3963    1.3963    1.3963    1.3963    1.3963    1.0472    0.6981
  1.3963    1.3963    1.3963    1.3963    1.0472    0.6981    0.3491
  1.3963    1.3963    1.3963    1.0472    0.6981    0.3491         0
  1.3963    1.3963    1.0472    0.6981    0.3491         0   -0.3491
```

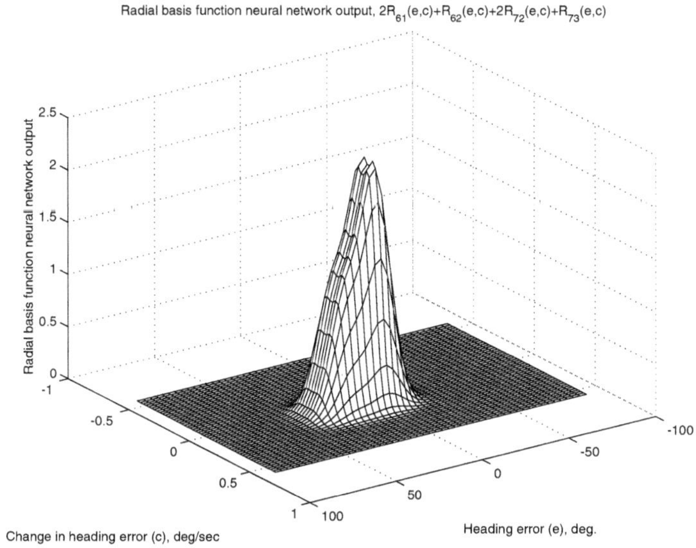

Figure 4.22: Scaling and addition of several receptive field units (i.e., $2R_{61}(e,c) + R_{62}(e,c) + 2R_{72}(e,c) + R_{73}(e,c)$).

1.3963	1.0472	0.6981	0.3491	0	-0.3491	-0.6981
1.0472	0.6981	0.3491	0	-0.3491	-0.6981	-1.0472
0.6981	0.3491	0	-0.3491	-0.6981	-1.0472	-1.3963
0.3491	0	-0.3491	-0.6981	-1.0472	-1.3963	-1.3963
0	-0.3491	-0.6981	-1.0472	-1.3963	-1.3963	-1.3963

Columns 8 through 11

1.0472	0.6981	0.3491	0
0.6981	0.3491	0	-0.3491
0.3491	0	-0.3491	-0.6981
0	-0.3491	-0.6981	-1.0472
-0.3491	-0.6981	-1.0472	-1.3963
-0.6981	-1.0472	-1.3963	-1.3963
-1.0472	-1.3963	-1.3963	-1.3963
-1.3963	-1.3963	-1.3963	-1.3963
-1.3963	-1.3963	-1.3963	-1.3963
-1.3963	-1.3963	-1.3963	-1.3963
-1.3963	-1.3963	-1.3963	-1.3963

Notice the pattern of elements in the matrix. For instance, for R_{61}, the receptive field unit in the center of the grid, we have a strength $b_{61} = 0$. Why? Because at this point $e = c = 0$ so the ship is on the proper heading and it is not deviating from that heading; hence, we do not make any corrections to the rudder angle. It is a useful exercise for you to consider another element in the above matrix and convince yourself that it is a good choice via relating its choice to what the controller should do for a particular (e, c) combination.

4.5.2 Stimulus-Response Characteristics

The stimulus-response characteristics of the radial basis function neural network $F_{rbf}(e, c)$ that we just designed are shown in Figure 4.23 in the form of a control surface, similar to how we illustrated the mapping for the multilayer perceptron (note that here the inputs are different).

Different inputs and neural networks lead to different stimulus-response characteristics for the controller.

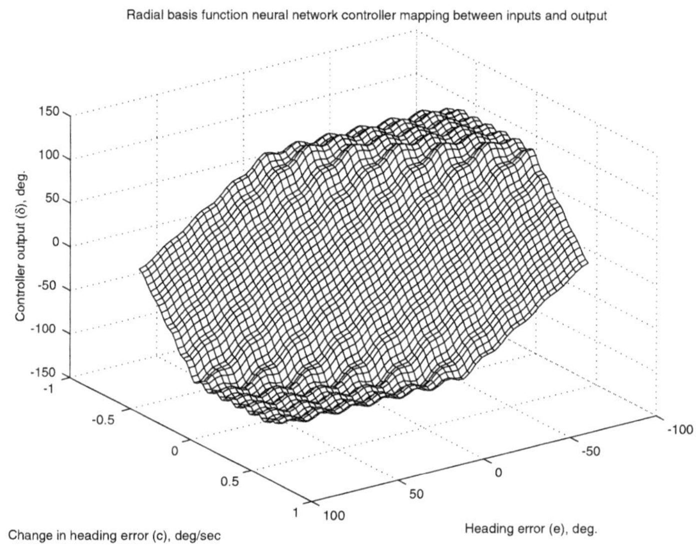

Figure 4.23: Stimulus-response characteristics of the radial basis function neural network for tanker ship heading regulation.

Note that the plot nicely summarizes the "decisions" that the neural network will make. Notice that if $e = c = 0$, the ship is heading in the proper direction and it is not deviating from that direction; hence, the controller sets $\delta = 0$. If, however, the ship heading error e is near 90 deg., with positive values of ψ and ψ_r, we know that the heading ψ is pointed about 90 deg. counterclockwise of the desired heading ψ_r. If along with this condition for e, we have that c is positive and near a value of 0.5 deg./sec., then the heading is moving to become even worse than it currently is. In this situation, the neural network will choose the largest possible *negative* rudder angle so that the heading will move clockwise towards the desired heading, counteracting the effects of having a rate of rotation in the wrong direction. For practice, it would be useful for you to consider other (e, c) values and explain why the decisions made by the neural controller are appropriate.

4.5.3 Behavior of the Ship Controlled by the Radial Basis Function Neural Network

To study how a radial basis function neural network can operate to regulate the ship heading, we will use simulation studies for a variety of operating conditions, the same ones as used in Section 4.3.

Closed-Loop Response, Nominal Conditions

If we use "nominal conditions," where we have "ballast" conditions, no wind, no sensor noise, and a speed of 5 meters/sec., we get a closed-loop response shown in Figures 4.24 and 4.25 when we use the radial basis function neural network developed in the last section in the control system in Figure 4.19. The ship heading ψ responds quickly, and while there is some overshoot past the desired value, the response settles to the proper value relatively quickly.

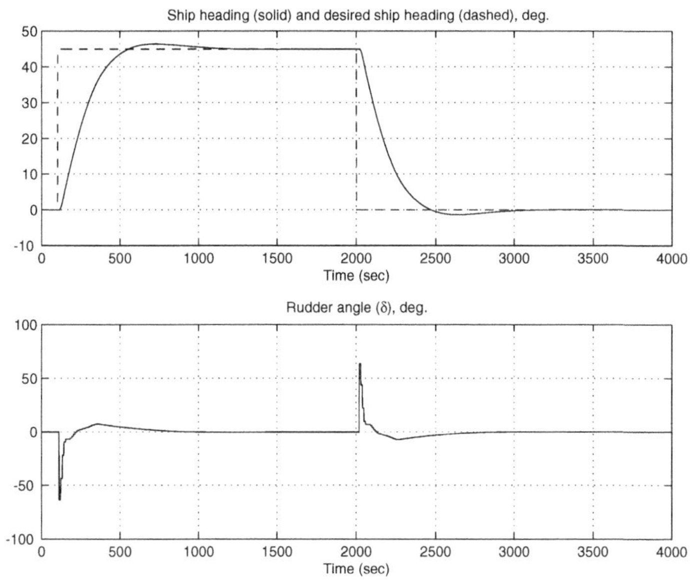

Figure 4.24: Closed-loop response resulting from using the radial basis function neural network for tanker ship steering.

While the response is generally superior to what we found for the multilayer perceptron, it is *not* appropriate to compare the two approaches. Why? Different inputs are used for the neural networks, there are far fewer parameters in the multilayer perceptron (how many were used in each case?), and we may have simply gotten lucky in our tuning for the radial basis function neural network.

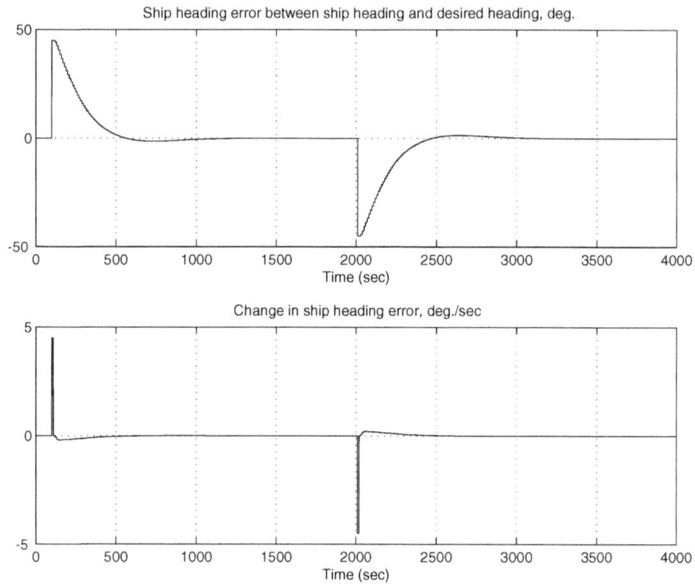

Figure 4.25: Closed-loop response resulting from using the radial basis function neural network for tanker ship steering.

Effects of Wind, Speed Changes, Sensor Noise, and Weight Changes on Heading Regulation

Next, consider the effects of a wind disturbance on the ship. In this case, we get the response in Figure 4.26. We see that the wind affects our ability to achieve very good steady-state regulation of the ship heading.

Next, consider the effect of a speed change on our ability to steer the ship. If we use a speed of $u = 3$ meters/sec. (i.e., a decreased speed compared to the previous simulations), then we get the response in Figure 4.27. We see that compared to the nominal conditions, the speed decrease causes more overshoot of the response since the rudder is not as effective in influencing the ship heading.

Using different inputs and a different neural network stimulus-response characteristic, we generally obtain different closed-loop responses.

As before, the sensor noise has little effect on the response. When there is a weight change and the ship is now full, we get the response in Figure 4.28. Notice that we get more overshoot than we did for nominal conditions; this is because a lighter ship is easier to steer so that the actions taken are too extreme and this results in the overshoot (you can think of the rudder as being more effective at steering for a light ship; hence, it generally needs smaller rudder inputs for a full ship). While the multilayer perceptron performed quite poorly for this condition, the radial basis function neural network performs reasonably well; however, just like for the nominal conditions above, it would be inappropriate to draw many conclusions from a comparison without more study. It would be

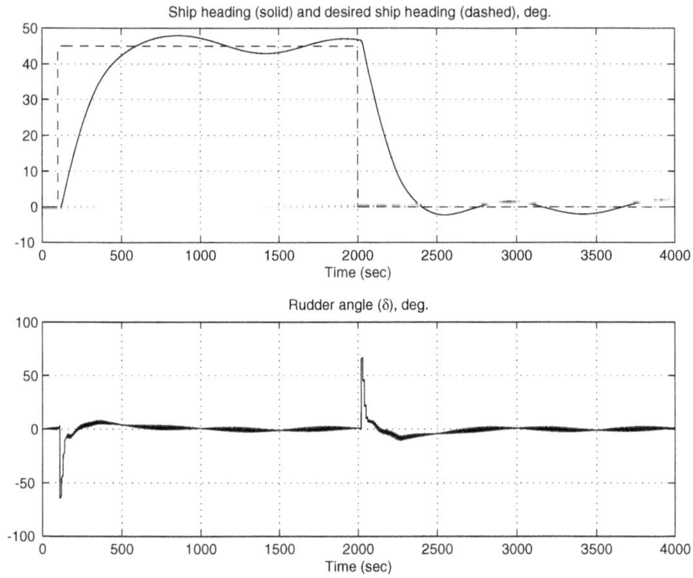

Figure 4.26: Closed-loop response resulting from using the radial basis function neural network for tanker ship steering, with wind.

especially inappropriate to try to conclude that the radial basis function neural network is generally superior to the multilayer perceptron.

4.6 Stability Analysis

For some applications, the designer is first concerned about investigating the stability properties of a control system, since it is often the case that if the system is unstable, there is no chance that any other performance specifications will hold. For example, if the control system for ship steering is unstable, you would be more concerned with the possibility of unsafe operation than with how well it regulates the heading to the desired angle. Fortunately, there has been significant attention given to the mathematical analysis of stability of nonlinear control systems, and certain results from that theory apply here. Here, we overview Lyapunov's direct method. For more complete introductions to stability analysis, see the "For Further Study" section at the end of this part.

Lyapunov stability analysis is an approach to verifying the correct operation of a control system.

4.6.1 Differential Equations and Equilibria

Suppose that a dynamic system is represented with

$$\dot{x}(t) = f(x(t)) \tag{4.13}$$

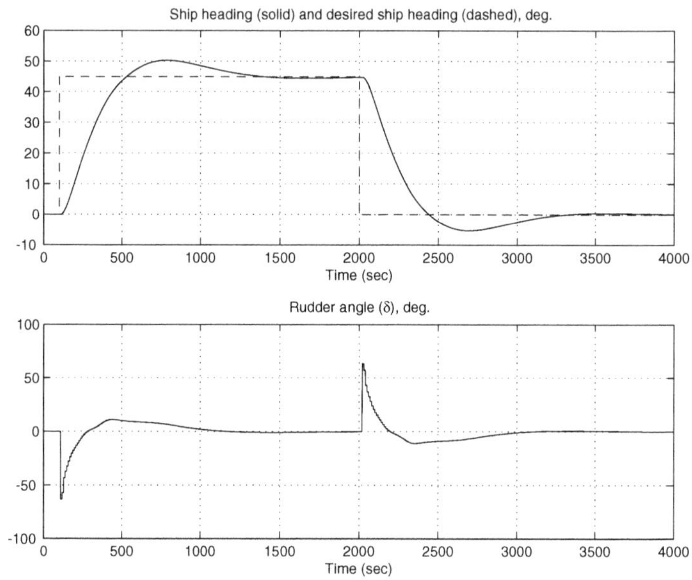

Figure 4.27: Closed-loop response resulting from using the radial basis function neural network for tanker ship steering, speed of 3 meters/sec.

where $x \in \Re^n$ is an n vector and $f : D \to \Re^n$ with $D = \Re^n$ or $D = B(h)$ for some $h > 0$ (h here is not to be confused with the integration step size used in the Runge-Kutta method) where

$$B(h) = \{x \in \Re^n : |x| < h\}$$

is a ball centered at the origin with a radius of h and $|\cdot|$ is a norm on $\Re^n$ (e.g., $|x| = \sqrt{(x^\top x)}$). If $D = \Re^n$, then we say that the dynamics of the system are defined globally, while if $D = B(h)$, they are only defined locally. Assume that for every x_0, the initial value problem

$$\dot{x}(t) = f(x(0)), \quad x(0) = x_0 \tag{4.14}$$

possesses a unique solution $\bar{\phi}(t, x_0)$ that depends continuously on x_0 ($\bar{\phi}(t, x_0)$ is a "solution" of Equation (4.13) if $\dot{\bar{\phi}}(t, x_0) = f(\bar{\phi}(t, x_0))$ where $\bar{\phi}(0, x_0) = x_0$). A point $x_e \in \Re^n$ is called an "equilibrium point" of Equation (4.13) if $f(x_e) = 0$ for all $t \geq 0$. An equilibrium point x_e is an "isolated equilibrium point" if there is an $h' > 0$ such that the ball around x_e,

$$B(x_e, h') = \{x \in \Re^n : |x - x_e| < h'\}$$

contains no other equilibrium points besides x_e. As is standard, we will assume that the equilibrium of interest is an isolated equilibrium located at the origin of $\Re^n$. This assumption results in no loss of generality since if $x_e \neq 0$ is an

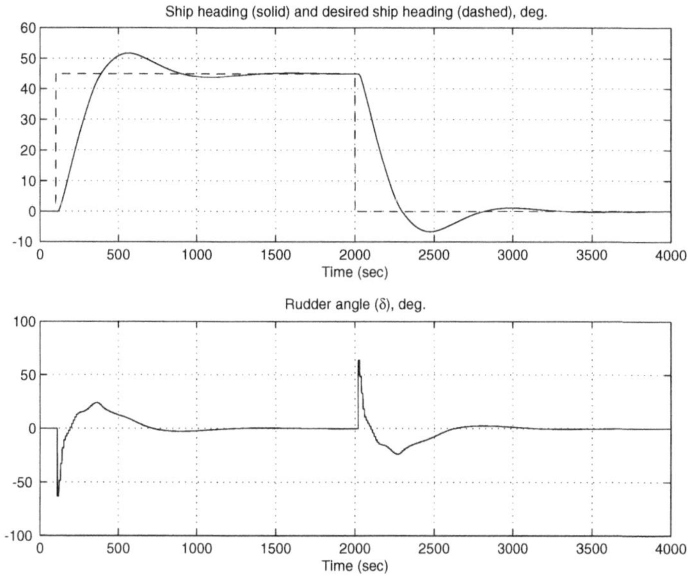

Figure 4.28: Closed-loop response resulting from using the radial basis function neural network for tanker ship steering, full rather than ballast conditions.

equilibrium of Equation (4.13) and we let $\bar{x}(t) = x(t) - x_e$, then $\bar{x} = 0$ is an equilibrium of the transformed system

$$\dot{\bar{x}}(t) = \bar{f}(\bar{x}(t)) = f(\bar{x}(t) + x_e)$$

To illustrate how to transform the equilibrium, we use a simple model of the pendulum shown in Figure 4.29 that is given by

$$\begin{aligned} \dot{x}_1 &= x_2 \\ \dot{x}_2 &= -\tfrac{g}{\ell}\sin(x_1) - \tfrac{k}{m}x_2 + \tfrac{1}{m\ell^2}T \end{aligned} \tag{4.15}$$

where $g = 9.81$, $\ell = 1.0$, $m = 1.0$, $k = 0.5$, x_1 is the angle (in radians) shown in Figure 4.29, x_2 is the angular velocity (in radians per second), and T is the control input.

If we assume that $T = 0$, then there are two distinct isolated equilibrium points, one in the downward position $[0, 0]^\top$ and one in the inverted position $[\pi, 0]^\top$. Suppose we are interested in the control of the pendulum about the inverted position; hence, we need to translate the equilibrium by letting $\bar{x} = x - [\pi, 0]^\top$. From this we obtain

$$\begin{aligned} \dot{\bar{x}}_1 &= \bar{x}_2 = \bar{f}_1(\bar{x}) \\ \dot{\bar{x}}_2 &= \tfrac{g}{\ell}\sin(\bar{x}_1) - \tfrac{k}{m}\bar{x}_2 + \tfrac{1}{m\ell^2}T = \bar{f}_2(\bar{x}) \end{aligned} \tag{4.16}$$

where if $T = 0$, then $\bar{x} = 0$ corresponds to the equilibrium $[\pi, 0]^\top$ in the original system in Equation (4.15), so studying the stability of $\bar{x} = 0$ corresponds to

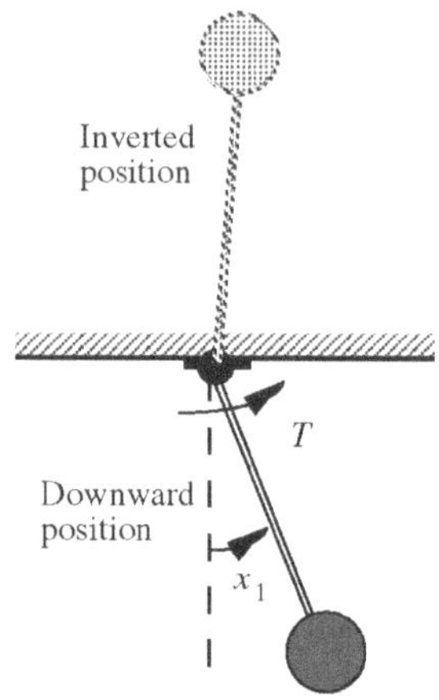

Figure 4.29: Pendulum.

studying the stability of the control system about the inverted position. Now, it is traditional to omit the cumbersome bar notation in Equation (4.16) and study the stability of $x = 0$ for the system

$$\begin{aligned} \dot{x}_1 &= x_2 = f_1(x) \\ \dot{x}_2 &= \tfrac{g}{\ell}\sin(x_1) - \tfrac{k}{m}x_2 + \tfrac{1}{m\ell^2}T = f_2(x) \end{aligned} \tag{4.17}$$

with the understanding that we are actually studying the stability of Equation (4.16).

4.6.2 Stability Definitions

The equilibrium $x_e = 0$ of Equation (4.13) is "stable" (in the sense of Lyapunov) if for every $\epsilon > 0$ there exists a $\delta(\epsilon) > 0$ such that $|\bar{\phi}(t, x_0)| < \epsilon$ for all $t \geq 0$ whenever $|x_0| < \delta(\epsilon)$ (i.e., it is stable if when it starts close to the equilibrium, it will stay close to it). The notation $\delta(\epsilon)$ means that δ depends on ϵ. A system that is not stable is called "unstable."

The equilibrium $x_e = 0$ of Equation (4.13) is said to be "asymptotically stable" if it is stable and there exists $\eta > 0$ such that $\lim_{t\to\infty} \bar{\phi}(t, x_0) = 0$ whenever $|x_0| < \eta$ (i.e., it is asymptotically stable, if when it starts close to the equilibrium, it will converge to it).

The set $X_d \subset \Re^n$ of all $x_0 \in \Re^n$ such that $\bar{\phi}(t, x_0) \to 0$ as $t \to \infty$ is called the "domain of attraction" of the equilibrium $x_e = 0$ of Equation (4.13). The equilibrium $x_e = 0$ is said to be "globally asymptotically stable" if $X_d = \Re^n$ (i.e., if no matter where the system starts, its state converges to the equilibrium asymptotically).

As an example, consider the scalar differential equation

$$\dot{x}(t) = -2x(t)$$

which is in the form of Equation (4.14). For this system, $D = \Re^1$ (i.e., the dynamics are defined on the entire real line, not just some region around zero). We have $x_e = 0$ as an equilibrium point of this system since $0 = -2x_e$. Notice that for any x_0, we have the solution

$$\bar{\phi}(t, x_0) = x_0 e^{-2t} \to 0$$

as $t \to \infty$ so that the equilibrium $x_e = 0$ is stable since, if you are given any $\epsilon > 0$, there exists a $\delta > 0$ such that if $|x_0| < \delta$, $|\bar{\phi}(t, x_0)| < \epsilon$. To see this, simply choose $\delta = \epsilon$ for any $\epsilon > 0$ that you choose. Also note that since for any $x_0 \in \Re^n$, $\bar{\phi}(t, x_0) \to 0$, the system is globally asymptotically stable. While determining if this system possesses certain stability properties is very simple since the system is so simple, for complex nonlinear systems it is not so easy. One reason why is that for complex nonlinear systems, it is difficult to even solve the ordinary differential equations (i.e., to find $\bar{\phi}(t, x_0)$ for all t and x_0). However, Lyapunov's direct method provides a technique that allows you to determine stability properties without solving the ordinary differential equations.

4.6.3 Lyapunov's Direct Method for Stability Analysis

The stability results for an equilibrium $x_e = 0$ of Equation (4.13) that we provide next depend on the existence of an appropriate "Lyapunov function"

$$V : D \to \Re$$

where $D = \Re^n$ for global results (e.g., global asymptotic stability) and $D = B(h)$ for some $h > 0$, for local results (e.g., stability in the sense of Lyapunov or asymptotic stability). If V is continuously differentiable with respect to its arguments, then the derivative of V with respect to t along the solutions of Equation (4.13) is

$$\dot{V}_{(4.13)}(x(t)) = \nabla V(x(t))^\top f(x(t))$$

where

$$\nabla V(x(t)) = \left[\frac{\partial V}{\partial x_1}, \frac{\partial V}{\partial x_2}, \ldots, \frac{\partial V}{\partial x_n} \right]^\top$$

is the gradient of V with respect to x. Using the subscript on $\dot{V}$ is sometimes cumbersome, so we will at times omit it with the understanding that the derivative of V is taken along the solutions of the differential equation.

Lyapunov's direct method is given by the following:

1. Let $x_e = 0$ be an equilibrium for Equation (4.13). Let $V : B(h) \to \Re$ be a continuously differentiable function on $B(h)$ such that $V(0) = 0$ and $V(x) > 0$ in $B(h) - \{0\}$, and $\dot{V}_{(4.13)}(x) \leq 0$ in $B(h)$. Then $x_e = 0$ is stable. If, in addition, $\dot{V}_{(4.13)}(x) < 0$ in $B(h) - \{0\}$, then $x_e = 0$ is asymptotically stable.

2. Let $x_e = 0$ be an equilibrium for Equation (4.13). Let $V : \Re^n \to \Re$ be a continuously differentiable function such that $V(0) = 0$ and $V(x) > 0$ for all $x \neq 0$, $|x| \to \infty$ implies that $V(x) \to \infty$, and $\dot{V}_{(4.13)}(x) < 0$ for all $x \neq 0$. Then $x_e = 0$ is globally asymptotically stable.

As an example, consider the scalar dynamical system

$$\dot{x} = -2x^3$$

that has an equilibrium $x_e = 0$. Choose

$$V(x) = \frac{1}{2}x^2$$

With this choice we have

$$\dot{V} = \frac{\partial V}{\partial x}\frac{dx}{dt} = x\dot{x} = -2x^4$$

so that clearly if $x \neq 0$, then $-2x^4 < 0$, so that by Lyapunov's direct method $x_e = 0$ is asymptotically stable. Notice that $x_e = 0$ is in fact globally asymptotically stable.

While Lyapunov's direct method has found wide application in conventional control, it is important to note that it is not always easy to find the "Lyapunov function" V that will have the above properties so that we can guarantee that the system is stable.

4.6.4 Stability of Discrete Time Systems

Consider the nonlinear discrete time system

$$x(k+1) = f(x(k)) \tag{4.18}$$

where k is the discrete time index, $x \in \Re^n$ is an n vector and $f : D \to \Re^n$ (with $D = \Re^n$ or $D = B(h)$ for some $h > 0$), and the equilibrium $x_e \in \Re^n$ is defined the same as in the continuous time case. Let $\bar{\phi}(k, x_0)$ denote a solution to the nonlinear discrete time system where $x_0 = x(0)$.

Stability in the sense of Lyapunov, (global) asymptotic stability, and regions of asymptotic stability are defined the same as in the continuous time case, except the time index "t" is replaced with the index "k."

Stability conditions for the discrete-time direct method of Lyapunov are slightly different from the continuous time case. We will only discuss asymptotic stability, as that property will be the one we are most interested in for our applications. According to Lyapunov's direct method, the equilibrium $x_e = 0$ of the system in Equation (4.18) is globally asymptotically stable if there exists a function $V(x)$ such that the following hold for all $x \in \Re^n$:

1. $V(x) \geq 0$ except at $x = 0$ where $V(x) = 0$,

2. $V(x) \to \infty$ if $|x| \to \infty$, and

3. $V(x(k+1)) - V(x(k)) < 0$.

If these conditions only hold locally, then we only obtain asymptotic stability. If they only hold on a region, that region is the region of asymptotic stability. Also, if x_e is an invariant set (i.e., where if we let $x_0 \in x_e$, $f(x_0) \in x_e$), then the same types of results hold (we will give an example of how to perform such analysis in the examples to follow).

As an example, consider

$$x(k+1) = ax(k)$$

where a is a fixed scalar and $x(k)$ is a scalar also. Notice that $x_e = 0$ is an isolated equilibrium. Suppose we want to find the conditions under which $x_e = 0$ is a globally asymptotically stable equilibrium. Choose $V = x^2$. Notice that the first two conditions above are satisfied for this choice. Next, notice that

$$V(x(k+1)) - V(x(k)) = x^2(k+1) - x^2(k) = a^2x^2 - x^2 = (a^2-1)x^2$$

Hence, if $a^2 - 1 < 0$, we have $V(x(k+1)) - V(x(k)) < 0$. In other words, if $a^2 < 1$, or if $a \in (-1, 1)$, then $x_e = 0$ is a globally asymptotically stable equilibrium.

4.6.5 Example: Stable Instinctual Neural Control

Suppose you are given the differential equation

$$\dot{x} = f(x) + gu$$

where $x(t)$ is a scalar, $g > 0$ is an unknown but fixed scalar (the following analysis works in a similar way if we know that $g < 0$), f is smooth (so solutions to the differential equation exist and are unique), and $f(0) = 0$. We will assume that while we do not know the exact form of $f(x)$, we do suppose that for some $\alpha > 0$,

$$|f(x)| < \alpha|x|$$

Lyapunov stability analysis is useful to verify the correct operation of a control system using an instinctual neural controller.

We emphasize, however, that there is *uncertainty* present in this control problem in the sense that we do not know the value of g and we do not know the specific form of the nonlinearity f, just that it satisfies the above inequality.

We seek to design a neural controller

$$u = F(x)$$

so that the equilibrium $x_e = 0$ is globally asymptotically stable. First, pick

$$V(x) = \frac{1}{2}x^2$$

so that

$$\dot{V} = x\dot{x}$$

and so

$$\dot{V} = xf(x) + gxu = xf(x) + gxF(x)$$

Notice that

$$\dot{V} \leq |x||f(x)| + gxF(x) \leq \alpha x^2 + gxF(x)$$

We want to design the neural controller $F(x)$ so that the second term in the above $\dot{V}$ equation is negative since then we will have $\dot{V} < 0$ for $x \neq 0$ and then $x_e = 0$ will be a globally asymptotically stable equilibrium. To do this, suppose that we design the controller so that $F(0) = 0$, $F(x)$ is smooth, and for some scalar $\beta > 0$,

$$\begin{aligned} F(x) &> -\beta x, \ x < 0 \\ F(x) &< -\beta x, \ x > 0 \end{aligned} \tag{4.19}$$

which simply constrains the nonlinear surface of the neural controller. Now, if $x > 0$, $F(x) < -\beta x$, so

$$\dot{V} \leq \alpha x^2 + gx(-\beta x) = (\alpha - g\beta)x^2$$

Also, if $x < 0$, $F(x) > -\beta x$, so once again

$$\dot{V} \leq \alpha x^2 - g\beta x^2 = (\alpha - g\beta)x^2$$

Hence, if we have $\alpha - g\beta < 0$ or $\beta > \alpha/g$, then $x_e = 0$ will be globally asymptotically stable.

So, intuitively, why does our neural controller stabilize this uncertain nonlinear plant? Basically, when $x > 0$, $F(x) < 0$ so the neural controller seeks to make the derivative $\dot{x}$ negative to get the state x to move toward $x_e = 0$. Similarly, if $x < 0$, $F(x) > 0$ so the neural controller seeks to make the derivative $\dot{x}$ positive to get the state to move toward $x_e = 0$. It should be clear that it is not necessary for $F(x)$ to be a neural controller to achieve the stabilization task; any controller that satisfies the conditions in Equation (4.19) (and the other constraints) will adequately perform the task.

All of this analysis is based on our ability to synthesize a neural controller so that Equation (4.19) is met. To do this, you would need to write out the mathematical form of $F(x)$ and prove that it satisfies Equation (4.19); perhaps in this simple case, you could use a somewhat heuristic graphical technique where you construct the neural controller and plot its surface to check Equation (4.19). Notice that for many neural controllers, the output saturates for some large magnitude values of x so that Equation (4.19) will often not be satisfied globally. In this case, the analysis is not global, but only for an interval of the x axis, so we can only conclude that x_e is asymptotically stable (i.e., a local property) or that there is some region of asymptotic stability.

4.7 Hierarchical Neural Networks

There are a variety of methods that can be employed to construct hierarchical neural networks. Here, we provide an example of how such hierarchies occur in nature, then discuss how multilayer perceptrons and radial basis function neural networks can be organized in a hierarchical fashion.

It is natural to view some neural networks as hierarchical.

4.7.1 Example: Marine Mollusc

In the marine mollusc, *Pleurobranchaea*, behaviors are organized hierarchically as dictated by the cellular arrangement of their neurons [312]. The arrangement (see Figure 472 in [312]), shows that the "swimming escape response" inhibits the other behaviors. Also, egg laying inhibits feeding, which in turn takes precedence over mating. The actual neural "circuitry" has been traced in these molluscs and this research has shown that when activated, command systems of neurons that are responsible for feeding and egg laying inhibit the neural networks dedicated to mating and locomotion.

The behavior of the mollusc is directly dictated by the underlying hierarchical organization of its neural network. Evolution has shaped a hierarchical arrangement in the neural network so that the behaviors that are exhibited increase the reproductive success of the mollusc.

4.7.2 Hierarchical Neural Structures

Here, we simply provide some ideas on how to structure neural networks in a hierarchical fashion. First, you could use a multilayer perceptron to turn on and off different parts of another multilayer perceptron. For example, the higher layer could simply output zeros and ones and these could multiply activation function outputs so that the lower level perceptron is reconfigurable based on different conditions.

For radial basis functions you may have a two-level hierarchical network with the higher layer defined on a coarse grid and the lower layer on a fine grid. Then, when a region is activated in the higher level network, that could activate a radial basis function neural network that is defined on a fine grid. This provides a type of "nesting" and focusing, and at times can provide for savings in computational complexity since only those radial basis functions with fine grids that are activated need to be stored in memory and computed.

4.8 Exercises and Design Problems

Exercise 4.1 (Building Multilayer Perceptrons): In this problem you will focus on constructing, "by hand," multilayer perceptrons to match certain functions.

(a) Construct two different multilayer perceptrons that try to match (approximate) the input-output properties of $y = f(x) = 2x$ where x and y are scalars over the range $x \in [-10, 10]$. In the first case you may use a linear activation function, and any combination of other neurons. In the second case, use a linear activation function in the output layer and a single hidden layer of no more than five logistic activation functions. Try to tune the parameters of the network *by hand* to make the mapping that is implemented by the neural network as close as possible to f over its entire domain. Do not use the neural network training methods that are introduced later in the book. Plot f vs. x and the mapping implemented by the neural network on the same plot in order to illustrate how close the network approximates the function.

(b) Repeat (a) but for $y = f(x) = 2x^2$. You may use any type of multilayer perceptron, with any number of neurons you would like.

(c) Repeat (a) but for $y = f(x) = 2\sin(x)$. You may use any type of multilayer perceptron, with any number of neurons you would like.

Exercise 4.2 (Building Radial Basis Function Neural Networks):

(a) Repeat Exercise 4.1 (a), but only construct one radial basis function neural network with no more than five receptive field units.

(b) Repeat Exercise 4.1 (b), but only construct one radial basis function neural network with no more than five receptive field units.

(c) Repeat Exercise 4.1 (c), but only construct one radial basis function neural network and you may use any number of receptive field units for it.

Exercise 4.3 (Lyapunov's Direct Method): Suppose that you are given the plant

$$\dot{x} = ax + bu$$

where $b > 0$ and $a < 0$ (so the system is stable) and x is a scalar. Suppose that you design an instinctual neural controller F that generates the input to the plant given the state of the plant (i.e., $u = F(x)$). Assume that you design the controller so that $F(0) = 0$ (so that $x = 0$ is an equilibrium) and so that $F(x)$ is continuous in x (so that a unique solution exists to the differential equation describing the closed-loop system).

(a) Use Lyapunov's direct method to show that if x and $F(x)$ always have opposite signs, then $x = 0$ is stable.

(b) What types of stability does $x = 0$ of the control system possess for part (a)? List all types of stability that it possesses.

(c) Design a (SISO) instinctual neural controller that satisfies the condition stated in (a) (and so that $F(0) = 0$ and $F(x)$ is continuous) and simulate the closed-loop system to help illustrate the stability of the neural control system. Choose the initial condition $x(0) = 1$, $a = -2$, and $b = 2$. Of course, the simulation does not *prove* that the closed-loop system is stable—it only shows that for one initial condition, the state appears to converge but cannot prove that it converges since the simulation is only for a finite amount of time.

Exercise 4.4 (Multilayer Perceptron for Tanker Ship Steering): Produce simulations to reproduce the results where we used a multilayer perceptron for tanker ship steering in the chapter (all the conditions). Add more comments to the code and produce a flowchart to demonstrate that you understand its operation.

Exercise 4.5 (Radial Basis Function Neural Network for Tanker Ship Steering): Produce simulations to reproduce the results where we used a radial basis function neural network for tanker ship steering in the chapter (all the conditions). Add more comments to the code and produce a flowchart to demonstrate that you understand its operation.

Design Problem 4.1 (Design of a Multilayer Perceptron for Tanker Ship Steering):

(a) Redesign the multilayer perceptron from Exercise (4.4) to improve performance of the closed-loop system for nominal conditions. Constrain the way that you perform the redesign to simply tuning of parameters, not changing the number of layers or neurons. Show plots to support your conclusions.

(b) Repeat (a) but design a multilayer perceptron that has two inputs, e and $\dot{e}$ (that you may approximate using an Euler approximation to the derivative), and one output δ. Hint: Build on the multilayer perceptron that was used in (a). Tune the multilayer perceptron so that it obtains "better" performance (you define precisely what this means for your study) than in (a). Plot the three-dimensional input-output map of the resulting tuned controller.

(c) Repeat (a) but you may use any type of multilayer perceptron (i.e., you choose the inputs, number of layers, and neurons). Try to achieve the best possible performance for all the different conditions considered in the chapter. You define what you mean by good performance, and you decide what an appropriate balance is in the quality of the results between the different conditions.

Design Problem 4.2 (Design of a Radial Basis Function Neural Network for Tanker Ship Steering):

(a) Redesign the radial basis function neural network from Exercise (4.5) to improve performance of the closed-loop system for nominal conditions. Constrain the way that you perform the redesign to simply tuning of parameters, not changing the number of receptive field units. Show plots to illustrate better performance. Plot the three-dimensional input-output map of the resulting tuned controller.

(b) Repeat (a), but you may use any radial basis function neural network (i.e., you choose the inputs and number of receptive field units).

(c) Repeat (a) with the modification in (b), but try to achieve the best possible performance for all the different conditions considered in the chapter. You define what you mean by good performance, and you decide what an appropriate balance is in the quality of the results between the different conditions.

Chapter 5

Rule-Based Control

Chapter Contents

One relatively complex task that humans can perform is a feedback control task. For instance, driving an automobile is a control task that humans regularly perform. There, the driver senses lane markers, vehicles, obstacles, and other cues to control the direction of travel by steering, and velocity via actuating the throttle and brakes. Humans perform many other control tasks when employed at, for example, chemical processing plants, manufacturing facilities, and in vehicular applications. In this chapter, we study rule-based control by studying the use of fuzzy and expert systems for control. These are probably the most popular intelligent control methods for automating feedback control tasks that have often been performed by humans in the past. The biomimicry here should be thought of as "human-mimicry" as it was explained in Part I, but clearly an accurate model of human reasoning and decision-making processes is neither sought, nor obtained.

The design example for the tanker ship serves to illustrate the heuristic nonlinear control design methodology that fuzzy and expert control allows. Here, there is a particularly important focus on design methodology for fuzzy controllers, as there have been certain problems in the literature with proper design methodology. We discuss effects of disturbances, noise, plant changes, stability, and limit cycles. It is emphasized that sound control engineering methodology, as outlined in Part I, should not be ignored.

5.1 Fuzzy Control

A block diagram of a fuzzy control system is shown in Figure 5.1. The fuzzy controller is composed of the following four elements:

1. *A rule base* (a set of If-Then rules), which contains a fuzzy logic quantification of the expert's linguistic description of how to achieve good control.

2. *An inference mechanism* (also called an "inference engine" or "fuzzy inference" module), which emulates the expert's decision-making in interpreting and applying knowledge about how best to control the plant.

3. *A fuzzification interface,* which converts controller inputs into information that the inference mechanism can easily use to activate and apply rules.

4. *A defuzzification interface,* which converts the conclusions of the inference mechanism into actual inputs for the process.

We introduce each of the components of the fuzzy controller for the simple problem of tanker ship heading regulation, as was shown in Figure 4.8.

5.1.1 Choosing Fuzzy Controller Inputs and Outputs

Consider a human-in-the-loop whose responsibility is to control the tanker ship (i.e., the ship captain), as shown in Figure 5.2. The fuzzy controller is to be designed to automate how a captain would control the system. First, the captain

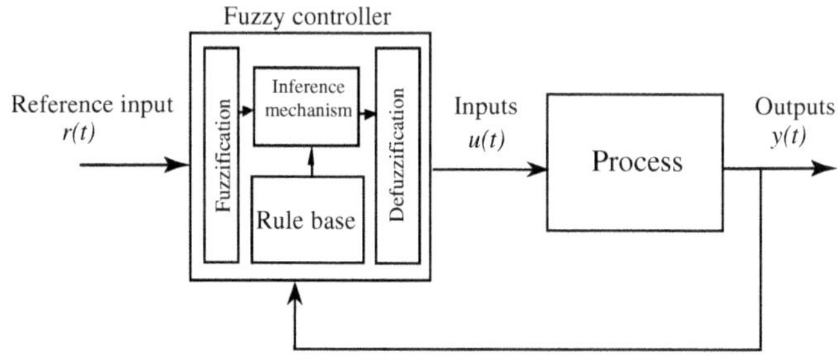

Figure 5.1: Fuzzy controller.

tells us (the designers of the fuzzy controller) what information she or he will use as inputs to the decision-making process. Suppose that for the tanker ship, the expert (this could be you, if you do not have the captain available) says that she or he will use

$$e(t) = \psi_r(t) - \psi(t)$$

and

$$\frac{de(t)}{dt} = \dot{e}(t)$$

as the variables on which to base decisions. Certainly, there are many other choices (e.g., the integral of the error e could also be used) but this choice makes good intuitive sense. Next, we must identify the controlled variable. For the tanker ship, it is assumed that we are only allowed to control the rudder so the input is δ (i.e., we do not consider the use of the ship speed for helping with steering).

Fuzzy controller input/output choice depends on what variables the expert uses and broadly affects the design of the controller.

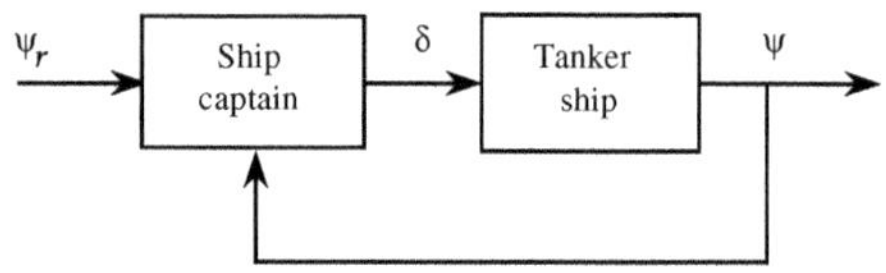

Figure 5.2: Human controlling a tanker ship.

For more complex applications, the choice of the inputs to the controller and outputs of the controller (inputs to the plant) can be more difficult. Essentially, you want to make sure that the controller will have the proper information available to be able to make good decisions and have proper control inputs to be able to move the system in the directions needed to be able to achieve high-performance operation. Practically speaking, access to information and the ability to effectively control the system often cost money. If the designer believes that proper information is not available for making control decisions, he or she may have to invest in another sensor that can provide a measurement of another system variable. Alternatively, the designer may implement some filtering or other processing of the plant outputs. In addition, if the designer determines that the current actuators will not allow for the precise control of the process, he or she may need to invest in designing and implementing an actuator that can properly affect the process. Hence, while in some academic problems you may be given the plant inputs and outputs, in many practical situations you may have some flexibility in their choice. These choices affect what information is available for making online decisions about the control of a process and hence affect how we design a fuzzy controller. Once the fuzzy controller inputs and outputs are chosen, you must determine the reference inputs. For the tanker ship, we will simply use step changes in ship heading. In general, the specification and generation of the reference input(s) can be more challenging.

After all the inputs and outputs are defined for the fuzzy controller, we can specify the fuzzy control system. The fuzzy control system for the tanker ship, with our choice of inputs and outputs, is shown in Figure 5.3. Now, *within this framework* we seek to obtain a description of how to control the process. We see then that the choice of the inputs and outputs of the controller places certain constraints on the remainder of the fuzzy control design process. If the proper information is not provided to the fuzzy controller, there will be little hope for being able to design a good rule base or inference mechanism. Moreover, even if the proper information is available to make control decisions, this will be of little use if the controller is not able to properly affect the process variables via the process inputs. It must be understood that the choice of the controller inputs and outputs is a fundamentally important part of the control design process for many practical applications.

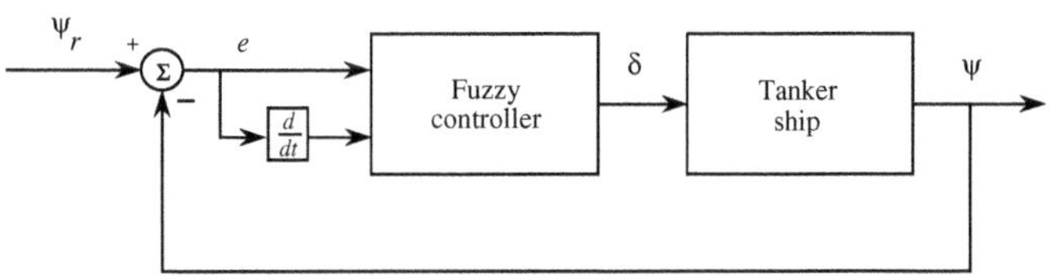

Figure 5.3: Fuzzy controller for a tanker ship steering problem.

5.1.2 Putting Control Knowledge into Rule Bases

Suppose that the human expert (captain) shown in Figure 5.2 provides a description of how best to control the plant in some natural language (e.g., English). Next, we characterize the expert's description with "linguistics."

Linguistic Descriptions

The linguistic description provided by the expert can generally be broken into several parts. There will be "linguistic variables" that describe each of the time-varying fuzzy controller inputs and outputs. For the tanker ship,

"error" describes $e(t)$
"change-in-error" describes $\frac{de(t)}{dt}$
"rudder-input" describes $\delta(t)$

Note that we use quotes to emphasize that certain words or phrases are linguistic descriptions, but emphasize that these variables do change over time. There are many possible choices for the linguistic descriptions for variables. Some designers like to choose them so that they are quite descriptive for documentation purposes. However, this can sometimes lead to long descriptions. Others seek to keep the linguistic descriptions as short as possible (e.g., using "$e(t)$" as the linguistic variable for $e(t)$), yet accurate enough so that they adequately represent the variables. Regardless, the choice of the linguistic variable has no effect on the way that the fuzzy controller operates; it is simply a notation that helps to facilitate the construction of the fuzzy controller via fuzzy logic.

Linguistic variables represent the key variables that the expert uses to make decisions.

Just as $e(t)$ takes on a value of, for example, 0.1 at $t = 2$ ($e(2) = 0.1$), linguistic variables assume "linguistic values." That is, the values that linguistic variables take on over time change dynamically. Suppose for the tanker ship example that "error," "change-in-error," and "rudder-input" take on the following values:

"neghuge"
"neglarge"
"negbig"
"negmed"
"negsmall"
"zero"
"possmall"
"posmed"
"posbig"
"poslarge"
"poshuge"

Note that we are using "negsmall" as an abbreviation for "negative small in size" and so on for the other variables. Such abbreviations help keep the linguistic descriptions short yet precise. For an even shorter description we could use integers:

"-5" to represent "neghuge"
"-4" to represent "neglarge"
"-3" to represent "negbig"
"-2" to represent "negmed"
"-1" to represent "negsmall"
"0" to represent "zero"
"1" to represent "possmall"
"2" to represent "posmed"
"3" to represent "posbig"
"4" to represent "poslarge"
"5" to represent "poshuge"

This is a particularly appealing choice for the linguistic values since the descriptions are short and nicely represent that the variable we are concerned with has a numeric quality. We are not, for example, associating "-1" with any particular number of radians of error; the use of the numbers for linguistic descriptions simply quantifies the sign of the error (in the usual way) and indicates the size in relation to the other linguistic values. We shall find the use of this type of linguistic value quite convenient when it comes to writing computer programs to simulate or implement fuzzy control systems and hence will give it the special name, "linguistic-numeric value."

The linguistic variables and values provide a language for the expert to express her or his ideas about the control decision-making process, in the context of the framework established by our choice of fuzzy controller inputs and outputs. Suppose that for the tanker ship $\psi_r(t) = 45$ deg. ($\psi_r(t) = \frac{45\pi}{180}$ rad.) and $e = r - y$ so that

$$e = \frac{45\pi}{180} - \psi$$

and

$$\frac{de}{dt} = -\frac{d\psi}{dt}$$

since $\frac{d\psi_r}{dt} = 0$. First, we will study how we can quantify certain dynamic behaviors with linguistics. In the next subsection we will study how to quantify knowledge about how to control the tanker ship using linguistic rules.

For the tanker ship, each of the following statements quantifies a different configuration of the ship (refer back to Figure 4.8 on page 117):

Linguistic statements characterize the status of the plant.

- The statement "error is poslarge" can represent the situation where the ship heading is at a significant angle counterclockwise to where it should be heading.

- The statement "error is negsmall" can represent the situation where the ship heading is just slightly clockwise of where it should be heading, but not too close to the reference heading ψ_r to justify quantifying it as "zero" and not too far away to justify quantifying it as "negmed."

- The statement "error is zero" can represent the situation where the ship heading is very near the desired heading (a linguistic quantification is not precise, hence we are willing to accept any value of the error around $e(t) = 0$ as being quantified linguistically by "zero" since this can be considered a better quantification than "possmall" or "negsmall").

- The statement "error is poslarge **and** change-in-error is possmall" can represent the situation where the ship heading is counterclockwise to where it should be and, since $\frac{d\psi}{dt} < 0$, the ship heading is moving *away* from the desired heading (note that in this case, the ship is moving counterclockwise).

- The statement "error is negsmall **and** change-in-error is possmall" can represent the situation where the ship heading is slightly clockwise of where it should be heading and, since $\frac{d\psi}{dt} < 0$, the ship heading is moving *toward* the desired heading (note that in this case, the ship is moving counterclockwise).

It is important for the reader to study each of the cases above to understand how the expert's linguistics quantify the current situation the ship is in (actually, each partially quantifies the ship's state).

Overall, we see that to quantify the dynamics of the process, we need to have a good understanding of the physics of the underlying process we are trying to control. While for the ship steering problem, the task of coming to a good understanding of the dynamics is relatively easy, this is not the case for many physical processes. Quantifying the process dynamics with linguistics is not always easy, and certainly a better understanding of the process dynamics generally leads to a better linguistic quantification. Often, this will naturally lead to a better fuzzy controller *provided* that you can adequately measure the system dynamics so that the fuzzy controller can make the right decisions at the proper time.

Rules

Next, we will use the above linguistic quantification to specify a set of rules (a rule base) that captures the expert's knowledge about how to control the plant. In particular, for the tanker ship in the three positions shown in Figure 5.4, we have the following rules (notice that we drop the quotes since the whole rule is linguistic):

Linguistic rules represent a description of the rules that the expert uses in control.

1. **If** error is negsmall **and** change-in-error is negsmall **Then** rudder-input is posmed

   This rule quantifies the situation in Figure 5.4(a) where the ship has a heading angle that is clockwise of the desired heading and is moving clockwise; hence, it is clear that we should apply a medium positive rudder angle so that we can get the ship moving in the proper direction.

2. **If** error is zero **and** change-in-error is possmall **Then** rudder-input is negsmall

   This rule quantifies the situation in Figure 5.4(b) where the ship is nearly moving in the proper direction (a linguistic quantification of zero does not imply that $e(t) = 0$ exactly) and is moving counterclockwise; hence, we should apply a small negative rudder angle to counteract the movement so that it moves toward zero (a positive rudder angle could result in the ship heading overshooting the desired angle).

3. **If** error is possmall **and** change-in-error is negsmall **Then** rudder-input is zero

   This rule quantifies the situation in Figure 5.4(c) where the ship is counterclockwise of the desired heading and is moving clockwise; hence, we apply a near zero rudder angle since the ship is already moving in the proper direction.

Specification of increasingly good rules generally requires increasingly good insights into the physics of the plant.

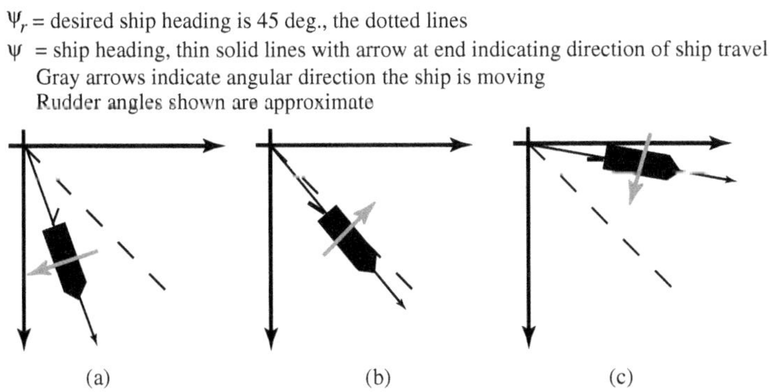

Figure 5.4: Tanker ship in various positions.

Each of the three rules listed above is a "linguistic rule" since it is formed solely from linguistic variables and values. Since linguistic values are not precise representations of the underlying quantities that they describe, linguistic rules are not precise either. They are simply abstract ideas about how to achieve good control that could mean somewhat different things to different people. They are, however, at a level of abstraction that humans are often comfortable with in terms of specifying how to control a process.

The general form of the linguistic rules listed above is

If premise **Then** consequent

As you can see from the three rules listed above, the premises (which are sometimes called "antecedents") are associated with the fuzzy controller inputs and are on the left-hand side of the rules. The consequents (sometimes called "actions") are associated with the fuzzy controller outputs and are on the right-hand side of the rules. Notice that each premise (or consequent) can be composed of the conjunction of several "terms" (e.g., in rule 3 above, "error is possmall **and** change-in-error is negsmall" is a premise that is the conjunction of two terms). The number of fuzzy controller inputs and outputs places an upper limit on the number of elements in the premises and consequents. Note that there does not need to be a premise (consequent) term for each input (output) in each rule, although often there is.

Rule Bases

Using the above approach, we could continue to write down rules for the ship steering problem for all possible cases (the reader should do this for practice, at least for a few more rules). Note that since we only specify a finite number of linguistic variables and linguistic values, there is only a finite number of possible rules. For the ship steering problem, with two inputs and eleven linguistic values for each of these, there are at most $11^2 = 121$ possible rules (all possible combinations of premise linguistic values for two inputs).

A convenient way to list all possible rules for the case where there are not too many inputs to the fuzzy controller (less than or equal to two or three) is to use a tabular representation. A tabular representation of one possible set of rules for the fuzzy controller for the ship is shown in Table 5.1. Notice that the body of the table lists the linguistic-numeric consequents of the rules, and the left column and top row of the table contain the linguistic-numeric premise terms. Then, for instance, the $(+1, -1)$ position (where the "+1" represents the row having "+1" for a numeric-linguistic value and the "−1" represents the column having "−1" for a numeric-linguistic value) has a 0 ("zero") in the body of the table and represents the rule

If error is possmall **and** change-in-error is negsmall **Then** rudder-input is zero

which is rule 3 above. Table 5.1 represents abstract knowledge that the expert has about how to control the tanker ship given the error and its derivative as inputs.

Table 5.1: Rule Table for the Tanker Ship

δ		$\dot{e}$										
		−5	−4	−3	−2	−1	0	1	2	3	4	5
e	−5	5	5	5	5	5	5	4	3	2	1	0
	−4	5	5	5	5	5	4	3	2	1	0	−1
	−3	5	5	5	5	4	3	2	1	0	−1	−2
	−2	5	5	5	4	3	2	1	0	−1	−2	−3
	−1	5	5	4	3	2	1	0	−1	−2	−3	−4
	0	5	4	3	2	1	0	−1	−2	−3	−4	−5
	1	4	3	2	1	0	−1	−2	−3	−4	−5	−5
	2	3	2	1	0	−1	−2	−3	−4	−5	−5	−5
	3	2	1	0	−1	−2	−3	−4	−5	−5	−5	−5
	4	1	0	−1	−2	−3	−4	−5	−5	−5	−5	−5
	5	0	−1	−2	−3	−4	−5	−5	−5	−5	−5	−5

Note that the other rules are also valid and take special note of the pattern of rule consequents that appears in the body of the table. Notice the diagonal of zeros. Viewing the body of the table as a matrix, we see that it has a certain symmetry to it. This symmetry that emerges when the rules are tabulated is no accident and is actually a representation of abstract knowledge about how to control the ship heading; it arises due to a symmetry in the system's dynamics. Similar patterns will often be found when constructing rule bases for other applications.

5.1.3 Fuzzy Quantification of Knowledge

Up to this point we have only quantified, in an abstract way, the knowledge that the human expert has about how to control the plant. Next, we will show how to use fuzzy logic to fully quantify the meaning of linguistic descriptions so that we may automate, in the fuzzy controller, the control rules specified by the expert.

Membership Functions

First, we quantify the meaning of the linguistic values using "membership functions." Consider, for example, Figure 5.5. This is a plot of a function μ versus $e(t)$ that takes on special meaning. The function μ quantifies the *certainty* that $e(t)$ can be classified linguistically as "possmall." In our discussion in this chapter, do not confuse the term "certainty" with "probability" or "likelihood." The membership function is not a probability density function, and there is no underlying probability space. By "certainty" we mean "degree of truth." The membership function does not quantify random behavior; it simply makes more accurate (less fuzzy) the meaning of linguistic descriptions.

To understand the way that a membership function works, it is best to perform a case analysis where we show how to interpret it for various values of $e(t)$:

Membership functions numerically quantify the meaning of linguistic statements by the expert.

- If $e(t) = -\frac{4\pi}{10}$, then $\mu(-\frac{4\pi}{10}) = 0$, indicating that we are certain that $e(t) = -\frac{4\pi}{10}$ is *not* "possmall" (indeed, it is negative).
- If $e(t) = \frac{2\pi}{20}$, then $\mu(\frac{2\pi}{20}) = 0.5$, indicating that we are halfway certain that $e(t) = \frac{2\pi}{20}$ is "possmall" (we are only halfway certain since it could also be "zero" with some degree of certainty—this value is in a "gray area" in terms of linguistic interpretation).
- If $e(t) = \frac{2\pi}{10}$, then $\mu(\frac{2\pi}{10}) = 1.0$, indicating that we are absolutely certain that $e(t) = \frac{2\pi}{10}$ is what we mean by "possmall."
- If $e(t) = \frac{8\pi}{10}$, then $\mu(\frac{8\pi}{10}) = 0$, indicating that we are certain that $e(t) = \frac{8\pi}{10}$ is not "possmall" (actually, we will soon see that we will quantify it as "poslarge").

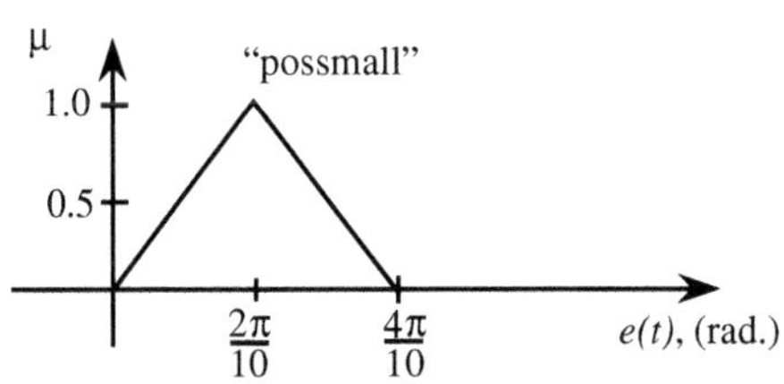

Figure 5.5: Membership function for linguistic value "possmall."

The membership function quantifies, in a continuous manner, whether values of $e(t)$ belong to (are members of) the set of values that are "possmall," and hence it quantifies the meaning of the linguistic statement "error is possmall." This is why it is called a membership function. It is important to recognize that the membership function in Figure 5.5 is only one possible definition of the meaning of "error is possmall;" you could use a bell-shaped function, a trapezoid, or many others, depending on what the expert means by "possmall."

For instance, consider the membership functions shown in Figure 5.6. For some applications someone may be able to argue that we are absolutely certain that any value of $e(t)$ near $\frac{2\pi}{10}$ is still "possmall" and only when you get sufficiently far from $\frac{2\pi}{10}$ do we lose our confidence that it is "possmall." One way to characterize this understanding of the meaning of "possmall" is via the trapezoid-shaped membership function in Figure 5.6(a). For other applications, you may think of membership in the set of "possmall" values as being dictated by the Gaussian-shaped membership function (not to be confused with the Gaussian probability density function) shown in Figure 5.6(b). For still other applications, you may not readily accept values far away from $\frac{2\pi}{10}$ as being "possmall," so you may use the membership function in Figure 5.6(c) to represent this. Finally, while we often think of symmetric characterizations of the meaning of linguistic values, we are not restricted to these symmetric representations. For instance, in Figure 5.6(d) we represent that we believe that as $e(t)$ moves to the left of $\frac{2\pi}{10}$, we are very quick to reduce our confidence that it is "possmall," but if we move to the right of $\frac{2\pi}{10}$, our confidence that $e(t)$ is "possmall" diminishes at a slower rate.

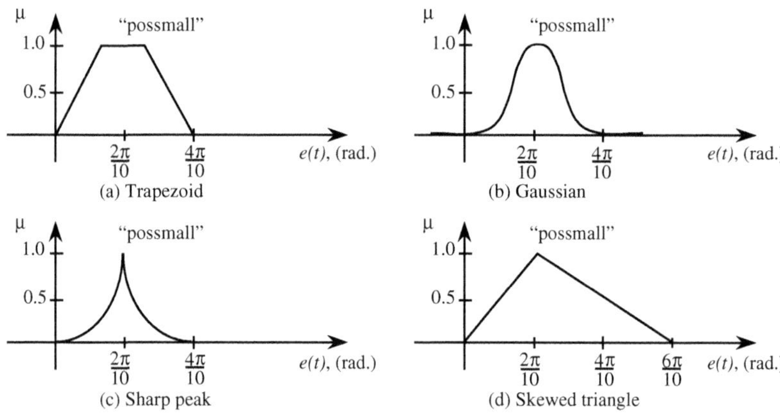

Figure 5.6: Some example membership function choices for representing "error is possmall."

Each of the membership functions in Figure 5.6 has a mathematical representation and these are useful in simulation and implementation of fuzzy controllers. For example, interval checking plus equations for lines can be used to implement the membership function in Figure 5.6(a) and a Gaussian function with an appropriate center, spread, and scale factor can implement the one in Figure 5.6(b).

In summary, we see that depending on the application and the designer (expert), many different choices of membership functions are possible. It is important to note here, however, that for the most part, the definition of a membership function is subjective rather than objective. That is, we simply quantify it in a manner that makes sense to us, but others may quantify it in a different manner.

The set of values that is described by μ as being "positive small" is called a "fuzzy set." Let A denote this fuzzy set. Notice that from Figure 5.5 we are absolutely certain that $e(t) = \frac{2\pi}{10}$ is an element of A, but we are less certain that $e(t) = \frac{2\pi}{40}$ is an element of A. Membership in the set, as specified by the membership function, is fuzzy; hence we use the term "fuzzy set." A "crisp" (as contrasted to "fuzzy") quantification of "possmall" can also be specified, but via the membership function shown in Figure 5.7. This membership function is simply an alternative representation for the interval on the real line $\frac{2\pi}{20} \leq e(t) \leq \frac{6\pi}{20}$, and it indicates that this interval of numbers represents "possmall." Clearly, this characterization of crisp sets is simply another way to represent a normal interval (set) of real numbers.

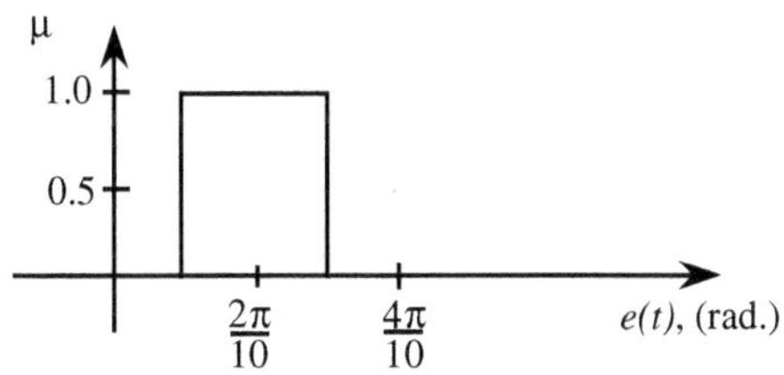

Figure 5.7: Membership function for a crisp set.

While the vertical axis in Figure 5.5 represents certainty, the horizontal axis is also given a special name. It is called the "universe of discourse" for the input $e(t)$ since it provides the range of values of $e(t)$ that can be quantified with linguistics and fuzzy sets. In conventional terminology, a universe of discourse for an input or output of a fuzzy system is simply the range of values the inputs and outputs can take on.

Now that we know how to specify the meaning of a linguistic value via a membership function (and hence a fuzzy set), we can easily specify the membership functions for all 33 linguistic values (eleven for each input and eleven for the output) of our ship steering example. See Figure 5.8 for one choice of membership functions.

For our later convenience, we list both the linguistic and linguistic-numeric values associated with each membership function. Hence, we see that the membership function in Figure 5.5 for "possmall" on the "error" universe of discourse is embedded among several others that describe other sizes of values (so that, for instance, the membership function to the right of the one for "possmall" is the

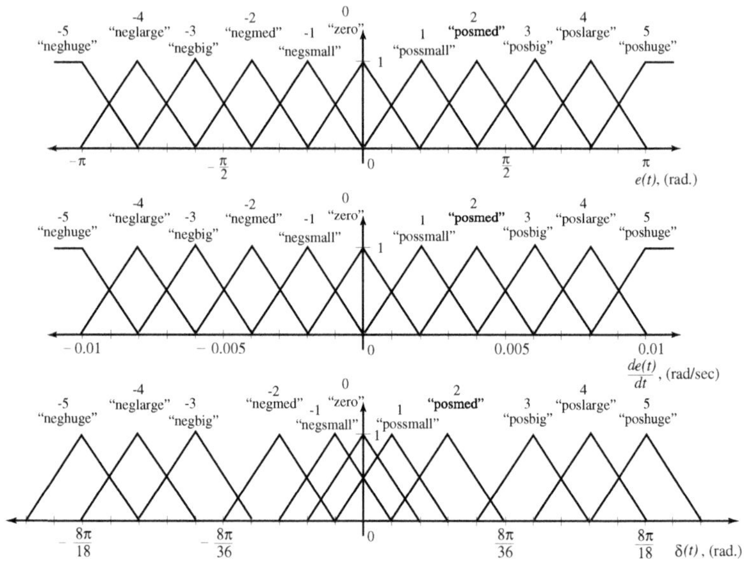

Figure 5.8: Membership functions for a ship steering example.

one that represents "error is posmed"). Note that other similarly shaped membership functions make sense (e.g., bell-shaped membership functions). The scale for the axis was chosen since $|e(t)|$ can be at most π radians since it is the difference between two angles and we use this fact to help define the relative sizes in our membership quantification of our linguistics. Notice that for the $\dot{e}$ universe of discourse, we use a set of membership functions similar to the ones on the e universe of discourse, but that the scale of the $\dot{e}$ axis is different. This scale was chosen since our expert captain felt that $|\dot{e}(t)| \geq 0.01$ radians per second (0.57 degrees per second) was a fast change in the ship heading. Notice then that the meaning of the linguistics on the $\dot{e}$ universe of discourse is different from those on the e universe of discourse.

The membership functions at the outer edges of the e and $\dot{e}$ universes of discourse in Figure 5.8 deserve special attention. For the inputs $e(t)$ and $\dot{e}$, we see that the outermost membership functions "saturate" at a value of one. This makes intuitive sense, as at some point the human expert would just group all large values together in a linguistic description such as "poshuge" (or "neghuge"). The membership functions at the outermost edges appropriately characterize this phenomenon since they characterize "greater than" (for the right side) and "less than" (for the left side). Study Figure 5.8 and convince yourself of this.

It is important to have a clear picture in your mind of how the values of the membership functions change as, for example, $e(t)$ changes its value over time. For instance, as $e(t)$ changes from $-\pi$ to π, we see that various membership functions will take on zero and nonzero values indicating the degree to which the corresponding linguistic value appropriately describes the current value of $e(t)$. For example, at $e(t) = -\pi$ we are certain that the error is "neghuge," and as the value of $e(t)$ moves toward $-8\pi/10$, we become less certain that it is "neghuge" and more certain that it is "neglarge." We see that the membership functions quantify the meaning of linguistic statements that describe time-varying signals.

Finally, note that often we will draw all the membership functions for one input or output variable on one graph; hence, we often omit the label for the vertical axis with the understanding that the plotted functions are membership functions describing the meaning of their associated linguistic values. Also, we will use the notation μ_{zero} to represent the membership function associated with the linguistic value "zero" and a similar notation for the others.

Next, consider the choice of membership functions for the "rudder-input" $\delta(t)$ universe of discourse in Figure 5.8. The horizontal scale was chosen since, as you may recall, the rudder input can only be moved between ± 80 degrees. Converting to radians, this means that it moves between $\pm 8\pi/18$ radians and this gives us the center values for the membership functions on the outer edges for the δ universe of discourse. Next, note that for the output δ, the membership functions at the outermost edges cannot be saturated for the fuzzy system to be properly defined (more details on this point will be provided at the end of Section 5.1.6 that starts on page 178). The basic reason for this is that in decision-making processes of the type we study, we seek to take actions that specify an exact value for the process input. We do not generally indicate to a process actuator, "any value bigger than, say, $\pm 8\pi/18$, is acceptable."

The Meaning of Membership Functions and Rules

Notice that the pattern of center positions (i.e., where the triangles peak at one) for the output membership functions in Figure 5.8 is not uniform as it is for the input universes of discourse. A uniform distribution (which with proper tuning can work for this ship steering example also) would imply that the captain would roughly make the rudder angle proportional to the error between the heading and desired heading, and the change in the heading error (except when the error and change in error are too big in magnitude, then she or he will simply move the rudder to its maximum deflection). To get a uniform distribution of output membership function centers you can choose the center values, which we denote by b_i where i is the linguistic-numeric index for the corresponding membership function, as

$$b_i = \frac{8\pi}{18}\left(\frac{i}{5}\right)$$

where $i = -5, -4, -3, -2, -1, 0, 1, 2, 3, 4, 5$ (and then use the same base widths and rule base). For practice, draw the resulting membership functions on the output universe of discourse.

The choice of having nonuniformly distributed membership functions in Figure 5.8 represents that for small heading errors, when the change in error is small also, the captain will not put in as big a rudder angle. Why? Through experience the captain has found that if the heading is close to where it should be and it is not moving away from where it should be fast, then small corrections are more effective in heading regulation. While the captain may have learned this through experience (or training), the basic reason for this arises from the physics of the process. For example, since the heading sensor measurement is noisy, for small heading errors this noise can have a relatively large impact on the error so that for small errors the noise can make it appear that there is a heading error when there is not. Now, for the case where, say, the error is "poshuge" but the change in error $\dot{e}$ is "neghuge" the captain also puts in a zero rudder angle input (see Table 5.1). The expert captain specified this rule since, while the heading is far from where it should be, the heading is moving very fast to correct this condition. If the error is "poshuge" and the change in error $\dot{e}$ is "neglarge," then the rudder input is "negsmall" and by Figure 5.8 this is only a small correction since the captain does not want to expend more rudder movement than necessary; the ship is moving to correct its own error in heading, why expend control energy (and wear out the rudder actuator) trying to do something that is already in the process of happening?

It is important to gain insights into the fuzzy logic quantification of the rule base to clearly understand what control expertise is being implemented by the fuzzy controller.

Next, note that from the pattern of output membership functions in the body of Table 5.1, we see that the captain will saturate the rudder either positive (upper left corner of the rule base) or negative (lower right corner of the rule base) if the error and change in error are too big in magnitude. The choice of when to saturate the rudder (i.e., to move it to its maximum deflection) is made through the captain's experience in heading regulation. If she or he is not willing to saturate it soon enough, larger heading deviations may occur. However, if the captain is too quick to saturate the rudder input, for example, even for relatively small errors, she or he will wear out the rudder actuator faster and will be continually moving the rudder.

With this discussion, it is important to note that the meaning of the linguistic rule base is not clear until the membership functions for the linguistic variables are all defined. The membership function definitions fully specify the meaning of the linguistics. Note that while on the e and $\dot{e}$ universes of discourse the meaning of the linguistics is similar, it is different by a scale factor on the horizontal axes (scaling the horizontal axis changes the meaning of the linguistics). Moreover, the meaning of the linguistics on the output universe of discourse is quite different from meaning of the linguistics on the input universes of discourse (e.g., for the membership functions at the outermost edges and in the nonuniform spacing of the output membership function centers).

Due to the lack of clarity of the meaning of control rules in the linguistic rule base shown in Table 5.1, schemes are often used which include membership function information in the rule base table. While many schemes are possible, a common one is shown in Table 5.2 where rather than listing the indices for the output membership functions, the centers of the appropriate output membership functions are listed, up to a scale factor, which in this case is $8\pi/18$ (i.e., to get the actual center from the rule base table you take the entry and multiply it by $8\pi/18$).

Coupled with our understanding of the meaning of the linguistic-numeric indices for the error and change in error, all the major components of the captain's

Table 5.2: Rule Table for the Tanker Ship (body of table holds the output membership function centers where each element should be multiplied by $8\pi/18$).

		$\dot{e}$										
		−5	−4	−3	−2	−1	0	1	2	3	4	5
e	−5	1	1	1	1	1	1	.8	.6	.3	.1	0
	−4	1	1	1	1	1	.8	.6	.3	.1	0	−.1
	−3	1	1	1	1	.8	.6	.3	.1	0	−.1	−.3
	−2	1	1	1	.8	.6	.3	.1	0	−.1	−.3	−.6
	−1	1	1	.8	.6	.3	.1	0	−.1	−.3	−.6	−.8
	0	1	.8	.6	.3	.1	0	−.1	−.3	−.6	−.8	−1
	1	.8	.6	.3	.1	0	−.1	−.3	−.6	−.8	−1	−1
	2	.6	.3	.1	0	−.1	−.3	−.6	−.8	−1	−1	−1
	3	.3	.1	0	−.1	−.3	−.6	−.8	−1	−1	−1	−1
	4	.1	0	−.1	−.3	−.6	−.8	−1	−1	−1	−1	−1
	5	0	−.1	−.3	−.6	−.8	−1	−1	−1	−1	−1	−1

knowledge of ship steering are directly evident from Table 5.2 in the following manner:

One good way to gain insights is to characterize the abstract patterns that often emerge when a rule base is constructed.

1. If the heading error and change in error are both too big (upper left and lower right corners of the rule base shown in Table 5.2), then use the appropriate maximum rudder input.

2. For zero e and $\dot{e}$, the rudder angle should be zero, but if e and $\dot{e}$ move positive, then the rudder should move negative (where if $\dot{e}$ moves significantly positive, then the rudder should move even more negative). Similar reasoning is used for e and $\dot{e}$ negative, where we then make the rudder angle positive. For the case where e and $\dot{e}$ have opposite signs and depending on the magnitude of the signals, we will make the rudder input either positive or negative.

3. For small e and $\dot{e}$, be conservative in making changes to the rudder position since such corrections may cause heading deviations instead (i.e., lower the "gain" of the controller near zero so that noise is not amplified). Also, if the ship's angular position is moving sufficiently fast to remove the heading error, then be conservative in using the rudder to help move it since this can require unnecessary control energy.

This provides a summary of the captain's knowledge about ship steering. The above three points can be thought of as "meta-rules," that is, abstract representations of control rules. Some designers use rules that are more abstract in the sense that they describe what control actions should occur whenever $e(t)$ and $\dot{e}(t)$ lie in a certain region. For example, the rule

If (error is neghuge **and** change-in-error is neghuge)
or (error is neghuge **and** change-in-error is neglarge)
or (error is neglarge **and** change-in-error is neghuge)
Then rudder-input is poshuge

represents what action should be taken if $e(t)$ and $\dot{e}(t)$ take on values such that any of the three rules in the upper left corner of the rule base in Table 5.2 are on. In this sense, the above rule represents three of the rules of the form discussed earlier. It achieves this apparently more compact representation via the use of the "disjunction" (or) in the premise of the above rule. If you were to use the above rule in the implementation you would use "maximum" to represent the disjunction; however, if you are concerned with implementation complexity, you must be careful to determine whether this approach is more or less complex than simply treating each rule separately.

Returning to our discussion on the tanker ship, we must emphasize that it is important in rule base construction that the control system designer can clearly list the expertise that is represented in the rule base. Lack of a clear understanding of the rule base is an indication that there is likely to be a later problem in simulation or implementation. Fuzzy control is not a methodology where you can haphazardly construct a rule base and expect in all cases for it to work well; you must put good control knowledge in to get good closed-loop system performance (it is not very often that you can get lucky and get good performance from a poorly constructed rule base).

In summary, the rule base of the fuzzy controller holds the linguistic variables, linguistic values, their associated membership functions, and the set of all linguistic rules (shown in Table 5.1 on page 162), so we have completed the description of the rule base for the ship steering problem. Next we describe the fuzzification process.

Fuzzification

It is actually the case that for most fuzzy controllers the fuzzification block in Figure 5.1 on page 156 can be ignored since this process is so simple. The reader should simply think of the fuzzification process as the act of obtaining a value of an input variable (e.g., $e(t)$) and finding the numeric values of the membership function(s) that are defined for that variable. For example, if $e(t) = 2\pi/10$ and $\dot{e}(t) = 0.001$, the fuzzification process amounts to finding the values of the input membership functions for these. In this case

$$\mu_{possmall}(e(t)) = 1$$

(with all others zero) and

$$\mu_{zero}\left(\dot{e}(t)\right) = \mu_{possmall}\left(\dot{e}(t)\right) = 0.5$$

Some think of the membership function values as an "encoding" of the fuzzy controller numeric input values. The encoded information is then used in the fuzzy inference process that starts with "matching."

5.1.4 Matching: Determining Which Rules to Use

Next, we seek to explain how the inference mechanism in Figure 5.1 on page 156 operates. The inference process generally involves two steps:

1. The premises of all the rules are compared to the controller inputs to determine which rules apply to the current situation. This "matching" process involves determining the certainty that each rule applies, and typically we will more strongly take into account the recommendations of rules that we are more certain apply to the current situation.

2. The conclusions (what control actions to take) are determined using the rules that have been determined to apply at the current time. The conclusions are characterized with a fuzzy set (or sets) that represents the certainty that the input to the plant should take on various values.

We will cover step 1 in this subsection and step 2 in the next.

Premise Quantification via Fuzzy Logic

To perform inference we must first quantify each of the rules with fuzzy logic. To do this, we first quantify the meaning of the premises of the rules that are composed of several terms, each of which involves a fuzzy controller input. Consider Figure 5.9, where we list two terms from the premise of the rule

If error is zero **and** change-in-error is possmall **Then** rudder-input is negsmall

Above, we had quantified the meaning of the linguistic terms "error is zero" and "change-in-error is possmall" via the membership functions shown in Figure 5.8. Now we seek to quantify the linguistic premise "error is zero **and** change-in-error is possmall." Hence, the main item to focus on is how to quantify the logical "and" operation that combines the meaning of two linguistic terms. While we could use standard Boolean logic to combine these linguistic terms, since we have quantified them more precisely with fuzzy sets (i.e., the membership functions), we can use these.

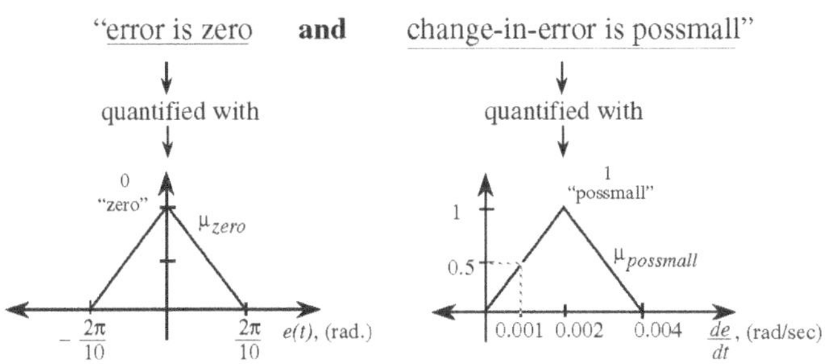

Figure 5.9: Membership functions of premise terms.

To see how to quantify the "and" operation, begin by supposing that $e(t) = \pi/10$ and $\dot{e}(t) = 0.0005$, so that using Figure 5.8 (or Figure 5.9) we see that

$$\mu_{zero}(e(t)) = 0.5$$

and

$$\mu_{possmall}\left(\dot{e}(t)\right) = 0.25$$

What, for these values of $e(t)$ and $\dot{e}(t)$, is the certainty of the statement

"error is zero **and** change-in-error is possmall"

that is the premise from the above rule? We will denote this certainty by $\mu_{premise}$. There are actually several ways to define it:

- *Minimum:* Define $\mu_{premise} = \min\{0.5, 0.25\} = 0.25$, that is, using the minimum of the two membership values.
- *Product:* Define $\mu_{premise} = (0.5)(0.25) = 0.125$, that is, using the product of the two membership values.

The premise of a rule is true to a certain degree and we think of rules that are "more true" as being more relevant to the current plant situation.

Do these quantifications make sense? Notice that both ways of quantifying the "and" operation in the premise indicate that you can be no more certain about the conjunction of two statements than you are about the individual terms that make them up (note that $0 \leq \mu_{premise} \leq 1$ for either case). If we are not very certain about the truth of one statement, how can we be any more certain about the truth of that statement "and" the other statement? It is important that you convince yourself that the above quantifications make sense. To do so, we recommend that you consider other examples of "anding" linguistic terms that have associated membership functions.

While we have simply shown how to quantify the "and" operation for one value of $e(t)$ and $\dot{e}(t)$, if we consider all possible $e(t)$ and $\dot{e}(t)$ values, we will obtain a multidimensional membership function $\mu_{premise}\left(e(t), \dot{e}(t)\right)$ that is a function of $e(t)$ and $\dot{e}(t)$ for each rule. For our example, if we choose the minimum operation to represent the "and" in the premise, then we get the multidimensional membership function $\mu_{premise}\left(e(t), \dot{e}(t)\right)$ shown in Figure 5.10 (and if we use product to represent the premise we get the premise membership function shown in Figure 5.11). Suppose that we use minimum to represent the conjunction in the premise. Notice that if we pick values for $e(t)$ and $\dot{e}(t)$, the value of the premise certainty $\mu_{premise}\left(e(t), \dot{e}(t)\right)$ represents how certain we are that the rule

If error is zero **and** change-in-error is possmall **Then** rudder-input is negsmall

is applicable for specifying the rudder input to the plant. As $e(t)$ and $\dot{e}(t)$ change, the value of $\mu_{premise}\left(e(t), \dot{e}(t)\right)$ changes according to Figure 5.10 (or Figure 5.11 if we use product to represent the rule), and we become less or more certain of the applicability of this rule.

In general we will have a different premise membership function for each of the rules in the rule base, and each of these will be a function of $e(t)$ and $\dot{e}(t)$

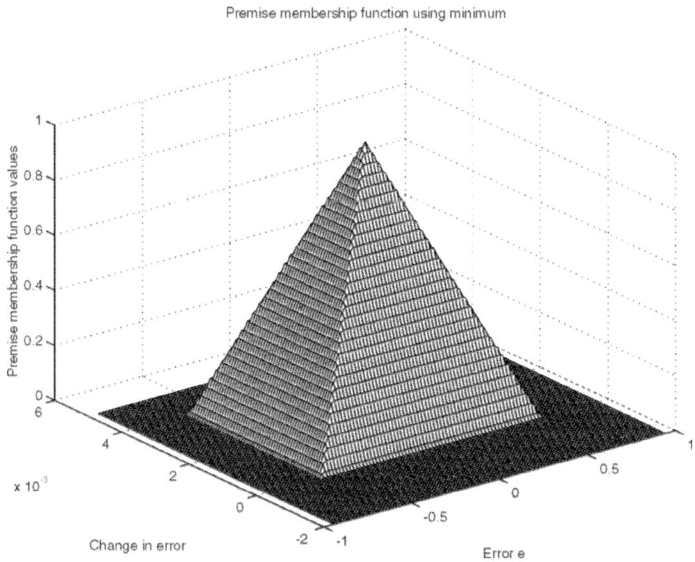

Figure 5.10: Membership function of the premise for a single rule using minimum to represent the conjunction.

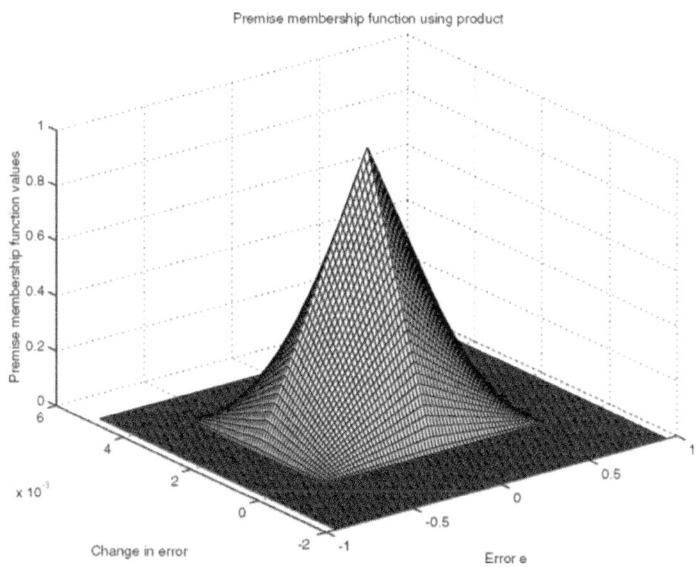

Figure 5.11: Membership function of the premise for a single rule using product to represent the conjunction.

so that given specific values of $e(t)$ and $\dot{e}(t)$, we obtain a quantification of the certainty that each rule in the rule base applies to the current situation. It is important you picture in your mind the situation where $e(t)$ and $\dot{e}(t)$ change dynamically over time. When this occurs, the values of $\mu_{premise}(e(t), \dot{e}(t))$ for each rule change, and hence the applicability of each rule in the rule base for specifying the rudder input to the ship, changes with time.

Determining Which Rules Are On

Determining the applicability of each rule is called "matching." We say that a rule is "on at time t" if its premise membership function $\mu_{premise}(e(t), \dot{e}(t)) > 0$. Hence, the inference mechanism seeks to determine which rules are on to find out which rules are relevant to the current situation. In the next step, the inference mechanism will seek to combine the recommendations of all the rules to come up with a single conclusion.

Generally, only a few rules are relevant to choosing the plant input at any one time.

Consider, for the ship steering example, how we compute the rules that are on. Suppose that

$$e(t) = 0$$

and

$$\dot{e}(t) = 0.0015$$

Figure 5.12 shows the membership functions for the inputs and indicates, with thick black vertical lines, the $e(t)$ and $\dot{e}(t)$ values. Notice that $\mu_{zero}(e(t)) = 1$ but that the other membership functions for the $e(t)$ input are all "off" (i.e., their values are zero). For the $\dot{e}(t)$ input we see that $\mu_{zero}(\dot{e}(t)) = 0.25$ and $\mu_{possmall}(\dot{e}(t)) = 0.75$ and that all the other membership functions are off. This implies that rules that have the premise terms

"error is zero"
"change-in-error is zero"
"change-in-error is possmall"

are on (all other rules have $\mu_{premise}(e(t), \dot{e}(t)) = 0$). So, which rules are these? Using Table 5.1 on page 162, we find that the following rules are on:

1. **If** error is zero **and** change-in-error is zero **Then** rudder-input is zero

2. **If** error is zero **and** change-in-error is possmall **Then** rudder-input is negsmall

Note that since for the ship steering example we have at most two membership functions overlapping, we will never have more than four rules on at one time (this concept generalizes to many inputs). Actually, for this system we will either have one, two, or four rules on at any one time. To get only one rule on choose, for example, $e(t) = 0$ and $\dot{e}(t) = 0.002$. In this example, only rule 2 above is on. What values would you choose for $e(t)$ and $\dot{e}(t)$ to get four rules on? Why is it impossible, for this system, to have exactly three rules on?

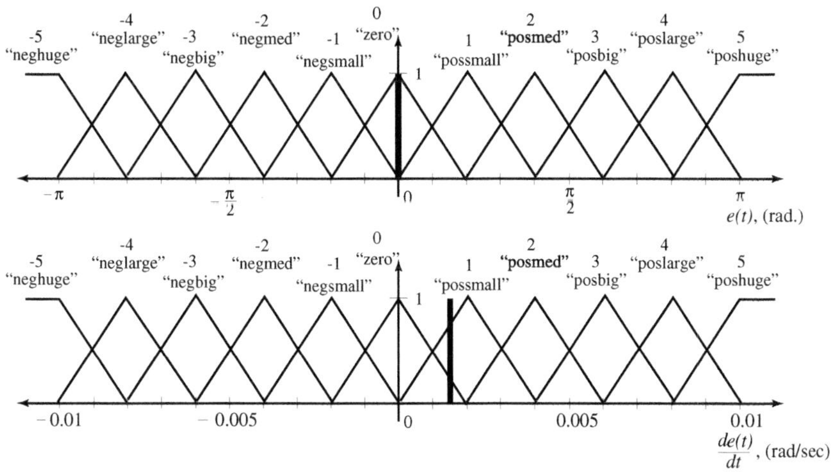

Figure 5.12: Input membership functions with input values.

It is useful to consider pictorially which rules are on. Consider Table 5.3, which is a copy of Table 5.2 on page 169 with boxes drawn around the consequents of the rules that are on (notice that these are the *same* two rules listed above). Notice that since $e(t) = 0$ ($e(t)$ is directly in the middle between the membership functions for "possmall" and "negsmall"), both of these membership functions are off. If we perturbed $e(t)$ slightly positive (negative), then we would have the two rules below (above) the two highlighted ones on also. With this, you should picture in your mind how a region of rules that are on (that involves no more than four cells in the body of Table 5.3, due to how we define the input membership functions) will dynamically move around in the table as the values of $e(t)$ and $\dot{e}(t)$ change. This completes our description of the "matching" phase of the inference mechanism.

5.1.5 Inference Step: Determining Conclusions

Next, we consider how to determine which conclusions should be reached when the rules that are on are applied to deciding what the rudder input to the ship should be. To do this, we will first consider the recommendations of each rule independently. Then later we will combine all the recommendations from all the rules to determine the rudder input to the tanker ship.

Recommendation from One Rule

Consider the conclusion reached by the rule

If error is zero **and** change-in-error is zero **Then** rudder-input is zero

Table 5.3: Rule Table for the Tanker Ship with Rules That Are "On" (highlighted). (Body of table holds the output membership function centers where each element should be multiplied by $8\pi/18$.)

		$\dot{e}$										
		−5	−4	−3	−2	−1	0	1	2	3	4	5
e	−5	1	1	1	1	1	1	.8	.6	.3	.1	0
	−4	1	1	1	1	1	.8	.6	.3	.1	0	−.1
	−3	1	1	1	1	.8	.6	.3	.1	0	−.1	−.3
	−2	1	1	1	.8	.6	.3	.1	0	−.1	−.3	−.6
	−1	1	1	.8	.6	.3	.1	0	−.1	−.3	−.6	−.8
	0	1	.8	.6	.3	.1	**0**	**−.1**	−.3	−.6	−.8	−1
	1	.8	.6	.3	.1	0	−.1	−.3	−.6	−.8	−1	−1
	2	.6	.3	.1	0	−.1	−.3	−.6	−.8	−1	−1	−1
	3	.3	.1	0	−.1	−.3	−.6	−.8	−1	−1	−1	−1
	4	.1	0	−.1	−.3	−.6	−.8	−1	−1	−1	−1	−1
	5	0	−.1	−.3	−.6	−.8	−1	−1	−1	−1	−1	−1

which for convenience we will refer to as "rule (1)." Using the minimum to represent the premise, we have

$$\mu_{premise_{(1)}} = \min\{1, 0.25\} = 0.25$$

(the notation $\mu_{premise_{(1)}}$ represents $\mu_{premise}$ for rule (1)) so that we are 0.25 certain that this rule applies to the current situation. The rule indicates that if its premise is true, then the action indicated by its consequent should be taken. For rule (1) the consequent is "rudder-input is zero" (this makes sense, for here the ship is headed in the proper direction, so we should not apply a rudder input that is different from zero since this would tend to move the ship heading away from the desired heading). The membership function for this consequent is shown in Figure 5.13(a). The membership function for the conclusion reached by rule (1), which we denote by $\mu_{(1)}$, is shown in Figure 5.13(b) and is given by

$$\mu_{(1)}(\delta) = \min\{\mu_{premise_{(1)}}, \mu_{zero}(\delta)\}$$

(where $\mu_{premise_{(1)}} = 0.25$ as determined above). This membership function defines the "implied fuzzy set"[1] for rule (1) (i.e., it is the conclusion that is implied by rule (1)). The justification for the use of the minimum operator to represent the implication is that *we can be no more certain about our consequent than our*

[1] This term has been used in the literature for a long time; however, there is no standard terminology for this fuzzy set. Others have called it, for example, a "consequent fuzzy set" or an "output fuzzy set" (which can be confused with the fuzzy sets that quantify the consequents of the rules).

premise. You should convince yourself that we could use the product operation to represent the implication also (in Section 5.1.6 we will do an example where we use the product).

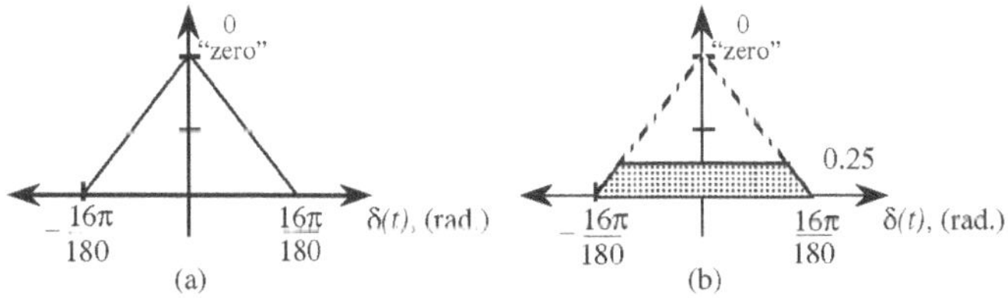

Figure 5.13: (a) Consequent membership function and (b) implied fuzzy set with membership function $\mu_{(1)}(\delta)$ for rule (1).

Notice that the membership function $\mu_{(1)}(\delta)$ is a *function* of δ and that the minimum operation will generally "chop off the top" of the $\mu_{zero}(\delta)$ membership function to produce $\mu_{(1)}(\delta)$. For different values of $e(t)$ and $\dot{e}(t)$ there will be different values of the premise certainty $\mu_{premise_{(1)}}(e(t), \dot{e}(t))$ for rule (1) and hence different *functions* $\mu_{(1)}(\delta)$ obtained (i.e., it will chop off the top at different points).

We see that $\mu_{(1)}(\delta)$ is in general a time-varying function that quantifies how certain rule (1) is that the force input δ should take on certain values. It is most certain that the force input should lie in a region around zero (see Figure 5.13(b)), and it indicates that it is certain that the force input should not be too large in either the positive or negative direction—this makes sense if you consider the linguistic meaning of the rule. The membership function $\mu_{(1)}(\delta)$ quantifies the conclusion reached by only rule (1) and only for the current $e(t)$ and $\dot{e}(t)$. It is important that the reader be able to picture how the shape of the implied fuzzy set changes as the rule's premise certainty changes over time.

Recommendation from Another Rule

Next, consider the conclusion reached by the other rule that is on:

If error is zero **and** change-in-error is possmall **Then** rudder-input is negsmall

which, for convenience, we will refer to as "rule (2)." Using the minimum to represent the premise, we have

Fuzzy control is "democratic" in that in deciding what input to put into the plant, it listens to the recommendation from each rule, to a degree specified by "how true" the premise of that rule is.

$$\mu_{premise_{(2)}} = \min\{1, 0.75\} = 0.75$$

so that we are 0.75 certain that this rule applies to the current situation. Notice that we are much more certain that rule (2) applies to the current situation than rule (1) does. For rule (2) the consequent is "rudder-input is negsmall" (this makes sense, for here the ship is heading in the proper direction but is moving in the counterclockwise direction with a small velocity). The membership function for this consequent is shown in Figure 5.14(a). The membership function for the conclusion reached by rule (2), which we denote by $\mu_{(2)}$, is shown in Figure 5.14(b) (the shaded region) and is given by

$$\mu_{(2)}(\delta) = \min\{\mu_{premise_{(2)}}, \mu_{negsmall}(\delta)\}$$

(where $\mu_{premise_{(2)}} = 0.75$ as determined above). This membership function defines the implied fuzzy set for rule (2) (i.e., it is the conclusion that is reached by rule (2)). Once again, for different values of $e(t)$ and $\dot{e}(t)$ there will be different values of $\mu_{premise_{(2)}}(e(t), \dot{e}(t))$ for rule (2) and hence different *functions* $\mu_{(2)}(\delta)$ obtained. The reader should carefully consider the meaning of the implied fuzzy set $\mu_{(2)}(\delta)$. Rule (2) is quite certain that the control output (process input) should be a small negative value. This makes sense since if the ship has some counterclockwise velocity, then we would want to apply a negative rudder angle input. As rule (2) has a premise membership function that has higher certainty than for rule (1), we see that we are more certain of the conclusion reached by rule (2).

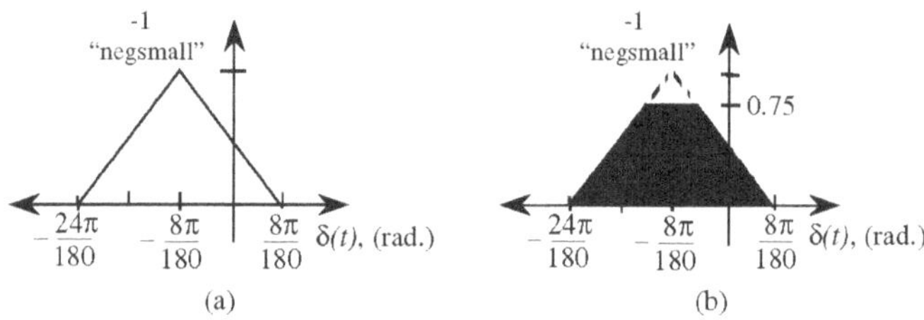

Figure 5.14: (a) Consequent membership function and (b) implied fuzzy set with membership function $\mu_{(2)}(\delta)$ for rule (2).

This completes the operations of the inference mechanism in Figure 5.1 on page 156. While the input to the inference process is the set of rules that are on, its output is the set of implied fuzzy sets that represent the conclusions reached by all the rules that are on. For our example, there are at most four conclusions reached since there are at most four rules on at any one time (and even some of these implied fuzzy sets may have a membership function that is zero for all values of δ so that we may ignore it).

Converting decisions to actions entails combining the recommendations of all the relevant rules.

5.1.6 Converting Decisions into Actions

Next, we consider the defuzzification operation, which is the final component of the fuzzy controller shown in Figure 5.1 on page 156. Defuzzification operates on the implied fuzzy sets produced by the inference mechanism and combines their effects to provide the "most certain" controller output (plant input). Some think of defuzzification as "decoding" the fuzzy set information produced by the inference process (i.e., the implied fuzzy sets) into numeric fuzzy controller outputs.

To understand defuzzification, it is best to first draw all the implied fuzzy sets on one axis as shown in Figure 5.15. We want to find the one output, which we denote by "δ^{crisp}," that best represents the conclusions of the fuzzy controller that are represented with the implied fuzzy sets. There are actually many approaches to defuzzification. We will consider two here.

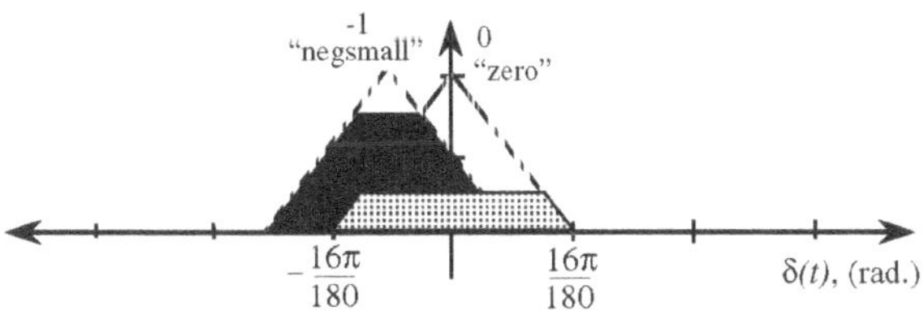

Figure 5.15: Implied fuzzy sets.

Combining Recommendations

Due to its popularity, we will first consider the "center of gravity" (COG) defuzzification method for combining the recommendations represented by the implied fuzzy sets from all the rules. Let b_i denote the center of the membership function for the implied fuzzy set for the ith rule (i.e., where the membership function for the ith rule reaches its peak for our example since the output fuzzy sets are all symmetric about their peaks). For our example we have

$$b_1 = 0.0$$

and

$$b_2 = -0.1\left(\frac{8\pi}{18}\right)$$

as shown in Figure 5.15. Let

$$\int \mu_{(i)}$$

denote the area under the membership function $\mu_{(i)}$. The COG method computes δ^{crisp} to be

$$\delta^{crisp} = \frac{\sum_i b_i \int \mu_{(i)}}{\sum_i \int \mu_{(i)}} \tag{5.1}$$

This is the classical formula for computing the center of gravity. In this case it is for computing the center of gravity of the implied fuzzy sets. Three items about Equation (5.1) are important to note:

1. Practically, we cannot have output membership functions that have infinite area since even though they may be "chopped off" in the minimum operation for the implication (or scaled for the product operation), they can still end up with infinite area. This is the reason we do not allow

infinite area membership functions for the linguistic values for the controller output (e.g., we did not allow the saturated membership functions at the outermost edges as we had for the inputs shown in Figure 5.8 on page 166).

2. You must be careful to define the input and output membership functions so that the sum in the denominator of Equation (5.1) is not equal to zero no matter what the inputs to the fuzzy controller are. Essentially, this means that we must have some sort of conclusion for all possible control situations we may encounter.

3. While at first glance it may not appear so, $\int \mu_{(i)}$ is easy to compute for our example. For the case where we have symmetric triangular output membership functions that peak at one and have a base width of w, simple geometry can be used to show that the area under a triangle "chopped off" at a height of h (such as the ones in Figures 5.13 and 5.14) is equal to

$$w\left(h - \frac{h^2}{2}\right)$$

Given this, the computations needed to compute δ^{crisp} are not too significant (note that if w is the same for every output membership function, then it cancels in Equation (5.1)).

We see that the property of membership functions being symmetric for the output is important since in this case no matter whether the minimum or product is used to represent the implication, it will be the case that the center of the implied fuzzy set will be the same as the center of the consequent fuzzy set from which it is computed. If the output membership functions are not symmetric, then their centers, which are needed in the computation of the COG, will change depending on the membership value of the premise. This will result in the need to recompute the center at each time instant.

Using Equation (5.1) with Figure 5.15, we have

$$\delta^{crisp} = \frac{(0)\left(0.25 - \frac{(0.25)^2}{2}\right) + \left(-0.1\frac{8\pi}{18}\right)\left(0.75 - \frac{(0.75)^2}{2}\right)}{\left(0.25 - \frac{(0.25)^2}{2}\right) + \left(0.75 - \frac{(0.75)^2}{2}\right)} = -0.0952$$

as the input to the ship for the given $e(t)$ and $\dot{e}(t)$.

Does this value for a force input (i.e., -5.4545 degrees) make sense? Consider Figure 5.16, where we have taken the implied fuzzy sets from Figure 5.15 and simply added an indication of what number COG defuzzification says is the best representation of the conclusions reached by the rules that are on. Notice that the value of δ^{crisp} is roughly in the middle of where the implied fuzzy sets say they are most certain about the value for the force input. In fact, recall that we had

$$e(t) = 0$$

and

$$\dot{e}(t) = 0.0015$$

so the ship is at the desired heading at this time instant but is moving counter-clockwise with a small velocity; hence, it makes sense to apply a small negative rudder input, and the fuzzy controller does this.

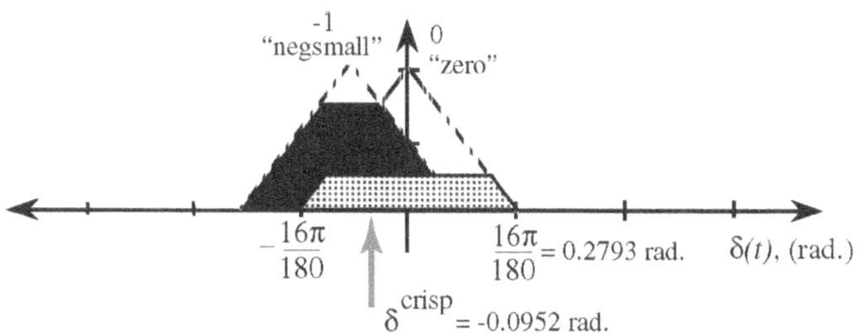

Figure 5.16: Implied fuzzy sets.

It is interesting to note that for our example it will be the case that

$$-\frac{8\pi}{18} \leq \delta^{crisp} \leq \frac{8\pi}{18}$$

To see this, consider Figure 5.17, where we have drawn the output membership functions. Notice that even though we have extended the membership functions at the outermost edges past $-8\pi/18$ and $+8\pi/18$ (see the shaded regions), the COG method will never compute a value outside this range.

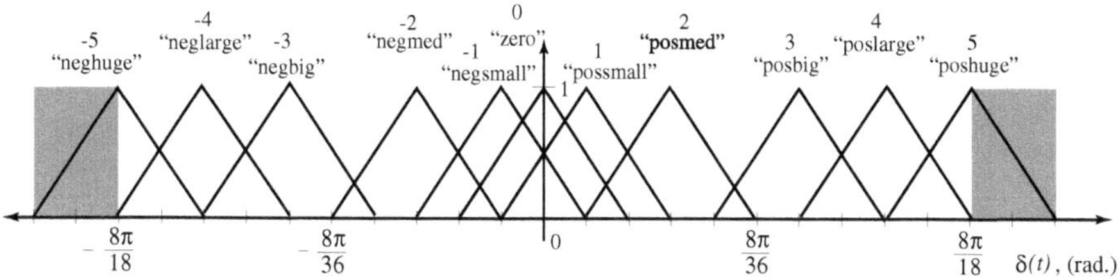

Figure 5.17: Output membership functions.

The reason for this comes directly from the definition of the COG method in Equation (5.1). The center of gravity for these shapes simply cannot extend beyond $-8\pi/18$ and $+8\pi/18$. Practically speaking, this ability to limit the range of inputs to the plant is useful in real applications since all real plant inputs are limited to lie in a specific range. The other conclusion that we would reach from this discussion is that in defining the membership functions for the fuzzy controller, we must take into account what method is going to be used for defuzzification.

Other Ways to Compute and Combine Recommendations

As another example, it is interesting to consider how to compute, by hand, the operations that the fuzzy controller takes when we use the product to represent the implication or the "center-average" defuzzification method.

First, consider the use of the product. Consider Figure 5.18, where we have drawn the output membership functions for "negsmall" and "zero" as dotted lines. The implied fuzzy set from rule (1) is given by the membership function

$$\mu_{(1)}(\delta) = 0.25\mu_{zero}(\delta)$$

shown in Figure 5.18 as the shaded triangle; the implied fuzzy set for rule (2) is given by the membership function

$$\mu_{(2)}(\delta) = 0.75\mu_{negsmall}(\delta)$$

shown in Figure 5.18 as the dark triangle. The computation of the COG is easy since we can use $\frac{1}{2}wh$ as the area for a triangle with base width w and height h (and the factor $\frac{1}{2}w$ cancels in Equation 5.1). When we use product to represent the implication, we obtain

$$\delta^{crisp} = \frac{(0)(0.25) + \left(-0.1\frac{8\pi}{18}\right)(0.75)}{0.25 + 0.75} = -0.1047$$

which also makes sense.

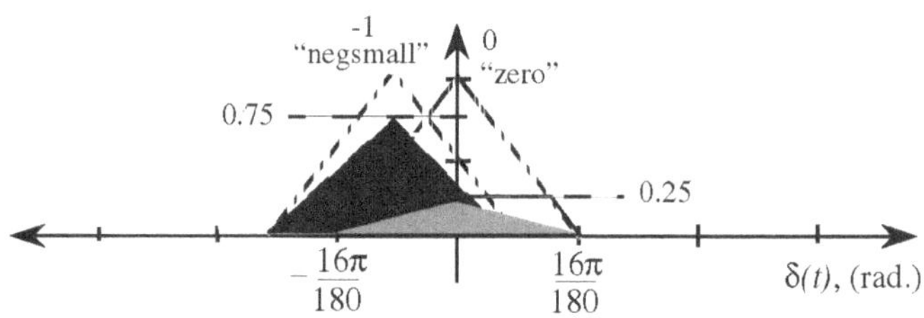

Figure 5.18: Implied fuzzy sets when the product is used to represent the implication.

Next, as another example of how to combine recommendations, we will introduce the "center-average" method for defuzzification. For this method we let

$$\delta^{crisp} = \frac{\sum_i b_i \mu_{premise_{(i)}}}{\sum_i \mu_{premise_{(i)}}} \tag{5.2}$$

where b_i once again denotes the center of the membership function for the implied fuzzy set for the ith rule (i.e., where the membership function for the ith rule reaches its peak for our example since the output fuzzy sets are all symmetric about their peaks). To compute the $\mu_{premise_{(i)}}$ we use, for example, minimum. We call it the "center-average" method since Equation (5.2) is a weighted average of the center values of the membership functions of the implied fuzzy sets (and output membership function centers). Basically, the center-average method replaces the areas of the implied fuzzy sets that are used in COG with the values of $\mu_{premise(i)}$. This is a valid replacement since the area of the implied fuzzy set is generally proportional to $\mu_{premise(i)}$ since $\mu_{premise(i)}$ is used to chop the top off (minimum) or scale (product) the triangular output membership function when COG is used for our example. For the above example, we have

$$\delta^{crisp} = \frac{(0)(0.25) + \left(-0.1\frac{8\pi}{18}\right)(0.75)}{0.25 + 0.75} = -0.1047$$

which is the same value as above (for this special case). Some like the center-average defuzzification method because the computations needed are generally simpler than for COG because when the output membership functions are symmetric (the usual case), they are easy to store since the only relevant information they provide is their center values (b_i) (i.e., their shape does not matter, just their center value, so this is all that needs to be stored). Moreover, the areas of the implied fuzzy sets do not have to be computed.

Notice that while both values computed for the different inference and defuzzification methods provide reasonable command inputs to the plant, it is difficult to say which is best without further investigations (e.g., simulations or implementation). This ambiguity about how to define the fuzzy controller actually extends to the general case and also arises in the specification of all the other fuzzy controller components, as we discuss below. Some would call this "ambiguity" a design flexibility, but unfortunately there are not too many guidelines on how best to choose the inference strategy and defuzzification method, so such flexibility is of questionable value.

Graphical Depiction of Fuzzy Decision Making

For convenience, we summarize the procedure that the fuzzy controller uses to compute its outputs given its inputs in Figure 5.19. Here, we use the minimum operator to represent the "and" in the premise and the implication and COG defuzzification. The reader is advised to study each step in this diagram to gain a fuller understanding of the operation of the fuzzy controller. To do this, develop a similar diagram for the case where the product operator is used to represent the "and" in the premise and the implication, and choose values of $e(t)$ and $\dot{e}(t)$ that will result in four rules being on. Then, repeat the process when center-average defuzzification is used with either minimum or product used for the premise. Also, learn how to picture in your mind how the parameters of this graphical representation of the fuzzy controller operations change as the fuzzy controller inputs change.

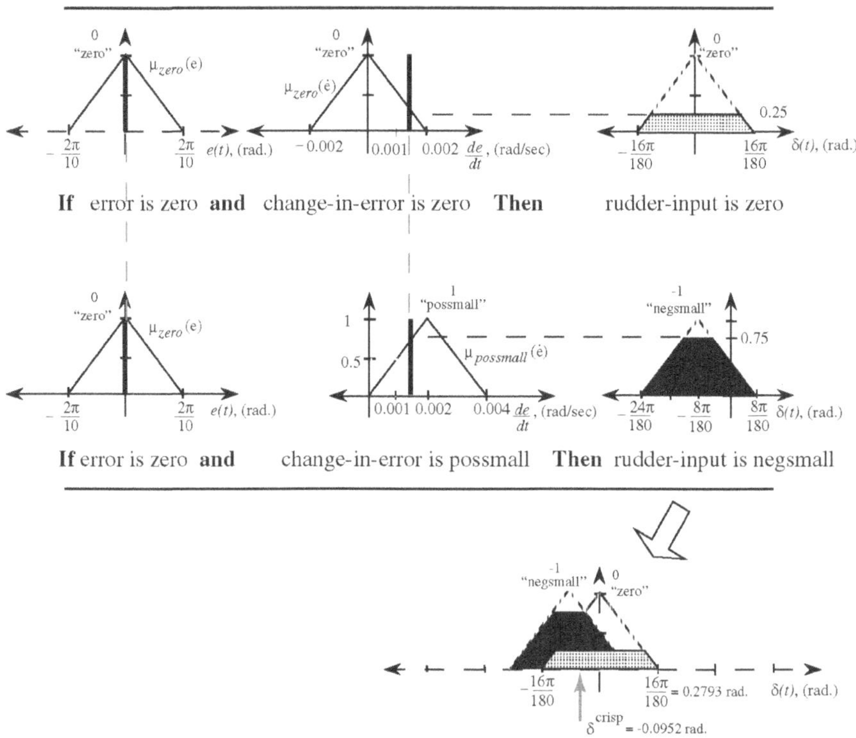

Figure 5.19: Graphical representation of fuzzy controller operations.

5.2 General Fuzzy Systems

In this section we introduce multi-input multi-output fuzzy systems, Takagi-Sugeno fuzzy systems, and then show how to develop mathematical representations for these.

5.2.1 Multiple Input Multiple Output Fuzzy Systems

A fuzzy system is a static nonlinear mapping between its inputs and outputs (i.e., it is not a dynamic system). Some people include the preprocessing of the inputs to the fuzzy system (e.g., differentiators or integrators) in the definition of the fuzzy system and thereby obtain a "fuzzy system" that *is* dynamic. In this book, we adopt the convention that such preprocessing is not part of the fuzzy system, and hence the fuzzy system will always be a memoryless nonlinear map.

A general multiple input multiple output (MIMO) fuzzy system with inputs

u_i, $i = 1, 2, \ldots, n$ and outputs y_j, $j = 1, 2, \ldots, m$ is shown in Figure 5.20. The inputs and outputs are "crisp;" that is, they are numeric values. The fuzzification block converts the crisp inputs to fuzzy sets (i.e., it converts them to "singleton" fuzzy sets, ones that have membership functions with zero width and a unit pulse at the value of the input; an example singleton membership function is shown in Figure 5.21). The inference mechanism uses the fuzzy rules in the rule base to produce fuzzy conclusions (e.g., the implied fuzzy sets), and the defuzzification block converts these fuzzy conclusions into the crisp outputs. In this subsection we explain how to define a MIMO fuzzy controller.

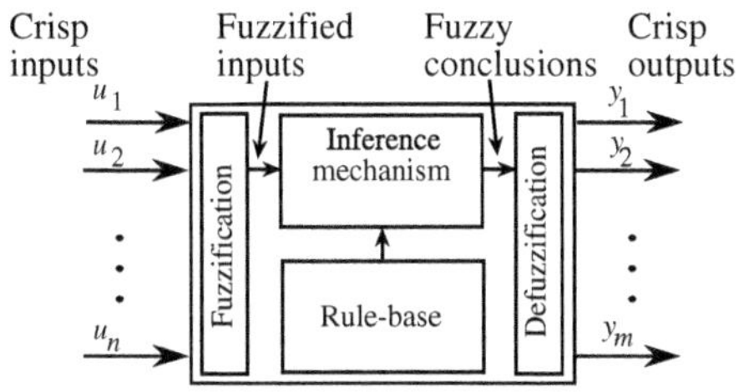

Figure 5.20: Fuzzy system (controller).

First, note that to define a MIMO fuzzy system you simply specify m multiple input single output (MISO) fuzzy systems, where the output of the jth fuzzy system is y_j, $j = 1, 2, \ldots, m$. We already know how to specify a MISO fuzzy system with $n = 2$ inputs so all we need to do is explain how to define a MISO fuzzy system with $n > 2$ inputs (then the case for $n = 1$ will be clear). To do this, note that for each input we define membership functions as we did for the e and $\dot{e}$ universes of discourse for the ship example. You form rules using the n inputs in n premise terms. Next, fuzzification is the same as earlier—you just compute the membership values on all the input universes of discourse. Next, we need to compute the fuzzy logic quantification of the conjunction between n premise terms rather than just two. To do this we take the same approach as before, but take the minimum (or product) of n membership function values to represent the conjunction of n premise terms. This will give us $\mu_{premise_{(i)}}$ for the ith rule and we will compute this for all the rules. From this point on the process is exactly the same as the two-input case since after the matching process, the inference mechanism computations of the implied fuzzy sets and the defuzzification computations only depend on $\mu_{premise_{(i)}}$. Therefore, the effect of additional inputs to the fuzzy system is on the premise and hence the computations needed to find $\mu_{premise_{(i)}}$ for all the rules.

The computations necessary for MISO fuzzy systems will be reviewed in Section 5.2.3 where we explain how to develop mathematical representations of fuzzy systems for n input MISO fuzzy systems.

5.2.2 Takagi-Sugeno Fuzzy Systems

The fuzzy systems discussed in the previous sections will be referred to as a "standard fuzzy system," regardless of the particular choices for premise representation, inference, defuzzification, etc. In this subsection we will define a "functional fuzzy system," of which the Takagi-Sugeno fuzzy system is a special case. For the functional fuzzy system, we use singleton fuzzification and the premise is defined the same as it is for the rule for the standard fuzzy system. The consequents of the rules are different, however. Instead of a linguistic term with an associated membership function, in the consequent we use a *function* $b_i = g_i(\cdot)$ (hence the name "functional fuzzy system") that does not have an associated membership function (or you can think of it as a singleton membership function whose position changes as specified by the function g_i for the ith rule). Notice that often the argument of g_i contains the fuzzy system inputs that are used in the premise of the rule, but other variables may also be used. The choice of the function depends on the application being considered. Below, we will discuss linear and affine functions but many others are possible. For instance, you may want to choose

$$b_i = g_i(\cdot) = a_{i,0} + a_{i,1}(u_1)^2 + \cdots + a_{i,n}(u_n)^2$$

or

$$b_i = g_i(\cdot) = \exp\left[a_{i,1} sin(u_1) + \cdots + a_{i,n} sin(u_n)\right]$$

Virtually any function can be used (e.g., a neural network mapping or another fuzzy system), which makes the functional fuzzy system very general.

Let R denote the number of rules. For the functional fuzzy system we can use an appropriate operation for representing the premise (e.g., minimum or product), and defuzzification may be obtained using

$$y = \frac{\sum_{i=1}^{R} b_i \mu_i(z)}{\sum_{i=1}^{R} \mu_i(z)} \tag{5.3}$$

where $\mu_i(z)$ is the premise membership function (rather than $\mu_{premise_{(i)}}$ which was used in our earlier discussion). It is assumed that the functional fuzzy system is defined so that no matter what its inputs are, we have $\sum_{i=1}^{R} \mu_i(z) \neq 0$. The vector z can be chosen in several ways. One common choice is to use $z = [u_1, u_2, \ldots, u_n]^\top$; however, sometimes z might hold other variables, or only a subset of the u_i values (with only a subset of the values, complexity of the mapping generally decreases since the computations needed to find $\mu_i(z)$ are simplified).

A Takagi-Sugeno fuzzy system is an interpolator between linear mappings.

In the special case where

$$b_i = g_i(\cdot) = a_{i,0} + a_{i,1}u_1 + \cdots + a_{i,n}u_n$$

(where the $a_{i,j}$ are fixed real numbers) the functional fuzzy system is referred to as a "Takagi-Sugeno fuzzy system."

If $a_{i,0} = 0$, then the $g_i(\cdot)$ mapping is a linear mapping and if $a_{i,0} \neq 0$, then the mapping is called "affine." Often, however, as is standard, we will refer to the affine mapping as a linear mapping for convenience. Overall, we see that the Takagi-Sugeno fuzzy system performs a nonlinear interpolation between linear mappings. In control applications, the linear mappings can each represent a different linear controller and the Takagi-Sugeno fuzzy system interpolates between these and applies combinations of the linear controller outputs (similar in some cases to what is called "gain scheduled control" in conventional control).

5.2.3 Mathematical Representations of Fuzzy Systems

Notice that each formula for defuzzification in the previous sections provides a mathematical description of a fuzzy system. There are many ways to represent the operations of a fuzzy system with mathematical formulas. Next, we clarify how to construct and interpret such mathematical formulas for the case where center-average defuzzification is used for n-input MISO fuzzy systems. Similar ideas apply for other defuzzification strategies, MIMO fuzzy systems, and Takagi-Sugeno fuzzy systems.

Two Different Approaches

Rules and Membership Functions: To represent linguistic rules, let $\tilde{u}_i$, $i = 1, 2, \ldots, n$, and $\tilde{y}$ denote the linguistic variables that describe u_i, $i = 1, 2, \ldots, n$, and y, respectively. Let $\tilde{A}_i^j$ denote the jth linguistic value for the ith input universe of discourse (here, suppose that $i = 1, 2, \ldots, n$, but that j can, for instance, take on values that are equal to the linguistic-numeric values). Similarly, let $\tilde{B}^p$ denote the pth linguistic value on the output universe of discourse that has linguistic variable $\tilde{y}$. With this, a linguistic rule may be described mathematically by

$$\begin{array}{c}\textbf{If } \tilde{u}_1 \text{ is } \tilde{A}_1^j \\ \textbf{and } \tilde{u}_2 \text{ is } \tilde{A}_2^k \\ \textbf{and } \cdots \\ \textbf{and } \tilde{u}_n \text{ is } \tilde{A}_n^l \\ \textbf{Then } \tilde{y} \text{ is } \tilde{B}^p\end{array}$$

Suppose that there are R such rules.

Next, consider the mathematical quantification of membership functions. Clearly, many other choices for the shape of the membership function are possible than the ones discussed so far, and these will each provide a different meaning for the linguistic values that they quantify. See Figure 5.21 for a graphical illustration of a variety of membership functions and Tables 5.4 and 5.5 for a mathematical characterization of the triangular and Gaussian membership functions, including the membership functions that are often used at the outermost edges of the input universe of discourse when the "center" membership functions are used at various positions along the input universe of discourse (other membership functions can be characterized with mathematics using a similar approach). For practice, you should sketch the membership functions that are described in Tables 5.4 and 5.5. Notice that for Table 5.4, c^L specifies the "saturation point" and w^L specifies the slope of the nonunity and nonzero part of μ^L. Similarly, for μ^R. For μ^C notice that c is the center of the triangle and w is the base width. Analogous definitions are used for the parameters in Table 5.5. In Table 5.5, for the "centers" case note that this is the traditional definition for the Gaussian membership function. This definition is clearly different from a standard Gaussian probability density function, in both the meaning of c and σ, and in the scaling of the exponential function. Recall that it is possible that a Gaussian probability density function has a maximum value at a value other than one; the standard Gaussian membership function always has its peak value at one.

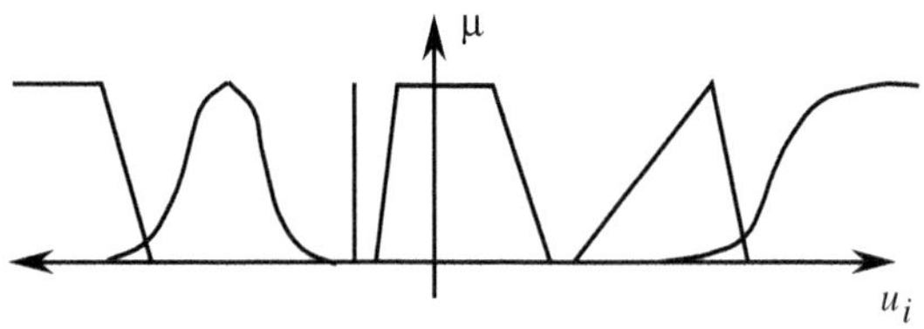

Figure 5.21: Some typical membership functions.

Table 5.4: Mathematical Characterization of Triangular Membership Functions

	Triangular and related membership functions
Left	$\mu^L(u) = \begin{cases} 1 & \text{if } u \leq c^L \\ \max\left\{0, 1 + \frac{c^L - u}{0.5w^L}\right\} & \text{otherwise} \end{cases}$
Centers	$\mu^C(u) = \begin{cases} \max\left\{0, 1 + \frac{u-c}{0.5w}\right\} & \text{if } u \leq c \\ \max\left\{0, 1 + \frac{c-u}{0.5w}\right\} & \text{otherwise} \end{cases}$
Right	$\mu^R(u) = \begin{cases} \max\left\{0, 1 + \frac{u - c^R}{0.5w^R}\right\} & \text{if } u \leq c^R \\ 1 & \text{otherwise} \end{cases}$

Approach 1: Given Membership Functions, All Possible Rules: Assume that we use center-average defuzzification so that the formula describing how to compute the output is

$$y = \frac{\sum_{i=1}^{R} b_i \mu_i}{\sum_{i=1}^{R} \mu_i} \tag{5.4}$$

where for convenience we use μ_i to represent the premise certainty for the ith rule (rather than $\mu_{premise_{(i)}}$ which was more descriptive for our earlier discussion, but a bit cumbersome).

Table 5.5: Mathematical Characterization of Gaussian Membership Functions

	Gaussian and related membership functions
Left	$\mu^L(u) = \begin{cases} 1 & \text{if } u \leq c^L \\ \exp\left(-\frac{1}{2}\left(\frac{u-c^L}{\sigma^L}\right)^2\right) & \text{otherwise} \end{cases}$
Centers	$\mu^C(u) = \exp\left(-\frac{1}{2}\left(\frac{u-c}{\sigma}\right)^2\right)$
Right	$\mu^R(u) = \begin{cases} \exp\left(-\frac{1}{2}\left(\frac{u-c^R}{\sigma^R}\right)^2\right) & \text{if } u \leq c^R \\ 1 & \text{otherwise} \end{cases}$

To be more explicit in Equation (5.4), we need to first define the premise membership functions μ_i in terms of the individual membership functions that describe each of the premise terms. Suppose that we use product to represent the conjunctions in the premise of each rule. Suppose that we use the triangular membership functions in Table 5.4 where we suppose that $\mu_j^L(u_j)$ ($\mu_j^R(u_j)$) is the "left-" ("right-") most membership function on the jth input universe of discourse. In addition, let $\mu_j^{C_i}(u_j)$ be the ith "center" membership function for the jth input universe of discourse. In this case, to define $\mu_j^L(u_j)$ we simply add a "j" subscript to the parameters of the "left" membership function from Table 5.4. In particular, we use c_j^L and w_j^L to denote the jth values of these parameters. We take a similar approach for the $\mu_j^R(u_j)$, $j = 1, 2, \ldots, n$. For $\mu_j^{C_i}(u_j)$ we use c_j^i (w_j^i) to denote the ith triangle center (triangle base width) on the jth input universe of discourse.

Suppose that we use all possible combinations of input membership functions to form the rules, and that each premise has a term associated with each and every input universe of discourse. A more detailed description of the fuzzy system in Equation (5.4) is given by

$$y = \frac{b_1 \prod_{j=1}^n \mu_j^L(u_j) + b_2 \mu_1^{C_1}(u_1) \prod_{j=2}^n \mu_j^L(u_j) + \cdots}{\prod_{j=1}^n \mu_j^L(u_j) + \mu_1^{C_1}(u_1) \prod_{j=2}^n \mu_j^L(u_j) + \cdots}$$

The first term in the numerator is $b_1\mu_1$ in Equation (5.4). Here, we have called the "first rule" the one that has premise terms all described by the membership functions $\mu_j^L(u_j)$, $j = 1, 2, \ldots, n$. The second term in the numerator is $b_2\mu_2$ and it uses $\mu_1^{C_1}(u_1)$ on the first universe of discourse and the leftmost ones on the other universes of discourse (i.e., $j = 2, 3, \ldots, n$). Continuing in a similar manner, the sum in the numerator (and denominator) extends to include all possible combinations of products of the input membership functions, and this fully defines the μ_i in Equation (5.4).

Overall, we see that because we need to define rules resulting from all possible combinations of *given* input membership functions, of which there are three kinds (left, center, right), the explicit mathematical representation of the fuzzy system is somewhat complicated. To avoid some of the complications, we first specify a single function that represents all three types of input membership functions. Suppose that on the jth input universe of discourse we number the input membership functions from left to right as $1, 2, \ldots, N_j$, where N_j is the number of input membership functions on the jth input universe of discourse. A single membership function that represents all three in Table 5.4 is

$$\mu_j^i(u_j) = \begin{cases} 1 & \text{if } (u_j \leq c_j^1, i = 1) \text{ or } (u_j \geq c_j^{N_j}, i = N_j) \\ \max\left\{0, 1 + \frac{u_j - c_j^i}{0.5 w_j^i}\right\} & \text{if } u_j \leq c_j^i \text{ and } (u_j > c_j^1 \text{ and } u_j < c_j^{N_j}) \\ \max\left\{0, 1 + \frac{c_j^i - u_j}{0.5 w_j^i}\right\} & \text{if } u_j > c_j^i \text{ and } (u_j > c_j^1 \text{ and } u_j < c_j^{N_j}) \end{cases}$$

A similar approach can be used for the Gaussian case in Table 5.5.

Suppose we use the shorthand notation

$$(j, k, \ldots, l; p)_i$$

to denote the ith rule shown above. In this notation, suppose the indices in (the "tuple") $(j, k, \ldots, l)$ range over $1 \leq j \leq N_1$, $1 \leq k \leq N_2$, ..., $1 \leq l \leq N_n$, and specify which linguistic value is used on each input universe of discourse. Correspondingly, each index in the tuple $(j, k, \ldots, l)$ also specifies the linguistic-numeric value of the input membership function used on each input universe of discourse.

Let

$$b^{(j,k,\ldots,l;p)_i}$$

denote the output membership function (a singleton) center for the ith rule. Note that we use "i" in the notation $(j, k, \ldots, l; p)_i$ simply as a label for each rule (i.e., we number the rules in the rule base from 1 to R, and i is this number). Hence, when we are given i, we know the values of j, k, ..., l, and p. Because of this, an explicit description of the fuzzy system in Equation (5.4) is given by

$$y = \frac{\sum_{i=1}^{R} b^{(j,k,\ldots,l;p)_i} \mu_1^j \mu_2^k \cdots \mu_n^l}{\sum_{i=1}^{R} \mu_1^j \mu_2^k \cdots \mu_n^l} \tag{5.5}$$

This formula clearly shows the use of the product to represent the premise. Notice that since we use all possible combinations of input membership functions to form the rules there are

$$R = \prod_{j=1}^{n} N_j$$

rules, and hence it takes

$$\sum_{j=1}^{n} 2N_j + \prod_{j=1}^{n} N_j \tag{5.6}$$

parameters to describe the fuzzy system since there are two parameters for each input membership function and R output membership function centers.

For some applications, however, all the output membership functions are not distinct. For example, consider the ship steering example where eleven output membership function centers are defined, and there are $R = 121$ rules. To define the center positions $b^{(j,k,\ldots,l;p)_i}$ so that they take on only a fixed number of given values, that is less than R, one approach is to specify them as a function of the indices of the input membership functions. What is this function for the ship steering example?

Approach 2: Parameterization in Terms of Rules: A different approach to avoiding some of the complications encountered in specifying a fuzzy system mathematically is to use a different notation, and hence a different definition for the fuzzy system. For this alternative approach, for the sake of variety, we will use Gaussian input membership functions. In particular, for simplicity, suppose that for the input universes of discourse we only use membership functions of the "center" Gaussian form shown in Table 5.5. For the ith rule, suppose that the input membership function is

$$\exp\left(-\frac{1}{2}\left(\frac{u_j - c_j^i}{\sigma_j^i}\right)^2\right)$$

for the jth input universe of discourse. Hence, even though we use the same notation for the membership function, these centers c_j^i are different from those used above, both because we are using Gaussian membership functions here, and because the "i" in c_j^i is the index for the rules, not the membership function on the jth input universe of discourse. Similar comments can be made about the σ_j^i, $i = 1, 2, \ldots, R$, $j = 1, 2, \ldots, n$. If we let b_i, $i = 1, 2, \ldots, R$, denote the center of the output membership function for the ith rule, use center-average defuzzification, and product to represent the conjunctions in the premise, then

$$y = \frac{\sum_{i=1}^{R} b_i \prod_{j=1}^{n} \exp\left(-\frac{1}{2}\left(\frac{u_j - c_j^i}{\sigma_j^i}\right)^2\right)}{\sum_{i=1}^{R} \prod_{j=1}^{n} \exp\left(-\frac{1}{2}\left(\frac{u_j - c_j^i}{\sigma_j^i}\right)^2\right)} \tag{5.7}$$

is an explicit representation of a fuzzy system. Note that we do not use the "left" and "right" versions of the Gaussian membership functions in Table 5.5 as this complicates the notation.

It is possible to write down the complete mathematical description of the mapping between the input and output of the fuzzy system.

There are nR input membership function centers, nR input membership function spreads, and R output membership function centers. Hence, we need a total of

$$R(2n + 1)$$

parameters to describe this fuzzy system.

Now, while the fuzzy systems in Equations (5.5) and (5.7) are in general different, it is interesting to compare the number of parameters needed to describe a fuzzy system using each approach. In practical situations, we often have $N_j \geq 3$ for each $j = 1, 2, \ldots, n$, and sometimes the number of membership functions on each input universe of discourse can be 10 or more. From Equation (5.6) we can clearly see that large values of n will result in a fuzzy system with many parameters (there is an exponential increase in the number of rules). On the other hand, using the fuzzy system in Equation (5.7), the user specifies the number of rules. This, coupled with the number of inputs n, specifies the total number of parameters. There is not an exponential growth in the number of parameters in Equation (5.7) in the same way as there is in the fuzzy system in Equation (5.5), so you may be tempted to view the definition in Equation (5.7) as a better one. Such a conclusion, can, however be erroneous for several reasons.

First, the type of fuzzy system defined by Equation (5.5) is sometimes more natural in control design when you use triangular membership functions since you often need to make sure that there will be no point on any input universe of discourse where there is no membership function with a nonzero value (why?). Of course, if you are careful, you can avoid this problem with the fuzzy system also represented by Equation (5.7). Second, suppose that the number of rules for Equation (5.7) is the same as that for Equation (5.5). In this case, the number of parameters needed to describe the fuzzy system in Equation (5.7) is

$$\left(\prod_{j=1}^{n} N_j \right) (2n + 1)$$

Now, comparing this to Equation (5.6) you see that for many values of N_j, $j = 1, 2, \ldots, n$, and number of inputs n, it is possible that the fuzzy system in Equation (5.7) will require many more parameters to specify it than the fuzzy system in Equation (5.5). Hence, the inefficiency in the representation in Equation (5.5) lies in having all possible combinations of output membership function centers, which results in exponential growth in the number of parameters needed to specify the fuzzy system. The inefficiency in the representation in Equation (5.7) lies in the fact that, in a sense, membership functions on the input universes of discourse are not reused by each rule. There are new input membership functions for every rule.

Generally, it is difficult to know which is the best fuzzy system for a particular problem. In this book, we will sometimes use the mathematical representation in Equation (5.7) because it is somewhat simpler, and possesses some properties that we will exploit. At other times we will be implicitly using the representation in Equation (5.5) because it will lend to the development of certain techniques.

Finally, we would like to recommend that you practice creating mathematical representations of fuzzy systems. For instance, it is good practice to create a mathematical representation of the fuzzy controller for ship steering of the form of Equation (5.5), and then also use Equation (5.7) to specify the same fuzzy system. Comparing these two approaches, and resolving the issues in specifying the output centers for the Equation (5.5) case, will help clarify the issues discussed in this section.

5.2.4 Relationships Between Neural and Fuzzy Systems

There are two ways in which there are relationships between fuzzy systems and neural networks. First, techniques from one area can be used in the other. Second, in some cases the functionality (i.e., the nonlinear function that they implement) is identical. Some label the intersection between fuzzy systems and neural networks with the term "fuzzy-neural" or "neuro-fuzzy" to highlight that techniques from both fields are being used. Here, we avoid this terminology and simply highlight the basic relationships between the two fields.

The multilayer perceptron should be viewed as a nonlinear network whose nonlinearity can be tuned by changing the weights and biases. The fuzzy system is also a tunable nonlinearity whose shape can be changed by tuning, for example, the membership functions. Since both are tunable nonlinearities, it is possible to use the methods of Part III to train either one (e.g., least squares, or gradient methods can be used to train both fuzzy and neural systems). While multilayer perceptron networks can take on a similar role to that of a fuzzy system in performing the function of being a tunable nonlinearity, an advantage that the fuzzy system may have, however, is that it often facilitates the incorporation of heuristic knowledge into the solution to the problem, which can, at times, have a significant impact on the quality of the solution.

Some radial basis function neural networks are *equivalent* to some standard fuzzy systems in the sense that they are functionally equivalent (i.e., given the same inputs, they will produce the same outputs). To see this, suppose that in Equation (4.12) we let $n_R = R$ (i.e., the number of receptive field units equal to the number of rules), let the receptive field unit strengths be equal to the output membership function centers, and choose the receptive field units as

$$R_i(x) = \mu_i(x)$$

(i.e., choose the receptive field units to be the same as the premise membership functions). In this case we see that the radial basis function neural network is *identical* to a certain fuzzy system that uses center-average defuzzification. This fuzzy system is then given by

$$y = F_{rbf}(x, \theta) = F_{fs}(x, \theta) = \frac{\sum_{i=1}^{R} b_i \mu_i(x)}{\sum_{i=1}^{R} \mu_i(x)}$$

where θ holds the membership function parameters for the fuzzy system or strengths and receptive field unit parameters for the radial basis function neural network.

The equivalence between this type of fuzzy system and a radial basis function neural network shows that *all the techniques in this book for the above type of fuzzy system work in the same way for the above type of radial basis function neural network.*

Due to the above relationships between fuzzy systems and neural networks, some would like to view fuzzy systems and neural networks as identical areas. This is, however, not the case for the following reasons:

- There are classes of neural networks (e.g., dynamic neural networks) that may have a fuzzy system analog, but if so, it would have to include not only standard fuzzy components but some form of a differential equation component.
- There are certain fuzzy systems that have no clear neural analog. Consider, for example, certain "fuzzy dynamic systems." We can, however, envision how you could go about designing a neural analog to such fuzzy systems.
- The neural network has traditionally been a "black box" approach where the weights and biases are trained (e.g., using gradient methods like backpropagation) using data, often without using extra heuristic knowledge we often have. In fuzzy systems you can incorporate heuristic information and use data to train them. This last difference is often quoted as being one of the advantages of fuzzy systems over neural networks, at least for some applications.

In Part III we will show how to train both neural networks and fuzzy systems and will *try* to provide some insights into which is best to use for a particular application.

5.3 Design Example: Fuzzy Control for Tanker Ship Steering

As there is no general systematic procedure for the design of fuzzy controllers that will definitely produce a high-performance fuzzy control system for a wide variety of applications, it is necessary to learn about fuzzy controller design via examples. Here, we continue with the ship steering example to provide an introduction to the typical procedures used in the design (and redesign) of a fuzzy controller. First, however, we discuss how to code the fuzzy controller for the tanker ship.

5.3.1 Simulation of a Fuzzy Controller

Often, before you implement a fuzzy controller, there is a need to perform a simulation-based evaluation of its performance. To perform a simulation, we will need a model of the plant and a computer program that will simulate the fuzzy control system (i.e., a program to simulate a nonlinear dynamic system). We explained in the last chapter how to simulate a nonlinear system; hence, all we need to do here is explain how to simulate the fuzzy controller.

Fuzzy Controller Arrays and Subroutines

The fuzzy controller can be programmed in C, Fortran, Matlab, or virtually any other programming language. There may be some advantage to programming it in C since it is then sometimes easier to transfer the code directly to an experimental setting for use in real-time control. At other times it may be advantageous to program it in Matlab since plotting capabilities and other control computations may be easier to perform there. Here, rather than discussing the syntax and characteristics of the multitude of languages that we could use to simulate the fuzzy controller, we will develop a computer program "pseudocode" that will be useful in developing the computer program in virtually any language. For readers who are not interested in learning how to write a program to simulate the fuzzy controller, this section will provide a nice overview of the steps used by the fuzzy controller to compute its outputs given some inputs.

Normalization and Scaling: We will use the ship steering example to illustrate the basic concepts on how to program the fuzzy controller. In particular, we will explain how to simulate the fuzzy control system shown in Figure 5.22. Notice that here we have added the gains g_1 and g_2 at the inputs to the fuzzy controller and g_0 at the output of the fuzzy controller. The reason for adding these is that they are often useful in tuning since they scale the horizontal input and output axes of the fuzzy controller. Hence, to simulate the fuzzy control system developed in the last section, we first "normalize" the input and output universes of discourse. For this example, this means that we simply change the membership functions to those shown in Figure 5.23 (i.e., normalize to an interval ± 1). With the indicated scaling gains in Figure 5.23 (i.e., the ones in the boxes) that are implemented as shown in Figure 5.22, we implement the membership functions shown in Figure 5.8.

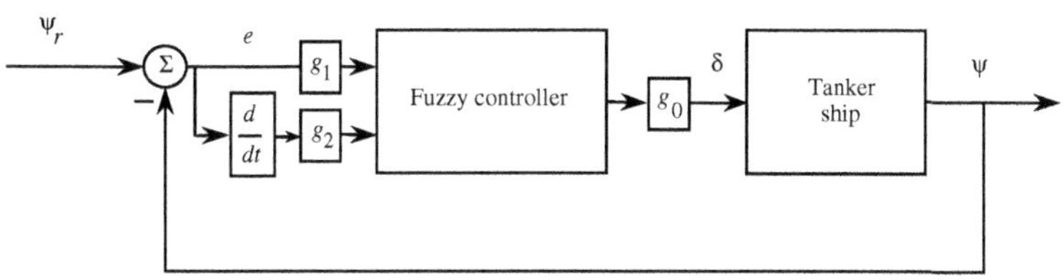

Figure 5.22: Fuzzy controller for tanker ship with scaling gains g_0, g_1, and g_2.

It is important to notice that a scaling gain g_1 on the input is equivalent to scaling the horizontal axis of the e universe of discourse by $1/g_1$ (yes, it is $1/g_1$; think about the fact that increasing g_1 changes, for instance, the meaning of "possmall" so that it quantifies *smaller* values of the error input that is passed through the gain g_1). In more detail, the scaling gain g_1 has the following effects:

- If $g_1 = 1$, there is no effect on the membership functions and there is no effect on the meaning of the linguistic values.
- If $g_1 < 1$, the membership functions are uniformly "spread out" by a factor of $1/g_1$ (notice that multiplication of each number on the e universe of discourse of Figure 5.23 by π which is $1/g_1$, gives you Figure 5.8 on

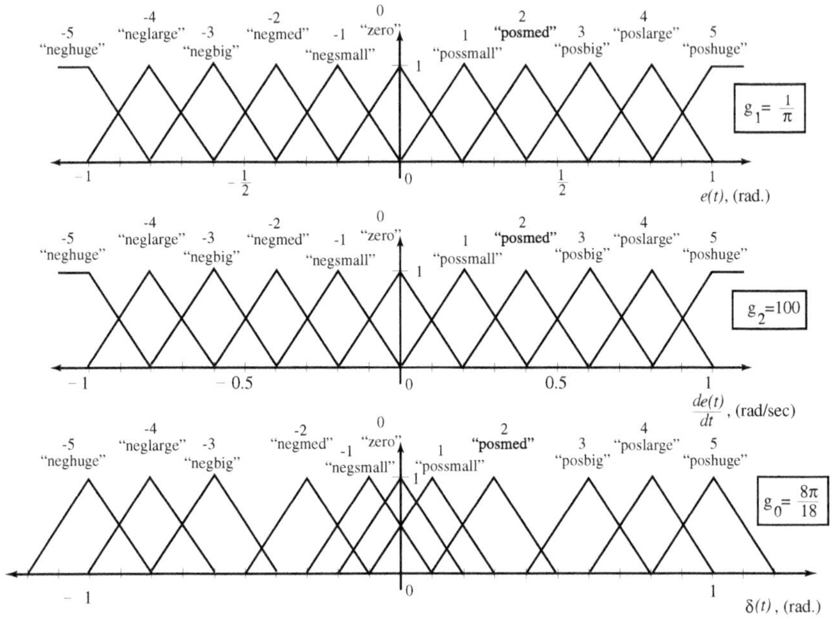

Figure 5.23: Normalized universes of discourse for fuzzy controller for tanker ship (and boxed values of the scaling gains give the original membership functions shown in Figure 5.8).

page 166). This changes the meaning of the linguistics so that, for example, "poslarge" is now characterized by a membership function that represents larger numbers.

- If $g_1 > 1$, the membership functions are uniformly "contracted." This changes the meaning of the linguistics so that, for example, "poslarge" is now characterized by a membership function that represents smaller numbers.

The scaling gain g_2 has similar effects, but for the $\dot{e}$ universe of discourse. However, for the output universe of discourse, the scaling is such that multiplying the output by the gain g_0 is the same as multiplying the horizontal δ axis by g_0.

Here, we will implement the membership functions in Figure 5.23 with the understanding that to get the membership functions in Figure 5.8 on page 166, all we need to do is multiply by scaling gains

$$g_1 = \frac{1}{\pi}, g_2 = 100, g_0 = \frac{8\pi}{18}$$

We will use the minimum operation to represent both the "and" in the premise and the implication (it will be obvious how to switch to using, for example, the product). We will use center of gravity defuzzification. At first we will make no attempt to code the fuzzy controller so that it will minimize execution time or minimize the use of memory. However, after introducing the pseudocode, we will address these issues.

Subroutines: First, suppose that for convenience we use a different set of linguistic-numeric descriptions for the input and output membership functions than we used up till now. Rather than numbering them

$$-5, -4, -3, -2, -1, 0, 1, 2, 3, 4, 5$$

we will renumber them as

$$1, 2, 3, 4, 5, 6, 7, 8, 9, 10, 11$$

so that we can use these as indices for arrays in the program. Suppose that we let the computer variable `x1` denote (notice that a different typeface is used for all computer variables) $e(t)$, which we will call the first input, and `x2` denote $\dot{e}(t)$, which we will call the second input. Next, we define the following arrays and functions:

- Let `mf1[i]` (`mf2[j]`) denote the value of the membership function associated with input 1 (2) and linguistic-numeric value `i` (`j`). In the computer program, `mf1[i]` could be a subroutine that computes the membership value for the ith membership function given a numeric value for the first input `x1` (note that in the subroutine we can use simple equations for lines to represent triangular membership functions). Similarly for `mf2[j]`.

- Let `rule[i,j]` denote the center of the consequent membership function of the rule that has linguistic-numeric value "`i`" as the first term in its premise and "`j`" as the second term in its premise. Hence `rule[i,j]` is essentially a matrix that holds the body of the rule base table shown in Table 5.2. In particular, for the tanker ship we have `rule[i,j]` as:

$$\begin{bmatrix}
1 & 1 & 1 & 1 & 1 & 1 & 0.8 & 0.6 & 0.3 & 0.1 & 0 \\
1 & 1 & 1 & 1 & 1 & 0.8 & 0.6 & 0.3 & 0.1 & 0 & -0.1 \\
1 & 1 & 1 & 1 & 0.8 & 0.6 & 0.3 & 0.1 & 0 & -0.1 & -0.3 \\
1 & 1 & 1 & 0.8 & 0.6 & 0.3 & 0.1 & 0 & -0.1 & -0.3 & -0.6 \\
1 & 1 & 0.8 & 0.6 & 0.3 & 0.1 & 0 & -0.1 & -0.3 & -0.6 & -0.8 \\
1 & 0.8 & 0.6 & 0.3 & 0.1 & 0 & -0.1 & -0.3 & -0.6 & -0.8 & -1 \\
0.8 & 0.6 & 0.3 & 0.1 & 0 & -0.1 & -0.3 & -0.6 & -0.8 & -1 & -1 \\
0.6 & 0.3 & 0.1 & 0 & -0.1 & -0.3 & -0.6 & -0.8 & -1 & -1 & -1 \\
0.3 & 0.1 & 0 & -0.1 & -0.3 & -0.6 & -0.8 & -1 & -1 & -1 & -1 \\
0.1 & 0 & -0.1 & -0.3 & -0.6 & -0.8 & -1 & -1 & -1 & -1 & -1 \\
0 & -0.1 & -0.3 & -0.6 & -0.8 & -1 & -1 & -1 & -1 & -1 & -1
\end{bmatrix}$$

  (recall that we will scale this matrix of centers by $g_0 = \frac{8\pi}{18}$ after we compute the output of the fuzzy controller).

- Let `prem[i,j]` denote the certainty of the premise of the rule that has linguistic-numeric value "`i`" as the first term in its premise and "`j`" as the second term in its premise given the inputs `x1` and `x2`.

- Let `areaimp[c,h]` denote the area under the output membership function with center `c` that has been chopped off at a height of `h` by the minimum operator. Hence, we can think of `areaimp[c,h]` as a subroutine that is used to compute areas under the membership functions for the implied fuzzy sets.

Fuzzy Controller Pseudocode

Using these definitions, consider the pseudocode for a simple fuzzy controller that is used to compute the fuzzy controller output given its two inputs:

```
1. Obtain x1 and x2 values
   (Get inputs to fuzzy controller)

2. Compute mf1[i] and mf2[j] for all i, j
   (Find the values of all membership functions given the values
   for x1 and x2)

3. Let num=0, den=0
   (Initialize the COG numerator and denominator values)

4. For i=1 to 11, For j=1 to 11,
   (Cycle through all areas to determine COG)

    prem[i,j]=min[mf1[i],mf2[j]]

    num=num+rule[i,j]*areaimp[rule[i,j],prem[i,j]]
       (Compute numerator for COG)

    den=den+areaimp[rule[i,j],prem[i,j]]
       (Compute denominator for COG)

5. Next i, Next j

6. Output ucrisp=num/den
   (Output the value computed by the fuzzy controller)

7. Go to Step 1.
```

To learn how this code operates, define each of the functions and arrays for the ship steering example and show how to compute the fuzzy controller output for the same (and some different) inputs used in the previous section. Following this, develop the computer code to simulate the fuzzy controller for the ship steering problem and verify that the computations made by the computer match the ones made by hand.[2]

We do not normally recommend that initially you use only the computer-aided design (CAD) packages for fuzzy systems since these tend to remove you from understanding the real details behind the operation of the fuzzy controller.

[2] One way to start with the coding of the fuzzy controller is to start with the code that is available for downloading at the Web site described in the Preface.

However, after you have developed your own code and fully understand the details of fuzzy control, we do advise that you use (or develop) the tools you believe are necessary to automate the process of constructing fuzzy controllers.

Aside from the effort that you must put into writing the code for the fuzzy controller, there are the additional efforts that you must take to initially type in the rule base and membership functions and possibly modify them later (which might be necessary if you need to perform redesigns of the fuzzy controller). For large rule bases, this effort could be considerable, especially for initially typing the rule base into the computer. While some CAD packages may help solve this problem, it is not hard to write a computer program to generate the rule base, because there are often certain regular patterns in the rule base.

Also notice that since there is a proportional correspondence between the input linguistic-numeric values and the values of the inputs, you will often find it easy to express the input membership functions as a nonlinear function of their linguistic-numeric values. Another trick that is used to make the adjustment of rule bases easier is to make the centers of the output membership functions a function of their linguistic-numeric indices.

Real-Time Implementation Issues

When it comes to implementing a fuzzy controller, you often want to try to minimize the amount of memory used and the time that it takes to compute the fuzzy controller outputs given some inputs. The pseudocode in the last section was not written to exploit certain characteristics of the fuzzy controller that we had developed for the ship; hence, if we were to actually implement this fuzzy controller and we had severe implementation constraints, we could try to optimize the code with respect to memory and computation time.

Computation Time: First, we will focus on reducing the amount of time it takes to compute the outputs for some given inputs. Notice the following about the pseudocode:

- We compute `prem[i,j]` for all values of `i` and `j` (121 values) when for our fuzzy controller for the ship, since there are never more than two membership functions overlapping, there will be at most four values of `prem[i,j]` needed (the rest will have zero values and hence will have no impact on the ultimate computation of the output).

- In a similar manner, while we compute `areaimp[rule[i,j],prem[i,j]]` for all `i` and `j`, we only need four of these values.

- If we compute only four values for `areaimp[rule[i,j],prem[i,j]]`, we will have at most four values to sum up in the numerator and denominator of the COG computation (and not 121 for each).

At this point, from the view of computational complexity, the reader may wonder why we even bothered with the pseudocode of the last section since it appears to be so inefficient. However, the code is only inefficient for the chosen form for the fuzzy controller. If we had chosen Gaussian-shaped (i.e., or some other bell-shaped) membership functions for the input membership functions, then no matter what the input was to the fuzzy controller, all the rules would be on so all the computations shown in the pseudocode were necessary and not too much could be done to improve on the computation time needed. Hence, if you are concerned with real-time implementation of your fuzzy controller, you may want to put constraints on the type of fuzzy controller (e.g., membership functions) you construct.

It is important to note that the problems with the efficiency of the pseudocode highlighted above become particularly acute when there are many inputs to the fuzzy controller and many membership functions for each input, since the number of rules increases exponentially with an increase in the number of inputs (assuming all possible rules are used, which is often the case). For example, if you have a two-input fuzzy controller with 21 membership functions for each input, you will have $21^2 = 441$ rules, and you can see that if you increase the number of inputs, this number will quickly increase.

To reduce computation time, most of which is used for finding which rules are on, it is important to recognize that only a few rules "near each other" are on at any one time.

How do we overcome this problem? Assume that you have defined your fuzzy controller so that at most two input membership functions overlap at any one point, as we had for the ship example. The trick is to modify your code so that it will compute only four values for the premise membership functions, only four values for areas of implied fuzzy sets, and hence, have only four additions in the numerator and denominator of the COG computation. There are many ways to do this. For instance, you can have the program scan `mf1[i]` beginning at position zero until a nonzero membership value is obtained. Call the index of the first nonzero membership value "`istar`." Repeat this process for `mf2[j]` to find a corresponding "`jstar`." The rules that are on are the following:

```
rule[istar,jstar]
rule[istar,jstar+1]
rule[istar+1,jstar]
rule[istar+1,jstar+1]
```

provided that the indicated indices are not out of range. If only the rules identified by the indices of the premises of these rules are used in the computations, then we will reduce the number of required computations significantly, because we will not be computing values that will be zero anyway (notice that for the ship example, there will be one, two, or four rules on at any one time, so there could still be a few wasted computations). Notice that even in the case where there are many inputs to the fuzzy controller *the problem of how to code efficiently reduces to a problem of how to determine the set of indices for the rules that are on.* So that you may fully understand the issues in coding the controller in an efficient manner, we challenge you to develop the code for an n-input fuzzy controller that will exploit the fact that only a hypercubical block of 2^n rules will be on at any one time (provided that at most two input membership functions overlap at any one point).

Memory Requirements: Next, we consider methods for reducing memory requirements. Basically, this can be done by recognizing that it may be possible to compute the rule base at each time instant rather than using a stored one. Notice that there is a regular pattern to the rule base for the ship; since there are at most four rules on at any one time, it would not be hard to write the code so that it would actually generate the rules while it computes the controller outputs. It may also be possible to use a memory-saving scheme for the output membership functions. Rather than storing their positions, there may be a way to specify their spacing with a function so that it can be computed in real-time. For large rule bases, these approaches can bring a huge savings in memory (however, if you are working with adaptive fuzzy systems where you automatically tune membership functions, then it may not be possible to use this memory-saving scheme). We are, however, gaining this savings in memory at the expense of possibly increasing computation time.

Finally, note that while we focus here on the real-time implementation issues by discussing the optimization of *software*, you could consider redesigning the *hardware* to make real-time implementation possible. Implementation prospects could improve by using a better microprocessor or signal processing chip. An alternative would be to investigate the advantages and disadvantages of using a "fuzzy processor" (i.e., a processor designed specifically for implementing fuzzy controllers). Of course, many additional issues must be taken into consideration when trying to decide if a switch in computing technology is needed. Not the least among these are cost, durability, and reliability.

5.3.2 Fuzzy Controller Tuning for the Tanker Ship

We will start out with the controller that we developed earlier and illustrate some basic ideas (from conventional control) that are often used to tune fuzzy controllers. In particular, note that increasing g_1 is analogous to increasing the proportional gain in a PD controller (i.e., it will often make the system respond faster, but may cause overshoot). Increasing the gain g_2 is analogous to increasing the derivative gain in a PD controller which tends to give the controller a better predictive capability and hence helps it avoid overshooting constant reference set points. Notice, also, that increasing g_0 has an effect of increasing the "gain in the loop" so it can be used to speed up the response.

Performance for the First Guess

First, consider the implementation of the fuzzy controller for ship steering developed in the previous sections which we will refer to as our "first guess." The closed-loop response, using the ship model specified in the previous section, is shown in Figure 5.24 (note that we use $g_0 = \frac{8\pi}{18}$, $g_1 = \frac{1}{\pi}$, and $g_2 = 100$ as scaling gains for our membership functions, which were normalized to the interval ± 1, to implement the membership functions in Figure 5.8). Note that while the response is at least tracking the step changes eventually, there is a significant amount of overshoot.

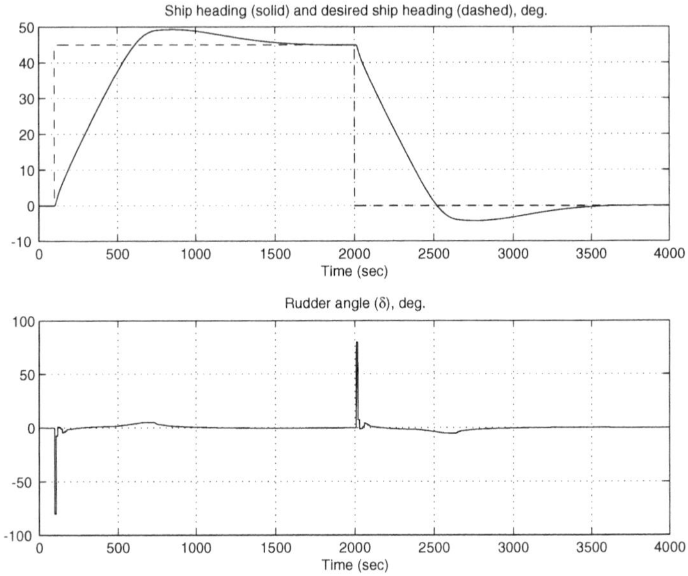

Figure 5.24: Response of fuzzy controller for tanker ship steering, $g_0 = \frac{8\pi}{18}$, $g_1 = \frac{1}{\pi}$, and $g_2 = 100$.

Tuning the Derivative Gain to Reduce Overshoot

Using standard ideas from tuning of conventional controllers (e.g., proportional-integral-derivative (PID) controllers), to reduce the overshoot, we should increase the gain on the derivative term (so that the controller gets more capability to "predict where the response is going"). To do this we choose $g_0 = \frac{8\pi}{18}$, $g_1 = \frac{1}{\pi}$, and $g_2 = 200$ and get the response in Figure 5.25, where we see that we have indeed reduced the overshoot. Unfortunately, however, this also reduced the response time of the system (i.e., it "slowed" the system).

We often use standard heuristic ideas from tuning conventional controllers for tuning fuzzy controllers.

Tuning the Proportional Gain to Decrease the Response Time: Finding "Good" Scaling Gains

Next, we seek to choose a good set of scaling gains by speeding up the response from the previous case. To do this we increase the gain on the proportional term so that we increase the speed of the response and hence reduce the response time. When we do this, however, this can cause some overshoot, so we also increase the gain on the derivative term to avoid that. In particular, choose $g_0 = \frac{8\pi}{18}$, $g_1 = \frac{2}{\pi}$, and $g_2 = 250$ to get a faster response with very little overshoot as seen in Figure 5.26. We take this set of gains as "good" values in that we consider the response that results from them to be good. Notice that we achieved all our tuning via the scaling gains, although this is certainly not possible in all applications.

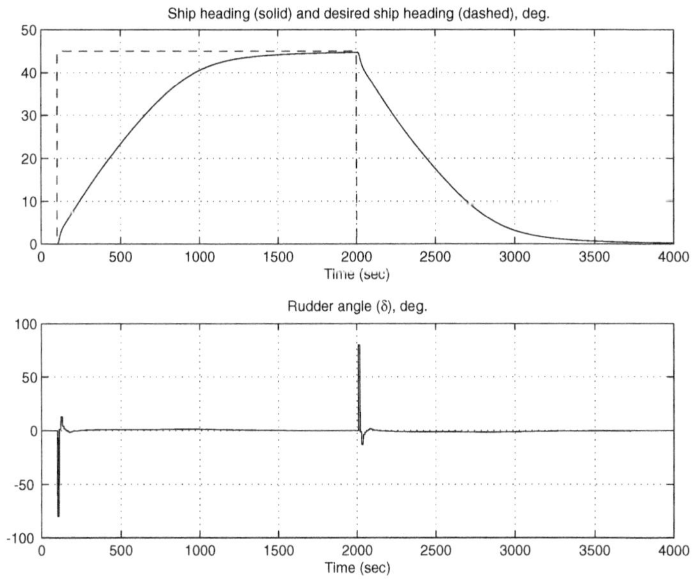

Figure 5.25: Response of fuzzy controller for tanker ship steering, $g_0 = \frac{8\pi}{18}$, $g_1 = \frac{1}{\pi}$, and $g_2 = 200$.

The Resulting Nonlinear Control Surface

To achieve this performance, the fuzzy controller implements a nonlinearity that is shown in Figure 5.27. Notice that this surface is another way to view the captain's expertise in ship steering (compare the list of the captain's steering expertise developed earlier to the shape of the surface; for instance, explain why the slope of the surface changes in the way it does).

The fuzzy controller implements a nonlinear input-output map. Rule construction and tuning shapes this map.

Note that the control surface for a simple proportional-derivative (PD) controller is a plane in three dimensions. With the proper choice of the PD gains, the linear PD controller can be made to have basically the same shape as the fuzzy controller near the origin. Hence, in this case the fuzzy controller will behave similarly to the PD controller provided its inputs are small. However, notice that there is no way that the linear PD controller can achieve a nonlinear control surface of the shape shown in Figure 5.27 (this is not surprising considering the complexity difference of the two controllers).

It is useful to notice that there is a type of interpolation that is performed by the fuzzy controller that is nicely illustrated in Figure 5.27. If you study the plot carefully, you will notice that the rippled surface is created by the rules and membership functions. For instance, if we kept a similar nonuniform distribution of membership functions for the input and outputs of the fuzzy system, but increased the number of membership functions, the ripples would correspondingly increase in number and the amplitude of the ripple would decrease. What is happening is that there is an interpolation between the rules. The output is

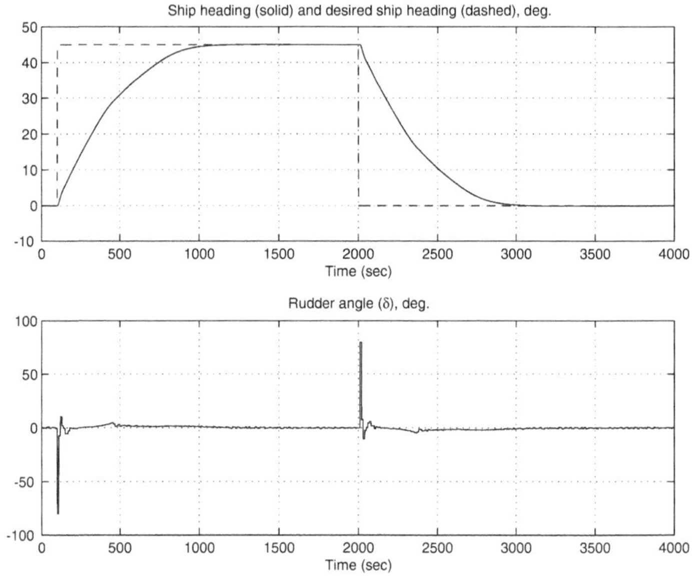

Figure 5.26: Response of fuzzy controller for tanker ship steering, $g_0 = \frac{8\pi}{18}$, $g_1 = \frac{2}{\pi}$, and $g_2 = 250$.

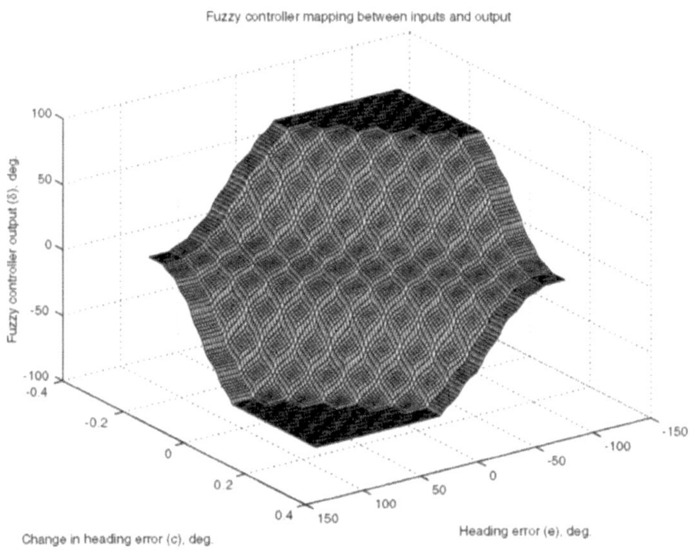

Figure 5.27: Nonlinear control surface implemented by the fuzzy controller, $g_0 = \frac{8\pi}{18}$, $g_1 = \frac{2}{\pi}$, and $g_2 = 250$.

an interpolation of the effects of the four rules that are on for the ship's fuzzy controller. For more general fuzzy controllers, it is important to keep in mind that this sort of interpolation is often occurring (but not always—it depends on your choice of the membership functions).

5.3.3 Design Concerns

While designing fuzzy controllers for practical problems you will encounter a whole variety of problems, not the least of which could be pragmatic issues in interfacing and communicating with the plant. In this section we outline some of these and offer ideas on how to solve them. Several of the design concerns we list are not specific to fuzzy control, but apply to any control system, including the ones considered later in this chapter. To illustrate several points we will use the tanker ship steering problem studied in the last section.

Understand the Control Problem

One key to developing a good solution to any problem is to make sure that you clearly understand the problem so that you are sure that you are solving the right problem! For control problems this means that you must do the following:

- *Obtain a good understanding of the plant:* It is critical that you gain a good understanding of the plant you are to control. Yes, this means understanding the physics of the problem and this may demand that you step outside your main area of expertise (e.g., to study thermodynamics, fluid mechanics, mechanics, circuit theory, etc.). Aside from returning to first principles, it may be beneficial to consult others who have operated the plant in the past or who have already developed a controller for it. It may be helpful to develop a simulation of the plant and study the effects of, for example, some inputs or disturbances on the output variables. Now, clearly one of the main advantages of fuzzy and expert control is that you do not explicitly need a mathematical model of a specific form to develop the controller; however, for some plants it is not too hard to develop an approximate mathematical model that can be very helpful in gaining an understanding of how to control the plant. Experience has shown that to develop good control systems you must use all the information you have about how to achieve good control. Some of this information may come in the form of rules from a human operator (or engineer) but other useful information can come from a mathematical model and this should not be ignored. Indeed, such a mathematical model will be needed for the implementation of a planning system.

- *Pay attention to plant constraints:* A particularly important part of the problem of obtaining a good understanding of the plant is to understand those plant characteristics that limit your ability to achieve high performance operation. Some typical limitations include the following:

  - Actuator saturation limits: All actuators have limits in which they can perform and these constrain the ways in which you can affect plant behavior. These limits come in the form of saturation limits on the magnitude of the input, limits on the rate of change of the input, etc.
  - Sensor noise: There is no perfect sensor. Better sensors cost more money. Noise on sensors limits the quality of information you can obtain about the plant and hence your ability to control the plant.
  - Plant dynamics: Unstable, highly nonlinear, nonminimum phase, and highly uncertain plants (i.e., those with significant noise and plant parameter variations), or plants with delays all provide unique challenges in control. The way each of these problems is manifested changes for different applications.

- *Develop appropriate specifications:* Sometimes the boss or customer is the one to provide the specifications of what they want in terms of performance. It is important to have a very clear understanding of the expectations for the plant in terms of typical measures of performance like the following:

  - Rise-time (amount of time for the output to get from 10% of the final value to 90% of the final value when there is a step input), overshoot (the amount the output increases above the final value of a step reference input), steady-state error (error between the plant output and commanded input as time goes to infinity).
  - Stability (e.g., for the ship steering problem, if you start the ship heading near a desired constant heading, will it move toward the reference heading and ultimately reduce the heading error to zero?), limit cycles (oscillations).
  - Performance robustness (e.g., how much can the plant be allowed to change before control system performance degrades significantly?), and stability robustness (e.g., how much can the plant be allowed to change before the system goes unstable?).

  If the requirements are unreasonable, you may have to return to the boss or customer and negotiate a reasonable set of specifications. If you find that more is possible than is being asked, then your company may have a competitive edge with the customer.

- *Consider if it is possible to redesign the plant:* For some control problems (e.g., aircraft control) there are significant efforts to design the plant so that it is easy to control. If you study the control problem (plant dynamics and control specifications) and find that the specifications cannot be met, another option may be to go back and redesign the plant so that the specifications can be met. This may entail adding a sensor, purchasing a better actuator, or even making structural changes to the plant to remove challenging nonlinearities.

- *Study similar problems:* There is an ever-expanding literature on the development of control systems and there may be some similar work done that is in the public domain that could be useful to you.

- *Try the simplest thing:* Trying the simplest thing first is good engineering practice and it may teach you something about the control problem. Fuzzy, expert, or planning systems-based control is probably not the first thing to try when implementing a control system. Even if you do not have a model, it is simple to develop a PID controller, or indeed a P controller (proportional controller), that can be tuned manually. The computations for simple proportional control involve, for instance, forming a difference between the reference input and plant output, and multiplying this difference by a gain. The numerical operations to implement a fuzzy controller are clearly more complex (although it is not always the case that they are more complex than a conventional controller).

Proper Rule Base Construction

Assuming that you are using fuzzy control, one of the most critical steps in the design process is the choice of the rule base. It is therefore very important to pay significant attention to this problem. The main sources of information for rule base construction are the following:

- Interviews of human plant operators (or learning how to operate the plant yourself).

- A good understanding of the plant, the constraints imposed by it, and the closed-loop specifications that you are trying to achieve.

- Modeling and simulation studies.

- Past development of controllers for the same plant (or similar ones).

- Controller implementation studies for controllers that ultimately do not adequately achieve the specifications (e.g., the controller that you are trying to replace in updating a control system to achieve higher performance).

There are several issues to pay attention to in rule base construction, including the following:

- *Conflicting rules:* Most often (but not always), the rules in the rule base should not conflict with one another (e.g., there should not be two rules that apply in the same situation that say to do two very different things). Note, however, that conflicting rules can be used in a fuzzy controller since, depending on how you define the inference mechanism, it will simply interpolate between the two different conclusions (e.g., in the ship steering fuzzy controller four different rules may come on that say to do somewhat different things and defuzzification combines the recommendations that are in a sense conflicting, if only mildly).

- *Completeness:* You must define the rule base so that there is at least one rule that is "on" at each time (and if there is, we will call it a "complete rule base"). This means that there is a sort of complete coverage of the input space of the fuzzy controller so that there is a premise membership function with nonzero certainty for all possible values of the inputs. For an expert controller, it must be the case that there is at least one rule with a premise term that evaluates to true at each time instant. Note that in the ship steering example, no matter what the values of e and $\dot{e}$ are, there is a premise membership function that has nonzero certainty so that there is always at least one rule on. If you do not have a complete rule base then, depending on how you define the fuzzy controller, it can be that the denominator in the defuzzification formula will have a zero value so that you will not be able to compute an explicit output (and your software will return a "divide by zero" error).

Reducing Controller Complexity

For simple academic problems, the complexity of the fuzzy controller is rarely a problem, especially when only simulation examples are considered. The problem is, however, that for real applications there are often limitations on computing power (memory and "throughput"), so it is important to carefully consider how to reduce the computations necessary for implementations. There are two fundamental reasons why complexity arises in fuzzy and expert controllers:

- *Complex nonlinear maps:* For challenging applications where you have spent a significant amount of time tuning the rule base, it is likely that the resulting controller surface has a very interesting and complex shape, and that this shape is critical in meeting the performance specifications. Complex nonlinear maps take significant computations to implement, so to get higher performance control you have to pay for it in controller complexity (you do not get something for nothing).

- *Exponential increase in number of rules:* Recall that in Section 5.2.3, we analyzed the number of parameters needed to define a fuzzy system for a given number of inputs and membership functions. We found that if you define rules for all possible combinations of linguistic values in the premises, then there is an exponential number of rules (similar analyses hold for expert controllers also). For example, for our ship steering problem with two inputs and eleven membership functions on each input universe of discourse there are $11^2 = 121$ possible rules. Hence, increasing the number of linguistic values or inputs causes large increases in the number of rules and hence the complexity of the fuzzy controller (e.g., going from using e and $\dot{e}$ as inputs to also using $\int_0^t e(\tau)d\tau$, with eleven membership functions on the $\int_0^t e(\tau)d\tau$ universe of discourse, would result in $11^3 = 1331$ rules).

Methods to reduce complexity are as numerous as there are applications since each is a special case and there are often methods to simplify the computations. There are, however, some general approaches to reducing complexity and these are listed next:

1. In some cases, upon further study, you may be able to determine that rule base completeness can be achieved with fewer rules simply because it may be the case that certain combinations of the inputs are not possible. In this case the corresponding rules can be removed since they will never be used anyway.

2. You can simply try to reduce the number of linguistic values so that the total number of possible rules is reduced. For example, for the tanker ship it is possible to get reasonably good performance (at least for nominal conditions) using only nine rules (three membership functions on each of the two input universes of discourse). Realize, however, that additional rules allow for the implementation of more complex nonlinear control surfaces, that can then result in higher performance operation. The key is to determine the minimum number of rules that still allows for the implementation of a control surface that can achieve adequate performance. In some cases you may have to go back to the customer and indicate that if you are only allowed a certain amount of computing power, then only a certain performance level is possible.

3. For the case of MIMO fuzzy controllers, study the problem carefully to determine if you truly need all the inputs for each of the fuzzy controllers for each plant input. Elimination of one input, for even one MISO controller, can result in significant savings.

4. Sometimes you may want to use some type of "multi-stage" fuzzy controller where, for example, there are two inputs to each of two controllers and their outputs are combined by a third fuzzy system that provides the input to the plant. In this case we will implement three two-input fuzzy controllers rather than one four-input fuzzy controller (which for some applications can make a big difference). This approach tends to be highly application specific but the principle is valid: try to reduce the number of inputs by cascading fuzzy controllers.

5. Another approach to reduction is to use one fuzzy controller to specify parameters in another. For instance, if you were to develop a controller for the ship that also took as an input the ship speed u, one approach would be to simply use a three-input fuzzy controller (where the rule base would indicate that for faster ship speeds a smaller rudder angle input is needed since the ship is easier to steer when it is moving fast). Another approach, one that avoids the implementation of a three-input fuzzy controller, is to use the two-input fuzzy controller we already developed for the ship but add another single-input single-output fuzzy controller with the ship speed as an input that specifies the amount of correction to the rudder

angle for different ship speeds. Now, this approach does not allow one to make coordinated control actions (e.g., different rudder corrections for different e and $\dot{e}$ and ship speeds), but it may be sufficient to solve the problem. Again, there are many approaches to reducing complexity using this approach since they are application-dependent.

Effects of Disturbances and Plant Changes

It is important to evaluate the performance of the fuzzy control system under adverse conditions.

Plant parameter variations, disturbances, speed changes, and sensor noise all affect our ability to achieve good control. In this section we will use the ship example to illustrate their effect on heading regulation performance when a fuzzy controller is used (this section parallels the simulation studies for the multilayer perceptron and radial basis function controllers for the ship in Sections 4.3 and 4.5; here, we use the same types of variations as we did in those sections). Our intent, however, is to alert the reader to these issues so that they can be taken into account in the design process.

First, we will consider the performance of the fuzzy controller when the ship is under "full" conditions. Figure 5.28 shows how the fuzzy control system, which was tuned for ballast conditions, performs for full conditions. We see that there now is a bit of overshoot in the ship heading since a lighter boat steers easier. We see that plant parameter variations can affect performance.

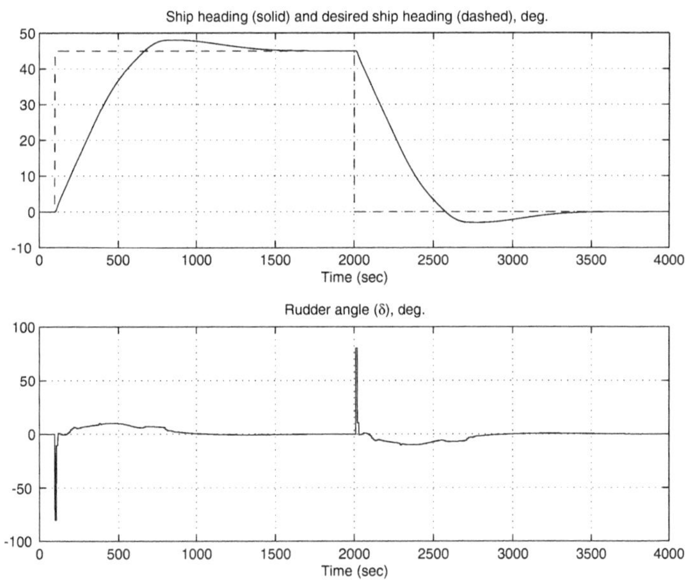

Figure 5.28: Response of fuzzy controller for tanker ship steering, "full" conditions, $g_0 = \frac{8\pi}{18}$, $g_1 = \frac{2}{\pi}$, and $g_2 = 250$.

Next, consider the effect of a wind disturbance on the ship. If we use $g_0 = \frac{8\pi}{18}$,

$g_1 = \frac{2}{\pi}$, and $g_2 = 250$ (i.e., the good tuned values), we get the response in Figure 5.29. We see that the wind affects our ability to achieve very good regulation of the ship heading since it causes a 1 to 2 degree variation in the tracking of the desired heading.

Adverse conditions generally degrade performance; however, good controller designs minimize such performance degradations.

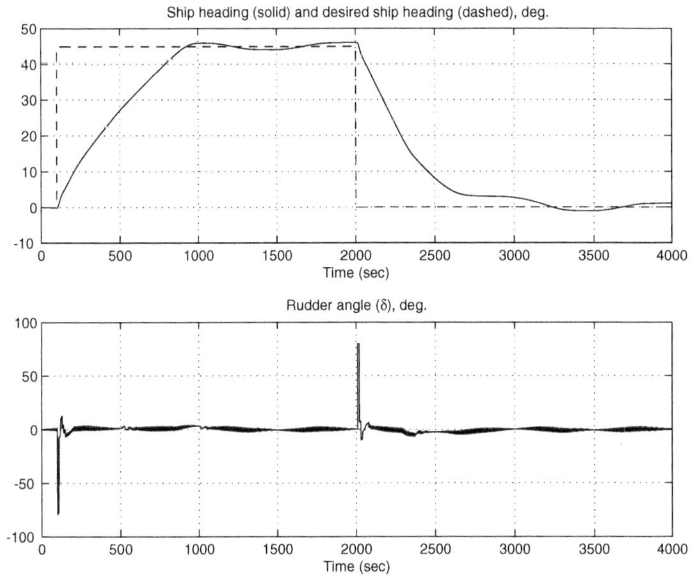

Figure 5.29: Response of fuzzy controller for tanker ship steering, wind disturbance, $g_0 = \frac{8\pi}{18}$, $g_1 = \frac{2}{\pi}$, and $g_2 = 250$.

If you use, for instance, an additive sensor noise uniformly distributed on $[-0.01, 0.01]$, there is little effect on the response so we do not show the plot (of course, if you get sensors with worse performance characteristics, then you will expect tracking errors to arise in an analogous manner to results for the wind).

Next, consider the effect of a speed change on our ability to steer the ship. If we use $g_0 = \frac{8\pi}{18}$, $g_1 = \frac{2}{\pi}$, and $g_2 = 250$ (i.e., the good tuned values), we get the response in Figure 5.30. We see that the speed decrease causes a significant overshoot in the response since the rudder is not as effective.

Tracking Error

Steady-state tracking error is the value

$$\lim_{t\to\infty} e(t)$$

and for most control problems we would like this to be as small as possible or zero when the reference input is, for example, a step change. Adding an integrator to the control loop is one approach that is often successful at reducing or eliminating steady-state error (since if the error is nonzero, the integrator's

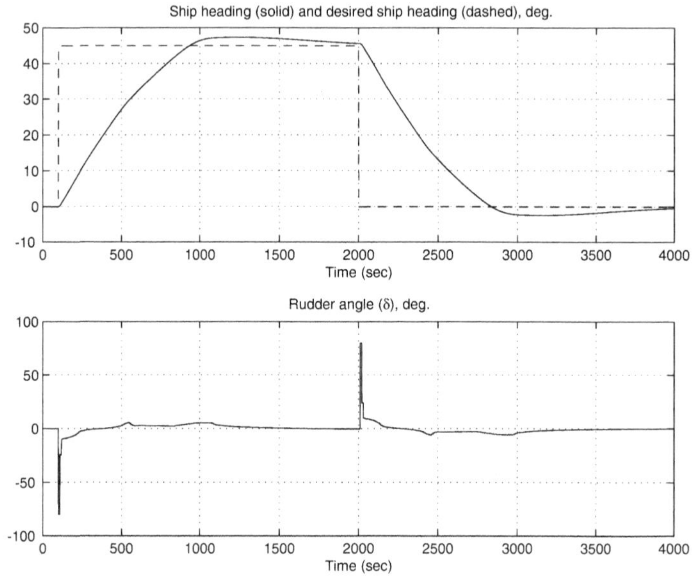

Figure 5.30: Response of fuzzy controller for tanker ship steering, speed decrease, $g_0 = \frac{8\pi}{18}$, $g_1 = \frac{2}{\pi}$, and $g_2 = 250$.

value gets larger, thus causing a larger control input to force the plant to move to reduce the error). In this ship example we did not need to add an integrator to the control loop by putting one in the controller since there is already an integration effect in the plant (note that if you hold the rudder angle input constant, the ship heading will tend off to infinity).

For fuzzy control design, since the plant and controller are nonlinear (you should not be trying to control a linear plant with a fuzzy controller since if the plant is truly linear all that is needed to succeed is a linear controller), we cannot be guaranteed that the addition of an integrator will help reduce steady-state error, but often it does. There are basically two ways to add an integrator to a fuzzy controller: as an input (to achieve, for instance, "PID fuzzy control," i.e., a fuzzy controller with P, I, and D inputs), or by adding an integrator to the output of the plant (which in some discrete time implementations some engineers do inherently by specifying that the output should be a change in the control variable, not an absolute value).

5.4 Stability Analysis

Here, we will be brief by simply providing some examples of how you can encounter limit cycles and instabilities for the tanker ship and a brief explanation of how to conduct stability analysis for fuzzy control systems.

5.4.1 Example: Stability and Limit Cycles in Ship Steering

Stability is often viewed as a fundamental property of a control system since if the system is unstable it is possible that the output response, and hence the tracking error (assuming a bounded reference input), grows without bound. For example, for the ship if you choose $g_0 = \frac{-8\pi}{18}$, $g_1 = \frac{2}{\pi}$, and $g_2 = 250$ (notice minus sign), you can get an unstable response. (In this case, the controller moves the rudder in the wrong direction to try to reduce a heading error and each time it does this, it creates a bigger error.) This is a rather simple mechanism that provides instability but there can be very complex ones.

Improper choices for the rule base can result in a closed-loop system with limit cycles and instability.

Another type of tracking error that can result (that is often considered to be a type of instability) is when $e(t)$ is oscillating, for example, when the reference input is a constant and the output of the plant is a sinusoid. In many applications this is an undesirable characteristic. If you pick the wrong values of the scaling gains in the ship steering problem, you can get such oscillatory behavior. For example, if you pick $g_0 = \frac{2000\pi}{18}$, $g_1 = \frac{2}{\pi}$, and $g_2 = 0.000001$ for the ship, you get the response shown in Figure 5.31 where we see that the oscillation characteristics are dependent on the magnitude of the reference input.

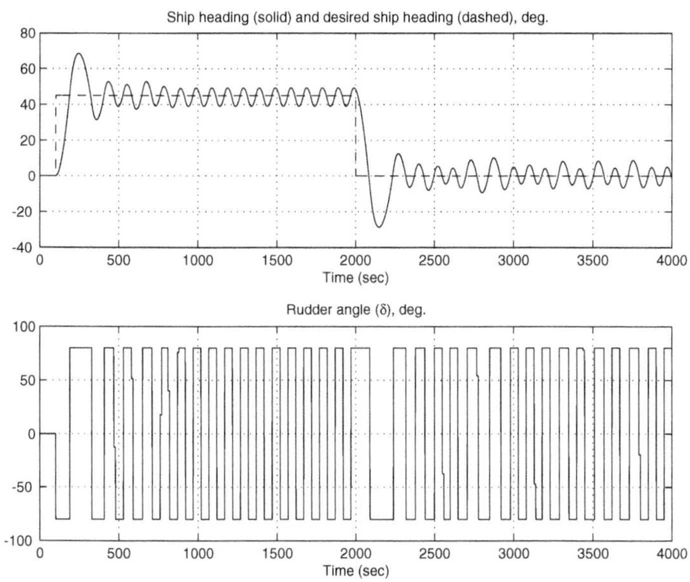

Figure 5.31: Response of fuzzy controller for tanker ship steering, $g_0 = \frac{2000\pi}{18}$, $g_1 = \frac{2}{\pi}$, and $g_2 = 0.000001$.

What is happening in the fuzzy control system to achieve this type of behavior? Usually, it is because some gains are set too large (or small) and the input or output signals are oscillating between their maximum values, forcing the plant to oscillate also (i.e., this is a "controller-induced oscillation" in this case). Clearly, the designer must be alerted to this possibility and try to avoid it. Methods to avoid this problem typically involve careful choice of rule bases and scaling gains.

5.4.2 Discussion: Lyapunov Stability Analysis of Fuzzy Control Systems

Lyapunov stability analysis can be useful for verification of fuzzy control systems.

The central issue in fuzzy controller design is to obtain good insights into how the plant behaves in order to determine how to shape the nonlinear function that is implemented by the fuzzy controller. Of course, this nonlinear function then affects the closed-loop dynamics. To characterize and analyze exactly how the nonlinearity affects the closed-loop stability properties, you can use mathematical stability analysis just as we did for the neural controller in the last chapter. Moreover, the reader may be interested to know that Lyapunov's first method (via linearization), absolute stability, and describing function analysis can be performed (see the "For Further Study" section at the end of this part for more information).

How exactly do you perform stability analysis via Lyapunov's direct method? Consider the simple example in Section 4.6.5 on page 147. Note that

$$u = F(x)$$

could be specified as a fuzzy controller so that $F(0) = 0$, $F(x)$ is smooth, and for some scalar $\beta > 0$,

$$\begin{aligned} F(x) &> -\beta x, \; x < 0 \\ F(x) &< -\beta x, \; x > 0 \end{aligned}$$

How do we construct a rule base so that this is the case? We will provide a problem of this type to the reader in Exercise 5.8. Basically, however, when you get familiar with the types of input-output mappings that are generated for certain choices of rule bases and membership functions, you will see how to construct nonlinear control surfaces with different shapes.

5.5 Expert Control

An expert system is a computer program that is designed to emulate a human's skills in a specific problem domain. If it is designed to emulate the expertise of a human in performing control activities, it is called an "expert controller." When the expert controller is connected to a plant, the closed-loop system is called an expert control system (see Figure 5.32). Traditionally, the expert system has been split into two components: the knowledge base and inference mechanism. The knowledge base is simply a generalization of the rule base in a fuzzy system where more general types of information can be characterized. Correspondingly, the inference mechanism is a generalization of the inference mechanism in a fuzzy controller that can incorporate other reasoning strategies. Hence, conceptually the expert controller is closely related to the fuzzy controller in its structure and function. Moreover, the design philosophy used to construct the expert controller is similar to that of the fuzzy controller. The main differences between the two approaches lie in the details of how the knowledge base and inference mechanism are constructed.

General representations of knowledge and inference can be used for emulating sophisticated control strategies.

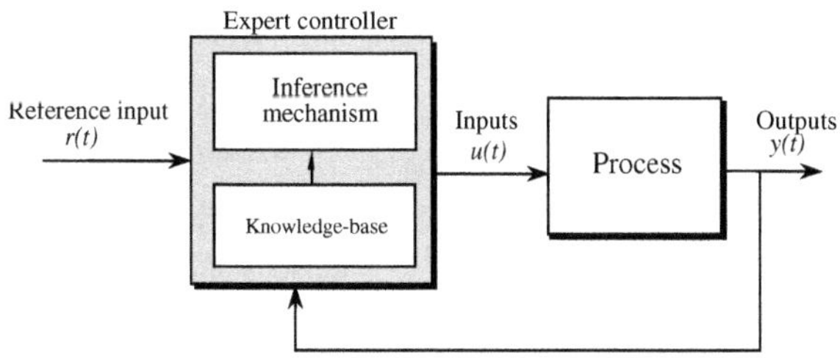

Figure 5.32: Expert control system.

5.5.1 General Knowledge Representations

The knowledge base in the expert controller could be a rule base, but is not necessarily so. It could be developed using other knowledge-representation structures, such as frames, semantic nets, causal diagrams, and so on (see the "For Further Study" section at the end of this part for information on these). Here, we will simply show how the form of the rules used in fuzzy controllers can be generalized and used in expert control.

The rule premises can be defined in much more general ways. For instance, any type of predicate logic can be used that can include any kind of Boolean logic, functions, relations, and existential quantifiers ("for all," and "there exists"). For example, a rule may have the form:

If $e(t) > 2$ **or** there exists a time over the last 10 sec. where $\frac{de(t)}{dt} \leq 0.5$ **Then** $u(t) = 2$.

Testing the validity of the premise can be defined in many ways, but normally the standard rules of logic are used (similar to how the premise part of a standard computer "if-then" statement is tested). Moreover, degrees of matching the premises to the current situation can be used in an analogous way to how it is in fuzzy systems.

The specific types of rules needed for control depend on the application being considered and often it requires significant expertise with the plant to develop an effective set of rules. Indeed, for practical applications this is typically an iterative trial-and-error process and may involve a team of process experts to test and develop the rule base. Conceptually, however, the synthesis of the rule base proceeds in basically the same way as for the fuzzy control methodology.

Finally, it is interesting to note that in practical control systems there are often rules used for "exception handling" and special situations. These rules sometimes override a currently operating controller (e.g., a PID controller) to take appropriate actions under special situations. The inclusion of such rules in controllers should not be viewed as a rare occurrence in the development of a practical control system; control rules are often present, and sometimes are significantly more difficult to develop than the "conventional" (e.g., PID) part of the overall control system.

5.5.2 General Inference Mechanisms

The inference mechanism in the expert controller is more general than that of the fuzzy controller. It can use more sophisticated matching strategies to determine which rules should be allowed to fire. Also, it can use more elaborate inference strategies. For instance, some expert systems use

- "Refraction," where if a rule has fired recently, it may not be allowed back into the "conflict set" (i.e., the set of rules that are allowed to fire).
- "Recency," where rules that were fired most recently are given priority in being fired again (sometimes a valid approach since such rules may be most relevant to the current situation).
- "Priority schemes" where certain rules are a priori given higher priority to fire if they are both in the conflict set. It is also possible to dynamically assign priority.

Verification of correct behavior of general reasoning systems used as feedback controllers is important and challenging.

It is in fact the case that an expert system is in a sense more general than a fuzzy system since it can be shown that a single rule in an expert controller can be used to represent an entire fuzzy controller. To see this, note that a single fuzzy controller can be represented with a single static input-output map. Then, a single rule in an expert controller can represent that mapping. If an entire set of fuzzy controllers is represented as a set of such rules, then the resulting expert controller will reason about how to successively apply fuzzy controllers at each time step.

5.5.3 Stability Analysis of Expert Control Systems

Just as for neural and fuzzy control systems, it is possible to analyze qualitative properties of expert control systems. For instance, a discrete time formulation can be used to study the following properties:

- Stability in the sense of Lyapunov that may characterize how well the expert system can stay focused on (attend to) a control task, or boundedness of plant variables in the closed-loop when an expert controller is used.
- "Reachability" properties where, for instance, search algorithms can be used to test if the expert controller can drive the plant into some state (e.g., the goal state).

- Cyclic properties where the expert system may get stuck in an infinite loop (circular reasoning) and hence not be able to achieve its goal.

Here, we will not develop the mathematical models and show explicitly how to conduct stability analysis for expert control systems since it basically follows the same conceptual approach as for neural or fuzzy control systems. We do, however, provide more references in the "For Further Study" section at the end of this part for the interested reader, and Design Problem 5.3 where the reader is asked to conduct stability analysis of a simple expert control system.

5.6 Hierarchical Rule-Based Control Systems

There are a variety of ways to construct hierarchical fuzzy or expert control systems. For instance, you could use the following approaches:

Knowledge and inference are sometimes conveniently represented via hierarchies.

- You could use a rule-based (fuzzy or expert) system as a supervisor for the operation of a rule-based controller. This supervisor could monitor certain plant conditions and modify the rules to try to maintain good performance. It may be more convenient to implement the system as two rule-based systems, rather than a single one that takes in all the inputs that the two systems do, and outputs the input to the plant (e.g., it may be more computationally efficient, or this may be the way that the human operator thinks about controlling the plant).

- You could use a rule-based system to supervise the operation of an adaptive control system. This possibility will be discussed in more detail in Section 9.4.5.

- Sometimes multiple layers of such supervision could be needed.

There are still other possibilities. For instance, you can think of the hierarchy in Figure 1.11 and suppose that each block is a fuzzy or expert system. The blocks at the low level may be standard fuzzy controllers. The blocks at the coordination level may contain fuzzy systems with rules about how to coordinate the operation of the fuzzy controllers at the execution level, and an expert system at the management level could supervise both the levels below it. In this context you may think of using rules at the higher levels to turn on appropriate rules at the low levels (some would think of this as pruning the rules at the lower levels).

5.7 Exercises and Design Problems

Exercise 5.1 (Defining Membership Functions: Single Universe of Discourse): In this problem you will study how to represent various concepts and quantify various relations with membership functions. For each part below, there is more than one correct answer. Provide one of these and justify your choice in each case.

(a) Draw a membership function (and hence define a fuzzy set) that quantifies the set of all people of medium height.

(b) Draw a membership function that quantifies the statement "the number x is near 10."

(c) Draw a membership function that quantifies the statement "the number x is less than 10."

Exercise 5.2 (Defining Membership Functions: Multiple Universes of Discourse): In this problem you will study how to represent various concepts and quantify various relations with membership functions when there is more than one universe of discourse. Use minimum to quantify the "and." For each part below, there is more than one correct answer. Provide one of these and justify your choice in each case. Also, in each case, provide the three-dimensional plot of the membership function.

(a) Draw a membership function (and hence define a fuzzy set) that quantifies the set of all people of medium height who are "tan" in color (i.e., tan *and* medium-height people). Think of peoples' colors being on a spectrum from white to black.

(b) Draw a membership function that quantifies the statement "the number x is near 10 and the number y is near 2."

Exercise 5.3 (Fuzzy Sets): There are many concepts that are used in fuzzy sets that sometimes become useful when studying fuzzy control. The following problems introduce some of the more popular fuzzy set concepts that were not treated earlier in the chapter.

(a) The "support" of a fuzzy set with membership function $\mu(x)$ is the (crisp) set of all points x on the universe of discourse such that $\mu(x) > 0$ and the "α-cut" is the (crisp) set of all points on the universe of discourse such that $\mu(x) > \alpha$. What is the support and 0.5-cut for the fuzzy set shown in Figure 5.5 on page 163?

(b) The "height" of a fuzzy set with membership function $\mu(x)$ is the highest value that $\mu(x)$ reaches on the universe of discourse on which it is defined. A fuzzy set is said to be "normal" if its height is equal to one. What is the height of the fuzzy set shown in Figure 5.5 on page 163? Is it normal? Give an example of a fuzzy set that is not normal.

(c) A fuzzy set with membership function $\mu(x)$ where the universe of discourse is the set of real numbers is said to be "convex" if and only if

$$\mu(\lambda x_1 + (1 - \lambda)x_2) \geq \min\{\mu(x_1), \mu(x_2)\} \tag{5.8}$$

for all x_1 and x_2 and all $\lambda \in [0, 1]$. Note that just because a fuzzy set is said to be convex does not mean that its membership function is a convex function in the usual sense. Prove that the fuzzy set shown in Figure 5.5 on page 163 is convex. Prove that the Gaussian membership function is convex. Give an example of a fuzzy set that is not convex.

(d) A linguistic "hedge" is a modifier to a linguistic value such as "very" or "more or less." When we use linguistic hedges for linguistic values that already have membership functions, we can simply modify these membership functions so that they represent the modified linguistic values. Consider the membership function in Figure 5.5 on page 163. Suppose that we obtain the membership function for "error is very possmall" from the one for "possmall" by squaring the membership values (i.e., $\mu_{verypossmall} = (\mu_{possmall})^2$). Sketch the membership function for "error is very possmall." For "error is more or less possmall" we could use $\mu_{moreorlesspossmall} = \sqrt{\mu_{possmall}}$. Sketch the membership function for "error is more or less possmall."

Exercise 5.4 (Fuzzy Logic): There are many concepts that are used in fuzzy logic that sometimes become useful when studying fuzzy control. The following problems introduce some of the more popular fuzzy logic concepts that were not treated earlier in the chapter or were treated only briefly.

(a) The complement ("not") of a fuzzy set with a membership function μ has a membership function given by $\bar{\mu}(x) = 1 - \mu(x)$. Sketch the complement of the fuzzy set shown in Figure 5.5 on page 163.

(b) There are other ways to represent the conjunction "and" using fuzzy sets, different from the minimum and product that were introduced in the chapter. Let μ^1 and μ^2 denote two specific membership function values. Then, to represent "and," we could use the "bounded difference" (i.e., $\max\{0, \mu^1 + \mu^2 - 1\}$) and "drastic intersection" (where its value is μ^1 when $\mu^2 = 1$, μ^2 when $\mu^1 = 1$, and zero otherwise). Consider the membership functions shown in Figure 5.8 on page 166. Sketch the membership function for the premise "error is zero **and** change-in-error is possmall" when the bounded difference is used to represent this conjunction (premise). Do the same for the case when we use the drastic intersection. Compare these to the case where the minimum operation and the product were used (i.e., plot these also and compare all four).

(c) Fuzzy logic can be used to represent the disjunction ("or") of, for example, two premise terms. While there are many ways to represent "or" in fuzzy logic, the most popular one seems to be to simply use the maximum of the membership values. Consider the membership functions shown in Figure 5.8 on page 166. Sketch the membership function for "error is zero **or** change-in-error is possmall" when the maximum is used to represent this disjunction.

Exercise 5.5 (Matching, Inference, and Defuzzification: Hand Calculations): Suppose that for the tanker ship you use the membership functions in Figure 5.8 on page 166 and the rule base in Table 5.1 on page 162. Also, suppose that we have

$$e(t) = \frac{\pi}{2}$$

and

$$\dot{e}(t) = -0.0045$$

at some time t. Assume that we use the rule base shown in Table 5.1 on page 162 and minimum to represent both the premise and implication.

(a) On Table 5.1, draw boxes around the centers of the output membership functions in the body of the table that correspond to the rules that are on.

(b) Draw all the implied fuzzy sets on the output universe of discourse.

(c) Find the output of the fuzzy controller using center-average defuzzification.

(d) Find the output of the fuzzy controller using COG defuzzification.

(e) Assume that we use the product to represent both the premise and implication. Repeat (b)–(d).

(f) Write a computer program to solve (b) and (c).

Exercise 5.6 (Graphical Depiction of Fuzzy Decision Making): Develop a graphical depiction of the operation of the fuzzy controller for the tanker ship similar to the one given in Figure 5.19 on page 184. For this, choose $e(t) = \frac{\pi}{2}$ and $\dot{e}(t) = -0.0045$, which will result in four rules being on. Be sure to show all parts of the graphical depiction, including an indication of the values for $e(t)$ and $\dot{e}(t)$, the implied fuzzy sets, and the final defuzzified value.

(a) Use minimum for the premise and implication and COG defuzzification.

(b) Use product for the premise and implication and center-average defuzzification.

Exercise 5.7 (Takagi-Sugeno Fuzzy Systems): In this problem you will study the way that a Takagi-Sugeno fuzzy system interpolates between linear mappings. In particular, as an example, suppose that $n = 1$, $R = 2$, and that we have rules

$$\textbf{If } \tilde{u}_1 \text{ is } \tilde{A}_1^1 \textbf{ Then } b_1 = 2 + u_1$$

$$\textbf{If } \tilde{u}_1 \text{ is } \tilde{A}_1^2 \textbf{ Then } b_2 = 1 + u_1$$

with the universe of discourse for u_1 given in Figure 5.33 so that μ_1 represents $\tilde{A}_1^1$ and μ_2 represents $\tilde{A}_1^2$. We have

$$y = \frac{b_1\mu_1 + b_2\mu_2}{\mu_1 + \mu_2} = b_1\mu_1 + b_2\mu_2$$

We see that for $u_1 > 1$, $\mu_1 = 0$, so $y = 1 + u_1$, which is a line. If $u_1 < -1$, $\mu_2 = 0$, so $y = 2 + u_1$, which is a different line. In between $-1 \leq u_1 \leq 1$, the output y is an interpolation between the two lines.

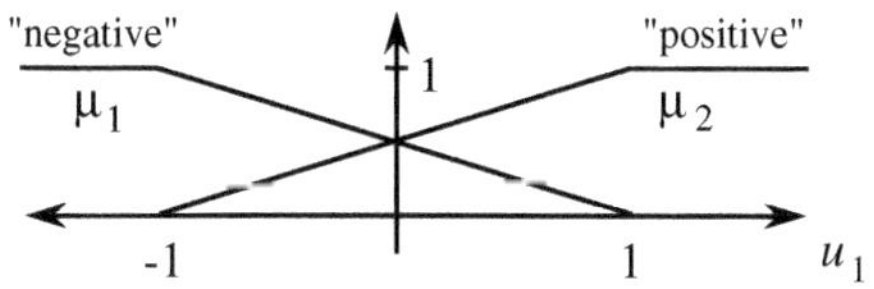

Figure 5.33: Membership functions for Takagi-Sugeno fuzzy system example.

(a) Show that the nonlinear mapping induced by this Takagi-Sugeno fuzzy system is given by

$$y = \begin{cases} 1 + u_1 & \text{if } u_1 > 1 \\ 0.5u_1 + 1.5 & \text{if } -1 \leq u_1 \leq 1 \\ 2 + u_1 & u_1 < -1 \end{cases}$$

(Hint: The Takagi-Sugeno fuzzy system represents three lines, two in the consequents of the rules and one that interpolates between these two.)

(b) Plot y versus u_1 over a sufficient range of u_1 to illustrate the nonlinear mapping implemented by the Takagi-Sugeno fuzzy system.

Exercise 5.8 (Lyapunov's Direct Method for Fuzzy Control Systems): Consider Exercise 4.3 but now suppose that you design $F(x)$ to be a fuzzy controller.

(a) Repeat parts (a)-(c) in Exercise 4.3.

(b) From the perspective of stability analysis, for this simple example, do you see any advantage of neural control over fuzzy control, or vice versa?

Design Problem 5.1 (Design of a Fuzzy Controller for Cargo Ship Steering): In this problem we study the development of fuzzy controllers for a cargo ship steering problem. Use the nonlinear model of the tanker ship provided in Equation (4.5) but with $K_0 = -3.86$, $\tau_{10} = 5.66$, $\tau_{20} = 0.38$, $\tau_{30} = 0.89$, and $l = 161$ meters [30]. Assume the rudder is saturated at ± 80 degrees as in the tanker case. Also, we will assume that the cargo ship is traveling in the x direction at a velocity of 5 meters/sec. Similar to the tanker ship, you should seek to get as good a steering response as possible.

(a) Develop a fuzzy controller for the cargo ship steering problem and simulate the closed-loop system to demonstrate its performance. Test the cases where there is a wind disturbance (assume it is modeled in the same way as for the tanker ship) and speed change.

(b) Develop a proportional-derivative (PD) controller for the cargo ship and test it under the same conditions as in (a).

(c) Compare the results in (a) and (b). Discuss.

Design Problem 5.2 (Design of a Fuzzy Controller that Balances an Inverted Pendulum): Consider the simple problem of balancing an inverted pendulum on a cart, as shown in Figure 5.34. Here, y denotes the angle that the pendulum makes with the vertical (in radians), l is the half-pendulum length (in meters), and u is the force input that moves the cart (in Newtons). We will use r to denote the desired angular position of the pendulum. The goal is to balance the pendulum in the upright position (i.e., $r = 0$) when it initially starts with some nonzero angle off the vertical (i.e., $y \neq 0$). This is a very simple and academic nonlinear control problem, and many good techniques already exist for its solution. Indeed, for this standard configuration, a simple PID controller works quite well, even in implementation. Here, you will develop a fuzzy controller for the inverted pendulum simply to gain practice in fuzzy control design.

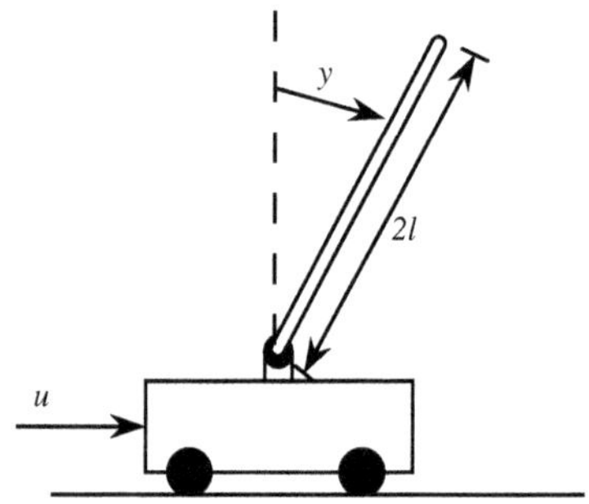

Figure 5.34: Inverted pendulum on a cart.

One model for the inverted pendulum shown in Figure 5.34 is given by

$$\begin{aligned} \ddot{y} &= \frac{9.8\sin(y) + \cos(y)\left[\frac{-\bar{u}-0.25\dot{y}^2\sin(y)}{1.5}\right]}{0.5\left[\frac{4}{3} - \frac{1}{3}\cos^2(y)\right]} \\ \dot{\bar{u}} &= -100\bar{u} + 100u. \end{aligned} \tag{5.9}$$

The first order filter on u to produce $\bar{u}$ represents an actuator. In the simulations of the fuzzy control system for balancing the inverted pendulum, be sure to use an appropriate numerical simulation technique for the nonlinear system and a small enough integration step size (e.g., a fourth-order Runge-Kutta method with an integration step size of $h = 0.001$). In your simulations, let the initial condition be $y(0) = 0.1$ radians ($= 5.73$ deg.), $\dot{y}(0) = 0$, and $\ddot{y}(0) = 0$ (this translates into an initial condition on the actuator state).

(a) Develop a fuzzy controller that uses $e = r - y$ and $\dot{e}$ as inputs, the minimum operator to represent both the "and" in the premise and the implication, and COG defuzzification. Simulate the closed-loop system and plot the output y and input u to demonstrate that your fuzzy controller can balance the pendulum. You should add scaling gains and tune the fuzzy controller as we did for the tanker ship steering problem.

(b) Repeat (a) for the case where you use product to represent the premise and implication and center-average defuzzification.

(c) Study the performance of the controllers in (a) and (b) for different initial conditions.

Design Problem 5.3 (Design and Stability Analysis of Expert Control Systems)*: This problem is based on a chapter in [410] that you should first obtain and read carefully before answering the following questions. You may also want to consult [338] for a related study.

(a) First, for the model of the tank provide a state transition diagram (circles for states, directed arrows between circles to represent plant changes for certain inputs) that represents the dynamics of the plant. Next, specify the state-transition diagram for the closed-loop system when the "seven-rule controller" is used. Also, draw the diagram for the case where the "three-rule" controller is used.

(b) Simulate the closed-loop system. In simulation, demonstrate that for each initial condition in the set $\{0, 1, 2, 3, 4, 5, 6, 7, 8, 9\}$ (initial liquid height) that the liquid level will converge to an appropriate set of values.

(c) Explain what "reachability" is, provide a mathematical definition for it, and analyze a reachability property of the tank system via simulation.

(d) Repeat the stability analysis shown in the chapter, providing full explanations at every step of the derivation, to illustrate mathematically that the closed-loop system processes the indicated stability properties (do this for both the seven-rule and three-rule controllers).

Chapter 6

Planning Systems

Chapter Contents

In the second chapter of this part we studied how fuzzy or expert systems could be used to represent human knowledge about how to perform control tasks. Essentially, they represent control software by emulating cognitive functionalities. Here, we focus on how to emulate the "software" functionality of more sophisticated reasoning strategies that use planning in order to decide how to control a plant. Since planning requires an ability to form representations (models) in the brain, exercise these representations to generate predictions about how the environment will react to various plans, choose among alternative plans, and execute a sequence of actions, it is only found in higher organisms (e.g., humans). While it certainly requires a neural network for implementation we do not focus on that; our focus here is on the functionalities basic to planning systems, and in particular, planning capabilities of humans as understood by psychologists.

Why is planning useful for control? Essentially, it is one approach that allows for more than simple reactions to what is sensed. It utilizes information about the problem and environment, often in the form of some type of model, and considers many options and chooses the best one to achieve the closed-loop control objectives. Planning provides for a very general and broadly applicable methodology and it has been exploited extensively in conventional control (e.g., in receding horizon control and model predictive control). As compared to the fuzzy and expert system approaches, it exploits the use of an explicit model to help it decide what actions to take. Like the fuzzy and expert system approaches, it is still, however, possible to incorporate heuristics that help to specify what control actions are the best to use. Hence, in a broad sense, planning approaches attempt to use both heuristic knowledge and model-based knowledge to make control decisions; this may be the fundamental reason for selecting a planning strategy over a simple rule-based one. It is often bad engineering practice to only favor the use of heuristics and ignore the information provided by a good mathematical model; planning strategies provide a way to incorporate this information.

6.1 Psychology of Planning

At an intuitive level, via introspection, you already understand what planning is. We plan our activities for the weekend, we plan a shopping trip, or plan how to solve a problem. A plan is a sequence of steps to achieve a goal, perhaps by performing tasks to achieve subgoals that then lead to the achievement of an overall goal.

6.1.1 Essential Features of Planning

Plans are typically hierarchical in that each task in sequence can often be viewed as a goal with a sequence of tasks to achieve it.

We often form "action plans" to try to achieve specific goals. For instance, consider Figure 6.1 where an "action hierarchy" is given as one type of action plan. Here, at the highest level there is the goal "eat lunch." Suppose that the person who is hungry, a professor who just got a job teaching at a university, develops a plan for how to achieve the goal to eat as the lunch hour approaches. The goal motivates the professor to pay *attention* to his hunger and pay attention to, and construct, a plan to meet his goal. This plan is formed using his past knowledge of what was successful for him when he was in graduate school, but modified somewhat due to his new role as a faculty member in a different university. In order to achieve his goal of eating lunch, he decides that he should consult the telephone directory given to him when he arrived since this may give him an idea of what restaurants are nearby. Next, since he is not familiar with any of the restaurants he found in the phone book, he asks some students and colleagues which restaurants have good food, are inexpensive, and yet have fast service. Notice then that some blocks in Figure 6.1 can be thought of as goals and tasks. Also, some of the blocks may need to be broken down further into tasks and goals. Next, the professor must pick a restaurant (based on personal tastes and priorities), find directions, choose a mode of transportation and route, and then travel to the restaurant. Hence, while a plan hierarchy may be conceptual, and as it is executed you may abstractly traverse the hierarchy, it may be that a subplan involves executing movements over a route that itself may be thought of as a planned path (e.g., the route to the restaurant). Clearly, there also may be a need for replanning, for example, if the planned route is unexpectedly blocked, or if the initial plan was in error due to someone providing bad directions. After arriving at the restaurant, the professor may execute a standard plan (a "script" available from his experience of eating at restaurants before) where he orders, eats, pays, and then returns to his office.

Learning and use of models for prediction is central to the activity of planning.

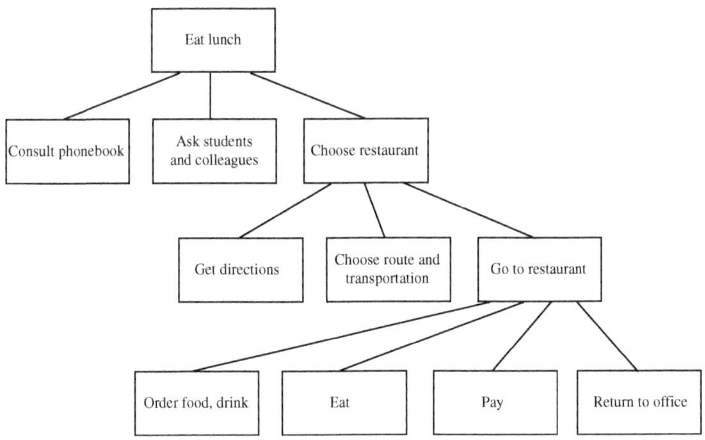

Figure 6.1: Action plan as an action hierarchy, an example.

Notice that the next day the professor's high-level goal may be the same near noon when he gets hungry, but he is likely to modify the plan based on his experience from the previous day. He may be inclined to return to the same restaurant if it was good, but may also want to sample others in order to learn whether others may be better (i.e., he may plan to try to learn more).

Clearly learning influences how we plan, and hence how the action hierarchy is formed and executed. For instance, as we learn the various routes to different restaurants, we essentially develop a "cognitive map" of the streets to get to the restaurants, and we use this map in the future (e.g., we plan over that map to minimize our travel time). We think of this "map" as a type of model that we learned that allows us to *predict* how a variety of plans will work, and hence it allows us to optimally achieve our goals by choosing the plan that minimizes travel time. This feature of exploiting past knowledge to predict (plan) ahead, and the process of choosing the "best plan," are essential features of successful planning, and flexible intelligent behavior. Moreover, the focus of attention is essential to planning, both the "internal focus" on traversing the action hierarchy, and the focus during execution of the plan to detect plan failures (e.g., observing a blocked street). Due to the hierarchical nature of the process, it seems that both planning and attention have hierarchical characteristics, and there is neurophysiological evidence of this intuition.

Optimization is essential to choose which plan is best.

6.1.2 Generic Planning Steps

While the above example serves to illustrate some of the essential features of how a human plans in one situation, it is useful to consider the following generic planning steps:

1. *Represent the problem ("planning domain"):* In order to plan, you must have some type of representation (model) of the problem that must be solved. This model could be in the form of a road map if you are trying to plan a route to get somewhere, or it could be a more conceptual map of the structural-functional characteristics of a problem. We generally think of these models as being acquired via experience (i.e., via learning), however, it is certainly the case that instincts (models passed to us via evolution) affect planning. For instance, we have certain "hard-wired" knowledge that can be thought of as aspects of models that influence planning (e.g., tendency to have a fear of snakes and some insects). Our performance in planning is critically dependent on our model of the problem. A poor model will generally lead to a bad plan, or at least to one that soon fails, thus requiring replanning. A high quality model that allows us to project far into the future (or down a hierarchy of tasks and subgoals), may lead to better plans. However, characteristics of the problem domain may make it impossible to specify a good model. For instance, time varying and stochastic features of some problem domains may make it impossible to predict into the future with any accuracy, and hence make it a waste of time to predict too far into the future. The difficulties in developing or generating a model include many of the same ones discussed in Part I for design and truth models. Differences arise however, since in planning we often learn the model as we plan.

2. *Set goal:* Setting goals is essential to planning, since without goals there is no purposeful behavior. Goals can be very different for different people,

environments, and times. Goals are driven by evolutionary characteristics (e.g., the goal of survival, the goal of reproduction), but in humans such goals can also be significantly affected by our values and ideals (e.g., ones set by culture). Goals can be learned, and can consist of a time-varying hierarchy or sequence of subgoals.

3. *Decide to plan:* Sometimes humans simply react to situations without considering the consequences of their actions. Other people decide to develop a plan since they may think that this will allow them to more successfully reach their goals. There are many issues that affect the decision of whether or not to plan (e.g., physiological and cultural). Many lower animals (e.g., some bacteria) cannot plan; they simply react to stimuli.

4. *Build a plan (select a strategy):* Normally the selection of a plan first involves projecting into the future using a model (e.g., in path planning on streets), and often involves considering a variety of sequences of tasks and subgoals to be executed (as in the action plan discussed above). In terms of a graph-theoretic view, you may think of this as a "tree" of plans where the nodes of the tree are tasks or subgoals, and links between these indicate plans (a path in the tree is a candidate plan). See Figure 6.2. How "deep" a tree to generate (e.g., how far to plan into the future) depends on the quality of the model, characteristics of the environment, and how much time or resources you have to plan. The second key component of selecting a plan is the solution of an optimization problem. For instance, suppose that the links on the "tree" that represents the set of possible plans are each labeled with integer values that represent the "cost" of performing the task represented by going in that direction in the tree. For instance, the cost may represent distance traveled or time to execute the task, and the characteristics of the cost are typically dictated by the goal. Next, suppose that the tree represents a finite number of possible plans, and that the cost of a plan is represented by summing the costs of each link that represents a step in the plan. We can then order the plans according to cost and perform minimization by picking the lowest cost plan (the "best" plan). Again, see Figure 6.2. For example, this may be the shortest route to the restaurant in the above example, if we are solving the subtask of route planning to the restaurant.

It is useful to view plan generation as forming a "tree" of possible behaviors for each plan. Plan selection involves ranking the quality of the behaviors and choosing the plan that produces the best behavior according to the model.

5. *Execute plan, monitor, and repair/replan:* After selecting a plan you must decide how to execute that plan. While we execute the plan, we monitor it by detecting deviations from what is expected to make sure that all is going well. Then, especially in an uncertain problem domain, it could be that there is a "plan failure" so that there is a need to repair the current plan, or to develop a completely new plan (the frequency of replanning is generally proportional to the amount of disturbances you have in the plant). The decision of whether to simply "tweak" the current plan, or develop a completely new one is difficult and can involve assessments of available resources (e.g., time), and the extent to which goals are being

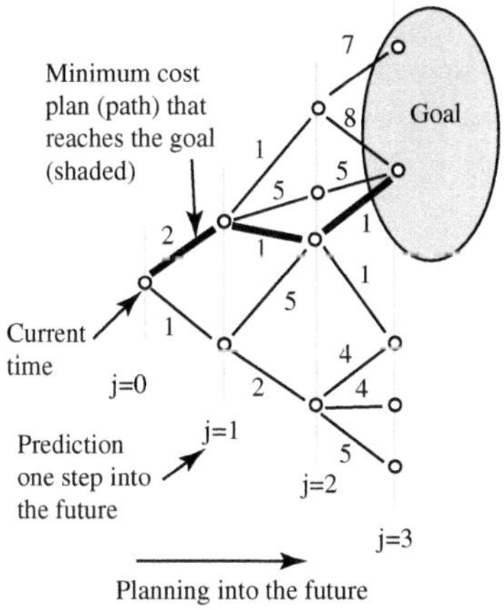

Figure 6.2: Tree representation of the alternative plans that can be considered at some point in time, along with the costs of executing such plans.

met. Some problem domains are particularly difficult to monitor and hence there may need to be a parallel process operating that estimates the "state" of the domain from available sensed information (this is sometimes called "situation assessment"). Our ability to do this depends on the "observability" properties of the problem domain (i.e., whether we can compute the state of the plant from measured inputs and outputs). When using such estimates, you may need to *guess* whether the plan is succeeding and subsequently replan.

Next, it is important to note that there are many cognitive factors that can influence how we plan. For instance, the amount of knowledge we have and our ability to learn is critical. Our current stress level, emotions, coping skills, personality, values, and self-confidence all affect our performance in planning. Moreover, the capabilities of our biological neural network in working memory affect the complexity of plans we can consider, and rate at which we can develop plans. Our attentional skills play a key role in ensuring that we stay focused on our goals, and on the most important planning task at hand.

6.2 Design Example: Vehicle Guidance

In this section we will develop a simple planning strategy for control of the position of an autonomous vehicle to move it toward a goal position (i.e., to guide the vehicle). This example only illustrates the first of several ways in which we use planning concepts for control in this book. It is primarily used to give intuitive insights into how planning strategies operate. In the next section we will explain more advanced concepts on how to design planning strategies for nonlinear dynamical systems. In Section 9.4.5 we will discuss how learning and planning can be combined in adaptive control. In Chapter 16.5 we will use basic ideas from planning systems to formulate an approach to evolutionary adaptive control. Finally, in Section 19.6 we will discuss how biomimicry of learning and planning of social foraging animals can be used in distributed coordination and control for vehicles.

6.2.1 Obstacle Course and Vehicle Characteristics

The particular type of vehicle guidance problem we will consider will be one where we seek to guide the vehicle from some initial position to a goal position while avoiding collisions with obstacles. For example, you might think of trying to guide a vehicle through the halls of your building without running into walls. We will assume that we have *perfect* information about where obstacles are, and for convenience we assume that the vehicle is in a rectangular room and that the obstacles are poles with known (x, y) positions. In particular, we consider a room such that the x-coordinate, $x \in [0, 30]$, and the y-coordinate, $y \in [0, 30]$, with the poles shown from a top view in Figure 6.3. We assume that the initial vehicle position is $(5, 5)$ and that the goal position is $(25, 25)$ as shown in the figure via the square and "$\times$" respectively.

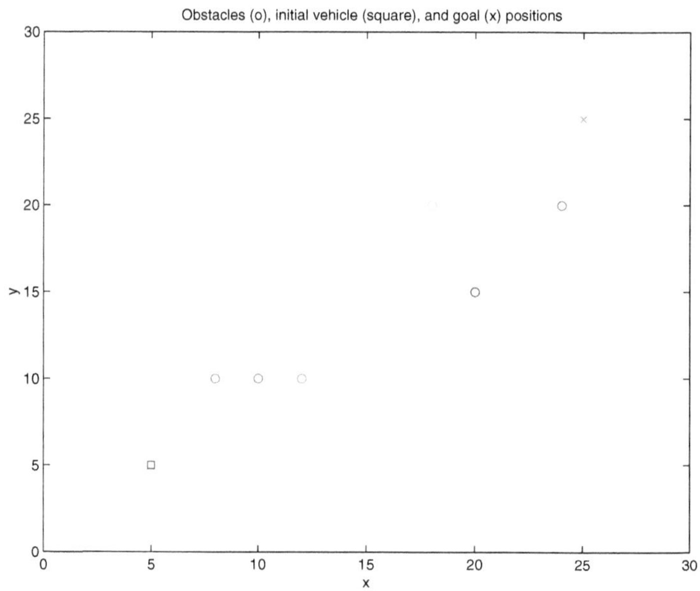

Figure 6.3: Initial vehicle position, goal position, and obstacles.

We assume that our vehicle is as shown in Figure 6.4 and that each side of the cubical vehicle measures 2.5 units so that it cannot fit in between the three poles shown in Figure 6.3 that are at positions $(8, 10)$, $(10, 10)$, and $(10, 12)$, but it can fit in between the other obstacles. We assume that the vehicle knows its own position (e.g., via an overhead computer vision system) and the goal position that it seeks to move to.

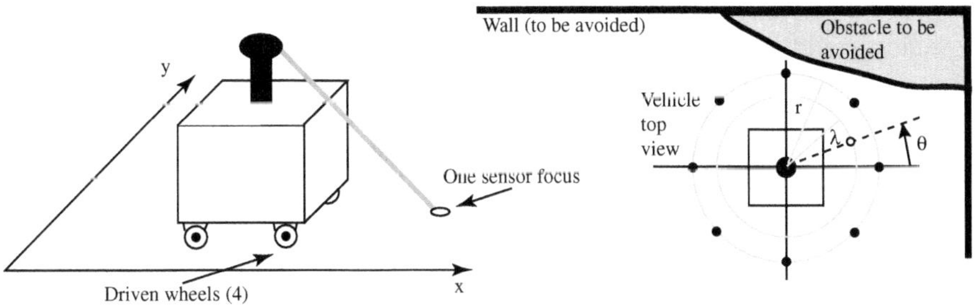

Figure 6.4: Autonomous vehicle guidance problem.

We will essentially ignore vehicle dynamics and assume that when a vehicle decides to move from one position to another position, it can approximately do so in one time step (but we do not put explicit units on distance or length of the time step). The "approximate" part is due to the fact that we assume it may not reach the precise desired position (e.g., due to inaccuracies in the vehicle drive system). In particular, if the vehicle's current position is $(x(k), y(k))$ and the onboard computer commands it to move at an angle θ a distance of λ (see Figure 6.4), it does so according to

$$\begin{bmatrix} x(k+1) \\ y(k+1) \end{bmatrix} = \begin{bmatrix} x(k) \\ y(k) \end{bmatrix} + \lambda \begin{bmatrix} \cos(\theta) \\ \sin(\theta) \end{bmatrix} + \Delta\lambda \begin{bmatrix} \cos(\Delta\theta) \\ \sin(\Delta\theta) \end{bmatrix}$$

where the sum of the first two terms on the right side of the equation represent the desired position. Here, we choose $\lambda = 0.1$. The last term is a noise term that represents effects of uncertainty that result in the vehicle not perfectly achieving the desired position. We choose $\Delta\lambda$ to be a random number chosen at each time step uniformly on $[-0.1\lambda, 0.1\lambda]$ representing that there is a 10% uncertainty in achieving the commanded radial movement. Also, we assume that $\Delta\theta$ is uniformly distributed on $[-\pi, \pi]$. Hence, when the vehicle is commanded to go to a particular position in one time step, all we know is that it ends up somewhere in a circular region of radius 0.1λ around the desired position. Notice that in order to make such movements, the vehicle needs to sample its own position at each time step. Hence, feedback control is used in the following way for guidance: the current position is sensed, and the command is made to move the vehicle to the new position. The vehicle may not end up where it was commanded to go, but at the next time instant, we will sense the vehicle's position and make adjustments from that point, and so on.

6.2.2 Path Planning Strategy

The assumption that we know exactly where all the obstacles are greatly simplifies the planning problem and allows us to focus only on some basic features of planning strategies; later we will remove this assumption and discuss how a vehicle could learn where obstacles are, plan based on that information, and even cope with moving obstacles. It should be clear that since we assume that we know where the vehicle and all the obstacles are, there is no need for a sensor that measures proximity to, or characteristics of obstacles. In a certain sense we have a perfect model of a *part* of our environment. We do not have a perfect model of the entire environment due to the uncertainty in reaching a desired commanded position that was discussed above.

Obstacle and Goal Functions

How can we represent and utilize the information given in Figure 6.3 about where the vehicle starts, where it should go, and where the obstacles are? First, since we are using a planning strategy, it is critical to realize that we need to formulate the path-finding problem as an optimization problem. To do this, we take the simple approach of constructing a surface (sometimes called a "potential field") that represents where the obstacles are. In particular, to represent the obstacles in Figure 6.3, we take Gaussian functions of unity height and center them at each of the obstacles and compute an "obstacle function" $J_o(x, y)$ that is the *maximum value* of each of those functions at each point (x, y) as shown in Figure 6.5 (the use of the maximum of the six Gaussian functions representing the six obstacles, rather than, for instance, simple addition of the six Gaussian functions, ensures that each obstacle position is represented independent of the others). In Figure 6.6 we show the contour plot of $J_o(x, y)$ along with the initial vehicle position and goal position. The contour nicely shows the "spreads" (variances) of the Gaussian functions and that there is a type of overlap such that values of $J_o(x, y)$ are at least a bit above zero for any position where the vehicle should not be in order to avoid collision with obstacles. Also, we will scale the obstacle function with a positive constant $w_1 > 0$ in our planning strategy below; however, here we choose $w_1 = 1$. Note that if you moved the vehicle about the environment in a way that the vehicle position is moved to points that try to *minimize* $J_o(x, y)$ (e.g., via hill climbing), then the vehicle will avoid the obstacles, due to the tails of the Gaussian functions. For many vehicle initial positions, the vehicle would move to the edge of the region, and when it arrives there, we always keep it on the edge.

Next, we show how to represent the goal of being at the position $(25, 25)$. To do this, suppose that we think of penalizing not being at this position by placing the minimum point of a quadratic (bowl) function

$$w_2 J_g(x, y) = w_2 \left[[x, y]^\top - [25, 25]^\top \right]^\top \left[[x, y]^\top - [25, 25]^\top \right]$$

where $w_2 > 0$ is a scale factor we choose as $w_2 = 0.0001$ (we will explain this below) that will multiply this function. The scaled function is shown in

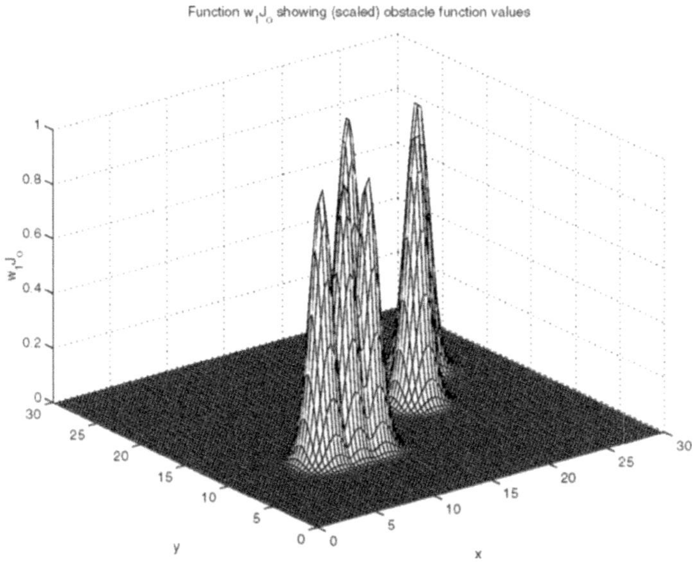

Figure 6.5: Obstacle function $J_o(x, y)$ (scaled by w_1).

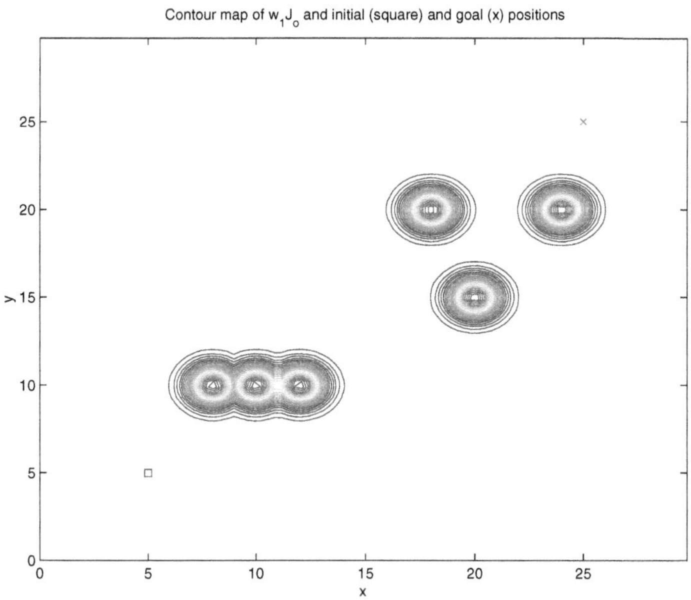

Figure 6.6: Obstacle function $J_o(x, y)$ (scaled), contour form, with initial vehicle position and goal position.

Figure 6.7 as a contour plot. If at each time step the vehicle moved to go down the surface, it will move toward the goal, but it may run into an obstacle.

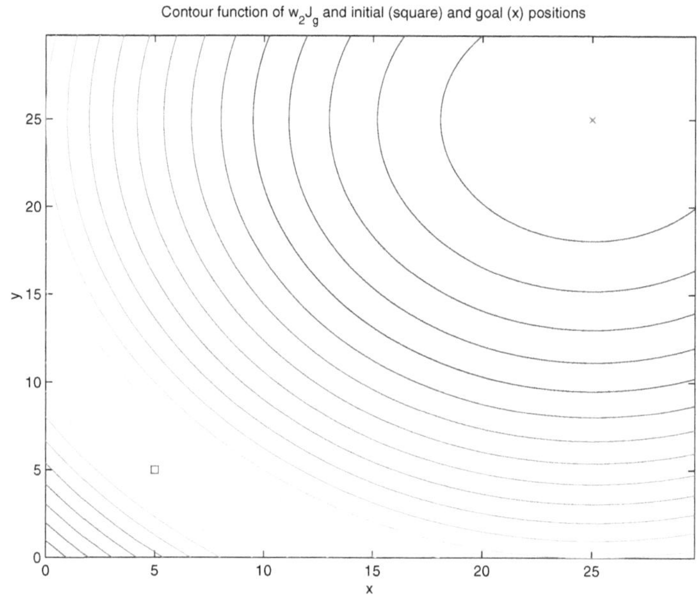

Figure 6.7: Goal function $w_2 J_g(x, y)$, contour form, with initial vehicle position and goal position.

Plan Generation and Selection

How does the planning strategy generate, evaluate, and select plans so that it can select which direction to move? To explain this, we first form our cost function for the planning strategy.

Multiobjective Cost Function: From the discussion in the previous subsection, it should be clear that if you commanded the vehicle to move a distance of λ in a direction θ that is chosen by simply moving in the "direction of steepest descent" on the function $J_o(x, y)$, then the vehicle would avoid obstacles but not reach the goal position and stay there. Similarly, if the direction was chosen to be the one with steepest descent for the $J_g(x, y)$ function, then it would move to the goal position but may collide with some obstacle for some initial vehicle positions.

Multiple goals can be represented by a multiobjective cost function.

To solve this problem we will use a "multiobjective cost function" (actually a special case where a "scalarization" approach is used to form a multiobjective cost, which is one of many ways to generate a Pareto cost)

$$J(x, y) = w_1 J_o(x, y) + w_2 J_g(x, y)$$

shown in Figure 6.8 where the weights w_1 and w_2 specify the relative importance of achieving obstacle avoidance and reaching the goal (but you must take into consideration the magnitudes of the values of each term in selecting these). Our choices of the weight values above represent that obstacle avoidance is important, but you must also keep moving toward the goal position. The choice of the weights will affect the shape of the trajectory that the vehicle will move on toward the goal position.

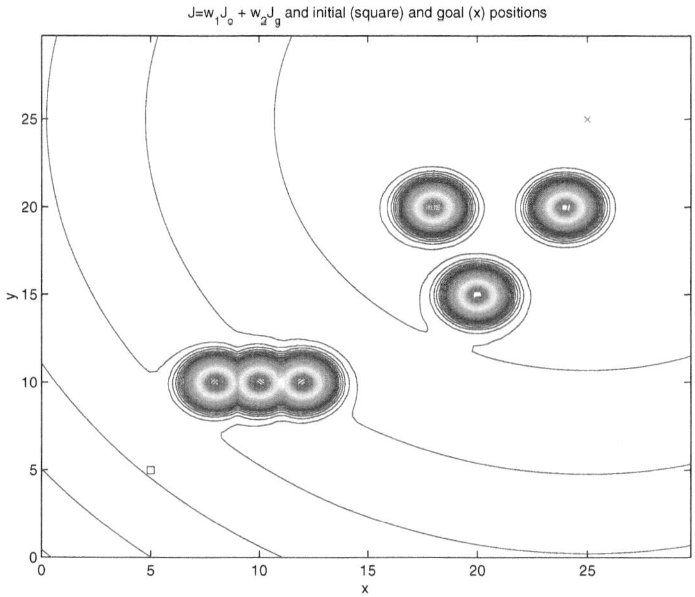

Figure 6.8: Multiobjective cost function $J(x, y)$ for evaluating plans.

Plan Generation and Selection: We take a very simple approach to plan generation and evaluation. If the vehicle is at a position (x, y), we simply compute the value of J at N_s values (x_i, y_i), $i = 1, 2, \ldots, N_s$, regularly spaced on a circle of radius r around the vehicle position (see Figure 6.4, where we have $N_s = 8$). Here, we will use $r = 1$ and $N_s = 16$. This generates 16 plans, where we "predict" one step ahead (clearly we could compute more values of J that are along other longer paths). We view the set of plans as "the vehicle is at (x, y), move it to (x_i, y_i)." We choose the plan to execute by finding a value i^* such that

$$J(x_{i^*}, y_{i^*}) \leq J(x_i, y_i), \; i = 1, 2, \ldots, N_s$$

(i.e., by finding the direction which will result in minimization of the multiobjective cost function). We then call this direction $\theta(k)$ and command the vehicle to take a step of length λ in the direction $\theta(k)$.

Notice that the above approach will approximate the "steepest descent approach" (hill-climbing) discussed above but we do not need analytical gradient information since we do not explicitly compute the gradient of the multiobjective cost function. Higher values of N_s cost more computations in plan generation and evaluation, but they also provide for more precise directional commands. Notice that using the above strategy, we expect that for any initial position on Figure 6.8, the vehicle will navigate so as to avoid the obstacles and move toward the goal by simply moving down the surface. Finally, notice that there is nothing special about the circular "pattern" of points that are evaluated on the J function. Other choices could work equally well. In fact, in Part V we will consider many other choices for the pattern of points that are used in deciding which direction to move to find the minimum point of a function (e.g., via pattern search methods), some of which are motivated by how animals search for food (a goal).

6.2.3 Simulation of the Guidance Strategy

Using the planning strategy, obstacle course, and vehicle, we get the trajectory shown in Figure 6.9. Clearly, the vehicle moves so as to avoid the obstacles (via the effect of J_o) but tries to stay on course to the goal (via the effect of J_g). The effects of the uncertainty in reaching commanded positions is seen by the small deviations on the trajectory that are "corrected" at each step since we assume that the vehicle gets a measurement of its own position at each time step. Other vehicle paths result from other choices of obstacle and goal functions and their scale factors (e.g., for this example, higher weight on the goal function tends to reduce deviations away from obstacles). Moreover, a different pattern of points where the multiobjective cost function is evaluated can result in a different path. For instance, using fewer points on the circular pattern results in trajectories that are not as smooth.

6.2.4 More Challenges: Complex Mazes, Mobile Obstacles, and Uncertainty

In this section we have studied a highly idealized planning problem. For instance, the assumption of *perfect* knowledge of the obstacle positions will not hold in any real obstacle avoidance problem. Removing the assumptions can quickly complicate the use of planning strategies, as we will see next.

Dead Ends and Circular Loops

Above, our type of obstacle course is quite simplistic. In some environments it is better to think of the obstacle course as a type of complex "maze" with many possible paths, many of which may not lead to the ultimate goal position (i.e., there may be "dead ends" or circular loops). See Figure 6.10(a). Suppose that we use the same basic approach as for our obstacle course in Figure 6.3 where we place functions that indicate that we should stay away from obstacles.

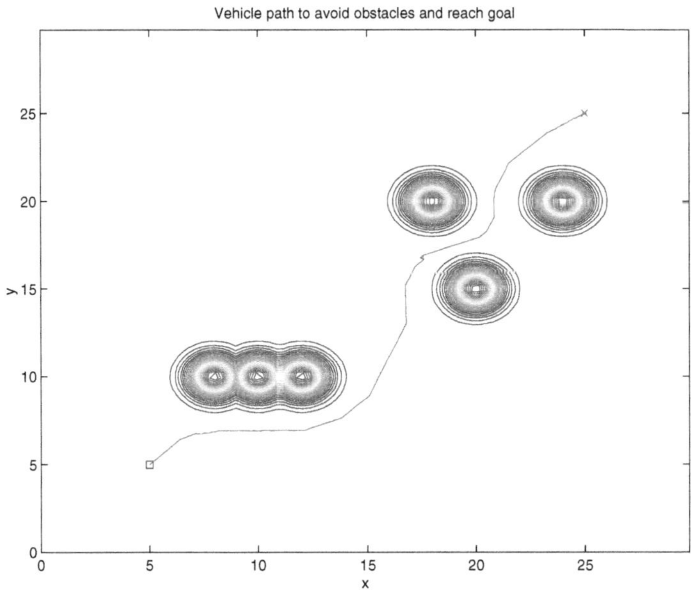

Figure 6.9: Vehicle path for obstacle avoidance and goal seeking.

How? While you could invent many types of functions, you could simply use a fine grid of appropriate Gaussian functions to get the proper shape for the $J_o(x, y)$ surface. A problem arises, however, with specifying the goal function and hence the multiobjective cost function. Suppose that we choose it as we did for the above example to be a simple quadratic function with a minimum point at the goal position. To see where the problem arises, suppose that we use the same planning strategy as in the previous subsection. In this case, it should be clear that for that obstacle course, with reasonable choices for the obstacle function and multiobjective cost function weights, the vehicle trajectory would move roughly diagonally (e.g., on paths 2, 3, or 4 in Figure 6.10(b)) toward the goal position similar to how it did in Figure 6.9 until it got to the curved wall in the "northeast" part of the maze. There, provided that r (the radius of the circular pattern of points where J is computed) is relatively small and we do not predict ahead more than one step, the vehicle will get stuck against the curved wall since it will listen to the goal function, but still try to avoid hitting the curved wall. It will get stuck at a "local minimum" on the multiobjective cost function. Notice that it does this even though if it could simply "see a little farther," it could navigate around the curved wall by going northwest, then back to the east to the goal position.

How can we solve this problem? One way is to use the a priori knowledge of the obstacle course and design the multiobjective cost function so that there is only one minimum, the global one, at the goal position. Another way is to design the obstacle and goal functions in a simplistic way as we did in the

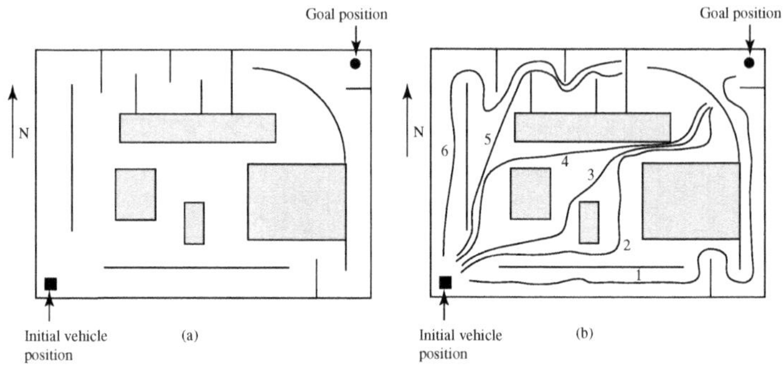

Figure 6.10: (a) Obstacle path viewed as a maze (notice dead ends), (b) Possible paths through the maze as computed by prediction in a planning strategy (numbered 1–6).

last section, then to exploit the "look-ahead" capabilities of planning strategies to find local minima on the multiobjective cost function that result in dead ends. To do this, suppose that at each step, the vehicle computer computes a tree of paths rooted at the current vehicle position (where it is assumed in the generation of that tree that the vehicle actually reaches the position desired on each move even though this will not be the case according to our model). For instance, suppose that it is constrained so that as shown in Figure 6.10(b), it computes six potential paths of the same length (length is not given by the physical length of the path, but by how many steps are taken, so in the figure each of the cases, 2–6, shows paths where the vehicle is stuck up against a wall for some time). It may come up with these potential paths by sampling the known multiobjective function, and some strategies even use minimization in the choice to limit the number of potential paths. For instance, in Figure 6.10(b) we show only six potential paths, not the many possible small deviations from these six. Next, we have to choose the best path. For this, we may use some method to detect when a plan will result in deadlock (no progress for a fixed number of steps), or we may try to minimize the number of required steps to get to the goal. The paths that are clearly unsuccessful can be eliminated from consideration and the first step suggested by the most successful plan can be taken. In the case where the maze is known perfectly and there is no uncertainty in reaching a desired position, there is no need for replanning at each step. You just follow the generated plan. However, in our model when we do not reach the commanded desired position, replanning (regeneration of plans and selection of new plans) is needed. How much replanning needs to be done? It depends on the magnitude of the uncertainty. Large uncertainty will lead to the need for frequent replanning.

Mobile Obstacles and Uncertainty

Next, note that if the obstacle environment is dynamic in the sense that, for instance, obstacles can move, our approaches require extensions. For instance, if some obstacle suddenly appeared at some position and we did not know about it, our vehicle can simply collide with it. Or if the walls and obstacles in Figure 6.10(a) moved in predictable ways, it should be clear that a "look-ahead strategy" may be needed. If the obstacles moved in unpredictable ways, then our model may not be able to accurately represent this so the vehicle will need to sense the environment while it navigates it and try to *learn* about obstacle positions and movements. Clearly this creates a very challenging obstacle avoidance problem.

6.3 Planning Strategy Design

Next, we distill the essential ideas from the psychology of planning in Section 6.1, some of which were explained via the path planning example of the last section, and show more clearly how they can be utilized in controllers for dynamical systems. Our focus here is on plants of the type that are typically considered in conventional control. First, we will think of planning systems as being computer programs that emulate the way experts plan to solve a control problem; notice the connection to how we thought of the heuristic design process for fuzzy and expert controllers. Note, however, there is an essential difference from how we thought of fuzzy and expert control: a planning system uses an explicit model of the plant. We will discuss several issues surrounding the choice of this model, plan generation, and selection. For simplicity, we will first ignore the hierarchical issues involved in planning and simply focus on how to plan at one "node" of an action hierarchy to achieve what might be a sequence of changing goals. Later in this section, however, we will discuss hierarchical planning.

6.3.1 Closed-Loop Planning Configuration

A generic planning system can be configured in the architecture of a standard control system as shown in Figure 6.11. In the context of human planning problems, the problem domain is the plant and environment. There are measured outputs $y(k)$ at step k (variables of the problem domain that can be sensed in real time), control actions $u(k)$ (the ways in which we can affect the problem domain), disturbances $d(k)$ (which represent random events that can affect the problem domain and hence the measured variable $y(k)$), and goal $r(k)$ (what we would like to achieve in the problem domain) which is called the reference input in conventional control terminology. There are closed-loop specifications that quantify performance specifications and stability requirements.

Planning (and replanning) often utilizes feedback to correct for prediction model errors.

The types of plants we consider in this section are those with

$$y(k+1) = f(x(k), u(k), d(k)) \tag{6.1}$$

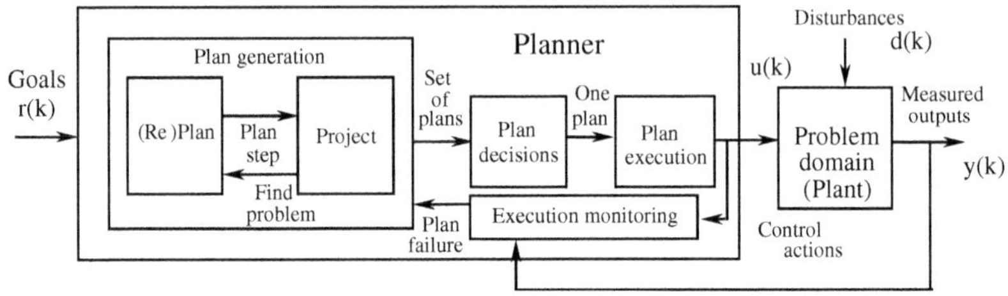

Figure 6.11: Closed-loop planning system.

where $y(k)$ is the measured output and f is a generally unknown smooth function of the state $u(k)$ and measurable state $x(k)$,

$$x(k) = [y(k), y(k-1), \ldots, y(k-n), u(k-1), u(k-2), \ldots, u(k-m)]^\top \quad (6.2)$$

Let

$$e(k+1) = r(k+1) - y(k+1)$$

be the tracking error. Generally, our objective here will be to make the tracking error as small as possible for all time, and in particular, we would like it to asymptotically approach zero so that the output follows the reference input.

Consider a plan to be a sequence of possible control inputs, where the ith plan of length N at time k is denoted by

$$u^i[k, N] = u^i(k, 0), u^i(k, 1), \ldots, u^i(k, N-1)$$

Our objective is to develop a controller that is based on a planning strategy. To do this, we will use a model and an optimization method to evaluate the quality of each plan. This will provide a ranking of the quality of the plans. After that we will choose the plan that is best (call it plan i^*), and let the control input at each time instant k be

$$u(k) = u^{i^*}(k, 0)$$

That is, at each time k we choose the best plan $u^{i^*}[k, N]$, then use the first input from the control sequence as the input to the plant. The process is repeated at each time instant. Clearly, you could use a lower frequency of replanning, where, for example, you could generate a new plan every other sampling instant, and execute the first *two* inputs from the optimal plan each time.

6.3.2 Models and Projecting into the Future

Good models lead to good plans; bad models can lead to unstable behavior and poor performance.

There are a wide range of possibilities for the type of model that is used, and the type depends on the problem domain, the capabilities of the planner to store and use the model, and also the goals. For instance, a model used for planning could be continuous or discrete (e.g., a differential or difference equation), and it could be linear or nonlinear. It may be deterministic, or it may contain an explicit representation of the uncertainty in the problem domain so that plans can be chosen taking the uncertainty into account. In industrial practice, in the so-called "model-predictive control" (MPC) method, linear models are often used for the plant and this approach is considered in Design Problem 6.2.

Just like the design model used for control design, it will *always* be the case that the model will not be a perfect representation of the plant and the environment. This implies that there will always be uncertainty in planning, and hence there will always be a bound on the amount of time that it makes sense to project (simulate the model) into the future. Projecting into the future too far becomes useless at some point since the predictions will become inaccurate at some point, and hence provide no good information on how to select the best plan. The difficulty is knowing how good your model is and how far to project into the future. Finally, note that you may actually want your model to be able to predict what goal is going to occur in the future since in the formulation in this section we could have a time-varying goal. If the goal can be predicted, contingencies can be developed, and earlier plans may be modified to try to ensure success for not only the current goals, but anticipated ones.

No model is perfectly accurate; hence, predictions based on it are always in error.

Here, we use a general nonlinear discrete time model

$$y_m(j+1) = f_m(x_m(j), u(j))$$

with output $y_m(j)$, state $x_m(j)$, and input $u(j)$ for $j = 0, 1, 2, \ldots, N-1$. Notice that this model can be quite general if needed; however, in practice, sometimes a linear model is all that is available and this may be sufficient. Let $y_m^i(k, j)$ denote the jth value generated at time k using the ith plan $u^i[k, N]$; similarly for $x_m(k, j)$. In order to predict the effects of plan i (project into the future) at each time k you compute for $j = 0, 1, 2, \ldots, N-1$,

$$y_m^i(k, j+1) = f_m(x_m(k, j), u^i(k, j))$$

At time k to simulate ahead in time, for $j = 0$ you initialize with $x_m(k, 0) = x(k)$. Then, generate $y_m(k, j+1)$, $j = 0, 1, 2, \ldots, N-1$, using the model (note that you will need to appropriately shift values in x_m at each step) and generate values of $u^i(k, j)$, $j = 1, 2, \ldots, N-1$, for each i.

6.3.3 Optimization Criterion and Method for Plan Selection

Next, the set of plans (strategies) is "pruned" to one plan that is the best one to apply at the current time (where "best" can be determined based on, e.g., consumption of resources). Hence, optimization is central to the activity of planning (just as we will later see that optimization is central to the activities of attention, learning, evolution, and foraging). The specific type of optimization approach that is used for plan selection depends on the goals, cost function, and type of model that is used to predict into the future. For instance, if the model of the plant is a finite automaton the optimization problem can in some cases be formulated as a "shortest path" problem where you choose the plan (sequence of actions) that results in minimizing a cost function (e.g., the sum of the costs for the steps of a candidate plan) in a manner similar to how we choose the best plan for Figure 6.2. Such a shortest path problem can be solved with a number of methods. For instance, you could use dynamic programming or standard combinatorial optimization methods. Alternatively, when the state space is large it may be advantageous to use some "heuristic search" methods, such as the A^* algorithm. For more details on such approaches, see the "For Further Study" section at the end of this part.

Criteria for Selecting Plans

We need a criterion to decide which plan is the best. Here, we will use a cost function $J(u^i[k,N])$ that quantifies the quality of each candidate plan $u^i[k,N]$ using the f_m model. First, assume that the reference input $r(k)$ is either known for all time, or at least that at time k it is known up till time $k+N$. Generally, you want the cost function to quantify over the next N steps how well the tracking objective is met. One cost function that we could use would be

$$J(u^i[k,N]) = w_1 \sum_{j=1}^{N} \left(r(k+j) - y_m^i(k,j)\right)^2 + w_2 \sum_{j=0}^{N-1} \left(u^i(k,j)\right)^2 \tag{6.3}$$

where $w_1 > 0$ and $w_2 > 0$ are scaling factors that are used to weight the importance of achieving the tracking error closely (first term) or minimizing the use of control energy (second term) to achieve that tracking error. Other cost functions could use the output of a "reference model" as we do for several adaptive control approaches in Part III (see "For Further Study" for more details), an error measure on the other past values of the inputs and outputs, or an error measure on some other system variable. The choice of the cost function for evaluating the quality of the plans is application-dependent. To specify the control at time k you simply take the best plan, as measured by $J(u^i[k,N])$, and call it plan $u^{i^*}[k,N]$ and generate the control using $u(k) = u^{i^*}(k,0)$ (i.e., the first control input in the sequence of inputs that was best).

Note that this specifies a variety of methods to achieve what is called "model predictive control" (or "receding horizon control") in conventional control theory. Clearly, different models, cost functions, and optimization methods will lead to different closed-loop system performance characteristics. It can be difficult to know which optimization method to choose for a particular application. Often, however, practical aspects of the problem govern many aspects of the choice as we discuss next.

Nonlinear Optimization for Plan Selection

The challenge is to pick an optimization method that will converge to the optimal plan, and one that can cope with the complexity presented by the large number of candidate plans. Why is this a "challenge"? First, focusing on the complexity aspect, note that the inputs and states for the plant under consideration can in general take on a continuum number of values (i.e., an infinite number of values, with as many possibilities as there are real numbers), even though in particular applications they may only take on a finite number of values. This is the case in analog control systems even considering actuator saturation. For digital control systems you may have a data acquisition system that results in a certain quantization and hence, theoretically speaking there are a finite number of inputs, states, and outputs, for the model specified by f_m since it is typically simulated on a digital computer. However, this number can be *very* large! There is then, in general, an infinite (continuum) number of possible plans that you must compute the cost of (in a brute-force approach) in order to form the ranking of plans according to cost, and select the best plan.

But, in the conventional model predictive control approach that is widely successful in industrial applications, this problem has been solved. How? There, most often linear plant models are used, a manageable size is chosen for the prediction horizon N (and perhaps a longer sampling interval is used for the model, than for the digital controller), and it then becomes feasible to specify an *analytical* solution to finding the optimal plan (sequence of inputs). The optimization approach can actually choose the best plan from the infinite set of plans. That analytical solution is the so-called "least squares" solution that is only possible due to the use of the linear model. But, of course, no real plant is linear (even though it may act as though it is almost linear in some situations).

Nonlinear or combinatorial optimization can be used for plan selection.

What if the nonlinear and uncertain characteristics dominate to the extent that a linear model is not sufficient for generating plans? Then, we could use a nonlinear model in the planner and try to employ some type of nonlinear optimization method where the "parameters" that are adjusted by the optimization method are ones that parameterize the infinite set of possible plans. Practically speaking, however, this can become problematic since if you use a nonlinear model for plan generation, you are confronted with a nonlinear optimization problem for which there is generally no analytical solution. There are, however, many algorithms that one could employ to try to solve this problem (e.g., steepest descent, Levenberg-Marquardt, etc., that are discussed in Part III). The problem is that none of these methods *guarantees* convergence to an optimal plan. They could even diverge and provide no solution, but typically they will converge to a local minimum. The plan that results from such a nonlinear optimization process cannot then be guaranteed to be the optimal one, and closed-loop performance can suffer. Having said all that, it is worthy to note, however, that in some practical industrial problems, engineers have managed to develop effective solutions via such a nonlinear optimization approach.

Brute-Force Approach to Plan Selection

Next, suppose that you do not want to take the standard nonlinear optimization approach, yet you want to use a nonlinear model since its use seems essential to represent the salient features of your plant. Is there another approach? One standard approach is to discretize the input, state, and output spaces, generate all possible plans and compute the cost of each of them explicitly (sometimes it is even possible to simultaneously generate plans and evaluate costs, and thereby greatly reduce the number of potential plans since ones that are of very high cost may not need to be generated). Creating such a discrete model is not a trivial exercise since you want it to be not only discrete in time, but also in space. The discretization typically (virtually always) leads to the creating of a less accurate model so that in taking this approach you are trading off complexity management and optimization ease with accuracy in evaluating the plans. Also, unless you use a very coarse quantization you may still end up with too many plans to consider. Why? Suppose that there are N_u possible input values obtained via discretization, and that the model is deterministic so that one control input leads to only one possible state, then there are

Creation and evaluation of all possible plans is often computationally prohibitive.

$$(N_u)^N$$

possible plans at each time k. Suppose that we simulate ahead in time $N = 100$ steps, and $N_u = 1000$ (not unreasonable considering the types of levels of discretization that could be accurate for many plants). Clearly, due to the exponential growth in the number of plans, we can quickly encounter problems with computational complexity if we take the brute-force approach of generating all possible plans. Moreover, even if we generate all the plans, we will also have to evaluate the cost of each one. And, this must be done at each sampling instant. Having said all that, however, there are classes of problems where a discrete model provides a reasonably good representation of the plant, even with a small N_u, and sometimes only a small N is needed to evaluate the quality of a plan. In this case, the brute-force approach may work very well. Besides, specific application-dependent characteristics often allow you to "prune" the tree of possible plans. For instance, if you have rate constraints on your plant, then typically for every state only certain inputs are possible, since the input cannot change too much from what it was the last time. Moreover, sometimes coarser quantizations in time and space may work adequately for some plants.

There are ways to trade off computational complexity for the quality of plan selection and ultimately, performance.

6.3.4 Planning Using Preset Controllers and Model Learning

Next, we will discuss another approach to solve the complexity and optimization challenges involved in plan generation and selection. This approach can be thought of as a method to prune the tree of possible plans that is generated at each sampling instant.

Planning Using Multiple Controllers

Consider a specific controller (a "preset" controller) applied to the current state and reference input to be a type of "plan template" in that it specifies one way to respond for a sequence of times into the future, but the precise manner in which it generates inputs depends on what occurs over time as the plan is implemented. There is an analogy with how humans plan. In some problem domains we may have learned a finite set of possible *approaches* to solve a problem and we start solving it, picking what seems to be the best approach at each step.

Suppose that there are S such plan templates, which have the form of functions F_u^i

$$u^i(k,j) = F_u^i(x(k,j), r(k+1)) \; i = 1, 2, \ldots, S$$

where we assume we can measure $r(k+1)$. Hence, at each step we take each of these S plans and project into the future how each will perform, pick the best one, then let the control input be $u^{i^*}(k,0)$ where i^* is the best plan as measured by some cost function. For some practical applications the value of S need not be too large, and hence, if we take the "brute-force" approach of the last section, we overcome the problems discussed there in complexity and optimization.

In a related approach, it is also possible to use planning systems as general supervisory controllers in a similar manner to how expert controllers are used for supervision. In this case, the planner will, for instance, coordinate the use of a set of controllers where different controllers are used for different operating conditions. We will discuss such methods in Sections 9.4.5 and 16.5.

Planning Using Multiple Models or Tuned Models

Suppose that upon entering some problem domain you know that it is best modeled by one of S models that you have learned. Suppose that as you begin taking actions in the problem domain, you gather information that tells you which model is most appropriate at the current time. If you enter it at a different time, a different model may be more appropriate. Also, some environments are dynamic in that their characteristics change over time so that as you are taking actions in the domain with one model, you continually monitor the quality of the predictions it makes, and if appropriate, you can switch to another model. How do you plan with the model possibly switching at each time? You can do it just the same as discussed above. You simply change the model that you predict with over time. You can think of this as *learning* the appropriate type of model and using it to plan (the optimization method employed to select the model is implementing a type of learning).

Other planning systems may perform "world modeling," where a model of the problem domain is developed or modified (tuned) in an online fashion (similar to online system identification), and "planner design" uses information from the world modeler to tune the planner (so that it makes the right plans for the current problem domain). The reader will, perhaps, think of such a planning system as a general adaptive controller. It integrates learning of models directly into the planning process, in a manner reminiscent of how humans learn while planning. While we will not illustrate the operation of such strategies in this chapter, in Part III and Part IV we will discuss how to use such strategies in adaptive control.

6.3.5 Hierarchical Planning Systems

First, suppose that there is a hierarchy of models available for generating plans. To provide a simple illustration of some key ideas in hierarchical planning, suppose that we are performing route planning for a mobile robot at an industrial complex that has several buildings. Moreover, suppose that we organize our planner according to the description of the hierarchy in Figure 1.11 where we have a higher-level management level, and lower-level coordination and execution levels.

Suppose that we have several models, detailed ones of each room in every building, maps of each building that simply show how the rooms are connected via hallways, and maps of possible connection routes between the buildings. Suppose that we want to plan how to move the mobile robots around the industrial complex. Suppose that at the highest level the human operator specifies that the robot should go to building 3, room 416, to deliver a part that is needed in some manufacturing process. A planner at the management level could generate a set of routes between buildings and pick the best one considering other traffic and minimization of time of travel. A planner at the coordination level could be used to plan how to move to the desired room once the building is reached, and the planner at the lower level could specify how to navigate the room.

There are other types of hierarchical planners that will use multiple planners at the coordination and execution levels. For instance, sometimes the goals specified by the human can be broken down into multiple sequences of tasks at the management level, each one representing a different way to reach the human-specified goal. One approach could be selected and passed to the coordination level. At the coordination level we could view the sequence of tasks chosen at the management level as a sequence of goals, and each planner at the coordination level may then develop sequences of operations to try to achieve those (sub)goals. Clearly this sets up a recursion and we can view the chosen coordination level as plans, and the execution level can view those as goals and develop plans to meet them. Implementation is achieved by executing the low level sequences that try to meet the subsubgoals, and thereby the subgoals, and hence the goal specified by the human.

There are many design issues involved in constructing such a hierarchical planning system. For instance, the accuracy of the models at the various levels and the form of the cost functions used will significantly affect the performance of the system. Computational complexity is affected by the choice of the planning horizons at the various levels, and the lengths of these horizons is in turn affected by the quality of the models we use in planning (and uncertainty in the environment). Moreover, one approach to coping with computational complexity in some planning applications is to split the planning problem into a hierarchical functionality, since sometimes this can simplify plan generation and evaluation. Finally, we note that it is possible to incorporate adaptation and learning into the planning processes at the various levels.

6.3.6 Discussion: Concepts for Stable Planning

It is possible to perform stability analysis of control systems whose controller uses a planning strategy; in such cases you may study, for example, convergence of tracking error. For instance, there has been extensive work on the study of stability conditions (e.g., in terms of horizon length) for conventional linear model predictive control methods. Moreover, there has been other work focusing on stability of planning systems for plants with a discrete event character (see Design Problem 6.3). These studies show that there are several essential characteristics that affect stability properties, several of which can be thought of in terms similar to the discussion in Section 6.2.4, where we discussed dead ends, circular loops, obstacle mobility, and obstacle position uncertainty:

Stability analysis of closed-loop planning strategies depends critically on model accuracy, plant uncertainty, and plant nonlinearities.

- *Model accuracy:* The accuracy of the model used to project into the future significantly affects the analysis. In most analysis it is assumed that a *perfect* model is known or that the model perfectly represents all possible ways that the plant will respond to inputs.
- *Navigating through uncertainty:* Your ability to achieve a goal state in a tree of possible paths that are simulated (e.g., as shown in Figure 6.2) depends on the uncertainty present in the plant. You can think of the uncertainty as a type of adversary, and that your objective is to keep moving in directions so that the uncertainty will not over time conspire to make it impossible for you to navigate to your goal state (in terms of Figure 6.2, the actual structure of the tree is random so at some points in time some paths may lead to the goal with a certain cost, while at other times the cost may increase/decrease, or may not even lead to the goal state). The planning strategy tries to navigate the tree in a way so that even though the plant may make unpredictable moves, it will not be able to make moves that will make it impossible to reach the goal. Clearly, the number of steps you project in the future can critically affect your ability to navigate through the uncertainty. If you do not look far enough into the future, for some plants it may be possible that you will enter a region of the state space such that the effects of the uncertainty dominate and there is no way to navigate out of that region and to the goal state (e.g., in Figure 6.2, note that there are some "dead-ends" in the tree that is shown). On the other hand, it may not make sense to project more than one or two steps into the future for some plants since longer projections may neither result in better plans, nor help navigate through the space.
- *Avoiding traps:* For some plants, without projecting far enough into the future, it may be possible to get "trapped" in a cycle where you repeatedly visit a finite sequence of states on a loop. Moreover, it is of course possible that such circular traps arise in a nondeterministic manner, essentially combining the concerns of the last point with those of this one (i.e., random dead-ends and cyclical traps can arise).

The above discussion is simply intended to provide the interested reader with some intuitions about some issues that significantly affect our ability to perform stability analysis of planning systems for some classes of plants. For further study on this topic, see Design Problem 6.3 and the "For Further Study" section at the end of this part.

6.4 Design Example: Planning for a Process Control Problem

In this section we will develop a planning strategy for a very simple yet representative process control problem. We begin by introducing the control problem and then we design and test a planning strategy.

6.4.1 Level Control in a Surge Tank

Consider the "surge tank," shown in Figure 6.12, that can be modeled by

$$\frac{dh(t)}{dt} = \frac{-\bar{d}\sqrt{2gh(t)}}{A(h(t))} + \frac{\bar{c}}{A(h(t))}u(t)$$

where $u(t)$ is the input flow (control input), which can be positive or negative (it can both pull liquid out of the tank and put it in); $h(t)$ is the liquid level (the output of the plant); $A(h(t)) = |\bar{a}h(t) + \bar{b}|$ is the cross-sectional area of the tank and $\bar{a} > 0$ and $\bar{b} > 0$ (their nominal values are $\bar{a} = 0.01$ and $\bar{b} = 0.2$); $g = 9.8$; $\bar{c} \in [0.9, 1]$ is a "clogging factor" for a filter in the pump actuator where if $\bar{c} = 0.9$, there is some clogging of the filter and if $\bar{c} = 1$, the filter is clean so there is no clogging (we will take $\bar{c} = 1$ as its nominal value); and $\bar{d} > 0$ is a parameter related to the diameter of the output pipe (and its nominal value is $\bar{d} = 1$). We think of all these plant parameters as being fixed (but unknown) for a particular surge tank; however, we could consider other values for these parameters and test the controller for these. This models the situation where you want to develop one controller for many different surge tanks.

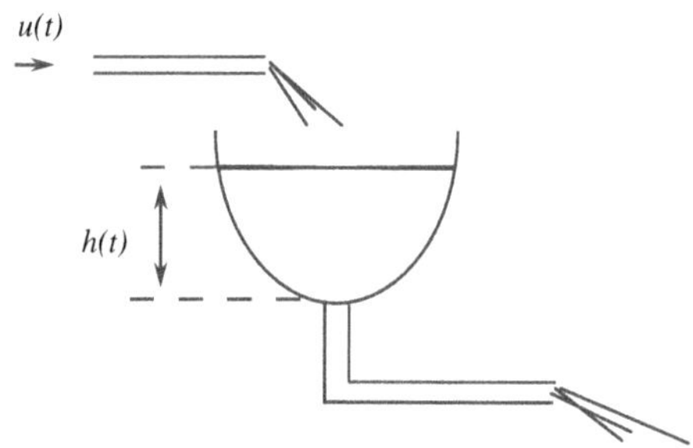

Figure 6.12: Surge tank.

Let $r(t)$ be the desired level of the liquid in the tank (the reference input) and $e(t) = r(t) - h(t)$ be the tracking error. Assume that you know the reference trajectory a priori and assume that $r(t) \in [0.1, 8]$ and that we will not have $h(t) > 10$. Assume that $h(0) = 1$.

To convert to a discrete-time approach, use an Euler approximation to the continuous dynamics to obtain

$$h(k+1) = h(k) + T\left(\frac{-\bar{d}\sqrt{19.6h(k)}}{|\bar{a}h(k)+\bar{b}|} + \frac{\bar{c}}{|\bar{a}h(k)+\bar{b}|}u(k)\right)$$

where $T = 0.1$. We assume that the plant input saturates at ± 50 so that if the controller generates an input $\bar{u}(k)$, then

$$u(k) = \begin{cases} 50 & \text{if } \bar{u}(k) > 50 \\ \bar{u}(k) & \text{if } -50 \leq \bar{u}(k) \leq 50 \\ -50 & \text{if } \bar{u}(k) < -50 \end{cases}$$

Also, to ensure that the liquid level never goes negative (which is physically impossible), we simulate our plant using

$$h(k+1) = \max\left\{0.001, h(k) + T\left(\frac{-\bar{d}\sqrt{19.6h(k)}}{|\bar{a}h(k)+\bar{b}|} + \frac{\bar{c}}{|\bar{a}h(k)+\bar{b}|}u(k)\right)\right\}$$

Note that all the simulations in this section will include these constraints.

6.4.2 Planner Design

Here, for the sake of illustration we will use a nonlinear discrete-time model for the nonlinear discrete-time plant (the "truth model"). We will generate candidate plans using this model using the "preset controllers" approach discussed in the last section.

Taking the model of the last subsection as the truth model for the plant, the model that we will use in our planning strategy will have

$$A(h(t)) = \bar{a}_m (h(t))^2 + \bar{b}_m$$

with $\bar{a}_m = 0.002$ and $\bar{b}_m = 0.2$. For the model we use the same nonlinear equations as given in the last section, but we do not assume that we know the values of $\bar{c}$ and $\bar{d}$, so for these we use $\bar{c}_m = 0.9$ and $\bar{d}_m = 0.8$. It is interesting to note that if you plot the cross-sectional area of the actual plant, and the one used in the model, you get Figure 6.13, so you can see that they are somewhat different so that our model is clearly not the same as the plant (model).

So, is the model accurate enough to be used in projection? To answer this question we develop a simple controller and test it on the plant and controller. We use a proportional integral (PI) controller as the "plan template." In particular, if $e(k) = r(k) - h(k)$, we use

$$u(k) = K_p e(k) + K_i \sum_{j=0}^{k} e(j) \tag{6.4}$$

Suppose that the goal is to get a reasonably fast response, with no overshoot in the tracking error $e(k)$.

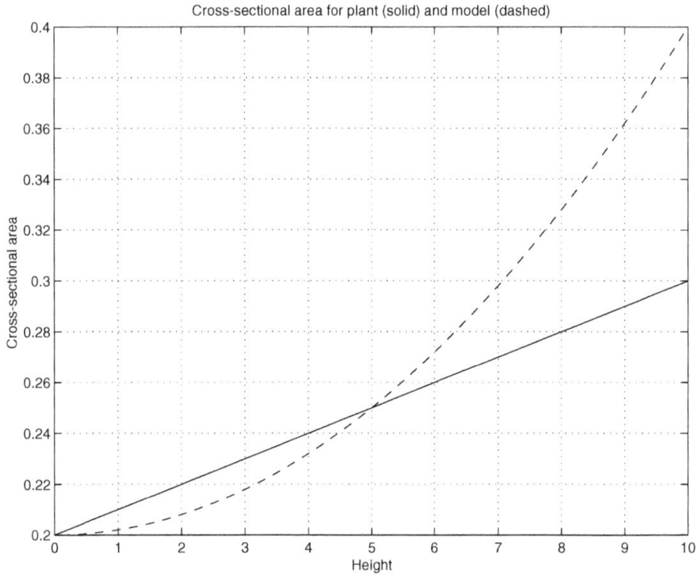

Figure 6.13: Cross-sectional area $A(h)$ for the plant (solid) and model to be used for projection (dashed).

Suppose that via experience in designing PI controllers for surge tanks with various cross-sectional areas, you know that typically

$$K_p \in [0, 0.2]$$

and

$$K_i \in [0.15, 0.4]$$

For instance, if you pick $K_p = 0.01$ and $K_i = 0.3$ and use the PI controller in Equation (6.4), you get the response in Figure 6.14. Notice that while the response is relatively fast, there is overshoot and that is undesirable.

You actually get a similar response if you use the same gains for the above model that will be used for projection in our planning strategy. To see this consider Figure 6.15, where we see that the difference between the regulated heights for the cases where we use the truth model for the plant, and where we use the projection model, are relatively small (there is more overshoot when the controller is used for the model rather than the plant). This gives us some confidence that our model is reasonably accurate; but of course to properly evaluate its accuracy, we need to consider how good a performance we can obtain when we use the model in a planning strategy for projection, and at the same time, use the truth model in the closed-loop.

We use the cost function in Equation (6.3) with $N = 20$ (for two seconds projection into the future), $w_1 = 1$, and $w_2 = 1$. Also, we assume at each

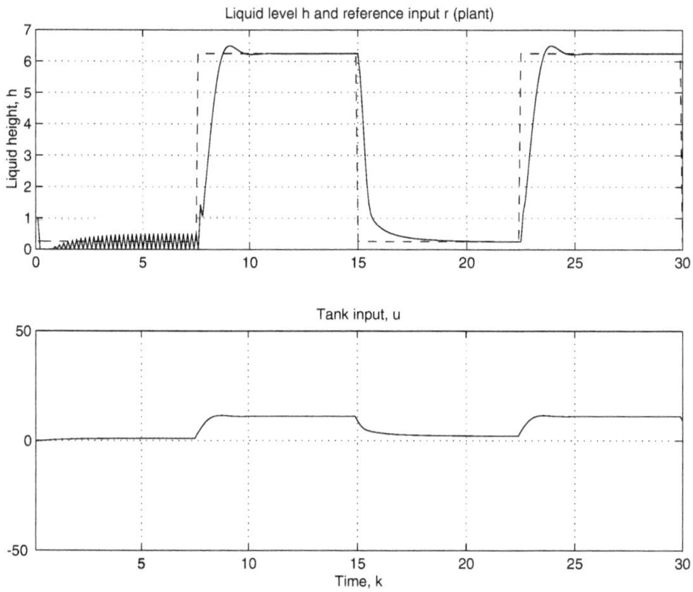

Figure 6.14: Closed-loop behavior of the surge tank using a PI controller.

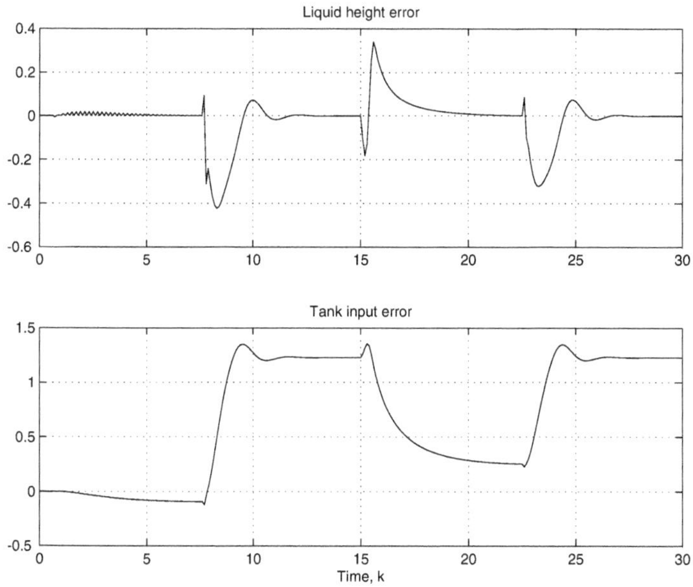

Figure 6.15: Error between cases where the truth model and projection model are used as the plant.

time instant that the reference input remains constant while we project into the future; this is equivalent to assuming that our evaluation of which controller is best is based on the reference input being constant.

Our plan templates are the PI controllers, with different values of K_p and K_i. In fact, we simply create a grid on the above ranges for the gains by considering all possible combinations of

$$K_p \in \{0, 0.05, 0.1, \ldots, 0.2\}$$

and

$$K_i \in \{0.15, 0.2, \ldots, 0.4\}$$

Hence, in this case there are $5 \times 6 = 30$ different plans (controllers) that are evaluated at each time step. To do this evaluation, we simulate using the projection model into the future two seconds for each PI controller. We initialize the simulations into the future with current error, and integral of the error.

6.4.3 Closed-Loop Performance

To see how the planning strategy operates, see Figure 6.16. Here, we see that we get a slower rise-time than in Figure 6.14 when we used the PI controller, but that we were able to tune the planning strategy (by adjusting w_1, w_2, and the grid on the PI gains) so that there is no overshoot, and still a reasonably good rise-time, and that was our main objective.

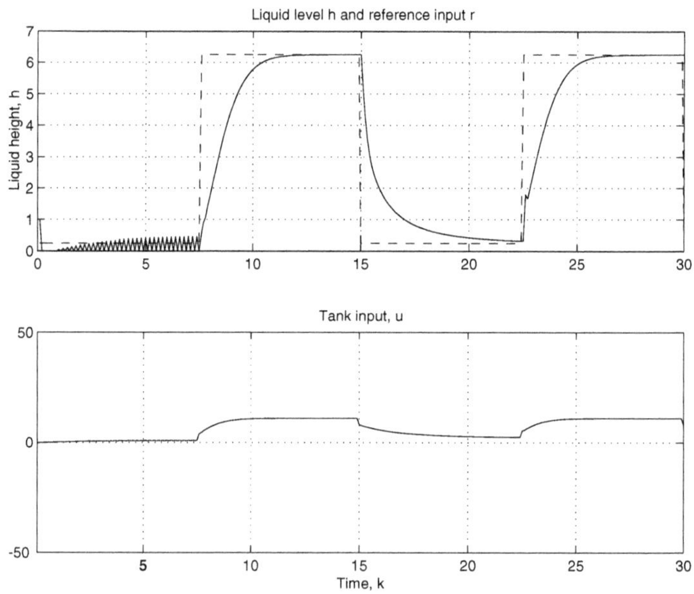

Figure 6.16: Closed-loop behavior of the surge tank using a planning strategy.

How does it achieve this performance? It switches controllers online and to see this, consider Figure 6.17. Note that we define the indices so that they are proportional in size to the K_p and K_i values (e.g., the $(1, 1)$ controller has $K_p = 0$ and $K_i = 0.15$) so that it seeks to increase the K_p value to reduce tracking error and get a good rise-time, and lowers the K_i value to try to reduce overshoot. If you choose different values of the planning horizon N, you will get different switching sequences.

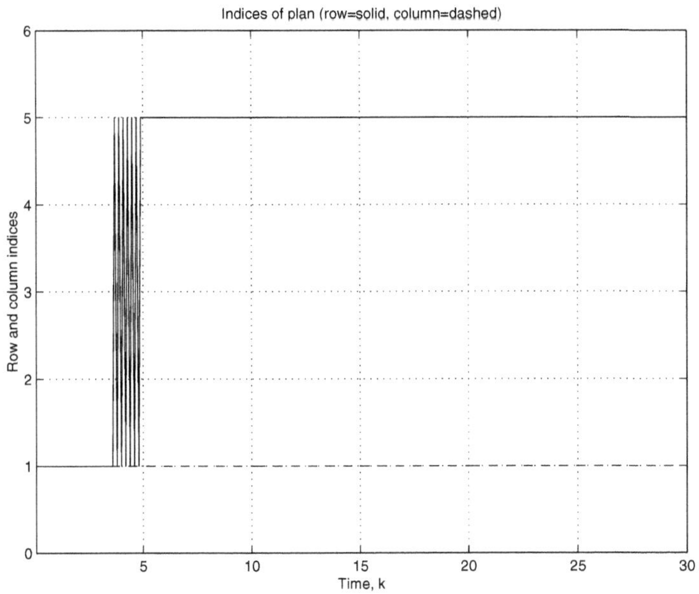

Figure 6.17: Indices of PI controllers that are used at each time step for the tank.

Similar performance to that shown above is found if you perturb some of the plant parameters. For example, if for the plant you let $\bar{c} = 0.8$ (representing more clogging), you get similar results to the above. Or, if you use the nominal value for $\bar{c}$ and use $\bar{a} = 0.05$, you get the cross-sectional area shown in Figure 6.18 and we get the closed-loop response in Figure 6.19.

Notice that while we still get an adequate rise-time, for this plant the planning strategy results in a small amount of overshoot; hence, you may want to tune the planner in order to improve the response. This shows that while the planning strategy may provide good performance for some plants, for some others the performance can degrade (not surprising). How robust is the controller to plant perturbations? It can be a challenging problem to design a single planning strategy that will perform adequately for all plants of a certain class (e.g., for a known set of structured perturbations about the nominal plant).

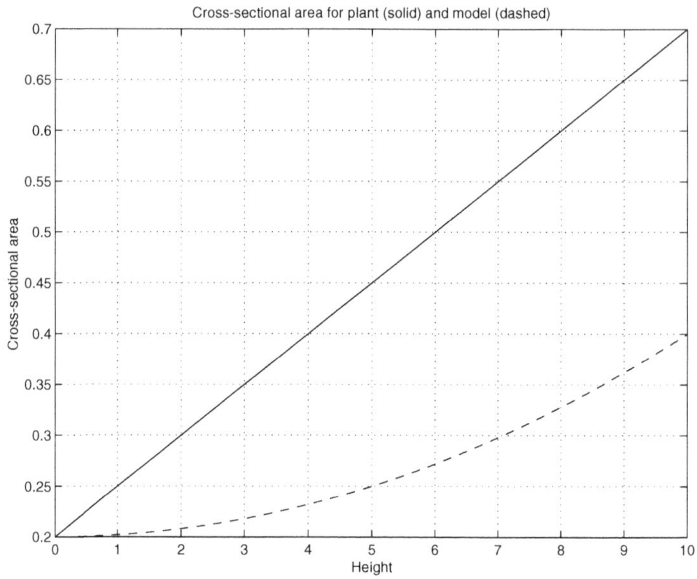

Figure 6.18: Cross-sectional area $A(h)$ for the plant (solid) and model to be used for projection (dashed).

6.4.4 Effects of Planning Horizon Length

Next, we return to using the parameters for the nominal plant and study the effect of changing the projection length N with all the same choices as in the previous subsection. In particular, we plot the tracking energy

$$\frac{1}{2}\sum_{k}(e(k))^2$$

and control energy

$$\frac{1}{2}\sum_{k}(u(k))^2$$

vs.

$$N \in \{1, 5, 10, 15, 17, 20, 25, 30, 33, 35, 36, 37, 38, 39, 40, 45, 50\}$$

as shown in Figures 6.20 and 6.21. This range of N was chosen by adding more points where the values of the tracking and control energy changed fast.

These plots show some justification for the choice of $N = 20$ in our earlier simulations. This choice did not cost too much computational complexity in projecting into the future, and yet gave a low tracking error (our main objective), with a reasonable amount of control energy. If you are not concerned about computational complexity, you may want to further increase the planning horizon, to get a similar value for the tracking energy, but with even lower

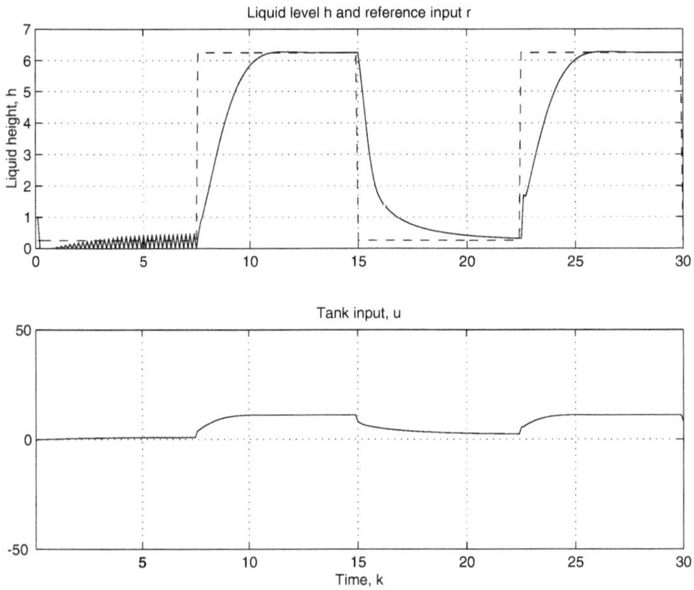

Figure 6.19: Closed-loop behavior of the surge tank using a planning strategy (different cross-sectional area).

control energy. Why did the values of the control energy change so quickly around the value of $N = 37$? Why does the tracking energy increase in the region from $N = 20$ to $N = 33$? Why is it the case that the control energy increases from $N = 1$ to $N = 10$? In general, how do you change the shape of the plots? Clearly, changing the w_1 and w_2 weights will change the shape, and hence, what choices you might make for what you call a "best" value of N. The model used for prediction, and the types of controllers that are simulated into the future will also change the shape. Moreover, the reference input can change it. Even though the generation of such plots can help you choose the planning horizon, it does not completely solve the problem. It simply provides insights.

Prediction horizon choice is difficult. Prediction too far into the future is computationally expensive and sometimes not useful due to plant uncertainty.

Finally, in some cases it is possible that longer planning horizons can actually *degrade* performance since the longer you simulate into the future with an inaccurate model, the less reliable the predictions tend to be. Hence, the optimization for plan choice can become inappropriate for selecting a good plan.

6.5 Exercises and Design Problems

Exercise 6.1 (Planning for Obstacle Avoidance):

(a) For the path planning problem in the chapter, use the simulation to generate plots that explain the effect of increasing the amount of uncertainty in where the vehicle ends up after a single step (i.e.,

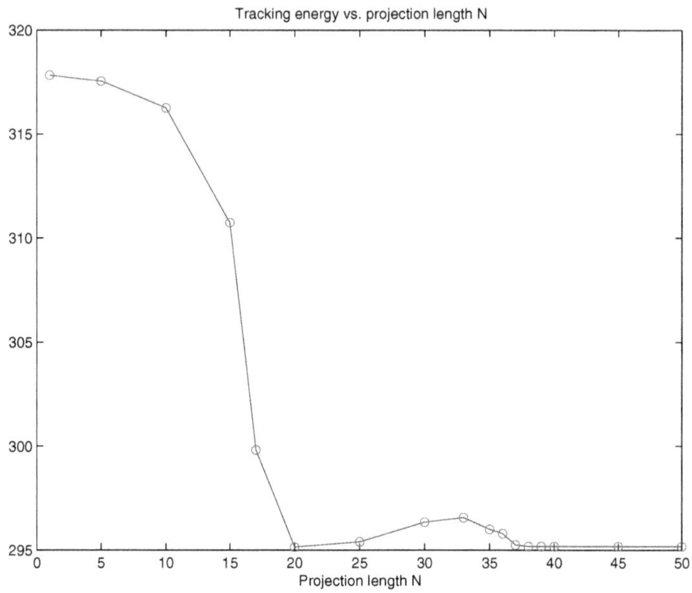

Figure 6.20: Tracking energy vs. projection length N.

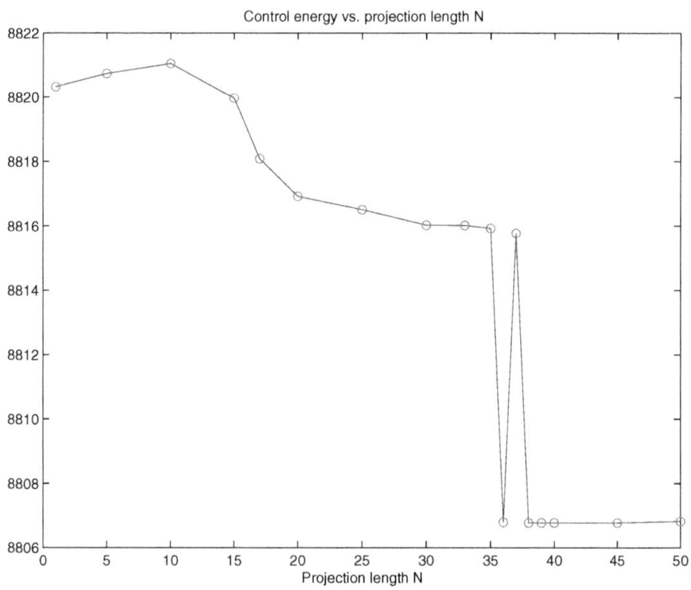

Figure 6.21: Control energy vs. projection length N.

explain the qualitative effects of the noise on the quality of planning). Can you choose the noise large enough so that the planning strategy fails to guide the vehicle to the goal after 500 steps?

(b) Next, study the effects of changing the value of r, the sensing radius. What happens if it is chosen smaller? Larger? If its value is too large, can the guidance algorithm fail? Illustrate your answer with a simulation.

(c) Illustrate the effects of changing N_s. What happens if $N_s = 4$? Illustrate with a simulation and explain how the movements of the vehicle change. What is the effect of using large values for N_s? Discuss smoothness of trajectories and computational complexity issues.

Exercise 6.2 (Model Predictive Control for a Simple Process Control Problem):

(a) For the MPC for the surge tank problem in the chapter, investigate the robustness of the strategy to measurement noise. To do this, you should precisely define what you mean by good performance, and investigate in simulation the effects of characteristics of noise (e.g., mean and standard deviation) on performance for a fixed MPC strategy.

(b) For the MPC for the surge tank problem in the chapter, investigate the robustness of the strategy to unknown characteristics of the tank cross-sectional area $A(h(t))$ (but for reasonable physical choices of the cross-sectional area). To do this, you should precisely define what you mean by good performance, and investigate in simulation the effects of characteristics of shape of the tank (e.g., if you characterize the shape with some nonlinear function, vary the parameters of the function) on performance for a fixed MPC strategy.

Design Problem 6.1 (Planning Ahead for Obstacle Avoidance):

(a) Simulate the obstacle avoidance problem in the chapter using the guidance algorithm defined there, but study a different placement of obstacles in the environment. Show a placement that the guidance strategy can successfully navigate, and one that it cannot successfully navigate.

(b) Repeat (a), but for a guidance strategy that predicts into the future multiple steps. Invent a placement of obstacles in the environment for which multistep prediction into the future allows successful navigation, where the strategy used in (a) does not. Hint: Consider a strategy that generates a tree of points with a root at the current position. For example, one approach would be to generate a circular pattern, pick the best point on that circle, then generate a circle of points around that point, and so on. The "best" plan is the path of best points found. You could experiment with different planning horizons and the frequency of replanning.

(c) Invent a placement of obstacles such that the strategy that you designed in (b) will fail in the sense that the robot will get stuck at a location other than the goal position. Redesign the look-ahead strategy so that it can successfully navigate it. Hint: Make the planning horizon vary with time in a way so that if it detects that it is "stuck," it lengthens its planning horizon until it finds its way around the obstacle (out of the local minimum). Illustrate the performance of the algorithm in simulation. Clearly explain your strategy and its operation. Discuss algorithm complexity.

Design Problem 6.2 (Model Predictive Controller Design for Tanker Ship Steering)*: In this problem you will study "model predictive control" (MPC) [192] for tanker ship steering. The tanker ship model that you will use as the truth model (to represent the plant) in all simulations should be the one given in Equation (4.5) that is simulated with a Runge-Kutta method in Section 4.3.1.

(a) For planning (prediction), use the linear model in Equation (4.4). Suppose this model is used with parameters specified for nominal conditions for the tanker ship; however, suppose that you use a discretized version of this model with a sampling period of $T = 1$. Hence, your discrete time model transfer function is

$$\frac{(-5.58e-05)z^3 - (5.635e-05)z^2 + (5.469e-05)z + 5.524e-05}{z^3 - 2.97z^2 + 2.939z - 0.9696}$$

which is obtained via a Tustin (bilinear) transformation when the nominal parameters ("ballast" conditions) are used and $T = 1$. Verify this. How accurate is this model? Simulate this linear discrete model and the nonlinear one. Highlight the similarities and differences in how the two models behave. In making this comparison induce disturbances in the nonlinear model (e.g., weight changes, wind, sensor noise, and speed variations) and explain how the plant differs from the linear discrete model in each case.

(b) Next, develop a method to project into the future and determine which sequence of inputs is best, then pick the first one to input to the plant for the current sampling instant. Let N denote the number of sampling instants that you simulate into the future. Suppose that you use a linear batch least squares approach (see Section 10.1 and in particular Equation (10.2)) to pick the best sequence of inputs to the plant, based on the linear discrete model. The key to solving this problem is to assume that the reference input trajectory is a constant, and to formulate the optimization problem as a linear least squares optimization problem. Show how to formulate the problem, and give an example where you code the example and actually find an optimal sequence of inputs to the plant. That is, simulate the closed-loop system when the MPC is used.

(c) Evaluate the performance of the MPC. Use similar reference inputs, and a similar sequence of investigations into the effects of disturbances, to what we did for the neural and fuzzy control methods. Develop a single numeric measure of performance (e.g., some quantification of tracking and control energy) and study how this measure changes for a range of values of the planning horizon N (make a plot). Repeat this for each disturbance condition. What is the best value to choose for N?

Design Problem 6.3 (Stability Analysis of Planning Systems)*: In this problem you will study stability analysis of planning strategies for two different plants that were studied in [196].

(a) Tank: Stability in the presence of uncertainty. Simulate, prove the strongest type of stability property possible.

(b) Load balancing in flexible manufacturing systems: Stability in the presence of traps. Repeat (a).

Chapter 7

Attentional Systems

Chapter Contents

Attention is a key component of all higher-level reasoning. A simplistic view is that the mechanism of attention seems to turn parts of the brain on and off in order to focus on what is currently important and ignore other things, but that does not tell the whole story. Characteristics of such attention processes depend on instincts, experiences and learning, and the human's goals and motivations. Attention helps us to cope with the large amount of information that can be acquired with our sensory systems in a short amount of time. In a sense, attention is one method that has evolved to ensure that we can succeed in the face of information overload. It helps us cope with complexity. Attention "filters out" less useful information from our senses (it "selects" the useful information), and thereby tries to optimally allocate cognitive resources. It also helps manage the complexity of internal reasoning (e.g., problem solving) by allowing us to focus on different internal representations, subproblems, and abstractions (i.e., it "selects" what to focus on when we are reasoning). Identifying components of attention is in fact complicated, as it closely intermixes with what are often considered other types of cognitive functions (e.g., planning and learning).

Due to its fundamental role in cognition, attention affects each type of control function that we have already considered in this part. On the other hand, control functions can affect attention since they dictate the behavior of dynamical attentional focusing. We may plan what to attend to, and have specific "attentional control rules" for how to focus. We can learn that certain stimuli are important to attend to since they help us reach our goals, or that such stimuli may play a significant detrimental role in our survival. We may learn that other stimuli can be ignored (i.e., learn that attending to some stimuli has no value to meeting our objectives). Indeed, we may even learn strategies for improving our attentional capabilities (e.g., how to concentrate better). We will not treat integrated attention-learning-planning in detail in this book. Instead, we will focus on the principles of dynamic focusing of attention, and analyze how control strategies can be used in attentional processes. We only briefly discuss how attentional strategies can be used in engineering applications for control and automation.

7.1 Neuroscience and Psychology of Attention

Attention is the process of focusing or concentrating. Often, we think of a hierarchy involving, in order of higher to lower levels, consciousness, sleeping, awareness, and attentiveness (e.g., you cannot be highly attentive when you are unconscious or asleep). At different points in our day we may turn off our attentional system. At others times our attentional system may be quite actively switching focus among different types of sensory data. For example, it may at one time disengage from one focus, move, and then engage on another focus. (Sometimes this is called "vigilance".) Attentional processes in the human brain are implemented with neural networks, but we will not consider this here (however, Design Problem 7.5 does request that you study the simulation of connectionist models of attentional systems).

We have certain types of attentional capabilities with all of our senses. For vision, we can pay attention to the object that we are looking at (e.g., focusing on these words as you read, while ignoring other peripheral visual stimuli or sounds). For auditory sensing, we may learn how to ignore background noise so that we are not distracted by it (e.g., if you have lived by a railroad track for a long time you may find yourself not even noticing a periodic passing train). For taste or smell, you can focus your attention on a certain spice in a food to try to identify it. For touch, we often ignore certain tactile senses (e.g., if you are holding this book, just an instant ago you probably did not notice the feeling of touching the book because you were probably attending to comprehension of the writing).

Attention allows us to amplify some sensory signals or internal thought processes and attenuate others.

A classical example of characteristics of our attentional system is given by the so-called "cocktail party effect." If you are at a party and there are many small groups of people talking, you have the useful ability to ignore (attenuate) what everyone is saying except for one person. The intriguing aspect, however, is that the person you are attending to (amplifying their signal) is not necessarily the one who is right next to you and talking the loudest. You may be able to virtually ignore this person to listen in from a distance on a quieter conversation that you are interested in (i.e., you may "eavesdrop").

In one famous experiment on human attention, "event-related potentials" (ERP) are measured via sensors on the scalp of a man via detection of electromagnetic waves. A specific ERP signal is the so-called "auditory N1 potential." The average voltage response for this ERP to an auditory stimulus that is attended to is relatively large in magnitude compared to an auditory stimulus that is not attended to. Some signals in the brain are amplified due to attention, and attenuated via lack of attention.

7.1.1 Dynamically Changing Focus

In the context of vision it is useful to think of our focus of attention as a type of "spotlight." This spotlight may coincide with where our eyes are focused ("overt" attention) or it may be that our eyes are focused at one point, and we attend to (shine our attentional spotlight) a different point ("covert" attention). Generally, we think of the spotlight as illuminating (amplifying) a region of sensory input. The dark region outside the spotlight is the region you are not attending to, and that visual sensory data are significantly attenuated.

There are two general types of control of attention, split according to what dictates the changes in attentional focus (i.e., what controls the dynamics of how the spot light moves). These are as follows:

Dynamic refocusing of attention can be driven by sensory data or explicit cognitive control.

- *Goal-driven (often "voluntary") attention reorientation:* Executive functions in the brain may reorient the focus of attention. This is thought of as a "top-down" refocusing that may be based on our problem-solving strategy and goals. For example, if you are reading and you decide to review a topic, you may go to the index of the book, find a key word, then go to another page and shift your focus of attention to another topic.

Typically, goal-driven reorientation of attention is slower and somewhat less "potent" than the stimulus-driven reorientation of attention that we discuss next.

- *Stimulus-driven (often "involuntary") attention reorientation:* Sensory signals can control the focus of attention in a "bottom-up" fashion. For example, we seem to have an instinct to pay attention to certain visual stimuli such as an object that is moving on a trajectory toward us, a bright flash of light (e.g., a fire), or blood (with evolutionary forces likely at work). Sensory inputs can achieve an *automatic* reorienting of attention, and often stimulus-driven attention reorientation is faster and more potent than goal-driven reorientation. For instance, if while reading this book, suddenly someone calls your name, yells in your ear, or your shirt catches on fire, it is likely that your attention will be diverted from this topic, no matter how interesting it is! Note, however, that if we repeatedly receive some external cue, and that cue does not indicate danger and we are not interested in it, we can typically learn to ignore it (i.e., learn not to allow sensory signals to reorient our attention). Hence, learning can play a key role in how our attentional dynamics operate.

Often, the two above methods to reorient attention are combined, or are interlaced over time. Clearly, both are influenced by knowledge acquired, and our instincts that have been established via evolution.

7.1.2 Multistage Processing: Filtering, Selection, and Resource Allocation

A functional model of the multistage attention process is given in Figure 7.1. There are sensory inputs that are "registered" (e.g., the receptor neurons detect sensory stimuli), then information is passed to the perceptual analysis and semantic encoding and analysis stages, where objects are recognized and processed for meaning. Information is then passed to executive functions, decision-making, memory, planning, etc. At the same time, there is feedback from executive functions that indicate what should be focused on (e.g., for voluntary control of attention).

Attention involves filtering out (discarding) some information.

There is evidence that at times, very early in the sensory processing process, there is selection of which stimuli are important, and which can be "filtered out." Evidence shows that in some situations this can be done before perceptual analysis or sensory encoding and analysis. For example, it seems that we have instinctual rules about certain types of stimuli that result in stimulus-driven reorientation of attention. It is this type of attentional control process that is involved in "early selection." On the other hand, "late selection" occurs in some situations, where more abstract analysis and processing of sensory signals (e.g., semantic encoding where meaning is determined) is conducted in order to reorient attention. For example, in some cases there might be some processing that determines whether the stimulus should gain full access to awareness, be

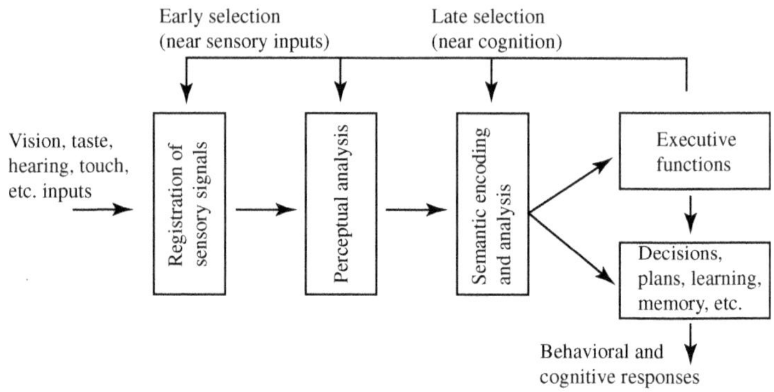

Figure 7.1: Multistage attention process.

encoded in memory, or result in some response. This type of processing may result in a goal-driven reorientation of attention.

Clearly, attention is a multistage process with feedback control paths. There is a cascaded filtering process that occurs where the most important information is focused on ("selected"), and less important information is ignored. Clearly, such a process is essential for high level cognitive functioning in humans. We have a finite amount of memory and processing power in our brain, and this naturally leads to "bottlenecks" in information processing. Attention allows us to allocate our cognitive resources to help us meet our goals. Hence, a key aspect of attention is the strategies used to allocate cognitive resources, especially in an "optimal" manner.

7.2 Dynamics of Attention: Search and Optimization Perspective

Here, we briefly discuss how to represent some of the underlying mechanisms of the dynamic focusing of attention as a search and optimization process. We will only focus on the dynamics of tracking objects in an attentional focus, how switching occurs from focusing on one object to another, and then the fine-tuning of the focus of attention after refocusing and during dynamic movement of an object. Our "model" is only based on the brief description in Section 7.1 of the psychology of attention, not neurophysiological studies, biophysics, or any of the other relevant underlying science. Hence, this is certainly of limited or possibly no value from a scientific perspective. Then, why provide such a model? First, the objective is to provide more detailed insight into the explanation of attention in Section 7.1. Second, we do not necessarily need a good model for the development of control and automation systems. The objective is to get the reader to think about dynamically focusing on information and hence ignoring other information. This is an essential feature of a complex automation problem where there is potential for "information overload" for the decision-making system, and hence the need to focus on the most important information.

We assume that there is a search component that finds objects in the "field of view" of the sensor and there is an optimization strategy at work that chooses the highest priority object and tracks it as it moves through the field of view. We do not focus on issues of the difference between where the sensor is directed versus where the focus of attention lies. Consider the model of attentional focusing provided in Figure 7.2. There, we have sensory data entering from the left into a block that processes these data to recognize objects, which we label with $i = 1, 2, \ldots, N$ (we assume a finite number of objects are in the field of view). For convenience, we assume in our discussion that the data are sensed about objects in an (x, y) plane. Next, the objects are prioritized by assigning a number $p_i > 0$, $i = 1, 2, \ldots, N$, where an object that is more important to focus on is given a higher value of p_i. Next, we assume that an "attention map" $J_a(x, y, t)$ is adjusted to represent the object positions and priorities at time t. Then, the priorities and attention map are input to a module that controls the focus point (i.e., where the focus is located in the (x, y) plane). To achieve control, it first compares the priorities to each other and picks the object $i^*(t)$ to focus on at time t that has the highest priority. That is, it lets

Dynamic focusing of attention can be modeled as optimization of a time-varying cost function.

$$i^*(t) = \arg \max_{i=1,2,\ldots,N} \{p_i\}$$

(arg max is simply the notation for finding the *index* of the priority that has the maximum value). Next, to pick the focus point, which we call (x_a, y_a), it considers which object should be focused on, where the current focus is relative to that object, and updates the focus point. It is assumed that it cannot move the focus point arbitrarily fast when it is trying to maintain focus on a particular object (e.g., as it moves across the plane), but that it can switch focus from one object to another very fast.

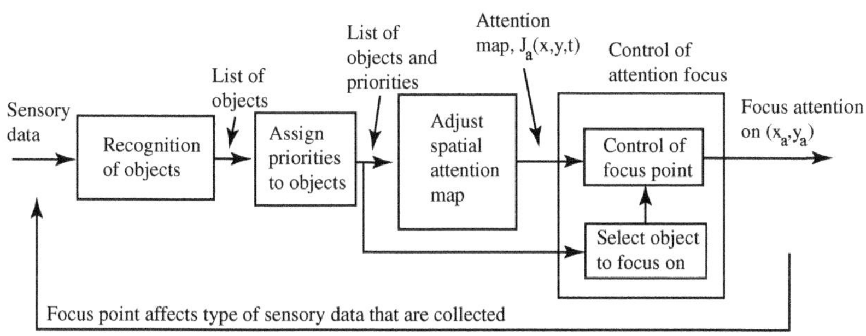

Figure 7.2: Functions involved in dynamically focusing attention.

7.2.1 Attentional Map

The key to our model of the dynamic focusing of attention lies in the definition of the attentional map $J_a(x, y, t)$. Here, we think of this map being generated internally (e.g., via pattern recognizer/semantic analysis), and assume that it is being used to indicate where it is important to focus on. In particular, we will define it as a continuous surface with $J_a(x, y) \in [-1, 0]$, where the point

$$(x_a, y_a) \in \{(x^*, y^*) : J_a(x^*, y^*) \leq J_a(x, y) \; \forall x, y\}$$

is a minimum point on the surface (note that there could be more than one such point, representing the possibility of a demand for split attention between equal priority points).

An example attentional map is shown in Figure 7.3. Here, we show an example attentional map that represents that there are two objects in the field of view: one that is high priority (the deeper valley) and one that is not as important (the shallow valley). The point that we want to focus on is the one defined by the point where the minimum is achieved on this map; that is, where the highest priority object is located.

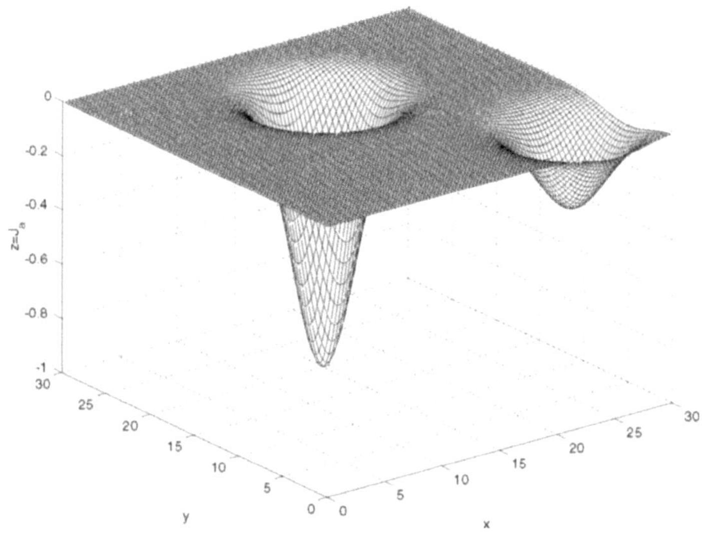

Figure 7.3: Example attentional map.

It is important to note that the map shown in Figure 7.3 is not static. It changes in several ways. First, if the objects move, the valleys move dynamically about the field of view. Also, if the field of view of the sensor is changed, the positions of the valleys change. The shape of the valleys may change (e.g., the widths of the valleys) depending on object positions. Moreover, the p_i priorities of the objects may change, which would dynamically change the depths of the valleys. For example, it may be possible that as relative positions of the objects changes, the shallow valley gets deeper and the deeper valley more shallow. If this happens, we would want the attentional focus to change from one object to another. If the valleys move, we want the attentional system to "track" the object that is of highest priority. Finally, note that as the field of view changes, and new objects appear and some disappear, it is possible that the number of valleys changes dynamically over time.

7.2.2 Optimization/Search Process for Focusing

It should be clear that in order to implement an attentional strategy using the attentional map, one could take an optimization/search perspective that has the following two components:

- *High priority object tracking:* Suppose that the current focus of attention (x_a, y_a) is located at the global minimum of the attentional map $J_a(x, y, t)$ (all other points on the J_a map are strictly above this point). Suppose that the field of view is constant, that objects do not leave the field of view, and that priorities of objects in the field of view stay constant. Suppose, however, that all the objects are moving and that some cognitive process keeps the attentional map up to date by dynamically adjusting the map. This will result in the centers of the valleys moving about the field of view dynamically. How does the attentional system work with the attentional map in order to maintain focus directly on the highest priority target? We could use a hill-climbing algorithm to continually climb down the attentional map at each step (e.g., it could move the focus of attention point in steps according to how someone would walk down a hill, moving in directions at each step toward the most significant decrease in the attentional map). Then, if the map does not move too fast, and the hill-climbing algorithm can keep up, it will tend to keep the focus of attention near the center of the valley that corresponds to the highest priority object. As the object moves about the field of view, the algorithm will tend to track the object.

- *Changing focus:* Next, suppose that the objects move about the field of view, and their priorities change dynamically. In this situation, the attentional tracking algorithm may track the highest priority object for a period of time, but its priority may decrease, and the priority of another object may increase. At the point where the global minimum of the attention function changes to correspond to the object with increasing priority, it should be the case that the strategy can switch focus from one object to another. How can this be achieved? Well, if the minimum points are known, switching is easy via a simple monitoring of the values of the minimum points of the attention function, ranking those values, and choosing to focus on the smallest one (a simple type of optimization). If those minimum points are not known, then one would need some type of "global" optimization procedure to determine when to switch. One approach would be to have N tracking algorithms of the type described above, and simply select for focusing the one that achieves the lowest value. There is some evidence that an analogous strategy is used in some cases in some biological attentional systems.

In summary, we see that one way to view the attentional process is as an optimization process for a cost function that is time-varying. Such an optimization problem can be very difficult to solve, but ideas from the optimization methods discussed in Part III and Part V provide many approaches to the problem.

7.3 Attentional Strategies for Multiple Predators and Prey

Consider an organism that is in some environment with multiple predators, and it is trying to attend to all of them to maintain as accurate a picture of its environment as possible in order that it can defend itself. Moreover, we assume that in the same environment, there are multiple prey that the organism would like to pay attention to in case it decides to pursue one of these to kill and eat. How should the organism dynamically focus its attention on the predators and prey to ensure its success in foraging and surviving? In this section we will model such a problem and introduce a variety of attentional ("scheduling") strategies for focusing attention. Hence, we think of needing to schedule our cognitive resources in order to maintain an accurate view of the environment. We will simulate the strategies and discuss issues in their design.

7.3.1 Cognitive Resource Allocation Model

We will assume that there is a recognizer for predators and prey that provides information to our attention strategy, so that it simply needs to decide what to focus on (cognitively process). The focus here is on the selection process that can be occurring in either early or late selection, or both. The key is that there is a "limited channel" or one resource that must be shared, and the attention strategy must decide how it is shared. We ignore issues of the possible differences in where the organism's sensor is pointed (e.g., where its vision is directed), versus where the center of the focus of attention is.

Quantifying Length of Time Predators/Prey Are Ignored

Suppose that we assume that the number of predators and prey is constant and that we number them and denote the set of predators and prey as

$$P = \{1, 2, \ldots, N\}$$

Let t denote time. Let

$$T_i(t), i \in P, t \geq 0$$

denote the *last time at which predator/prey i was detected* (later you will see that this is defined by the instant t' when, by focusing on predator/prey i, we get $T_i(t') = 0$). By "detected" we mean that the organism has focused its attention on the predator or prey, and has identified it and its characteristics (e.g., its position).

Generally, we will view the attentional strategies as "controllers" that take as inputs the $T_i(t)$, $i \in P$, and choose which predator/prey to focus on next. This is shown in Figure 7.4. We will assume that there is a cognitive tracking mechanism that is trying to estimate where predators/prey are moving, and that it has a certain level of accuracy in achieving this task. We will not require perfect accuracy in tracking multiple predators/prey; we will allow them to be "lost" for a period of time. Loss of tracking could result from predators or prey hiding (e.g., behind a tree), due to the sensor having only a limited "field of view," or from possible additional (but finite) time required to reacquire tracking when attentional focus is shifted. We will discuss how we model such issues in a moment.

An organism seeks to schedule its cognitive resources over time to enhance its chance of survival.

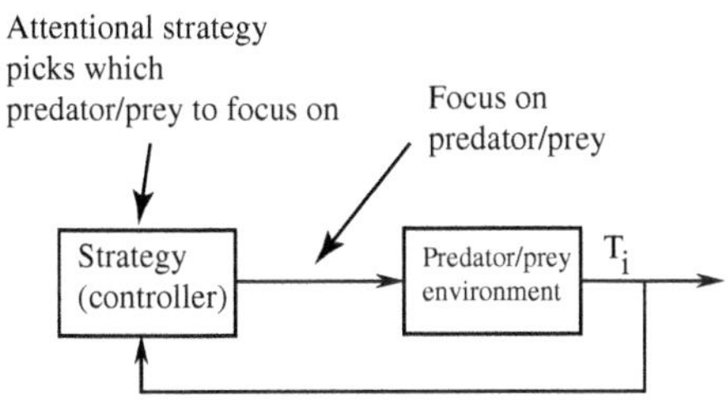

Figure 7.4: Attentional strategy viewed as a controller.

Suppose that initially

$$T_i(0) = 0, i \in P$$

so that we act as though initially we had simultaneously detected all the predators/prey, which is clearly physically impossible. Note, however, that this is a good initialization considering the fact that our attentional strategies will make decisions about which predator/prey to focus on based on the sizes of the $T_i(t)$, $i \in P$ (i.e., based on how long they have been ignored). Basically, for many strategies this initialization indicates that at $t = 0$, there is no priority to seek one predator/prey rather than any other one. For many strategies, an initialization with $T_i(0) > T_{i'}(0)$ for $i \neq i'$ would indicate an initial preference to first focus on the ith predator/prey over predator/prey i'.

Note that if the organism was not actively engaged in paying attention to its environment (e.g., it was sleeping or doing something else), then clearly

$$T_i(t) \to \infty, i \in P, t \to \infty$$

since it will never detect a predator/prey. The goal of the attentional strategy is to try to avoid $T_i(t) \to \infty$ for any $i \in P$ and indeed it will try to keep the $T_i(t)$

values *as small as possible* since this represents that the organism has recently detected each predator/prey and hence has good information about the predators/prey. It is assumed that each predator/prey will *persistently* periodically "appear" (i.e., not be occluded by some object, or lost due to poor cognitive tracking) so that there is a finite amount of time between predator/prey appearances to the attentional strategy; this is assumed since, if some predator/prey $i \in P$ only appears for a finite amount of time, and never appears again, then at some point it will clearly be impossible to detect it again so that $T_i(t) \to \infty$ as $t \to \infty$.

The organism wants to minimize the amount of time it ignores any predator/prey to ensure it has accurate information about its environment.

Environmental and Cognitive Delays Affecting Attentional Switching

Let $\delta(t) > 0$ denote a "processing delay" that may represent the delay from the environment (e.g., due to a predator being occluded for a brief period of time) and a "cognitive processing delay." The cognitive processing delay may be used to represent the amount of time that it takes for the organism to switch from paying attention to one predator/prey i to another predator/prey j, $j \neq i$. We will call this type of delay $\delta_{i,j}$ and assume it is a fixed known delay (if it were unknown but bounded, then the attentional strategies and analysis still hold). For convenience, we will assume that these attentional switching delays are all the same and will denote that value by $\delta_s = \delta_{i,j}$ for all $i, j \in P$.

The variable $\delta(t)$ may also incorporate delays in being able to detect a predator/prey. For instance, each predator/prey has a type of frequency of appearance that is driven by a variety of characteristics such as how effectively the prey can hide in the current environment, or how fast a predator can run. Suppose that for a known predator/prey type i, there is some bound δ^i on the amount of time that it would take for the organism to first realize that the predator/prey may be at some location, if that was the only predator/prey that the organism focused on (clearly, this would depend on the predator/prey appearance period). Getting the first indication of the presence of a predator or prey does not correspond to achieving a detection of a predator/prey. Suppose that $\delta_e(t)$ denotes the delay incurred by the organism in first getting an indication of the presence of a predator/prey, from the time that it gets switched to focus on that predator/prey. It could be that many characteristics contribute to this delay, including cognitive tracking mechanisms and environmental characteristics. Note that if we let

$$\bar{\delta} = \max_i \left\{\delta^i\right\}$$

then $\delta_e(t) \leq \bar{\delta}$. Let

$$\delta(t) = \delta_s + \delta_e(t)$$

For convenience, we let δ denote a constant that is the least upper bound on $\delta(t)$ so that $\delta(t) \leq \delta$ (i.e., we simply remove the time index to denote the least upper bound on the variable).

To summarize, when the attentional strategy issues a command to focus on predator/prey i, there is a delay to switch to the attention to focus on it, and then there is an additional (time-varying) delay since the predator/prey may not have appeared. This additional delay is shorter than $\bar{\delta}$. After these two types of delays occur we assume that the organism knows that a predator/prey is where it is focusing (but we do *not* assume that the organism has identified all the characteristics of the predator/prey and hence, has not yet "detected" it).

Rate of Cognitive Processing

We will suppose that the organism may take additional time to detect a predator/prey that it has not detected for a long period of time. That is, we think of the organism as having successively more difficult times finding a predator/prey that it has not found for longer periods of time since it, in a sense, becomes "desynchronized" with that predator/prey and cannot easily determine when or where it will appear, or its other characteristics. To quantify this phenomenon, we will use parameters

$$a_i, i \in P$$

where $1/a_i$ represents a "rate" at which the organism cognitively processes information about predators/prey in order to detect them. These a_i parameters require further explanation. Consider the case where there is only one predator/prey ($N = 1$), named "predator/prey 1." Suppose that at some time t', the amount of time that has elapsed since the last time predator/prey 1 was detected is $T_1(t') > 0$ as shown in Figure 7.5.

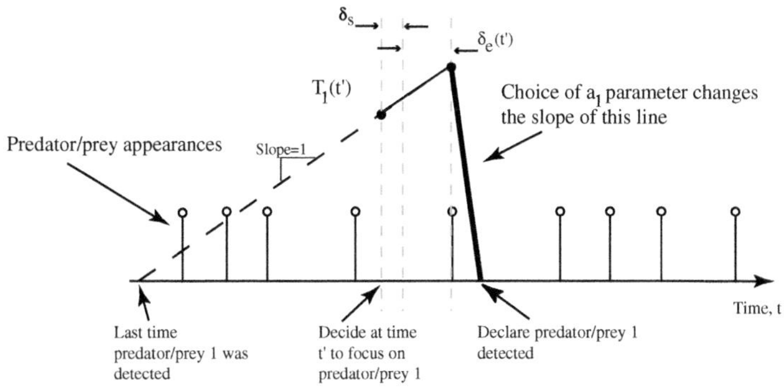

Figure 7.5: Illustration of timing of organism decision-making and predator/prey appearances (note that pulses represent the first times that predators/prey appear).

At time $t' + \delta_s$, the organism has switched its focus to predator/prey 1. So, starting at $t' + \delta_s$, the organism is looking for predator/prey 1 and before $t' + \delta_s + \delta^1$, we know that a predator/prey appearance will occur. Name the delay between achieving a switch in focus to the time where a predator/prey appearance is first found $\delta_e(t')$. Then, at time $t' + \delta_s + \delta_e(t')$, the organism initiates the completion of the "detection" of predator/prey 1 and the amount of time that it takes to do that is dictated by the a_1 parameter. (As you will see below, smaller values of a_1 correspond to it taking shorter amounts of time to fully detect the predator/prey.) We declare predator/prey 1 "detected" at the time at which T_1 is decreased to zero. Next, we need to further clarify the meaning of the a_i parameters by explaining how they produce the slope of the bold line in Figure 7.5 and hence quantify how long it takes to detect a predator/prey. Also, we need to explain how the organism chooses which predator/prey to focus on. To do this, we will introduce a specific attentional strategy and explain how to interpret the a_i, $i \in P$ parameters.

7.3.2 Focus on a Predator/Prey Ignored for the Longest Time

First, let D_{k_r} denote the time at which the attentional strategy chooses a predator/prey to focus on (i.e., it is the *decision time*), and suppose that $D_1 = 0$. An attentional strategy that focuses on the predator/prey that was ignored for the longest time makes choices of which predator/prey to focus on such that at D_{k_r}, the attentional strategy chooses to focus on predator/prey $i^*(k_r)$ such that

$$T_{i^*(k_r)}(D_{k_r}) \geq T_i(D_{k_r}), \forall i \in P \tag{7.1}$$

and focuses on it until it detects it. If there is more than one maximizer, then the attentional strategy will simply choose one of these at random.

Decision-Timing for Attentional Switches

First, notice that the actual time when focusing starts for predator/prey $i^*(k_r)$ occurs after some delay, and then it may take some additional (but finite time) for the predator/prey to appear ($\delta_e(D_{k_r}) \leq \bar{\delta}$), and still more time based on how long it has been since the predator/prey was last detected (i.e., the effect of the a_i). Note also that while the delays occur, the time since the last detection is still increasing. Hence, the times when the attentional strategy makes decisions are given by

$$D_{k_r+1} = D_{k_r} + \delta(D_{k_r}) + a_{i^*(k_r)} T_{i^*(k_r)}(D_{k_r}) + (D_{k_r+1} - D_{k_r})\, a_{i^*(k_r)} \tag{7.2}$$

Here, the next decision point D_{k_r+1} is the time when the detection of the last predator/prey that was focused on is detected and this formula gives the time D_{k_r+1} when the next decision will be made. The value of D_{k_r+1} is given by the sum of four terms. The first term is simply the last decision point D_{k_r}. The second term is the delay $\delta(D_{k_r})$ where

$$\delta(D_{k_r}) = \delta_s + \delta_e(D_{k_r})$$

Third, the term $a_{i^*(k_r)} T_{i^*(k_r)}(D_{k_r})$ is the amount of time it takes to detect predator/prey $i^*(k_r)$ that arises due to the fact that we have not detected it for some time. (Note the proportionality—if it has not been detected for a long time, then it will take more time to find it and this represents that predators/prey that have not been detected for a long time become more difficult to detect.) Finally, the fourth term quantifies that additional time is needed to detect the predator/prey simply because during the time that the cognitive processing for the predator/prey is occurring, even when it is focused on, the length of time since the last detection continues to increase (we do not consider a predator/prey $i^*(k_r)$ fully detected until $T_{i^*(k_r)}(D_{k_r+1}) = 0$).

Using simple algebra to rearrange Equation (7.2), we get

$$D_{k_r+1} = D_{k_r} + \frac{\delta(D_{k_r}) + a_{i^*(k_r)} T_{i^*(k_r)}(D_{k_r})}{1 - a_{i^*(k_r)}} \tag{7.3}$$

Notice that as expected, the delay δ directly influences the rate at which we can switch attentional focus. Also, this equation shows us that the length of time between decisions can be lengthened if a particular predator/prey has been ignored for too long due to the effects of the a_i parameters.

The Cognitive Capacity Constraint

In fact, using Equation (7.2), it is now possible to complete the explanation of Figure 7.5 and further explain how to interpret the a_i parameters. What is the effect of the a_i parameters on how fast a predator/prey is detected? Notice that we incur the delay $\delta(t)$, and from Figure 7.5 we see that the slope of the bold line dictates then how fast we achieve detection. What is the slope of the bold line in Figure 7.5? We use simple geometry to determine this. First, notice that the peak value

$$T_{i^*(k_r)}(D_{k_r} + \delta_s + \delta_e(D_{k_r})) = T_{i^*(k_r)}(D_{k_r}) + \delta_s + \delta_e(D_{k_r})$$

since the slope of the dashed line in Figure 7.5 is unity. Next, notice that Equation (7.3) gives the amount of time between the decision time D_{k_r} and time of detection D_{k_r+1} so that the slope of the bold line in Figure 7.5 is

$$-\left\{ \frac{T_{i^*(k_r)}(D_{k_r}) + \delta_s + \delta_e(D_{k_r})}{\frac{\delta_s + \delta_e(D_{k_r}) + a_{i^*(k_r)} T_{i^*(k_r)}(D_{k_r})}{1 - a_{i^*(k_r)}} - (\delta_s + \delta_e(D_{k_r}))} \right\}$$

which with some simple algebra reduces to

$$-\frac{(1 - a_{i^*(k_r)})}{a_{i^*(k_r)}} \tag{7.4}$$

In a moment you will see that it is necessary that $a_{i^*(k_r)} < 1$. Using this fact, Equation (7.4) indicates how fast detection occurs as shown in Figure 7.6 (i.e., how fast cognitive processing occurs). With small values of a_i (high values of $1/a_i$, the rate of processing by the organism in trying to detect) we get fast detection, and with larger ones we get slower detection. So, how do we interpret the a_i parameters? They are parameters used to model how difficult

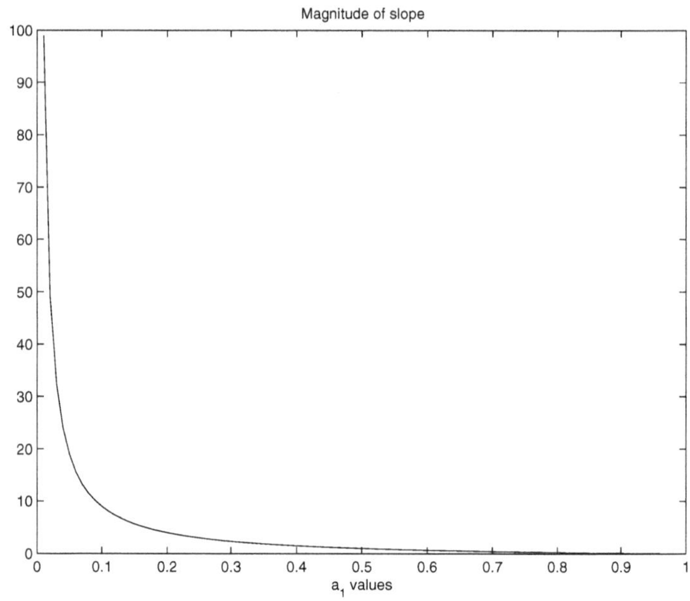

Figure 7.6: Magnitude of the slope of the bold line in Figure 7.5 for various values of a_1.

it is to detect a predator/prey, where if a predator/prey has not been detected for a long period of time, it can become more difficult to detect.

Clearly, it is *necessary* that the "capacity condition"

$$\rho = \sum_{i=1}^{N} a_i < 1 \tag{7.5}$$

be satisfied in order for *any* attentional strategy to ensure that the values of $T_i(t)$, $i \in P$, remain bounded. How should this capacity condition be interpreted? Intuitively, it says that it must be the case that even if the predators/prey can become more difficult to detect if they have not been detected for a long time, the organism must be able to operate "fast enough" to be able to find them. For instance, Equation (7.5) is satisfied if for each $i \in P$,

Cognitive capacity quantifies when an environment presents too large of an attentional load for an organism so that it will miss important information.

$$a_i < \frac{1}{N}$$

This shows us that as the number of predators/prey grows, it is possible that the cognitive capacity of the organism is overwhelmed and it is being given too much work, so that there is no way that it can keep up, so it will end up being the case that $T_i \to \infty$ for at least some $i \in P$ (or more than one i).

Equation (7.5) can be used to gain insight into the operation of attentional strategies by using the ideas in [418]. First, note that you can think of a_i as the amount of "load" (or the number of time units of "work") that is brought to the organism for the attentional task at each time instant by predator/prey i. Hence, if the organism is to succeed, on average the organism can only afford to spend a portion $(1-\rho)$ of its total time being idle. If you assume that the delay $\delta(t)$ is a constant δ, then each decision time when we switch from focusing on one predator/prey to another costs δ time units of idle time; hence, the average frequency of decision times is bounded above by

$$\frac{1-\rho}{\delta}$$

Now, if ρ is very close to 1 (representing an organism that is heavily loaded), $(1-\rho)\delta^{-1}$ is very small so the frequency of switching attentional focus between different predators/prey is low (which means that it can take a long time for the organism to find each predator/prey, so the organism will tend to have large $T_i(t)$ values and hence will not perform as well).

7.3.3 Additional Attentional Strategies

There are a wide variety of possible attentional strategies. Next, we consider one that is more general than the one of the previous subsection given in Equation (7.1), in the sense that at each decision point D_{k_r} it could make *exactly* the same decision as it did there, but could also make other choices.

Focus on a Predator/Prey Ignored More Than the Average One

The particular strategy is given by choosing the predator/prey to focus on that has been ignored more than the average time that all the predators/prey have been ignored. In particular, at D_{k_r}, the attentional strategy chooses to focus on predator/prey $i^*(k_r)$ such that

Attentional strategies are feedback controllers that dynamically refocus.

$$T_{i^*(k_r)}(D_{k_r}) \geq \frac{1}{N}\sum_{i=1}^{N} T_i(D_{k_r}) \tag{7.6}$$

and focuses on it until it detects it (in a similar way to the strategy of the last section). Note that Equation (7.3) also holds for this strategy, and that of course the capacity condition Equation (7.5) must hold.

Note that for this strategy, *any* predator/prey that has been ignored for more time than the average predator/prey has can be focused on. How does the strategy choose which particular predator/prey to focus on? One simple approach is to randomly choose one. However, more sophisticated criteria are possible. For instance, it could try to optimize some other system quantity, or it may use Equation (7.6) to provide a set of possible predators/prey to choose and then use "predator/prey priorities" (some indication of which predator/prey is most important) to choose the one to focus on. In the simulations of the next section, when we study this strategy, we will assume that predator/prey i has priority i and higher values of i correspond to higher priorities.

Focus on a Predator/Prey That May be Most Difficult to Find

An attentional strategy that focuses on the most difficult to find predator/prey makes choices of which predator/prey to focus on such that at D_{k_r}, the attentional strategy chooses to focus on predator/prey $i^*(k_r)$ such that

$$a_{i^*(k_r)}T_{i^*(k_r)}(D_{k_r}) \geq a_i T_i(D_{k_r}), \forall i \in P \tag{7.7}$$

and focuses on it until it detects it. If there is more than one maximizer, then the attentional strategy will simply choose one of these at random.

Clearly, this is similar to the attentional strategy that focuses on the predator/prey that has been ignored for the longest time that was given in Equation (7.1). Here, however, we have the scalings by the a_i parameters and this changes the attentional strategy. Intuitively, since a_i is the amount of "load," you can think of this attentional strategy as choosing the the most difficult one to find predator/prey to focus on.

Focus on a Predator/Prey Expected to be Most Difficult to Detect

The strategy to be developed next is motivated by the above strategy and is modeled after the one in [418] that has been found to be very effective in a different class of resource allocation problems. Recall from our earlier analysis that if you pick predator/prey $i^*(k_r)$ to focus on,

$$T_{i^*(k_r)}(D_{k_r}) + \delta_s + \delta_e(D_{k_r})$$

is the peak that $T_{i^*(k_r)}(D_{k_r})$ reaches before the predator/prey is detected. Note that in general we do not know $\delta_e(D_{k_r})$ since it depends on how the organism decision times are aligned with the predator/prey appearance times. A known bound, however, on the peak value is given by

$$T_{i^*(k_r)}(D_{k_r}) + \delta_s + \delta_e(D_{k_r}) \leq T_{i^*(k_r)}(D_{k_r}) + \delta_s + \delta^{i^*(k_r)}$$

Hence, predators/prey with larger δ^i values (i.e., ones with possibly lower frequency appearances) can be considered on average more difficult to detect in this framework. Also, the a_i parameters, which model another characteristic of the difficulty of predator/prey detection, will also affect how soon detection can occur.

Consider choosing predator/prey $i^*(k_r)$ to focus on at time D_{k_r} if

$$i^*(k_r) = \arg\max_i \left\{ w_i \left(\frac{T_i(D_{k_r}) + \delta_s + \delta^i}{\frac{(1-a_i)}{a_i}} \right) \right\} \tag{7.8}$$

where $w_i > 0$, $i \in P$ are weighting factors. Notice that in this formula, the numerator is the bound on the peak value and the denominator is the magnitude of the slope of the bold line in Figure 7.5 given by Equation (7.4). Why divide by the slope in the above formula? If the slope is greater in magnitude (smaller a_i value), this corresponds to an easier-to-detect predator/prey and this will result in Equation (7.8) with a *reduced* emphasis on focusing on that predator/prey. Hence, the strategy picks the predator/prey to focus on that is *expected* to be the most difficult to detect in the sense that it estimates which predator/prey will take the longest time to detect and selects it (assuming $w_i = 1$ for all i). To see this geometrically, notice via Figure 7.5 that the numerator $T_i(D_{k_r}) + \delta_s + \delta^i$ in Equation (7.8) should be thought of as an estimate of where the peak occurs and we divide it by the slope; hence, this value is the length of time that elapses from the time that the peak occurs, until detection.

The weighting factors w_i can be chosen to force the organism to focus on some predators/prey more than others. Equal weighting would correspond to the choice of $w_i = 1$ for all $i \in P$. If $w_i >> w_j$, $i \neq j$, then Equation (7.8) will tend to choose i rather than j to focus on. This may be useful in some predator/prey environments since it provides a way to indicate which predator/prey should be focused on. Another possibility is to weight predators more than prey so that the organism always focuses on those more. While the weighting factors provide an opportunity to tune the strategy, there is no guarantee that this strategy will be better than any of the others introduced above according to typical performance measures. Generally, you would want to choose the weights so as to make the attentional strategy perform as successfully as possible (where you define what is meant by "successfully").

7.3.4 Attentional Strategies Based on Predator/Prey Priority

In the last subsection, we introduced two ways to incorporate priorities of predators/prey into scheduling strategies. First, in Equation (7.6) we used priority as a "secondary" selection mechanism to choose from the set of predators/prey that has been ignored longer than the average one. Second, in Equation (7.8) we introduced the weighting factors w_i which allow us to emphasize the processing of one predator/prey more than another (and this will be illustrated in the simulation examples in Section 7.4). In this subsection we will introduce yet another priority scheme, but one that integrates the consideration of predator/prey priorities so that predator/prey priority is neither a secondary consideration nor set by secondary weighting parameters that have loose connections with the predator/prey priorities.

To do this, we introduce a set of parameters $p_i > 0$, $p_i \in \Re$, $i \in P$, that represent the predator/prey priorities (larger values correspond to higher priorities). We allow the designer to take two different views of the priority parameters:

Attentional strategies can include information on which predators/prey are most important to pay attention to.

1. *Predator/prey environment information:* You can assume that the values of the parameters p_i, $i \in P$, are set a priori and remain constant throughout the activity (e.g., foraging) of the organism. Hence, you can view them as part of the a priori information about the predator/prey environment.

2. *Design parameters:* Alternatively, you may view the priority parameters as design parameters that can be tuned (e.g., via extensive simulations of the

predator/prey environment) before an organism engages in the attentional task.

How do we integrate the priority parameters into *each* of the strategies defined in the previous subsections? For example, how can we use them to modify the strategy in Equation (7.1) where we chose to focus on the predator/prey that was ignored the longest. Here, we simply *scale* T_i by p_i, $i \in P$ in each of the cases and then make all decisions based on the same formulas as above, but with T_i replaced by $p_i T_i$, $i \in P$. What is the effect of such a scaling? It serves to scale the lengths of times that the predators/prey have been ignored, with higher weights given to predators/prey with higher priorities. Thereby, it biases the attentional strategy toward higher priority predators/prey.

For such strategies to be stable, it is clearly necessary that we modify our capacity condition. With priorities, we require that

$$\rho_p = \sum_{i=1}^{N} p_i a_i < 1 \tag{7.9}$$

be satisfied to ensure that the values of $T_i(t)$, $i \in P$, remain bounded.

How does the scaling affect the behavior of the strategies? While it is clear that predators/prey $i \in P$ with T_i scaled by higher values of p_i will have $p_i T_i$ grow faster (the slope of the line representing the growth is p_i), the behavior is also affected by the range of values that you allow for the priorities. For instance, if you dictate that your priorities $p_i \in (0, 1]$, $i \in P$, then if you were given some a_i values that satisfied Equation (7.5), the p_i and a_i values would also satisfy Equation (7.9). Hence, if you use a proper range of values for the priority parameters, any strategy that satisfies the capacity condition without priorities will satisfy Equation (7.9). Note that there is really no reason why you cannot make the choice of $p_i \in (0, 1]$, $i \in P$, since the parameters are simply used to rank order the predators/prey. It is also interesting to note that if you repeat the analysis in Sections 7.3.1 and 7.3.2, the result in Equation (7.4) still holds (due to cancellations of the priority parameters in the algebra); hence, simulation of the class of priority strategies discussed here is quite similar to the earlier strategies.

To summarize, *you can embed the priority parameters into any of the above strategies.* For instance, Equation (7.1), when converted to a priority scheme using this approach, becomes one where the attentional strategy chooses to focus on predator/prey $i^*(k_r)$ such that

$$p_{i^*(k_r)} T_{i^*(k_r)}(D_{k_r}) \geq p_i T_i(D_{k_r}), \forall i \in P \tag{7.10}$$

In this way you can have a strategy that selects predators/prey based on both priorities and how long they have been ignored. The scheduling strategies in earlier subsections are modified in a similar manner. Finally, note that you can still use the two priority schemes we discussed earlier in conjunction with this priority scheme.

7.3.5 Viewpoint of Attention Scheduling as Online Optimization

Next, note that we can provide an interpretation of the above attentional strategies in terms of optimization. The key is to think of the attentional decision-making in terms of *optimizing* a cost function J_p, and that J_p is the result of a computation made in the scheduler (controller) so that it can make scheduling decisions. With this view, we have the following:

Attentional strategies make decisions that optimize some short-term performance measure in hopes of optimizing a long-term one.

- *Focus on a Predator/Prey Ignored for the Longest Time:* Here, for the strategy in Equation (7.1), we have

$$J_p = -\max\{T_i(D_{k_r}) : i = 1, 2, \ldots, N\}$$

  and hence in trying to maximize J_p, we try to minimize the longest time that the organism ignores any predator/prey. In this way, the scheduler tries to focus on predators/prey so as to keep the values of $T_i(t)$ low so that the organism has good information about the predators/prey.

- *Focus on a Predator/Prey Ignored More Than the Average One:* Here, for the strategy in Equation (7.6), we have

$$J_p = -\sum_{i=1}^{N} T_i(D_{k_r})$$

  and hence in trying to maximize J_p, we try to minimize the average time that the organism ignores any predator/prey (it attempts this even though there is not a single maximizer at each decision time). Again, the scheduler tries to focus on predators/prey so as to keep the values of $T_i(t)$ low so that the organism has good information about the predators/prey. Here, however, it makes decisions in a different manner since it tries to maximize a different J_p.

Using this same approach, it is simple to specify J_p measures for the other strategies we defined above. For instance, for the strategy in Equation (7.7), we have $J_p = -\max\{a_i T_i(D_{k_r}) : i = 1, 2, \ldots, N\}$ and hence, in trying to maximize J_p, we try to minimize the longest time that the organism ignores any predator/prey, but scaled by the "load" of the predator/prey. For Equation (7.8), our J_p would quantify the desire to keep the peaks of the $T_i(t)$ as low as possible (which may or may not result in a lower average delay). Clearly, if you embed a priority scheme via the priority parameters p_i, $i \in P$, the same concepts hold.

Note that the above J_p measures should not be thought of as measures of attentional success over the long term, but as instantaneous measures that are used to guide decisions about which predator/prey to focus on. Achieving an instantaneous optimization does not necessarily result in making optimal decisions to try to ensure that the organism gets the best information over the long term.

7.4 Design Example: Attentional Strategies

In this section, we will simulate the attentional strategies of the last section in order to provide insights into their operation. Moreover, we will discuss several issues in how to design attentional strategies.

7.4.1 Simulation Approach and Performance Measures

For convenience, we simulate the predator/prey environment and organism as a discrete-time system. We will use a sampling period of $T_s = 0.01$ and in all our simulations we will have $N = 4$ predators/prey. Each predator/prey will be characterized by a sequence of appearances, which we simply model as unity height signal at some sampling instant. When there is no appearance, the signal height is zero. For instance, for all our simulations below we will have the predator/prey appearance sequences shown in Figure 7.7. We use different frequencies of appearance for different predators/prey, but for simplicity we keep the appearance frequencies constant (for predators/prey $i = 1, 2, 3, 4$ we have them appear every 1, 1.1, 1.2, and 1.3 sec.).

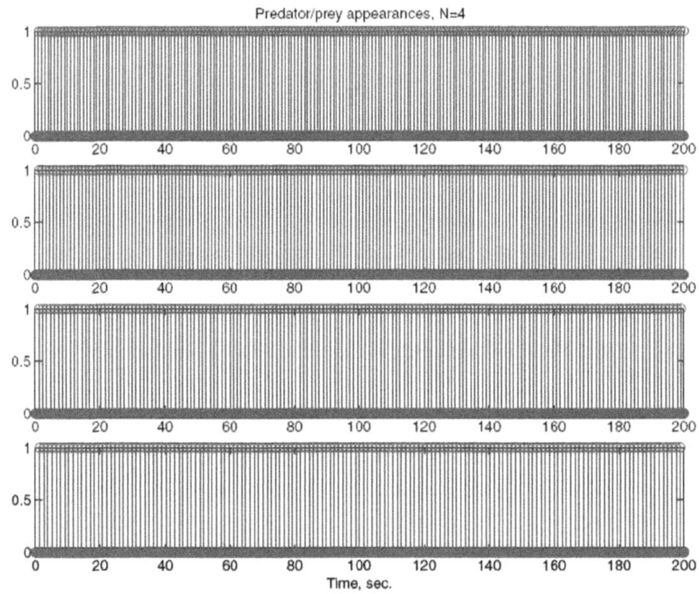

Figure 7.7: Predator/prey appearance sequences, for $N = 4$ predators/prey (predator/prey $i = 1$ is the top plot, $i = 2$ is the next one down, $i = 3$ is below that, and $i = 4$ is the bottom plot).

Suppose that we know that the bounds on the spacing between appearances are

$$\delta^1 = 1.05,\ \delta^2 = 1.15,\ \delta^3 = 1.25,\ \delta^4 = 1.35$$

Notice that these are simply bounds for periods given in Figure 7.7. We choose $\delta_s = 0.03$. To model detection difficulty, and in order to satisfy the capacity condition, we choose

$$a_1 = 0.1,\ a_2 = 0.2,\ a_3 = 0.3,\ a_4 = 0.1$$

This gives $\sum_{i=1}^{4} a_i = 0.7$, which represents that the organism will be quite busy in detecting predators/prey (lower values of this sum correspond to light loads).

There are several ways to measure performance of the attentional strategies. Here we will compute the average of the length of time since any predator/prey has been detected

$$\frac{1}{N}\sum_{i=1}^{N} T_i(k)$$

at each step k. We will also compute the time average of this quantity (i.e., the time average of the average values) and the maximum average value achieved over the entire simulation run. We will compute the maximum time that any predator/prey has been ignored at each time step k

$$\max_i\{T_i(k)\}$$

We will also compute the time average of this quantity (i.e., the time average of the maximum values) and the maximum of the maximum values achieved over the entire simulation run. In order to measure how well we have focused on higher priority predators/prey, we will use

$$\frac{1}{N}\sum_{k} i^*(k)$$

where $i^*(k)$ is the predator/prey chosen as step k. Clearly, higher values of this measure will correspond to the case where on average, higher priority predators/prey were focused on, in the case where we use i to both label the predators/prey and as a priority parameter.

7.4.2 Attentional Strategy Behavior: Focus on Longest Ignored

Here, we will illustrate the performance of the attentional strategy in Equation (7.1) that chooses the predator/prey to focus on that has not been detected for the longest period of time.

First, consider Figure 7.8 where the top plot shows $i^*(t)$, the predator/prey being focused on at each time. The plot below it shows $T_1(t)$, and the bottom plot shows $T_2(t)$. From the top plot it is interesting to note that the sequence of predators/prey that is focused on is: $1, 2, 3, 4, 1, 2, 3, 4, 1, 2, 3, 4, \ldots$. But, the lengths of time that each is focused on is different, due to how the organism decision times happen to line up with the predator/prey appearances and due to the a_i values. Notice the periodic behavior of the $T_1(t)$ and $T_2(t)$ plots (due to the switching from focusing on one predator/prey to another). Figure 7.9 shows a similar plot, but for predators/prey 3 and 4. Notice that the periodic behavior of T_3 and T_4 is different from those shown in Figure 7.8. Ultimately, the pattern of the behavior of the $T_i(t)$ depends on the pattern of predator/prey pulses, the a_i values, the delay values, and how the predator/prey appearances align with the decision times.

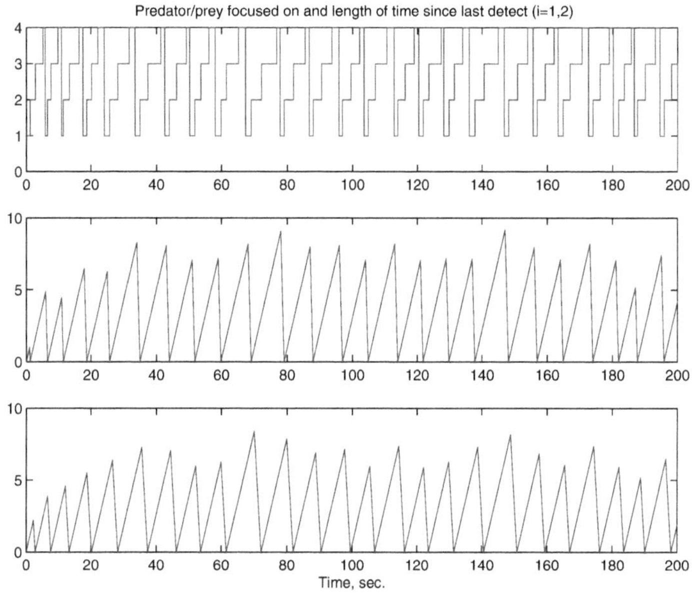

Figure 7.8: Attention scheduler decisions, and $T_i(t)$ for predators/prey 1 and 2.

Figure 7.10 shows a summary view of the dynamics of the attentional scheduling process. There, in the top plot, we also plot the average of the priorities of the predators/prey (assuming that priorities are defined by the i indices). The bottom plot shows the dynamics by showing all the $T_i(t)$ functions on one plot so that you can see the pattern of switching, and the maximum amount of time that the organism ignores any predator/prey. In Figure 7.11, we plot the performance measures of the average length of time since the last detection and maximum length of time since the last detection (and their average values as the straight lines).

Next, the program outputs some numeric values of the performance measures: (i) The time average of the priorities is 2.5670, (ii) the time average of the average values of the lengths of times waited is 3.4066, (iii) the maximum of the average values of the lengths of times waited is 5.7949, (iv) the time average of the maximum values of the lengths of times waited is 5.8297, and (v) the maximum of the maximum values of the lengths of times waited is 9.6199.

The time average of the average values is 3.4066, and this provides a good measure of scheduler performance. What does this value mean? It means that

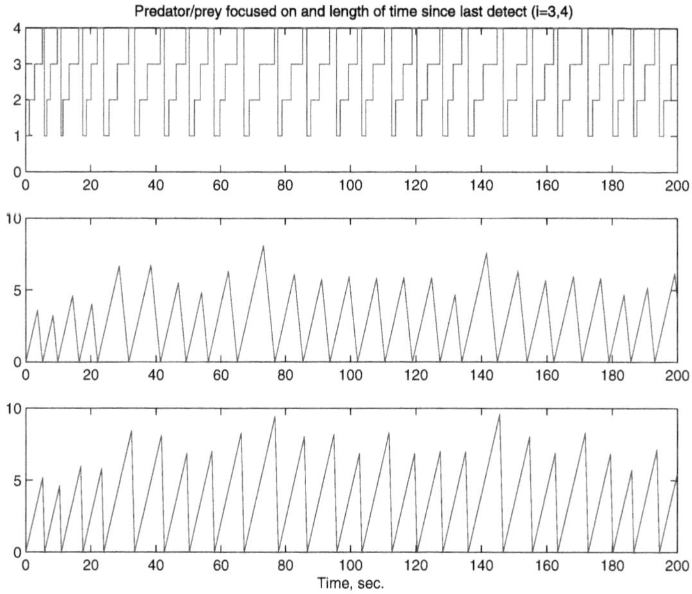

Figure 7.9: Scheduler decisions, and $T_i(t)$ for predators/prey 3 and 4.

on average, the organism detected each predator/prey every 3.4066 seconds. Is this good performance? Notice that the predator/prey appearances occurred every $1, 1.1, 1.2,$ and 1.3 seconds (predators/prey $i = 1, 2, 3, 4$ respectively). Considering the relative low rates of processing to detect the predators/prey, and the delays in switching and waiting for appearances, this appears to be reasonably good performance. Clearly, the performance could go up or down if the frequency or timing of the predators/prey appearances changed.

7.4.3 Effect of Focusing on Higher Priority Predators/Prey

Next, we use the strategy in Equation (7.6) that picks the predator/prey that has been ignored longer than the average one. For the set of predators/prey that has been ignored longer than the average one, we choose the one that has highest priority (i.e., predator/prey i with the greatest value of i). In this way, we study how priorities enter into attentional strategies by augmenting the strategy with a priority scheme. In this case, we get Figures 7.12 and 7.13. We see in Figure 7.12 that the sequence of predators/prey that is focused on is different from the previous strategy, and that the sequence is not periodic in the same way (e.g., it is not a simple $1, 2, 3, 4$ sequence). Also, we see that the average value of the priority of the predator/prey that is focused on is a bit higher, as we would expect. The bottom plot in Figure 7.12 shows quite a different behavior than the bottom plot in Figure 7.10; notice that here there is not an equal "balance" in focusing, since we see that the average values of

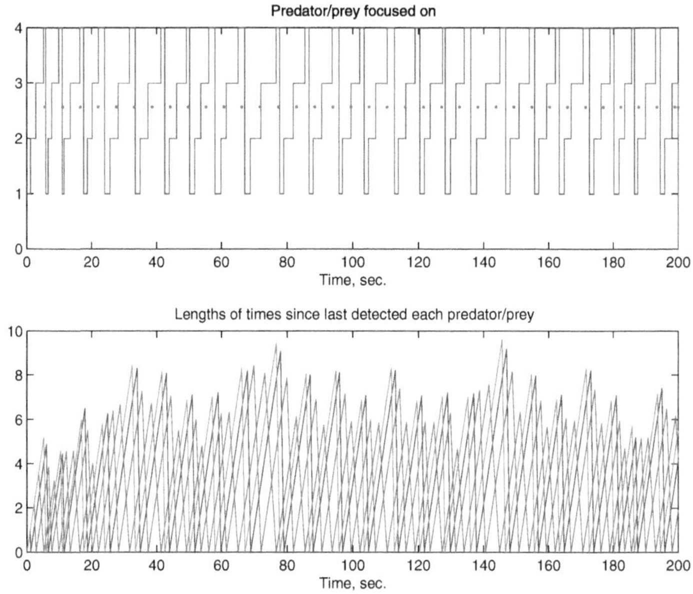

Figure 7.10: Attention scheduler decisions, and $T_i(t)$ for predators/prey $i = 1, 2, 3, 4$.

the $T_i(t)$ are quite different (e.g., see the occasional peaks). Next, notice that in Figure 7.13 we get poorer performance than that shown in Figure 7.11.

Frequent focusing on high priority predators/prey generally requires you to ignore others for longer periods of time.

To quantify the performance further, notice that the numeric performance measures are: (i) the time average of the priorities is 2.6896, (ii) the time average of the average values of the lengths of times waited is 3.8204, (iii) the maximum of the average values of the lengths of times waited is 6.4525, (iv) the time average of the maximum values of the lengths of times waited is 7.6574, and (v) the maximum of the maximum values of the lengths of times waited is 15.7399. This clearly shows that while we get *slightly* better focusing on higher priority predators/prey, we get poorer performance for all the other performance measures. We have paid a price in focusing on high priority predators/prey by ignoring other predators/prey for longer periods of time.

7.4.4 Tuning Attentional Strategy Parameters

As we saw with Equations (7.8) and (7.10), there are ways to define attentional strategies in terms of a set of parameters that specify how they make decisions (e.g., weights or priorities that modify J_p). For instance, we could specify the w_i weights such that there is a high emphasis on focusing on one predator or prey. To do this, you simply make one w_i value much larger than the others. This will result in frequent focusing on the corresponding predator/prey. Suppose that we are not concerned with predator/prey priority, or that all the predators/prey

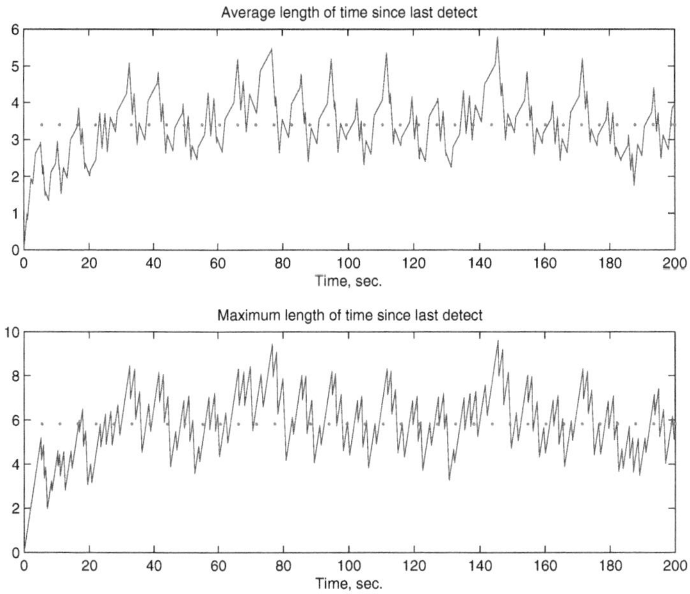

Figure 7.11: Performance measures (average and maximum times since last detection) and the time averages of their values.

have the same priority.

Can we tune the w_i values in Equation (7.8) in order to try to improve the performance measures? That is, can we use the parameters to simply try to improve performance, rather than emphasize focusing on a particular high priority predator/prey? The answer is yes, and to illustrate this, we ran a few simulations, tuning the w_i values with a focus on trying to minimize time average of the average values of the lengths of times waited. We obtained $w_1 = 4$, $w_2 = 2$, $w_3 = 1$, and $w_4 = 4$ and we get the performance in Figure 7.14. The tuning strategy used was to try a set of w_i values and look at the $T_i(t)$ plots. Then the value of w_i was increased a bit for the predator/prey that had higher peak values in order to try to make the strategy focus on that predator/prey more heavily.

The performance for this new set of w_i values is quantified via the following: (i) the time average of the priorities is 2.5802, (ii) the time average of the average values of the lengths of times waited is 3.2755, (iii) the maximum of the average values of the lengths of times waited is 5.3599, (iv) the time average of the maximum values of the lengths of times waited is 5.6423, and (v) the maximum of the maximum values of the lengths of times waited is 9.0899.

Notice that compared to the result in Section 7.4.2, we have tuned the w_i values to get a better value for time average of the average values of the lengths of times waited (there we obtained 3.4066). Is there further room to improve the performance of the scheduler? This seems likely, as the tuning process used

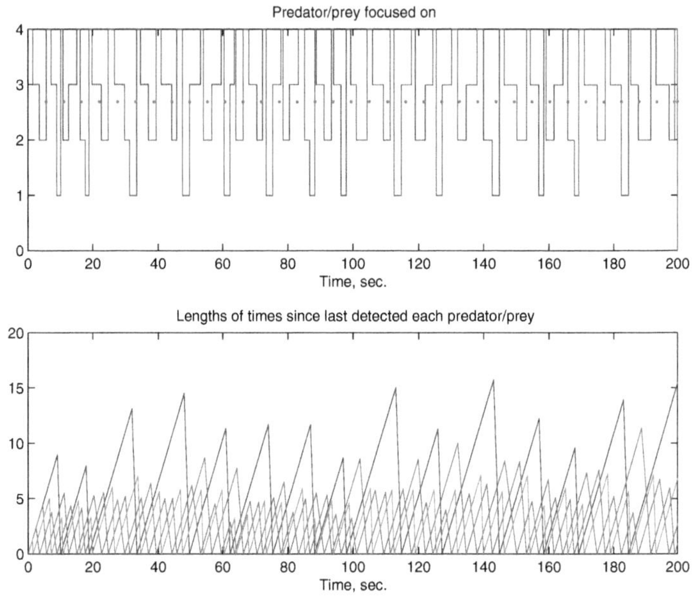

Figure 7.12: Attention scheduler decisions, and $T_i(t)$ for predators/prey $i = 1, 2, 3, 4$.

did not involve consideration of too many values of the parameters. It should be clear that the tuning problem can be quite difficult, especially if there are many predators/prey.

7.5 Stability Analysis of Attentional Strategies

In this section, the first three attentional strategies defined earlier will be proven to be stable, given that the capacity condition in Equation (7.5) holds. Stability of the strategy defined in Equation (7.8) can be studied using a similar proof procedure. Moreover, it is simple to extend the analysis below to the case where priority parameters are added as discussed in Section 7.3.4. At the end of this section, we will explain how to design a strategy that will stabilize *any* scheduling strategy, such as the ones that we will discuss in the next section.

7.5.1 Stability Properties of Attentional Strategies

We begin with the strategies defined in Equations (7.1) and (7.6).

Theorem 1: Assume that Equation (7.5) holds. The attentional strategies where the predator/prey that was ignored the longest time, or one that has been ignored longer than the average one, as defined in Equations (7.1) and (7.6),

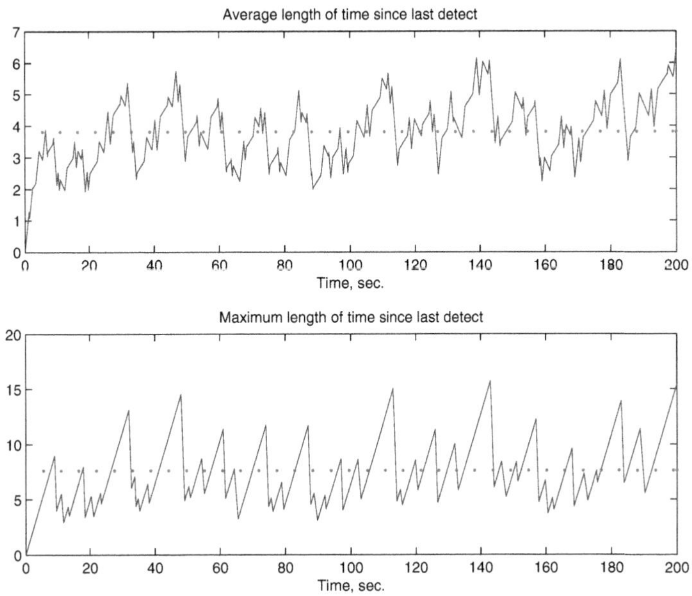

Figure 7.13: Performance measures (average and maximum times since last detection) and the time averages of their values.

have the following properties: They are *stable* in that

$$\sup_{t \geq 0}\{T_i(t)\} < B_i, i \in P$$

for some $B_i > 0$, $i \in P$ so that they will not ignore any predator/prey for too long. A specific bound on the ultimate longest time that the organism will ignore any predator/prey is given by

$$\lim_{t\to\infty} \sup \sum_{i=1}^{N} T_i(t) \leq \delta \left[\frac{\sum_{i=1}^{N} a_i}{\underline{a}} + \frac{\bar{a}N}{\underline{a}\left(1 - \sum_{i=1}^{N} a_i\right)} \max_i \left\{ \frac{-a_i + \sum_{i=1}^{N} a_i}{a_i} \right\} \right]$$

where $\underline{a} = \min_i\{a_i\}$ and $\bar{a} = \max_i\{a_i\}$.

Proof: Let

$$V(t) = \sum_{i=1}^{N} a_i T_i(t)$$

be a "Lyapunov-like" function (strictly speaking it is not a Lyapunov function because $T_i(t)$ is not the state of the system, e.g., due to the presence of the delays). You can think of $V(t)$ as the amount of work that the organism needs to do at time t in order to obtain perfect information about all the predators/prey.

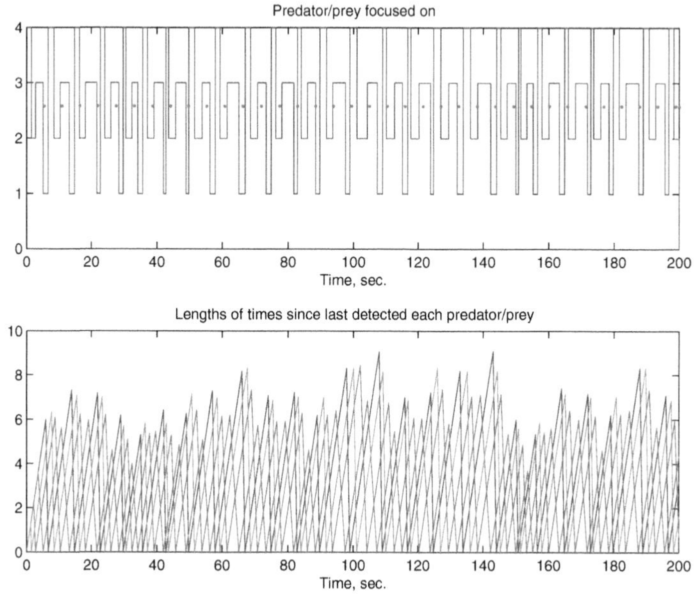

Figure 7.14: Attention scheduler decisions, and $T_i(t)$ for predators/prey $i = 1, 2, 3, 4$.

The proof to follow focuses on the strategy where the predator/prey is chosen that has been ignored longer than the average one; however, a special case of this is when the one that is ignored the longest is chosen at each decision point so the above bounds hold for that attentional strategy also.

Note that since $T_{i^*(k_r)}(D_{k_r+1}) = 0$ ($i^*(k_r)$ was the predator/prey that was just detected),

$$V(D_{k_r+1}) = \sum_{i=1}^{N} a_i T_i(D_{k_r+1}) = \sum_{i \neq i^*(k_r)}^{N} a_i T_i(D_{k_r+1})$$

Also,

$$\sum_{i \neq i^*(k_r)}^{N} a_i T_i(D_{k_r+1}) = \sum_{i \neq i^*(k_r)}^{N} a_i \left(T_i(D_{k_r}) + (D_{k_r+1} - D_{k_r})\right)$$

since when the organism is focusing on predator/prey $i^*(k_r)$, the amount of time that all other predators/prey are ignored increases by $(D_{k_r+1} - D_{k_r})$ for each i, $i \neq i^*(k_r)$. Rearrange this equation to obtain

$$V(D_{k_r+1}) = V(D_{k_r}) - a_{i^*(k_r)} T_{i^*(k_r)}(D_{k_r}) + (D_{k_r+1} - D_{k_r}) \sum_{i \neq i^*(k_r)}^{N} a_i \quad (7.11)$$

Now, use Equation (7.3) to obtain

$$V(D_{k_r+1}) \leq V(D_{k_r}) - \alpha(i^*(k_r))T_{i^*(k_r)}(D_{k_r}) + \beta(i^*(k_r)) \tag{7.12}$$

where

$$\alpha(i) = \frac{a_i\left(1 - \sum_{j=1}^{N} a_j\right)}{1 - a_i}$$

and

$$\beta(i) = \delta\frac{\left(-a_i + \sum_{j=1}^{N} a_j\right)}{1 - a_i}$$

Note that $\alpha(i) > 0$ and $\beta(i) > 0$ for all $i \in P$. To understand how Equation (7.12) is found, using Equation (7.3), note that since $\delta(D_{k_r}) \leq \delta$

$$\begin{aligned} V(D_{k_r+1}) &\leq V(D_{k_r}) - a_{i^*(k_r)}T_{i^*(k_r)}(D_{k_r}) \\ &+ \left[\sum_{j\neq i^*(k_r)}^{N} a_j\right](1 - a_{i^*(k_r)})^{-1}\left[\delta + a_{i^*(k_r)}T_{i^*(k_r)}(D_{k_r})\right] \end{aligned}$$

The term due to δ creates $\beta(i^*(k_r))$. For the remaining terms, besides $V(D_{k_r})$, by grouping we get

$$\begin{aligned} &-a_{i^*(k_r)}\left(1 - \frac{\sum_{j\neq i^*(k_r)}^{N} a_j}{1 - a_{i^*(k_r)}}\right)T_{i^*(k_r)}(D_{k_r}) = \\ &-a_{i^*(k_r)}\left(\frac{1 - \sum_{j=1}^{N} a_j}{1 - a_{i^*(k_r)}}\right)T_{i^*(k_r)}(D_{k_r}) \end{aligned}$$

and this is used to define $\alpha(i^*(k_r))$.

Next, notice that due to the definition of *either* attentional strategy

$$\alpha(i^*(k_r))T_{i^*(k_r)}(D_{k_r}) \geq \alpha(i^*(k_r))\frac{1}{N}\sum_{i=1}^{N} T_i(D_{k_r})$$

and due to the definition of $\bar{a}$,

$$\alpha(i^*(k_r))\frac{1}{N}\sum_{i=1}^{N} T_i(D_{k_r}) \geq \alpha(i^*(k_r))\frac{1}{N}\bar{a}^{-1}\sum_{i=1}^{N} a_i T_i(D_{k_r})$$

(since $\frac{a_i}{\bar{a}} \leq 1$). But notice that

$$\alpha(i^*(k_r))\frac{1}{N}\bar{a}^{-1}\sum_{i=1}^{N} a_i T_i(D_{k_r}) = \alpha(i^*(k_r))\frac{1}{N}\bar{a}^{-1}V(D_{k_r}) \tag{7.13}$$

Combine this with Equation (7.12) to get

$$V(D_{k_r+1}) \leq \left[1 - \bar{a}^{-1}N^{-1}\alpha(i^*(k_r))\right]V(D_{k_r}) + \beta(i^*(k_r)) \tag{7.14}$$

Subtract $\bar{a}N \max_i \frac{\beta(i)}{\alpha(i)}$ from both sides of Equation (7.14) and after a bit of algebra, you get

$$
\begin{aligned}
V(D_{k_r+1}) &- \bar{a}N \max_i \frac{\beta(i)}{\alpha(i)} \\
&\leq \left[V(D_{k_r}) - \bar{a}N \max_i \frac{\beta(i)}{\alpha(i)}\right]\left[1 - \bar{a}^{-1}N^{-1}\alpha(i^*(k_r))\right] \\
&+ \beta(i^*(k_r))\left[1 - \frac{\alpha(i^*(k_r))}{\beta(i^*(k_r))} \max_i \frac{\beta(i)}{\alpha(i)}\right]
\end{aligned}
$$

Focus for a moment on the last term in this equation, and notice that

$$
\beta(i^*(k_r))\left[1 - \frac{\alpha(i^*(k_r))}{\beta(i^*(k_r))} \max_i \frac{\beta(i)}{\alpha(i)}\right] \leq 0
$$

How do you get the last inequality? Note that $\beta(i) > 0$. If the $\max_i \frac{\beta(i)}{\alpha(i)}$ term is maximized at some particular value j, then clearly this value divided by any value considered in the maximization will be greater than or equal to 1.

Now, we have

$$
\begin{aligned}
&\left[V(D_{k_r+1}) - \bar{a}N \max_i \frac{\beta(i)}{\alpha(i)}\right] \\
\leq &\left[V(D_{k_r}) - \bar{a}N \max_i \frac{\beta(i)}{\alpha(i)}\right]\left[1 - \bar{a}^{-1}N^{-1}\alpha(i^*(k_r))\right] \qquad (7.15)
\end{aligned}
$$

But, notice that the second term on the right-hand side of this equation

$$
\begin{aligned}
\left[1 - \bar{a}^{-1}N^{-1}\alpha(i^*(k_r))\right] &\leq 1 - \bar{a}^{-1}N^{-1} \min_i \left\{\frac{a_i\left(1 - \sum_{j=1}^N a_j\right)}{1 - a_i}\right\} \\
&\leq 1 - \bar{a}^{-1}N^{-1}\left[\frac{\underline{a}\left(1 - \sum_{j=1}^N a_j\right)}{1 - \underline{a}}\right]
\end{aligned}
$$

Notice that

$$
0 < \frac{\left(1 - \sum_{j=1}^N a_j\right)}{1 - \underline{a}} < 1
$$

and

$$
0 < \frac{\underline{a}}{\bar{a}} < 1
$$

so that

$$
0 < 1 - \bar{a}^{-1}N^{-1}\alpha(i^*(k_r)) < 1
$$

which makes the mapping in Equation (7.15) contractive so that

$$
\lim_{k_r \to \infty} \sup \left\{V(D_{k_r}) - \bar{a}N \max_i \frac{\beta(i)}{\alpha(i)}\right\} = 0
$$

But this (ultimate) bound is in terms of only the decision points D_{k_r}, $k_r = 1, 2, 3, \ldots$ Due to the delay δ, the $T_i(t)$ values can rise higher at times t not at the decision points. However, for $D_{k_r} \leq t \leq D_{k_r+1}$

$$V(t) \leq V(D_{k_r} + \delta)$$

But notice that

$$V(D_{k_r} + \delta) = \sum_{i=1}^{N} a_i T_i(D_{k_r} + \delta) = \sum_{i=1}^{N} a_i T_i(D_{k_r}) + \delta \sum_{i=1}^{N} a_i = V(D_{k_r}) + \delta \sum_{i=1}^{N} a_i$$

This gives us

$$\lim_{t\to\infty} \sup V(t) \leq \delta \sum_{i=1}^{N} a_i + \bar{a} N \max_i \frac{\beta(i)}{\alpha(i)}$$

and since

$$\lim_{t\to\infty} \sup \sum_{i=1}^{N} T_i(t) \leq \frac{1}{\underline{a}} \lim_{t\to\infty} \sup V(t)$$

we know

$$\lim_{t\to\infty} \sup \sum_{i=1}^{N} T_i(t) \leq \frac{\delta \sum_{i=1}^{N} a_i}{\underline{a}} + \frac{\bar{a} N}{\underline{a}} \max_i \frac{\delta \left(-a_i + \sum_{j=1}^{N} a_j\right)}{a_i \left(1 - \sum_{j=1}^{N} a_j\right)}$$

which gives the desired result. ■

Note that since the above bound for Theorem 1 may be conservative for some situations, it would be of interest to specify "tight" bounds since this would provide good guarantees for bounding the maximum time that a predator/prey is ignored.

Next, we will study the stability properties of the other strategy defined in the last section where we get a different bound on the maximum length of time that a predator/prey will be ignored by the organism. The analysis, is however, only slightly different and depends on the above proof.

Theorem 2: Assume that Equation (7.5) holds. The attentional strategies defined in Equation (7.7) have the following properties: It is *stable* in that

$$\sup_{t\geq 0}\{T_i(t)\} < B_i, i \in P$$

for some $B_i > 0$, $i \in P$ so that it will not ignore any predator/prey for too long. A specific bound on the ultimate longest time that the organism will ignore any predator/prey is given by

$$\begin{aligned} \lim_{t\to\infty} \sup V(t) &\leq \frac{\delta(N-1)}{1 - \sum_{i=1}^{N} a_i} \left(-\underline{a} + \sum_{i=1}^{N} a_i\right) + \delta \sum_{i=1}^{N} a_i \\ &\leq \delta \left[\sum_{i=1}^{N} a_i\right] \frac{N - \sum_{i=1}^{N} a_i}{1 - \sum_{i=1}^{N} a_i} \end{aligned}$$

where $\underline{a} = \min_i\{a_i\}$ and $\bar{a} \max_i\{a_i\}$.

Proof: Use the ideas from the proof for Theorem 1 and note that Equation (7.13) in this case is

$$\alpha(i^*(k_r))T_{i^*(k_r)}(D_{k_r}) = \frac{\alpha(i^*(k_r))}{a_{i^*(k_r)}} a_{i^*(k_r)} T_{i^*(k_r)}(D_{k_r})$$

$$\geq \frac{\alpha(i^*(k_r))}{(N-1)a_{i^*(k_r)}} \sum_{i=1}^{N} a_i T_i(D_{k_r}) = \frac{\alpha(i^*(k_r))}{(N-1)a_{i^*(k_r)}} V(D_{k_r})$$

The $N-1$ factor appears, rather than N, for one i, $T_i = 0$. To complete the proof, simply take the same approach as in the remainder of the proof of Theorem 1, below Equation (7.13). ■

So, do these bounds give an indication of which of the three strategies is "best"? Unfortunately, they generally do not since the bounds can be conservative. It is for this reason that simulation analysis is generally needed to analyze the *performance* of particular strategies and determine which is best for a particular predator/prey environment.

7.5.2 Stabilizing Mechanism for Attentional Strategies

At times there is significant knowledge about the predator/prey environment and organism that is relevant to the design of attentional strategies. There is then a natural tendency to incorporate this information in the specification of the attentional strategy, often in the form of "scheduling heuristics." We will briefly discuss two such approaches in the next section. The problem with this approach, however, is that the resulting strategies may end up being somewhat nonstandard and there may be concerns about whether they will be stable.

Fortunately, the approach in [290] to specifying a "universal stabilizing mechanism" (USM) for any scheduling strategy actually holds for the attention scheduling problem. (Actually, the approach in [290] was developed for a fixed size delay and we have a time-varying but bounded delay; however, the proofs there can be directly extended to our case with no difficulty.) This mechanism can then be applied to *any* heuristically constructed attentional strategy, and you will be ensured that the overall strategy will be stable. In this section, we introduce the USM from [290]. In the next two sections, we introduce two types of schedulers that exploit predator/prey domain information to try to enhance scheduler performance, and which can be stabilized by the USM introduced here.

The USM allows the designer to focus on improving performance of the attentional strategy.

The key fact is that for the resource allocation problems we consider here, as long as the capacity condition is satisfied, it is possible to define a USM which, when used to supervise a scheduling strategy, will always result in stable operation. To define the USM, let Q denote a first-come first-serve (FCFS) priority queue for predators/prey that have been ignored for a long time. For instance, if predator/prey's T_i value becomes too large, we will have criteria for entering the queue at some time t'. If predators/prey i and j are in the priority queue, and j entered it before i,

$$Q = (\ldots, i, j, \ldots)$$

then when this priority queue is serviced, predator/prey j will be taken off the queue and focused on before predator/prey i (the "tail" of the queue is the first predator/prey listed after the "(" and the "head" of the queue is the predator/prey listed just before the ")" in the definition of Q above). We need some additional parameters to specify the USM. Let $L > 0$ be a large number satisfying

$$L > \frac{N\delta}{1-\rho}$$

where ρ is specified in Equation (7.5), N is the number of predators/prey, and δ is the bound on the maximum delay. Next, let

$$H_i > 0, \; i \in P$$

denote a set of parameters, the interpretation of which will become clear as we define the USM.

The USM is implemented by the following set of rules:

1. *Truncation rule:* The organism can process no predator/prey i longer than La_i time units. This means that if at time $t'+\delta_s+\delta_e(t')$ the organism starts to try to detect predator/prey i, then it can only try to detect it no longer than up to the time $t' + \delta_s + \delta_e(t') + La_i$. If detection occurs before that time, then the strategy acts as usual and selects another predator/prey to focus on. If, however, it has not yet detected predator/prey i by this time, it is forced to make a new decision (which could entail switching predators/prey). Note that if it does switch to another predator/prey, we assume that the progress it had made on predator/prey i is used, but that the time since it was last detected, T_i, begins to increase again.

2. *Rule for entering Q:* Predator/prey i enters the tail of the priority queue Q at time t if we have not just decided to focus on i or are currently focusing (cognitively processing) to detect i, and $T_i(t) > H_i$ (hence, the H_i are thresholds for when a predator/prey is placed in the queue).

3. *Predator/prey selection rule:* If Q is not empty when the organism has finished focusing on a predator/prey (either by achieving $T_i = 0$ or via rule 1 above), then the predator/prey at the head of the priority queue Q (i.e., FCFS) is chosen.

4. *Rule for leaving Q:* A predator/prey i leaves Q at the time $t' + \delta_s + \delta_e(t')$ where t' is the time point when predator/prey i was selected by rule 3.

5. *Rule for processing-time for a predator/prey from Q:* If predator/prey i from Q is chosen to be focused on, then beyond the time t' defined in rule 4, it is processed for La_i time units unless it is detected (i.e., $T_i = 0$) before this time elapses.

Notice that this is not simply another attentional strategy. It actually defines a "supervisor" for any attentional strategy (e.g., ones that exploit heuristic information from the problem domain) that ensures it will result in stable operation. If you have constructed a stable strategy, and you choose L and the H_i, $i \in P$, large enough, then the USM will *never* intervene. The USM simply truncates the processing of predators/prey that are not found fast enough, and via Q makes sure that predators/prey that have been ignored for too long will get attention. How do we pick L and the H_i parameters? If you pick $H_i = 0$, $i \in P$, then the USM simply enforces a type of FCFS strategy on predators/prey with $T_i > 0$, but it stops processing any predator/prey that is focused on too long. In this case, the USM always intervenes. As you increase the size of L and the parameters H_i, $i \in P$, the USM intervenes less frequently.

What is the value of the USM? In a sense, it frees the designer of attentional strategies from being concerned about the stability of the myriad possible attentional strategies (but of course, the stability analysis of Section 7.5 is still useful, particularly if the analysis helps to clarify how to design the strategy to achieve high performance operation). You can adopt a design philosophy where you construct a very complicated attentional strategy, possibly exploiting heuristic ideas about how to achieve the best performance. Then, you can augment such strategies with the USM and be assured that you will obtain stable operation. Essentially, the USM allows the designer to focus on the design of attentional strategies to improve attention scheduling *performance*. To illustrate this point, in the next section we will briefly discuss the design of two heuristic strategies, ones based on our intuitions about the problem domain.

7.5.3 Planning and Attention

In this section, we discuss two ways to use planning concepts from Chapter 6 in attentional strategies. Intuitively, this should make sense. We can plan how to pay attention to a set of predators and prey if we have some idea of how the environment might behave, and if we consider alternative predators/prey to focus on based on predictions about how they might behave. We consider the alternatives and choose what we think is the best one to focus on based on these predictions. As an example, per our discussion in Sections 7.3.3 and 7.3.5, it should be clear that even our earlier strategies used a type of online optimization to choose which predator/prey to focus on. Moreover, for the strategy in Equation (7.8), we used a type of prediction in determining which was the best predator/prey to focus on (there we predicted which T_i would be highest after a delay, scaled that prediction, and then used it to decide which predator/prey to focus on). For that strategy, the information we used was quite simple, and only incorporated some information about delays in the organism and environment.

If environmental or organism information is available, it can be used to plan what to attend to.

Hence, in a limited way we have already considered the use of planning concepts in attentional strategies. Here, however, we will consider two explicit ways to incorporate more detailed information about the environment. We invite the reader to evaluate the performance of these strategies in Design Problem 7.6.

Attentional Strategies that Use Predator/Prey Behavioral Characteristics

The strategies considered up to this point do not incorporate a significant amount of a priori information that may be available about the likely timing of predator/prey appearances. For example, if the organism has identified the predator/prey *type*, it may have a good guess of when the next appearance time will be, or if it has observed a fixed pattern of appearances in the past, it may have a guess of when it will appear again. Without using such a priori information, the attentional strategies may focus on a predator/prey even though it is unlikely that it will appear or be found for some period of time, and during this time, the organism could more profitably search for and detect other predators/prey.

How can such a priori information be incorporated? We simply provide a few ideas here. First, suppose that we use a "certainty of appearance" function

$$C_i^k(t, t_i^k)$$

for each predator/prey i, which is defined along the time-line $t \geq t_i^k$ starting from the time t_i^k when predator/prey i was last detected (i.e., from the time that the predator/prey was detected for the kth time). Suppose that this function has values in the range of $[0, 1]$, with 0 representing that it is unlikely that there will be an appearance, 0.5 representing uncertainty about whether there will be an appearance, and 1 representing that you are certain that there will be an appearance (based on a priori information). Now, suppose we define a strategy that at each decision point simply picks the predator/prey to focus on that is most likely to appear (and perhaps taking into account any delay in switching focus to a different predator/prey). See Figure 7.15.

In Figure 7.15, notice that there are appearance certainty functions for four predators/prey. Predator/prey 1 is predicted to appear with a higher frequency, and the width of each of the humps quantifies the certainty of occurrence of appearance; hence, appearances are most certain at the peaks. Notice that predators/prey 2 and 3 are predicted to have similar (lower) frequency appearances, but the precise timing of the appearances is not as certain and this is quantified via the spreads of the humps being larger. Predator/prey 4 is predicted to be a lower frequency illuminator, but the certainties of when the appearances will occur is similar to that specified for predator/prey 1. Note that the parameters defining the $C_i^k(t, t_i^k)$ functions (e.g., the points where the peaks occur and the spreads) could be estimated in some situations by some other cognitive subsystem, and then the $C_i^k(t, t_i^k)$ functions used by the attention scheduler could be changed. Finally, note that these certainties could be scaled by predator/prey "priorities" so that the strategy could choose to focus on the highest priority predator/prey that is likely to produce an appearance.

Will this result in a stable strategy? No, not if that is all that is used in the attentional strategy. It could be that you have bad a priori information so that bad guesses are made and the appearances are never found for a predator/prey and so the length of time that it is ignored goes to infinity (representing that

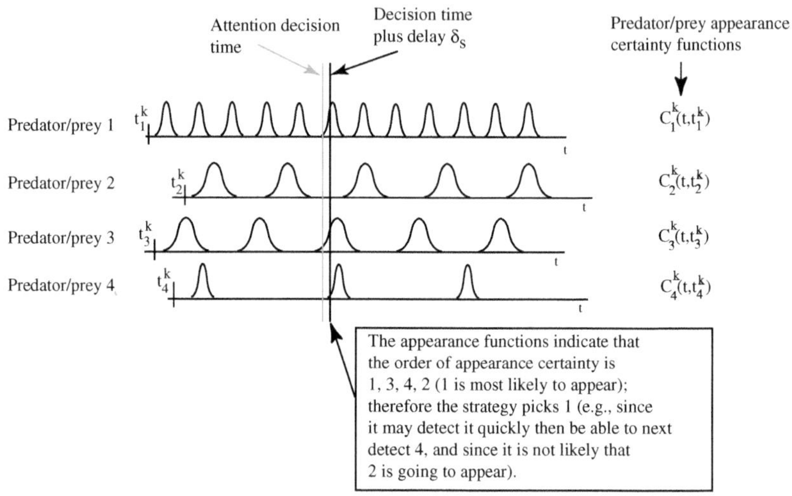

Figure 7.15: Attention strategy that exploits information about likely times of appearance of predators/prey.

ultimately it knows nothing about the predator/prey). We can, however, use the USM of the previous section to ensure stable operation. Moreover, if the $C_i^k(t, t_i^k)$ represent good predictions about the predators and prey, it is possible that very good scheduling performance can be achieved.

Attentional Strategies Based on Model Predictive Control

In most engineering applications we can simulate, to a reasonable degree of accuracy, the domain in which we make decisions. For instance, in this chapter we have simulated the predator/prey environment in Section 7.4. The actual predator/prey environment is certainly somewhat different from what our simulations would lead us to believe. Let us suppose, however, that we can simulate the predator/prey environment reasonably well, at least in its broad characteristics. Furthermore, suppose that the organism can simulate this model of the predator/prey environment *in real-time* in some cognitive module. Would such a simulation provide useful information to help decide which predator/prey to focus on? Below, we study this question by providing one way to incorporate a simulated predator/prey environment into a scheduling strategy.

Suppose that we use the model of the predator/prey environment to predict how the organism will perform using different strategies or orders of focusing on predators/prey. Suppose that we use the model to predict M different behaviors that result from M different candidate sequences of predators/prey to focus on of length N_h (a specification of a sequence of N_h predators/prey to focus on). The strategy is shown in Figure 7.16.

As shown in the figure, for the MPC strategy we rank order the M different

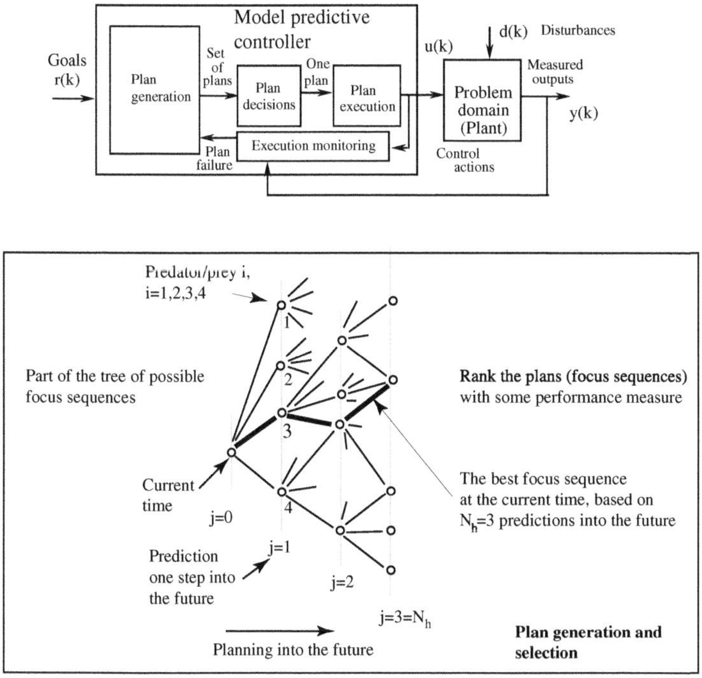

Figure 7.16: Model predictive control (MPC) for use in an attentional strategy.

predator/prey focus sequences and choose the best one, and then we focus on the predator/prey specified as the first one to focus on in the best predator/prey sequence. The process repeats at the next decision point, i.e., when that predator/prey is detected. You can think of the MPC attentional strategy as a more sophisticated version of the attentional strategy discussed in the last section. We think of N_h as specifying a "receding horizon" or length of time we predict ahead in time. For a very uncertain predator/prey environment it typically does not make sense to make N_h very large since the predictions will typically become more inaccurate as we predict farther ahead in time. If, however, your model is good and you have sufficient computational resources you may want to predict into the future for longer periods of time so that the best possible predator/prey is chosen to focus on.

Clearly, if information was gathered online, you may be able to profitably update the model that is used in the MPC strategy (this would then result in the incorporation of learning and planning into attention). Moreover, it is not difficult to incorporate a predator/prey priority scheme. Will MPC-type strategies result in stable scheduling? Probably not. However, once again we can use the USM to ensure that we obtain stable operation.

7.6 Attentional Systems in Control and Automation

In this section we will overview how attentional systems and multisensor integration can be used in control and automation. For more information on each of these topics, see the "For Further Study" section at the end of this part.

7.6.1 Attentional Strategies for Control

In this section, we briefly explain how to augment the control strategies considered so far with attentional mechanisms. Later, in Chapter 9, we briefly discuss relationships between learning and attention and in Section 9.4.5, we discuss how to augment adaptive (learning) controllers with attentional mechanisms.

At the neural level, attentional mechanisms can be implemented by neurons so that an organism focuses on the most important aspects of its environment in achieving a control task (e.g., stimulus-driven attention reorienting that is implemented in a network of neurons). There has been a variety of neural network models introduced for attentional systems, and some of these have been experimentally validated to a certain extent. Some of the models have incorporated the hierarchical aspects of attention, while others have illustrated how attention is integrated with visual processing such as object recognition. Here, we do not investigate neural network models for attention, but in Design Problem 7.5 we provide some references and invite the reader to do so.

Attentional strategies can be employed in rule-based planning and learning controllers.

Typically, the central issue in augmenting a fuzzy or expert controller with an attentional mechanism is to add a mechanism that manages the matching process since that is typically the most complex part of those systems, and the part where sensory data are processed to determine how they should be used. The attentional system in this case could try to prune the number of rules that are on at any one time based on contextual information that is gathered. For instance, suppose that you have a controller with many inputs (e.g., 1000 or more). In this case, you could define priorities for your control objectives and then you could only consider inputs that help you to meet those objectives, or you could process the inputs to capture the essential features. This would be a supervisory strategy that managed the flow of input information so that the computational complexity is reduced. This strategy is shown in Figure 7.17.

To achieve "attentive planning," the ideas for integrating planning and attention in the last section could be useful, or, the attentional system could prune projections into the future (as in Figure 6.2) since that is often the most computationally complex part of the planning process. This is pictured in Figure 7.18. Goals, hard constraints, and other inputs may provide the information for how to prune. Attention can make the complex problem of predicting the many ways that the system can behave in reaction to different sequences of inputs, but it could result in a performance degradation in control performance. Essentially, attention tries to reduce complexity to a manageable level, without sacrificing too much performance.

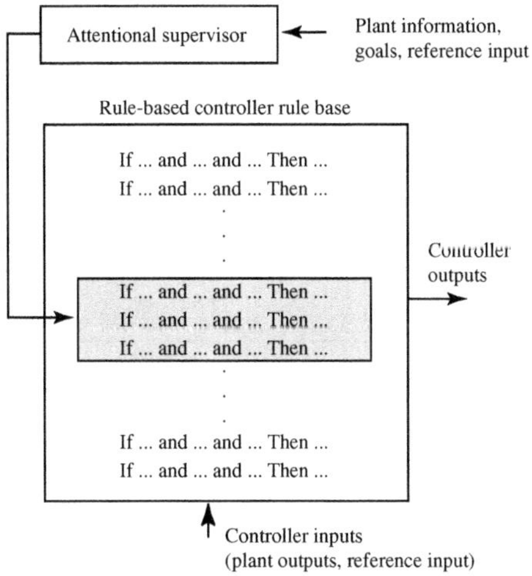

Figure 7.17: Attentional strategy for rule pruning for rule-based control.

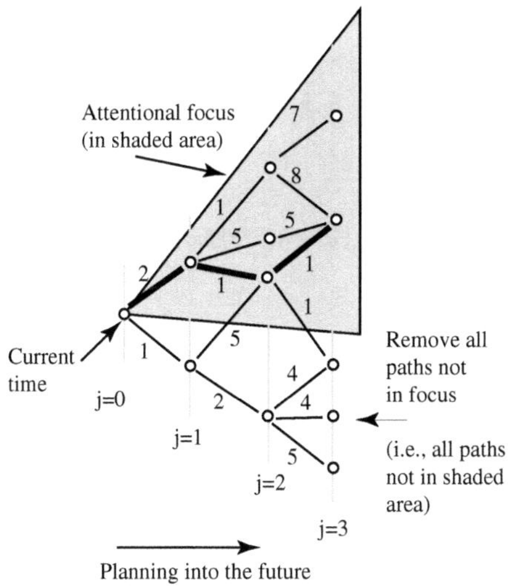

Figure 7.18: Attentional strategy for plan pruning.

7.6.2 Filtering and Focusing: Multisensor Integration

In complex highly automated systems, it is often necessary to use multiple types of sensors for obtaining information about the environment (plant). For instance, a mobile robot may need sensors for velocity, acceleration, yaw, etc. It may also need a vision system for obstacle avoidance, coupled perhaps with radar or an ultrasonic sensor for reliability in achieving obstacle avoidance. The robot must decide how to combine this information for object recognition, decision-making, and other tasks. For some tasks it may ignore some sensor data, and pay attention to other data. For other tasks it may "fuse" data from two or more different sensors. The general task for a "multisensor integration system" is to distill the most useful information from the suite of sensors. A general sensor integration system is shown in Figure 7.19.

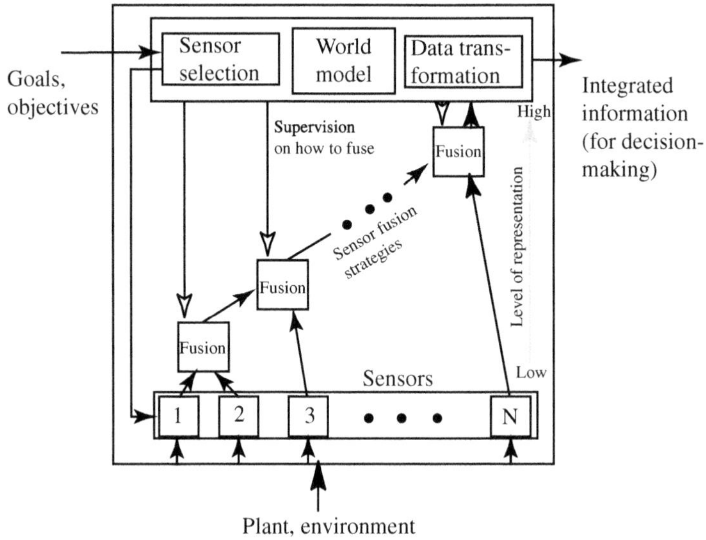

Figure 7.19: Multisensor integration and fusion (adapted from [339], © IEEE, used with permission).

Sensor fusion and integration are closely related to key functionalities in attentional systems.

Here, we see that there are N sensors, each possibly of a different type, or copies of one type of sensor (e.g., for reliability purposes). There is a sensor selector that decides which sensors should be enabled. Then, there is a sensor fusion strategy. In Figure 7.19, we show one strategy where information from sensor 1 is fused with information from sensor 2, and then that fused information is fused with the information from sensor 3, and so on. Other strategies are also possible (e.g., having two fusion strategies combine information from two sensors each, then you could have another fusion approach for the fused information from those). There are a wide variety of methods for multisensor fusion (e.g., Kalman filtering, Bayesian estimation, etc.), world modeling, sensor selection, and data transformation. The interested reader should consult the references in the "For Further Study" section at the end of this part.

The fusion strategies may have guidance from the higher level functionality. The fused information is passed to the higher level and may be stored in a "world model" (a representation of aspects of the environment that are useful for decision-making to reach the goals, but which may also help guide the overall strategy as to how to fill information that is needed). To achieve world modeling and ultimately sensor integration, we will often have to also perform data transformations. Ultimately then, multisensor integration systems do possess some key features of learning, and that topic is covered in Part III. Next, note that there is a general process of filtering that is naturally involved in sensor fusion where some information is discarded and other information is derived by combining data. This results in a "low" level of representation early in the fusion process and a "high" level of representation at the end where the most useful information has been obtained.

There are several possible advantages to integrating information from multiple sensors. For instance, some sensors may provide *redundant* information which can reduce overall uncertainty about what is being sensed, or it can provide for fault tolerance in case a sensor fails. Sometimes information is *complementary* in the sense that it may allow, via appropriate processing, for the perception of some objects that could not be perceived otherwise. Sometimes, multisensor integration can speed up the overall process of decision-making by providing the proper information faster. Other times, it may be possible that using multisensor integration strategies will result in a less-expensive system.

Finally, we note that the focus in this chapter is largely *not* on attentional strategies for control or on multisensor integration, but on how to use control concepts (scheduling for resource allocation) for attentional strategies.

7.7 Exercises and Design Problems

Exercise 7.1 (Simulation of Attentional Strategies):

(a) Simulate all the attentional strategies in Section 7.4, reproducing the results found there.

(b) Let $N = 3$ and use

$$\delta^1 = 0.9,\ \delta^2 = 1,\ \delta^3 = 1.2$$

Synthesize sequences similar to those shown in Figure 7.7 that satisfy these constraints (make the appearances periodic). Choose $\delta_s = 0.03$ and

$$a_1 = 0.3,\ a_2 = 0.2,\ a_3 = 0.1$$

Simulate the three attentional strategies studied in Sections 7.4.2 and 7.4.3 and evaluate the performance of each attentional strategy. Tune the w_i parameters to obtain as good performance as you can via manually tuning these parameters.

Exercise 7.2 (Stability Analysis of Priority-Based Attentional Strategies): Prove that the policy defined by embedding priorities via the approach in Section 7.3.4 into the strategy defined by Equation (7.1) is stable if appropriate conditions are met (state these, and show each step in your proof). Provide explicit ultimate bounds on the T_i values.

Exercise 7.3 (Stability Analysis of an Attentional Strategy): Prove that the attentional strategy defined in Equation (7.8) is stable if appropriate conditions are satisfied (specify the conditions and show each step in your proof). Provide explicit ultimate bounds on the T_i values.

Design Problem 7.1 (Tuning Attentional Strategies with Priorities): This problem focuses on how to tune attentional strategies to improve their performance.

(a) Specify priority parameters $p_i \in P$ and explain how to embed a priority scheme into the attentional strategy in Section 7.3.2 using the ideas in Section 7.3.4. For a specific set of priority parameters, develop a simulation of the priority attentional strategy and evaluate its performance for the scheduling problem defined in Section 7.4. Tune the priority parameters to try to improve performance, where you measure performance by the time average of the average values of the lengths of times waited.

(b) Next simulate the strategy given by Equation (7.8) and tune the w_i parameters to obtain better performance, as measured by the time average of the average values of the lengths of times waited, than what we obtained in the chapter. Compare the performance that you obtained to that which you obtained in (a).

Design Problem 7.2 (Stable Attentional Strategy Design):

(a) Suppose that you consider the set P of labels for the predators/prey as specifying the sequence that they should be focused on (and suppose that this sequence is fixed a priori by the labeling). Suppose that you define a policy that at each decision point simply picks predator/prey $1, 2, \ldots, N$ in sequence, and after it finishes with predator/prey N, it returns to predator/prey 1 and repeats the process. Will this result in a stable attentional strategy? Why? Why not? Can you generate a counterexample to stability, or provide a proof of stability that does not use the USM?

(b) Can you define an attentional strategy that will result in stable operation, but is different from the others discussed in this chapter and does not use the USM? Specify the strategy and prove stability.

Design Problem 7.3 (Design of Universal Stabilizing Mechanisms): For the scheduling problem in Section 7.4, employ the attentional strategy defined in Design Problem 7.2(a). Augment the strategy with the USM. Simulate the strategy for various choices of USM parameters and explain the effects of these parameters on attentional strategy behavior and performance (measure performance by the time average of the average values of the lengths of times waited). Be sure to simulate the attentional strategy for a sufficient period of time so that the performance measures represent the long-term performance of the attentional strategy. To get accurate performance measures, do you need to repeat the simulation many times with different sequences of choices of predators/prey to focus on?

Design Problem 7.4 (Design of "Optimal" Attentional Strategies): This problem builds on Design Problem 7.1 by exploring systematic ways to pick the best attentional strategy parameters. Choose a *stable* attentional strategy (you may use the USM) that seems to have the potential to obtain a better value for the time average of the average values of the lengths of times waited than the one in Section 7.4.2. One approach to this is to tune the priority or weighting parameters for the attentional strategies in Design Problem 7.1 to try to obtain better performance. Tune the parameters of the attentional strategy with a goal of obtaining a better value for the time average of the average values of the lengths of times waited than the one in Section 7.4.2. Hint: You may want to produce a systematic approach to tuning the parameters of the scheduler rather than just manually tuning them. One approach to do this is to use ideas from the "response surface methodology" discussed in Chapter 15. A simple version of this approach is to simply create a grid of attentional strategy parameters and simulate the strategy for each point on the grid (which can take significant computational resources) and pick the parameters that correspond to the best performance. Another approach would be to use the "simultaneous perturbation stochastic approximation" algorithm that is studied in Chapter 15.

Design Problem 7.5 (Neural Models of Attentional Systems)*: There is research in the literature on how to develop neural network models of attentional mechanisms and this problem studies the simulation of attentional systems via such models.

(a) For background reading, read the article [372]. Search the literature on this topic to supplement this study.

(b) Implement code necessary to study the attentional system and reproduce the simulations shown in [372]. Focus on the simulation of the "spotlight" view of attention.

(c) Explain how such an attentional mechanism may be useful in a control system. Identify at least two ways in which it can be used.

Design Problem 7.6 (Attentional Strategies Based on Planning and Learning)*: In Section 7.5.3 we introduced two ways to use planning concepts in attentional strategies. Here, you will completely specify such a strategy, simulate it, and evaluate its performance.

(a) Using the ideas in Section 7.5.3, develop an attentional strategy that incorporates planning concepts. You do not have to precisely follow the methodologies specified earlier; you can invent your own method. Specify the attentional strategy, explain what environmental/organism information it needs in order to predict how the attentional strategy will operate, explain what cost function will be used to select a single sequence of predator/prey focuses (plan) from the set that is generated, and explain how the overall approach seeks to improve attentional performance. Specify the strategy in a way that will ensure that it is stable (you may use the USM).

(b) Specify a single performance measure that you would like to optimize. Develop a simulation of the attentional strategy that you specify in (a) and tune the strategy to try to optimize your chosen performance measure. You should use a scheduling problem and performance measure similar to the ones in Section 7.4.

(c) Next, expand on your strategy by incorporating a method to learn the model that is used by your planning strategy to predict. Repeat (b) for this strategy.

Design Problem 7.7 (Cooperative Attentional Systems)*: Suppose that there are M agents, each with an attentional system given by the model used in the chapter. Suppose that they are seeking predators and prey, but that they do so cooperatively in the sense that they identify N predators or prey and then cooperate on paying attention to them. With cooperation we expect that there will be an increased "capacity" to pay attention.

(a) Define two cooperative attention strategies. For instance, suppose that the $M < N$ agents act autonomously but share an "unattended" set of things that are not paid attention to at the current time. There is then a corresponding set of predators/prey that the group of M agents is attending to. A decision strategy can be defined in terms of what each agent does at its decision times. For instance, it may "check out" (using a mutual exclusion strategy) the unattended set and pick a particular predator/prey to focus on; then it can "return" the new unattended set to the others. It can then focus on that predator/prey until it is detected. The agents would then make all their decisions asynchronously. What predator/prey should be chosen from the unattended set? Mathematically define two strategies for the agents to make these choices.

(b) Simulate the cooperative attentional strategy and show plots as we did in the chapter to evaluate their performance (e.g., relative to the $M = 1$ case).

(c) Augment your strategies with the learning/planning methods you studied in Design Problem 7.6 and then evaluate their performance in simulation.

(d) Find conditions under which the strategies of (a) will result in stability in the sense that it was studied in the chapter.

Chapter 8

For Further Study

To deepen your understanding of the wide variety of neural network methods and applications, see [238], and for neural bio-foundations, see [268, 269, 558, 130]. More details on fuzzy control theory and applications are given in [412]. For more details on planning for autonomous robots, see [278] and a tutorial on model predictive control is given in [192]. The approach to the study of attentional systems was based on viewing the problem as a resource allocation problem and there is a large amount of literature on this topic. Here, the development was based on [418].

Neural Networks and Motor Control: We must emphasize that there are many topics in the area of neural networks that are not covered here. For instance, we do not discuss associative memories and recurrent networks, or Boltzmann machines. We refer the reader to [238, 262, 356] for treatment of these topics, or for references to sources where these topics can be studied. A recent theoretical treatment of modeling and analysis methods for theoretical neuroscience is in [130]. There, the authors cover a variety of topics including encoding and decoding information, neuron and neural network models, and adaptation and learning. By studying [130], the reader can gain a better understanding of "firing rate models," and "tuning curves," of neurons and hence, how accurate our neural network models are.

There has been extensive work on neurophysiological studies of motor control (e.g., on the hierarchy of the controls) in [206, 268, 269, 558], and the interested reader is recommended to see [274] and the current literature where close connections to control system methodology are found.

A method that has been popular in the control of robots has been the cerebellar model articulation controller (CMAC), which was first introduced in [9], and later applied in a different form in [362, 286]. Other neural network applications to robotic systems are contained in [279, 207].

Fuzzy and Expert Control: A recent, and relatively detailed, overview of the literature and methods of fuzzy control is given in [412]. You may also want to consider [155, 531, 262, 134]. Fuzzy logic is covered in [350, 280, 559, 440] and other places (here, we do not emphasize fuzzy logic formalisms for their own sake, but only use ideas from fuzzy logic needed for control systems, and in Section 11.6, for clustering methods in pattern recognition). For a detailed overview of how to perform stability analysis (absolute stability and via Lyapunov's first and second methods), how to study limit cycles via describing function analysis, and how to analyze tracking error for fuzzy control systems, see [263]. For an analysis of complexity of fuzzy systems and problems of the curse of dimensionality, see [235].

Expert systems are covered in many different books, so it is best, perhaps, to start with a general book on artificial intelligence, such as [444, 387, 136, 97, 82, 166]. The reader interested in studying stability analysis of expert control systems should consult [410, 338].

Planning Systems: The section on psychology and cognitive neuroscience of planning was based on [188, 206], and the work in [408] that was developed by using conventional control-theoretic concepts together with how the field of artificial intelligence views planning. Physiological foundations, particularly formation of cognitive maps for planning, are discussed in [239]. Planning systems are also discussed in [444, 387, 136] (a nice discussion on hierarchical and adaptive planning is in [136]) and in [137, 189]. Methods of anticipating the future (prediction) in microorganisms are discussed in [161]. These involve "circadian rhythms" and biological "clocks" [541]. Clearly, the use of regularly appearing events can be used to predict and hence react (early) quickly to stimuli, and these ideas are all related to the predictive nature of planning.

Planning in robotic applications has been studied extensively, and some papers you may want to study are [76, 77, 78, 44, 550], or the books [11, 258, 520]. For more information on path planning for robots, see [278, 90] where the "potential field method" is described (basically the method we introduce in this chapter where we use optimization over functions to guide a robot through a maze). In fact, some other functions that can be useful to build "obstacle functions" are given in [278]. A bibliography for heuristic search which has been used in planning is provided in [493] and an introductory treatment is given in [413]. The reader should be aware that there is a very large literature on combinatorial optimization methods, including many good books [400, 218] that provide methods to select a plan (e.g., by searching trees or graphs). One investigation on neural substrates for planning is given in [453].

A survey of model predictive control (which is also studied in Design Problem 6.2), what you can think of as the existing body of knowledge on planning methods in conventional control, is given in [192]. References on scheduling theory, which has many relevant concepts and techniques to planning, are provided below.

Attentional Systems: Discussions in cognitive neuroscience, clinical neuropsychology, and computational studies of attentional systems are contained in [404, 206, 522, 401]. The section in the chapter was developed using primarily [404, 206]. For other mathematical models of attentional systems that (unlike the one here) have been experimentally validated, at least to some extent, see [85, 482]. There is also a large literature on attention deficit disorder (a common disorder in children) that may provide insights into the operation and modeling of the human attentional system.

There are many books that treat the topic of scheduling and the topics treated there are relevant to both planning systems and attentional systems. Two books to consider are [419, 210]. Building on basic ideas in sequencing and scheduling [40, 109, 186], the approach to the development of attentional strategies here depends critically on the work in [418, 290] by using the time-based policies that were first studied in [81] and a discrete-event system [95, 244] theoretic framework for stability analysis that is described in detail in [411, 409].

Recently, some discussion on the use of attentional systems for control appeared in [11]. Earlier, it was shown in [296] how to augment adaptive controllers with attentional mechanisms.

The attentional systems approach of this part has been extended to the case where there are multiple agents cooperatively paying attention to multiple predators/prey in [212]. Also, the approach in [211] shows how such ideas can be used to allocate the focusing of multiple vehicle activities.

The section on multisensor integration is based on [339, 556, 158]. There are, however, many other papers on the theory and particularly the application of multisensor integration and management ideas. For more information on "world modeling," see [11].

Bayesian Belief Networks: One method that we did not introduce here, since to date it has found little use in control, is that of "Bayesian belief networks." This method has, however, found some use in a variety of engineering applications, such as diagnostic systems and decision-support applications. Moreover, it has potential for use in developing and implementing expert systems that reason under uncertainty and act as controllers. The reader interested in this method should consult [414, 96, 444] and the book [383] on learning Bayesian networks from data.

Part III

LEARNING

Part Contents

Sequence of Essential Concepts

- Learning theories from psychology and neuroscience form the foundations of biomimicry for incorporating learning into control and automation systems. The key underlying theories are classical conditioning, operant conditioning (reinforcement learning), and function approximation.

- Learning can be represented as an optimization process (e.g., gradient method) that involves sensing aspects of the environment and forming the associations or representations in memory that are best for trying to maximize performance and hence, survival chances. The type of association or representation sought depends on the situation. For instance, sometimes an organism learns to predict a stimulus from the occurrence of another stimulus (classical conditioning); other times, it learns the action that will result in a reward in a certain situation (operant conditioning, reinforcement learning).

- Neural networks and fuzzy systems can serve as tunable function approximators (interpolators) that hold associations or representations for making control decisions. We think of them as "approximators" for an unknown ideal mapping, one that we view as the target of our optimization process for incrementally learning the mapping. The neural or fuzzy systems can be trained (tuned) online to control a plant via reinforcement learning. For this, control decisions that lead to more reward (good closed-loop performance) are reinforced, and others are not. Under certain conditions, this iterative reinforcement leads to an appropriately shaped controller mapping after a long time period. If later the plant changes, then earlier "good actions" may not lead to rewards but other actions may lead to rewards. Then, the iterative reinforcement process reshapes the controller mapping in response to plant changes. Reinforcement learning control leads to what we call "adaptive control" for the plant.

- The key feature of using neural or fuzzy systems as tunable mappings for adaptive control is how to train them from data. There are a wide variety of training methods that you can use to learn functions from data. In linear least squares methods, we focus on tuning only a subset of the parameters of the approximator that enter linearly. Batch least squares methods focus on processing of data gathered offline, and recursive least squares methods focus on incrementally adjusting the approximator mapping as data are gathered in real time. The gradient methods that we discuss provide ways to tune all the parameters in an approximator structure, including the ones that enter in a nonlinear fashion, either in a batch or online mode.

- There are several fundamental issues involved in training approximators. The information in the training data is best in a specific form, but for control applications, it is often beyond the direct control of the training algorithm. The choice of the approximator structure, its complexity, and which parameters to tune has a significant impact on the quality of approximation and hence learning. The training method, initialization of the approximator parameters, and the parameters used to specify it (e.g., step size, termination method) can significantly affect learning performance. Moreover, there are fundamental issues to pay attention to in the training and testing process, including "generalization" (the ability of the trained approximator to respond similarly for similar inputs, or to discriminate between inputs that you would like it to), "local learning" (the ability of the method to learn the shape of the function in one region, and not to disturb what it has learned there when it learns in some other region), approximator "complexity" (while increases in approximator complexity generally give you an ability to approximate more complex functions, if the approximator is too complex, it can lead to poor generalization), and "overfitting" and "overtraining" (so that the approximator tries to match noise in the data, or tries to match the data gathered so closely that it does a poor job at generalization).

- The view of "learning as optimization" can be exploited to show how online optimization methods can be used to tune approximator structures to achieve adaptive control. The focus in such approaches is how to use data gathered online to shape functions (i.e., how to perform online function approximation). In the "indirect adaptive control" approach, the focus is on tuning approximators to match the nonlinear plant dynamics, and then using the approximations to specify the control inputs (using a "certainty equivalence approach"). In the "direct adaptive control" approach, the focus is on directly (i.e., without an approximation to the plant dynamics) tuning an approximator so that it approximates a controller that will achieve adaptive control. Optimization methods arising from learning (and foraging or evolutionary theory as studied later in this book) provide ways to adjust parameters for either the indirect or direct approaches. Since gradient-based adjustments reflect at least some biological learning/adaptation processes, online optimization approaches can also be viewed as biologically motivated. However, here we will depart somewhat from this focus to concentrate on what conventional optimization and approximation theory teaches us about the functionality and operation of adaptation mechanisms.

- Stability characterizes, for instance, how well a controller can achieve tracking of a desired reference input. Stable adaptive methods focus on how to construct, for example, online approximation-based controllers that will achieve stable and robust operation. While at the foundation of such approaches is Lyapunov stability theory, the methods essentially seek to minimize an instantaneous energy-based characterization of tracking performance. Hence, there is a close relationship to the online optimization methods. In stable adaptive control the focus is, however, on conditions under which the online optimization method will result in stable closed-loop control.

Chapter 9

Learning and Control

Chapter Contents

Learning seems to be an essential characteristic needed to achieve what we normally think of as highly intelligent behavior. Learning requires some method to store information and change behavior based on that stored information. A wide range of animals exhibit varying degrees of learning capabilities. Learning, in the forms of "classical" and "operant conditioning," has been studied for many years in psychology and neuroscience; actually, these both seem to be relatively mature fields (e.g., compared to planning and attention), at least for some types of animals (e.g., rats and pigeons). Classical and operant conditioning provide a foundation of concepts for biomimicry of learning processes for use in control and automation.

Here, we will introduce bio-motivated "heuristic" methods for adaptive control. In our neural control method, we will use a reinforcement learning approach to tune a neural network to act as a controller. In our adaptive fuzzy control method, we use "human mimicry" of adaptation heuristics to specify an adaptive control rule tuning method. Both methods use the basic features of learning via embedded online function approximators to approximate an unknown controller mapping (the one that succeeds in controlling the plant). In the next two chapters on least squares and gradient methods, we will discuss a number of offline and online function approximation methods. In the final chapter of this part, we build on the least squares and gradient methods by showing how they can form a foundation for optimization-based methods for function approximation in adaptive control. Finally, at the end of the final chapter of this part, we show how to achieve stable adaptive control with online function approximation methods. Hence, this chapter establishes bio-foundations for this entire part, and heuristic adaptive control methods that are directly based on the bio-foundations. The later three chapters in this part depart somewhat from the bio-foundations to focus more on engineering applications and methodology.

9.1 Psychology and Neuroscience of Learning

Learning can be defined as "any process through which experience at one time can alter an individual's behavior at a future time" [223] (from a system-theoretic view, it seems then that any system with memory has the potential to be a learning system). Alternatively, in [152] the author writes that "learning is an enduring change in the mechanisms of behavior involving specific stimuli and/or responses that results from prior experience with similar stimuli and experiences." In control engineering, many (including the author) have often thought of learning as the process of the organism interacting with its environment and using that experience to modify its behavior so that it is more successful in its environment in the future. But, does learning always imply performance improvement? Actually, many other factors affect performance (e.g., sensory and motor capabilities) and moreover, we may learn something at one time and it may not affect performance until much later. It is the case, however, that in control engineering, performance is typically measured via metrics over the entire lifetime (or long time periods), and we construct learning systems with the objective of improving performance, so it is sometimes useful to think of learning as leading to performance improvement. It is useful, however, to note that even if performance improvement is the objective of our learning system, and even if performance improvements are achieved, it does not mean that our systems have learned what you think they need to learn in order to get the performance improvement (e.g., you will see that in indirect adaptive control, it may be that a very poor model of the plant is learned but we still achieve good performance in terms of tracking).

In control systems, learning often *involves improving performance by interacting with the environment. Memory is necessary for learning.*

Generally speaking, there are two types of learning: (i) learning aspects of the environment and storing facts, relations, characteristics, etc., in "explicit memory" (i.e., memory available to our consciousness so we can deliberately recall it); and (ii) learning how to do things (e.g., when you acquire motor or perceptual skills) that we store in "implicit memory" (i.e., a type of memory that is unavailable to consciousness so it generally cannot be recalled). Some types of learning involve both explicit and implicit memory. In fact, the process of "explication" is when an expert manages to explain precisely what she knew implicitly (something that a nonexpert cannot recognize) and where she proceeds to improve the process (e.g., a pro explaining how to swing a golf club). While we can prove that we learned something that is stored in explicit memory simply by recalling it, to prove that we learned something that was stored in implicit memory, we must demonstrate improved performance in some task (e.g., in learning a motor skill like swinging a golf club). Generally, however, you must be careful in measuring the extent of what is learned. To quantify how much is learned, psychologists generally give a task to one group in some environment, and a similar task to a second (control) group in an identical environment. Then, the groups are tested in identical conditions and the difference in performance between the two groups is a measure of the amount of learning by the first group.

We acquire information via our senses, store information in several types of memory, and exploit this information later via explicit recall or via motor actions that were modified via physiological changes due to the interaction with the environment.

Before discussing specific types of learning in organisms, it is useful to point out that there is an area of learning theory called the "ecological perspective" where it is hypothesized that learning in animals is a process of "fill in the blanks" in their species-typical behavior. As one example of a conclusion from this viewpoint, it seems that animals are much more intelligent when given problems similar to those in their natural environment (e.g., birds that achieve a specialized type of place learning for storing food in thousands of locations, and then later retrieving it). Evolution shaped the animal to be most capable to learn how to solve problems in its natural environment. Moreover, it is important to recognize that evolution has shaped the sensory processes and the ability to perceive stimuli, an organism's basic physiology and hence constraints to the generation of actions, and basic goals (reward structure) of learning. (For example, evolution is the process that resulted in a pigeon being able to see a seed and retrieve it. The result was that the seed was found to be edible and nourishing so it is a reward.) Effects of evolution on learning seem fundamental to learning processes. Both evolution and learning are adaptive processes, and as we will see later in Part IV, each can affect the other. For now, we only briefly discuss evolutionary aspects as we discuss learning.

Evolution "invented" and shapes all aspects of learning processes.

9.1.1 Habituation and Sensitization

Learning can be broken down into associative methods where associations (e.g., between different stimuli) are learned and nonassociative methods. In this section we will discuss nonassociative learning and then later we discuss classical and operant conditioning, which are associative methods. In nonassociative methods relations between different stimuli are not learned. Only a single type of stimulus is repeatedly presented, and this brings about changes in how the organism responds to it.

In "habituation," after having repeated encounters with the same information, the animal may respond differently than when it first encountered the information (e.g., consider the human's ability to attenuate certain strong smells or loud sounds that are encountered over a long period of time). Basically, in habituation, we witness a decrease in a response to a benign stimulus. At the neural level, this seems to be implemented by a decrease in synaptic transmission, and sufficiently strong (long) habituation seems to, in some cases, involve the actual "pruning" (removal) of synaptic connections.

Habituation involves learning to ignore benign stimuli. Sensitization involves learning to react to important stimuli.

In "sensitization," once an animal has encountered a certain stimulus, it becomes more sensitive to it at later times (e.g., consider a human's ability to pick a familiar face out of a crowd). Basically, in sensitization, we witness an increase in response to important stimuli. At the neural level, this seems to be implemented by an increase in synaptic transmission, and with sufficient sensitization, new synaptic connections may be formed in some cases. In some organisms, sensitization can undo the effects of habituation (this is called "dishabituation"). In fact, according to the "dual process theory," habituation and sensitization both occur simultaneously and the net effect is in the direction of the one that has the strongest underlying process. Note, however, that there is a "stimulus specificity" characteristic to habituation (e.g., some animals can only be habituated to certain stimuli) that is generally not present for sensitization (many animals can become sensitized to almost any stimuli that they can sense). Hence, while you can broadly think of habituation and sensitization as "duals," they do have different characteristics. Also, note that there are other related "homeostatic" (feedback) theories such as the "opponent process theory," which is used to explain affective dynamics (neurophysiological mechanisms involved in emotional behavior that serve to maintain emotional stability). A feedback control theorist would also, perhaps, be interested in the homeostatic "compensatory-response model" [152].

9.1.2 Classical Conditioning: Learning Associations Between Stimuli

Behaviorism is a branch of psychology that attempts to understand human behavior generally by only considering observable inputs (stimuli) and outputs (responses). Classical conditioning is a behaviorist approach to characterize learning processes.

The Classical Conditioning Process

Suppose you are given an organism that has a natural (instinctual) reflexive-type response (called the "unconditioned response," UR) to some stimulus (called the "unconditioned stimulus," US) that could be neutral, rewarding, or noxious. Suppose you also know of a stimulus (called the "conditioned stimulus," CS) that will not instinctually elicit this same response. Next, a learning (training) process is conducted where the unconditioned and conditioned stimuli are "paired" by presenting the conditioned stimulus somewhat before the unconditioned stimulus to the organism. The time separation between the presentation of the two stimuli is called the inter-stimulus interval (ISI). After repeating this experiment several times, if the stimuli and the ISI are chosen properly, the organism will actually evoke the unconditioned response, UR (sometimes also called the "conditioned response," CR, especially if it is slightly different from the UR) when only the conditioned stimulus is applied (something that it would not do instinctually). It learned to pair (associate) the conditioned and unconditioned stimuli so that even when the unconditioned stimulus is not present, the conditioned stimulus can evoke the response.

Classical conditioning involves learning to associate two stimuli and hence, learn a reflex between a stimulus and response.

Classical conditioning can be viewed as learning to predict one stimulus from another.

Basically, you may think of classical conditioning as the learning of reflexes from instinctual reflexes. It is also useful to think of classical conditioning as *learning how to predict* the unconditioned stimulus by observing the conditioned stimulus that precedes the unconditioned stimulus. In this way we can see how an organism can learn to predict events in its environment. It has been found that for some animals, if the CS and US are "novel" ("surprising"), then learning is faster. Moreover, some animals seem to have a genetic predisposition to associating the CS and US (i.e., using one to predict the other).

An early experiment in classical conditioning involved Pavlov's dog where the dog was observed to salivate (unconditioned response) when food was placed in its mouth (unconditioned stimulus), but the dog did not have the instinctive response to salivate when a bell rang. In his experiments (which were quite extensive, studying many aspects of the learning process), Pavlov paired the bell ringing (conditioned stimulus) with placing food in the dog's mouth (unconditioned stimulus). After a sufficient number of such pairings, the dog would salivate even if only the bell rang. It learned to associate the conditioned stimulus with the unconditioned response.

Conditioned learning is related to Aristotle's "law of association by contiguity" (contiguity means closeness in space or time): "If a person experiences two environmental events (stimuli) at the same time or one right after the other (contiguously), those events will become associated in the person's mind, such that the thought of one will, in the future, tend to elicit the thought of the other" [223]. Of course, such a law is more difficult to verify for general learning of thoughts and concepts because it is more difficult to measure the responses than it is in an experiment like Pavlov's.

There are a number of characteristics of classical conditioning. We describe some of these next to provide a deeper understanding of the training process.

Blocking Phenomena

There is a phenomenon called "blocking" that can occur in classical conditioning. To understand it, consider Figure 9.1. Suppose that you have two conditioned stimuli, CS_1 and CS_2, and in step 1 of a two stage training process, you perform classical conditioning so CS_1 will evoke the UR as shown. Next, for the organism trained in step 1, train again at step 2 but now with the US paired with CS_1 and CS_2 (e.g., via having CS_1 and CS_2 occur simultaneously; also imagine what would happen if they are not simultaneous, but still precede the US). Now, while as expected, CS_1 will still evoke the UR, CS_2 *will not.* Learning to predict the US via CS_2 was "blocked" by CS_1. Essentially, the amount of conditioning depends on how "surprising" the UR is (more surprising URs result if there is new information in the set of conditioned stimuli). In this case, CS_1 adequately predicts the UR so CS_2 is not needed for this. The blocking phenomenon may show the potential to evaluate an organism's instinctual encoding of Shannon's entropy measure of how surprising information is.

Organisms tend to use the minimal amount of information to predict an event.

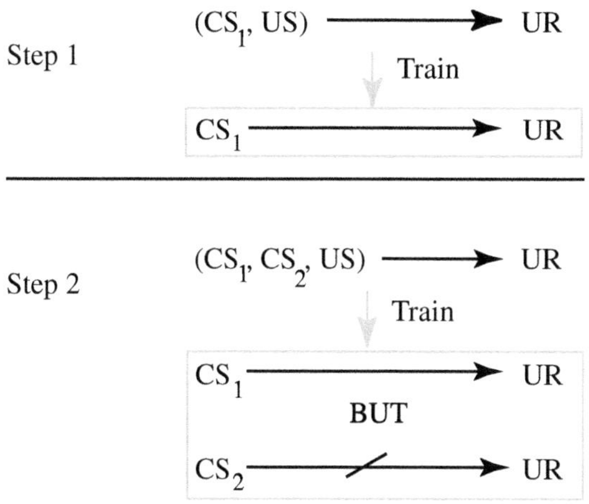

Figure 9.1: Blocking phenomenon in classical conditioning.

Extinction

Next, there is the issue of the permanence of the "conditioned reflex" (e.g., will the dog, for the rest of its life, always salivate when it hears a bell ring?). In fact, "extinction" of the conditioned reflex occurs if the bell rings a number of times without the unconditioned stimulus (food). However, it has been discovered that a type of "spontaneous recovery" occurs where, after extinction, the conditioned reflex is partially strengthened, and that, after extinction, a new sequence of training can more easily evoke the conditioned reflex (hence, Pavlov concluded that the conditioned reflex is only suppressed, but not forgotten by extinction). Extinction should not be thought of as forgetting, but as a process

Learning new associations can inhibit past associations that were learned.

of learning something new (e.g., that something will not occur). Some think of extinction and spontaneous recovery as an influence of habituation on classical conditioning.

Generalization and Discrimination

Particularly relevant to the studies on learning in this part are phenomena in conditioned learning associated with "generalization," where after training with CS_1, other stimuli (CS_i, $i \neq 1$) that are similar to the conditioned stimulus will actually evoke the conditioned reflex in the same way as the conditioned stimulus. A process of "discrimination training" can be used to reduce the effects of generalization. For instance, suppose that two stimuli, CS_1 and CS_2, are similar and hence after training an organism, will produce the conditioned reflex in response to either one. Then, suppose that in another training sequence, CS_1 is again presented with the unconditioned stimulus, but CS_2 is presented repeatedly without the unconditioned stimulus. The conditioned reflex between CS_2 and the unconditioned response will become extinct, and hence the organism will learn different responses to the two stimuli (we say that it has learned to discriminate between the two).

With sufficient sensory capabilities, some organisms can learn to discriminate between similar stimuli.

Some think of the process of sensitization as having basic influences on generalization, and so discrimination training is a process of making the organism more sensitive to (able to discriminate between) slightly different stimuli. It is interesting to note that this concept has formed the basis of quantifying the quality of sensory processes in animals and infants (if they can learn to discriminate between two similar stimuli, they must be able to perceive the difference).

9.1.3 Hebbian/Gradient Model of Neural-Level Classical Conditioning

Here, we give an example of classical conditioning processes at the neural level and provide a simple model of the learning process that is motivated by extensions of Hebb's classical learning rule modeled as a gradient optimization method.

Neural Mechanisms of Learning in *Aplysia*

Here, we will consider neural mechanisms of learning in a shell-less sea-dwelling mollusc called a sea slug or sea hare (referred to as *Aplysia*). *Aplysia* only have about 20,000 neurons and this makes them easier to study than the nervous system of mammals, for example. Moreover, some of the neurons are quite large and this makes the neural mechanisms even easier to study. It has been found that several of its natural behaviors can be modified by learning and some of these behaviors are only affected by as few as 100 neurons. A behavior of this type is the so called "gill-withdrawal" reflex where if the *Aplysia* is touched anywhere on its skin, it pulls its gill into its body as if it were protecting against an attack.

In [267], Kandel and his colleagues report finding that the gill-withdrawal reflex can be forced to occur in response to a stimulus that would normally not elicit it. In this case, the unconditioned response is the normal gill-withdrawal reflex in response to a touch to the skin. Unlike many studies in conditioned learning, they actually determined how the underlying neural mechanisms operate to achieve classical conditioning, and this is shown diagrammatically in Figure 9.2. A stimulation of sufficient strength anywhere on the skin excites the sensory neurons, which then excite the motor neurons that signal muscles to withdraw the gill. At the same time, the sensory neurons signal "modulatory (facilitating) interneurons" but these are only activated when there is a particularly strong stimulus (e.g., an electric shock, or in nature when bitten by a predator). When the modulatory interneurons are active, they release a chemical substance called a "neuromodulator" at some "slow" synapses onto axon terminals of sensory neurons. When these modulatory interneurons repeat this many times, the neuromodulator chemicals have the capability to start a chain reaction in the sensory neurons where they grow new synaptic connections onto motor neurons and "strengthen" existing ones. This makes the motor neurons more sensitive to inputs from the sensory neurons so that a weak stimulus that normally would not cause a gill reflex becomes capable of evoking it (this demonstrates the process of sensitization). In fact, the sensory neurons are affected more significantly by the neuromodulator if they have just been activated recently, within the ISI; this provides a possible mechanism for classical conditioning.

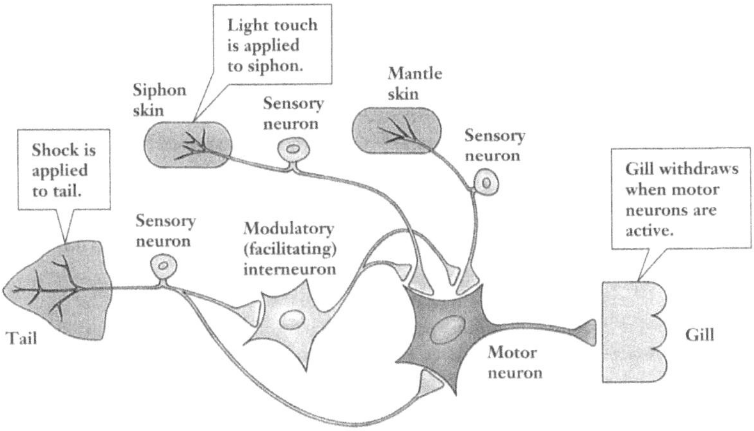

Figure 9.2: Neural level learning for gill-withdrawal learning in *Aplysia* (figure taken from [223], © 1991, 1994, and 1999 by Worth Publishers Inc., and used with permission).

Here, an electrical shock to the tail will serve as the unconditioned stimulus and a very weak stimulus to the skin (in particular, the "siphon") can serve as the conditioned stimulus. Then, after several times where the conditioned and unconditioned stimuli are paired at an appropriate ISI, if only the conditioned stimulus is applied, it will evoke the gill-withdrawal reflex. This is due to the fact that the motor neurons become more sensitive to the sensory neurons so that a light touch can evoke the gill-reflex that normally will only occur in response to a stronger stimulus to the skin.

The relationship between ISI and the strength of conditioning (i.e., how much is learned) is shown in Figure 9.3. The figure shows that the *Aplysia* can best learn to predict events that are spaced at about 0.5 sec. If the events are simultaneous, it can learn nothing about how to predict one stimulus from another. If they are spaced too far apart, then it cannot learn how to predict the US from the CS. Essentially, the ISI affects the rate and extent of learning. Even though biological evidence for it has not been found (to my knowledge), it is tempting to hypothesize that evolution can explain why *Aplysia* learn to predict best with this ISI (how would you explain this using concepts related to the speed of predators, physiology of *Aplysia*, and the stochastic nature of the environment?).

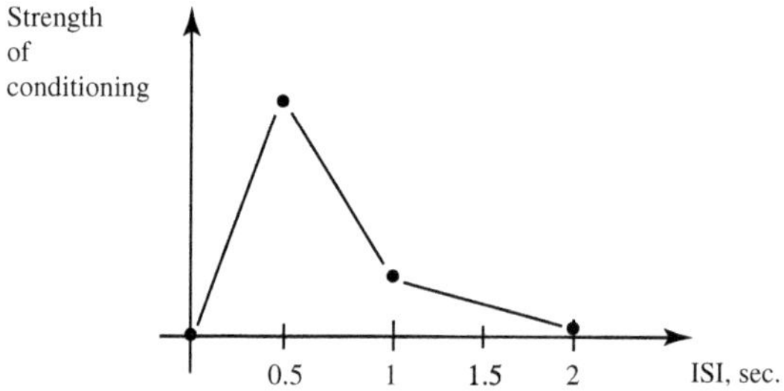

Figure 9.3: Effect of inter-stimulus interval on strength of conditioning for *Aplysia* (data taken from [268]; however, only the general shape is plotted here).

Hebbian Learning Modeled as a Gradient Method

Here, we will produce a very simple model of classical conditioning fashioned after general aspects of how classical conditioning occurs in *Aplysia*. To do this, consider Figure 9.4. Here, we have two inputs: the unconditioned stimulus x_1 that represents an electrical shock to the tail and the conditioned stimulus x_2 that represents a weak stimulus to the skin (particularly, the siphon). Recall that instinctually the *Aplysia* will respond to the US with a gill-withdrawal response, but it will not do this for the CS. Here, we model the sensory neurons in Figure 9.2 as generating signals x_1 and x_2 that are passed through synapses modeled as w_1 and w_2 to the motor neuron. The function f is the activation function of the motor neuron and suppose that it is linear so

$$y(k) = w_1 x_1(k) + w_2(k) x_2(k) + b$$

where k is the index of the conditioning step and w_1 and b are fixed and hence, not influenced by learning. We will suppose for our example here that $w_1 = 1$ and $b = -0.5$. We consider the UR to occur if $y \geq 0$ and not to occur if $y < 0$. Suppose that we let $x_1(k) = 1$ represent that the US is applied at conditioning step k and $x_2(k) = 0.1$ represent that the CS is applied at step k, and in both cases, these are set to zero if the corresponding stimulus is not applied at step k. The conditioning process results in learning that is modeled by changing $w_2(k)$.

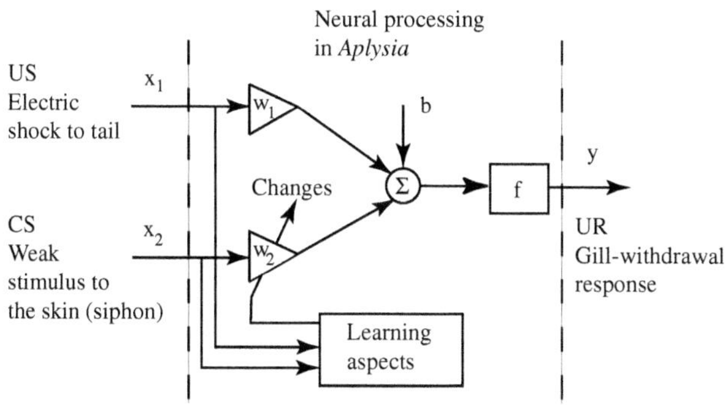

Figure 9.4: Neural network model for *Aplysia.*

Suppose that initially, before conditioning, $w_2(0) = 0$ and consider the influence of stimuli at this point. If $x_1(0) = x_2(0) = 0$, representing that no stimuli are applied, then $y(0) = -0.5 < 0$ so that the UR will not occur. The US will invoke the gill-withdrawal response since $y(0) > 0$ if $x_1(0) = 1$ and either $x_2(0) = 0$ or $x_2(0) = 0.1$ (i.e., whether the CS is present or not). If, however, $x_1(0) = 0$ and $x_2(0) = 0.1$ representing that only the CS is present, due to $b = -0.5$, $y(0) < 0$ representing that the UR will not occur.

Notice that we have not yet modeled the modulatory interneuron in Figure 9.2 and the learning process. Here, we will model the learning process with a synaptic connection (weight) update formula for $w_2(k)$ as shown in Figure 9.4. In particular, we will assume that the neurons operate so as to adjust w_2 in order to minimize the "cost function" (a characterization of how much learning needs to be done)

$$J(w_2) = \frac{1}{2}\left(\alpha x_1 - w_2 x_2\right)^2 \tag{9.1}$$

Here, $\alpha > 0$ is defined to represent the modulatory interneuron as a linear mapping. Suppose that $\alpha = 1$. We suppose that conditioning amounts to minimizing J according to a gradient method called "steepest descent." In particular, we assume that the w_2 weight is adjusted by the formula

$$w_2(k+1) = w_2(k) - \lambda \left.\frac{\partial J}{\partial w_2}\right|_{w_2 = w_2(k)} \tag{9.2}$$

where $\frac{\partial J}{\partial w_2}$ is the gradient of J with respect to w_2 and $\lambda > 0$ is the "step size." This is the steepest descent gradient method. Why is it called this? Notice that it updates $w_2(k)$ along the direction of the negative gradient to move in the direction of maximal decrease of J (this amounts to moving down the cost function surface, like climbing down a hill, taking step at each conditioning step of size λ). Equation (9.2) represents the change in the weight due to one pairing of the US with the CS at time k with the appropriate ISI (so it actually represents the physical presentation of the CS and US at two different times). For this example,

$$\frac{\partial J}{\partial w_2} = -\left(\alpha x_1 - w_2 x_2\right) x_2$$

so that

$$w_2(k+1) = w_2(k) + \lambda\left(\alpha x_1(k) - w_2(k) x_2(k)\right) x_2(k) \tag{9.3}$$

represents the learning process where the weight is adjusted to minimize how much learning needs to be done as characterized by Equation (9.1).

Next, consider the effects of long sequences of various values of $x_1(k)$ and $x_2(k)$, $k \geq 0$, when $w_2(0) \geq 0$ takes on a fixed positive value. We have the following:

1. *No US or CS:* Here $x_1(k) = x_2(k) = 0$, $k \geq 0$, so via Equation (9.3), $w_2(k) = w_2(0)$, $k \geq 0$. The weight value stays the same, representing that there is no learning or extinction.

2. *No US, but CS present; Extinction:* Here, $x_1(k) = 0$ and $x_2(k) = 0.1$, $k \geq 0$, so via Equation (9.3),

   $$w_2(k+1) = w_2(k) - \lambda w_2(k) x_2^2(k) = \left(1 - \lambda x_2^2(k)\right) w_2(k)$$

   Assume that $0 < 1 - \lambda x_2^2(k) < 1$, which we can always achieve for a fixed size $x_2(k) > 0$ by choosing $\lambda > 0$. With such a choice, $w_2(k) \to 0$ as $k \to \infty$ for any $w_2(0)$ value. This represents the case where there is extinction since after a certain point, $w_2(k)$ will be small enough so that $y(k) < 0$ and the CS will not elicit the UR if the US is not present.

3. *US present, but no CS:* Here, $x_1(k) = 1$ and $x_2(k) = 0$, $k \geq 0$, so via Equation (9.3), $w_2(k+1) = w_2(k)$, $k \geq 0$, and there is no learning since the CS is not present.

4. *US and CS present; Learning:* Here, $x_1(k) = 1$ and $x_2(k) = 0.1$, $k \geq 0$, so via Equation (9.3),

   $$w_2(k+1) = \left(1 - \lambda x_2^2(k)\right) w_2(k) + \lambda \alpha x_1(k) x_2(k)$$

   The first term arises in case 2 above and we have $0 < 1 - \lambda x_2^2(k) < 1$ so that $w_2(k)$ will be bounded, and will take on a positive value since $\lambda \alpha x_1(k) x_2(k) \geq 0$. Also, for our case, each conditioning step will result in an increasing size to $w_2(k)$ which represents that learning occurs.

As an example, suppose that $\lambda = 20$, then the weight trajectory for $w_2(0) = 0$ is shown in Figure 9.5(a) and the UR output y is shown in Figure 9.5(b) for each $w_2(k)$ value (we think of the output y as being proportional to the strength of the UR). For this simulation in the first 20 conditioning steps, $x_1 = 1$ and $x_2 = 0.1$ but in the remainder of the steps, $x_1 = 0$ and $x_2 = 0.1$. Hence, in the first phase we see that with the US present, even at the first step the UR occurs, and then after a sufficient number of steps, the UR would occur without the US present, only via the presence of the CS. Why? Notice that with no US but a CS present

$$y(k) = 1(0) + w_2(k)0.1 - 0.5$$

What weight value $w_2(k)$ must be present before the CS will by itself result in $y(k) \geq 0$? Note that by step k such that $w_2(k) \geq 0.5$, it has achieved conditioning. Next, during the second phase when the US is not present but the CS is, $w_2(k)$ decreases and once it is such that $w_2(k) < 0.5$, then $y(k) < 0$ and extinction has been achieved.

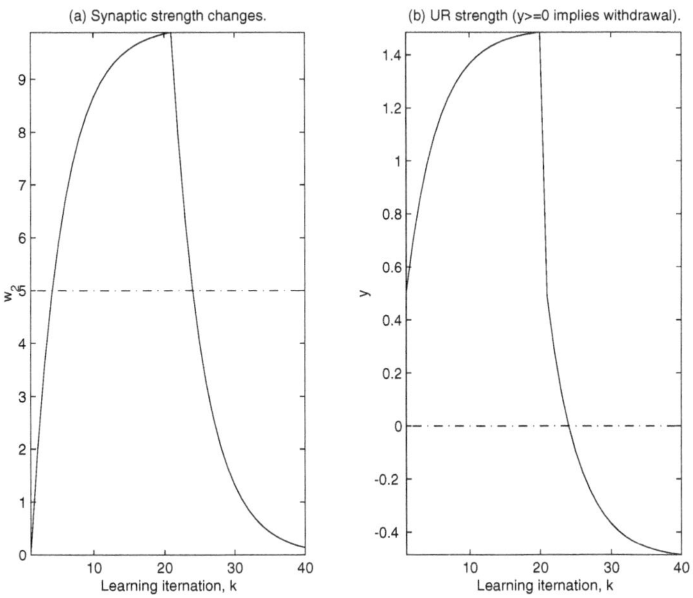

Figure 9.5: Gradient/Hebbian learning in *Aplysia*, synaptic strength changes during classical conditioning (a) and motor neuron output (b).

While the above model of classical conditioning at the neural level seems to represent some characteristics of learning, is there a neuro-scientific basis for the model? To explain how learning takes place at the neural level, D.O. Hebb hypothesized that if an axon of, say, neuron 1, repeatedly helps fire neuron 2, then the connection between the neurons is modified so that in the future, neuron 1 more easily fires neuron 2. Hence, the "connection" provides a mechanism for memory, and it is due to Hebb's learning rule hypothesis that we often think of modifying neural network weights as learning since they model the synaptic connections between neurons. Now, using the view in [206], "associative long-term potentiation" is an extension of Hebb's rule that says: if a weak and strong input act on a cell at the same time, the weak synapse becomes stronger. This means that the above weight update formula that results from viewing learning as a gradient method is a form of Hebbian learning.

It must be emphasized, however, that this model of part of the neural processing in *Aplysia* only represents the gross characteristics of its neurophysiology and learning. There are many aspects of the neural network and learning that are not modeled. For instance, the feature of extinction where learning will occur faster during the second training phase is not represented. The shapes of the curves and training rates are not based on physiological data. Also, there are a number of other aspects of classical conditioning characteristics that are not represented. Our intent, however, was not to create an accurate model of learning in this biological system; it was only to illustrate the *plausibility* of modeling learning as a gradient method to motivate the use of gradient methods. For more work on modeling classical conditioning, see the "For Further Study" section at the end of the part.

Supervised Vs. Unsupervised Learning Perspectives

In "supervised learning," there is a "teacher" who provides stimulus-response pairs repeatedly to a "student" whose goal is to learn the association between the given stimulus and the response that the teacher provides (like rote memorization). Classical conditioning is often thought of as "unsupervised learning" since, while stimuli are provided (e.g., the conditioned and unconditioned stimuli), the response is not since it is generated by the organism (student), not the teacher. For example, for the *Aplysia* example, the US and CS stimuli can be provided but the teacher cannot specify the gill response.

There is, however, another way to view classical conditioning where it can also be thought of as supervised learning (training). Recall that classical conditioning can be viewed as learning to predict that a stimulus will occur when another stimulus is present. Can we view the CS as the stimulus and the US as the response (that occurs later than the CS by the ISI) so that the organism (student) then learns to associate the CS to the US? It seems so. To do this, note that presentation of the US changes the internal state of the organism in some way so that it will elicit the UR. The CS changes the internal state of the organism in a way that does not result in the UR. Classical conditioning can be viewed as changing the internal state generated by the CS so that it elicits the UR. The changes could be so that the state produced by the CS triggers the state produced by the US even when the US is not present. Or, it could be that the changes result in the CS producing the same state as the US. Regardless, we can view the CS as the stimulus and the internal state produced by the US as the response that the teacher wants. Repeated presentations of the paired CS and internal state produced by the US should result in the formation of an association between the CS and the internal state produced by the CS so that when the CS is presented without the US, the UR occurs.

Consider the *Aplysia* example. Let the training pairs be $(x_2(k), x_{us}(k))$ where $x_2(k)$ is the CS and the internal state generated by the US is

$$x_{us}(k) = \alpha x_1(k)$$

which is generated after the CS by a time interval specified by the ISI. So, part of what the teacher is demanding as a response is an internal characteristic of the network that is being trained. Recall that $\alpha = 1$ and $x_1(k) = 1(0)$ represents the presence (absence, respectively) of the US. A cost function for supervised learning measures how well the student produces the pairs presented by the teacher. One way to do this for one training pair is via Equation (9.1) with $x_{us}(k) = \alpha x_1(k)$, since this is a measure of how well the output of the sensory neuron connected to the motor neuron (what we call $w_2 x_2(k)$) produces a signal that is the same as the one produced by the US (what we call $x_{us}(k)$). Clearly, if we use the steepest descent gradient method to model the supervised learning process we get the same general form for the weight update equation as in Equation (9.2), but since $x_{us}(k) = \alpha x_1(k)$,

$$w_2(k+1) = w_2(k) + \lambda \left(x_{us}(k) - w_2(k) x_2(k) \right) x_2(k) \tag{9.4}$$

where $(x_2(k), x_{us}(k))$ are specified at each conditioning step. Clearly this is the *same* as Equation (9.3); all we have done is change our interpretation of what is happening in the conditioning process. Now, we think of conditioning as teaching to associate two stimuli spaced by an ISI (or later, in some cases, we think of learning a function that represents the associations; i.e., we think of it from a "learning as function approximation" perspective). Extinction involves presenting $(x_2(k), x_{us}(k)) = (0.1, 0)$ repeatedly so it can be thought of as teaching the organism not to associate the two stimuli. Due to the view of modeling Hebbian learning via a gradient method discussed above, Equation (9.4) is sometimes called a model of supervised Hebbian learning.

So, why concern ourselves with the different viewpoints of unsupervised and supervised learning? In training of artificial neural networks, the distinction is sometimes important. Here, however, the key reason is that later, in our function approximation approaches, we will use some supervised learning approaches and the above discussion clarifies that gradient learning is a plausible model for this case also. A particular case we will be interested in is when the "response" in the stimulus-response pair is some value that is specified in order to lead to a reward for the organism. In this way, the supervised learning model of classical conditioning can be viewed as closely connected to operant conditioning.

9.1.4 Operant Conditioning: Learning to Predict Consequences of Actions

Generally, the consequences of actions (operations) that an organism takes in its environment increase or decrease the likelihood that those actions are taken again depending on whether they helped the organism achieve its goals (e.g., getting food). The effectiveness of the actions can be measured on a continuous scale by some type of performance measure that quantifies how well the actions help the organism achieve its goals. Psychologists (e.g., E.L. Thorndike and B.F. Skinner) constructed environments for training animals where if the animals learned how to take the proper actions, they would be rewarded and if they did not, they would not be rewarded (e.g., via food). Skinner called the training process "operant conditioning" (others refer to it as "instrumental conditioning" or "trial-and-error" learning).

Operant conditioning involves learning which actions are most likely to lead to goal achievement.

It is interesting to note that some view habituation, sensitization, and classical conditioning as "building blocks" for learning in the sense that some hypothesize that it would be logical for evolution to build more complex learning strategies from existing ones in simpler organisms. From this viewpoint, you will see shadows of the learning strategies we have covered earlier in this treatment of operant conditioning.

Example: Operant Conditioning for Pigeons

As an example, recall that Skinner trained pigeons to peck at images of boats projected onto a screen by giving them a food reward each time that they pecked at the correct position (review Part I, Section 2.3.2 on page 70). You may think of operant conditioning as the training process depicted in Figure 9.6. In the figure, at the start of the training process, the pigeons took actions that did not lead them to a reward (note that the thick black lines are used to represent more likely actions). Then, they discovered how to take actions that led them to rewards (e.g., by pecking at various locations). After further training, however, they discovered that pecking on images of boats produced immediate rewards so after several trials, the likelihood that they took that action increased significantly. Notice the change of line thickness through the training process, where the pigeon incrementally modifies the likelihood of taking the actions and biases these modifications in the direction of actions that are likely to provide a reward.

During operant conditioning, there is a typical response dynamic studied by Skinner where once the pigeon figures out the proper response to get the reward, it will tend to increase the frequency at which it performs that action and this produces a type of feedback that allows the pigeon to quickly increase the likelihood of producing the proper action. Note, however, that this is for a constant environment and reward system. Clearly, if there is a dynamically changing environment and reward system, the operant conditioning will have to "track" the rewards, at least by strengthening/weakening the likelihood of some actions and possibly by inventing new actions to find the rewards. Some general theory that studies the rates of responding and rates of reinforcement is found in the study of the "matching law" and self-control [152].

Next, we provide more details on some characteristics of the operant conditioning process.

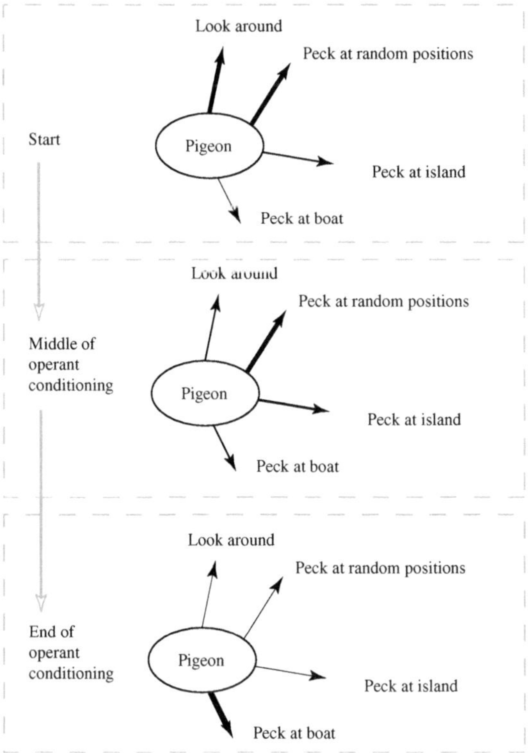

Figure 9.6: Depiction of reinforcement in operant conditioning (Thorndike's "law of effect").

Learning to Predict Consequences of Actions

In an analogous manner to how we viewed classical conditioning as learning how to predict stimulus events, we can view operant conditioning as *learning to predict consequences of actions.* We think of operant conditioning as a process where the organism modifies its frequency of taking various actions to optimize the likelihood of getting rewards. Operant conditioning is constrained by the interval of time between action and reward (but see the section on "chaining" below) in an analogous way to how the ISI works in classical conditioning. Also, environmental context can affect operant learning so that the organism can learn to perform certain actions in specific environments. From an evolutionary perspective, there may be a strong selective advantage for organisms that have operant learning capabilities rather than just classical conditioning capabilities. For instance, operant learning generally gives the organism an ability to shape its own environment so it is best suited for survival, rather than just trying to cope with a given environment. Moreover, instincts can affect what is learned and how it is learned.

Operant conditioning can be viewed as learning how to predict consequences of actions.

Shaping, Partial Reinforcement, Extinction, Reinforcer Control

Concepts related to "generalization" in classical conditioning have been studied for operant conditioning. In some environments, an organism may never find the action that will lead to a reward, and hence the training process can be viewed as a failure. To solve this problem for specific training exercises, experimentalists have used a concept called "shaping." For instance, note that in the pigeon example, if the pigeon pecks at the screen in random locations (a likely possibility due to the natural behavior of the pigeon), this type of action will be rewarded occasionally since the pigeon may get lucky and hit a boat image. The random pecking response is "shaped" so that it is similar to the proper response in the sense that it will give a "partial reward" (i.e., not as much as for the correct action for the amount of effort expended). You may think of the random pecking response as leading the pigeon to the correct response. Skinner set the pigeons up for success since he already understood their capabilities and tendencies.

Just as with classical conditioning, the operantly conditioned response becomes extinct if the organism does not get a reward for performing it for a long period of time. As with classical conditioning, the response is not actually forgotten, it is just inhibited, and "spontaneous recovery" and higher-rate relearning is normal.

Skinner introduced the term "reinforcer" to use instead of goal or reward (and this will later lead to "reinforcement learning control" in Section 9.4). Then, there is the possibility of having a "partial reinforcer" (periodic or lower magnitude reward), "positive reinforcement" (a process that increases the likelihood that a response will occur), and "negative reinforcement" (when the removal of a stimulus after a response makes the response more likely to occur). For instance, once positive reinforcement has been used to train the pigeons, it would be normal for them to continue pecking at boat images, even if each time they did it properly you did not give them the positive reinforcer. Instead, you could give it to them only periodically. Essentially, they learn to be persistent to get their reward.

There are also methods of training that typically result in making it more difficult for a response to become extinct. Researchers have used different types of *schedules* for providing rewards. For example, there are *fixed* schedules where reinforcement is given periodically in time or where reinforcement is given after n occurrences of the correct action. Alternatively, some study *variable* schedules where you vary the frequency of reinforcement randomly about a mean value or number of occurrences of the correct action about some mean value. One interesting effect from such studies, sometimes called the "partial reinforcement learning effect," is that the variable schedule training methods are typically more resistant to extinction. This should not be surprising since the animal also learns that it has to be patient.

Next, we briefly note that some animals can learn that some reinforcer is controllable or uncontrollable, and such concepts are often studied in the control of aversive stimuli. For example, there is the "learned helplessness effect" where an animal can essentially learn that it cannot avoid an aversive stimulus, and thereby this can adversely affect its ability to later learn. Finally, note that it is possible to learn to pay attention to some stimulus, but along similar conceptual lines to the learned helplessness effect, it is also possible to learn to have an attention deficit (e.g., when an animal receives a bombardment of stimuli and cannot choose actions to direct behavior at getting rewards fast enough due to other physiological limitations and hence, it then learns to ignore the stimuli). This provides some connections between learning and attention.

Discrimination Training and Chaining

In "discrimination training" you can train an animal to recognize the *situation* that it is currently in and to only take actions when in that situation. You reinforce a response when some situation ("contextual information") is present, and extinguish the response when that situation is absent. Via discrimination training, animals can be trained to perform sequences of events. To do this, the key is to note that after discrimination training, the situation is associated with receiving a reinforcer so the situation itself acquires some reinforcing value. The situation is sometimes said to be a "secondary reinforcer" (for humans, money is a typical secondary reinforcer). Hence, if you set up an additional operant training sequence with the situation as the reinforcer, it can learn the action it needs to get to the situation as follows:

Training via chaining results in an ability to predict sequences of events.

$$\begin{gathered} \text{action } 1 \rightarrow \text{situation } 1 \text{ (secondary reinforcer 1)}, \\ \cdots \\ \text{action } n-1 \rightarrow \text{situation } n \text{ (secondary reinforcer } n) \rightarrow \\ \text{action } n \rightarrow \text{goal (reinforcer)} \end{gathered}$$

Essentially, the training can occur in a "backward manner" in the learning process. You train using operant conditioning so that the animal finds action n so that it gets a reward. Via this process, it associates the situation it was in while learning (situation n) with getting a reward so situation n gets some reinforcing value, and thereby can become the reinforcer for an operant conditioning process for the animal to learn action $n-1$ so that it obtains the secondary reinforcer of situation n. The process then repeats. Long "chains" of sequences have been taught to animals, and clearly human behavior is affected by such processes also. Foraging in a variety of animals can be viewed as exploiting a type of operant conditioning that involves chaining.

9.1.5 Control System Model of Behavioral-Level Operant Conditioning

Consider the block diagram representation of the operant conditioning process shown in Figure 9.7. The organism is placed in some environment (just as in control systems, even though we break the organism and environment into two boxes, *the organism is in the environment*). The organism can take actions on the environment and if it takes the proper actions, it will change the environment in a way so that the environment "gives" it a reward. Operant conditioning occurs if the organism can find an action that will change the environment in a way so that it gets a reward, and if the likelihood of the organism taking that action again increases.

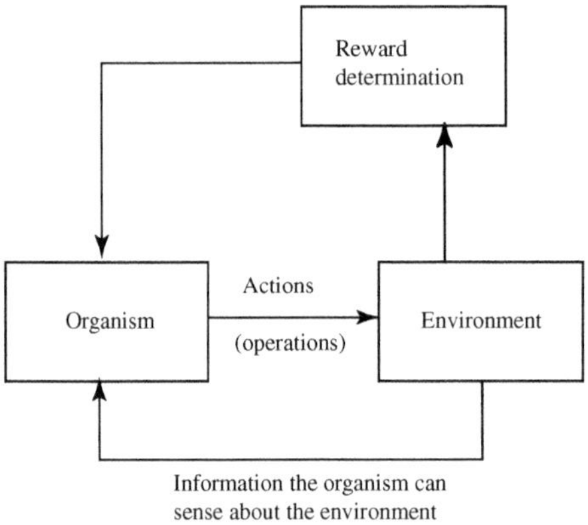

Figure 9.7: The operant conditioning process.

Feedback Control System Model of Operant Conditioning

In order to model the operant conditioning process, we will think of the organism as having an ability to act as a controller and the environment as being our "plant" (process to be controlled). Taking a control-engineering approach, we will begin by providing an example of how we can model the environment. In particular, for illustration purposes, suppose that the environment can be represented by

$$\begin{aligned} x(k+1) &= f(x(k), w(k), u(k)) \\ y(k) &= h(x(k), w(k), u(k)) \end{aligned} \tag{9.5}$$

General operant conditioning processes can be represented as a feedback control process.

where $x \in \Re^n$ is the state, $w \in \Re^{n_w}$ is a disturbance or noise, $u \in \Re^{n_u}$ is the input, and $y \in \Re^{n_y}$ is the output. Here, the state $x(k)$ at time k is the "situation" in the environment at time k, $u(k)$ represents the actions the organism takes on the environment at time k, $w(k)$ is an uncertain influence (e.g., sometimes taking a specific action in some state may lead to an unpredictable situation in the environment), and $y(k)$ represents variables that the organism can sense (generally, due to sensory processing limitations, it is not possible for any one organism to sense every aspect of its environment and this forces the organism to make decisions under uncertainty). The functions f and h can generally be assumed to satisfy certain properties (e.g., they are generally "smooth" in their arguments) but they are normally unknown by the organism, at least when the organism first encounters the environment (however, this ignores what instinctual information the organism may have about the environment it is to live in that has been "encoded" via evolution).

Reward determination differs for every type of organism. Suppose, however, that the organism is rewarded if it can make

$$y(k) = y_d(k)$$

for all $k \geq T$, for some finite T, where $y_d(k)$ is a vector representing measurable aspects of the environment that, if the organism can steer the environment to them, the organism will get a reward (e.g., it may represent that the organism is in a specific location and has moved some lever so that the environment changes in a way to make a reward available). Note that if we let

$$e(k) = y_d(k) - y(k)$$

then

$$J(k) = e^\top(k)e(k)$$

(the sum of the squares) represents a measure of how close the organism is to a reward so that J represents a way to provide partial reinforcement (note that small rewards are given for large values of J, and bigger ones are given for smaller values of J).

To model one type of operant conditioning, suppose that Equation (9.5) is at some state $\bar{x}$, that we seek a reward $\bar{y}_d$ (a constant), and that there is no influence from the disturbance $w(k)$ (represented with $w = 0$). Moreover, suppose that any action the organism takes results in the next state being $\bar{x}$ (so it is put back in the same situation to try to find a more rewarding action again). The organism experiments with different $u(k)$ values to try to minimize J (to try to get a big reward). In particular, during operant conditioning the organism *solves an optimization problem* by generating the actions $u(i)$, $i = 1, 2, \ldots, M$, and

$$\begin{aligned} \bar{x} &= f(\bar{x}, 0, u(i)) \\ \bar{y}(i) &= h(\bar{x}, 0, u(i)) \end{aligned}$$

with rewards

$$J(i) = e^\top(k)e(k) = (\bar{y}_d - \bar{y}(i))^\top(\bar{y}_d - \bar{y}(i))$$

After training, given that it is placed in the same environmental situation $\bar{x}$, the organism will simply choose the input $u(i)$, say $u(i^*)$, that gave it the most reward (i.e., we think of learning a function relating x and u to rewards, which is a function approximation perspective on learning). Mathematically,

$$i^* = \arg\min \{J(i) : i = 1, 2, \ldots, M\}$$

In fact, the organism's strategy for finding the highest reward when it is in the situation $\bar{x}$ can be viewed as an *optimization strategy* (controller) that picks a sequence of actions to maximize $J(i)$ using feedback information from the environment (both the situation that the actions take the environment to and the resulting amount of reward). What types of optimization strategies might an organism use to find the action that will provide the most reward and thereby learn that action? This depends on the physiology of the organism and characteristics of its environment. Here, we will not concern ourselves with modeling actual organisms using the above formalism for representing operant conditioning. Instead, we will explain how it can represent characteristics of operant conditioning, and then discuss how operant conditioning can be viewed as a relatively general building block for learning in decision-making systems that operate as feedback control systems.

Modeling Characteristics of Operant Conditioning

If you think of long sequences of events as forming a tree where at some twig there is a reward, it should be clear that "chaining" is a way to train the organism by starting at the reward (twig) and tracing backward to a situation you want to start the organism in, and be able to find the twig. Characteristics of shaping, extinction, and discrimination training have clear analogies. How?

It seems that in some ways, the specific way that we modeled operant conditioning in the last subsection represents a set of essential assumptions of the *pure* behavioralist view. From the "cognitive perspective," we should view the organism itself as a stochastic dynamical system

$$\begin{aligned} x^c(k+1) &= f^c(x^c(k), w^c(k), y^c(k)) \\ u^c(k) &= h^c(x^c(k), w^c(k), y^c(k)) \end{aligned} \tag{9.6}$$

Here, x^c is an internal state of the organism (state of the brain and physiological condition), w^c represents unmodeled influences and noise, u^c represents the actions it takes on the environment (e.g., $u^c = u$), and y^c represents inputs from the environment (e.g., $y^c = y$). Note that f^c and h^c can represent many aspects of cognition and physiology, but it will certainly represent aspects of the reward structure (e.g., the J function discussed above).

Equation (9.6) has memory, and possibly parts of the mappings f^c and h^c that influence learning (e.g., via existing information in memory) but are not changed by it (e.g., physiological constraints). The functions f^c and h^c can actually be used to represent "internal mental models" of the environment that can help it succeed (e.g., a cognitive map). In fact, operant conditioning can be viewed as a process of model building where the organism learns to predict the consequences of its actions.

9.2 Function Approximation as Learning

Recall that in Chapter 4 we modeled two types of neurons: one with a tuning curve in the form of a sigmoidal function that led to the multilayer perceptron, and another with a tuning curve in the form of a Gaussian function that led to the radial basis function neural network. In Chapter 4 it was also explained how such networks could be constructed by hand to implement complex input-output mappings. Here, we study how to construct these neural networks by using supervised learning to train the networks to approximate an unknown function that is represented by a set of input-output data pairs (i.e., more general neural networks and approaches to training them than what we considered in our simple *Aplysia* model). Hence, the function approximation viewpoint here can be thought of as having bio-foundations in classical and operant conditioning (e.g., by viewing the teacher's demanded response as arising from a stimulus, a reward, or specific value that leads to a reward), and later we will show how to achieve the training via least squares and gradient methods that have their bio-foundations in Hebbian learning. You will see a number of characteristics from classical and operant conditioning arise in this and the remaining sections of this chapter. However, you will also see a general departure from neural bio-foundations and an increasing focus on mimicry of the behavioral level and engineering applications.

First, we provide a mathematical definition of a function approximation problem. Let

$$F(x, \theta)$$

denote a tunable nonlinear function that we will use as a "function approximator." The input, which is known, is $x = [x_1, x_2, \ldots, x_n]^\top$ and the parameter vector $\theta = [\theta_1, \theta_2, \ldots, \theta_p]^\top$ (i.e., there are n inputs to the approximator and p parameters which can be changed to modify the functional mapping that F implements). The function F could be a standard fuzzy system which we denoted with $F_{fs}(x, \theta)$, a Takagi-Sugeno fuzzy system, which we denote by $F_{ts}(x, \theta)$, a multilayer perceptron which we denote by $F_{mlp}(x, \theta)$, a radial basis function neural network $F_{rbf}(x, \theta)$, or some other tunable nonlinear function (e.g., a polynomial). The function F is the "approximator structure" and θ is the parameter vector which holds the set of tunable parameters for the approximator structure. The value of p will be called the "size" of the approximator. For example, the size of the multilayer perceptron is given by the number of tunable weights and biases. In the following chapters, we view the parameters θ as the values that are learned (tuned to shape the nonlinearity $F(x, \theta)$); however, "structure learning" provides another way to achieve function approximation and this will be discussed briefly in Section 9.4.6.

9.2.1 Using Functions to Represent Mappings in Data

Let

$$y = G(x, z)$$

where its input is $x = [x_1, x_2, \ldots, x_n]^\top$, $z = [z_1, z_2, \ldots, z_{n_z}]^\top$ is an unknown "auxiliary variable," and its output is the scalar y. Here, n is the number of inputs, and n_z is the number of auxiliary inputs. If $n_z = 0$, then this means that $G(x, z)$ is not a function of z and in this case we will denote it by $G(x)$. We assume that $G(x, z)$ is a function for which we do not have an explicit mathematical form. Suppose, however, that we do have an ability to *learn* about its form by performing experiments and gathering input and output data from the function.

Suppose that for the ith experiment, we let the input data be

$$x(i) = [x_1(i), x_2(i), ..., x_n(i)]^\top$$

the auxiliary variable

$$z(i) = [z_1(i), z_2(i), ..., z_{n_z}(i)]^\top$$

and the output data be

$$y(i) = G(x(i), z(i))$$

(hence, $x_j(i)$ is the jth element of the ith data vector so it has a specific value and is not a variable like x_j; similarly for $z(i)$). Typically, in practice, our experiments are constrained so that we know that

$$x(i) \in X \subset \Re^n$$

for some (bounded) set X that we know a priori; similarly for

$$z(i) \in Z \subset \Re^{n_z}$$

While for the ith experiment we know the value of $x(i)$, it is assumed that we do not know $z(i)$ but that we can obtain $y(i) = G(x(i), z(i))$ from our experiment. For example, $x(i)$ may be the (known) input value to a function, $z(i)$ may result from some noise or other values that we cannot measure, and $y(i)$ is the subsequent output that is generated and that we can measure; of course, the function $G(x, z)$ could generate other unmeasurable outputs but we are not interested in these.

We will call the pair $(x(i), y(i))$ an input-output data pair and each such pair can be used as a piece of "training data." We call the set of input-output data pairs the *training data set* and denote it by

$$G = \{(x(1), y(1)), \ldots, (x(M), y(M))\} \tag{9.7}$$

where M denotes the number of input-output data pairs contained in G. Hence, G is the data set of input-output pairs that is gathered to gain information about the unknown function $G(x, z)$.

Function approximation involves constructing an interpolator for data so it properly represents the function from which the data were gathered.

If you perform many experiments, the set G will contain a significant amount of information about the mapping that is inherent between the $x(i)$ vectors and the $y(i)$. In some ideal situation where we could perform an infinite number of experiments in a way so that all of X is covered (i.e., we pick all possible values in $X \subset \Re^n$), then we would still not have complete information about $G(x, z)$ because of the influence of the unknown variable z on the shape of the function.

The function approximation problem is the problem of how to pick the value for the parameter vector θ in $F(x, \theta)$, a function whose explicit form we know, so that

$$G(x, z) = F(x, \theta) + e(x, z) \tag{9.8}$$

where the approximation error $e(x,z)$ *is as small as possible* for all $x \in \Re^n$ and $z \in \Re^{n_z}$, even at x such that $(x,y) \notin G$ (which is quite challenging if we know nothing of the function $G(x,z)$ besides what is in the training data G). If $e(x,z)$ is small for all $x \in \Re^n$ and $z \in \Re^{n_z}$, we say that $F(x,\theta)$ does a good job at approximating the function $G(x,z)$; that is, $F(x,\theta)$ does a good job at representing the mapping that is inherent in the data set G.

9.2.2 Choosing the Training Data Set

The fact that classical conditioning can be viewed as a type of function approximation relies on being able to form a training data set G. The choice of how to structure G so as to represent different types of learning problems, particularly adaptive estimation and control problems, is an underlying theme of this part. In this section we will discuss how to pick the *particular* data pairs after you have formulated the function approximation problem by specifying what the (x,y) data pairs in G represent.

While the method for adjusting the parameters θ of $F(x,\theta)$ is critical to the overall success of the approximation method, there is virtually no way that you can succeed at having $F(x,\theta)$ approximate $G(x,z)$ if there is not appropriate information present in the training data set G. Basically, we would like G to contain as much information as possible about $G(x,z)$. Unfortunately, most often the number of training data pairs is relatively small, or it is difficult to use too much data since this affects the computational complexity of the algorithms that are used to adjust θ. The key question is the following: How would we like the limited amount of data in G structured so that we can adjust θ so that $F(x,\theta)$ matches $G(x,z)$ very closely?

Quality of function approximation depends critically on whether the training data set provides good information about the unknown function.

Uniform Coverage May Help

There are several issues involved in answering this question. Intuitively, if we can manage to spread the data over the input space uniformly (i.e., so that there is a regular spacing between points and not too many more points in one region than another) and so that we get coverage of the whole input space, we would often expect that we may be able to adjust θ properly, provided that the space between the points is not too large. This is because we would then expect to have information about how the mapping $G(x,z)$ is shaped in all regions so we should be able to approximate it well in all regions (assuming small influences from z). The accuracy will generally depend on the slope of $G(x,z)$ in various regions. Assuming the influence of z is small, in regions where the slope is high, we may need more data points to get more information so that we can do good approximation. In regions with lower slopes, we may not need as many points. This intuition, though, may not hold for all methods of adjusting θ. For some methods, you may need just as many points in "flat" regions as for those with ones that have high slopes. It is for this reason that we seek data sets that have uniform coverage of the X space. If you feel that more data points are needed, you may want to simply add them more uniformly over the entire space to try to improve accuracy.

We Often Cannot Control What Is in the Data Set

While the above intuitive ideas do help give direction on how to choose G for many applications, they cannot always be put directly into use. The reason for this is that for many applications, we cannot directly pick the data pairs in G. For instance, most often in "system identification," you cannot directly pick the data pairs. Recall that in system identification the objective is to use data from the plant (and perhaps other information) to construct a mathematical model of that system. It is then a type of function approximation problem. Notice that our input portion of the input-output training data pairs (i.e., x) typically contains *both* the inputs and the outputs of the system (i.e., x is often a regression vector) since the system has memory and so past inputs and outputs can affect the current output. It is for this reason that it is not easy to pick an input to the system that will ensure that the outputs will have appropriate values so that we get x values that uniformly cover the space X (basically it is an issue of controllability of the system, which for the nonlinear case can quickly become complicated).

Another situation that is commonly encountered where you cannot pick the input to the system is in some "online" function approximation problems. For instance, in adaptive control, the input may be chosen by a controller whose prime objective is to achieve tracking, and this may be in conflict with its other objective of providing a persistently exciting signal.

Similar problems may exist for other applications (e.g., parameter estimation), but for some applications this may not be a problem. For instance, in constructing a fuzzy controller (a valid approximation structure) from human decision-making data, we may be able to ensure that we have the human provide data on how to respond to a whole range of input data (i.e., we may have full control over what the input portion of the training data in G is).

Relationships to Persistent Excitation

It is interesting to note that there are fundamental relationships between a data set that has uniform coverage of X and the idea of "sufficiently rich" signals in system identification (i.e., "persistency of excitation" in adaptive systems). Intuitively, for system identification we must choose an input signal to "excite" the dynamics of the system so that we can "see," via the plant input-output data, what the dynamics are that generated the output data (i.e., we can see inside the "black box"). Normally, constraints from conventional linear system identification will require that, for example, a certain number of sinusoids be present in the input signal to be able to estimate a certain number of parameters. The idea is that if we excite more modes of the system, we will be able to identify these modes. Following this line of reasoning, if we use white noise for the input signal, then we should excite all frequencies of the system—and therefore, we should be able to better identify the dynamics of the plant.

Excitation with a noise signal (or a random binary signal) will have a tendency to place points in X over a whole range of locations; however, there is no guarantee that uniform coverage will be achieved for nonlinear identification problems with standard ideas from conventional linear identification. Hence, it is a difficult problem to know how to pick the input signal so that G is a good data set for solving a function approximation problem. Sometimes we will be able to make a choice for the input signal that makes sense for a particular application. For other applications, excitation with noise may be the best choice that you can make since it can be difficult to pick the input signal that results in a better data set G; however, sometimes putting noise into the system is not really a viable option due to practical considerations.

Other "Experiment Design" Issues

In system identification, the issue of how to design the experiments to collect data is a well-studied but complicated problem that offers ideas on how to solve the problem of choice of the data set G for the function approximation problem. In practical problems, there can be many issues that arise when working with data from a physical system. For instance, there is the issue of what to measure and when to measure it (i.e., what sampling period to use). There is often a need to preprocess the data to eliminate high frequency effects or noise. Sometimes there are certain outliers or missing data (discontinuous data records) that must be dealt with. The interested reader is referred to the "For Further Study" section at the end of this part for more information.

Data Scaling

There are times when scaling the data can be helpful in the sense that the algorithms that are used to process the data to find θ can sometimes perform better if the data are scaled. One simple way to scale the data is to simply multiply by a number that will force all the data values to be between -1 and $+1$. Such scaling can help with numerical issues, and can speed convergence of some training methods. Again, the interested reader is referred to the "For Further Study" section at the end of this part for more information.

9.2.3 Example: Collecting Data for Function Approximation

The process of collecting the training data pairs in the data set G from the function $G(x, z)$ generally differs for each application; however, there are common characteristics in the process and we will use a simple problem here to illustrate them. The particular example we consider has an unknown function $G(x, z)$ where x is a scalar ($n = 1$) that we can pick and at first, we assume that $n_z = 0$ so that z does not influence $G(x, z)$ in any way. We collect $M = 7$ pieces of training data as shown by the circles in Figure 9.8. Here, the $x(i)$ values are given on the horizontal axis and the data values $y(i)$ are shown on the vertical. We assume that we know the domain of the training conditions is $X = [-6, 6]$, which is simply an interval on the real line. Note that while we show evenly spaced input values $x(i)$, it is often the case that in practical applications their values are not uniformly spaced (and in fact, there may be significant regions in X where there may be no data). The function approximation problem amounts to finding a function $F(x, \theta)$ by manipulating θ so that $F(x, \theta)$ fits these data as closely as possible. Notice that there is some interesting nonlinear behavior that is exhibited by the training data. For instance, for values near $x = -6$, the function appears to smooth out. As x increases, there seem to be different slopes to the function. Finally, near $x = 6$, the function appears to be increasing quickly.

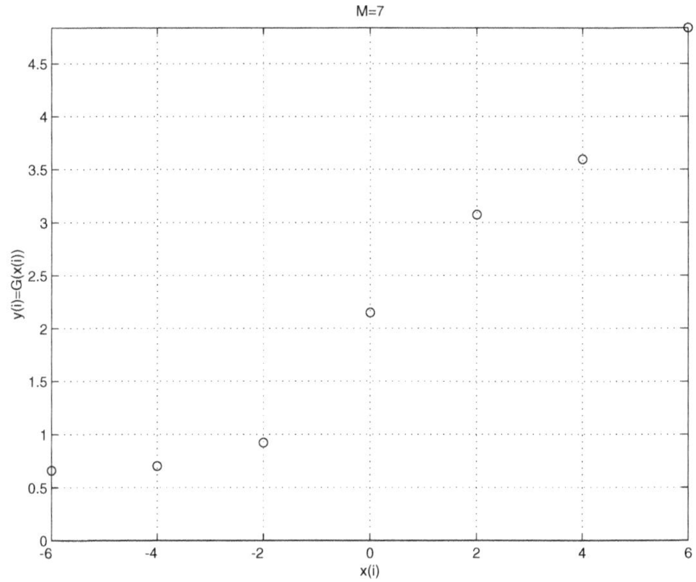

Figure 9.8: The training data G generated from the function $G(x, z)$, $M = 7$, $n_z = 0$.

Sparse data sets generally do not show how a function is shaped. "Dense" data sets can provide more information, but it is generally impossible to be perfectly confident that you have enough information.

Generally, more training data gives us more information about the underlying function $G(x, z)$. If $M = 1$, then little is known about the function. For larger values of M, you generally get more information about the function. Unfortunately, in practice, you are either constrained in the size of M, the ranges over which x can be generated, and often you cannot explicitly pick values for x—they are chosen for you. As an example of how more data gives us more information, we use $M = 121$ evenly spaced data points for the same function as above and we get the data pairs shown in Figure 9.9. With this amount of data, we now more clearly see the shape of the unknown function. The higher frequency oscillations were not seen before since our grid size was too large;

however, it may be that our grid size is still not small enough since with even higher values of M, we may find even higher frequency oscillations between the known data points. We see that without additional information about the unknown function (e.g., the maximum slope of the function), it is quite difficult to know when you have enough data to have a good representation of the function.

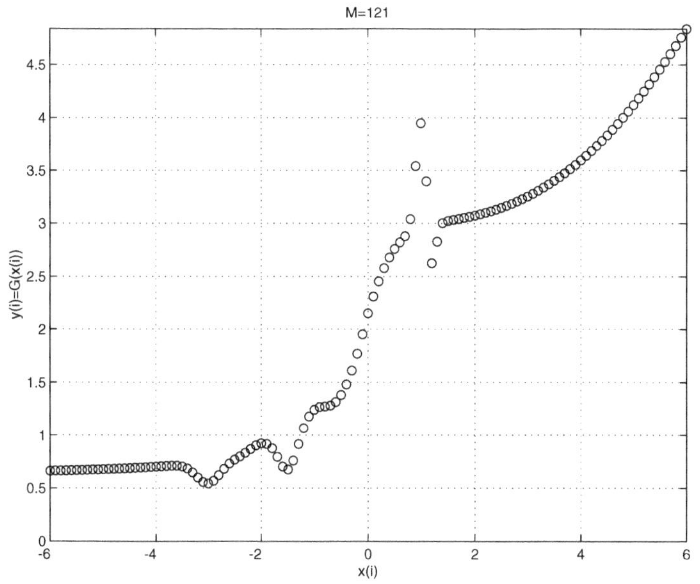

Figure 9.9: The training data G generated from the function $G(x, z)$, $M = 121$, $n_z = 0$.

Moreover, if we considered the auxiliary variable z, there would in essence be a set of functions that we would like our single function approximator $F(x, \theta)$ to represent. Why? Because we could be given one random value of z and generate a whole set of data showing the shape of the function. Then, if a different value of z was chosen and another set of data was gathered, a different shape could be produced. Clearly, if z has a significant impact on the shape of $G(x, z)$, it will be very difficult to find a single $F(x, \theta)$ to closely approximate it.

As an example of how the z variable can complicate the function approximation problem, consider the case when $n_z = 1$. There are many types of influences that z can have on $G(x, z)$. It could be that z is simply fixed but unknown, z could be additive white Gaussian noise, z could be other types of noise that influence $G(x, z)$ in complex nonlinear ways, or z could contain time as one of its components so that $G(x, z)$ is nonlinear, stochastic, and time-varying. Here, for this example, we will not indicate the type of influence that z has, besides to show the training data that were gathered. When we collect training data for the $M = 121$ case, we get the data shown in Figure 9.10. Now we see that $G(x, z)$ appears to be a very complex function to approximate accurately. First, it appears that there could be very high frequency information in the function when in actuality, this is not the influence of useful information, but a type of noise (that has in fact masked some of the behavior of the function that we saw in Figure 9.9). Second, even though we have gathered a lot of data, it is not clear how much data should be gathered, since these data might not be providing useful information, even though they do tell more about the noise characteristics of the process. Regardless, it is clear that the effects of the unknown auxiliary variable can be significant and can greatly complicate the function approximation problem.

Noise often corrupts training data and masks the true form of the underlying nonlinearity.

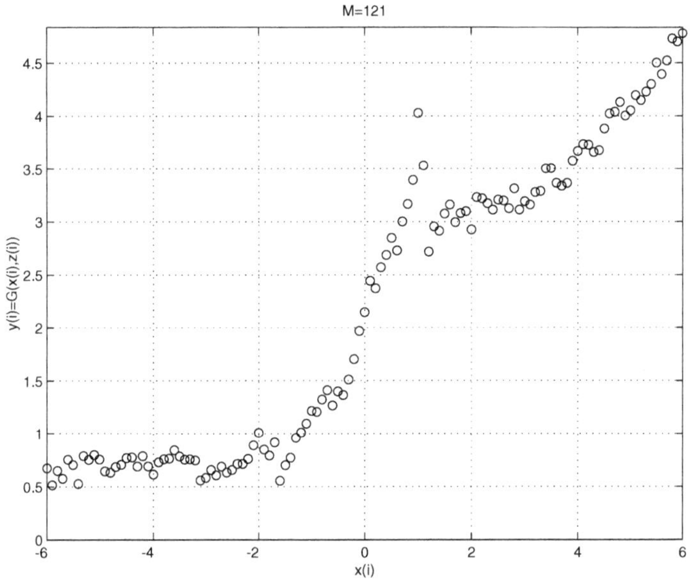

Figure 9.10: The training data G generated from the function $G(x, z)$, $M = 121$, and influences of the auxiliary variable z are shown.

9.2.4 Measuring Approximation Accuracy: Using a Test Set

How do we evaluate how closely the function $F(x, \theta)$ approximates the function $G(x, z)$ for a given θ? Notice that

$$W = \sup_{x \in X, z \in Z} \{|G(x, z) - F(x, \theta)|\} \tag{9.9}$$

is a bound on the approximation error $e(x, z)$ (if it exists) and

$$W^* = \inf_{\theta} \sup_{x \in X, z \in Z} \{|G(x, z) - F(x, \theta)|\} \tag{9.10}$$

is the "ideal approximation error" for a given approximator structure. Interpreting the mathematics in Equation (9.9), the ideal approximation error is the value of W when θ is chosen to make W as small as possible. The value (or values) of θ that achieves W^* is called the "ideal parameter value" and it is denoted by θ^*.

Unfortunately, in practice, specification of such a bound like W requires that the function $G(x, z)$ be completely known; however, as stated above, we know only a part of $G(x, z)$ given by the finite set G. Therefore, in practice we are only able to evaluate the accuracy of approximation by evaluating the error between $G(x, z)$ and $F(x, \theta)$ at certain points $x \in X$ given by *available* input-output data. We call this set of input-output data the *test set* and denote it as Γ, where

$$\Gamma = \{(x(1), y(1)), \ldots, (x(M_\Gamma), y(M_\Gamma))\} \tag{9.11}$$

Here, M_Γ denotes the number of known input-output data pairs contained within the test set. You can use error measures like

$$e_\Gamma = \frac{1}{M_\Gamma} \sum_{(x(i),y(i))\in\Gamma} (y(i) - F(x(i), \theta))^2 \tag{9.12}$$

(the mean squared error) or

$$e_\Gamma = \sup_{(x(i),y(i))\in\Gamma} \{|y(i) - F(x(i), \theta)|\} \tag{9.13}$$

(the maximum error for all points in the test set) to measure the approximation error. Accurate function approximation requires that some expression of this nature be small; however, this clearly does not guarantee perfect representation of $G(x, z)$ with $F(x, \theta)$, since most often, we cannot test that $F(x, \theta)$ matches $G(x, z)$ over all possible input points $x \in X$ and auxiliary variable values $z \in Z$.

It is important to note that the input-output data pairs $(x(i), y(i))$ contained in Γ may not be contained in G, or vice versa. It also might be the case that the test set is equal to the training set ($\Gamma = G$); however, this choice is often not a good one. Most often you will want to test the system with quite a bit of data that were *not* used to construct $F(x, \theta)$ since this will often provide a more realistic assessment of the quality of the approximation. In fact, one way to see if you have chosen M large enough is to use a test set with $M_\Gamma >> M$ (i.e., significantly bigger than M) and find the error that results for the training data set and the test set and compare them. If the two resulting error measures are close, then it is likely that you have chosen the training data set size (i.e., M) to be large enough.

Test data generally should contain training data that were not used in training to properly evaluate how well the approximator works (e.g., generalizes).

9.3 Approximator Structures as Substrates for Learning

The type of function $F(x, \theta)$ that you choose to adjust to fit $G(x, z)$ can have a significant impact on the ultimate accuracy of the approximator. For instance, it may be that a radial basis function neural network or a Takagi-Sugeno (or functional) fuzzy system will provide a better approximator than a standard fuzzy system, or a multilayer perceptron may provide an even better approximation, for a particular application. We think of $F(x, \theta)$ as a "structure" for an approximator that is parameterized by θ. In this section, we will summarize the approximator structures we will regularly use in this book. We will outline properties of these approximators, explain how they can be used on-line to approximate functions as data are gathered, and then in the next two chapters, study several methods for picking θ so that $F(x, \theta)$ does a good job of approximation.

9.3.1 Linear and Polynomial Approximator Structures

Linear approximator structures are a special case of polynomial approximator structures and they have been used in many capacities in estimation and control.

Linear Approximators

Linear mappings are tunable approximators that can be adjusted to perfectly match linear relationships if there is no noise.

A simple approximator structure is a linear or affine function. In this case, the approximator is given by

$$y = F_l(x, \theta) = \theta_1 x_1 + \theta_2 x_2 + \cdots + \theta_n x_n + \theta_{n+1}$$

(notice that the last term is simply a constant) or in vector notation if we let $\phi = \left[x^\top, 1\right]^\top$

$$y = F_l(x, \theta) = \theta^\top \phi(x) \tag{9.14}$$

where $x = [x_1, x_2, \ldots, x_n]^\top$, $\theta = [\theta_1, \theta_2, \ldots, \theta_n, \theta_{n+1}]^\top$, and we have $p = n + 1$. If $\theta_{n+1} \neq 0$, then F_l is called an "affine" function while if $\theta_{n+1} = 0$, it is called a "linear" function (although many refer to an affine function as being linear).

Linear approximators only provide for perfect representation of a class of functions that are linear (likewise for affine). If you know that the underlying function $G(x, z)$ generating the training data G is linear, then this type of approximator is all that is needed and the tuning of θ simply involves fitting the line to the data (in the case where $n = 1$ and if $n_z = 0$ so that z has no influence, and $G(x)$ is truly linear, you would only need two training data pairs to fully define the line and you would get a perfect fit). Generally, however, you do not know many properties of the underlying function $G(x, z)$ that is generating the training data and you may simply try to fit a linear function to the data to see how accurate it is. As an example, consider Figure 9.11, where we have fit a line to the data in Figure 9.10. Here, to fit the data, we simply guessed at the values of the slope and intercept of the line. Clearly this is not the "best" fit for a line to data (in Chapter 10, we will show how to fit a line to data by minimizing the sum of the squared errors in the distance from the data to the line).

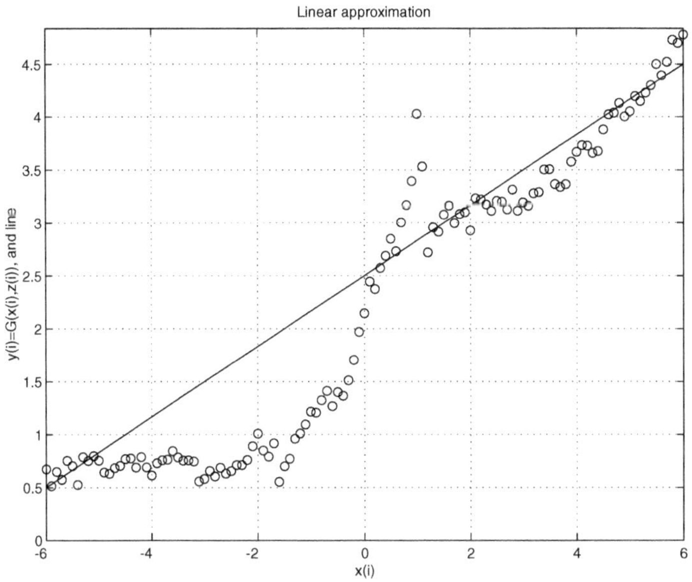

Figure 9.11: Using a line to approximate data.

Polynomial Approximators

Next, consider a polynomial approximator

$$\begin{aligned} F_{poly}(x,\theta) &= \theta_1 x_1 + \theta_2 x_2 + \cdots + \theta_n x_n + \theta_{n+1} + \cdots \\ & +\theta_{n+2} x_1 x_2 + \theta_{n+3} x_1 x_3 + \cdots \\ & +\theta_{n+k} x_1^2 x_2 + \cdots \end{aligned} \tag{9.15}$$

where the parameters θ_i that we adjust are used to scale terms that are products of the x_i (to any finite order), and notice that F_l is a special case of F_{poly}. We can, of course, pick θ so that $F_{poly}(x,\theta)$ is a linear function of θ (to do this, simply load in the coefficients of the polynomial) or a nonlinear function of θ (in this case, in addition to the coefficients, load in the powers of the x_i).

Just like for the linear approximator structure, we can pick the vector θ (that holds all the θ_i) so that $F_{poly}(x,\theta)$ best approximates the data. As an example, consider the data in Figure 9.10. To construct F_{poly}, you must first decide on the structure of the approximator. For polynomial approximators, the structure is determined by which terms you use (e.g., do you use the $x_1^3 x_2^5 x_3$ term?). Suppose that for our example you choose the polynomial approximator structure

Polynomial approximators can represent a wider class of functions than linear approximators; they have a more flexible approximator structure.

$$F_{poly}(x,\theta) = \theta_1 + \theta_2 x^2$$

(i.e., a parabola) that is a linear (affine) function of the parameters. What values would you choose for θ_1 and θ_2 to make this function best fit the data?

Figure 9.12 shows one choice (but not the best one) and the resulting fit to the data. Does this polynomial fit the data better than the line shown in Figure 9.11? Can we improve approximation accuracy by using additional terms (e.g., the x^3 term) in the approximator? In this part, we will explain how to pick the parameter vector θ of a variety of approximators using a variety of methods, we will show how to evaluate approximation accuracy, and we will give some ideas on how to pick the approximator structure (including some automated procedures). Ultimately, however, approximator *structure* choice is still a difficult problem, and is likely to be so for a long time.

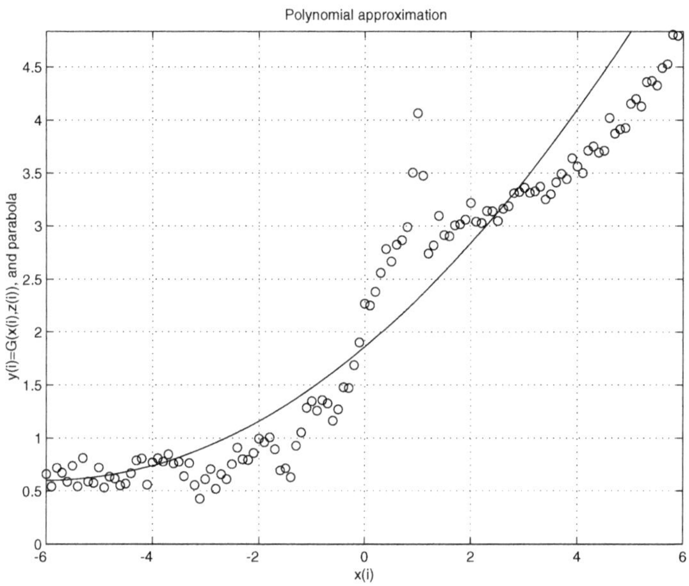

Figure 9.12: Using a polynomial to approximate data.

9.3.2 Neural and Fuzzy System Approximator Structures

In Chapter 5 we introduced standard and functional fuzzy systems, each of which can take on several forms depending on, for example, the types of membership functions, inference, and defuzzification strategies used. In Chapter 4 we described two classes of neural networks, the multilayer perceptron and the radial basis function neural network, each of which can take on several forms depending on, for example, the number of layers and type of activation function, or receptive field unit. Overall, we see that there are a wide range of possible neural and fuzzy system structures that are candidates for approximators. It is for this reason that we will pick the following three representative structures and focus on their construction in the remainder of this part (extension to other approximator structures is straightforward using the ideas in this part).

Multilayer Perceptron, Two Hidden Layers

The full equations for the two hidden layer case are given in Chapter 4. Suppose that we denote the output of this particular neural network by

$$y = F_{mlp}^{(2)}(x, \theta) \tag{9.16}$$

where θ holds all the weights and biases. We will assume that for $F_{mlp}^{(2)}$ we use hyperbolic tangent activation functions in the two hidden layers, but in the output layer, all the activation functions are linear.

Multilayer Perceptron, One Hidden Layer

In this case, the network is shown in Figure 9.13. Notice that we only consider the case where there is $m = 1$ output so there is only one linear activation function in the output layer; however, it is straightforward to treat the multi-output case by simply constructing m such networks.

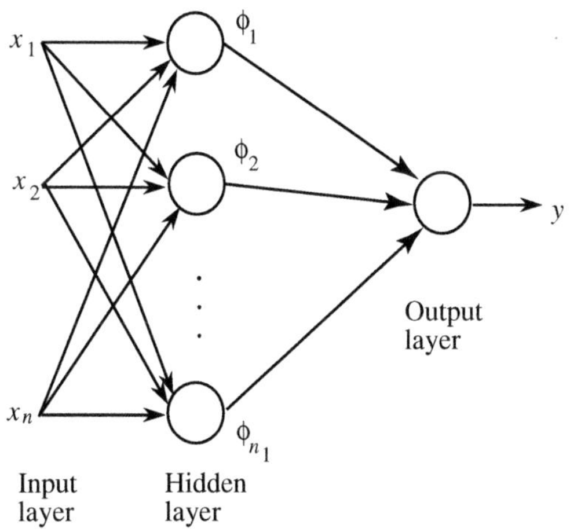

Figure 9.13: Multilayer perceptron with one hidden layer.

For our single hidden layer network, let ϕ_j, $j = 1, 2, \ldots, n_1$ denote the output of the jth neuron in the hidden layer, let b_j be the bias, and let

$$w^j = [w_{1,j}, w_{2,j}, \ldots, w_{n,j}]^\top$$

Hence, we have

$$\phi_j = f(b_j + (w^j)^\top x)$$

where f is the activation function (we could have different activation functions for each neuron, but for simplicity we let them all be the same). We will assume that the neurons in the hidden layer use nonlinear activation functions (any of the nonlinear activation functions we discussed earlier, except for the linear one).

Let w_j, $j = 1, 2, \ldots, n_1$ denote a weight in the output layer and let b be the bias. Let

$$w = [w_1, w_2, \ldots, w_{n_1}]^\top$$

With this, and the choice of using a linear activation function in the output layer, the mathematical formula describing Figure 9.13 is

$$y = b + \sum_{j=1}^{n_1} w_j \left(f(b_j + (w^j)^\top x) \right) = b + \sum_{j=1}^{n_1} w_j \phi_j$$

Parameters of an approximator can be classified as entering in a linear or nonlinear fashion.

Now, if we choose

$$\phi = [\phi_1, \phi_2, \ldots, \phi_{n_1}, 1]^\top$$

so that

$$y = F_{mlp}(x, \theta) = \left[w^\top, b \right] \phi \tag{9.17}$$

(note that we could use $F_{mlp}^{(1)}$ to denote this single hidden layer perceptron, but we omit the "(1)" for simplicity).

We will consider two different choices for the parameter θ:

- Nonlinear in the parameters: Choose

  $$\theta = [(w^1)^\top, b_1, (w^2)^\top, b_2, \ldots, (w^{n_1})^\top, b_{n_1}, w^\top, b]^\top$$

  so that we seek to tune all the weights and biases of the perceptron. Notice that the output y is a nonlinear function of the parameters w^j and b_j, $j = 1, 2, \ldots, n_1$, due to the choice of having nonlinear activation functions in the hidden layer.

- Linear in the parameters: Suppose that you know the values of the w^j and b_j, $j = 1, 2, \ldots, n_1$, so they do not need to be tuned. In this case, we choose

  $$\theta = [w^\top, b]^\top$$

  Notice that if the w^j and b_j, $j = 1, 2, \ldots, n_1$, are known, then the ϕ_j, $j = 1, 2, \ldots, n_1$ are known once the input x is specified, so ϕ is known. For this choice of θ,

  $$y = F_{mlp}(x, \theta) = \theta^\top \phi$$

  so y is a linear function of the parameter vector θ.

Clearly, the nonlinear in the parameter case is more general since the added parameters can be tuned to shape the nonlinear mapping implemented by the perceptron in more complex ways than when only the parameters that enter linearly are tuned. We will see, however, that we have better methods to adjust linear in the parameter approximators.

As an example of what types of nonlinear functions this neural network can implement, consider the case where $n = 1$, $n_1 = 2$ (i.e., two neurons in the hidden layer and only one input, x_1), and let each neuron be the logistic (sigmoidal) nonlinearity. We have

$$w^1 = [w_{1,1}]^\top$$
$$w^2 = [w_{1,2}]^\top$$

as the weights and b_1 and b_2 as the biases for the hidden layer. For the output layer, we have

$$w = [w_1, w_2]^\top$$

where w_j, $j = 1, 2$, are weights and b is the bias for the neuron in the output layer. All these weights and biases specify the shape of the nonlinearity that is implemented by the neural network.

It is useful to know how to manually construct an approximator to gain insight. Some parameter adjustment methods only apply to the parameters that enter linearly and others profit from good initializations of parameters.

Suppose that we want to shape the nonlinearity of the network to match the function in Figure 9.10. While later we will examine several methods (e.g., least squares and gradient techniques) for tuning the parameters, here we will study how to use simple heuristic ideas to specify the weights and biases. Our overall intent is to provide insights into how the choices of various parameters affect the shape of the nonlinearity implemented by the neural network.

To pick the parameters, first study Figure 4.6, where we show the shape of the logistic function nonlinearity that is given by

$$f(\bar{x}) = \frac{1}{1 + \exp(-\bar{x})}$$

For this nonlinearity, note that its slope is given by

$$\frac{\partial f(\bar{x})}{\partial \bar{x}} = 2f(\bar{x})(1 - f(\bar{x})) \tag{9.18}$$

Note that at $\bar{x} = 0$, $f(0) = 0.5$, and hence, the slope of the logistic function is 0.5. Also note that, as $\bar{x}$ moves to increasingly large values in the negative direction $f(\bar{x}) \to 0$, while if it moves to increasingly large values in the positive direction $f(\bar{x}) \to 1$. As shown in Figure 4.6, by the time that $\bar{x}$ reaches -5 in the negative direction ($+5$ in the positive direction), $f(\bar{x})$ is quite close to 0 (1).

To construct the network with two neurons in the hidden layer, we will proceed by first constructing a network with one neuron in the hidden layer. First, note that we are trying to approximate the function in Figure 9.10 with a network that has a single sigmoid, which you may think of as a "smooth step function" where you can shift the step horizontally (i.e., to the left and right) by changing the value of the bias b_1, change the steepness of the step by adjusting $w^1 = w_{1,1}$, scale the size of the step by changing $w = w_1$, and offset the step vertically by adjusting the (output) bias b.

To manually construct a multilayer perceptron, it is useful to view the neurons as "smooth step" building blocks for the interpolator.

Notice that part of the function in Figure 9.10 (the part for $x < 0.5$) actually takes on a shape like a sigmoid, at least to a rough approximation. Now, notice there is really no hope of perfectly approximating the peaks that appear to occur at about $x = 1$, or the apparently linearly increasing behavior for larger values of x. Hence, we will seek to achieve a gross approximation of the function in

Figure 9.10 by using the sigmoid where it will be most effective. To do this, first note that if we pick $b = 0.6$, that will shift the whole function up so that for small negative values of x, we should get a close match (see discussion above on how fast the nonlinearity goes to zero for negative values). Next, notice that viewed as a smooth step, the unknown nonlinearity increases in magnitude about 2.5 to 3 units so we pick $w = w_1 = 3$. Next, by inspection we can see that we will want the value of the neural network to be about 2.1 at $x = 0$. With our current choices, we will have a value of $0.6 + 3f(b_1 + w^1 x)$ at $x = 0$, so we want $f(b_1) = 0.5$, so we choose $b_1 = 0$. We pick $w_{1,1} = 1.5$ by imagining drawing a line on Figure 9.10 through $x = 0$ that tries to approximate the slope of the function at that point, and we use this value of the slope as the value for $w_{1,1}$. This completes our heuristic tuning process for the parameters and Figure 9.14 shows the resulting approximation.

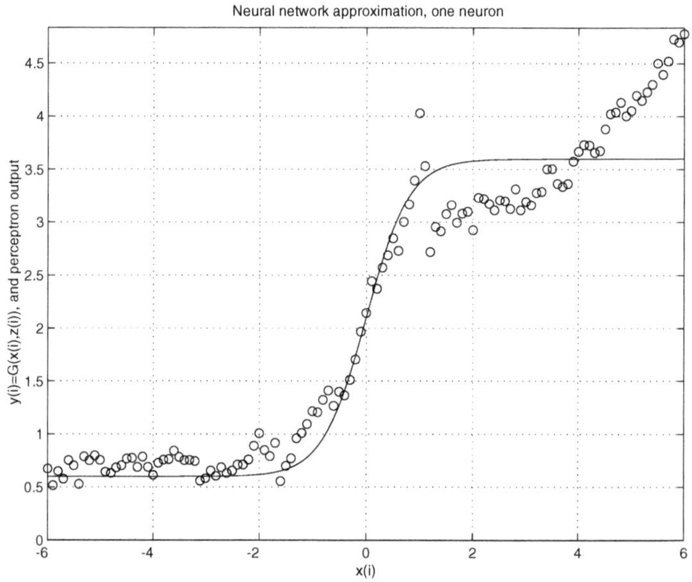

Figure 9.14: Neural network approximation, one neuron.

Next, we will add a neuron to the hidden layer to obtain a hidden layer composed of two neurons. Now, we must accept that with only two neurons, we will only get the ability to have two smooth steps, but these can be scaled and translated (note that the steps can be either up or down since we can use negative values) to try to improve approximation accuracy. To keep things simple, we will simply try to place a neuron so that the increasing part of the smooth step is in the region near $x = 5$ so that the increasing part of the unknown function can be approximated more accurately. To achieve this goal, we first choose $w = [3, 1]^\top$ so that we scale the output of the second neuron with unity (somewhat arbitrary since we do not need to specify how much this neuron contributes for very large x), and $b = 0.6$ to get matching for large negative values of x as for the single neuron case. Next, we notice that the slope of the increasing part of the function on the right of the plot is slightly lower than that near $x = 0$, so we pick $w_{1,2} = 1.25$ and then we tune by shifting the step to the left (by tuning b_2 to $b_2 = -6$) until we get an improvement in approximation accuracy. The result is shown in Figure 9.15. We emphasize that our objective is not to achieve the best possible approximation; it is simply to show how, by tuning the neural network appropriately, we can get improvements in accuracy. Indeed, it is very difficult to tune these functions by hand (and repeated computer simulations) and this is the reason we will study how to use least squares and gradient methods for tuning neural networks later in this part.

Additional structure, defined properly, can lead to more accurate approximation.

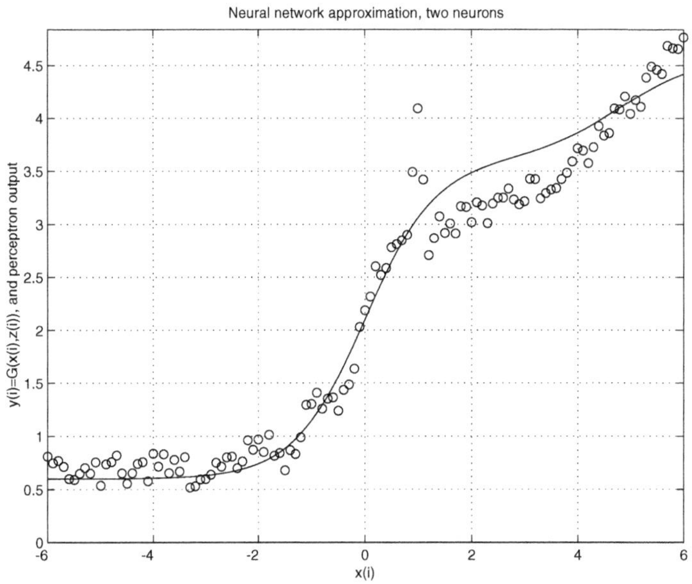

Figure 9.15: Neural network approximation, two neurons.

Takagi-Sugeno Fuzzy Systems

We will consider two types of Takagi-Sugeno fuzzy systems: ones that are linear in the parameters and ones that are not. We consider a Takagi-Sugeno fuzzy system that is given by

$$y = F_{ts}(x, \theta) = \frac{\sum_{i=1}^{R} g_i(x)\mu_i(x)}{\sum_{i=1}^{R} \mu_i(x)}$$

where if $a_{i,j}$ are constants,

$$g_i(x) = a_{i,0} + a_{i,1}x_1 + \cdots + a_{i,n}x_n$$

Also, for $i = 1, 2, \ldots, R$,

$$\mu_i(x) = \prod_{j=1}^{n} \exp\left(-\frac{1}{2}\left(\frac{x_j - c_j^i}{\sigma_j^i}\right)^2\right)$$

where c_j^i is the point in the jth input universe of discourse where the membership function for the ith rule achieves a maximum, and $\sigma_j^i > 0$ is the relative width of the membership function for the jth input and the ith rule. Clearly, we are using center-average defuzzification and product for the premise and implication. Notice that the outermost input membership functions do not saturate as is the usual case in control. Also, note that by considering Takagi-Sugeno fuzzy systems, we are also inherently considering the class of standard fuzzy systems that is defined by letting all the $a_{i,j} = 0$ except for the cases where $j = 0$ (i.e., by only leaving the affine terms we get a standard fuzzy system that uses center-average defuzzification).

We will consider two different choices for the parameter vector θ:

- Nonlinear in the parameters: One choice for θ above is to let

$$\begin{aligned}\theta &= [c_1^1, \ldots, c_n^R, \sigma_1^1, \ldots, \sigma_n^R, \\ &\quad a_{1,0}, a_{2,0}, \ldots, a_{R,0}, a_{1,1}, a_{2,1}, \ldots, a_{R,1}, \ldots, a_{1,n}, a_{2,n}, \ldots, a_{R,n}]^\top\end{aligned}$$

  For this choice, y is a nonlinear function of θ, particularly the premise membership functions parameters.

- Linear in the parameters: Next, we develop a choice for θ so that y is a linear function of θ. Note that

$$y = \frac{\sum_{i=1}^R a_{i,0}\mu_i(x)}{\sum_{i=1}^R \mu_i(x)} + \frac{\sum_{i=1}^R a_{i,1}x_1\mu_i(x)}{\sum_{i=1}^R \mu_i(x)} + \cdots + \frac{\sum_{i=1}^R a_{i,n}x_n\mu_i(x)}{\sum_{i=1}^R \mu_i(x)}$$

  We see that the first term is the standard fuzzy system as we discussed above. Let

$$\begin{aligned}\phi &= [\xi_1(x), \xi_2(x), \ldots, \xi_R(x), x_1\xi_1(x), x_1\xi_2(x), \ldots, x_1\xi_R(x), \ldots, \\ &\quad x_n\xi_1(x), x_n\xi_2(x), \ldots, x_n\xi_R(x)]^\top\end{aligned}$$

  and

$$\theta = [a_{1,0}, a_{2,0}, \ldots, a_{R,0}, a_{1,1}, a_{2,1}, \ldots, a_{R,1}, \ldots, a_{1,n}, a_{2,n}, \ldots, a_{R,n}]^\top$$

  and

$$\xi_j = \frac{\mu_j(x)}{\sum_{i=1}^R \mu_i(x)}$$

  $j = 1, 2, \ldots, R$, so that

$$y = F_{ts}(x, \theta) = \theta^\top \phi(x)$$

represents the Takagi-Sugeno fuzzy system. This choice of θ will not allow us to tune the fuzzy system to match as complex mappings as the nonlinear in the parameter Takagi-Sugeno fuzzy systems above; however, we will develop several good methods to tune the approximators for the linear in the parameter case.

As an example, consider a Takagi-Sugeno fuzzy system with one input and $R = 4$ rules that we will use to approximate our same unknown function in Figure 9.10. In this case, we have to pick the parameters for

To manually construct a Takagi-Sugeno fuzzy system, it is useful to think of the building blocks as local linear mappings that are connected.

$$\mu_i(x) = \exp\left(-\frac{1}{2}\left(\frac{x_1 - c_1^i}{\sigma_1^i}\right)^2\right)$$

$i = 1, 2, 3, 4$ and the parameters of the corresponding g_i functions. Notice that for the $n = 1$ case, we have

$$g_i(x) = a_{i,0} + a_{i,1}x_1$$

which are lines. Hence, in this case, the Takagi-Sugeno fuzzy system uses the functions μ_i to interpolate between lines. Each μ_i will specify the region in which we want to use the line given by g_i. Hence, our overall strategy will be to examine a plot of the training data, pick a number of lines (i.e., pick R) that will appear to give reasonable approximation accuracy, sketch lines on the data to get the slopes $a_{i,1}$ and intercepts $a_{i,0}$ for each g_i line, then pick the parameters of μ_i to specify where we will use each line in the approximation.

Consider Figure 9.16, where we plot the training data and sketch lines (with circled numbers 1 to 4) that we will use as a basis for constructing our approximator. In particular, if you use the lines shown and find their slopes and intercepts, by inspection we get

$$a_{1,0} = 1, a_{1,1} = \frac{0.5}{6}$$

$$a_{2,0} = 2.25, a_{2,1} = \frac{4.4}{4}$$

$$a_{3,0} = 2.9, a_{3,1} = \frac{1}{12}$$

$$a_{4,0} = 1.3, a_{4,1} = \frac{4.8}{8}$$

Next, we need to specify the functions μ_i and to do this, we will explain how to specify it for the $i = 2$ rule; then it will be clear how to specify them for the other three cases. Note that

$$\mu_2(x) = \exp\left(-\frac{1}{2}\left(\frac{x_1 - c_1^2}{\sigma_1^2}\right)^2\right)$$

so we need to specify values for the "center" c_1^2 and "spread" σ_1^2. Notice from Figure 9.16 that the approximate range of validity for rule 2 (and line 2) is

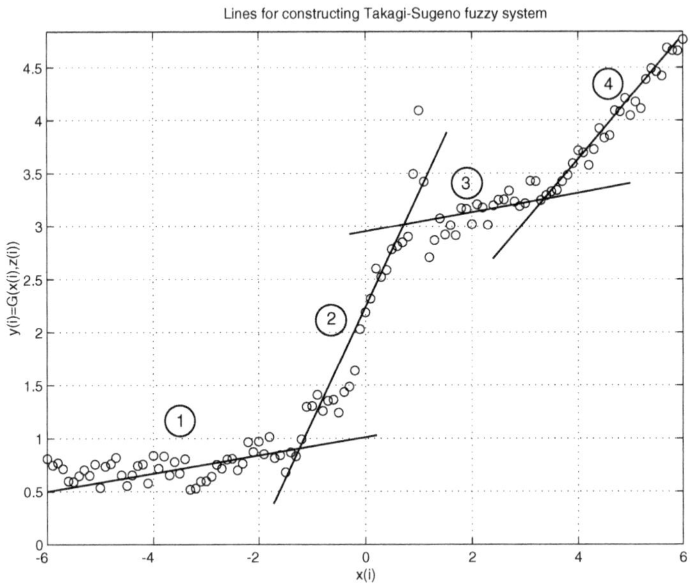

Figure 9.16: Training data and lines used in the Takagi-Sugeno fuzzy system approximator.

when $-1.2 \leq x \leq 0.7$ (this range is simply chosen by inspection). Hence, in this region, we want μ_2 to be "on" (i.e., have a value greater than zero) and the other μ_i, $i \neq 2$ to be near zero. Pick $c_1^2 = -0.25$ to be the middle of this range and choose $\sigma_1^2 = 0.6$ so that μ_2 rolls off fast enough so that it is close to zero outside the range $-1.2 \leq x \leq 0.7$. Continuing in a similar manner, we end up with

$$c_1^1 = -3.5, \sigma_1^1 = 0.8$$
$$c_1^2 = -0.25, \sigma_1^2 = 0.6$$
$$c_1^3 = 2, \sigma_1^3 = 0.4$$
$$c_1^4 = 4.5, \sigma_1^4 = 0.8$$

which are shown in Figure 9.17. Notice that the premise membership functions turn on in the regions where we want to use the corresponding lines for approximation.

To be more precise, however, notice that the lines g_i are activated by the "basis functions"

$$\xi_j = \frac{\mu_j(x)}{\sum_{i=1}^{4} \mu_i(x)}$$

$j = 1, 2, 3, 4$. A plot of these functions is shown in Figure 9.18. This clearly shows when each g_i line will be used, and where the transition between using two different lines occurs. The premise membership functions specify the regions,

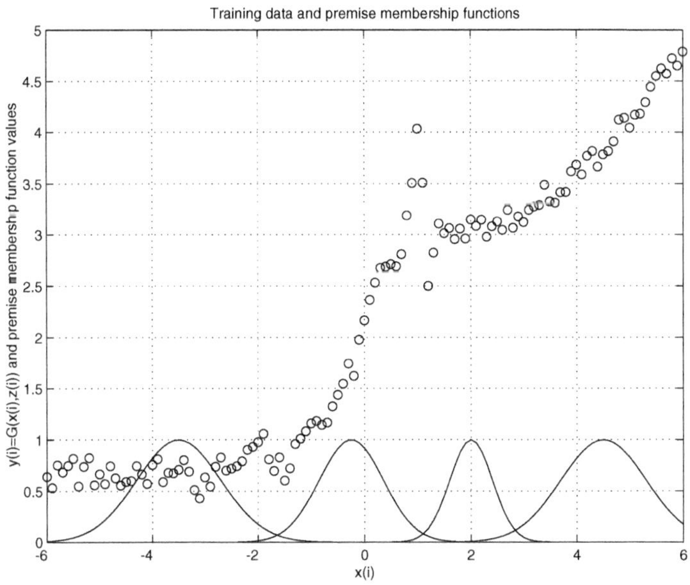

Figure 9.17: Training data and membership functions used in the Takagi-Sugeno fuzzy system approximator.

but notice two important things that the basis functions add. First, at the outermost edges, the basis functions are constant and the other basis functions are essentially zero; this ensures that the associated lines (i.e., g_1 and g_4) are used for extrapolation at the outer edges. Second, notice that for the two middle basis functions (i.e., g_2 and g_3), the functions flatten out in their middles and the others go to zero to ensure that only their associated lines are used in the interpolation. Also, notice that

$$\sum_{j=1}^{4} \xi_j(x) = 1$$

This is the reason why near the outer edges the functions go to unity, and why the transitions are shaped as they are, crossing at 0.5 in each case.

With this, we have specified the entire approximator and its accuracy is illustrated in Figure 9.19. Notice that the approximator performs as expected, using lines to approximate regions that have different slopes. The transition regions between the lines have curves whose shapes are partially dictated by the choice for the μ_i and hence ξ_i. For instance, the shape near $x = 1$ could be changed by moving the c_1^3 to the right.

Clearly, we have not achieved the best possible approximation for the unknown function $G(x, z)$; however, we have improved it over the cases considered earlier. This improvement came, however, with an increase in approximator

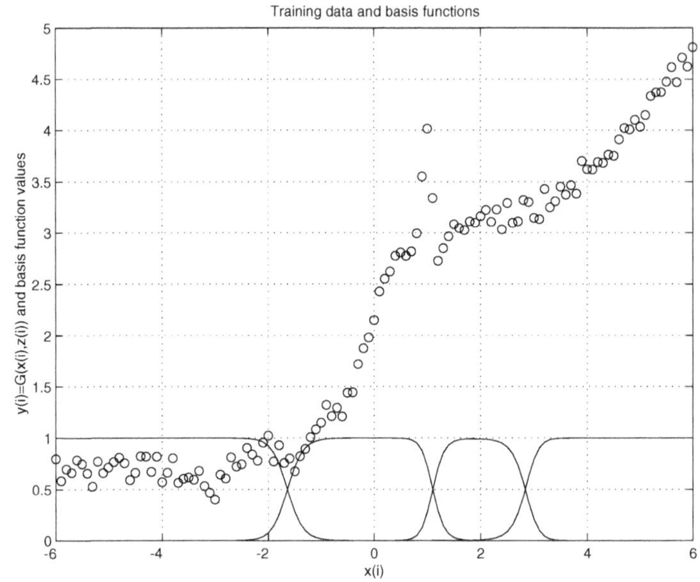

Figure 9.18: Training data and basis functions used in the Takagi-Sugeno fuzzy system approximator.

complexity. Notice that for the linear and polynomial approximators, we only needed two parameters. For the neural network approximator, we only used 7 parameters (but of course we could have chosen to use more and could have then gotten better approximation accuracy there also). For the Takagi-Sugeno fuzzy system, we used 16 parameters and this allowed us to get the better approximation accuracy shown in Figure 9.19. You will often find such a general relationship between approximator complexity and its ability to achieve good approximation; however this will not always be the case. Ultimately, the achievable accuracy depends on how you tune the approximator structure, the flexibility of the approximation structure, and on the form of the underlying unknown nonlinear function.

9.3.3 Universal Approximation Property and Substrate Capabilities

Neural networks and fuzzy systems have very strong functional capabilities. That is, if properly constructed, they can perform very complex operations and implement very complex nonlinear mappings (e.g., much more complex than those that can be implemented by a linear mapping). Actually, many neural networks and fuzzy systems are known to satisfy the "universal approximation property."

To study this idea, let $\mathcal{F}$ denote the set of all possible approximator struc-

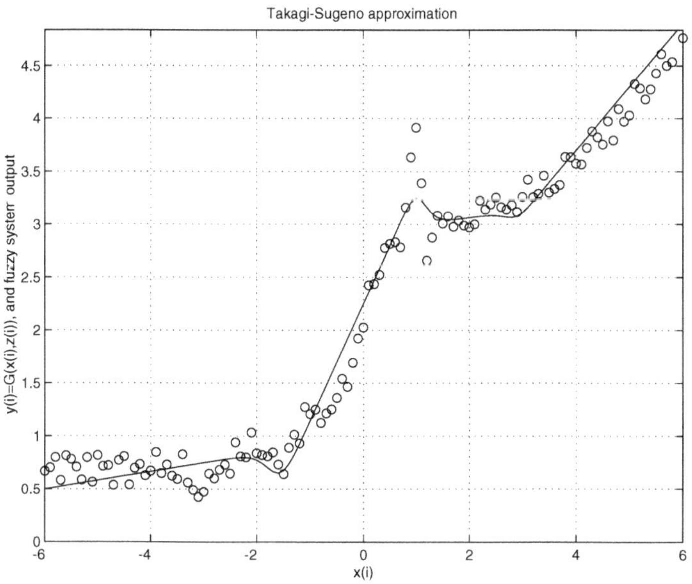

Figure 9.19: Training data and Takagi-Sugeno fuzzy system approximator.

tures of type $F(x, \theta)$. For example, if $F(x, \theta)$ is a multilayer perceptron with a single hidden layer, then $\mathcal{F}$ would contain an infinite number of approximator structures, each one with different interconnections, nonlinear activation functions, and numbers of neurons in the various layers and values for the tunable parameters θ.

Assume in this section that $n_z = 0$ so that there is no influence from the auxiliary variable z. If $G(x)$ is any real valued continuous function defined on a closed and bounded set $X \subset \Re^n$ and for an arbitrary $\epsilon > 0$, there exists an approximator structure $F(x, \theta) \in \mathcal{F}$ such that

$$\sup_{x \in X} |G(x) - F(x, \theta)| < \epsilon$$

then the approximator structure $F(x, \theta)$ is said to satisfy the "universal approximation property." Using the Stone-Weierstrass theorem, it is easy to show that multilayer perceptrons, radial basis function neural networks, and standard and Takagi-Sugeno fuzzy systems all satisfy the universal approximation property. Clearly, however, the linear approximator structure $F_l(x, \theta)$ does not satisfy the universal approximation property.

An approximator that is a universal approximator is flexible enough to be able to represent many functions; however, this "flexibility" may require a large approximator structure and extraordinary parameter tuning capabilities.

Satisfaction of the universal approximation property guarantees that there exists a way to define the particular approximator structure $F(x, \theta)$ and its parameters θ to represent the unknown nonlinearity as accurately as you would like. It does *not* say how to find the particular $F(x, \theta)$, which can, in general, be very difficult (i.e., it does not say how many neurons should be in the hidden layer, how many rules should be in the fuzzy system, or anything about how to pick θ to get good accuracy). Furthermore, for arbitrary accuracy you may need an arbitrarily large number of parameters (i.e., for $\theta = [\theta_1, \ldots, \theta_p]^\top$ you may need p to be arbitrarily large).

The value of the universal approximation property is simply that it shows that if you work hard enough at structure choice, are willing to use a large structure, and do good parameter tuning, you should be able to make the neural networks and fuzzy systems achieve what you are trying to get done. For estimation, this means that there is great flexibility in working with the neural and fuzzy approximator structures. If you choose enough neurons in the hidden layers of the multilayer perceptron, or enough rules and membership functions in a fuzzy system, there is a way to tune the neural or fuzzy system so that it will perform its estimation task very well. For control, practically speaking, it means that there is great flexibility in tuning the nonlinear function implemented by, e.g., the fuzzy controller or a neural network.

To summarize, recall that W is the bound on the representation error of the unknown function $G(x)$ with the approximator $F(x, \theta)$. For a *given* approximator structure $F(x, \theta)$ that satisfies the universal approximation property, all we know is that the bound on the approximation error $W > 0$ exists. We typically do not know how small it is. The universal approximation property simply says that we may increase the size of the approximation structure and properly define the parameters of the approximator to achieve any desired accuracy (i.e., to make W as small as we want); it does not say how big the approximator must be or if you fix the structure $F(x, \theta)$ how small W is.

9.3.4 Approximator Complexity Vs. Substrate Flexibility

Approximator complexity refers to the complexity of implementing the approximator structure. Clearly, the best approximator is a "flexible" one that can be tuned to accurately approximate many types of nonlinear functions (e.g., one that satisfies the universal approximation property) yet only requires minimal memory and processing time to implement. Generally, there is a proportional relationship between approximator flexibility and complexity; however, this is not always the case since, if bad choices are made for the structure, you may have added significant complexity without providing additional flexibility. Hence, one goal in approximation structure choice is to make sure that any additions to the structure (e.g., more neurons in a hidden layer or rules to a rule base) are done in a way that improves approximator flexibility, hopefully, in a significant and useful way.

Generally, for neural networks and fuzzy systems, implementation amounts to coding certain nonlinear functions (e.g., activation functions or membership functions) and sums, differences, products, or the division of such functions with other constants and other nonlinear functions. For some applications, complexity, in terms of the actual number of, for example, multiplications that are required for implementation, is very important to reduce as much as possible. How much can complexity be reduced? This depends on the application and often the only way to answer this question is to experiment with a variety of approximators of varying complexities. Often, such analysis can lead you to a quantification of the trade-off between performance and complexity, leading you to conclude that performance costs money (not a surprising conclusion).

Next, we identify a few key features of neural and fuzzy approximator structures as they relate to complexity: First, for multilayer perceptrons, you should be careful in adding layers to the network since more than two layers may not add much tuning flexibility, but it certainly adds more complexity. Next, be careful with creating grids of, for example, membership functions of a fuzzy system (or receptive field units of radial basis function neural networks) on the input space, since a grid with N membership functions on each of the n input dimensions results in N^n membership functions (i.e., we get an exponential growth in approximator complexity). While in this case finer and finer grids will tend to improve the potential for good approximation accuracy, in practical applications if n is large, complexity may be prohibitive for implementation. This does not prohibit the use of fuzzy systems (or radial basis function neural networks). It just points to the importance of making careful choices for the approximator structure. For example, in the case outlined above, if the input x truly spans the whole n-dimensional input space, then the complexity may be warranted. In practical applications, however, it is often the case that large regions of the input space will never be visited by x so that portions of the approximator structure are wasted. There are two approaches to solve this problem. First, you could simply fix portions of the approximator structure on the portions of the input space that you know will be visited. Second, if you do not know which portions will be visited, then you can tune the structure to have it focus on the portion of the input space that has been visited. Finally, it is important to note that these comments also hold for some commonly used multilayer perceptrons since we can think of allocating activation functions (which provide for an increased flexibility in tuning) to input dimensions in a similar way.

Generally, more complex approximators are more flexible in that they can represent more functions.

Regardless of which approximator structure is used, this discussion should show that it is possible that nonlinear in the parameter approximators (the ones that allow for tuning of the grid of membership functions or receptive field units, or positions of activation functions) provide the potential to save complexity, at the expense of proper tuning of the parameters that enter in a nonlinear fashion. This is quantified and discussed further in the next section.

9.3.5 Linear Vs. Nonlinear in the Parameter Approximators: Substrate Tunability

Do we want to use linear or nonlinear in the parameter approximators? Results from approximation theory (see [45]) give us some clues as to how to answer this question:

- It has been shown that for a nonlinear in the parameter approximator (like the multilayer perceptron shown in Figure 9.13 that uses sigmoids for the activation functions in the hidden layer and a linear activation function in the output layer) and for a certain class of functions (with certain smoothness properties) that we would like to approximate, if we tune the parameters of the approximator properly, then the integral squared error over the approximation domain is less than

Nonlinear in the parameter approximators have the potential to provide for low approximation errors with a relatively small approximator structure; however, in practice it may be difficult to tune them so that such gains are realized.

$$\frac{C}{n_1}$$

where n_1 is the number of neurons in the hidden layer. The value of C depends on the size of the domain over which the approximation takes place (it increases for larger domains), and how oscillatory the function is that we are trying to approximate (roughly speaking, with more oscillations C increases). For certain general classes of functions, C can increase exponentially as the input dimension n increases, but for a fixed n and a fixed domain size, the results show that by adding more sigmoidal functions and *if we tune the parameters properly,* we will get a definite decrease in the approximation error (and with only a linear increase in approximator size, i.e., a linear increase in the number of parameters). In summary, what this says is that if, in this case, the parameters that enter in a nonlinear fashion are allowed to depend on the function to be approximated (and the only way they can depend on the function for our problem formulation is via the training data and hence how the parameters are tuned), then the size of n_1 to achieve a certain level of approximation accuracy is much less than if these parameters are fixed a priori. In the approaches to follow, we will consider both the case where we fix the parameters that enter in a nonlinear fashion (making it a linear in the parameter approximator), and the case where we try to tune them.

- For linear in the parameter approximators, for the same type of functions to be approximated as in the nonlinear case discussed above, it has been shown that there is *no way* to tune the parameters that enter linearly (for a given fixed number of "basis functions," i.e., an approximator structure with the part where the parameters enter in a nonlinear fashion fixed) so that the approximation error is better than

Linear in the parameter approximator structure complexity can increase exponentially if you try to get low approximation errors; however, there exist good methods to tune them.

$$\frac{C_L}{n_1^{2/n}}$$

Here, C_L has similar dependencies as the nonlinear in the parameter case discussed above. Note, however that there is a dependence on the dimension n in the bound so that for high dimensional function approximation you may need many parameters n_1 to achieve good approximation accuracy, so you probably want a nonlinear in the parameter approximator to avoid the "curse of dimensionality." Also, you will want to be very careful with the choice of the nonlinear part of the structure or you may add more approximator structure and not gain any more ability to reduce the approximation error. We will study the case where the linear in the parameter approximator is tuned and discuss issues in the choice of the size of the approximator structure on approximation accuracy.

To summarize, it is desirable to use approximators that are nonlinear in their parameters, since a nonlinear in the parameters approximator can be simpler than a linear in the parameters one (in terms of the size of its structure and hence number of parameters) yet achieve the same approximation accuracy (i.e., the same W above). The general problem: we know how to tune linear in the parameter approximators, but in certain cases they may not be able to reduce approximation error very well, and we do not know as much about how to tune nonlinear in the parameter approximators, but we know that if we can tune them properly, we can definitely reduce the approximation error. We emphasize, however, that finding the best approximator structure is a difficult problem that normally requires trial-and-error in practical applications.

The fundamental guideline that you can use is to view more complex approximators as more "flexible" in terms of the functions that they can be tuned to approximate. You do not, however, want to use the most flexible (complex) approximator, since a simpler solution is often the best (e.g., due to implementation complexity, and issues with "generalization" that will be discussed later).

9.3.6 Online Function Approximation: Dynamic Learning

Above, we discussed the case of approximating functions when G is given a priori and then we train. In control, however, another common situation is where we obtain the training data in a sequence over time as the system operates and we need to use it immediately. In this section, we briefly outline general issues of using training data gathered online to perform "online" function approximation. The basic ideas of this section form a foundation on which we will later build a number of adaptive controllers. Assume that each experiment is performed at the click of a clock, $k = 0, 1, 2, \ldots$ and we get an infinite sequence of training data pairs

$$(x(k), y(k))$$

$k = 0, 1, 2, \ldots$ (notice that we switch the index of the training data to the time index k). Suppose that each time we get a new training data pair, we want to update the parameter vector $\theta(k)$ of the approximator $F(x, \theta(k))$ to try to make it match $G(x, z)$ closer. It is hoped that the new training data gives information about the shape of the $G(x, z)$ nonlinearity, and we want to immediately use it to update the shape; this will result in $F(x, \theta(k))$ being a time-varying nonlinear function that is searching for an appropriate shape. This "online function approximation" scheme is shown in Figure 9.20, where we illustrate the tuning of $\theta(k)$ (illustrated via the diagonal arrow through the box containing the approximator) using $(x(k), y(k))$ information.

Approximator mappings can be dynamically tuned as data are gathered.

The goal is to get $\epsilon(k) = y(k) - \hat{y}(k) \rightarrow 0$ as $k \rightarrow \infty$. This is, however, a challenging objective for the following reasons:

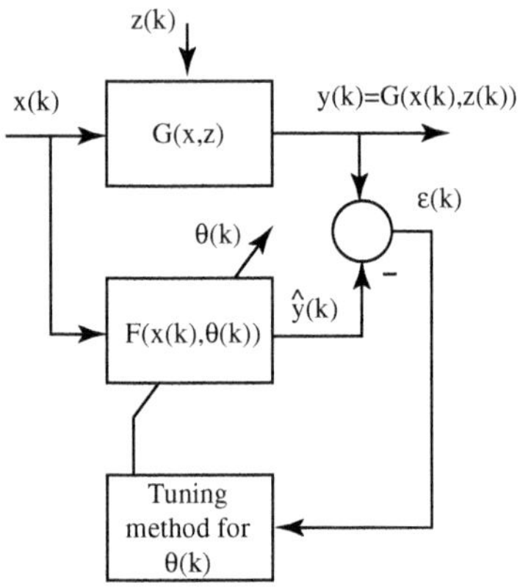

Figure 9.20: Online function approximation scheme.

1. In practical applications, we often cannot pick the explicit values of $x(k)$ (e.g., they may be chosen by some system) so we cannot ensure that we have training data in all regions of X, so we can do very poor function approximation in some regions where we do not get enough data.

2. Often, we do not know z or whether it has spanned the space Z, and often we do not have a clear picture of how z affects the shape of $G(x, z)$.

3. Choosing the best approximator structure can be difficult. In fact, if you do not make a good choice for the approximator structure, repeated retuning of θ may be needed, for example, if the input data $x(k)$ switches between regions of X. In this case, with repeated encounters of $x(k)$, all in one region of X, it may tune $F(x, \theta(k))$ to model the shape of $G(x, z)$ in that region very well. If, however, not enough parameters are used in the structure, and if repeated encounters are made with $x(k)$ in a different region of X, the tuning method will need to *forget* what it learned about the first region and focus on retuning the parameters for the second region. In this way, it is possible that a continual reshaping of $F(x, \theta(k))$ is needed to try to obtain the highest accuracy possible for the current operating region. Clearly, it would be better if there were enough approximator structure available so that continual retuning is not needed.

4. The quality of the estimate $\hat{y}(k)$ depends on how good our training method is for $\theta(k)$. We may not have a good tuning method for some approximator structures that may be able to achieve high approximation accuracy.

5. Perhaps most important, however, is the fact that as time goes on and the tuning method shapes the nonlinearity, due to the presence of z, there is in general a potential need for retuning, even if there is enough approximator structure and the same $x(k)$ is encountered because $z(k)$ may have

changed so that $G(x(k), z(k))$ is different. Hence, via repeated encounters with a single x value or x values all in one region, we may *learn* the approximate shape of $G(x, z)$ only for some value of z. We see that it is important that the method can forget about past $x(k)$ information it has encountered and retune the nonlinearity. Clearly, for some applications where we have many measurements, good control over $x(k)$, and very little influence from $z(k)$, we can expect to shape $F(x, 0)$ in different regions of X and as we encounter new regions, we can shape the nonlinearity for those regions too. In this case, if we return to a region of X in which we have already learned the shape, very little retuning will be needed and we can think of continually building a better and better function approximator. Unfortunately, in practical applications, there is always a need for retuning, either because it is impossible to perfectly tune the given approximator structure to match the function as discussed above or because of the unknown influences of z as we discussed here. In fact, in practice it is often difficult to separate the adverse influences of poor approximator structure choice from influences of z, since poor approximator structure choice can in some ways be modeled by the presence of z and because both problems always seem to be present to some degree and have similar effects on the need for retuning.

Later in this part we will study a variety of methods to tune parameters for online approximators. These include recursive least squares, steepest descent gradient methods, and Levenberg-Marquardt methods.

9.4 Biomimicry for Heuristic Adaptive Control

While there are some control tasks that seem to be instinctual for the human (e.g., balancing while standing), there are many others that are learned during our lifetime, some of which then seem to become instinctual once they are learned (e.g., riding a bike), and others which require continual practice and learning to be able to maintain skill, perhaps because the learning environment changes somewhat over time (e.g., the operator who manually controls a process that changes over time due to changing process conditions and process redesigns by the engineering team). In control engineering, we often think of a controller as being an artificial decision-maker and hence we naturally think that during its "lifetime," it should be able to learn to perform better control by gaining experience in controlling the plant. Here, we will investigate two approaches to such adaptive decision-making for control. We will first consider the use of a neural network to learn the nonlinear plant input-output mapping so it can be used to specify a control input; we will refer to the general approach as "neural control" (hence, neural control refers to methods where adaptation is used to implement learning). Basically, this shows how to augment the instinctual neural controllers in Chapter 4 with learning capabilities. This will endow them with the capability to learn how to control a plant, and thereby we will be free from having to perform the tuning of the shape of the control surface that they implement. Following this, we will study how automated rule training in fuzzy systems can provide fuzzy controller synthesis and tuning in response to plant changes (adaptive fuzzy control).

To a certain extent, you can view both the approaches in this section as examples of the direct application, via biomimicry, of principles of classical and operant conditioning to constructing adaptive controllers. In the neural control approach, we think of modeling neural-level learning processes (especially reinforcement learning), whereas in the adaptive fuzzy control approach, we consider modeling human adaptation expertise. Hence, in a sense we view one as a cognitive hardware emulation and the other as a software emulation.

It is important to further clarify how you should view these heuristic adaptive control methods. First, they are being used here to provide an accessible introduction to several important concepts in adaptive control of nonlinear systems. For this reason, extended discussions are provided to give intuition into the construction and subsequent dynamical learning processes. You will see similar concepts emerge in methods in subsequent chapters in this part (e.g., when we discuss stable adaptive control methods in Chapter 12). Second, there are many possible different heuristic strategies that can be used; here, we provide representative methods and discuss the principles of their operation. Third, the methods of this section have been successfully employed in a variety of applications; however, the heuristic focus should not mislead the reader into a sloppy design and evaluation procedure where the most you could hope for is to get lucky that it will work properly in an implementation.

Finally, note that to connect the learning concepts with the methods of Part II, we close this section with a brief discussion on the role of learning in expert, planning, and attentional systems. Moreover, we include a brief discussion on development and plasticity and how biomimicry for these might be useful in control and automation.

9.4.1 Reinforcement Learning for Neural Control

This will be the first of several adaptive control methods that we will study in this part. Since this is the first, we will try to keep it very simple. We will design a reinforcement learning method for a radial basis function neural network from Chapter 4.

In "reinforcement learning control," the controller learns via interactions with its environment (via generating plant inputs and measuring plant outputs). The learning controller quantifies the relative success or failure of its actions. If the action it took tended to lead it closer to its goal, then it strengthens the tendency to pick that action again (i.e., it reinforces the action), and may in fact augment that good action in a direction that seems as if it will lead to more performance improvements in the future. If, on the other hand, the action that was taken was unsuccessful, it will weaken the tendency to select that action. After iteratively interacting with its environment, the controller should learn what leads to success. The reader should see a clear connection to concepts in operant conditioning (what are they?).

Reinforcement learning controllers self-adjust to obtain rewards that correspond to good performance.

To design a particular reinforcement learning controller, we need to choose the underlying controller that is to be adjusted (here, as an example, we use the radial basis function neural network), the reinforcement learning strategy, which entails picking a reinforcement learning signal, and a method to adjust the controller based on values of that signal.

Reinforcement Function

The key components of the learning mechanism for a reinforcement learning approach to neural control are the "reinforcement signal" and how we use this signal to adjust the neural network. The reinforcement signal is generated via a (typically scalar) "reinforcement function," which we will denote with J_R, that uses data that are gathered during the operation of the control system. The function J_R is typically generated using characterizations of how you would like the closed-loop system to behave that are based on the real time operation of the system. One way to do this is to use a reference model, as we discuss next.

Use of a Reference Model for a Reinforcement Function Input: A "reference model" is used to quantify the desired performance of the closed-loop system. It does this by generating a reference model trajectory that specifies where the output of the plant should be at each time instant. Basically, for the reference model, you want to specify a desirable performance, but also a reasonable one. If you ask for too much, the controller will not be able to deliver it. Certain characteristics of real-world plants place practical constraints on what performance can be achieved. It is not always easy to pick a good reference model, since it is sometimes difficult to know what level of performance we can expect, or because we have no idea how to characterize the performance for some of the plant output variables.

In general, the reference model may be discrete or continuous time, linear or nonlinear, time-invariant or time-varying, and so on. For example, suppose that we would like to have the response track the output of the continuous time model

$$G(s) = \frac{1}{s+1}$$

Suppose that for your discrete-time implementation, you use $T = 0.1$ sec. Using a bilinear (Tustin) transformation to find the discrete equivalent to the continuous-time transfer function $G(s)$, we replace s with $\frac{2}{T}\frac{z-1}{z+1}$ to obtain

$$\frac{Y_m(z)}{R(z)} = H(z) = \frac{\frac{1}{21}(z+1)}{z - \frac{19}{21}}$$

where $Y_m(z)$ and $R(z)$ are the z-transform of $y_m(kT)$ and $r(kT)$, respectively. Now, for a discrete-time implementation, we would choose

$$y_m(kT+T) = \frac{19}{21} y_m(kT) + \frac{1}{21} r(kT+T) + \frac{1}{21} r(kT)$$

This choice would then represent that we would like our output $y(kT)$ to track a smooth, stable, first-order-type response of $y_m(kT)$. A similar approach can be used to, for example, track a second-order system with a specified damping ratio ζ and undamped natural frequency ω_n.

The performance of the overall system is computed with respect to the reference model by the learning mechanism by generating an error signal

$$y_e(kT) = y_m(kT) - y(kT)$$

Given that the reference model characterizes design criteria such as rise-time and overshoot and the input to the reference model is the reference input $r(kT)$, the desired performance of the controlled process is met if the learning mechanism forces $y_e(kT)$ to remain very small for all time no matter what the reference input is or what plant parameter variations occur. Hence, the error $y_e(kT)$ provides a characterization of the extent to which the desired performance is met at time kT. If the performance is met (i.e., $y_e(kT)$ is small), then the learning mechanism should not make significant modifications to the controller. On the other hand, if $y_e(kT)$ is big, the desired performance is not achieved and the learning mechanism must adjust the controller.

Use of Other Variables: In practical applications, there are often many ways to characterize how the closed-loop system is performing at any given time step. These might include using measures of sizes of *trajectories*. Along these lines, one common approach is to use the first derivative (or higher order derivatives) of the error between the reference model output and the plant output. For instance, you could use

$$y_c(kT) = \frac{y_e(kT) - y_e(kT - T)}{T}$$

This signal measures whether the plant output is moving in a direction to increase the error between the reference model output and the plant output. Clearly this can be a useful signal to decide how much to adjust the underlying controller.

Quantifying When the Response Is Good Enough: Generally, in designing a reinforcement signal, it is good to add a feature where, if the plant output response is quite close to where it should be, you do not adjust the controller. If you make adjustments for each small possible deviation, then as time progresses, these changes can accumulate and force unnecessarily large inputs to the plant. Essentially, the deviations conspire over time to make the underlying controller have an increasingly "high gain effect" that results in large plant inputs, and if there is a positive feedback effect, the small adjustments to increase the gain can make the system go unstable (a key observation that forms the basis for robust adaptive control).

Here, to avoid these problems, we will design the reinforcement function so that when its inputs are small (indicating that the plant response is close to where we want it to be), the reinforcement signal will be zero, indicating that the underlying controller should not be adjusted. Below, for the ship steering application, we will show one simple choice to turn off the adaptation when the response is close to what we want.

Adjusting the Neural Network

There are many methods to adjust the parameters, or even structure of the neural network given the reinforcement signal. Here, we will only consider the adjustment of the receptive field unit strengths b_i, $i = 1, 2, \ldots, n_R$. Consider the adjustment approach where we choose

Adjustments to the controller are made to try to obtain better closed-loop responses.

$$b_i(kT) = b_i(kT - T) + J_R(kT)R_i(kT - T)$$

where $R_i(kT)$ is the output of the ith receptive field unit (we drop the input arguments of this function and replace them with the time index). Suppose that inputs to the plant affect the output of the plant in one time step (if it took d time steps, then we would simply use $b_i(kT - dT)$ and $R_i(kT - dT)$ rather than $b_i(kT-T)$ and $R_i(kT-T)$ above). The input to the plant at $kT-T$ (determined by the $b_i(kT - T)$ and $R_i(kT - T)$) is then the input that generated the current $y_e(kT)$, $y_c(kT)$, and hence $J_R(kT)$. The above formula is designed to modify the input that *was* generated so that if the same conditions are encountered again, a more appropriate input will be generated for the plant.

Notice that since the functions R_i have a type of "local support" (i.e., for values of their inputs far from the center of the receptive field unit, their outputs will be near zero), we view the adjustment mechanism as only making local adjustments to the overall mapping that is implemented by the radial basis function neural network. This is important so that it learns the shape of the function in the region where it is operating and this does not destroy what was learned before (i.e., so that it will not forget). We say that the learning approach makes local rather than global adjustments.

Finally, note that there are many possible modifications to the adjustment rule given above. Notice that using the above rule every b_i, $i = 1, 2, \ldots, n_R$, is adjusted at each step. It would be easy to specify a threshold $\epsilon > 0$ such that we would only modify b_i if $R_i(kT - T) > \epsilon$ and it will have little impact on the behavior of the adaptation (if ϵ is chosen small enough), but will simplify computations since only a subset of the b_i would be updated at each step. Another useful modification is possible when you know a range of values for the b_i values. For example, as in gradient optimization when we know a range on the parameter values and use a projection method to keep the parameters in the proper bounds, and then if the above adjustment formula indicates to put a b_i value outside the known set, it simply keeps it on the boundary. However, if it indicates it should stay on the boundary or move to the interior of the allowable range, then it uses the update indicated by the formula.

9.4.2 Design Example: Neural Control for the Tanker Ship

To improve fuel efficiency and reduce wear on ship components, autopilot systems have been developed and implemented for controlling the directional heading of ships. Often, the autopilots utilize simple control schemes such as PID control. However, the capability for manual adjustments of the parameters of the controller is added to compensate for disturbances acting upon the ship such as wind and currents. Once suitable controller parameters are found manually, the controller will generally work well for small variations in the operating conditions. For large variations, however, the parameters of the autopilot must be continually modified. Such continual adjustments are necessary because the dynamics of a ship vary with, for example, speed, trim, and loading. Also, it is useful to change the autopilot control law parameters when the ship is exposed to large disturbances resulting from changes in the wind, waves, current, and water depth. Manual adjustment of the controller parameters is often a burden on the crew. Moreover, poor adjustment may result from human error. As a result, it is of interest to have a method for automatically adjusting or modifying the underlying controller.

Here we will develop a reinforcement learning strategy for a neural controller and use simulation studies to evaluate its performance in achieving heading regulation when there are plant changes and disturbances. In particular, we will tune the radial basis function neural network that we developed in Chapter 4 using a reinforcement learning strategy as shown in Figure 9.21. This will provide us with a way to quantify, for this application, possible advantages of using adaptive rather than fixed controllers.

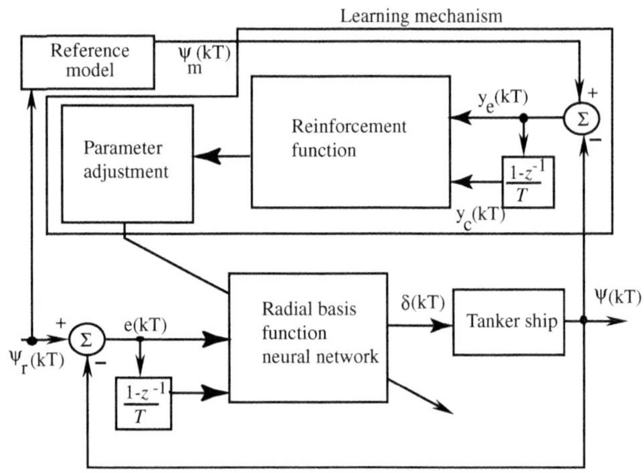

Figure 9.21: Reinforcement learning control strategy for ship heading regulation.

Learning Mechanism Design

We will use the radial basis function neural network designed in Section 4.5 on page 133. In particular, we will use the same $n_R = 121$ and the grid of receptive field units with the same center values and spreads used there. Here, we will consider the tuning of only the receptive field unit strengths b_i, $i = 1, 2, \ldots, n_R$. In each case below, we initialize the radial basis function receptive field unit strengths to be zero (i.e., $b_i = 0$, $i = 1, 2, \ldots, n_R$) to try to represent that the neural network knows little about how to control the ship heading. Of course, another option would be to initialize the b_i values with the ones we manually chose in Chapter 4.

To define the reinforcement signal, first define a reference model

$$G(s) = \frac{1/150}{s + 1/150}$$

This choice indicates that we want a smooth first order response for changes in the desired ship heading. It represents a reasonable but not trivial performance request for the control system for ship heading. We will discretize this transfer function in the simulations below, and hence, compute its output $\psi_m(kT)$.

We use

$$y_e(kT) = \psi_m(kT) - \psi(kT)$$

and

$$y_c(kT) = \frac{y_e(kT) - y_e(kT - T)}{T}$$

as inputs to the reinforcement function. Choose

$$\bar{J}_R(y_e(kT), y_c(kT)) = \eta\left(-\eta_e y_e(kT) - \eta_c y_c(kT)\right)$$

to help define the reinforcement function. Moreover, when the arguments of this function are small, we will set the output of the reinforcement signal to zero. The final reinforcement signal is then generated by the function

$$J_R(y_e(kT), y_c(kT)) = \begin{cases} \bar{J}_R(y_e(kT), y_c(kT)) & if\ |\bar{J}_R(y_e(kT), y_c(kT))| \geq \alpha \\ 0 & if\ |\bar{J}_R(y_e(kT), y_c(kT))| < \alpha \end{cases}$$

where we choose the design parameter $\alpha = 0.005$. Here, we consider η, η_e, and η_c to be design parameters. The parameters η_e and η_c are adjusted to indicate the importance of achieving tracking and deviations in tracking, respectively. The parameter η is the "adaptation gain"; if you choose it small, adaptation will be slow, while if you choose it large, it will try to adapt very fast (which can sometimes lead to instabilities). Here, after a bit of tuning via simulations, we choose $\eta = 1$, $\eta_e = 1$, and $\eta_c = 20$.

How do we adjust the receptive field unit strengths b_i, $i = 1, 2, \ldots, n_R$? Here, we simply use

$$b_i(kT) = b_i(kT - T) + J_R(kT) R_i(kT - T)$$

(note that for the ship, $d = 1$ so that the input δ affects the next value of the ship heading).

It is important to recognize the general approach to learning that is being used here. When the plant output is not big enough, the receptive field unit strengths that led to the control input that resulted in it not being big enough are increased. This *concept* generally holds, whether the signals are positive or negative.

Closed-Loop Response, Tuned Mapping, Nominal Conditions

Adaptive control can achieve initial controller synthesis and later tuning as it becomes necessary.

We will follow the same sequence of simulations that we used in Chapter 4 in Section 4.5.3 on page 139. If we use nominal conditions, where we have ballast conditions, no wind, no sensor noise, and a speed of 5 meters/sec., we get a closed-loop response shown in Figure 9.22. Notice that due to the lack of initial knowledge about how to control the ship heading, there is significant overshoot of the desired response early in the simulation. Then, as it learns from the data that are gathered online (the size of the reinforcement signal shown on the bottom of Figure 9.22 can be thought of as being proportional to the amount of learning that is taking place at any particular time in that it represents the magnitude of the local adjustments to the mapping), it improves the response, giving less and less overshoot until ultimately it very closely tracks the desired heading ψ_m. You can think of this as providing for synthesis of the stimulus-response characteristics of the neural network. To achieve this, it simply tunes a portion of the parameters of the neural network, and not the structure of the network (e.g., the number of receptive field units). It is interesting to note that by the end of the simulation, we could consider the response that is obtained to be superior to the one obtained in Figures 4.24 and 4.25, where we had manually constructed the instinctual neural controller for this application.

Adaptive control may be able to use online information to construct a better controller than is possible via a priori manual tuning since it can exploit information that was not known a priori.

The stimulus-response characteristics of the synthesized neural network can be seen by considering the controller surface that is achieved by the end of the simulation in Figure 9.22. This learned input-output mapping is shown in Figure 9.23 (although since in Figure 9.22 the reinforcement signal has not converged to zero, the mapping is still being shaped). First, it is interesting to compare this mapping that was constructed using online data to the one we manually synthesized for this application shown in Figure 4.23. There are three important things to notice. First, the mapping in Figure 9.23 only has certain regions that have been tuned since the heading error and change in error values only visited certain regions. When they visit the regions where the mapping is zero, those regions will also be learned. Second, the general shape of the mapping in Figure 9.23, in some regions, resembles the one that we had manually constructed in Figure 4.23 (which makes sense if it is to succeed). Third, the online tuning approach does not necessarily converge to one that you construct manually; it may find a better mapping for the current conditions, or the learning mechanism may not be as effective as you are in specifying a good controller mapping. (Clearly, different learning mechanisms will result in different tuned mappings.)

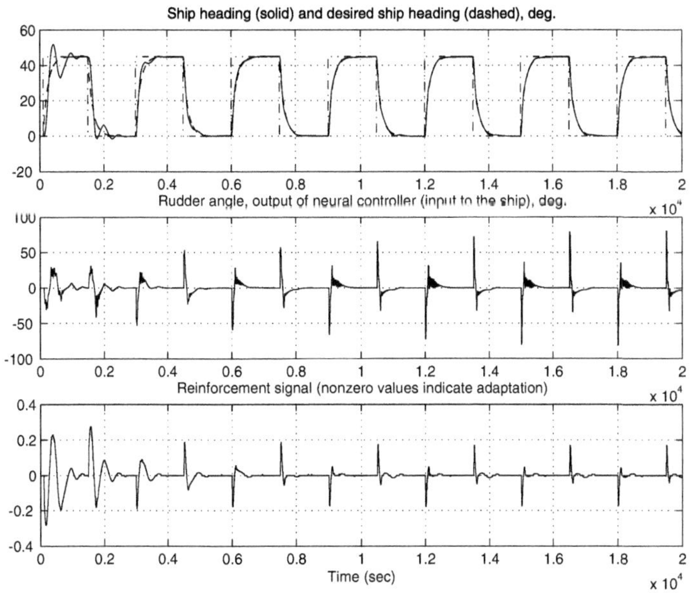

Figure 9.22: Closed-loop response resulting from using the neural controller for tanker ship steering.

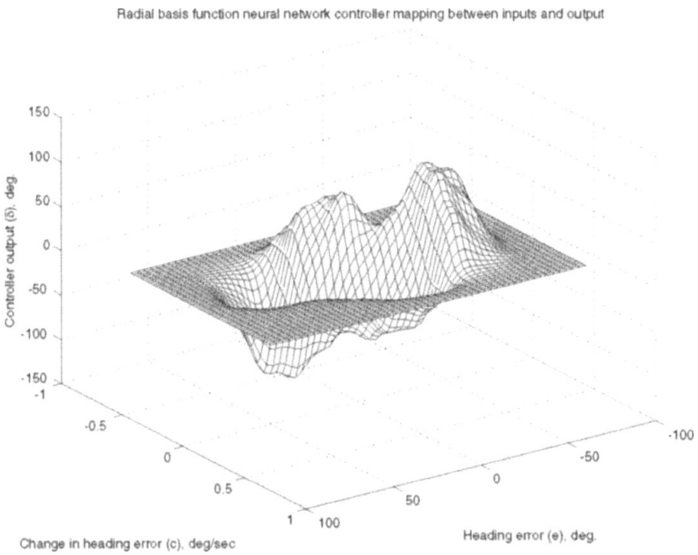

Figure 9.23: Final tuned mapping for the neural network for tanker ship steering.

Effects of Plant Changes and Disturbances

Next, consider the effects of a wind disturbance on the ship. In this case, we get the response in Figure 9.24. While in Figure 4.26 we found that wind affects our ability to achieve very good regulation of the ship heading, here we see that the neural controller learns to compensate for its effects. The map that is synthesized by the end of the simulation is similar, but not identical, to the one synthesized for the nominal conditions. The neural controller figures out how to shape the mapping to reduce the effects of the wind disturbance, even though the precise characteristics of the wind disturbance are not known, but are only inferred via the data that are gathered online.

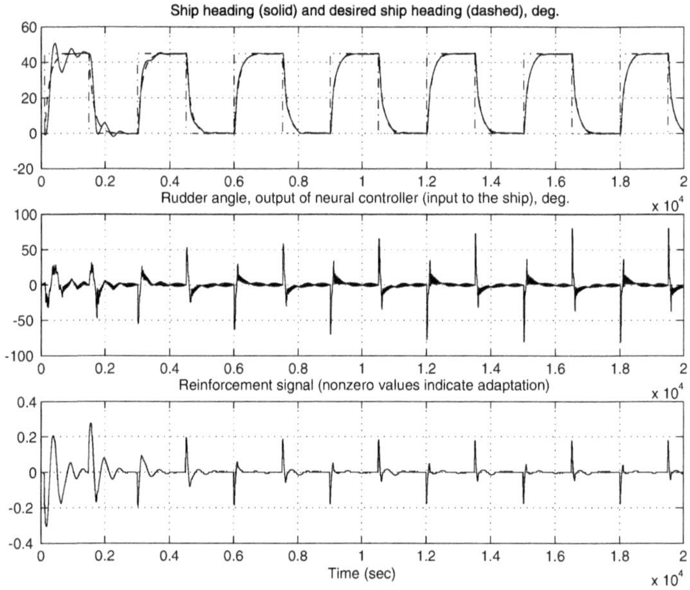

Figure 9.24: Closed-loop response resulting from using the neural controller for tanker ship steering, with wind.

Next, consider the effect of a speed change on our ability to steer the ship. Here, for the first 9000 seconds we will operate the ship at a speed of $u =$ 5 meters/sec., then abruptly at $t = 9000$ sec. we will change the speed to $u = 3$ meters/sec. (the "abrupt" change is certainly physically unreasonable; however, it seems likely that similar behavior will result for a slower change in speed). In this case we get the response in Figure 9.25, and the final tuned controller surface shown in Figure 9.26. Of course, we get the same response as for nominal conditions for the first 9000 sec. Then, it adapts to the speed change after it happens, first with a slightly degraded transient response, which it later improves. Notice that the final tuned controller mapping in Figure 9.26 is different from the one that resulted from nominal conditions (or when there was wind) since adapting to the speed change is different from adapting to those conditions.

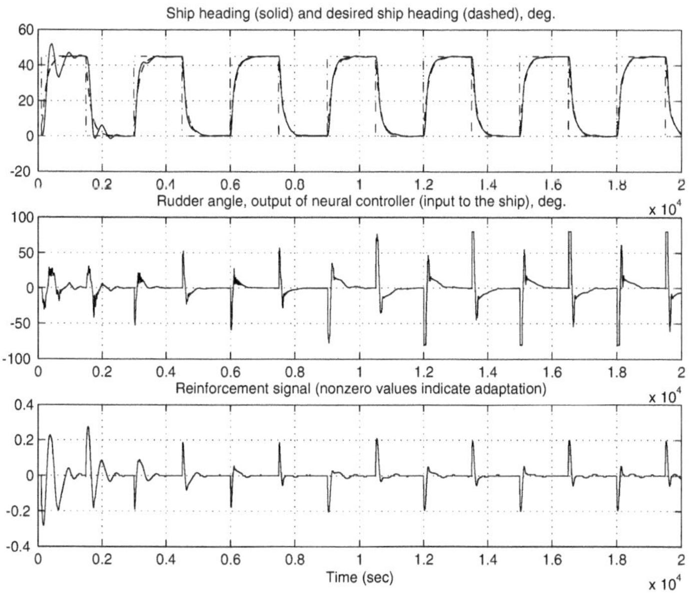

Figure 9.25: Closed-loop response resulting from using the neural network for tanker ship steering, speed change from 5 to 3 meters/sec. at $t = 9000$ sec.

If you use, for instance, an additive sensor noise uniformly distributed on $[-0.01, 0.01]$, just like the nonadaptive case, there is little effect on the response so we do not show the plot (of course, if you get sensors with worse performance characteristics, then you will expect tracking errors to arise in an analogous manner to results for the wind). Next, we will consider the case of how the ship steers when at $t = 9000$ sec. we switch from ballast to full conditions abruptly (again, this is physically unreasonable, but the resulting responses are representative of what would happen if the weight was changed in a physically reasonable manner). In this case, we get the response in Figure 9.27 (compare with the nonadaptive case in Figure 4.28). Notice that while immediately after the weight change the transient response degrades, soon afterward the neural controller learns how to compensate for the weight change (essentially it must "lower the gain" of the controller; why?). Of course, the final tuned controller mapping that results in this case is different from the previous ones, and its shape will be discussed in the next subsection.

Controller Map Shape Changes

Next, we will return to the case where there is a weight change from ballast to full conditions, but show how the map shape changes when the controller adapts

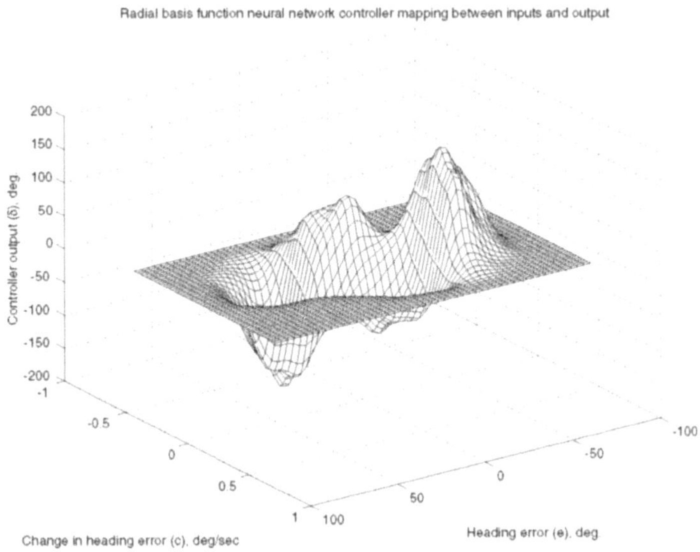

Figure 9.26: Final tuned mapping for the neural network for tanker ship steering, speed change from 5 to 3 meters/sec. at $t = 9000$ sec.

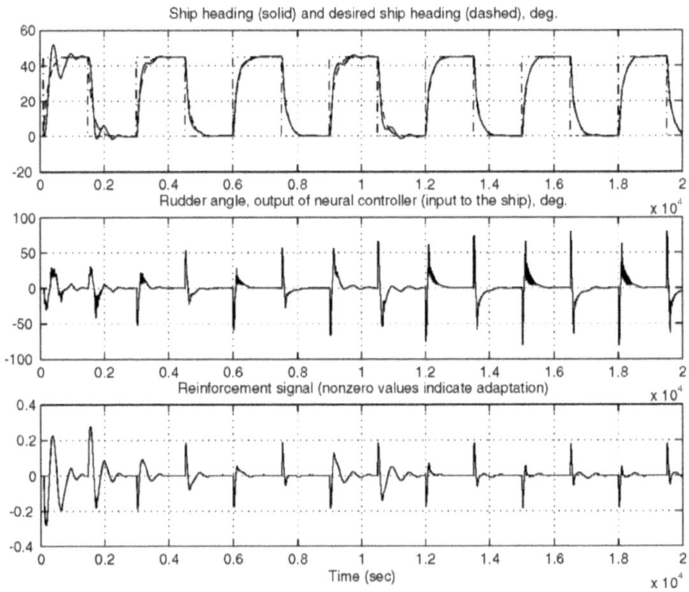

Figure 9.27: Closed-loop response resulting from using the neural controller for tanker ship steering, switch from ballast to full conditions at $t = 9000$ sec.

from a ballast condition to a full one. First, consider the controller map shown in Figure 9.28 that is obtained by $t = 8999$ sec. when nominal conditions are used (this is similar to the map in Figure 9.23, but different since the lengths of the simulations are different). Of course, keep in mind that the mapping shapes are still changing some near the points where we show them, just as with Figure 9.23.

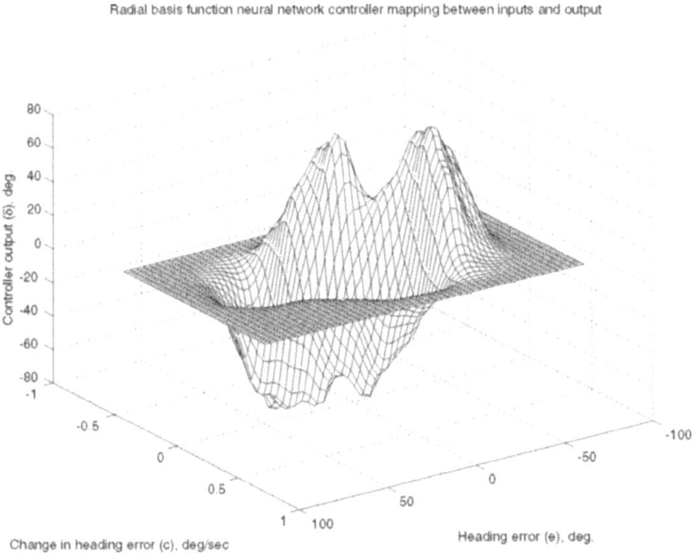

Figure 9.28: Controller mapping of the radial basis function neural network obtained by $t = 8999$ sec., nominal conditions (before ship weight changes to full).

At $t = 9000$ sec. the ship weight abruptly changes from ballast to full conditions. The controller adapts to this situation where the ship suddenly becomes *easier* to steer. In particular, the mapping change after the simulation ends at $t = 20,000$ sec. is shown in Figure 9.29. Notice that the mapping shape has changed compared to Figure 9.28 and it is the change in this map shape that summarizes what the effects of the adaptation were. Notice that with the continual changes that are still occurring to the map (see the bottom plot in Figure 9.27), the peaks in the map in Figure 9.29 have generally been increased. This can actually cause a problem in that the parameters may be adjusted so that as $t \to \infty$, the peaks can go to infinity. There are several solutions to this problem. First, you could use the "projection" method for the parameters so that no region of the map will increase too much. Second, by adjusting the reinforcement function, you can generally avoid having the map diverge (e.g., you may change the value of the α threshold and the other gains). Generally, if your adaptation is too aggressive, you can encounter problems of parameter divergence; here, rather than focus on making sure that we end up with boundedness, we will simply defer the issue to later chapters of this part where we will use projection, or will design the learning mechanism so that we will be ensured that stable closed-loop operation will be achieved.

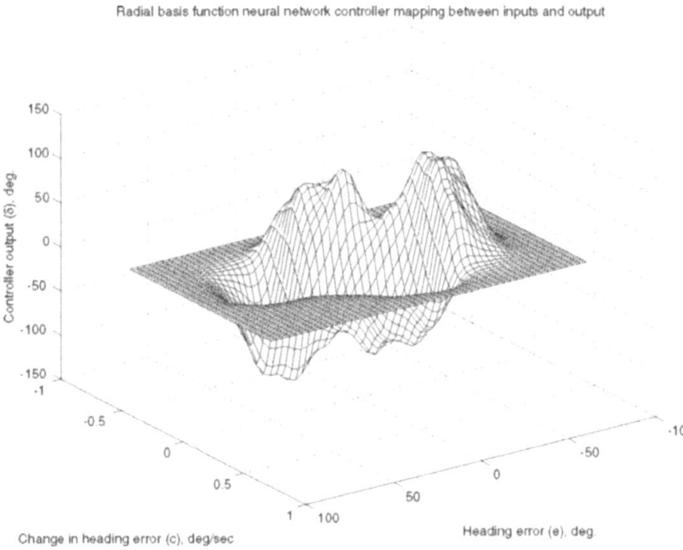

Figure 9.29: Controller mapping of the radial basis function neural network obtained by $t = 20,000$ sec., nominal ballast weight conditions followed by a weight change to full at $t = 9000$ sec.

Next, it is interesting to study the *change* in the map shapes between $t = 8999$ and $t = 20,000$ sec. To do this, consider Figure 9.30. The positive peaks indicate areas where the mapping has increased by the end of the simulation, and the negative ones indicate regions where the mapping has decreased relative to the shape at the middle of the simulation. To gain more insight into the shape changes, consider the contour plot in Figure 9.31 where the gray color indicates *increases* in the map. Notice that the changes are local, not global, to where the system was operating. For example, if there were changes in one region of the contour plot, there were not changes in another region, provided that the two regions were far enough away from each other, with "far" defined by the particular values used for the spreads for the receptive field units.

Localized Learning, Forgetting, and Adaptation

Here, we will consider the effects of using a simpler neural network than in the last subsection for the ship heading regulation. Suppose that we think of radial basis functions with lower values of n_R as being simpler. The parameter n_R controls the size of the parameter update vector, and hence, quantifies how much *flexibility* the underlying neural network has for learning. Here, we will study

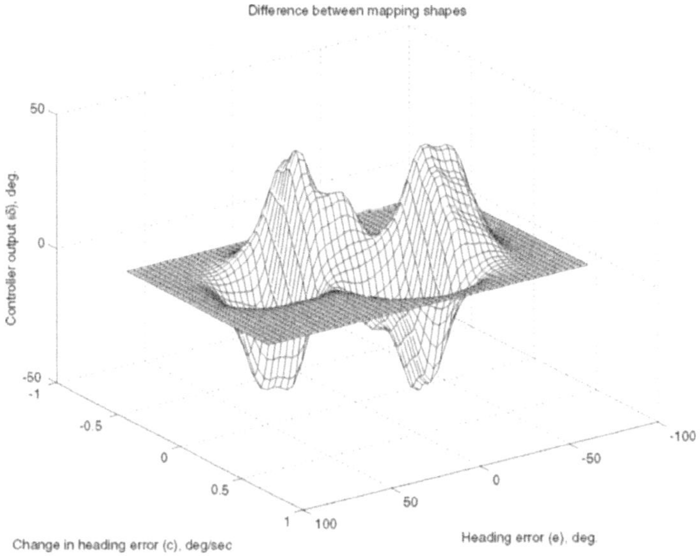

Figure 9.30: Difference of controller mappings of the radial basis function neural network (one obtained by subtracting the one at $t = 20,000$ sec. from the one at $t = 8999$ sec.).

the effects of reducing the size of n_R. Clearly, there can be practical advantages to reducing n_R since the resulting algorithm will be computationally simpler.

Adjustments to the controller are made locally, allowing learning without forgetting.

Consider using a radial basis function neural network with $n_R = 9$ and a 3×3 grid for the centers of the receptive field units as shown in Figure 9.32. For the spreads σ_j^i, we use

$$\sigma_1^i = 0.7 \frac{\pi}{\sqrt{n_R}}$$

and

$$\sigma_2^i = 0.7 \frac{0.02}{\sqrt{n_R}}$$

for $i = 1, 2, \ldots, 9$ where we consider e to be our first input and c to be our second. A plot of one of the receptive field units is shown in Figure 9.33.

We use exactly the same reinforcement learning strategy as above, but with different parameters; in particular, after a bit of tuning we use $\eta = 1$, $\eta_e = 0.5$, and $\eta_c = 100$. We use $\alpha = 0.005$, as above.

Simpler is Better—Achieving Similar Performance: It is basic to engineering design that "simpler is better." Here, roughly the same type of performance can be achieved as compared to the case where we had $n_R = 121$, which represented a much more complex controller. For instance, consider the case similar to the one illustrated in Figure 9.27 where, at the middle of the simulation, we switch from ballast to full conditions. For the new controller,

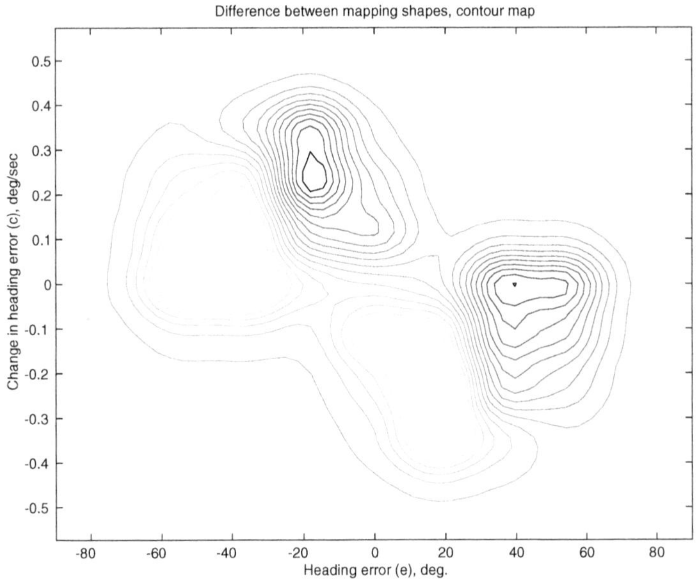

Figure 9.31: Contour map of difference of controller mappings of the radial basis function neural network (one obtained by subtracting the one at $t = 20,000$ sec. from the one at $t = 8999$ sec.).

we get the results shown in Figure 9.34. Notice that it performs similar (not the same, notice the oscillations) to the one for $n_R = 121$ shown in Figure 9.27, yet the computational complexity of the controller here is much simpler. Next, while we will not show the plots, we note that we also get similar results to the $n_R = 121$ case if we have a speed decrease, sensor noise, and wind. While this is certainly not an extensive test for comparing the performance of the simpler strategy relative to the $n_R = 121$ one, it does give us some confidence that we could succeed with a less complex design. Moreover, it illustrates an important engineering principle.

High neural network complexity can allow for highly flexible and local learning, and less forgetting.

Low Neural Network Complexity Causes More Forgetting and Hence, Demands More Adaptation: We do, however, pay a price for the simplifications of going to a $n_R = 9$ controller. We pay in that we end up with a less flexible learning strategy in that the underlying control surface that is tuned can only result from changes to larger regions of the (e, c) plane compared to the $n_R = 121$ case. To illustrate this, consider the mapping obtained by $t = 8999$ sec. (just before the change from ballast to full) in Figure 9.35. Notice that the overall map shape is much smoother; this is due to the low number of receptive field units (we are only adding up 9 scaled Gaussian functions to produce this plot, compared to 121 in the earlier case). At $t = 9000$ sec. the ship weight abruptly changes from ballast to full conditions. The mapping after the simu-

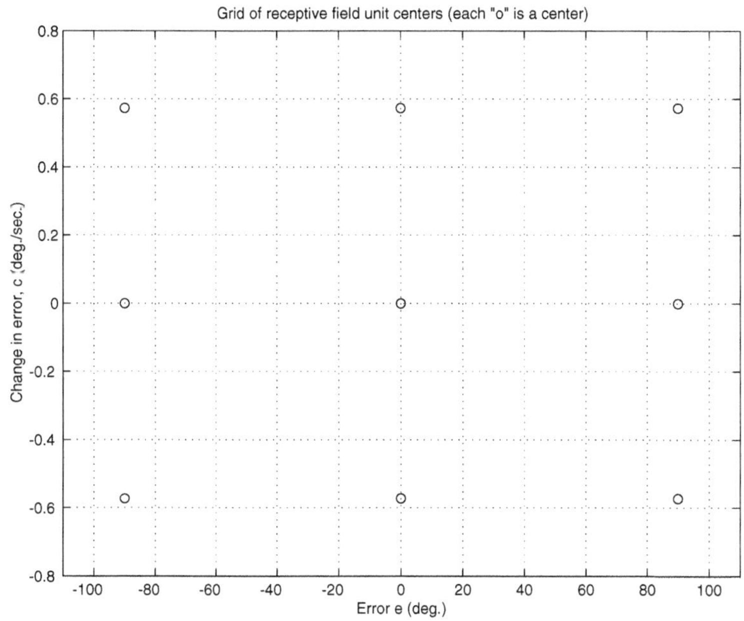

Figure 9.32: Grid of 9 center points for receptive field units.

lation ends at $t = 20,000$ sec. is shown in Figure 9.36.

Next, it is interesting to further study the *change* in the map shapes between $t = 8999$ and $t = 20,000$ sec. To do this, consider Figure 9.37. The positive peaks indicate areas where the mapping has increased by the end of the simulation, and the negative ones indicate regions where the mapping has decreased relative to the shape at the middle of the simulation. To gain more insight into the shape changes, consider the contour plot in Figure 9.38, where the gray color indicates *increases* in the map. Notice that the changes are local, not global, to where the system was operating. However, compare this plot to Figure 9.31. Notice that while broadly speaking the gross changes in the maps are similar, there are many differences. In particular, note that in Figure 9.38, we can make less precise changes in the map since we have a much *coarser* grid of receptive field unit centers, and since we have spreads that are larger so the receptive field unit "supports" (i.e., where it is above some small positive number) cover much larger regions of the (e, c) space. With this larger coverage, changes in the b_i result in "more global" changes in the map shape (i.e., over the whole support of receptive field unit R_i). With the *fine* grid of receptive field unit centers for the $n_R = 121$ case, the effect of the changes in the b_i are much more local.

Notice that there are some fundamental principles here. First, with the finer grid ($n_R = 121$ case), we get more flexibility in learning, and an ability to learn and not to forget, but this costs computations. With the coarse grid ($n_R = 9$ case), we save computations, but when we change one b_i, it affects a

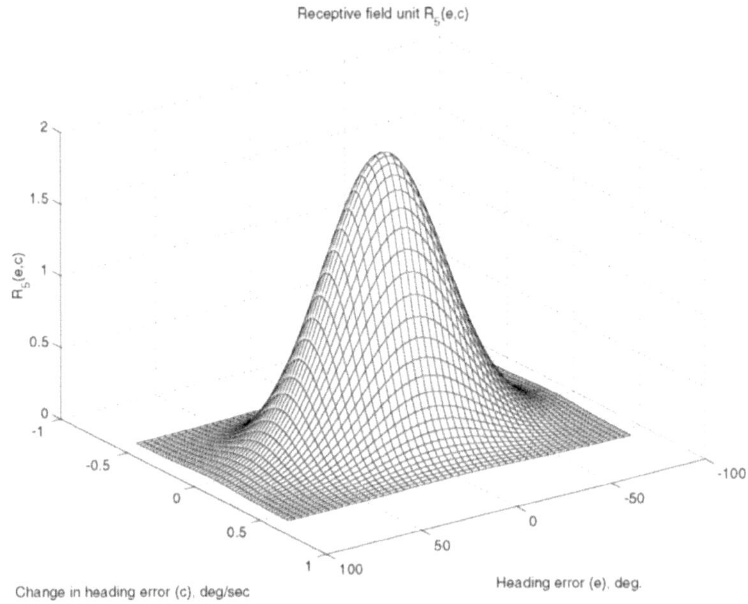

Figure 9.33: Receptive field unit mapping when 9 receptive field units are used in the radial basis function neural network.

larger region. We think of this as having possibly adverse effects. It can result in "forgetting" the shape of the map in a region where we had already shaped it (since the spreads are relatively large), and this then requires that we "re-learn" the map shape if we return to the region where we had shaped it (this is actually what typically causes the additional oscillations that resulted when we used $n_R = 9$). Clearly, choice of the grid partition can be very difficult for a particular application; however, keeping the above principles in mind can help guide the design. You should be reminded of Section 9.3, where we studied how to improve approximation accuracy via the use of more complex approximators; here, we consider similar ideas, but for adaptive control problems.

9.4.3 Adaptive Fuzzy Control: Emulating Adaptation Expertise

Humans often have expertise in how to learn about how to control a process.

In this section, as one example of an adaptive fuzzy controller where we emulate operator intuition about adaptation, we introduce the "fuzzy model reference learning controller" (FMRLC), which is a (direct) model reference adaptive controller. You will find several similarities between neural control via reinforcement learning and this adaptive fuzzy control approach. Here, however, we provide additional insights into adaptation mechanism design, some of which are more conveniently discussed in the context of fuzzy systems (e.g., controller initialization).

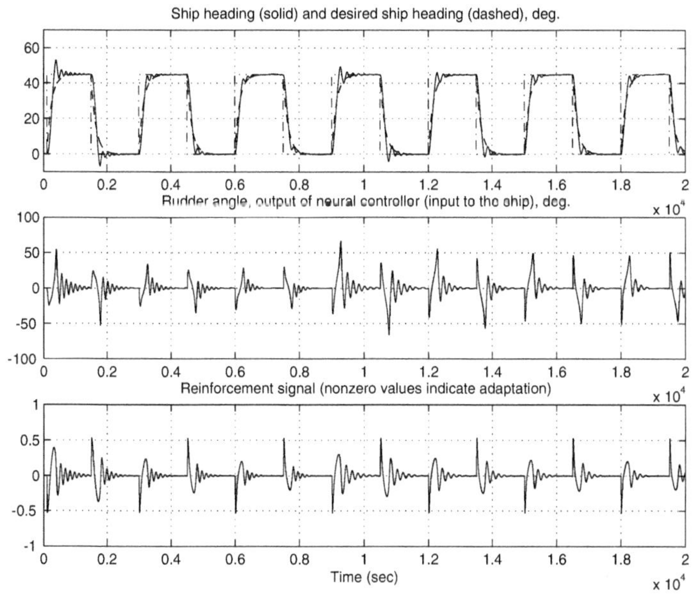

Figure 9.34: Closed-loop response resulting from using the neural controller for tanker ship steering, switch from ballast to full conditions at $t = 9000$ sec., using only 9 receptive field units.

The functional block diagram for the FMRLC is shown in Figure 9.39. It has four main parts: the plant, the fuzzy controller to be tuned, the reference model, and the learning mechanism (an adaptation mechanism). The FMRLC uses the learning mechanism to observe numerical data from a fuzzy control system (i.e., $r(kT)$ and $y(kT)$ where T is the sampling period). Using these numerical data, it characterizes the fuzzy control system's current performance and automatically synthesizes or adjusts the fuzzy controller so that some given performance objectives are met. These performance objectives (closed-loop specifications) are characterized via the reference model shown in Figure 9.39. The learning mechanism seeks to adjust the fuzzy controller so that the closed-loop system (the map from $r(kT)$ to $y(kT)$) acts like the given reference model (the map from $r(kT)$ to $y_m(kT)$). Basically, the fuzzy control system loop (the lower part of Figure 9.39) operates to make $y(kT)$ track $r(kT)$ by manipulating $u(kT)$, while the upper-level adaptation control loop (the upper part of Figure 9.39) seeks to make the output of the plant $y(kT)$ track the output of the reference model $y_m(kT)$ by manipulating the fuzzy controller parameters.

Next, we describe each component of the FMRLC in more detail for the case where there is one input and one output from the plant (the same concepts are easy to extend to the multi-input multi-output case).

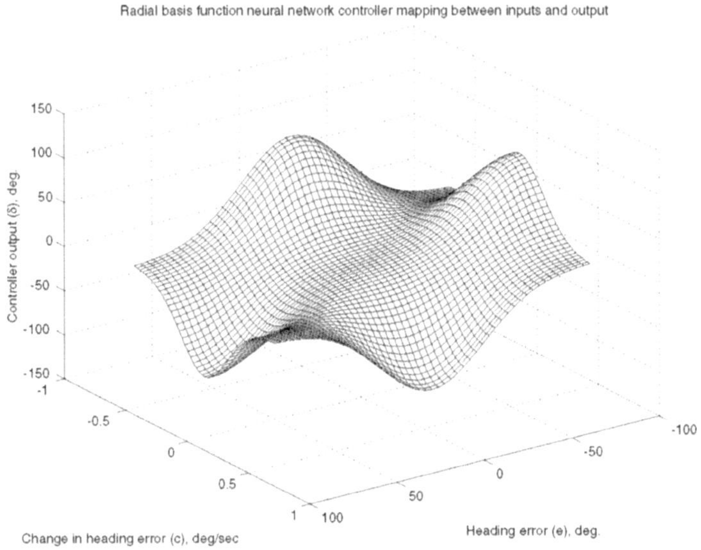

Figure 9.35: Controller mapping of the radial basis function neural network obtained by $t = 8999$ sec., nominal conditions (before ship weight changes to full).

The Fuzzy Controller

Most often, the inputs to the fuzzy controller are generated via some function of the plant output $y(kT)$ and reference input $r(kT)$. Figure 9.39 shows a simple example of such a map that has been found to be useful in some applications. For this, the inputs to the fuzzy controller are the error $e(kT) = r(kT) - y(kT)$ and change in error

$$c(kT) = \frac{e(kT) - e(kT - T)}{T}$$

(i.e., a PD fuzzy controller). There are times when it is beneficial to place a smoothing filter between the $r(kT)$ reference input and the summing junction. Such a filter is sometimes needed to make sure that smooth and reasonable requests are made of the fuzzy controller (e.g., a square wave input for $r(kT)$ may be unreasonable for some systems that you know cannot respond instantaneously). Sometimes, if you ask for the system to perfectly track an unreasonable reference input, the FMRLC will essentially keep adjusting the "gain" of the fuzzy controller until it becomes too large. Generally, it is important to choose the inputs to the fuzzy controller, and how you process $r(kT)$ and $y(kT)$, properly; otherwise, performance can be adversely affected and it may not be possible to maintain stability.

Returning to Figure 9.39, we use scaling gains g_e, g_c, and g_u for the error $e(kT)$, change in error $c(kT)$, and controller output $u(kT)$, respectively. A first guess at these gains can be obtained in the following way: the gain g_e can be

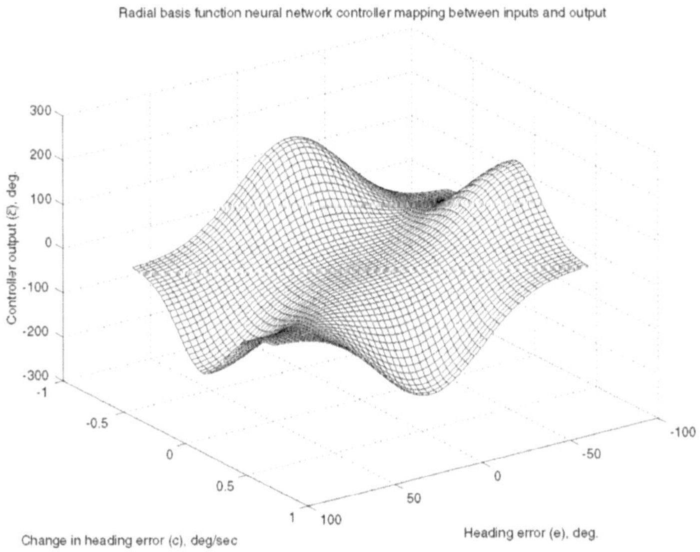

Figure 9.36: Controller mapping of the radial basis function neural network obtained by $t = 20,000$ sec., nominal ballast weight conditions followed by a weight change to full at $t = 9000$ sec. with 9 receptive field units.

chosen so that the range of values that $e(kT)$ typically takes on will not make it so that its values result in saturation of the corresponding outermost input membership functions. The gain g_c can be determined by experimenting with various inputs to the fuzzy control system (without the adaptation mechanism) to determine the normal range of values that $c(kT)$ will take on. Using this, we choose the gain g_c so that normally encountered values of $c(kT)$ will not result in saturation of the outermost input membership functions. We can choose g_u so that the range of outputs that are possible is the maximum range possible, yet still so that the input to the plant will not saturate (for practical problems, the inputs to the plant will always saturate at some value). Clearly, this is a very heuristic choice for the gains and hence, may not always work. Sometimes, tuning of these gains will need to be performed when we tune the overall FMRLC.

Rule Base: The rule base for the fuzzy controller has rules of the form

$$\textbf{If } \tilde{e} \text{ is } \tilde{E}^j \textbf{ and } \tilde{c} \text{ is } \tilde{C}^l \textbf{ Then } \tilde{u} \text{ is } \tilde{U}^m$$

where $\tilde{e}$ and $\tilde{c}$ denote the linguistic variables associated with controller inputs $e(kT)$ and $c(kT)$, respectively, $\tilde{u}$ denotes the linguistic variable associated with the controller output u, $\tilde{E}^j$ and $\tilde{C}^l$ denote the jth (l^{th}) linguistic value associated with $\tilde{e}$ ($\tilde{c}$), respectively, and $\tilde{U}^m$ denotes the consequent linguistic value associated with $\tilde{u}$. Hence, as an example, one fuzzy control rule could be

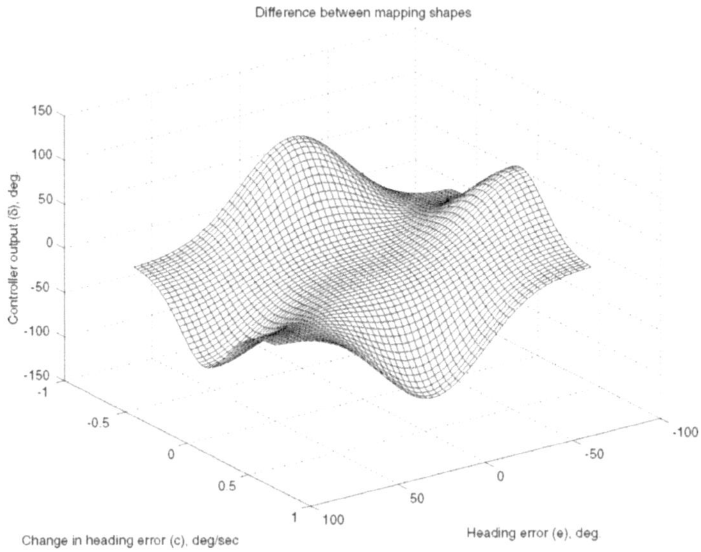

Figure 9.37: Difference of controller mappings of the radial basis function neural network (one obtained by subtracting the one at $t = 20,000$ sec. from the one at $t = 8999$ sec.) with 9 receptive field units.

If error is positive-large **and** change-in-error is negative-small
Then plant-input is positive-big

(in this case $\tilde{e}$ = "error", $\tilde{E}^4$ = "positive-large", etc.). We use a standard choice for all the membership functions on all the input universes of discourse, such as the ones shown in Figure 9.40. Hence, we would simply use some membership functions similar to those in Figure 9.40, but with a scaled horizontal axis, for the $c(kT)$ input.

We will use all possible combinations of rules for the rule base. For example, we could choose to have 11 membership functions on each of the two input universes of discourse, in which case we would have $11^2 = 121$ rules in the rule base. At first glance it would appear that the complexity of the controller could make implementation prohibitive for applications where it is necessary to have many inputs to the fuzzy controller. However, we must remind the reader of the results in Section 5.3.1 on page 199, where we explain how implementation tricks can be used to significantly reduce computation time when there are input membership functions of the form shown in Figure 9.40.

We use minimum or product to represent the conjunction in the premise and the implication (in this book we will use minimum for the FMRLC unless otherwise stated) and the standard center-of-gravity defuzzification technique. As an alternative, we could use appropriately initialized singleton output membership functions and center-average defuzzification since there are certain computational advantages to this approach.

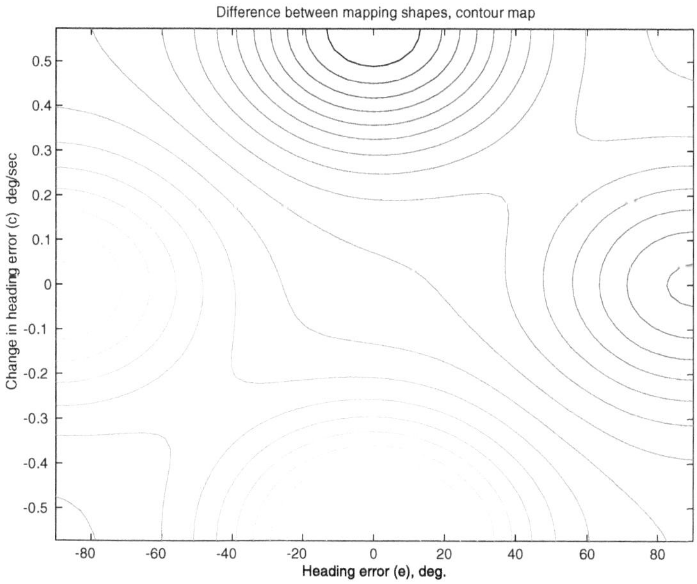

Figure 9.38: Contour map of difference of controller mappings of the radial basis function neural network (one obtained by subtracting the one at $t = 20,000$ sec. from the one at $t = 8999$ sec.) with 9 receptive field units.

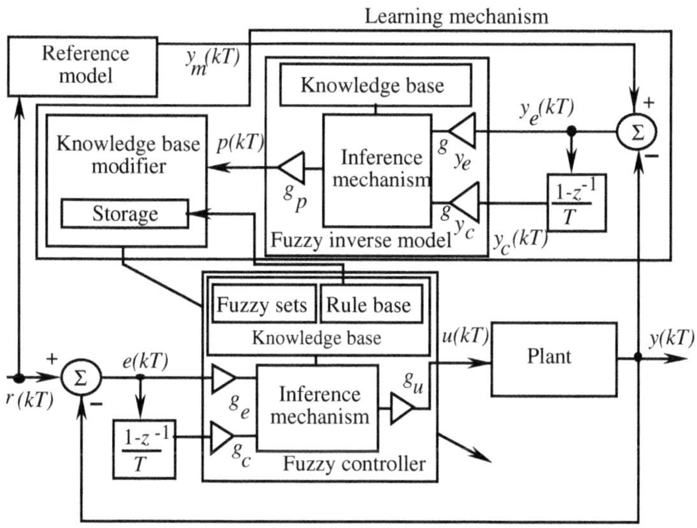

Figure 9.39: Fuzzy model reference learning controller (figure taken from [301], © IEEE, and used with permission).

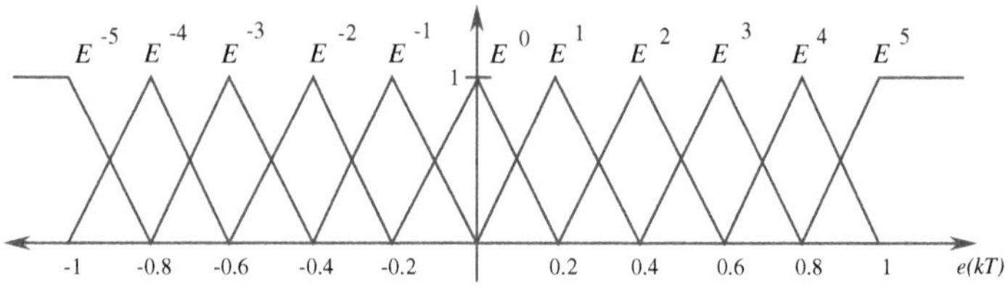

Figure 9.40: Membership functions for input universe of discourse (figure taken from [301], © IEEE, and used with permission).

The fuzzy controller holds the best guess at how a fixed controller should behave.

Rule Base Initialization: The input membership functions are defined to characterize the premises of the rules that define the various situations in which rules should be applied. The input membership functions are left constant and are not tuned by the FMRLC (although we do discuss some approaches to tuning them below). The membership functions on the output universe of discourse are assumed to be unknown. They are what the FMRLC will automatically synthesize or tune. Hence, the FMRLC tries to fill in what actions ought to be taken for the various situations that are characterized by the premises.

We must choose initial values for each of the output membership functions. For example, for an output universe of discourse $[-1, 1]$, we could choose triangular-shaped membership functions with base widths of 0.4 and centers at zero. This choice represents that the fuzzy controller initially knows nothing about how to control the plant so it inputs $u = 0$ to the plant initially (well, really it does know something, since we specify the remainder of the fuzzy controller a priori). Of course, one can often make a reasonable best guess at how to specify a fuzzy controller that is "more knowledgeable" than simply placing the output membership function centers at zero. For example, we could pick the initial fuzzy controller to be the best one that we can design for the nominal plant. Notice, however, that this choice is not always the best one. Really, what you often want to choose is the fuzzy controller that is best for the operating condition that the plant will begin in (this may not be the nominal condition). Unfortunately, it is not always possible to pick such a controller, since you may not be able to measure the operating condition of the plant, so making a best guess or simply placing the membership function centers at zero are common choices.

Learning, Memorization, and Controller Input Choice: For some applications you may want to use an integral of the error or other preprocessing of the inputs to the fuzzy controller. Sometimes the same guidelines that are used for the choice of the inputs for a nonadaptive fuzzy controller are useful for the FMRLC. We have found, however, times where it is advantageous to replace part of a conventional controller with a fuzzy controller and use the FMRLC to tune it. In these cases the complex preprocessing of inputs to the fuzzy controller is achieved via a conventional controller. Sometimes there is also the need for postprocessing of the fuzzy controller outputs.

Generally, however, choice of the inputs also involves issues related to the learning dynamics of the FMRLC. As the FMRLC operates, the learning mechanism will tune the fuzzy controller's output membership functions. In particular, in our example, for each different combination of $e(kT)$ and $c(kT)$ inputs, it will try to learn what the best control actions are. In general, there is a close connection between what inputs are provided to the controller and the controller's ability to learn to control the plant for different reference inputs and plant operating conditions. We would like to be able to design the FMRLC so that it will learn and remember different fuzzy controllers for all the different plant operating conditions and reference inputs; hence, the fuzzy controller needs information about these. Often, however, we cannot measure the operating condition of the plant, so the FMRLC does not know exactly what operating condition it is learning the controller for. Moreover, it then does not know exactly when it has returned to an operating condition (you can think of this as an online function approximation problem, where the unknown function is a controller that can meet the specifications and there is the influence of the auxiliary variable whose value we do not know). Clearly, then, if the fuzzy controller has better information about the plant's operating conditions, the FMRLC will be able to learn and apply better control actions. If it does not have good information, it will continually adapt, but it will not properly remember.

Controller behavior is parameterized via its inputs, so with more inputs it can learn how to behave differently in a wider range of situations.

For instance, for some plants $e(kT)$ and $c(kT)$ may only grossly characterize the operating conditions of the plant. In this situation, the FMRLC is not able to learn different controllers for different operating conditions; it will use its limited information about the operating condition and continually adapt to search for the best controller. It degrades from a learning system to an adaptive system that will not properly remember the control actions. (This is not to imply, however, that there will automatically be a corresponding degradation in performance.)

Generally, we think of the inputs to the fuzzy controller as specifying what conditions we need to learn different controllers for. This should be one guideline used for the choice of the fuzzy controller inputs for practical applications. A competing objective is, however, to keep the number of fuzzy controller inputs low due to concerns about computational complexity. In fact, to help with computational complexity, we will sometimes use multiple fuzzy controllers with fewer inputs to each of them rather than one fuzzy controller with many inputs; then we may, for instance, sum the outputs of the individual controllers.

The Learning Mechanism

The learning mechanism tunes the rule base of the direct fuzzy controller so that the closed-loop system behaves like the reference model (here, we use the same reference model as for the neural controller). These rule base modifications are made by observing data from the controlled process, the reference model, and the fuzzy controller. The learning mechanism consists of two parts: a "fuzzy inverse model" and a "rule base modifier." The fuzzy inverse model performs the function of mapping $y_e(kT)$ (representing the deviation from the desired behavior), to changes in the process inputs $p(kT)$ that are necessary to force $y_e(kT)$ to zero. The rule base modifier performs the function of modifying the fuzzy controller's rule base to produce the needed changes in the process inputs. We explain each of these components in detail next.

Engineers often know how to tune their decision-making strategy since they generally understand how to change inputs to move the plant output in a desired direction.

Fuzzy Inverse Model: Using the fact that most often a control engineer will know how to roughly characterize the inverse model of the plant (i.e., how the input should be changed to get a certain change in the output), we use a fuzzy system to map $y_e(kT)$, and possibly, functions of $y_e(kT)$ such as $y_c(kT) = \frac{1}{T}(y_e(kT) - y_e(kT - T))$ (or any other closed-loop system data), to the necessary changes in the process inputs $p(kT)$. This fuzzy system is sometimes called the "fuzzy inverse model" since information about the plant inverse dynamics is used in its specification. Some, however, avoid this terminology and simply view the fuzzy system in the adaptation loop in Figure 9.39 to be a controller that tries to pick $p(kT)$ to reduce the error $y_e(kT)$. A neural control viewpoint would highlight an analogy between the inverse model and the reinforcement function (what is it?).

Note that similar to the fuzzy controller, the fuzzy inverse model shown in Figure 9.39 contains scaling gains, but now we denote them with g_{y_e}, g_{y_c}, and g_p. We will explain how to choose these scaling gains below. Given that $g_{y_e}y_e$ and $g_{y_c}y_c$ are inputs to the fuzzy inverse model, the rule base for the fuzzy inverse model contains rules of the form

$$\textbf{If } \tilde{y}_e \text{ is } \tilde{Y}_e^j \textbf{ and } \tilde{y}_c \text{ is } \tilde{Y}_c^l \textbf{ Then } \tilde{p} \text{ is } \tilde{P}^m$$

where $\tilde{Y}_e^j$ and $\tilde{Y}_c^l$ denote linguistic values and $\tilde{P}^m$ denotes the linguistic value associated with the mth output fuzzy set. In this book, we often utilize membership functions for the input universes of discourse as shown in Figure 9.40, symmetric triangular-shaped membership functions for the output universes of discourse, minimum to represent the premise and implication, and COG defuzzification. Other choices can work equally well. For instance, we could make the same choices, except use singleton output membership functions and center-average defuzzification.

While the choice of a particular fuzzy inverse model is application-dependent, there are, however, some general guidelines for the choice of the fuzzy inverse model. First, we note that for a variety of applications, we find that the specification of the fuzzy inverse model is not much more difficult than the specification of a direct fuzzy controller. In fact, the fuzzy inverse model often takes on a form that is quite similar to a direct fuzzy controller. For instance, the rule base often has some typical symmetry properties. "First guess" values for the input scaling gains for the inverse model can be often found using a similar approach to that used for a standard fuzzy controller. Second, to pick g_p it is probably best to pick $g_p = 0$ and tune the fuzzy controller to get a reasonable performance for the nominal system (sometimes, however, this is not possible and the adaptation mechanism is needed to get reasonable performance levels). If this is not possible, and no good guess is known for the fuzzy controller, then start with a small value of g_p (the "adaptation gain") and experiment with choices for the other values using standard ideas from tuning conventional controllers as a guideline. Then, slowly increase g_p until adequate adaptation speeds are achieved (perhaps retuning each time you pick a new value for g_p).

Rule Base Modifier: Given the information about the necessary changes in the input, which are represented by $p(kT)$, to force the error y_e to zero, the rule base modifier changes the rule base of the fuzzy controller so that the previously applied control action will be modified by the amount $p(kT)$. Consider the previously computed control action $u(kT-T)$, and assume that it contributed to the present good or bad system performance (i.e., it resulted in the value of $y(kT)$ such that it did not match $y_m(kT)$). Hence, for illustration purposes we are assuming that in one step the plant input can affect the plant output; in Section 9.4.3 we will explain what to do if it takes d steps for the plant input to affect the plant output. Note that $e(kT-T)$ and $c(kT-T)$ would have been the error and change in error that were input to the fuzzy controller at that time. By modifying the fuzzy controller's rule base, we may force the fuzzy controller to produce a desired output $u(kT-T)+p(kT)$, which *we should have put in at time* $kT-T$ *to make* $y_e(kT)$ *smaller.* Then, the next time we get similar values for the error and change in error, the input to the plant will be one that will reduce the error between the reference model and plant output.

Assume that we use symmetric output membership functions for the fuzzy controller, and let b_m denote the center of the membership function associated with $\tilde{U}^m$. Rule base modification is performed by shifting centers b_m of the membership functions of the output linguistic value $\tilde{U}^m$ that are associated with the fuzzy controller rules that contributed to the previous control action $u(kT-T)$. This is a two-step process:

1. Find all the rules in the fuzzy controller whose premise certainty

$$\mu_i(e(kT-T), c(kT-T)) > 0 \tag{9.19}$$

   and call this the "active set" of rules at time $kT-T$. We can characterize the active set by the indices of the input membership functions of each rule that is on (since we use all possible combinations of rules, there will be one output membership function for each possible rule that is on).

2. Let $b_m(kT)$ denote the center of the mth output membership function at time kT. For all rules in the active set, use

$$b_m(kT) = b_m(kT-T) + p(kT) \tag{9.20}$$

   to modify the output membership function centers. Rules that are not in the active set do not have their output membership functions modified.

Notice that for our development, when COG is used, this update will guarantee that the previous input would have been $u(kT-T)+p(kT)$ for the same $e(kT-T)$ and $c(kT-T)$ (to see this, simply analyze the formula for COG to see that adding the amount $p(kT)$ to the centers of the rules that were on will make the output shift by $p(kT)$). For the case where the fuzzy controller has input membership functions of the form shown in Figure 9.40, there will only be at most four rules in the active set at any one time instant (i.e., four rules with $\mu_i(e(kT-T), c(kT-T)) > 0$ at time kT). Then we only need to update at most four output membership function centers via Equation (9.20). Note the similarities between this parameter update scheme and the one used in neural control earlier in this chapter.

Example Rule Parameter Update: As an example of the rule base modification procedure, assume that all the scaling gains for both the fuzzy controller and the fuzzy inverse model are unity. Suppose that the fuzzy inverse model produces an output $p(kT) = 0.5$, indicating that the value of the output to the plant at time $kT - T$ should have been $u(kT-T) + 0.5$ to improve performance (i.e., to force $y_e \approx 0$). Next, suppose that $e(kT-T) = 0.75$ and $c(kT-T) = -0.2$ and that the membership functions for the inputs to the fuzzy controller are given in Figure 9.40. Then rules

$$\mathcal{R}_1\text{: } \textbf{If } E^3 \textbf{ and } C^{-1} \textbf{ Then } U^1$$

and

$$\mathcal{R}_2\text{: } \textbf{If } E^4 \textbf{ and } C^{-1} \textbf{ Then } U^2$$

are the only rules that are in the active set (notice that we chose to use the indices for the rule "1" and "2" simply for convenience). In particular, from Figure 9.40, we have $\mu_1 = 0.25$ and $\mu_2 = 0.75$, so rules $\mathcal{R}_1$ and $\mathcal{R}_2$ are the only ones that have their consequent fuzzy sets (U^1, U^2) modified. Suppose that at time $kT-T$ we had $b_1(kT-T) = 1$ and $b_2(kT-T) = 3$. To modify these fuzzy sets, we simply shift their centers according to Equation (9.20) to get

$$b_1(kT) = b_1(kT-T) + p(kT) = 1 + 0.5 = 1.5$$

and

$$b_2(kT) = b_2(kT-T) + p(kT) = 3 + 0.5 = 3.5$$

Learning, Memorization, and Inverse Model Input Choice: Notice that the changes made to the rule base are only *local* ones. That is, the entire rule base is not updated at every time step, just the rules that needed to be updated to force $y_e(kT)$ to zero. Notice that this local learning is important since it allows the changes that were made in the past to be remembered by the fuzzy controller. Recall that the type and amount of memory depends critically on the inputs to the fuzzy controller. Different parts of the rule base are "filled in" based on different operating conditions for the system (as characterized by the fuzzy controller inputs), and when one area of the rule base is updated, other rules are not affected. Hence, if the appropriate inputs are provided to the fuzzy controller so that it can distinguish between the situations in which it should behave differently, the controller adapts to new situations and also remembers how it adapted to past situations.

Rules for how to adapt that have more inputs can behave differently in a wider range of situations.

Just as the choice of inputs to the fuzzy controller has a fundamental impact on learning and memorization, so does the choice of inputs to the inverse model. For instance, you may want to choose the inputs to the inverse model so that it will adapt differently in different operating conditions. In one operating condition, we may want to adapt more slowly than in another. In some operating conditions, the direction of adjustment of the output membership function centers may be the opposite of that in another. If there are multiple fuzzy controllers, you may want multiple inverse models to adjust them. This can sometimes help with computational complexity, since we could then be using fewer inputs per fuzzy inverse model.

The choice of inputs to the fuzzy inverse model shown in Figure 9.39, indicates that we want to adapt differently for different errors and error rates between the reference model and plant output. The inverse model may be designed so that, for example, if the error is small, then the adjustments to the fuzzy controller should be small, and if the error is small but the rate of error increase is high, then the adjustments should be larger. It is rules such as these that are loaded into the fuzzy inverse model.

Alternative Rule Base Modifiers

Here, we discuss a variety of options for how to define the rule base modifier.

Plants with Delays: Recall that we had assumed that the plant input $u(kT)$ would affect the plant output in one time step so that $y(kT+T)$ would be affected by $u(kT)$. To remove this assumption and hence generalize the approach, let d denote the number of time steps that it takes for an input to the plant $u(kT)$ to first affect its output. That is, $y(kT + dT)$ is affected by $u(kT)$. To handle this case, we use the same approach but we go back d steps to modify the rules. Hence, we use

$$\mu_i(e(kT - dT), c(kT - dT)) > 0 \tag{9.21}$$

to form the "active set" of rules at time $kT - dT$. To update the rules in the active set, we let

$$b_m(kT) = b_m(kT - dT) + p(kT) \tag{9.22}$$

(when $d = 1$, we get the case in Equations (9.19) and (9.20)). This ensures that we modify the rules that actually contributed to the current output $y(kT)$ that resulted in the performance characterization $y_e(kT)$. For applications we have found that we can most often perform a simple experiment with the plant to find d (e.g., put a short-duration pulse into the plant and determine how long it takes for the input to affect the output), and with this choice we can often design a very effective FMRLC. However, this has not always been the case. Sometimes we need to treat d as a tuning parameter for the rule base modifier.

Thresholds for Entry to the Active Set: There are several alternatives to how the basic rule base modification procedure can work that can be used in conjunction with the d-step back approach. For instance, note that an alternative to Equation (9.19) would be to include rules in the active set that have

$$\mu_i(e(kT - dT), c(kT - dT)) > \alpha$$

where $0 \leq \alpha < 1$. In this case, we will not modify rules whose premise certainty is below some given threshold α. This makes some intuitive sense, since we will then not modify rules if the fuzzy system is not too sure that they should be on. However, one could argue that any rule that contributed to the computation of $u(kT - dT)$ should be modified. This approach may be needed if you choose to use Gaussian membership functions for the input universes of discourse, since it will ensure that you will not have to modify all the output centers at each time step, and hence the local learning characteristic is maintained.

Update Formula Alternatives, Parameter Constraints: There are also alternatives to the center update procedure given in Equation (9.20). For instance, we could choose

$$b_m(kT) = b_m(kT - dT) + \mu_m(e(kT - dT), c(kT - dT))p(kT)$$

so that we scale the amount we shift the membership functions by the μ_m certainty of their premises. Intuitively, this makes sense since we will then change the membership functions from rules that were on more by larger amounts, and for rules that are not on as much, we will not modify them as much. Notice that for our example in this section we have that

$$\sum_{i=1}^{R} \mu_i = 1$$

where $R = 121$ is the number of rules (this is due to how we overlap the membership functions, and relies on maintaining this type of overlap if you adjust the input membership functions over time). We see that with this the premise membership functions are the same as the basis functions

$$\xi_i = \frac{\mu_i}{\sum_{i=1}^{R} \mu_i} = \mu_i$$

and the update formula bears some similarity to the gradient update formulas studied in Chapter 11.

This approach has proven to be more effective than the one in Equation (9.20) for some applications; however, it is difficult to determine a priori which approach to use. We usually try the scaled approach if the one in Equation (9.20) does not seem to work well, particularly if there are some unwanted oscillations in the system that seem to result from excessive modification of output membership function center positions. Sometimes, using too few rules in the fuzzy controller causes such problems.

Another modification to the center update law is also necessary in some practical applications to ensure that the centers stay in some prespecified range. For instance, you may want the centers to always be positive so that the controller will never provide a negative output. Other times you may want the centers no larger than some prespecified value to ensure that the control output will become no larger than this value. In general, suppose that we know a priori that the centers should be in the range $[b_{min}, b_{max}]$ where b_{min} and b_{max} are given scalars. We can modify the output center update rule to ensure that if the centers start in this range, they will stay in the range by adding the following two rules after the update formula:

$$\textbf{If } b_m(kT) < b_{min} \textbf{ Then } b_m(kT) = b_{min}$$

$$\textbf{If } b_m(kT) > b_{max} \textbf{ Then } b_m(kT) = b_{max}$$

In other words, if the centers jump over the boundaries, they are set equal to the boundary values. This is same "projection" approach as used in gradient methods to constrain the update of parameters.

Hybrid Methods, Heuristic Robustification: Notice that you could combine the above alternatives to rule base modification so that we set a threshold for including rules in the active set, scale the updates to the centers, bound the updates to the centers, and use any number of time steps back to form the active set. There are yet other alternatives that can be used for rule base modification procedures. For instance, parts of the rule base could be left intact (i.e., we would not let them be modified). This can be useful when we know part of the fuzzy controller that is to be learned, we embed this part into the fuzzy controller that is tuned, and do not let the learning mechanism change it.

Overaggressive pursuit of closed-loop objectives can require continual adjustment of parameters that can lead to instability.

In the section on stable adaptive control we will show how "robustification" schemes can be used to ensure stable operation of the adaptive system for uncertain plants. Here, we simply discuss some heuristic methods that try to achieve robustification. For instance, for many practical applications it is necessary to define the inverse model so that when the response of the plant is following the output of the reference model very closely, the fuzzy inverse model turns off the adaptation. In this way, once the inputs to the fuzzy inverse model get close to zero, the output of the fuzzy inverse model becomes zero. We think of this as forcing the fuzzy inverse model to be satisfied with the performance as long as the plant output is quite close to the reference model; there is no need to make it exact in many applications. Designing this characteristic into the fuzzy inverse model can sometimes help ensure stability of the overall closed-loop system. Another way to implement such a strategy is to directly modify the output of the fuzzy inverse model by using the rule:

$$\textbf{If } |p(kT)| < \epsilon_p \textbf{ Then } p(kT) = 0$$

where $\epsilon_p > 0$ is a small number that is specified a priori. For typical fuzzy inverse model designs (i.e., ones where the size of the output of the fuzzy inverse model is directly proportional to the size of the inputs to the fuzzy inverse model), this rule will make sure that when the inputs to the fuzzy inverse model are in a region of zero, its output will be modified to zero. Hence, for small fuzzy inverse model inputs, the learning mechanism will turn off. If, however, the error between the plant output and the reference input grows, then the learning mechanism will turn back on and it will try to reduce the error. Such approaches to modifying the adaptation online are related to "robustification" methods in conventional adaptive control.

As another alternative, when a center is updated, you could always wait d or more steps before updating the center again. This can be useful as a more "cautious" update procedure. It updates, then waits to see if the update was sufficient to correct the error y_e before it updates again. We have successfully used this approach to avoid inducing oscillations when operating at a set point.

9.4.4 Design Example: Rule-Tuning for the Tanker Ship

Here, we consider the same tanker ship heading regulation problem as we did for both the neural control earlier in this chapter, and the instinctual neural controller and fixed fuzzy controller in Part II. This will provide us with the opportunity to compare the results to a variety of other cases, and particularly in the context of fuzzy control, show why adaptation can be valuable.

FMRLC Design for the Tanker Ship

We had chosen the fuzzy controller inputs to be

$$e(kT) = \psi_r(kT) - \psi(kT)$$

$$c(kT) = \frac{e(kT) - e(kT - T)}{T}$$

where $\psi_r(kT)$ is the desired ship heading (we chose $T = 10$ seconds). We will use this same choice for the FMRLC. The controller output is the rudder angle, $\delta(kT)$, of the ship. We use 11 uniformly spaced triangular membership functions for each controller input as shown in Figure 9.40 (i.e., on normalized universes of discourse). We chose $g_e = \frac{2}{\pi}$, $g_c = 250$, and $g_u = \frac{8\pi}{18}$ (the "good" tuned values). The output membership functions are assumed to be symmetric and triangular shaped with a base width of 0.4 (on a normalized output universe of discourse), and centered via the tuning of the fuzzy controller for the "nominal plant" in Part II.

The reference model is chosen to be

$$\frac{\psi_m(s)}{\psi_r(s)} = \frac{\frac{1}{150}}{s + \frac{1}{150}}$$

where $\psi_m(t)$ specifies the desired system performance for the ship heading $\psi(t)$.

The fuzzy inverse model inputs are chosen to be

$$\psi_e(kT) = \psi_m(kT) - \psi(kT)$$

$$\psi_c(kT) = \frac{\psi_e(kT) - \psi_e(kT - T)}{T}$$

We use 11 fuzzy sets defined with symmetric and triangular shaped membership functions, which are evenly distributed on the appropriate universes of discourse (the same as shown in Figure 9.40).

For fuzzy inverse model design, note that for a tanker ship, an increase in the rudder angle $\delta(kT)$ will generally result in a *decrease* in the ship heading angle (see Figure 4.8). This is the information about the inverse dynamics of the plant that we use in the fuzzy inverse model rules. Specifically, we will use rules of the form

$$\textbf{If } \tilde{\psi}_e \text{ is } \tilde{\Psi}_e^i \textbf{ and } \tilde{\psi}_c \text{ is } \tilde{\Psi}_c^j \textbf{ Then } \tilde{p} \text{ is } \tilde{P}^m$$

Our output membership function centers are $c_{i,j}$ (i^{th} membership function for $\tilde{\psi}_e$ and j^{th} membership function for $\tilde{\psi}_c$). We use the rule base shown in Table 9.1 for the fuzzy inverse model.

Table 9.1: Rule base for the tanker ship fuzzy inverse model.

$c_{i,j}$		Ψ_c^j										
		−5	−4	−3	−2	−1	0	1	2	3	4	5
	−5	1	1	1	1	1	1	.8	.6	.4	.2	0
	−4	1	1	1	1	1	.8	.6	.4	.2	0	−.2
	−3	1	1	1	1	.8	.6	.4	.2	0	−.2	−.4
	−2	1	1	1	.8	.6	.4	.2	0	−.2	−.4	−.6
	−1	1	1	.8	.6	.4	.2	0	−.2	−.4	−.6	−.8
Ψ_e^i	0	1	.8	.6	.4	.2	0	−.2	−.4	−.6	−.8	−1
	1	.8	.6	.4	.2	0	−.2	−.4	−.6	−.8	−1	−1
	2	.6	.4	.2	0	−.2	−.4	−.6	−.8	−1	−1	−1
	3	.4	.2	0	−.2	−.4	−.6	−.8	−1	−1	−1	−1
	4	.2	0	−.2	−.4	−.6	−.8	−1	−1	−1	−1	−1
	5	0	−.2	−.4	−.6	−.8	−1	−1	−1	−1	−1	−1

In Table 9.1, Ψ_e^i denotes the ith fuzzy set associated with the error signal ψ_e, and Ψ_c^j denotes the jth fuzzy set associated with the change in error signal ψ_c. The entries of the table represent the center values of symmetric triangular membership functions $c_{i,j}$ with base widths 0.4 for output fuzzy sets P^m on the normalized universe of discourse.

To see that this rule base captures our intuitive knowledge about inverse model design, we consider a few example rules. First, notice that if $i = j = 0$, then we see from the table that $c_{i,j} = c_{0,0} = 0$ (this is the center of the table). This cell in the table represents the rule that says "if $\psi_e = 0$ and $\psi_c = 0$, then y is tracking y_m perfectly so you should not update the fuzzy controller." Hence, the output of the fuzzy inverse model will be zero. If, on the other hand, $i = 2$ and $j = 1$, then $c_{i,j} = c_{2,1} = -0.6$. This rule indicates that "if ψ_e is positive (i.e., ψ_m is greater than ψ) and ψ_c is positive (i.e., $\psi_m - \psi$ is increasing), then change the input to the fuzzy controller that is generated to produce these values of ψ_e and ψ_c by decreasing it. This is because we want ψ to increase so we want to decrease δ to get this (see Figure 4.8). We see that the inverse model indicates that whatever the input was in this situation, it should have been less so it subtracts some amount (the amount affected by the scaling gain g_p). Other rules can be explained similarly.

We choose the fuzzy inverse model scaling gains as $g_{\psi_e} = \frac{2}{\pi}$, $g_{\psi_c} = 10$, and $g_p = 0.4$, respectively, according to the ideas for tuning these gains presented earlier.

We use Equation (9.20) to implement the parameter update method. We use $d = 1$ and add a heuristic robustification scheme where if

$$|p(kT)| < 0.01 g_p$$

then we let $p(kT) = 0$. This way, with the choice of the inverse model, we will tend to turn off the adaptation when the error between the reference model (and its change in error) are small. We make the threshold value depend on g_p simply for convenience in tuning since this way, we can think of not updating when the size of the suggested update is smaller than some percentage of the adaptation gain.

Performance Using a "Good Guess" Initialization

First, consider the "good" set of gains that we found in Part II when we tuned the direct fuzzy controller, where $g_1 = 2/\pi$, $g_2 = 250$, and $g_0 = 8\pi/18$. As shown in Figure 9.41, when we start with a good initial guess at the fuzzy controller, very little tuning occurs and we get good tracking performance. We will, in fact, use these same gains for the fuzzy controller for the rest of the simulations in this section.

Also, consider Figure 9.42, which shows the fuzzy controller surface at the end of the simulation. Notice that the shape of the surface is similar to the one in Figure 5.27 from the nonadaptive case; this makes sense since we started with a good guess at the fuzzy controller, so the FMRLC does not adjust it too much. Compare this to the neural control case and note the difference in the shapes of the maps due to initialization and adjustment approach differences.

Note that from Figure 9.41, you can see that there are still adjustments being made to the fuzzy controller; hence, this surface has a shape that is still changing at the end of the simulation (to see how you could plot the shape over several time steps). Indeed, in some cases the shape will continually change for

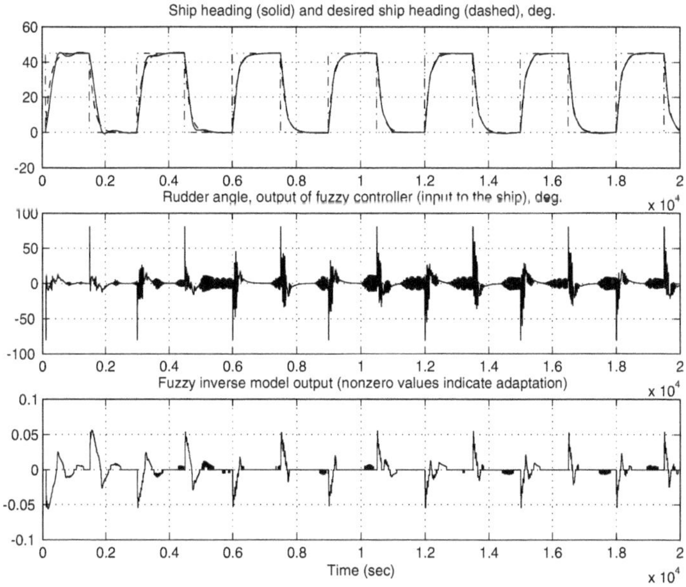

Figure 9.41: Tanker FMRLC response, good initial controller.

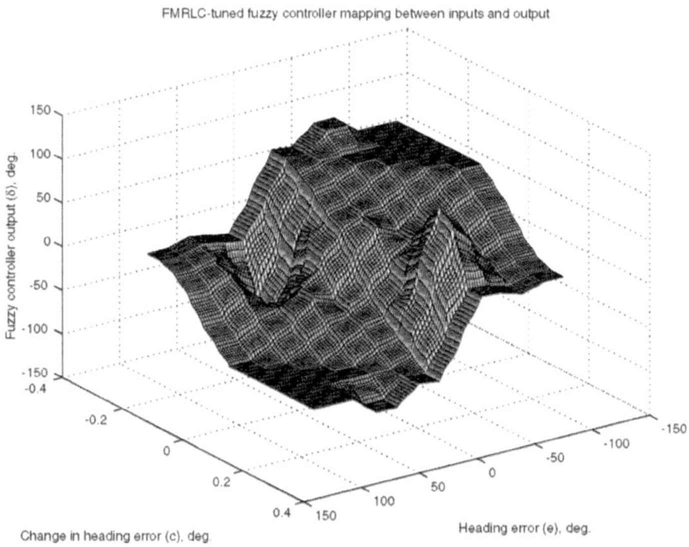

Figure 9.42: Tuned fuzzy controller surface.

all time for some reference inputs. Whether it stops changing shape depends on several things. First, it depends on the reference input. Second, it depends on the adaptation method. Third, it depends on the plant dynamics and what information is available from the plant, and how that information is used as inputs to the fuzzy controller and inverse model.

Effects of Plant Changes and Disturbances

Here, we change from "ballast" to "full" conditions at $t = 9000$ sec. so we appropriately change the plant parameters (suddenly) at this time to represent this weight change. (It is of course unreasonable to suddenly change the weight of the ship; here, we are simply thinking of the simulation as showing what happens when the FMRLC is used in the nominal conditions for a while, the ship unloads ballast, and then the ship continues with a lighter load using the same controller.) As shown in Figure 9.43, while the initial transient performance just after $t = 9000$ sec. is not as good, the FMRLC does tune the FMRLC to achieve good performance by the end of the simulation. Of course, it has tuned differently to achieve this type of performance so that the shape of the tuned controller nonlinearity will be different.

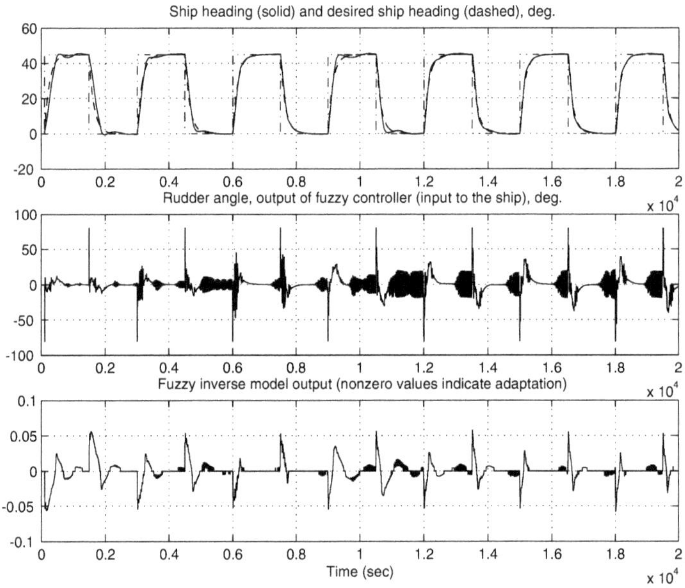

Figure 9.43: Tanker FMRLC response, plant condition change, ballast to full.

Next, we leave everything else the same as in the last simulation, but do not change the ship weight and add a wind disturbance (the same one described in Part II). As shown in Figure 9.44, the FMRLC rejects the disturbance (recall that for the nonadaptive controller, the wind disturbance had a noticeable adverse impact on the response as shown in Figure 5.29). Next, we study the effect of changing the speed of the ship from $u = 5$ to $u = 3$ meters/sec. As shown in Figure 9.45, the FMRLC also performs reasonably well for this case, certainly much better than in the nonadaptive case shown in Figure 5.30.

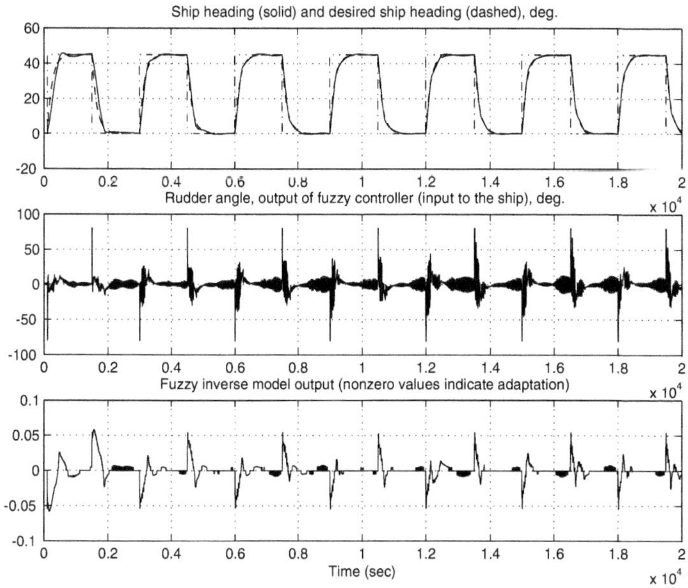

Figure 9.44: Tanker FMRLC response, wind disturbance.

Notice that overall we get similar performance to the neural control case. In fact, the slight performance differences cannot be used to conclude anything about the superiority of one approach over the other. This is especially the case due to the difference in complexity of the two controllers and the fact that we certainly cannot be sure that both are tuned for "optimal performance," however you would define that.

9.4.5 Expert, Planning, and Attentional Systems for Adaptive Control

There are several ways that expert, planning, and attentional systems can be used in an adaptive controller. Here, we outline a few heuristic approaches to use such methods.

Expert Control for Adaptation Strategies

Rules can be used to tune adaptive strategies.

Expert systems can be used to incorporate high-level heuristic ideas on how to tune adaptation mechanisms online. For instance, in a neural controller or the FMRLC discussed in the last section, an expert controller may monitor plant information and the reference input and may subsequently tune the adaptation gain or reference model. There are times when it is known that in certain

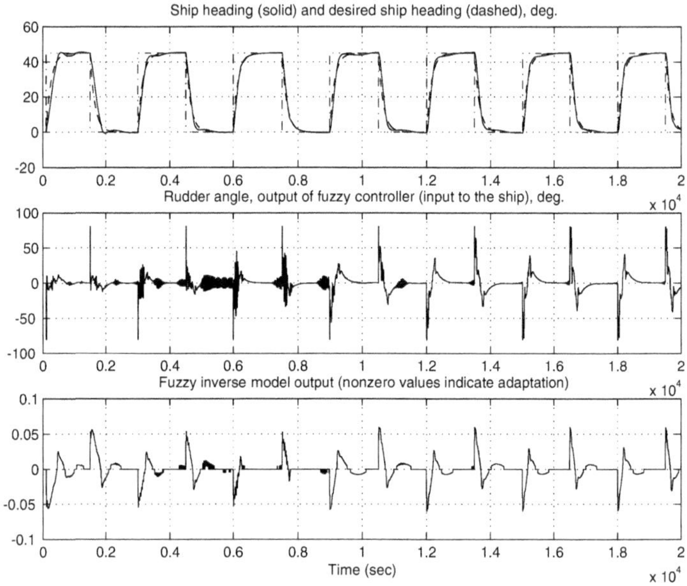

Figure 9.45: Tanker FMRLC response, speed decrease to 3 meters/sec.

conditions (e.g., if there are indications that there is a large plant change), more significant changes should be made to the controller so that the closed-loop system maintains high performance operation. One application where this can happen is in fault tolerant control, where if you receive an indication that there is a failure, you may want to quickly adapt the controller to recover from the failure quickly. Then, when some reasonable performance level is achieved, you may want to slow the changes (i.e., lower the adaptation gain) to help ensure stable operation. In such a situation it may also be beneficial to tune the reference model so that it always reflects reasonable performance requirements for the plant. If there is a failure, then you may want the expert controller to pick a reference model that reflects less demanding performance specifications. Then, if the failure is fixed, the expert system should return the reference model to its nominal choice. Such a scheme could be said to be "performance adaptive" in the sense that it would try to adapt to achieve the best possible performance, but one that is reasonable for the current operating conditions. Overall, you may view the tuning of adaptive controller parameters as adding another level of adaptation and hence as providing a method *to learn how to learn.*

Expert systems can also be used to incorporate ideas about how to coordinate the application of a variety of conventional or intelligent controllers. For instance, in some applications you may know that under certain conditions you may want to use one type of controller, where in others a different type of controller may be better. Often, this can be quantified with rules in an expert system. You may think of this as a general sort of "gain scheduling" approach where the expert controller supervises the application of different controllers by picking which controller to apply in each plant operating condition.

Adaptive Planning

Next, after providing some bio-foundations, we explain how learning and planning can be combined for control.

Place Learning: In other work that has taken the "cognitive perspective," E. Tolman and his colleagues performed research that led them to conclude that animals learn a "cognitive map" of their environment in order to learn about places. This cognitive map can be thought of as a mental representation of the spatial layout of its environment, and they found that this map is learned solely through exploration, whether an animal finds a reward or not. Clearly, acquisition of such cognitive maps has great value when used to predict, and hence it should be clear that it is reasonable to hypothesize that this ability could evolve from simpler learning processes. Clearly, the idea of learning cognitive maps has relevance to adaptive planning system development. For this, see more discussion below, and also the idea of using a "surrogate model method" for combined learning and planning that is discussed in Section 19.6 and Design Problem 19.5.

Humans have an ability to learn cognitive maps of their environment and to learn by observation.

Planning Mechanisms in Adaptive Model Predictive Control: "Model predictive control" (MPC) (in some contexts called "receding horizon control") is a very popular method in the process control industries, as it has been shown to provide effective solutions to complex control problems (see Chapter 6). In MPC, a model of the plant is used at each step to predict into the future how the plant will respond to a sequence of control inputs. The length of time of the simulation into the future is called the "prediction horizon" and often it is a design parameter for the controller. Next, of all the possible control input sequences, the control sequence is chosen that is predicted to best satisfy the performance requirements. Then, typically the first control input of this sequence is applied to the plant and the process repeats. In the linear case, typically a quadratic cost criterion is used and there is a unique optimal sequence that can be found using least squares methods. It should be clear that a type of planning mechanism is used in MPC.

Models can be used to predict reactions to inputs in order to plan how to choose the best input.

One of the problems with MPC is that if the model is not good, then the predictions about performance are poor so that the choice of control inputs can be less than optimal. Methods have been investigated to solve this problem by using an online model identification method to specify the model used in MPC. If good identification is performed, and this model is used in the standard MPC strategy, sometimes a good controller can be obtained. The direct genetic adaptive control scheme (to be covered in Chapter 16, Section 16.5) employs a type of planning mechanism, and when integrated with an indirect genetic adaptive scheme, adaptive model predictive control is achieved.

To clarify how adaptive model predictive control can be achieved, it is interesting to consider how the ideas can be used to develop an adaptive fuzzy model predictive control (FMPC) method. To do this, you can simply use a Takagi-Sugeno fuzzy system model of the plant as the model in the FMPC method and, if formulated properly, you can solve an optimization problem that will provide an "optimal" input at each step. Several of the methods of this part can be used to adapt the Takagi-Sugeno model of the plant. See the "For Further Study" section at the end of the part for more details.

Finally, to directly use biomimicry of the cognitive map idea at the beginning of this section, the idea of using a "surrogate model method" for combined learning and planning is discussed in Section 19.6 and Design Problem 19.5.

Attentional Mechanisms for Adaptation

It is clear that humans can learn how to pay attention to a task. In a similar way, you can design adaptive controllers that seek to pay attention to a control task. If you augment a controller with an attentional mechanism, this mechanism can play a special role in adaptation. To make the ideas concrete, we will discuss one type of attentive mechanism for the FMRLC that was introduced earlier. It should then be clear how the methods could be generalized and used in other methods (e.g., in neural control).

Attentional strategies can augment or serve as adaptive strategies.

Human Attentional Mechanisms for Control Tasks: Learning controllers are often designed to mimic the manner in which a human in the control loop would learn how to control a system while it operates. Some characteristics of this human learning process may include the following:

1. A natural tendency for the human to *focus* his learning by paying particular attention to the current operating conditions of the system since these may be most relevant to determining how to enhance performance.

2. After the human has learned how to control the plant for some operating condition, if the operating conditions change, then the best way to control the system may have to be relearned.

3. A human with a significant amount of experience at controlling the system in one operating region should not forget this experience if the operating condition changes.

To mimic these types of human learning behavior, in this section we discuss three strategies that can be used to dynamically focus a learning controller onto the current operating region of the system. The subsequent "dynamically focused learning" (DFL) can in some situations be used to enhance the performance of the FMRLC.

Motivation for Dynamically Focused Learning: There are several motivations for using DFL. First, with DFL we will be tuning not only the centers of the output membership functions, but also the input membership functions of the rules. In this sense, we will have more tuning flexibility and hence, it is possible that we can do a better job at matching the underlying nonlinearities. We can think of this as using DFL as a way to allocate approximator structure to the places where it is needed to improve approximation accuracy; hence, it provides a way to tune the parameters of the approximator that enter in a nonlinear fashion.

Second, thinking of DFL as an attentional system, there is the possibility that the amount of needed computations can be decreased. For instance, note that in the FMRLC, it may be that many of the rules of the fuzzy controller are never visited so that their corresponding output membership function centers are never modified. In this case, these rules are never used and are only a waste of computational resources (e.g., memory for their storage). In other cases, the FMRLC may only learn a portion of the rule base by a certain point in time, but later, the other portions may be learned when there are different reference inputs or plant changes, and in a sense, there is not a waste of memory resources.

Third, you can think of how allocation of structure and computational resources are related. Note that if you can continually reallocate approximator structure, it may be possible that you can learn (with a "fine-grained" set of rules) how to control in one region, then when the system operating condition changes, you can simply reallocate the structure of the approximator to where it is needed. This can save computational resources, but of course, something about how to control is forgotten in the process (but this may be satisfactory if relearning can occur quickly and computational resources are at a premium). It is for reasons such as these that we turn to DFL strategies. Next, we outline a few such strategies.

Auto-Attentive Mechanisms: In some FMRLC designs, the reference input sequence does not excite the whole range of the designated input universes of discourse. Instead, the rule base learned for the input sequence only covers the center part of the rule base. Hence, to achieve an adequate number of rules to enhance the granularity of the rule base near the center, it would be necessary to design the rule base so that it is located at exactly where most of the rules are needed. However, we would like to ensure that we can adapt the fuzzy rule base should a different input sequence drive the operation of the system out of this center region.

To do this, we can use an "auto-tuning" mechanism that simply focuses the rule base about the origin by adjusting the two scaling gains on the fuzzy controller (you might think of the region where the rules' premises are not saturated as where a "spotlight" is shining, i.e., where we are currently focusing our attention). Note that an increase in these two scaling gains results in sharper focusing, while a decrease in the values results in coarse rule distribution and hence less-sharp focusing. One scheme that has been investigated to adjust each of the gains is to simply make the gains the inverse of the maximum absolute value achieved at each input over a fixed time window. In this way, we avoid letting the fuzzy controller have its inputs saturate on the input universes of discourse. Of course, adjusting the focus will destroy (or at least modify) what was learned in the past; however, if you learn faster than you focus, then this problem can be overcome and you can achieve a focusing of the computational resources by proper allocation of the structure of the approximator.

Next, note that this method will only allow for focusing where the center of the focus is the origin of the input space of the fuzzy controller. It is of course possible to use more general auto-attentive strategies where you move the spotlight (the region where the premise membership functions are not saturated) across the input space of the fuzzy controller. In this way, if the inputs are only in one region of the input space, the rule base can be focused there. Clearly, the use of auto-tuning the size of the spotlight and its position in the input space can both be used to get the best attentional approach.

Note, however, that in the auto-attentive DFL strategy, every shift of the rule base (spotlight) will create a new unexplored region. Having to learn the new regions from scratch after every movement of the rule base can cause degradations in the performance of the auto-attentive FMRLC, since it will require the learning mechanism to fill in the unknown rules (i.e., additional time for learning will be needed). For example, if an auto-attentive FMRLC has been operating for a long time on an input sequence, then at some time instant a disturbance affected the controller inputs and forced the rule base to leave its current position, some of the rules are lost and replaced by new rules that will accommodate the disturbance. When the temporary disturbance is stopped and the rule base returns to its initial position again, its previous experience is lost and it has to "relearn" everything about how to control in a region where it actually has gained a significant amount of experience.

One way to avoid this problem is to add memory to the attentive mechanism. One way to do this is to simply add another fuzzy system, where when the FMRLC does a good job of learning in one region, before a shift to another region, some information about the shape of the local learned surface is transferred to the other fuzzy system. Then, when it moves to a new region, it initializes its learning with what was previously stored (or initialized). Actually, this can occur as a continuous process where, as the spotlight moves into new regions, it loads in information that was gathered via experience, and uses it at the edges of the rule base that is being learned. Clearly, as the spotlight is moved, it must be the case that learning is occurring at a higher rate than the spotlight is moved, or at least it must be the case that a good initialization was used so that even if the spotlight moves quickly, there will be reasonably good control inputs that can be generated.

9.4.6 Development, Plasticity, and Control

We briefly discuss ideas for using biomimicry of development and plasticity for control and automation.

Development and Plasticity

Neural networks grow in animals and humans and incrementally acquire information (e.g., rules or plans on how to control). Hence, often learning occurs not only via parameter tuning, but also via adjustments (e.g., additions or deletions) to the neural network (e.g., the number of neurons in a neural network). In this section we briefly overview how development influences learning, and we identify the concept of plasticity. In Design Problem 11.2 you will be asked to design methods for "approximator structure synthesis" whose biomimicry foundations partly lie in the concepts in this section.

It is difficult to separate development from learning since they occur simultaneously, at least during our childhood, and some evidence suggests that even new neural connections can be made at least until we are 20 years old. It is for this reason that in addition to the standard learning theories, it is important to consider the effects of development.

Developmental psychologists often view the infant as an "explorer." Via "habituation," infants have been found to look at novel objects, scenes, faces, etc., longer than ones that they have already looked at for some period of time. It is thought that they learn after a time and that there is a fundamental drive to explore to learn more. Moreover, it has been found that infants are much more interested in the parts of their environment that they can control themselves. Infants quickly learn how to explore not only with their eyes also but with their hands and mouths. They are typically quite curious and have been shown to understand many of the basic laws of physics at a very early age (e.g., "object permanence," where if you cover up an object, they know that it did not vanish but is just hidden). There is even a debate as to how much is learned and how much of the knowledge of physics is genetically determined (i.e., is instinctual).

Mental development is driven by interactions with the environment.

J. Piaget and many subsequent researchers have studied how reasoning develops. One fundamental idea that grew from Piaget's research is that mental development occurs via a child's actions on his own environment. Piaget pointed out that infants at first seem to only react to events via instinctual reflexes. Then later they incrementally gain voluntary control of their actions as they construct mental representations of the types of actions that they can perform on different classes of objects. Piaget named the mental representations "schemes," and you should think of these as "blueprints" for what actions can be taken on different objects in the world. Piaget felt that part of mental development involved the growth of schemes in two ways. First, it uses "assimilation," which is a process where new experiences are incorporated into existing schemes (e.g., a baby uses its sucking scheme to explore the shapes of his blocks). Second, it uses "accommodation," where existing schemes change to accommodate a new object or event (e.g., extend the sucking scheme to learn how to drink out of a glass). This assimilation and accommodation process results in a child's learning about his environment and how to succeed in it. Piaget proposed that children are most interested in experiences that can be assimilated into existing ones, but not easily, so that they use accommodation and over time incrementally try to maximize their mental growth.

Next, Piaget proposed that to grow out of infancy, the actions that most contribute to mental development are "operations," which he defines as "reversible actions" (e.g., turning a switch on and off). By exploring with operations, children gradually develop "operational schemes," which are blueprints for how to think about simple actions that they take in their environment. Later stages of development depend on building on these operational schemes. In fact, Piaget went on to propose stages of mental development. First, there is the "sensorimotor stage" where the child acts on objects that are present, but does not think about ones that are not present. Objects are assimilated into what you might call instinctual schemes (e.g., via exploration); after learning properties of objects, they learn to think of these properties even when they are not present. Next, in the "preoperational stage," the child can symbolize objects and events that are not present. Then, in the "concrete-operational stage," via their explorations, they gain the capability to think about the reversible consequences of actions. Finally, they enter the "formal-operational stage," where they think about similarities of actions that can be performed on different objects and hence are able to reason more abstractly.

In addition to Piaget's perspective on mental development, which generally focused on how the brain grew as a single entity, there is an "information processing perspective" in psychology where researchers focus on how the interacting components of the brain can develop somewhat independently based on what they encounter and how this can affect cognitive development. For this, researchers view the brain as an information processor like a computer and try to explain mental development in terms of, essentially, hardware and software development (e.g., they may view the brain's functionality as being analogous to that of a computer operating system). For instance, they have shown that mental development is affected by improved abilities in attention, an increased capacity of working memory (think of this as analogous to RAM in a computer), and a faster speed of processing as a child develops. From a software perspective, they propose that we obtain strategies and rules for solving specific problems over time and this contributes to our mental development. Moreover, one set of studies proposes that a portion of mental development depends on our ability to turn "implicit memories" or procedures (ones that we are not conscious of but use regularly) into "explicit" ones that we have knowledge of and can then modify to improve them. This process of "explication" or "proceduralization" gives us creative problem solving capabilities.

How can our neural network be flexible enough (plastic) to learn, yet be "stable" so that when we learn new information, old information is not forgotten? In some organisms researchers have hypothesized that infants learn easier, but then as they grow up, it becomes more difficult to learn so that stability is maintained. Others have hypothesized that the degradation in learning capabilities as we age is due to evolutionary selective pressures (presumably the old have already reproduced so they do not need the selective advantage of learning that the young do). Regardless of which scientific explanation turns out to be true, it is well known that our brains have a variety of plasticity characteristics where the actual topology of our neural network is adjusted to enable learning.

Finally, it should be noted that evolution has certainly had significant impacts on development. For instance, it seems reasonable to hypothesize that evolution must have created the drive in infants to explore. Significant research activities are under way, however, to further clarify the synergistic effects between evolution and development.

Development and Plasticity in Control and Automation?

Is there any value to using biomimicry of development and plasticity ideas in control and automation? To do this, it is profitable to think of development and plasticity as achieving a very general level of adaptation during an organism's lifetime. Do the ideas provide for a general view of achieving high levels of adaptability in control and automation systems? Consider the following questions:

Development can be viewed as a very flexible adaptation strategy.

- Could a control system recognize a major structural change in a plant and then rewrite its own code to expand itself to cope (e.g., by adding or deleting neurons or rules) with the new part of the system?
- If there were a failure of part of the controller, could the code reconfigure itself to maintain performance, even if the change demanded more than simple parameter adjustments?
- How would such high levels of adaptation affect the role of learning in expert, planning, and attentional systems?

Additional relevant discussion is in Section 12.8. Clearly, here we are only providing some general ideas for using biomimicry of development and plasticity. Relatively little research has been done on biomimicry of development and plasticity for control and automation.

9.5 Exercises and Design Problems

Exercise 9.1 (Neural Control for a Tanker Ship):

(a) Simulate the neural controller ($n_R = 121$) described in the chapter and produce all the plots shown for the nominal, plant change, and disturbance conditions given there.

(b) Repeat (a) but for the case where $n_R = 9$.

(c) Repeat (a) and (b) but for the case where you choose a threshold $\epsilon > 0$ for testing whether a particular radial basis function parameter should be updated. Justify your choice via simulations.

Exercise 9.2 (Neural Control for Cargo Ship Steering):

(a) Develop an $n_R = 121$ reinforcement learning radial basis function based neural controller for the cargo ship from Design Problem 9.1.

Simulate the closed-loop system to demonstrate its performance for the nominal, plant change, and disturbance cases considered in the chapter. You should be able to tune the controller to achieve qualitatively similar performance to the tanker ship case, when comparing to the reference model in each case.

(b) Repeat (a) but use $n_R = 9$.

Design Problem 9.1 (FMRLC for Tanker and Cargo Ship Steering): In this problem, we will study the use of the FMRLC for steering various ships.

(a) Verify the results in the chapter for the tanker ship by simulating the FMRLC under the same conditions.

(b) The key difference in the code for the FMRLC compared to the standard fuzzy controller lies in the rule base update mechanism, since the fuzzy inverse model is simply coded in a similar manner to a standard direct fuzzy controller. In particular, the code fragment that implements the rule base update for the code that produces the plots for the tanker ship in the chapter is

```
for k=(meme_int(d)-meme_count(d)+1):meme_int(d)
for l=(memc_int(d)-memc_count(d)+1):memc_int(d)
  rules(k,l)=rules(k,l)+p(index);
end
end
```

Using the code provided, explain what the `e_int` and `e_count` variables are used for (in your explanation, draw the rule base table and show an example of what the `e_int` and `e_count` variables could be for one particular value of e and c inputs). The variables `c_int` and `c_count` are used in a similar manner for the c universe of discourse. Next, explain what the variables `meme_int(d)` and `meme_count(d)` are for. The variables `memc_int(d)` and `memc_count(d)` are used in a similar manner for the c universe of discourse. Explain why the particular ranges

```
k=(meme_int(d)-meme_count(d)+1):meme_int(d)
```

and

```
l=(memc_int(d)-memc_count(d)+1):memc_int(d)
```

are used in the rule base update formula given above (do not ignore the "+1"). To do this, use the rule base table drawing you produced and give an example for $d = 1$ of what values `meme_int(d)`, `meme_count(d)`, `memc_int(d)`, and `memc_count(d)` could hold. (You do not need to do the actual calculations, just use values to explain what these variables would hold to demonstrate that you understand the code.)

(c) Design a 9-rule fuzzy controller that uses center-average defuzzification for the tanker ship. To start, use the code that generates the plots in the chapter (i.e., from (a)) and let $g_p = 0$ specify the membership functions and rule base, and tune the 9-rule (nonadaptive) fuzzy controller to get performance similar to what we got for the 121-rule nonadaptive case (i.e., the "good" response we got in Part II that has little overshoot). Plot the responses for the closed-loop system and fuzzy controller surface. Then, for your tuned values, show the response of the closed-loop system for a tanker with a different weight (i.e., when it is under "full" conditions rather than "ballast" ones). To do this, simply change the plant parameters and rerun the simulation (do not attempt to tune for the case where the ship is full; "optimize" your design by tuning only for the ballast case and simply test the resulting design on the full case).

Next, design an FMRLC for the tanker that will tune the 9-rule fuzzy controller that you just developed. Begin with the code that was used to generate the plots in the chapter. However, use a rule base update scheme where you multiply the update suggested by the inverse model by the premise certainty d steps ago (i.e., use $b_m(kT) = b_m(kT - dT) + \mu_m(e(kT - dT), c(kT - dT))p(kT))$. Use $d = 1$ as we had in the chapter. First, try to tune the FMRLC by leaving the scaling gains on the fuzzy controller alone and focus on the tuning of the inverse model gains (if, however, this does not work, you may have to go back and change the scaling gains on the fuzzy controller, then retune the inverse model gains). To study performance, test both the nominal case and the case where the weight of the ship changes at $t =$ 9000 sec. from ballast to full conditions. You should tune the FMRLC so that it overcomes the plant changes and gets good tracking (at least asymptotically), much better than the tracking you found above in the nonadaptive case where there was a plant condition change. Plot the closed-loop responses for the nominal and weight change cases, and the tuned fuzzy controller surface in each case and discuss. Hints: (i) start with the "good" tuned values for the scaling gains for the controller and fuzzy inverse model developed in the chapter for the 121-rule case, and initialize the fuzzy controller with the one you developed above for the nominal case; (ii) do not change the number of rules in the inverse model, or any other aspect of its design, except for possible tuning of its input-output scaling gains; (iii) note that when you change the number of membership functions, you must check the remainder of the code to make sure that the change will work (it will not—you must also change the specification of the base widths and membership function centers on the e and c universes of discourse, and the way that the certainties are computed from the membership functions on the outermost edges of the e and c universes of discourse); and (iv) the more interesting change relates to making the code compute $\mu_m(e(kT-dT), c(kT-dT))$ for use in the rule base update $b_m(kT) = b_m(kT-dT) + \mu_m(e(kT-dT), c(kT-dT))p(kT)$ where you must modify the code to save the four values of the premise membership functions that were on d steps ago when you make the computations for the fuzzy controller, and then modify the knowledge update procedure to use these in the rule base update.

(d) In this part, you will study the steering of a "cargo ship" defined in [30]. Use the nonlinear model of the tanker ship provided in Equation (4.5) but with $K_0 = -3.86$, $\tau_{10} = 5.66$, $\tau_{20} = 0.38$, $\tau_{30} = 0.89$, and $l = 161$ meters. Assume the rudder saturates at ± 80 deg. and that we have the same types of disturbances as studied for the tanker ship (wind and speed changes). Also, we will assume that the cargo ship is traveling in the x direction at a velocity of 5 meters/sec. You should seek to get as good steering performance as possible, in an analogous way to how we did this in the chapter for the tanker ship (i.e., design an FMRLC that will achieve good performance even if there are significant disturbances and plant changes). Use a reference model

$$\frac{\psi_m(s)}{\psi_r(s)} = \frac{\frac{1}{100}}{s + \frac{1}{100}}$$

so that we are requesting a faster response from the cargo ship (reasonable since it is a smaller ship). Use the same reference input as was used for the tanker ship.

Begin by turning off the adaptation mechanism and tuning the nonadaptive fuzzy controller so that it gets a closed-loop response that is similar to that specified by the reference model (use the fuzzy controller developed for the tanker ship and tune its scaling gains). Show the closed-loop response and fuzzy controller surface for this case. Next, develop an FMRLC that performs well for nominal conditions, but also performs well without retuning when the wind disturbance is introduced, or when there is a speed change from 5 to 3 meters/sec. at $t = 4000$ sec. Show the closed-loop response in each case and the tuned fuzzy controller surface. Compare to the control surface for the nonadaptive case and discuss.

Design Problem 9.2 (Generic Adaptive Control for Ship Steering)*: Suppose that we consider an adaptive controller to be "generic" if it is meant to be able to achieve good performance for a whole class of applications, and in particular, in the case where a manufacturer will produce one type of product that requires a controller, and they want the same controller to be able to adapt and perform well if it is applied to a product they later design. For example, suppose that you work for a company that has designed controllers for ship heading for many years, and every few years they introduce a new ship, and so you have to go through the costly process of controller redesign each time. Design a generic adaptive control strategy that will achieve high performance operation, whether it is installed on a cargo ship or tanker. Suppose that you first design the controller for the cargo ship, but that you intend to later use it when you design a controller for the tanker (that will be released in a few years so you do not have a model when you are designing the cargo ship controller and hence, you cannot evaluate its performance for the tanker ship when you are designing the cargo ship controller). You may only tune the controller on the cargo ship, but must be able to show that if it is later used on the tanker ship, it will adapt and achieve high performance operation. You may assume that you have a sensor that tells you that the controller is implemented on a different ship, but you cannot assume it gives you any more information than that (e.g., the weight or length of the new ship). Specify a manner to evaluate the performance of your generic adaptive controller, and illustrate its performance and operation in simulation. Hint: You may want to consider using a supervisory scheme to tune an adaptive controller.

Chapter 10

Linear Least Squares Methods

Chapter Contents

In this chapter, we introduce batch and recursive least squares methods for tuning approximator structures where the parameters that will be tuned enter linearly. In particular, we study the tuning of the $p \times 1$ vector θ for the linear in the parameters approximator

$$F_{lip}(x, \theta) = \theta^\top \phi(x)$$

where $\phi(x)$ is a known specified $p \times 1$ vector function. For the tuning, we use the given set of training data $G = \{(x(i), y(i)) : i = 1, 2, \ldots, M\}$.

Section 9.3 outlined several approximator structures that fit this form, including the linear approximator $F_l(x, \theta)$; the polynomial approximator $F_{poly}(x, \theta)$ where the coefficients are tuned; the multilayer perceptron $F_{mlp}(x, \theta)$ with one hidden layer, a linear activation function at the output, and known activation functions in the hidden layer; and the Takagi-Sugeno fuzzy system $F_{ts}(x, \theta)$ with known premise membership functions. In each case, the function $\phi(x)$ is known once x is specified, and the form of θ depends on which approximator structure you use.

The batch least squares method can be used to find approximator parameters that enter linearly when all training data is given a priori.

In this chapter, we simply focus on tuning of θ and will not concern ourselves (except in the examples) with which of the approximator structures is used to implement the approximator (i.e., we will not focus on the construction of $\phi(x)$).

10.1 Batch Least Squares

First, we derive the least squares solution to the approximation problem. Then we provide a simple example where we fit a line to data, and a more interesting example where we train a multilayer perceptron and Takagi-Sugeno fuzzy system to match the function in Figure 9.10.

10.1.1 Batch Least Squares Derivation

In the batch least squares method, we define

$$Y(M) = [y(1), y(2), \ldots, y(M)]^\top$$

to be an $M \times 1$ vector of output data where the $y(i)$, $i = 1, 2, \ldots, M$ come from G (i.e., $y(i)$ such that $(x(i), y(i)) \in G$). We let

$$\Phi(M) = \begin{bmatrix} \phi^\top(x(1)) \\ \phi^\top(x(2)) \\ \vdots \\ \phi^\top(x(M)) \end{bmatrix}$$

be an $M \times p$ matrix that is constructed by stacking the $1 \times p$ $\phi^\top(x(i))$ vectors into a matrix (i.e., the $x(i)$ are such that $(x(i), y(i)) \in G$). Let $\epsilon(i) = y(i) - F_{lip}(x(i), \theta) = y(i) - \theta^\top \phi(x(i))$, which is the same as

$$\epsilon(i) = y(i) - \phi^\top(x(i))\theta$$

be the error in approximating the data pair $(x(i), y(i)) \in G$ where θ is used in the approximation structure. Define

$$E(M) = [\epsilon(1), \epsilon(2), \ldots, \epsilon(M)]^\top$$

so that

$$E = Y - \Phi\theta$$

Choose

$$J(\theta, G) = \frac{1}{2} E^\top E$$

to be a measure of how good the approximation is for all the data in G for a given θ. $J(\theta, G)$ is the sum of the squares of the errors in approximation for each of the training data pairs. We want to pick θ to minimize $J(\theta, G)$ and that is why we use the term "least squares." It is "linear" least squares since our approximator is linear in the parameters.

Notice that $J(\theta, G)$ is convex in θ so that a local minimum is a global minimum. Next, we seek to find the value of θ that will achieve the global minimum. Using basic ideas from calculus, if we take the partial derivative of J with respect to θ and set it equal to zero, we get an equation for θ, the best estimate (in the least squares sense) of the unknown θ^*. Leaving this approach to the derivation (which depends on the use of vector calculus) to a homework exercise, we take a simple (matrix) algebraic approach to the minimization by noting that

$$2J = E^\top E = Y^\top Y - Y^\top \Phi\theta - \theta^\top \Phi^\top Y + \theta^\top \Phi^\top \Phi\theta$$

Then, we "complete the square" by assuming that $\Phi^\top \Phi$ is invertible and letting

$$\begin{aligned} 2J &= Y^\top Y - Y^\top \Phi\theta - \theta^\top \Phi^\top Y + \theta^\top \Phi^\top \Phi\theta \\ &\quad + Y^\top \Phi(\Phi^\top \Phi)^{-1} \Phi^\top Y - Y^\top \Phi(\Phi^\top \Phi)^{-1} \Phi^\top Y \end{aligned}$$

(where we are simply adding and subtracting the same terms at the end of the equation). Hence,

$$\begin{aligned} 2J &= Y^\top (I - \Phi(\Phi^\top \Phi)^{-1} \Phi^\top) Y \\ &\quad + (\theta - (\Phi^\top \Phi)^{-1} \Phi^\top Y)^\top \Phi^\top \Phi (\theta - (\Phi^\top \Phi)^{-1} \Phi^\top Y) \end{aligned} \tag{10.1}$$

The parameters computed via batch least squares minimize the sum of the squared error between the approximator output and the training data outputs; however, it only adjusts the parameters that enter linearly to achieve this minimization.

The first term in this equation is independent of θ, so we cannot reduce $J(\theta, G)$ via this term, so it can be ignored. Hence, to get the smallest value of $J(\theta, G)$, we choose θ so that the second term is zero. We will denote the value of the parameters that achieves the minimization of J by θ, and we notice that

$$\theta = (\Phi^\top \Phi)^{-1} \Phi^\top Y \tag{10.2}$$

since the smallest we can make the last term in the above equation is zero (since it is positive). This is the equation for batch least squares that shows we can directly compute the least squares estimate θ from the "batch" of data that are taken from G and loaded into Φ and Y. If we pick the inputs to the system so that it is "sufficiently excited" [331], then we will be guaranteed that $\Phi^\top \Phi$ is invertible; if the data come from a linear mapping with p (the number of parameters in the linear in the parameters approximator) as the number of underlying linear terms in the nonlinear function, then for sufficiently large M we will achieve perfect estimation of the plant parameters.

In "weighted" batch least squares, we use

$$J(\theta, G) = \frac{1}{2} E^\top W E \tag{10.3}$$

where, for example, W is an $M \times M$ diagonal matrix with its diagonal elements $w_i > 0$ for $i = 1, 2, \ldots, M$ and its off-diagonal elements equal to zero. These w_i can be used to weight the importance of certain elements of G more than others. For example, we may choose to have it put less emphasis on older data by choosing $w_1 < w_2 < \cdots < w_M$ when $x(2)$ is collected after $x(1)$, $x(3)$ is collected after $x(2)$, and so on. One way to select the weights in this case is to suppose that $0 < \lambda \leq 1$, then let $w_i = \lambda^{M-i}$, $i = 1, 2, \ldots, M$. In any case, the resulting parameter estimates can be shown to be given by

$$\theta_{wbls} = (\Phi^\top W \Phi)^{-1} \Phi^\top W Y \tag{10.4}$$

To show this, simply use Equation (10.3) and proceed with the derivation in the same manner as above.

10.1.2 Numerical Issues in Computing the Estimate

In practical problems, numerical issues often arise in computing the inverse

$$(\Phi^\top \Phi)^{-1}$$

needed to compute the batch least squares solution due to $\Phi^\top \Phi$ being "ill-conditioned." Such issues can arise even for relatively simple "academic" problems. For example, these issues arise in the examples to be considered in this book where we typically use the Matlab "backslash" operation to compute the least squares estimate as

```
theta = Phi \ Y
```

where `theta` is θ, `Phi` is Φ, and Y is `Y`. Basically, most view the inverse in Equation (10.2) as a statement of how the least squares estimate is found theoretically. In practice, direct computation of the inverse is generally not used.

In practical applications, numerical issues in computing the least squares estimate must be confronted.

To avoid numerical issues you have several options. First, if you can select $x(i)$ explicitly (which you often cannot, either due to physical limitations of the mapping you are trying to learn, or because you cannot pick $x(i)$ because it it provided by another system), then you can avoid the problems. To do this, basically you want to choose the $x(i)$ so that the $\phi(x(i))$ that are loaded into Φ have values that are aligned in such a way that the inverse can be computed (i.e., so that $\Phi^\top \Phi$ is positive definite, with a good "condition number"). Without getting into details, one way to get "rich" enough data so that the inverse is computable is to use noise as the components of $x(i)$. Of course, this is not always possible, so we often have to turn to other methods.

For instance, a common approach to solve numerical problems with computing the least squares solution is to use a "square root" method. The details of the variety of possible methods and their advantages and disadvantages are beyond the scope of this discussion; however, if you run into numerical problems, you can basically proceed in four ways. First, you can rely on an existing software package to provide a numerically sound solution (i.e., perhaps you should not just employ a direct method to computing the inverse but use more sophisticated methods). Second, you can see the "For Further Study" section at the end of this part to find references to learn more about how to overcome numerical problems. Third, you could turn to an RLS (or gradient) approach to process the data sequentially (e.g., by cycling several times through the data set G) as we explain in the next sections. Fourth, you could use the singular value decomposition approach that we discuss next, whose solution has interesting and useful properties.

It is possible to provide the least squares solution whether or not $\Phi^\top \Phi$ is invertible. The common approach to doing this is to use the singular value decomposition (SVD) method. If Φ is an $M \times p$ matrix and U and V are $M \times M$ and $p \times p$ "unitary" matrices, respectively (i.e., $U^\top U = I$ and $V^\top V = I$ so $U^{-1} = U^\top$ and $V^{-1} = V^\top$), then the SVD of Φ is

$$U^\top \Phi V = \begin{bmatrix} \Sigma & 0 \\ 0 & 0 \end{bmatrix} = S$$

where S is $M \times p$, the "0" elements in S are, in general, matrices (what are their dimensions?),

$$\Sigma = diag(\sigma_1, \sigma_2, \ldots, \sigma_r)$$

where

$$\sigma_1 \geq \sigma_2 \geq, \cdots, \geq \sigma_r > 0$$

are the "singular values" and $r = rank(\Phi)$.

The least squares estimate is then

$$\theta = (\Phi^\top \Phi)^{-1} \Phi^\top Y = V \begin{bmatrix} \Sigma^{-1} & 0 \\ 0 & 0 \end{bmatrix} U^\top Y$$

Note that the matrix

$$\begin{bmatrix} \Sigma^{-1} & 0 \\ 0 & 0 \end{bmatrix}$$

in this computation is a $p \times M$ matrix. Also, note that the SVD computes $(\Phi^\top \Phi)^{-1} \Phi^\top$, which is the "pseudoinverse" of matrix Φ. To see that this is a valid computation for the least squares estimate, recall that

$$J(\theta, G) = \frac{1}{2} E^\top E = \frac{1}{2} (Y - \Phi\theta)^\top (Y - \Phi\theta)$$

and since U and V are unitary,

$$J(\theta, G) = \frac{1}{2}\left[U^\top (Y - \Phi V^\top V\theta)\right]^\top \left[U^\top (Y - \Phi V^\top V\theta)\right]$$

Now, let

$$V^\top \theta = \bar{v} = \begin{bmatrix} \bar{v}_1 \\ \bar{v}_2 \end{bmatrix}$$

where $\bar{v}_1$ is $r \times 1$ and $\bar{v}_2$ is $(n-r) \times 1$, and

$$U^\top Y = \bar{u} = \begin{bmatrix} \bar{u}_1 \\ \bar{u}_2 \end{bmatrix}$$

where $\bar{u}_1$ is $r \times 1$ and $\bar{u}_2$ is $(M-r) \times 1$. Note that since we can choose θ to minimize $J(\theta, G)$, we can choose $\bar{v}$. Since

$$U^\top \Phi V = \begin{bmatrix} \Sigma & 0 \\ 0 & 0 \end{bmatrix}$$

we know that

$$\begin{aligned} J(\theta, G) &= \left[\begin{bmatrix} \bar{u}_1 \\ \bar{u}_2 \end{bmatrix} - \begin{bmatrix} \Sigma & 0 \\ 0 & 0 \end{bmatrix}\begin{bmatrix} \bar{v}_1 \\ \bar{v}_2 \end{bmatrix}\right]^\top \left[\begin{bmatrix} \bar{u}_1 \\ \bar{u}_2 \end{bmatrix} - \begin{bmatrix} \Sigma & 0 \\ 0 & 0 \end{bmatrix}\begin{bmatrix} \bar{v}_1 \\ \bar{v}_2 \end{bmatrix}\right] \\ &= \begin{bmatrix} \bar{u}_1 - \Sigma\bar{v}_1 \\ \bar{u}_2 \end{bmatrix}^\top \begin{bmatrix} \bar{u}_1 - \Sigma\bar{v}_1 \\ \bar{u}_2 \end{bmatrix} \end{aligned}$$

To get $J(\theta, G)$ as small as possible, choose

$$\bar{v}_1 = \Sigma^{-1}\bar{u}_1$$

and also choose $\bar{v}_2 = 0$ (since we can choose it to be anything we would like). We have $V^\top \theta = \bar{v}$ so

$$\theta = V\bar{v} = V\begin{bmatrix} \Sigma^{-1}\bar{u}_1 \\ 0 \end{bmatrix} = V\begin{bmatrix} \Sigma^{-1} & 0 \\ 0 & 0 \end{bmatrix}\begin{bmatrix} \bar{u}_1 \\ \bar{u}_2 \end{bmatrix} = V\begin{bmatrix} \Sigma^{-1} & 0 \\ 0 & 0 \end{bmatrix} U^\top Y$$

In the weighted batch least squares case, with W a diagonal matrix with all positive numbers on its diagonal, if you let

$$W = \sqrt{W}\sqrt{W}$$

we know that $\sqrt{W} = \sqrt{W}^\top$. Hence, $\theta_{wbls} = (\Phi^\top \sqrt{W}\sqrt{W}\Phi)^{-1}\Phi^\top \sqrt{W}\sqrt{W}Y$, and if we let $\bar{\Phi} = \sqrt{W}\Phi$ and $\bar{Y} = \sqrt{W}Y$, we have $\theta_{wbls} = (\bar{\Phi}^\top \bar{\Phi})^{-1}\bar{\Phi}^\top \bar{Y}$ and so you can use the same approach as above.

Finally, it is interesting to note that even in the case where $M < p$, where we have the "underdetermined case," the singular value decomposition will provide a solution even though in this case, there are an infinite number of θ solutions. Actually, out of the infinite number of possible solutions to the linear least squares problem in this case, the θ computed via the singular value decomposition is the one solution such that $\theta^\top \theta$ has the smallest possible size (so sometimes it is a reasonable choice).

10.1.3 Example: Fitting a Line to Data

As an example of how batch least squares can be used, suppose that we would like to use this method to fit a line to a set of data. Suppose that $n = 1$. In this case, our parameterized linear (polynomial) approximator is

$$y = F_{lip}(x, \theta) = \theta^\top \phi(x) = \theta^\top [\phi_1(x), 1]^\top = \theta_1 x_1 + \theta_2 \tag{10.5}$$

which is an equation for a line (note that the 1 in the second row of $\phi(x) = [\phi_1(x), 1]^\top$ is used to include the affine term θ_2). Suppose that the data that we would like to fit the line to is given by

$$G = \{(1, 1), (2, 1), (3, 3)\}$$

and that these data were generated from an unknown function $G(x, z)$ (we assume that they are numbered from left to right, so that $(x(1), y(1)) = (1, 1)$). Notice that $M = 3$.

We will use Equation (10.2) to compute the parameters for the line that best fits the data (in the sense that it will minimize the sum of the squared distances between the line and the data). First, let

$$Y = \begin{bmatrix} 1 \\ 1 \\ 3 \end{bmatrix}$$

Next, form the $\phi(x(i))$, $i = 1, 2, 3$, and let

$$\Phi = \begin{bmatrix} 1 & 1 \\ 2 & 1 \\ 3 & 1 \end{bmatrix}$$

With this,

$$\theta = (\Phi^\top \Phi)^{-1} \Phi^\top Y = \left(\begin{bmatrix} 14 & 6 \\ 6 & 3 \end{bmatrix} \right)^{-1} \begin{bmatrix} 12 \\ 5 \end{bmatrix} = \begin{bmatrix} 1 \\ -\frac{1}{3} \end{bmatrix}$$

Hence, the line

$$y = x_1 - \frac{1}{3}$$

best fits the data in the least squares sense.

To see that the line fits the data, consider Figure 10.1, where we plot both the data in G and the line $F_{lip}(x, \theta)$. Clearly, the data were not generated by a linear mapping. The least squares method tries to overcome this problem, and results in a good fit to the data, the best in the least squares sense that is possible for a linear approximator. Notice that the line is raised up toward the two points above it, balancing out the error that is created in the approximation, considering that there is only one point below it.

The same general approach works for larger data sets. The reader may want to experiment with weighted batch least squares to see how the weights w_i affect

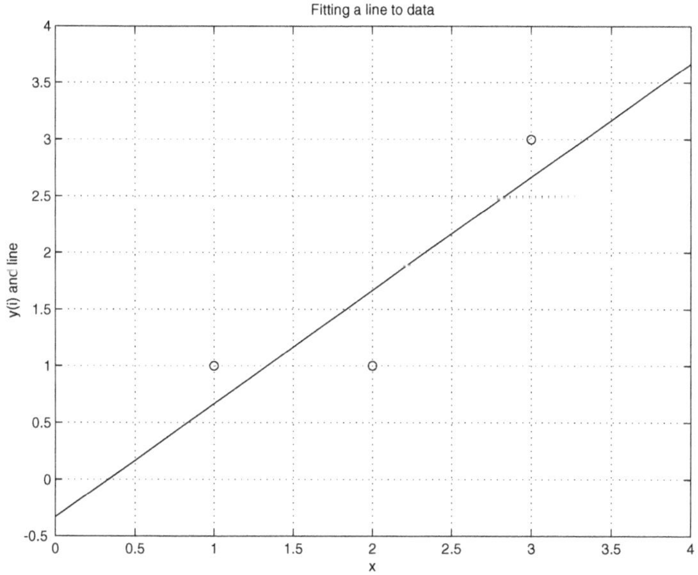

Figure 10.1: Training data and line that is the best fit to the data.

the way that the line will fit the data (making it more or less important that the data fit at certain points). In doing this, you will see that you can, by various choices of the weighting factors, move the line so that it more closely matches any point that you put a relatively high weight value on.

10.2 Example: Offline Tuning of Approximators

In this section we will show how to use batch least squares to tune a multilayer perceptron and Takagi-Sugeno fuzzy system to match the training data shown in Figure 9.10 (this defines G and in our case, we have $M = 121$). In particular, we will first use the same multilayer perceptron and Takagi-Sugeno fuzzy system as studied in Section 9.3 and compare the approximation accuracy when a least squares approach is used to tune the parameters that enter linearly to the approximation accuracy that we obtained via manual tuning.

It is good practice to first try a linear in the parameter approximator, or even one that is linear in its inputs.

10.2.1 Multilayer Perceptrons

Improved Accuracy Over the Manually Tuned Neural Network

Recall that we were using the perceptron with a single hidden layer shown in Figure 9.13 with $n_1 = 2$ neurons in the hidden layer. This is represented by

$$y = F_{mlp}(x, \theta) = \theta^\top \phi(x) = [w_1, w_2, b][\phi_1(x), \phi_2(x), 1]^\top$$

where via our heuristic approach, we used $f(\bar{x}) = \frac{1}{1+\exp(-\bar{x})}$ and had chosen

$$\phi_1(x) = f(b_1 + w_{1,1}x)$$

with $b_1 = 0$ and $w_{1,1} = 1.5$, and

$$\phi_2(x) = f(b_2 + w_{1,2}x)$$

with $b_2 = -6$ and $w_{1,2} = 1.25$. We had chosen $\theta = [3, 1, 0.6]^\top$. We will use the batch least squares approach to see how it can pick a better θ.

Batch least squares is often a very effective method for computing the parameters that enter linearly; however, it relies on your choice of the parameters that enter nonlinearly.

To do this, we simply form the matrices Y and Φ and use the batch least squares formula to find

$$\theta = [2.5747, 1.6101, 0.7071]^\top$$

which, when we use these values for the approximator parameters, results in the approximator shown with the training data in Figure 10.2. The approximation accuracy is clearly better than what we obtained via manual tuning (see Figure 9.15) and the batch least squares method provided an automatic method to pick some of the parameters, in particular, w_1, w_2, and b. For the other parameters we relied on our heuristic tuning discussed earlier.

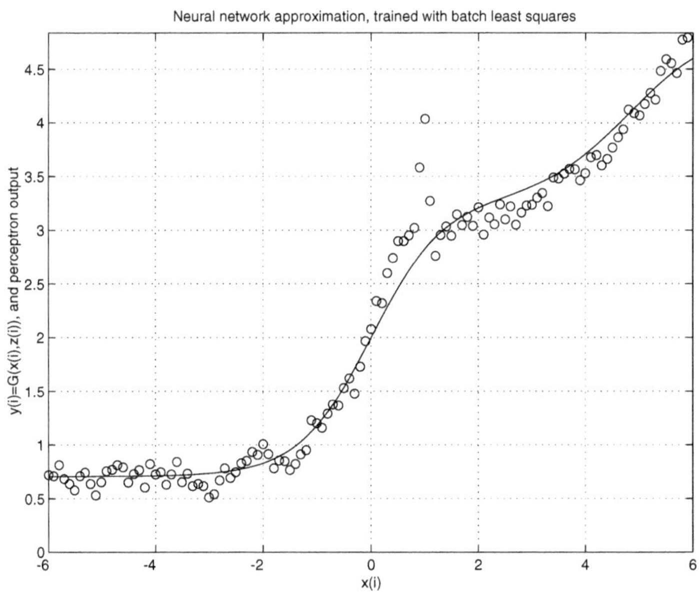

Figure 10.2: Multilayer perceptron approximator trained with batch least squares, 2 neurons.

Seeking More Approximation Accuracy: Increasing the Number of Hidden Neurons

The main reason for not considering more neurons in the hidden layer when we were considering manual tuning of the perceptron for this example was that the tuning can become complicated due to interactions between the tuning parameters. With the assistance of batch least squares, however, we can easily tune approximators with more parameters. Generally, you want to use much more training data than parameters (to avoid what is called "overfitting" below) so since we use $M = 121$, we will now consider $n_1 = 11$ neurons in the hidden layer (for a total of $11(2) + 11 + 1 = 34$ parameters).

Notice, however, that we need a scheme to pick the weights and biases of the hidden layer. To do this, we will use a simple heuristic approach (others are possible, some suggested by the application at hand). To pick the biases, we choose them to be evenly spaced over -5 to 5, so that $b_1 = -5$, $b_2 = -4$, all the way to $b_{11} = 5$. This should help spread the points where the activation functions turn on across the input space. The choice for the w^j, $j = 1, 2, \ldots, 11$, is more difficult if you take the view that we did in the manual tuning of the perceptron. Notice that there we assumed that we could examine the training data and pick off slopes to set these values. This is often unrealistic for complex real world problems. Here, we will exploit the fact that the scaling factors in w are used to modify the slopes to what we will need, so we simply pick $w^j = 1$, $j = 1, 2, \ldots, 11$ (for applications where $n > 1$, this scheme may not be as effective; in those cases, you will want the weights to take on values that will allow for a range of slopes). This completes the specification of the hidden layer.

Next, we use batch least squares to tune the 12 parameters in $\theta = [w^\top, b]^\top$. We get

$$\begin{aligned}\theta &= [2.7480, 2.0120, -11.9865, 34.7556, -69.6968, 93.4042,\\ &\quad -80.8496, 57.0819, -34.6710, 15.8048, -3.9398, 0.8087]\end{aligned}$$

For this case we get the approximation shown in Figure 10.3, which is a significant improvement over Figure 10.2, where we used $n_1 = 2$ neurons in the hidden layer and Figure 9.15, where we tuned the approximator manually. Notice that in the vector θ, we have both positive and negative values. The negative ones help to implement the parts of the nonlinearity where the slope goes negative. Clearly, it would be quite difficult to tune the approximator manually to get this kind of accuracy.

Fine-Tuning to Capture High Frequency Behavior

Next, to illustrate what can happen if you use even more parameters in your approximator, we use $n_1 = 25$ neurons (to get a total of $25(2) + 25 + 1 = 76$ parameters). We choose the biases in a similar fashion to the above, but spread them over the whole range -6 to 6 to get $b_1 = -6$, $b_2 = -5.5$, all the way to $b_{25} = 6$. As above, we pick all the weights in the hidden layer to be unity. We use batch

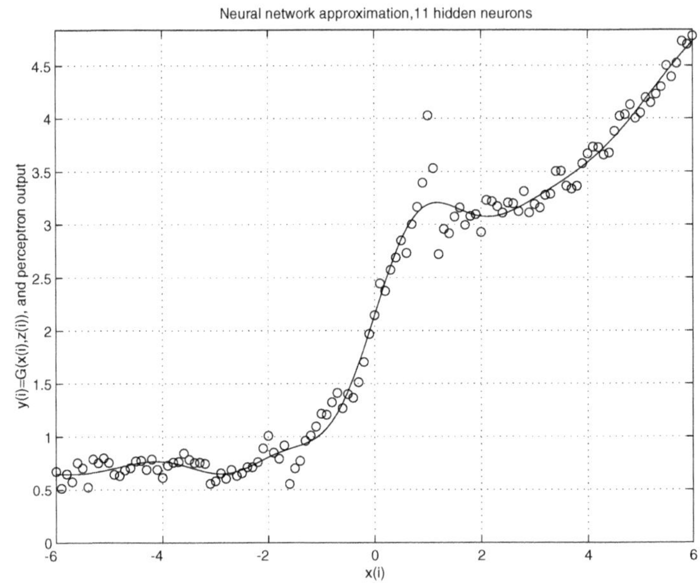

Figure 10.3: Multilayer perceptron approximator trained with batch least squares, 11 neurons.

Generally, using a larger approximator structure can improve approximation accuracy; however, if you use a structure that is too complex, it can be too aggressive in trying to represent the noise (overfitting) rather than seeking to achieve a good interpolation.

least squares to tune the 26 parameters in $\theta = [w^\top, b]^\top$. For this case, we get the approximation shown in Figure 10.4, which is an improvement over Figure 10.3, where we used $n_1 = 11$ neurons (notice that the approximator is starting to find some of the structure of the underlying function that is illustrated in Figure 9.9; least squares is particularly good at finding this structure, *in this case*, due to how the noise on z enters). Also, notice that with more neurons we are able to approximate more and more of the "high frequency" behavior in the function (with even more neurons, perhaps concentrated in the region around 1, we can get an even more accurate approximation of the peaking behavior found in that region).

Overfitting Where the Approximator Seeks to Model Noise

Next, we show that this approach of continually increasing n_1 can be taken too far. Suppose that we choose $n_1 = 121$ (for a total of $121(2) + 121 + 1 = 364$ parameters), $b_1 = -6$, $b_2 = -5.9$, all the way to $b_{121} = 6$, and the weights in the hidden layer as all unity. In this case, we have θ as a 122×1 vector so that we have more parameters to tune than data pairs. We use batch least squares to train the network and the result is shown in Figure 10.5. Notice that in this plot, we have also plotted approximator nonlinearity on top of the function $G(x)$ (i.e., where we have removed the effects of the noise z). This illustrates a very important fact: if you use too many parameters, you may start trying

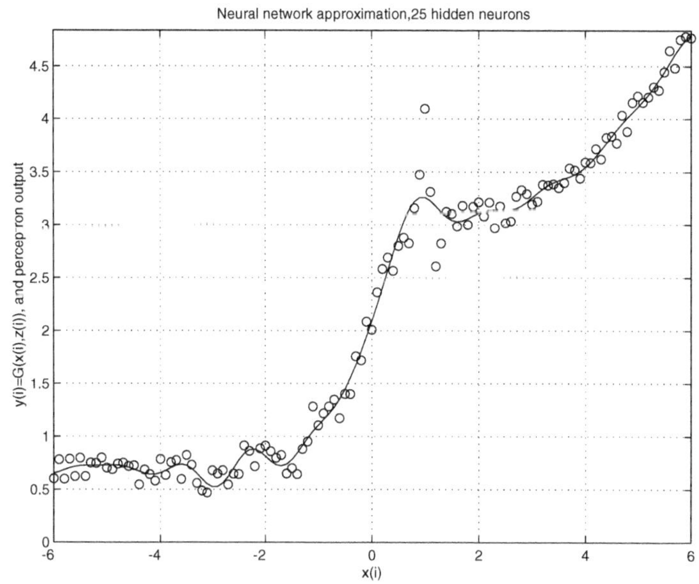

Figure 10.4: Multilayer perceptron approximator trained with batch least squares, 25 neurons.

to approximate characteristics of the noise, and not the underlying function. Even without the presence of z, it is possible to get similar "overfitting" where in between the training data, the approximator moves far away from where it should be (but for this example, if you train without the influence of z in the data, the approximator will do a very good job at approximating the function and does not exhibit this problem). Basically, this highlights the fact that there are often situations where it is desirable to capture some of the higher frequency behavior, but not behavior that is too high a frequency since this may represent uncertainty (noise) in the system.

10.2.2 Takagi-Sugeno Fuzzy Systems

In this section, we study how to tune the Takagi-Sugeno fuzzy system to match the function in Figure 9.10. Here, however, we will not consider the many different cases as we did for the neural network in the last section since the same basic ideas apply (least squares offers a nice automated method for tuning, additional parameters can be used to achieve improved accuracy, and if you use too many parameters, you can get a type of overfitting). Instead, our focus will simply be on how to construct the premise membership functions.

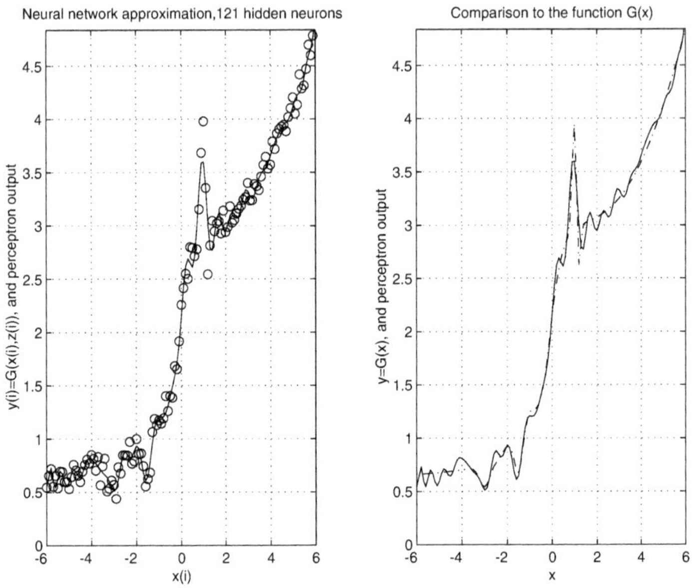

Figure 10.5: Multilayer perceptron approximator trained with batch least squares, 121 neurons, plus comparison to $G(x)$.

Getting Similar Accuracy to the Neural Network

Suppose we use $R = 20$ rules so that we will have $4R = 80$ parameters to tune, a number close to the 76 parameters used for the neural network above, with 25 neurons in the hidden layer. It is interesting to note that we will tune 40 values of the Takagi-Sugeno fuzzy system compared to 26 for the perceptron with $n_1 = 25$ neurons in the hidden layer. For this reason, we will have fewer parameters to tune manually (i.e., the function ϕ for the Takagi-Sugeno fuzzy system takes fewer parameters to specify than the one for the neural network).

Comparisons between approximator structure types must include complexity of the structure, ease of training, and approximation accuracy.

For the Takagi-Sugeno fuzzy system, we have to pick the parameters for

$$\mu_i(x) = \exp\left(-\frac{1}{2}\left(\frac{x_j - c_j^i}{\sigma_j^i}\right)^2\right)$$

where $j = 1$ (since $n = 1$) and $i = 1, 2, \ldots, R$. A logical strategy is to space the c_1^i points on a uniform grid across the x axis (especially in cases where you do not know the form of the underlying function; for this example, since we can easily examine the data, it would make more sense to use a nonuniform distribution of the c_1^i points, with more concentrated where there is more high frequency behavior). To do this, for convenience, we choose to spread the 20 c_1^i points across the range $[-5.4, 6]$ in increments of 0.6. Next, we pick all the $\sigma_1^i = 0.1$. This gives us the ξ_i functions shown in Figure 10.6 and the approximator shown in Figure 10.7. Notice that with our choice of $\sigma_1^i = 0.1$, we get very steep slopes between the basis functions so that they switch somewhat abruptly from one line for the approximator to the next. This results in the somewhat erratic behavior in the plot.

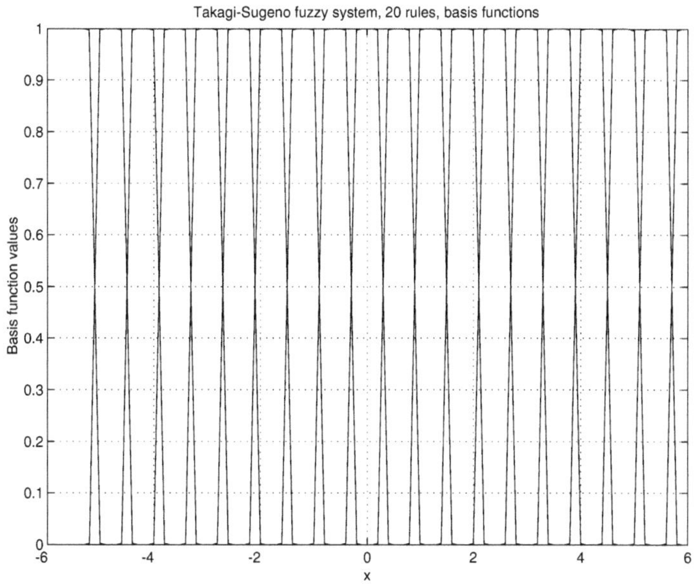

Figure 10.6: Takagi-Sugeno basis functions, $R = 20$, $\sigma_1^i = 0.1$ case.

Manually Tuning the Nonlinear Part of the Approximator

If we pick $\sigma_1^i = 1$, we get the ξ_i functions shown in Figure 10.8 (why are they not perfectly symmetric?) and the approximator shown in Figure 10.9. This shows that the value of $\sigma_1^i = 1$ provides for a much smoother transition between basis functions, which results in smoother transitions between the lines used for approximation. Overall, in terms of approximator accuracy, we obtain results similar to those obtained for the perceptron with $n_1 = 25$ neurons in the hidden layer; however, this may not always be the case. Sometimes, one approximator will be able to achieve better accuracy with fewer parameters.

There exist good intuitive ideas on how to manually tune the nonlinear part of the approximator structure.

Overall, this shows some ideas on how to tune the premise membership functions (that extend to the more general case where $n > 1$). As a final note, we caution against using this discussion—and that given in the last section—to draw general conclusions about which approximator structure to use. In general, different applications will dictate the need for different approximator structures, numbers of parameters to tune, and methods to tune them.

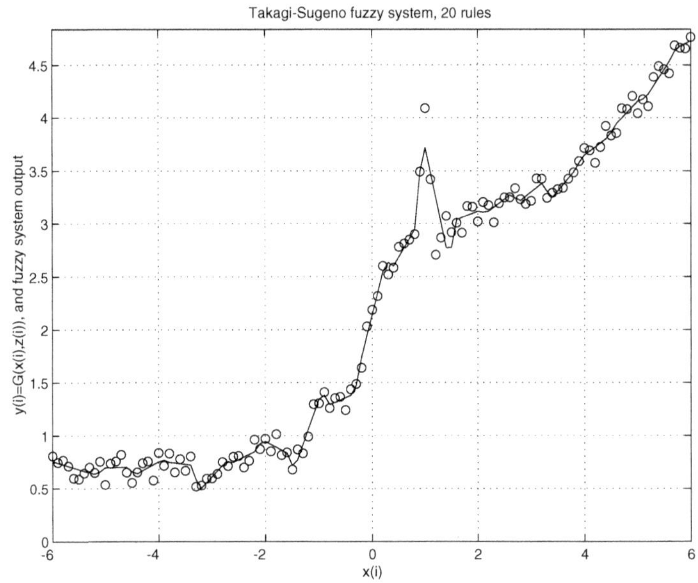

Figure 10.7: Takagi-Sugeno approximator, $R = 20$, $\sigma_1^i = 0.1$ case.

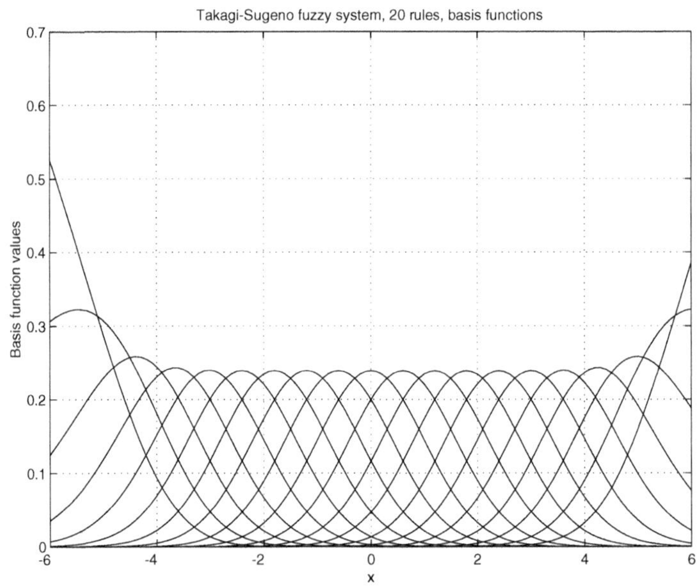

Figure 10.8: Takagi-Sugeno basis functions, $R = 20$, $\sigma_1^i = 1$ case.

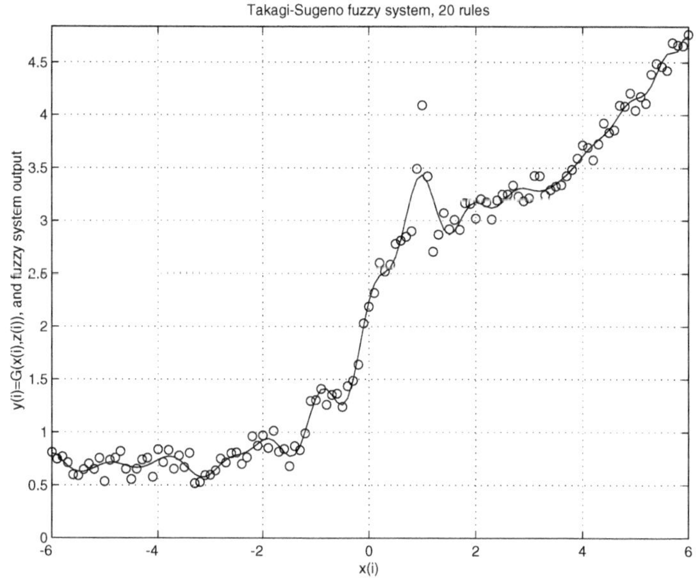

Figure 10.9: Takagi-Sugeno approximator, $R = 20$, $\sigma_1^i = 1$ case.

10.3 Design Example: Rule Synthesis Using Operator Data

In this problem, taken directly from [498] (where other estimation methods are studied), suppose you are given data from how a human operator controls a chemical plant (see the Web site for this book to get the data set). See Figure 10.10, where we suppose that the operator has measurements of monomer concentration (u_1), change in monomer concentration (u_2), monomer flow rate (u_3), some local temperatures in the plant (u_4, u_5), and with these makes decisions on how to select the set point for the monomer flow rate (y). The actual value of the monomer flow rate to be put into the plant is controlled by a PID controller and the value of y is the set point for that controller.

It is possible to construct a fuzzy (or neural) controller from a set of numeric examples of how an expert human would solve the problem. This offers another nonmodel-based strategy to construct a nonlinear controller.

In Chapter 5 we studied how to construct a fuzzy controller using heuristic ideas about how the plant behaves. Here, we take a different approach where we gather plant data (that actually represents the heuristic control ideas of the operator about how to control the plant) and create an interpolator for these data using a fuzzy controller. After appropriate testing, this controller could then be put into operation either to provide advice to novice operators or to completely replace the expert operator.

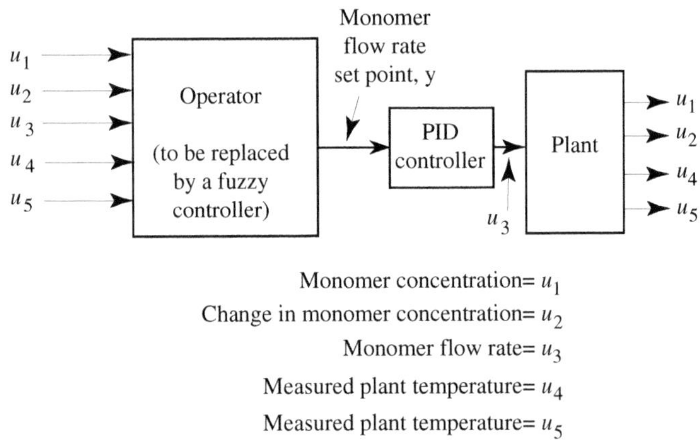

Figure 10.10: Operator for controlling a plant.

10.3.1 Data Analysis, Correlation Analysis, and Controller Input Selection

There are $M = 70$ data pairs that were obtained by monitoring how the operator performs this task. In the data set, the first 5 columns hold $u_i(k)$, $i = 1, 2, 3, 4, 5$, and each row corresponds to a different time k. The last column holds the corresponding set point values $y(k)$ that are determined by the operator. The data are shown in Figure 10.11. In practical applications, it is often good to plot the data and examine them. Here, it is interesting to note that we clearly may not have enough data to perform a good approximation over a wide range of values of the inputs since we do not have output settings for a very wide range of input combinations. Moreover, by examining the plots more carefully you may suspect that the operator is not using all five data values to make decisions (the operator is the expert, so while the data might be available, the operator may not use it to make decisions since the operator may have found a few key variables are the important ones to consider).

From our examination of the data, we begin by performing some data analysis to study how the operator makes decisions. In particular, using the approach in [343], we calculate the correlation coefficients between each input and the output (and in fact, between all the different variables) and we show this in Figure 10.12. Now, while this is a linear analysis, it does give an indication of which inputs are important to the operator in making decisions. Notice that $u_4(k)$ and $u_5(k)$ do not have a high correlation with the output $y(k)$ (the magnitudes of the correlation coefficients are less than about 0.2 for both cases), so this leads us to suspect that the operator is ignoring these inputs in his decision-making process (perhaps the operator could be asked if this is the case). Moreover, $u_2(k)$ has a correlation coefficient of only about 0.33 so it does not seem to be a key variable for decision-making either. Notice, however, that $u_1(k)$ and

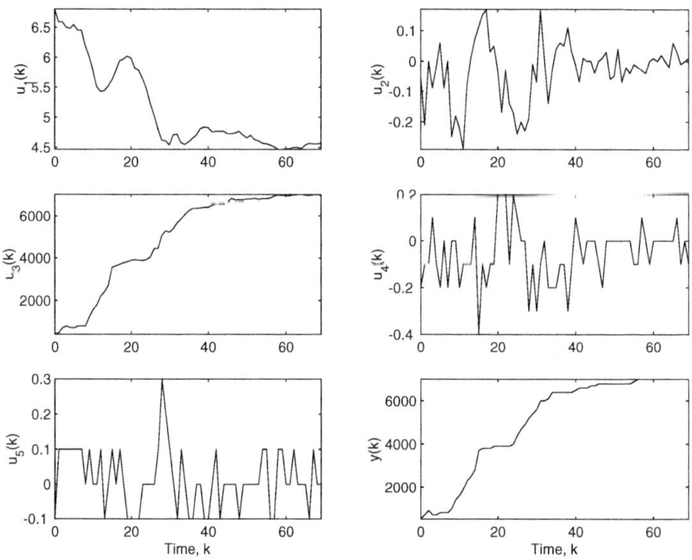

Figure 10.11: Data that indicate how the operator selects the set point for the monomer flow rate.

$u_3(k)$ have correlation coefficients that are close to 1 in magnitude and this indicates that each of these variables seems to be important in the decision-making process. Recall that $u_3(k)$ is the monomer flow rate (the output of the PID controller) so we expect a correlation with $y(k)$, the set point for the controller (the correlation indicates that the PID controller is successful in forcing the actual monomer flow rate to be equal to the one that is commanded by the operator). The input $u_1(k)$ is the monomer concentration and seems to be a key variable for decision-making.

While the above analysis is instructive, it is also important to consider the cross-correlation between the inputs that we decide to keep as inputs to the controller. If one input is significantly correlated to another one that you want to keep, then it may be that they are carrying basically the same information so it might be possible to remove one of them. For instance, the correlation coefficient between $u_1(k)$ and $u_3(k)$ is -0.9381 so since its magnitude is near 1, it seems that removing one of these inputs is possible. From the physics of the problem, it does not make sense to *only* use the input $u_3(k)$ as an input to the controller since it is the actual value of the monomer flow rate that is input to the plant, and its value is directly dictated by the monomer flow rate set point that is set by the controller to be constructed. Hence, when we only want to use one input variable, we will consider the case where we remove $u_3(k)$ and hence, only use $u_1(k)$ as the input to the controller. We will, however, also consider the use of other inputs as you will see below.

Before proceeding, however, note that there are other methods for selecting

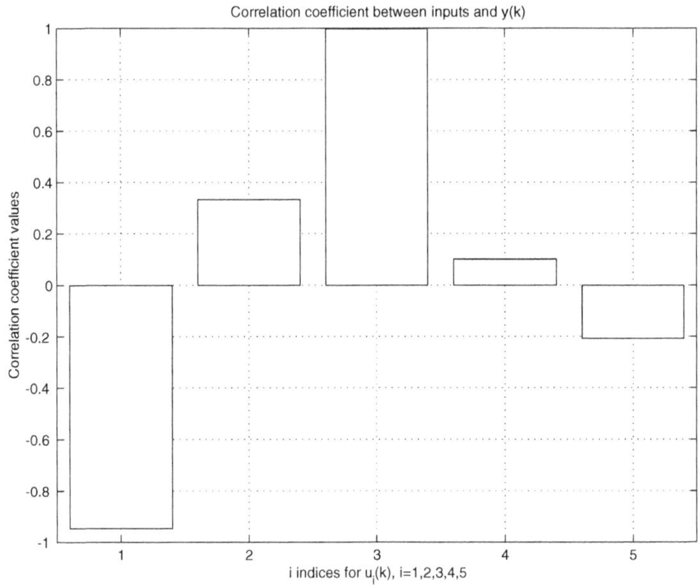

Figure 10.12: Correlation coefficients between each input, $u_i(k)$, $i = 1, 2, 3, 4, 5$, and the output $y(k)$.

inputs to the controller and that these of course also apply to general function approximation problems (note that we are essentially trying here to pick the regressor vector length and composition). While here we use the approach in [343], you could, in some situations (e.g., when you have plenty of training data and not too many inputs), simply use an exhaustive approach where you train approximators for all possible combinations of input variables. Another approach is to normalize the data so that they all lie between -1 and 1, and then construct a linear least squares estimator between the inputs and the output and consider the magnitude of the regressor coefficients. Then you can discard regressor components that have coefficients that are small in magnitude. Note that even though this is also a linear analysis approach, it can lead to different conclusions from the correlation analysis above. Moreover, all this analysis is complicated by the fact that the conclusions that you reach can depend on the controller that you end up constructing (e.g., you may have two sets of inputs to choose between, and your analysis may say that one is better than the other, but you may not be able to construct a nonlinear approximator that performs better for that set of inputs).

10.3.2 Determine if a Linear Controller Is Sufficient

We start by trying to use a linear (actually affine) mapping to fit the operator data so that we get a linear (affine) controller. We do this first for two reasons.

First, if the linear controller performs reasonably well, then we will guess that the linear correlation analysis of the last subsection is valid. Second, if the linear controller works well, then we will want to use it since it is simpler to implement than a fuzzy controller.

Due to the lack of a significant amount of training data, we will train the approximator using all $M = 70$ data pairs (this specifies G). This approach, however, creates problems with validating the accuracy of the approximation since we can only test at the data that the approximator was trained at. Here, since we cannot access the plant to generate more data, we will artificially generate a test data set. To do this, we simply create data points in between each of the given data points by taking the average value of two adjacent points (i.e., average value of each component), and associating it with the average value between two output data points. This will give us $M_\Gamma = 69$ test data pairs in our test set Γ. We will treat these values as if they were actually generated in an experimental setting.

Using a linear least squares method to train an affine approximator structure, we get the results shown in Figure 10.13. For this, if $F(x, \theta)$ is the affine approximator mapping with θ chosen using batch least squares and we use all the inputs so

$$x(k) = [u_1(k), u_2(k), u_3(k), u_4(k), u_5(k)]^\top$$

we get a mean squared error at the training data of

$$\frac{1}{M} \sum_{(x,y) \in G} ((y - F(x, \theta))^2 = 1.1142 \times 10^4$$

and we get a mean squared error at the test data of

$$\frac{1}{M_\Gamma} \sum_{(x,y) \in \Gamma} ((y - F(x, \theta))^2 = 8.6598 \times 10^3$$

Note that the mean squared error values at the training and testing data are similar, but in this case the training error is higher (this is a bit atypical; normally the test error is slightly higher).

Notice that we achieve reasonable approximation accuracy, but there are several points at which there are significant deviations between what the operator did and what the linear controller does (suppose that the operator feels that the errors are "significant"). We could conclude from this, however, that a linear estimator does reasonably well, so we place more confidence in our earlier correlation analysis. From this, we suspect that we may be able to remove input variables and achieve similar approximation accuracy.

10.3.3 Study the Effects of Removing Input Variables

We could study the performance of the approximator by successively removing more input variables. Here, we will trust the earlier correlation analysis and first consider a two-input linear controller that only uses $u_1(k)$ and $u_3(k)$ as

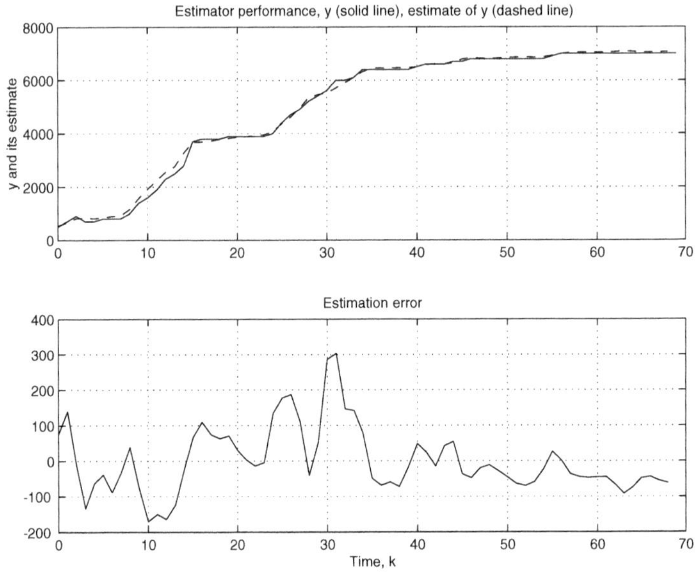

Figure 10.13: Operator settings, linear controller settings, and error between these.

inputs (the case for using inputs $u_1(k)$, $u_2(k)$, and $u_3(k)$ is similar, with just a slightly worse approximation error than the case where we use all the inputs). After that we will consider the case where we only use the input $u_1(k)$.

Using a linear least squares method to train an affine approximator structure with only two inputs, we get the results shown in Figure 10.14. For this, if $F(x, \theta)$ is the affine approximator mapping with θ chosen using batch least squares and

$$x(k) = [u_1(k), u_3(k)]^\top$$

we get a mean squared error at the training data of

$$\frac{1}{M} \sum_{(x,y)\in G} ((y - F(x, \theta))^2 = 1.1669 \times 10^4$$

and we get a mean squared error at the test data of

$$\frac{1}{M_\Gamma} \sum_{(x,y)\in \Gamma} ((y - F(x, \theta))^2 = 9.1087 \times 10^3$$

Notice that our mean squared error did not increase drastically even though we removed three inputs.

Notice, however, that if we only use $u_1(k)$ as an input, then using a linear least squares method to train an affine approximator structure with only one

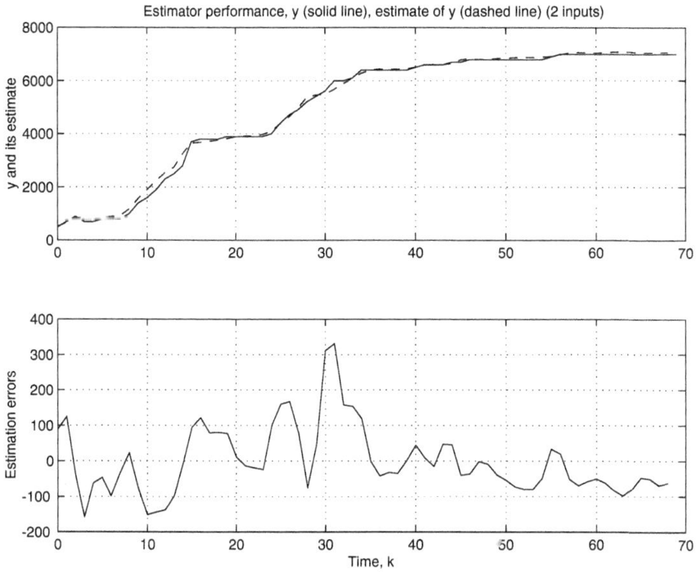

Figure 10.14: Operator settings, linear controller settings, and error between these, $u_1(k)$ and $u_3(k)$ as inputs.

input, we get the results shown in Figure 10.15. For this, if $F(x, \theta)$ is the affine approximator mapping with θ chosen using batch least squares and

$$x(k) = [u_1(k)]^\top$$

we get a mean squared error at the training data of

$$\frac{1}{M} \sum_{(x,y) \in G} ((y - F(x, \theta))^2 = 5.2090 \times 10^5$$

and we get a mean squared error at the test data of

$$\frac{1}{M_\Gamma} \sum_{(x,y) \in \Gamma} ((y - F(x, \theta))^2 = 5.0309 \times 10^5$$

Notice that in this case, we get a significant degradation in performance. You could be led to several different conclusions. First, you may think, via the earlier correlation analysis, that even though the cross-correlation between $u_1(k)$ and $u_3(k)$ was high, there was still some important information in the $u_3(k)$ input that we are now ignoring. In that case it would seem that the operator is primarily looking at two inputs to make decisions. However, there is a second possibility that is important to consider. It is possible that the linear approach is failing. In particular, it could be that the errors for the single-input

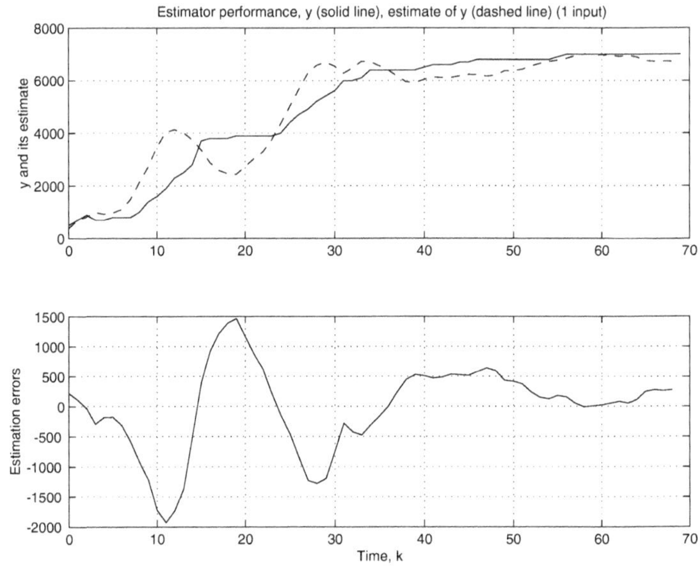

Figure 10.15: Operator settings, linear controller settings, and error between these, $u_1(k)$ as an input.

case are arising due to the fact that there is a nonlinearity in the underlying operator decision-making that the linear mapping is not suited to represent. It is for this reason that we turn to a nonlinear approximator, the fuzzy system, and try to use only one input (of course you could also construct a neural network in an analogous manner). We keep in mind, however, that if we do not succeed in this approach, we will try to add a second input, $u_3(k)$.

10.3.4 Construct a Fuzzy Controller from Operator Data

Next, we attempt to reduce the approximation errors so that the decisions made are closer to those of the operator than what we were able to obtain with a linear approximator, no matter how many inputs were used. We first construct a single-input fuzzy controller. We will, in fact, construct a Takagi-Sugeno fuzzy system with Gaussian input membership functions and affine consequent functions, both with only one input. When this is done, using $R = 9$ rules and one choice for the membership function parameters, we do reduce the approximation error (we get a mean squared testing error of 3.0702×10^5), but not significantly, and it is *worse* than the cases in the last section, where we also used $u_2(k)$ and $u_3(k)$ as inputs. Now, you could increase the number of membership functions on the input universe of discourse to try to improve accuracy; however, we will take a different approach here since the development of the linear approximators indicates that the inputs $u_2(k)$ and $u_3(k)$ do carry some information.

First Attempt: Problems with Overfitting

Since we have found via the correlation analysis that $u_1(k)$ seems to be the most informative input variable, we will use it as an input to the premise membership functions; however, for the consequent membership functions we will use either $u_1(k)$, $u_2(k)$, and $u_3(k)$, or all the inputs to try to get as much information from the inputs as possible. Notice that this will increase the complexity of the approximator, but for R rules there are only $R(n+1)$ consequent parameters (n is the number of inputs to the consequent membership functions). For R rules, with only one input to the premise membership functions, there are only $2R$ parameters needed to define the membership functions. Notice that if you used all the inputs to the premise membership functions, and all possible combinations of rules (that results from gridding the input space with N membership functions on each input dimension), then we need $2R$ parameters but in this case, $R = N^n$ so that the approximator can easily become very complex. (If we used $n = 5$ inputs with $N = 3$ membership functions on each input universe of discourse, then there would be $R = 3^5 = 243$ rules which would be defined with 486 parameters, which is far greater than M, not even considering the additional parameters needed for the consequent functions.)

Here, we will consider two cases. In both cases we will use one input to the premise membership function and $R = 9$ rules, so we will need 18 parameters to define the input membership functions. We will, however, consider different numbers of inputs to the consequent functions. First, we will consider using 3 inputs to the consequent functions ($u_1(k)$, $u_2(k)$, and $u_3(k)$); hence, with $n = 3$ we will need $R(n+1)$ parameters for a total of $18 + 9(4) = 54$ parameters in the approximator. In the second case, we will consider using all the inputs to the consequent functions so that there will be $18 + 9(6) = 72$, which is greater than $M = 70$; hence, in the second case we must be especially concerned that the approximator will "overfit" the data and hence not generalize well in between the data.

We use a grid on the input space of 9 input membership functions so we get $R = 9$ (we omit the actual values of the centers and spreads of the Gaussian input membership functions and invite the reader to solve this problem in a design problem at the end of the chapter). We use a linear least squares method to train the Takagi-Sugeno fuzzy system approximator. In the first case, we use $u_1(k)$, $u_2(k)$,and $u_3(k)$ as inputs to the consequent functions and get the results shown in Figure 10.16. For this, if $F(x, \theta)$ is the Takagi-Sugeno fuzzy system approximator mapping with θ chosen using batch least squares, we get a mean squared error at the training data of

$$\frac{1}{M} \sum_{(x,y) \in G} ((y - F(x, \theta))^2 = 2.2077 \times 10^3$$

and we get a mean squared error at the test data of

$$\frac{1}{M_\Gamma} \sum_{(x,y) \in \Gamma} ((y - F(x, \theta))^2 = 1.1329 \times 10^4$$

Note that the testing error is significantly higher than the training error. This result shows that there is some overfitting occurring (in between the training data there are some excursions where the interpolation is not performing very well). If we proceeded according to our plan and used all the inputs in the consequent functions, we find that this overfitting problem gets significantly worse so we do not present those results (we get a mean squared training error of 594.1489 and a mean squared testing error of 6.3452×10^9). Clearly, the fact that we have more parameters to tune than training data is causing a significant problem in this last case. This example clearly shows the importance of using both training and testing sets; if you only used the training data set you would think that you had significantly improved approximation accuracy when in fact all you have done is match the training data very well. While in operation, when data different from the training data are encountered, the controller could provide very unreasonable inputs.

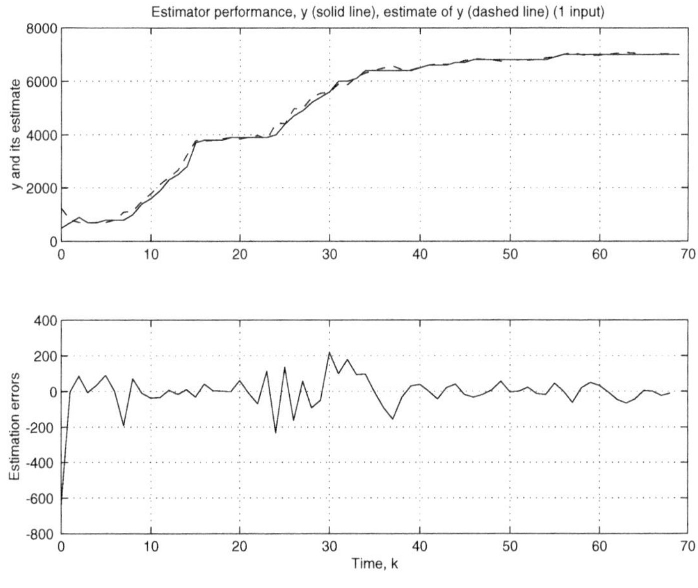

Figure 10.16: Operator settings, fuzzy controller settings, and error between these (one input to the premise membership functions, three to the consequent functions).

Before we continue with the design process for the approximator, consider Figure 10.17 where, using ideas from [343], we see that by including the $u_1(k)$, $u_2(k)$, and $u_3(k)$ inputs in producing the result in Figure 10.16, they are uncorrelated with the estimation error (notice that while this correlation analysis is again linear, it does take into account the nonlinear mapping implemented by the Takagi-Sugeno fuzzy system). Notice also that the $u_4(k)$ and $u_5(k)$ inputs are only a bit correlated with the approximation error so that we expect that if we add these inputs, they probably would not help much with approximation accuracy (but this is just a guess based on the linear analysis).

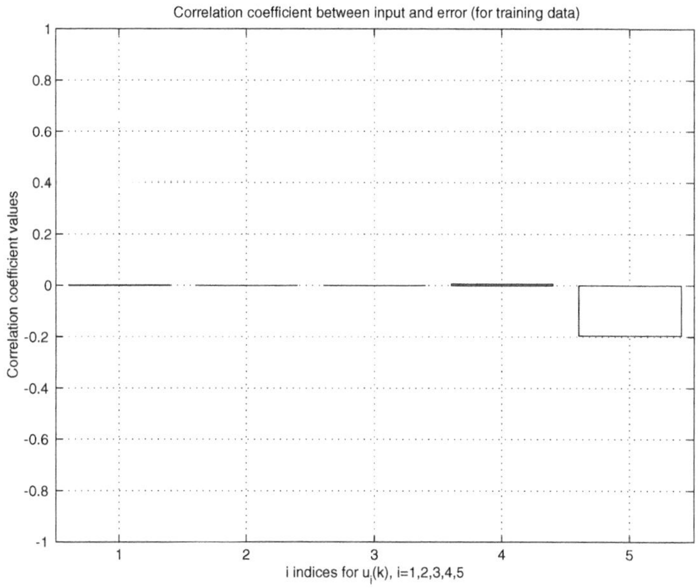

Figure 10.17: Correlation coefficients between each input, $u_i(k)$, $i = 1, 2, 3, 4, 5$, and the output approximation error, using the training data (one input to the premise membership functions and three to the consequent functions).

Second Attempt: A Good Controller

Notice that we have not improved the approximation accuracy over the previous cases by using a nonlinear approximator. How do we improve the accuracy? We could certainly try to improve the accuracy by tuning the premise membership function parameters or adding more rules (but we are limited in this last approach by the small amount of training data). Another approach would be to use more inputs to the premise membership functions. Recall from the past section that if we add such inputs, we can quickly increase the number of rules and hence the number of parameters in the approximator; therefore, we only add one more input, $u_3(k)$. This approach may make sense since then it will provide for a nonlinear map between two variables that the operator seems to be using in decision-making.

In this case, we grid the membership functions on the input space and get the results shown in Figure 10.18. For this, if $F(x, \theta)$ is the Takagi-Sugeno fuzzy system approximator mapping with θ chosen using batch least squares, we get a mean squared error at the training data of

$$\frac{1}{M} \sum_{(x,y)\in G} ((y - F(x,\theta))^2 = 4.0401 \times 10^3$$

and we get a mean squared error at the test data of

$$\frac{1}{M_\Gamma} \sum_{(x,y)\in \Gamma} ((y - F(x,\theta))^2 = 3.1657 \times 10^3$$

which is significantly better than any of the controllers that we have constructed so far.

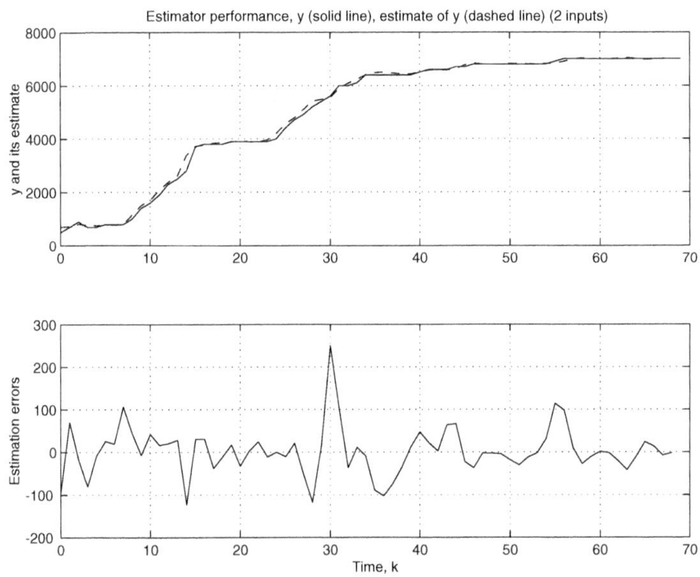

Figure 10.18: Operator settings, fuzzy controller settings, and error between these (u_1 and u_3 inputs to the premise membership functions and consequent functions).

Before we continue with the design process for the approximator, consider Figure 10.19 where, using ideas from [343], we see that by including the $u_1(k)$ and $u_3(k)$ inputs in producing the result in Figure 10.18, they are uncorrelated with the estimation error. Notice also that the $u_2(k)$, $u_4(k)$ and $u_5(k)$ inputs are only a bit correlated with the approximation error so that we guess that if we add these inputs, they probably would not help much with approximation accuracy.

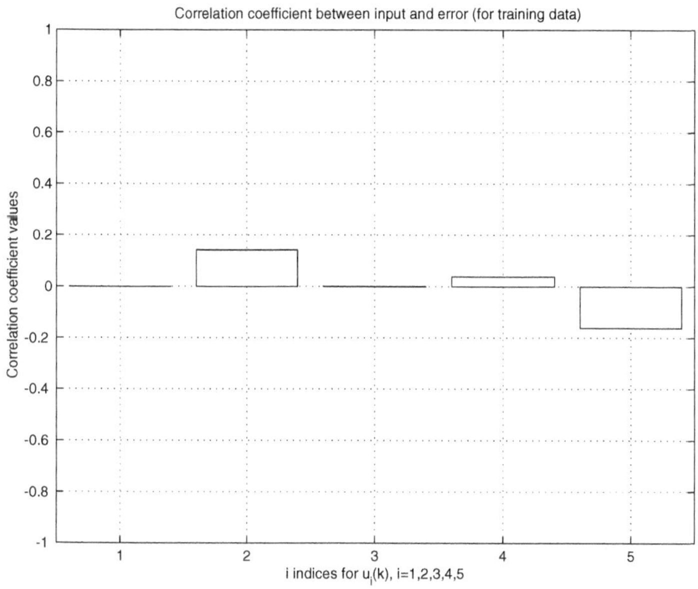

Figure 10.19: Correlation coefficients between each input, $u_i(k)$, $i = 1, 2, 3, 4, 5$, and the output approximation error, using the training data (one input to the premise membership functions and three to the consequent functions).

Third Attempt: No, Nothing Better

At this point, we return to our original correlation analysis and evaluate if there are other possibilities for improving on the performance. Recall that the analysis indicated that if we kept u_1 as an input, then we may not need to keep u_3 since these are correlated quite strongly. Earlier, in the construction of the linear estimators, we found that we could eliminate all the variables but these and we would still do pretty well, and that if we eliminated u_3, and hence only used u_1, there was a significant decrease in performance. This led us to conclude that there must be some useful information in the u_3 variable.

There is, however, a different line of reasoning that can be used. Note that u_1 is certainly a useful variable and so suppose that we use it as an input. Then, based on constraints due to approximator complexity in relation to the size of the training data set, and the problems we encountered in overfitting, we could try to pick a second variable that has the highest correlation with the output, but the lowest correlation with u_1. Which variable is this? Recall that u_2 had the third highest correlation with the output (a value of 0.3343), but it has correlation with u_1 of -0.1906. Note that u_4 has a relatively low correlation with the output of only 0.1012 (the u_5 variable had a correlation of -0.2068) but a correlation with u_1 of only -0.0283 (and the u_5 variable had a correlation with u_1 of 0.0837). To take a different approach, suppose that due to the low correlation between u_4 and u_1, we try to use u_4 as an input (u_5 may also work).

We will simply take the same approach as in our "second attempt," but we will use u_4 as an input, rather than u_3. We try different numbers of inputs, but do not find better results than earlier. Hence, while the line of reasoning above made sense, it did not end up helping to improve approximator accuracy for this approximator, and this training strategy. It may have been a good approach if we had used a different approximator or different training method. Why even discuss a case that does not work well? It helps to illustrate the normal process that you encounter in a real-world problem. Generally, you need to establish a logical approach to construction or improvement of an approximator, and try out the approach. It may or may not work better than your previous approach. Sometimes you win, sometimes you lose!

10.3.5 Methods to Test Generalization/Extrapolation and Controller Validity

While the best Takagi-Sugeno fuzzy system (as measured by the mean squared error) that we constructed in the last subsection gave a good approximation error, it could be that in between the training and testing data there are large excursions in the fuzzy controller mapping that intuitively may not be reasonable interpolations. Similarly, there could be large excursions in certain regions where there is a need to extrapolate since there is not good data in those regions. For low dimensional cases, it is possible to test the validity of the approximator by visually inspecting the approximator mapping (or perhaps a few dimensions can be studied at a time). Another alternative is to analytically determine the maximum slope of the approximator mapping. Or, as an approximation, you could numerically determine the maximum slope of the function on a fine grid of input data. In particular, in this approach, you would numerically compute an approximation to

$$\frac{\partial F(x, \theta)}{\partial x}$$

for the value of θ that was used in the approximator. You may want to study the data and analyze two different cases, one where the x is in a region where we had training data (to test generalization), and the other where x is in a region where there were no training data (to test extrapolation).

It is important to note that often an integral part of the validation of the controller will be to consult the operator and ask if it is reasonable. To do this, it may be convenient to convert the $R = 9$ rules that were trained into a type of linguistic equivalent. To do this, you could first assign linguistic values to the input membership functions that were specified for the Takagi-Sugeno fuzzy system. Now, if you used a standard fuzzy system you could assign linguistics to the output membership functions; then the rules would be simple to explain to the operator to get their "approval" of the rules. When we use a Takagi-Sugeno fuzzy system, you need to discuss this with the operator as being a "smooth switching" between the use of different linear (affine) functions of the inputs. See more details in [498].

10.4 Recursive Least Squares

While the batch least squares approach has proven to be very successful for a variety of applications, it is by its very nature a "batch" approach (i.e., all the data are gathered, then processing is done). For small M, we could clearly repeat the batch calculation for increasingly more data as they are gathered, but the computations can become prohibitive due to the computation of the inverse of $\Phi^\top \Phi$ and due to the fact that the dimensions of Φ and Y depend on M. Next, we derive a recursive version of the batch least squares method that will allow us to update our θ estimate each time we get a new data pair, without using all the old data in the computation and without having to compute the inverse of $\Phi^\top \Phi$. This "recursive least squares" approach allows us to implement an online function approximator, as we will illustrate in our examples in the next section.

The recursive least squares method can be used to tune the approximator parameters that enter linearly, with adjustments to the approximator mapping made online as each new training data pair is obtained.

10.4.1 Recursive Least Squares Derivation

Since we will be considering successively increasing the size of G, and we will assume that we increase the size by one at each time step, we assume that $(x(i), y(i))$ as gathered at time $k = i$. At time $k = 0$ we have no data. Suppose that $k = M$ so that you have gathered M pieces of training data and for $k \geq 1$, let the $p \times p$ matrix

$$P(k) = (\Phi^\top \Phi)^{-1} = \left(\sum_{i=1}^{k} \phi(x(i))\phi^\top(x(i)) \right)^{-1} \tag{10.6}$$

($P(k)$ is called the "covariance matrix"). We will define $P(0)$ when we explain how to initialize the recursive least squares algorithm. Assume that $\Phi^\top \Phi$ is nonsingular for all k. We have

$$P^{-1}(k) = \Phi^\top \Phi = \sum_{i=1}^{k} \phi(x(i))\phi^\top(x(i))$$

so we can pull the last term from the summation to get

$$P^{-1}(k) = \sum_{i=1}^{k-1} \phi(x(i))\phi^\top(x(i)) + \phi(x(k))\phi^\top(x(k))$$

and hence

$$P^{-1}(k) = P^{-1}(k-1) + \phi(x(k))\phi^\top(x(k)) \tag{10.7}$$

Now, the least squares estimate for k pieces of training data is $\theta(k)$, which using Equation (10.2), is

$$\theta(k) \quad = \quad (\Phi^\top \Phi)^{-1} \Phi^\top Y$$

$$
\begin{aligned}
&= \left(\sum_{i=1}^{k} \phi(x(i))\phi^\top(x(i))\right)^{-1} \left(\sum_{i=1}^{k} \phi(x(i))y(i)\right) \\
&= P(k)\left(\sum_{i=1}^{k} \phi(x(i))y(i)\right) \\
&= P(k)\left(\sum_{i=1}^{k-1} \phi(x(i))y(i) + \phi(x(k))y(k)\right) \qquad (10.8)
\end{aligned}
$$

Hence, from the second to last equation, if we shift the time index back one, we have

$$\theta(k-1) = P(k-1)\sum_{i=1}^{k-1} \phi(x(i))y(i)$$

If we multiply both sides of this equation by $P^{-1}(k-1)$, we get

$$P^{-1}(k-1)\theta(k-1) = \sum_{i=1}^{k-1} \phi(x(i))y(i)$$

Now, replacing $P^{-1}(k-1)$ in this equation with the result in Equation (10.7), we get

$$(P^{-1}(k) - \phi(x(k))\phi^\top(x(k)))\theta(k-1) = \sum_{i=1}^{k-1} \phi(x(i))y(i)$$

Using the result from Equation (10.8), this gives us

$$
\begin{aligned}
\theta(k) &= P(k)(P^{-1}(k) - \phi(x(k))\phi^\top(x(k)))\theta(k-1) + P(k)\phi(x(k))y(k) \\
&= \theta(k-1) - P(k)\phi(x(k))\phi^\top(x(k))\theta(k-1) + P(k)\phi(x(k))y(k) \\
&= \theta(k-1) + P(k)\phi(x(k))(y(k) - \phi^\top(x(k))\theta(k-1)). \qquad (10.9)
\end{aligned}
$$

This provides a method to compute an estimate of the parameters $\theta(k)$ at each time step k from the past estimate $\theta(k-1)$ and the latest data pair that we received, $(x(k), y(k))$. Notice that $(y(k) - \phi^\top(x(k))\theta(k-1))$ is the error in predicting $y(k)$ using $\theta(k-1)$.

To update θ in Equation (10.9), we need $P(k)$, so we could use

$$P^{-1}(k) = P^{-1}(k-1) + \phi(x(k))\phi^\top(x(k)) \qquad (10.10)$$

But then we will have to compute an inverse of a matrix at each time step (i.e., each time we get another data pair $(x(k), y(k))$). Clearly, this is not desirable for real time implementation, so we would like to avoid this. To do so, recall that the "matrix inversion lemma" indicates that if A, C, and $(C^{-1} + DA^{-1}B)$ are nonsingular square matrices, then $A + BCD$ is invertible and

$$(A + BCD)^{-1} = A^{-1} - A^{-1}B(C^{-1} + DA^{-1}B)^{-1}DA^{-1}$$

We will use this fact to remove the need to compute the inverse of $P^{-1}(k)$ that comes from Equation (10.10) so that it can be used in Equation (10.9) to update θ. Notice that

$$\begin{aligned} P(k) &= (\Phi^\top(k)\Phi(k))^{-1} \\ &= (\Phi^\top(k-1)\Phi(k-1) + \phi(x(k))\phi^\top(x(k)))^{-1} \\ &= (P^{-1}(k-1) + \phi(x(k))\phi^\top(x(k)))^{-1} \end{aligned}$$

and that if we use the matrix inversion lemma with $A = P^{-1}(k-1)$, $B = \phi(x(k))$, $C = I$, and $D = \phi^\top(x(k))$, we get

$$\begin{aligned} &P(k) = P(k-1) - \\ &P(k-1)\phi(x(k))(1 + \phi^\top(x(k))P(k-1)\phi(x(k)))^{-1}\phi^\top(x(k))P(k-1) \end{aligned} \tag{10.11}$$

Now, using the fact that

$$P(k)\phi(x(k)) = \frac{P(k-1)\phi(x(k))}{1 + \phi^\top(x(k))P(k-1)\phi(x(k))}$$

(to see this, substitute $P(k)$ from Equation (10.11) into $P(k)\phi(x(k))$ on the left side of this equation). This gives us

$$\begin{aligned} \theta(k) &= \theta(k-1) + K(k)\left(y(k) - \phi^\top(x(k))\theta(k-1)\right) \\ K(k) &= \frac{P(k-1)\phi(x(k))}{1 + \phi^\top(x(k))P(k-1)\phi(x(k))} \\ P(k) &= \left(I - K(k)\phi^\top(x(k))\right)P(k-1) \end{aligned} \tag{10.12}$$

which is called the "recursive least squares (RLS) algorithm." Basically, the matrix inversion lemma turns a matrix inversion into the inversion of a scalar (i.e., the term $(1+\phi^\top(x(k))P(k-1)\phi(x(k)))^{-1}$ is a scalar). Note that $K(k)$ is sometimes viewed as a time-varying gain on the prediction error $y(k)-\phi^\top(x(k))\theta(k-1)$ that dictates how $\theta(k-1)$ is adjusted to get $\theta(k)$.

We need to initialize the RLS algorithm (i.e., choose $\theta(0)$ and $P(0)$). One approach to do this is to use $\theta(0) = 0$ and $P(0) = P_0$ where $P_0 = \alpha I$ for some large $\alpha > 0$. This is the choice that is often used in practice. Other times, you may pick $P(0) = P_0$, but choose $\theta(0)$ to be the best guess that you have at what the parameter values are.

10.4.2 Weighted Recursive Least Squares: Using a Forgetting Factor

There is a "weighted recursive least squares" (WRLS) algorithm also. Suppose that the parameters of the physical system vary slowly. In this case, it may be advantageous to minimize

$$J(\theta, G)|_{M=k} = \frac{1}{2}\sum_{i=1}^{k} \lambda^{k-i}\left(y(i) - \phi^\top(x(i))\theta\right)^2$$

where $0 < \lambda \leq 1$ is called a "forgetting factor" since it gives the more recent data higher weight in the optimization (to see this, consider the effect of the term λ^{k-i} in the above summation). See the discussion on weighted batch least squares in the previous section.

Using a similar approach to the RLS case, you can show that the equations for WRLS are given by

$$\begin{aligned} \theta(k) &= \theta(k-1) + K(k)\left(y(k) - \phi^\top(x(k))\theta(k-1)\right) \\ K(k) &= \frac{P(k-1)\phi(x(k))}{\lambda + \phi^\top(x(k))P(k-1)\phi(x(k))} \\ P(k) &= \frac{1}{\lambda}\left(I - K(k)\phi^\top(x(k))\right)P(k-1) \end{aligned} \tag{10.13}$$

(where when $\lambda = 1$, we get the equation for standard RLS given above).

10.4.3 Numerical Issues and Covariance Modifications

Briefly, we note that for practical problems, you can have numerical problems with the computation of the (weighted) recursive least squares update algorithm. One solution to this problem is to employ the "factorization" methods that are highlighted in the "For Further Study" section at the end of this part. Here, we will discuss other issues that arise in the use of the recursive least squares method.

RLS with Covariance Resetting

One particular problem that has been encountered in implementations of the online recursive least squares algorithm (e.g., in adaptive control) is when the matrix $P(k)$ has elements that become too small so that $P^{-1}(k)$ is difficult to compute. This can occur, for instance, when you do not get data that gives sufficient information about the underlying unknown mapping. To be more concrete, notice that for RLS ($\lambda = 1$), we had derived in Equation (10.7) that

$$P^{-1}(k) = P^{-1}(k-1) + \phi(x(k))\phi^\top(x(k))$$

Notice that $\phi(x(k))\phi^\top(x(k))$ is a $p \times p$ matrix with squared terms so that they are always positive. From this it is easy to see that it is possible that the elements of $P^{-1}(k)$ can grow unbounded.

In this situation, sometimes a "covariance resetting method" is used where at each time instant k, you check to see if

$$\lambda_{min}(P(k)) < \delta_1$$

(where $\lambda_{min}(P(k))$ is the minimum eigenvalue of $P(k)$) for some fixed $\delta_1 > 0$ and if it is, then you let

$$P(k+1) = \delta_2 I$$

where I is the identity matrix and $\delta_2 \geq \delta_1$ (e.g., you could choose $\delta_2 = \alpha > \delta_1$ where α was used to initialize $P(0)$). This ensures that we keep the minimum eigenvalue of $P(k)$ above some fixed value so that $P(k)$ is positive definite so the inverse $P^{-1}(k)$ exists (i.e., it is bounded). Now, this modification results in a method that is not a pure least squares method (of course, between the resets, it is); however, in adaptive control it is often found to be adequate to maintain certain desirable closed-loop properties.

WRLS with Covariance Modification

It is interesting to note that for the WRLS method where $0 < \lambda < 1$, we must proceed differently. In this case, you can show that

$$P^{-1}(k) = \lambda P^{-1}(k-1) + \phi(x(k))\phi^\top(x(k))$$

(indeed, this is one step in the derivation of the WRLS formula in Equation (10.13)) so that $P^{-1}(k)$ will stay bounded (you can think of the above equation as a stable discrete time system with a bounded input so long as $\phi(x(k))$ is bounded, which it often is simply by the choice of the structure of the approximator). However, in this case, $P(k)$ may have elements that grow without bound. To see this, recall that we had derived

$$P(k) = \frac{1}{\lambda}P(k-1) - \frac{1}{\lambda}\frac{P(k-1)\phi(x(k))\phi^\top(x(k))P(k-1)}{\lambda + \phi^\top(x(k))P(k-1)\phi(x(k))}$$

Notice that while

$$\lambda + \phi^\top(x(k))P(k-1)\phi(x(k)) > 0$$

since $\lambda > 0$, it could be that

$$P(k-1)\phi(x(k))\phi^\top(x(k))P(k-1) = 0$$

Now, since $\frac{1}{\lambda} > 1$, it is possible that elements of $P(k)$ can grow without bound.

To avoid this, we can modify the WRLS algorithm. To do this, we use Equation (10.13) as an update formula for $P(k)$ so long as

$$||P(k)||_2 \leq \delta_1$$

for some $\delta_1 > 0$ such that

$$||P(0)||_2 < \delta_1$$

If, however, at some time k,

$$||P(k)||_2 > \delta_1$$

we simply update $P(k)$ by letting $P(k) = P(k-1)$ (i.e., rather than using Equation (10.13)). This will ensure that all the elements of $P(k)$ will stay bounded. Here, note that

$$||P(k)||_2 = \left[\lambda_{max}\left(P^\top(k)P(k)\right)\right]^{\frac{1}{2}}$$

where $\lambda_{max}\left(P^\top(k)P(k)\right)$ is the maximum eigenvalue of $P^\top(k)P(k)$. But since $P(k)$ is symmetric and $P(k) \geq 0$ (i.e., it is positive semidefinite), we know that

$$||P(k)||_2 = \lambda_{max}\left(P(k)\right)$$

so that all we need to implement the modification to WRLS is to test eigenvalues.

10.4.4 Example: Fitting a Line to Data

As an example of how recursive least squares can be used, suppose that we would like to use this method to fit a line to data that are generated by a time-varying function. Suppose that $n = 1$. In this case our parameterized linear approximator is

$$y = F_{lip}(x, \theta) = \theta^\top \phi(x) = \theta^\top [\phi_1(x), 1]^\top = \theta^\top [x_1, 1]^\top = \theta_1 x_1 + \theta_2 \qquad (10.14)$$

which is an equation for a line. Suppose that the unknown function (we need to explicitly provide it here for the sake of illustration)

$$y(k) = G(x(k), z(k)) = \sin(0.01k)x_1(k) + 1$$

where $x(k) = x_1(k)$ and $z(k)$ captures the time-varying nature of the function (e.g., we could say that $z(k) = k$). Notice that we can think of the $\sin(0.01k)$ as a time-varying slope and the 1 as the intercept of a time-varying line. We assume that we have examined the data and determined that there is some type of time-varying behavior, so we decide to try to use recursive least squares. We emphasize, however, that we assume that we do not know the explicit structure of the unknown function.

First, we must specify the input $x(k)$. Here, we simply choose $x(k)$ to be a random value that is uniformly distributed on $[-1, 1]$. Next, we pick $\theta(0) = [0, 0]^\top$ and $P(0) = \alpha I$ where $\alpha = 100$. Then we show the performance of the estimator in Figure 10.20 for the case where $\lambda = 1, 0.98, 0.95, 0.7$. First, notice that in the nonweighted case ($\lambda = 1$), the estimate θ_1 quickly converges to the true value of 1 but that the estimate θ_2 is quite poor. The reason for this is that even though the underlying system has a parameter that changes over time, the estimation algorithm does not forget what the old data told it about how to do a good estimate. On the other hand, in the case where $\lambda = 0.98$, which represents that we want to forget *some* old information, the estimator does much better. Continuing with the tuning in this manner, we see that we get successive improvements, until the case where $\lambda = 0.7$ and we get a very good estimate. In this case, we tuned λ so that we are forgetting enough old information about the underlying function so that we are listening enough to what new data are saying about how the underlying function is shaped. It is clear then that when you have slowly varying changes in the underlying function, you probably want a large λ (i.e., near 1), where if the underlying system changes quickly, you probably want a smaller value of λ so that the algorithm quickly forgets old information; however, generally you do not want to pick λ to be too small since then it will quickly forget old information and may not perform well.

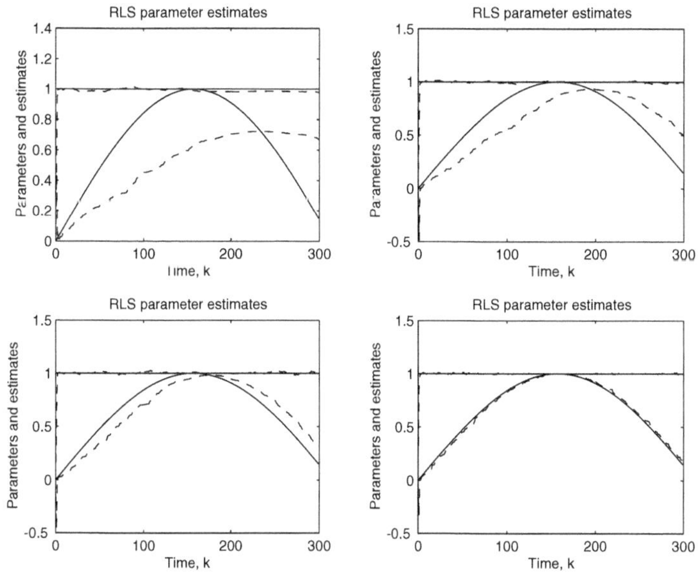

Figure 10.20: RLS parameter estimates, $\lambda = 1$ (upper left), $\lambda = 0.98$ (upper right), $\lambda = 0.95$ (lower left), $\lambda = 0.7$ (lower right).

10.5 Example: Online Tuning of Approximators

In this section, we continue with the examples considered in Section 10.2 where we considered the use of the batch least squares method to tune the approximator parameters.

10.5.1 Multilayer Perceptrons

Here, we consider the case where we had $n_1 = 25$ neurons and use all the same values of the biases and weights in the hidden layer that we developed in Section 10.2. Here, we use RLS to tune the neural network output layer weights and bias (these are stacked in the 26×1 vector θ). We will focus in particular on how the training data are presented to the algorithm and how this affects the accuracy of the approximator, how the forgetting factor affects the algorithm, and how initialization affects approximator accuracy.

Relatively Uniform Coverage of the Input Space

We will let the input x be uniformly distributed on $[-6, 6]$ and try to train the neural network to match the function in Figure 9.10. We let $\lambda = 1$ and initialize the algorithm with $\theta(0) = 0$ and $P(0) = \alpha I$ with $\alpha = 100$. To illustrate how the shape of the approximator nonlinearity evolves over time, we show the first 10 iterations of the RLS algorithm in Figure 10.21. Notice that for $k = 1$,

The performance of recursive least squares in constructing a mapping over the entire input space depends critically on the gathered data being properly spread over this space.

we actually have two data points shown since we generated one data pair at $k = 0$ simply in case you wanted to also evaluate the estimation error of the approximator at time $k = 0$. At this time, however, only one data point has been used to tune the approximator. Notice that with only one data point, the estimator mapping is far from providing a good approximation of the mapping in Figure 9.10 (not surprising since with only one data pair, it knows little about its shape). Notice, however, that as k increases, we get more and more training data and our representation becomes more and more accurate. Here, $x(k)$ provides a relatively uniform coverage of the input space so even after only 10 iterations, we get a relatively good approximation.

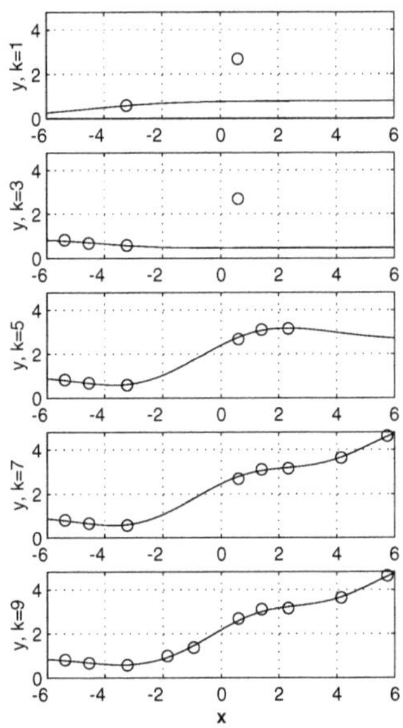

Figure 10.21: Neural network mappings generated using increasing amounts of training data ($k = 1$ to $k = 10$).

Nonuniform Coverage of the Input Space

Next, consider what happens when we do not get a good coverage of the input space. To do this, we simply run the program used to generate Figure 10.21 a few times, until by random chance it does not place any x data in one region of the input space. When this happens, we get the sequence of neural network mappings shown in Figure 10.22, where there are no input data in the region near $x = -6$ so we see that the approximation is poor in that region (until at $k = 10$, where it gets one more point and it improves the approximation). This is, of course, not surprising since, in this case the approximator is *extrapolating* for $k \leq 9$ so we cannot expect it to perform very well. It is also the case that if there are "holes" in the input space where there are no data, we will generally get poor approximation accuracy in that region. The general principle is that if you want to get good accuracy in any region, you need data to tell you what the shape of the function in that region is.

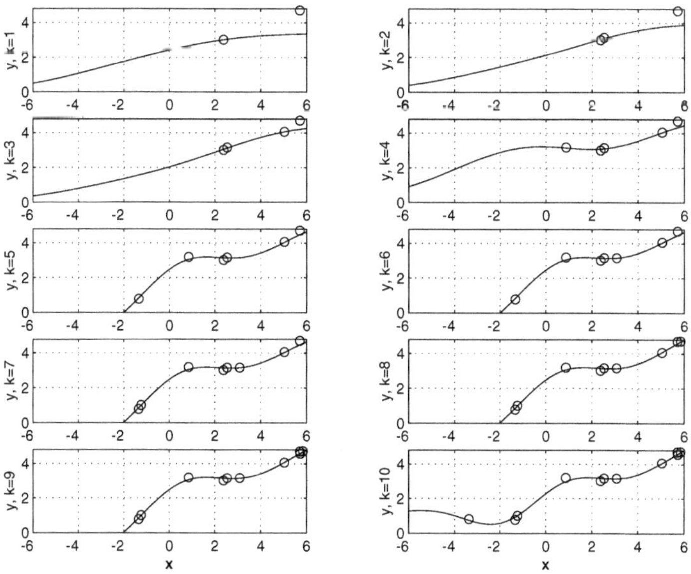

Figure 10.22: Neural network mappings generated using increasing amounts of training data (no data near $x = -6$ for $k \leq 9$).

Effects of Tuning the Forgetting Factor

In this case we will use a data set that has many input-output data pairs that are uniformly distributed across the input space. To do this, we will simply run the RLS algorithm for 300 steps, generating data at each of these steps (qualitatively, we obtain similar results if we use 1000 steps). To avoid showing 300 plots, we simply show how the last approximator performs compared to the training data. In particular, see Figure 10.23, and notice that this is a reasonably good approximation (we rely on random chance to get the uniform distribution of training data shown). Here, the approximator misses some of the structure inherent in the function, especially the "high frequency content" of the function (e.g., the peak at about $x = 1$ is considered to be an outlier since it is not encountered often). It seems to smooth out the approximator shape and does avoid being distracted by noise.

It is interesting to compare the accuracy of this approximator to that which was trained with batch least squares (see Figure 10.4). The one trained with batch least squares appears to be more accurate even though it was trained with

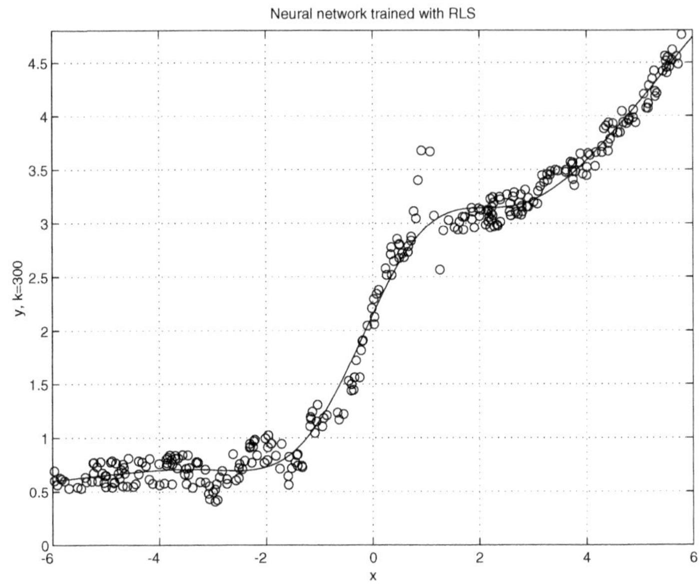

Figure 10.23: Neural network tuned with RLS for 300 steps.

fewer data (i.e., using less information about the unknown function). Notice, however, that the critical issue here is *how* the data were used. For batch least squares, all 121 data pairs were used at the same time to minimize the approximation error. For RLS, we provide the 300 data pairs in a sequence, and update the approximator each time we get a new data pair, placing equal importance on each piece of data obtained (i.e., we have $\lambda = 1$ here). Actually, by training with fewer steps (e.g., 121), you may get better overall accuracy (similar to that of the batch least squares).

Note, however, that by tuning λ, it is sometimes possible to get the approximator to do a better job at approximating the higher frequency content of the function. (Of course, another option is to increase the number of neurons in the hidden layer as we did in the batch least squares case.) For instance, if we let $\lambda = 0.95$, we get the plot shown in Figure 10.24, where we see that the approximator is trying to model the higher frequency behavior. Why? Well, the effect of $\lambda = 0.95$ is to place less significance on old data (i.e., data encountered for low values of k when we are at a higher value of k) so that different regions tend to become somewhat independent of each other so we can shape based on local data. However, this example is not to be overgeneralized. Sometimes you can pick λ to take on certain values and get disastrous results (where the approximator shape diverges from the shape it should have). The parameter λ offers the *potential* for performance improvements but cannot be guaranteed to provide these every time.

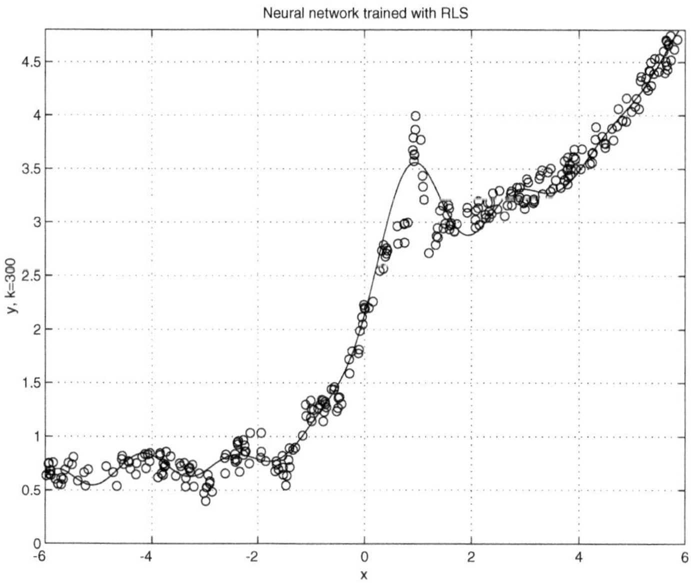

Figure 10.24: Neural network tuned with RLS for 300 steps, $\lambda = 0.95$.

Effects of Good Initialization of the Approximator

It should be intuitively clear that if we initialize, via choice of $\theta(0)$, the approximator close to the shape that it should ultimately be, that the RLS method should perform better. How do we initialize θ? One approach is to collect a set of training data and train with batch least squares first. If you do this for our example using the result of the batch least squares training studied in Section 10.2, then RLS ends up tuning the shape very little, even after 300 iterations (this is partly because the $(y(k) - \phi(x(k))^\top \theta(k-1))$ term in Equation (10.12), which is the error in predicting $y(k)$ using $\theta(k-1)$, is small for a good initialization, so the updates to $\theta(k)$ are small).

Generally, better initialization of the mapping results in higher quality approximators after training; however, training could decrease the quality of the initial guess.

We would like to show, however, that if you have a reasonably good initialization this can help the approximator accuracy, but that RLS also can improve on this accuracy by using more data to tune the approximator. So, how do we obtain a "reasonably good" initialization? One approach is to simply guess, but this can be very difficult when there are many parameters. Another approach is to use fewer training data in the batch least squares approach (in practical applications, this is typically the approach used). Here, however, simply for the sake of illustration, we will perturb the θ we had obtained from the batch least squares training by letting $\theta(0) = \theta + 0.25\theta$. In this case, we obtain the results shown in Figures 10.25 and 10.26. Notice that in Figure 10.25, the initial approximator shape is better than in, for instance, Figure 10.21, but that near $x = 6$, the approximation is not very good. As data are gathered, however, the RLS further improves the accuracy of the approximator and to see this, the final approximator is illustrated in Figure 10.26.

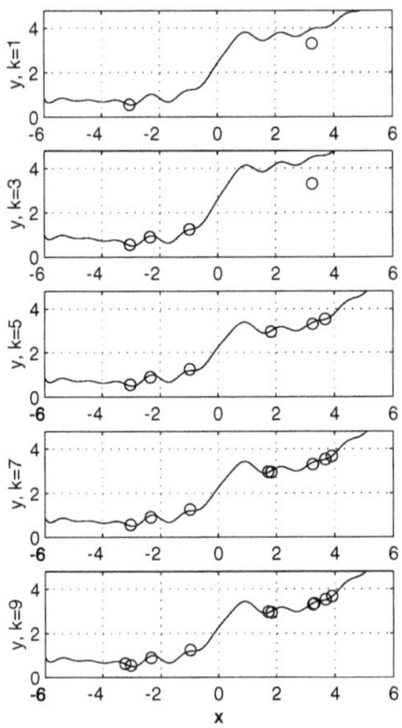

Figure 10.25: Neural network mappings generated using increasing amounts of training data (reasonably good initialization).

10.5.2 Takagi-Sugeno Fuzzy Systems

Here, we consider the case where we had $R = 20$ rules and use all the same values of the centers and spreads for the premise membership functions that we developed in Section 10.2. Here, we use RLS to tune the Takagi-Sugeno fuzzy system output function parameters (these are stacked in the 40×1 vector θ).

Input Space Coverage

Due to locality properties of the Takagi-Sugeno fuzzy system, it tends to adjust the mapping only where it gets data and tends not to destroy what it has learned in one area when making adjustments in another area.

In Figure 10.27 we show what happens in the first 10 steps of the algorithm for training the Takagi-Sugeno fuzzy system. Notice that in this case, we are initializing the parameter vector to be all zeros so that the mapping is zero initially. As it begins to get data, it shapes the mapping, but apparently in a very different way from what was done for the neural network (compare to Figure 10.21). Each time it gets another data point, it tries to pass the approximator mapping through that data point. In a sense, it trusts the initialization and does not adjust parameters to match in places where it does not know how to adjust. Of course, while this does not appear to be a good property in this example, for early steps of the approximator construction it can be quite beneficial since the approximator shape only changes locally.

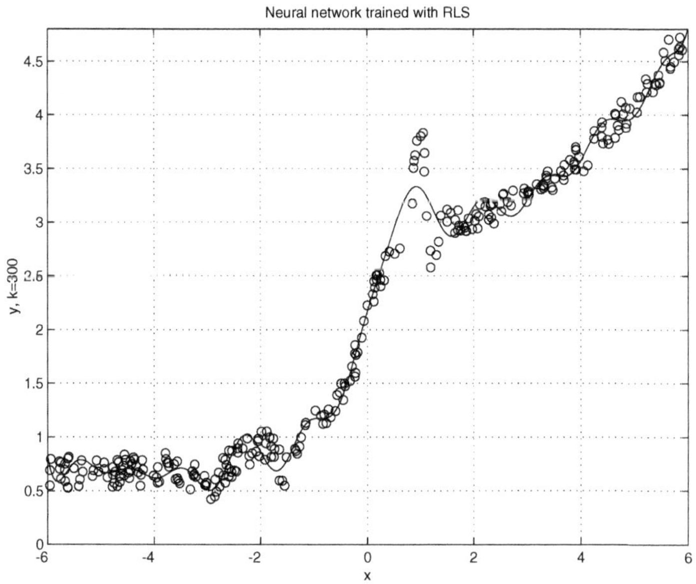

Figure 10.26: Neural network mappings generated using increasing amounts of training data, after 300 iterations.

Figure 10.27 dramatically illustrates issues of input space coverage. Bad coverage can clearly result in bad mappings, just as in the neural network case. Good coverage is obtained with more data. For example, if we use 300 pieces of training data, we get the mapping shown in Figure 10.28. The approximation error is somewhat lower compared to the neural network as you can see by comparing to Figure 10.23 (but do not draw any general conclusions from that). The tuning of the forgetting factor acts in a similar way (qualitatively) to that which was illustrated for the neural network.

Effects of Good Initialization of the Approximator

Here, we briefly illustrate the effects of using a good initialization for the parameters of the Takagi-Sugeno fuzzy system. We will follow the same approach as for the neural network in the last section. First, we show in Figure 10.29 what happens if you use the exact value of θ found by batch least squares. Notice that, as compared to Figure 10.27, the initial shapes are quite good due to the initialization (and if you run this for 300 steps, then it gets the result shown in Figure 10.30).

Next, we will perturb the good initialization by letting $\theta(0) = \theta + 0.25\theta$ where θ was found using the batch least squares method. The results for this case are shown in Figure 10.31. By studying the sequence of plots, you can see that as data are obtained in new regions, the approximator updates the shape

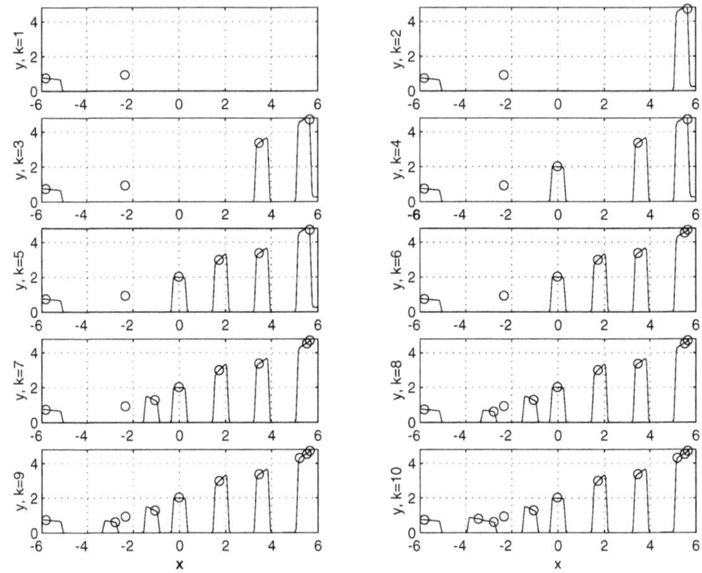

Figure 10.27: Takagi-Sugeno fuzzy system mappings generated using increasing amounts of training data ($k = 1$ to $k = 10$).

to more closely approximate the unknown function. In fact, after 300 iterations, the shape shown in Figure 10.32 is quite good and even similar to the case where a very good initialization was used. This shows how RLS can improve upon an initialization to provide good approximation accuracy.

10.6 Exercises and Design Problems

Exercise 10.1 (Batch Least Squares Derivation): Recall that for batch least squares, we had

$$J(\theta) = \frac{1}{2} E^\top E$$

(a) Using basic ideas from calculus, take the partial of J with respect to θ and set it equal to zero. From this, derive an equation for how to pick the least squares estimate. Compare it to Equation (10.2). Hint: If m and b are two $n \times 1$ vectors and A is an $n \times n$ symmetric matrix (i.e., $A = A^\top$), then $\frac{d}{dm} b^\top m = b$, $\frac{d}{dm} m^\top b = b$, and $\frac{d}{dm} m^\top Am = 2Am$.

(b) Repeat (a) for the weighted batch least squares approach and compare it to Equation (10.4).

Exercise 10.2 (Recursive Least Squares Derivation): In this problem you will derive the RLS method for two different cases.

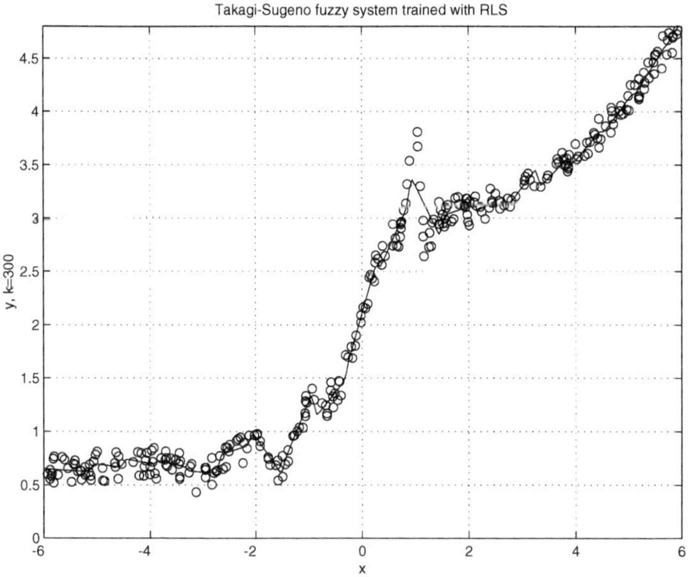

Figure 10.28: Takagi-Sugeno fuzzy system approximator mapping after 300 steps.

(a) Derive the update Equations (10.13) for the weighted recursive least squares approach (show every step of the derivations).

(b) In some applications we have a vector of measurements at each time instant (i.e., multiple measurements), rather than a single measurement. Derive the RLS equations for this case, and clearly identify the dimensions of all the matrices and vectors that you use.

Design Problem 10.1 (Gas Furnace CO_2 One-Step Ahead Prediction): See the Web page for the book and download the Box-Jenkins data for the gas furnace (it is in the form of a Matlab `.dat` file, but will download as a text file that you will remove the .txt extension from, before using in Matlab; of course, you can examine the file and easily use it in other programs).

(a) Develop an estimator using a batch least squares approach for $y(k)$, assuming that you use all the inputs to the estimator. You can think of this as a one-step ahead predictor since the estimate will depend on past inputs and outputs, in particular, $y(k-1)$, $y(k-2)$, $y(k-3)$, $y(k-4)$, $u(k-1)$, $u(k-2)$, $u(k-3)$, $u(k-4)$, $u(k-5)$, and $u(k-6)$. Use an affine approximator mapping (i.e. linear with bias). There are 290 data pairs in the "boxjenkins.dat" file. Use only the first 145 for training (i.e., $M = 145$). Use the last 145 for testing (i.e., $M_\Gamma = 145$). Provide values of the mean squared error for the approximator that

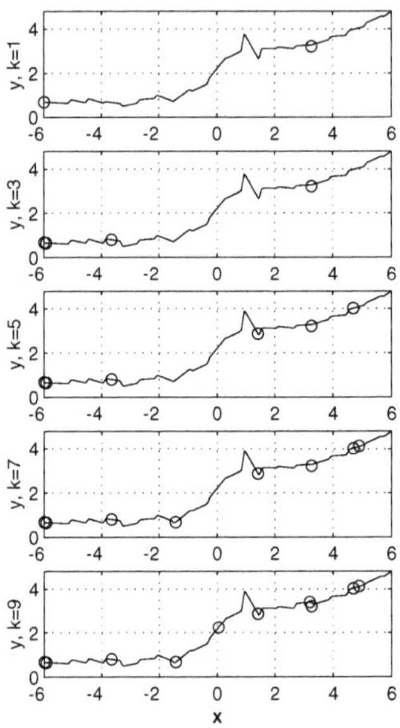

Figure 10.29: Takagi-Sugeno fuzzy system approximator mapping for the first 10 steps, good initialization.

you trained, both at the training data and the testing data. Plot $y(k)$ vs. k and the estimate of $y(k)$ vs. k on the same plot (use different line types for each) for $k = 1, 2, ..., 145, 146, ..., 290$, so that this plot will show how the estimator will perform at both the training and testing data (with the plots concatenated). Plot the error between the approximator estimate and the training data and testing data on the same type of plot (but on a different graph). Discuss the results.

(b) Suppose that you are constrained to only be able to use two inputs to your estimator. Pick the best two (with "best" defined by minimizing the mean squared error for the testing data). Moreover, specify a methodology for picking these. Use this methodology and repeat the process for three inputs. Compare, using the same plotting methodology and mean squared errors, to part (a).

(c) For (a), switch the training and test data. Repeat the design and testing of the estimator. Comment on the values of the mean squared error relative to the ones found in (a).

Design Problem 10.2 (Controller Construction from Process Operator Data):

In this problem you will follow the development in Section 10.3 to construct a controller using process operator data. (You can get the data at the Web site for the book.)

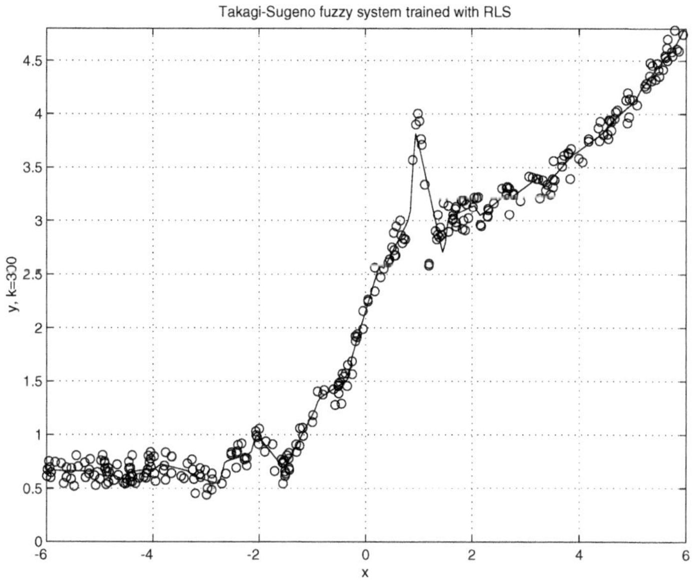

Figure 10.30: Takagi-Sugeno fuzzy system approximator mapping obtained after 300 iterations, good initialization.

(a) Verify the correlation analysis that was used to try to select inputs to the controller by reproducing Figure 10.12.

(b) Develop using a batch least squares approach linear (affine) controllers for the cases where you use all the inputs, 3 inputs, 2 inputs, and 1 input, as was done in the chapter. Verify the results there for the values obtained for the mean square errors in each case. In each case, produce the same types of plots to illustrate the performance of the estimator. Also, for each case, compute the correlation coefficients between the inputs and the estimation error and explain the resulting values using ideas from Section 10.3.

(c) Next, using ideas presented in Section 10.3, develop a fuzzy controller that obtains a lower mean squared error than you obtained for any case in (b). Produce the same types of plots as you did in (b) to illustrate the performance of the estimator and provide the value of the mean square error. You may use the same type of approach as given in the chapter (including the correlation analysis), or you may want to try to tune a neural network, or try a different tuning method. No matter which approach you take, be certain to pay attention to the number of parameters that you use in the approximator you are tuning. In fact, for each approximator structure you study, provide the number of parameters in the approximator and keep in mind that it is typically best to use the simplest approximator (where "simple"

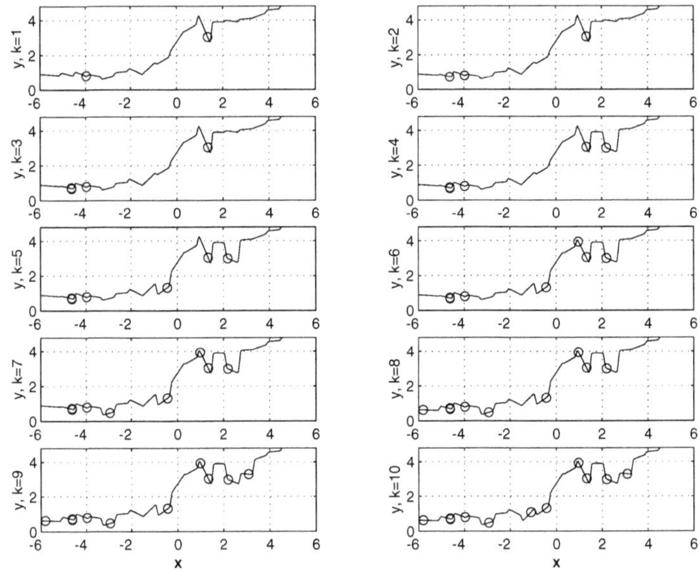

Figure 10.31: Takagi-Sugeno fuzzy system approximator mapping for the first 10 steps, reasonably good initialization.

may be measured by the size of the approximator, i.e., the number of parameters that are used in its definition).

(e) Optional: If you used a standard fuzzy controller in (c), develop linguistic rules from the ones that were constructed with data. You can pick linguistic values. Use these rules to explain some operating characteristics of the plant.

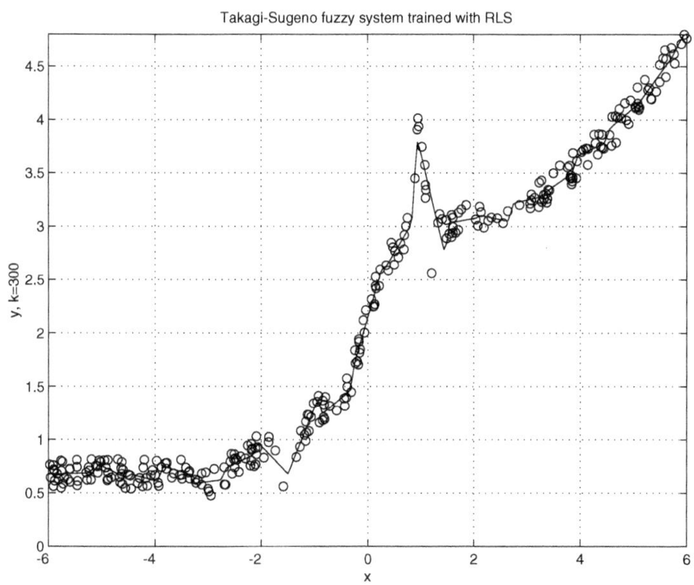

Figure 10.32: Takagi-Sugeno fuzzy system approximator mapping after 300 iterations, reasonably good initialization.

Chapter 11

Gradient Methods

Chapter Contents

Gradient techniques offer practical and effective methods to perform optimization. When applied to either the offline or online function approximation problems, they seek to find θ to minimize the function approximation error. The methods operate in an iterative fashion by successively improving on "guesses" (estimates) of the ideal parameter vector.

Gradient methods can be used in a batch or online manner to tune all parameters of the approximator. The basic approach is to iteratively adjust the parameters to minimize the approximation error.

Consider minimizing

$$J(\theta, G) = \frac{1}{2}\sum_{i=1}^{M} |y(i) - F(x(i), \theta)|^2 \tag{11.1}$$

by the choice of $\theta = [\theta_1, \theta_2, \ldots, \theta_p]^\top$ for a given training data set

$$G = \{(x(i), y(i)) : i = 1, 2, \ldots, M\}$$

(note that in several cases below, we will develop the theory for the case where $F(x(i), \theta)$ and $y(i)$ are $\bar{N} \times 1$ vectors so that it will be clear how to update multi-input multi-output approximators if you need to do that). In Equation (11.1), $|\bar{x}| = \sqrt{\bar{x}^\top \bar{x}}$ if $\bar{x}$ is an $\bar{N} \times 1$ vector. Clearly, if we can pick θ to minimize $J(\theta, G)$, we will have done a good job at function approximation, at least at the training data pairs in G (perhaps not at the test set Γ where $G \subset \Gamma$).

At the outset it is important to highlight the fact that, in general, the optimal solution to the function approximation problem is difficult to find. Why? Basically, the answer lies in the fact that $J(\theta, G)$ can be a very complex "landscape" with many hills and valleys, such as the one shown in Figure 11.1, which is also shown in Figure 11.2 as a contour map.

The methods may search for the global minimum of such a function (which in this case, by inspection, is at $(15, 5)$), but it can get distracted by the multiple local minima at other positions (e.g., at $(20, 15)$, which is easier to see on the contour plot), or the very flat regions of the surface where, if you only have a local view (i.e., not the perspective you have by looking at the plots from a bird's-eye view where you can see the peaks and valleys), it is difficult to know which direction to head to find the minimum. In fact, on such low slope portions of the surface, relatively large changes in the parameters make very little progress in minimizing the function (and such low slope regions are often found in function approximation problems where the approximator structure is "overparameterized," i.e., more approximator structure is used than is needed to get a low representation error and hence, large changes in some parameters may have little effect on the shape of the function and hence, the quality of the approximation).

Local minima arise since, in general, the cost function that characterizes approximation error is not convex. Getting trapped at a local minimum corresponds to not tuning the approximator in a way that could further reduce the approximation error.

In general, we will not know that we have converged to a global or local minima for the gradient methods studied here (except in special cases). The most we will be able to hope for is to converge to a "stationary point;" that is, a zero slope region of the surface such as a local minimum or a flat region on the landscape.

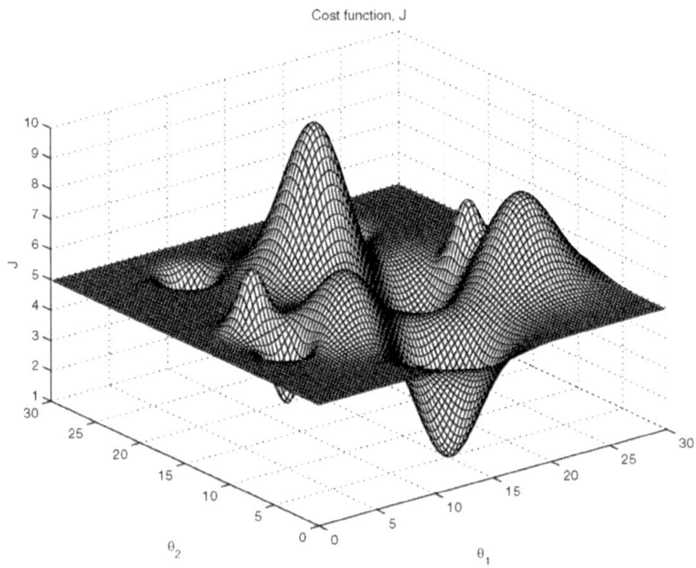

Figure 11.1: Candidate function for which we may seek to find the minimum.

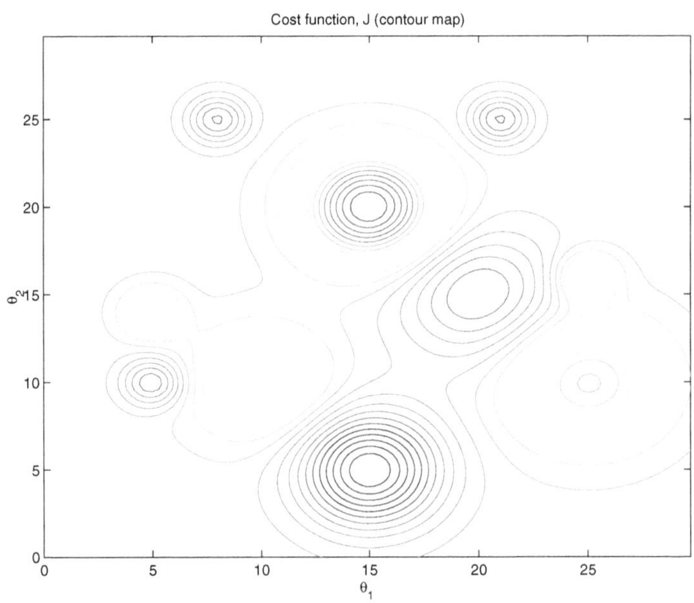

Figure 11.2: Candidate function for which we may seek to find the minimum (contour map).

11.1 The Steepest Descent Method

Let $\theta(j)$ be the current estimate of the parameter vector at iteration j (note that when we indexed θ with time we used k, but here j is used to emphasize that the index may simply be for the training data, not time).

11.1.1 Steepest Descent Parameter Updates

The basic form of the update using a gradient method to minimize the function $J(\theta, G)$ via the choice of $\theta(j)$, is

$$\theta(j+1) = \theta(j) + \lambda_j d(j) \tag{11.2}$$

where $d(j)$ is the $p \times 1$ "descent direction," and $\lambda_j > 0$ is a (scalar) positive "step size" that can depend on the iteration number j.

To see the rationale for the choice of this parameter update formula, consider the case where θ is a scalar and where we use the simple quadratic function

$$J(\theta, G) = \theta^2$$

in Figure 11.3, where we are searching for the point where the function reaches the minimum by picking the scalar θ. Name the point where the minimum is achieved θ^* and assume that it is unknown and that we want to find its value.

It is useful to think of gradient methods as "hill climbing" (here, climbing down hills).

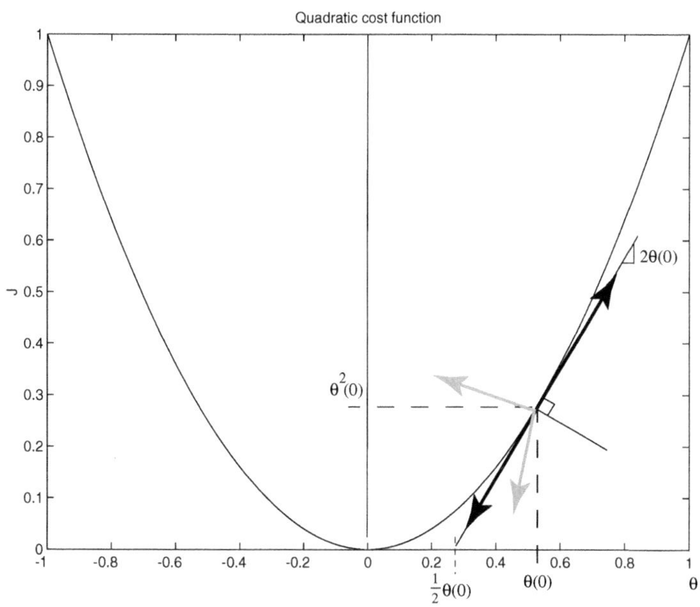

Figure 11.3: Scalar quadratic $J(\theta, G)$.

The update formula for this scalar case, where $\lambda_j = \lambda$ is a positive constant, is

$$\theta(j+1) = \theta(j) + \lambda d(j)$$

Notice that

$$d(j) = \frac{\theta(j+1) - \theta(j)}{\lambda} \tag{11.3}$$

With λ as a step size, we see that $d(j)$ is a descent direction in the sense that it is the direction in which the parameter is moving in order to try to minimize $J(\theta, G)$. What direction would we like this to be? We would like the parameter updates to always move in a direction that decreases $J(\theta, G)$ because, if it does this over a whole sequence of iterations (perhaps an infinite number of iterations), we may get $\theta(j) \to \theta^*$ as $j \to \infty$ (so $J(\theta^*, G) \leq J(\theta, G)$ for all other possible θ).

The above formula (Equation (11.3)) suggests that the direction should be the slope of the function $J(\theta, G)$ at $\theta = \theta(j)$. To see this, consider the example in Figure 11.3. Suppose that the initial (best) guess of θ^* is the $\theta(0)$ shown. Based on this guess, how would you next guess at θ? That is, how would you generate $\theta(1)$? Suppose that we can compute the slope (gradient) of $J(\theta, G)$ at $\theta(0)$. For our example, this gradient is

$$\left. \frac{\partial J(\theta, G)}{\partial \theta} \right|_{\theta=\theta(0)} = 2\theta|_{\theta=\theta(0)} = 2\theta(0) \tag{11.4}$$

and it is shown as the black arrow pointing up and to the right in Figure 11.3. The negative of this gradient is $-2\theta(0)$ and it is shown as the black arrow pointing down and to the left in Figure 11.3. These arrows indicate possible directions $d(j)$ to update the guess at $\theta(0)$. Clearly, to move *down* the function $J(\theta, G)$ to minimize it, one choice would be to use the direction

Steepest descent involves updating the parameters in a direction that appears to decrease the cost function the most.

$$d(j) = -\left. \frac{\partial J(\theta, G)}{\partial \theta} \right|_{\theta=\theta(j)} \tag{11.5}$$

(i.e., to move along the negative gradient). Intuitively, this choice is the "direction of steepest descent" (it corresponds to how a skier often moves down a snow-covered mountain) and hence, the parameter update formula for the steepest descent method, even for the p-dimensional case, is given by

$$\theta(j+1) = \theta(j) - \lambda_j \left. \frac{\partial J(\theta, G)}{\partial \theta} \right|_{\theta=\theta(j)} \tag{11.6}$$

11.1.2 Example: Convergence, Step Size, and Termination Issues

Using the hill climbing (or skiing) perspective, the step size indicates how big a step to move in the $d(j)$ direction. For instance, for $\lambda_j = \lambda = 0.1$ for all j, the update formula becomes

$$\begin{aligned}\theta(j+1) &= \theta(j) - 0.2\theta(j) \\ &= 0.8\theta(j)\end{aligned}$$

so clearly as $j \to \infty$, $\theta(j) \to \theta^* = 0$, the optimum point.

In optimization you do not know where the minimum point is to begin with, so you generally do not directly know if you are approaching it (or if you are at it), so it is difficult to know how to terminate the updates to $\theta(j)$. That is, we do not know when $\theta(j)$ has approached θ^* since θ^* is unknown. So, how do we terminate the gradient algorithm if we need a solution after a finite number of iterations? One way is to monitor

$$d(j) = -\left.\frac{\partial J(\theta, G)}{\partial \theta}\right|_{\theta=\theta(j)}$$

and if

$$d(j) = -\left.\frac{\partial J(\theta, G)}{\partial \theta}\right|_{\theta=\theta(j)} = 0$$

clearly Equation (11.6) will stop making changes to $\theta(j)$. In practice, we often simply terminate if

$$|d(j)| = \left| -\left.\frac{\partial J(\theta, G)}{\partial \theta}\right|_{\theta=\theta(j)} \right| \leq \epsilon$$

for some prechosen constant $\epsilon > 0$; however, there are other termination issues and methods that will be discussed later.

Next, it is important to further consider the effect of the choice of the step size, particularly on the convergence of the estimate to its true value. If you choose $\lambda_j = \lambda = 0.2$ for all j (i.e., double the step size compared to the choice above), the update formula becomes

$$\begin{aligned}\theta(j+1) &= \theta(j) - 0.4\theta(j) \\ &= 0.6\theta(j)\end{aligned}$$

so again as $j \to \infty$, $\theta(j) \to \theta^* = 0$, the optimum point. But notice that the "rate of convergence" is much faster in this case since this choice for λ corresponds to taking larger steps at each iteration. So this line of reasoning may lead one to believe that larger step sizes are generally better; this is not, however, the case. Notice that if you pick $\lambda_j = \lambda = 10$ for all j, then the update formula is

$$\begin{aligned}\theta(j+1) &= \theta(j) - 20\theta(j) \\ &= -19\theta(j)\end{aligned}$$

so that as j increases, the value of $\theta(j)$ oscillates between larger and larger values and does not converge (in terms of Figure 11.3 it climbs up the parabola). In this case, the step size is too big, so the algorithm is too aggressive and fails to converge to the minimum value of $J(\theta, G)$.

Generally, $\theta(j)$ may not have a limit point, it may diverge as in this example, or it may oscillate; however, gradient methods are generally able to find isolated stationary points (i.e., ones where you can draw a sphere around them and no other stationary points are in the sphere) if they start close to them (this is why initialization of the algorithms is so important). Sometimes, all we can hope to be able to show is that the parameters will remain bounded, and sometimes we can do this by appropriately choosing the step size so that at each iteration we are guaranteed to get descent. In this case, the parameter vector will belong to a bounded set and so it must have at least one limit point; however, it can be difficult to guarantee that the parameter vector will converge to a single limit point. Clearly, the direction of descent and step size are important parameters in the development of a gradient update formula, especially when $J(\theta, G)$ is very complex, as it often is in practical applications.

11.1.3 Descent Direction Possibilities and Momentum

The above simple example shows intuitively that the choice of

$$d(j) = -\left.\frac{\partial J(\theta, G)}{\partial \theta}\right|_{\theta=\theta(j)}$$

(i.e., the negative gradient) is the direction of steepest descent and that as long as an appropriate choice is made for the step size λ_j, the algorithm will converge for this example (since the function we are minimizing is convex and only has one minimum). There are, however, many other choices that can be made for the descent direction and these others can be useful in practical applications. (Indeed, in the case where $J(\theta, G)$ is quadratic, there are well-known methods for the solution to the optimization problem; in practical applications, it is most often not quadratic.) Notice that any direction $d(j)$ is a descent direction provided that the angle it makes with

Iterative update of the parameters in any direction that locally appears to decrease the cost results in viable gradient descent methods.

$$\left.\frac{\partial J(\theta, G)}{\partial \theta}\right|_{\theta=\theta(j)}$$

is more than 90°. Hence, the shaded arrows in Figure 11.3 are also descent directions for $\theta(0)$. The angle is greater than 90° if

$$\left(\frac{\partial J(\theta(j), G)}{\partial \theta(j)}\right)^\top d(j) < 0$$

As indicated, this formula also holds for the vector case. In the vector case, $d(j)$ is a $p \times 1$ vector and the gradient is a $p \times 1$ vector that is denoted by

$$\nabla J(\theta(j), G) = \frac{\partial J(\theta(j), G)}{\partial \theta(j)} = \begin{bmatrix} \frac{\partial J(\theta(j),G)}{\partial \theta_1(j)} \\ \frac{\partial J(\theta(j),G)}{\partial \theta_2(j)} \\ \vdots \\ \frac{\partial J(\theta(j),G)}{\partial \theta_p(j)} \end{bmatrix} \tag{11.7}$$

Clearly, the choice of

$$d(j) = -\left.\frac{\partial J(\theta, G)}{\partial \theta}\right|_{\theta=\theta(j)} = -\nabla J(\theta(j), G)$$

results in the satisfaction of this formula, but clearly many other choices do also.

One modification to the descent direction that has been found to be useful in some applications is to use a "momentum term." In this case the update formula is

$$\theta(j+1) = \theta(j) - \lambda_j \nabla J(\theta(j), G) + \beta\left(\theta(j) - \theta(j-1)\right) \tag{11.8}$$

where $0 \leq \beta < 1$ is a fixed gain and $\beta\left(\theta(j) - \theta(j-1)\right)$ is the momentum term. Basically, momentum accelerates progress of the update where the gradients $\nabla J(\theta(j), G)$ are pointing in the same direction, but restricts update sizes when successive gradients are roughly opposite in direction. This can tend to damp oscillations in the parameter vector and keep the parameter vector moving in the proper direction.

In the following sections we will consider other choices for the descent direction, ones that can be particularly effective in practical applications (e.g., for tuning approximators that are not linear in the parameters). First, however, we will consider the application of the steepest descent method to the tuning of linear in the parameter approximators.

11.1.4 The Linear in the Parameter Case

For a linear in the parameter approximator $y = F(x, \theta) = \theta^\top \phi(x)$ we have, in the case where $\bar{N} = 1$,

$$d(j) = -\left.\frac{\partial J(\theta, G)}{\partial \theta}\right|_{\theta=\theta(j)} = -\left.\frac{1}{2}\frac{\partial}{\partial \theta}\sum_{i=1}^{M}\left(y(i) - \theta^\top \phi(x(i))\right)^2\right|_{\theta=\theta(j)}$$

and if we use the notation from Chapter 10, this is expressed as

$$d(j) = -\left.\frac{1}{2}\frac{\partial}{\partial \theta}E^\top E\right|_{\theta=\theta(j)}$$

where

$$E = \left[\epsilon_1, \epsilon_2, \ldots, \epsilon_M\right]^\top = Y - \Phi\theta$$

with $\epsilon(i) = y(i) - \theta^\top \phi(x(i))$. Now, if we follow the derivation of the batch least squares estimate, we find

$$\begin{aligned} d(j) &= -\left.\frac{1}{2}\frac{\partial}{\partial \theta}(Y - \Phi\theta)^\top (Y - \Phi\theta)\right|_{\theta=\theta(j)} \\ &= -\frac{1}{2}\frac{\partial}{\partial \theta}\left(Y^\top (I - \Phi(\Phi^\top \Phi)^{-1}\Phi^\top)Y + \right. \end{aligned}$$

$$
\begin{aligned}
&(\theta - (\Phi^\top \Phi)^{-1}\Phi^\top Y)^\top \Phi^\top \Phi(\theta - (\Phi^\top \Phi)^{-1}\Phi^\top Y))\big|_{\theta=\theta(j)} \\
&= -\frac{1}{2}\frac{\partial}{\partial \theta}\left(\theta^\top \Phi^\top \Phi \theta - 2\theta^\top \Phi^\top \Phi (\Phi^\top \Phi)^{-1}\Phi^\top Y\right)\Big|_{\theta=\theta(j)} \\
&= -\frac{1}{2}\frac{\partial}{\partial \theta}\left(\theta^\top \Phi^\top \Phi \theta - 2\theta^\top \Phi^\top Y\right)\Big|_{\theta=\theta(j)} \\
&= -\frac{1}{2}\left(2\Phi^\top \Phi \theta - 2\Phi^\top Y\right)\big|_{\theta=\theta(j)} \\
&= \Phi^\top (Y - \Phi\theta)\big|_{\theta=\theta(j)} \\
&= \Phi^\top E\big|_{\theta=\theta(j)}
\end{aligned}
$$

Suppose that $\lambda_k = \lambda > 0$ is a constant so that the steepest descent update formula for the linear in the parameters case is

$$
\begin{aligned}
\theta(j+1) &= \theta(j) + \lambda \Phi^\top E = \theta(j) + \lambda \Phi^\top (Y - \Phi\theta(j)) \\
&= \theta(j) + \lambda \sum_{i=1}^{M} \phi(x(i))(y(i) - \theta^\top(j)\phi(x(i))) \qquad (11.9) \\
&= \lambda \left(I\frac{1}{\lambda} - \Phi^\top \Phi\right)\theta(j) + \lambda \Phi^\top Y
\end{aligned}
$$

Now, if the steepest descent approach converges (and it will, assuming certain technical conditions are satisfied, for instance, having a diminishing step size), we get

$$\theta(j+1) \to \theta(j) = \theta_{sd}$$

as $j \to \infty$ (where we call θ_{sd} the value that the steepest descent converges to). In this case, we have

$$\theta_{sd} = \lambda \left(I\frac{1}{\lambda} - \Phi^\top \Phi\right)\theta_{sd} + \lambda \Phi^\top Y$$

Now, notice that

$$
\begin{aligned}
\left(I - \lambda\left(I\frac{1}{\lambda} - \Phi^\top \Phi\right)\right)\theta_{sd} &= \lambda \Phi^\top Y \\
\lambda \Phi^\top \Phi \theta_{sd} &= \lambda \Phi^\top Y
\end{aligned}
$$

so that if the inverse exists

$$\theta_{sd} = \left(\Phi^\top \Phi\right)^{-1}\Phi^\top Y$$

and we see that (if it converges) it converges to the least squares solution that we found in Equation (10.2). Notice, however, that this is an analysis of the *asymptotic* behavior of the estimate, and sometimes it is better simply to use an appropriate software package to compute the least squares estimate.

This provided a comparison to the batch least squares approach in the case where the batch of data is used in the steepest descent gradient method. What if, instead, we proceeded as in the recursive least squares case and processed the data sequentially? To do this, define

$$G_k = \{(x(k), y(k))\}$$

to be the data set at time k. In this case, since we make an iteration of the gradient method at each step, our update formula is

$$\theta(k+1) = \theta(k) + \lambda_k d(k)$$

(i.e., we replace j with k) and

$$d(k) = -\left.\frac{\partial J(\theta, G_k)}{\partial \theta}\right|_{\theta=\theta(k)}$$

so with $\lambda_k = \lambda = 1$ for all k, using Equation (11.9) (with $M = 1$ piece of data, the piece in G_k), we have

$$\theta(k+1) = \theta(k) + \phi(x(k))\left(y(k) - \theta^\top(k)\phi(x(k))\right)$$

Compare this to the update formula for recursive least squares given in Equation (10.12) that can be used for online parameter adjustment. Notice that the two are not the same. The update formula for recursive least squares has the extra $P(k+1)$ term that multiplies the second term in the above equation. Which is the better approach? In adaptive control problems, the recursive least squares approach tends to converge faster but you pay for this faster convergence by having to compute $P(k)$ and include it in the update formulas. Hence, sometimes for simplicity we may use a steepest descent gradient approach (sometimes, with certain modifications), even for the linear in the parameter case.

11.1.5 Step Size Choice

While in the last section (and in several subsequent ones) we focus on the selection of the descent direction $d(j)$, here we will consider the choice of the scalar step size λ_j. Clearly, while we will only discuss scalar step sizes, it is possible to have a diagonal matrix of step sizes, so that different parameters are updated at different rates.

Step size choice affects rate of convergence, how the algorithm copes with local minima, and asymptotic behavior of the algorithm.

Constant Step Size

For some applications (e.g., in adaptive control), a fixed step size $\lambda_j = \lambda$ for all j can be sufficient. Other times, it can be difficult to select λ. For instance, see the example in Section 11.1, where for a simple example, we showed that it is possible to pick it so that convergence is not achieved. Generally, as we saw in that example, if λ is too small, we get slow convergence; if it is too large, then we can get divergence. Indeed, in many problems a constant step size can result in "limit cycling" (oscillations in parameter values) near a local minimum since, when you are in a region near a local minimum, you must often take successively smaller steps to ensure that you do not "overshoot" the solution. Next, we discuss the case where a successively smaller step size is used.

Diminishing Step Size

In this case, the step size converges to zero as j goes to infinity, according to some formula. That is, we pick an algorithm for forcing

$$\lambda_j \to 0$$

as $j \to \infty$. For instance, we may choose

$$\lambda_j = \frac{\lambda}{j+1}$$

where $\lambda > 0$ is a constant or we could pick

$$\lambda_j = e^{-\alpha j}$$

for some $\alpha > 0$. While these rules can be simple to implement, in some cases λ_j may be chosen so that it goes to zero too fast so that the algorithm slows prematurely, before it gets near a solution. It is for this reason that often it is required that

$$\sum_{j=0}^{\infty} \lambda_j = \infty$$

which, in effect, forces the step size to persistently update the parameters (provided the gradient is sufficiently large).

Another way that a diminishing step size is sometimes implemented in practical applications is to let

$$\lambda_j = \max\left\{\lambda_{min}, \frac{\lambda_{max}}{1+\alpha j}\right\}$$

where at $j = 0$ we have $\lambda_0 = \lambda_{max} > 0$ and as j increases λ_j decreases, with rate governed by the choice of $\alpha > 0$, to $\lambda_{min} > 0$. All these parameters would need to be tuned for a particular application. Generally, this approach offers a big step size early in the processing and the step size diminishes as time goes on, but no lower than the value λ_{min}. This ensures that the step size will not get too small so that updates are persistently made (but of course in this case, we do not get $\lambda_j \to 0$ as $j \to \infty$).

Regardless, the general problem with a diminishing step size approach for practical applications is that it often slows convergence too much, or does not provide a fast enough update early in the processing. Due to this, tuning of the step size rule is normally needed. This is why in some practical problems, many have turned to line minimization approaches and the Armijo step size rule that we discuss next.

Line Minimization Approaches

For this, pick a scalar $\lambda^0 > 0$ that is the largest step size you think is reasonable for the problem at hand (sometimes information from the problem domain can suggest a choice for λ^0, while other times you must guess at it and experiment with the performance of the algorithm to get a good choice). Then, you pick λ_j so that

$$J(\theta(j) + \lambda_j d(j), G) = \min_{\lambda \in [0, \lambda^0]} J(\theta(j) + \lambda d(j), G)$$

This is simply a one-dimensional "line" minimization problem. The resulting value of λ_j yields the greatest reduction in the cost function over all step sizes such that $\lambda_j \geq 0$ (to keep the updates moving in the proper direction) and $\lambda_j \leq \lambda^0$ as it is specified above. "Line search," or what are sometimes called "line minimization" approaches, are used to solve this problem.

There are many line minimization methods. Some, like Newton's approach, require the use of second derivatives, while methods like the "secant method" only require the use of first derivatives. Other approaches use interpolation with candidate points generated in $[0, \lambda^0]$, or "golden section search," which is an interval reduction method. See the "For Further Study" section for more details.

Armijo Step Size Rule

In practice it is often the case that the line minimization approaches are computationally expensive, so methods that are based on "successive step size reduction" are often used. One popular method of this type is the Armijo step size rule.

In this rule, first pick the following scalars:

1. λ^0, an initial guess at the size of the step (often, for applications that are properly "scaled" you can pick $\lambda^0 = 1$).
2. γ, $0 < \gamma < 1$, a "reduction factor" (often, for applications, $\frac{1}{10} \leq \gamma \leq \frac{1}{2}$).
3. σ, $0 < \sigma < 1$, a scale factor (often, for applications, $0.00001 \leq \sigma \leq 0.1$).

While this provides guidelines for the choice of these parameters, they may need to be tuned. The Armijo step size rule is actually an iterative process for finding the step size λ_j that proceeds as follows for each iteration j:

1. Let $m = 0$.
2. Let $\lambda_j = \gamma^m \lambda^0$.
3. If
$$J(\theta(j), G) - J(\theta(j) + \lambda_j d(j), G) \geq -\sigma \lambda_j \nabla J(\theta(j), G)^\top d(j)$$
then return λ_j as the step size to be used at iteration j. Otherwise, let $m = m + 1$ (i.e., increase m by one) and go to Step 2.

Suppose that we apply the Armijo step size rule to the steepest descent method so that $d(j) = -\nabla J(\theta(j), G)$. The test in Step 3 then evaluates if the advance in reducing $J(\theta, G)$ (i.e., the left side of the inequality, $J(\theta(j), G) - J(\theta(j+1), G)$) is greater than a scaled version of the size of the gradient at $\theta(j)$. The Armijo step size rule decreases the step size from the initial guess λ^0 by a factor of γ until it is small enough to ensure a certain amount of decrease in $J(\theta(j+1), G)$ relative to $J(\theta(j), G)$. The amount of reduction is governed by the parameter σ, while the rate of decrease of λ_j is governed by the choice of γ (if γ is chosen too large, it can take too many iterations to find a solution, while if it is chosen too small, it may miss a larger valid step size that could have resulted in more decrease in the value of $J(\theta(j+1), G)$). Note that if $d(j)$ is a descent direction, $\nabla J(\theta(j), G)^\top d(j) < 0$ so that the Armijo step size rule algorithm will be guaranteed to converge in a finite number of iterations. Basically, the Armijo step size rule tries to combine the positive effects of a constant step size rule and a diminishing step size rule. Generally, when the gradient is large, it will try to take a big step (the size limited by the choice of λ^0) since it knows that it is probably not near a local minimum. When the gradient is small, it assumes that it is near a local minimum, and it takes smaller steps (clearly there is the possibility that it could reduce the steps too much so that convergence is slowed).

Step Size Choice Via Normalizing the Gradient

While there are many other ways to pick the step size, we will close this section with a brief discussion on one approach that has been found to be useful for online function approximation problems. In particular, we will focus on picking the step size for the steepest descent case for linear in the parameter approximators by "normalizing" the gradient.

For a linear in the parameter approximator $y = F(x, \theta) = \theta^\top \phi(x)$ and the case where we process one training data pair at a time (i.e., recursive processing as is often done in online approximation) we have, following the development in Section 11.1.4 in the case where $\bar{N} = 1$,

$$\theta(k+1) = \theta(k) + \lambda_k \phi(x(k))(y(k) - \theta^\top(k)\phi(x(k)))$$

where k is the time index. Consider the choice of

$$\lambda_k = \frac{\lambda}{1 + \gamma \phi^\top(x(k))\phi(x(k))} \tag{11.10}$$

where we consider $\lambda > 0$ to be a type of constant step size and γ to be a tuning parameter. To see how this works, first, view the term

$$\left(y(k) - \theta^\top(k)\phi(x(k))\right)$$

in the above update formula as simply a scalar (approximation error) that indicates how close the approximator is to doing a good job at approximation at time k. If it does well, then this term is small while if it does poorly then this term is larger. It does not, however, contribute to the direction of the update (except in its sign), and only to its size so we will not consider it to be a part of the "update direction" $d(k)$ (you could think of it as part of the step size that says that if you are not doing good approximation at that point, then make a big update, but that if you are doing a good job, then make a small update). Now, we see that the *direction* of the update is dictated by the vector $\phi(x(k))$ and of course, depending on its form, it can also affect the magnitude of the update. Clearly the size of the elements of $\phi(x(k))$ set how big the updates will be for the corresponding components of $\theta(k)$. In fact, $\phi^\top(x(k))\phi(x(k))$ is a measure of the overall size of the update suggested by $\phi(x(k))$ ("big" updates result from a large $\phi^\top(x(k))\phi(x(k))$.

Now, with this in mind, and with the step size in Equation (11.10), we see that if the gradient size, as characterized by $\phi^\top(x(k))\phi(x(k))$, is big, $1 + \gamma\phi^\top(x(k))\phi(x(k))$ will be particularly large, so that the step size λ_k will be small. If the gradient size, as characterized by $\phi^\top(x(k))\phi(x(k))$, is small, $1 + \gamma\phi^\top(x(k))\phi(x(k))$ will be close to 1, so that the step size λ_k is close to λ. In this case, we see that the parameter γ scales our characterization of size of the gradient so that if, for example, γ is very small, then even larger gradients will still result in a step size near λ.

To summarize, we see that the value of λ_k will vary between 0 (for a very big gradient) and λ (for a very small gradient), with γ governing the rate of variation between the two extremes. Hence, the step size will "diminish" if the gradient is large, but it does not generally "diminish" to zero as time progresses since if a local minimum is approached, then the gradient size goes to zero (not to mention the approximation error) and the step size approaches λ. Notice, however, that it does have one characteristic that is similar to some of the other rules. It fixes a maximum step size (λ) no matter how big the gradient is (compare to the Armijo step size rule).

11.1.6 Parameter Initialization, Constraints, and Update Termination

In this section we will discuss the practical issues of how to initialize the gradient methods (i.e., to pick $\theta(0)$), how to incorporate constraints on the parameter values into the dynamics of the algorithm, and if it is needed, how to terminate the algorithm.

Parameter Initialization

The choice of $\theta(0)$ can obviously have a significant effect on the quality of a solution provided by the gradient method. While in general you do not know where the optimal solution θ^* is, it is clearly best to pick $\theta(0)$ as close to this desired value as possible. The examples in Section 10.5 illustrated this for the recursive least squares algorithm (for the linear in the parameter case) and the same general concepts hold here. There are, however, other practical issues in choosing $\theta(0)$ for gradient methods.

Initialization affects performance of the algorithm and ultimately the achieved approximation accuracy.

First, note that since the functions $J(\theta, G)$ that we seek to minimize are in general not necessarily bowl-shaped and can have many local minima, it is often wise to execute the gradient algorithm for several choices of $\theta(0)$. This can help ensure that the value you found is indeed a global minimum. However, simple tests with multiple $\theta(0)$ cannot, in general, guarantee that you have found a global minimum.

Second, for a neural network with sigmoid nonlinearities, if you pick $\theta(0)$ (which will also have weights and biases of a hidden layer in it) to have certain components that are all too large, then it could be that all (or many of) the sigmoids are saturated so that the gradient will be small and updates will be slow (at least at the beginning). It is for this reason that for neural networks, a good choice for the weights and biases is often random small values. For a fuzzy system, there can be similar issues in specifying $\theta(0)$ (e.g., if Gaussian shaped membership functions are used and all the centers are specified to be too large, and the consequent function parameters are set to be zero, then the gradient will be small, at least initially).

In addition, when you generate the update formulas for a function approximator, you may find that the parameters cannot be allowed to take on certain values (both initially and for $j > 0$) or there will be a divide-by-zero error or other numerical problems. Hence, after you derive the update formulas, you should examine them for such problems and initialize appropriately. Moreover, you will then also need to make sure that during its operation, the algorithm never moves the parameters to inappropriate values. We discuss this next.

Constraints on Parameters

Generally, there are several reasons why there are constraints on the parameters in the optimization problems we study:

1. If the parameters $\theta(j)$ take on certain values, there are numerical problems as we discussed above (e.g., division by zero).

2. Since the parameters correspond to physical values (e.g., parameters for a neural network), there are practical limitations to their size (even in software there can be overflows if the values are too large).

3. Sometimes we know a "feasible region" for the optimal parameter values and hence, impose constraints on the set of values that $\theta(j)$ may take on because searching outside this set is fruitless.

4. Sometimes, we have extra information about the form of the underlying function that we seek to approximate (e.g., by physical insights or by simple inspection of the training data), and this can be used to constrain the choices of the parameters. For instance, in some problems you may know where on the input domain the unknown function has more nonlinear or oscillatory behavior, and hence, where you would like to "allocate" more of the approximator structure, since this is where it is needed to get good

approximation accuracy. For instance, if there are some oscillations in the function in a certain region, then you may want more sigmoids or membership functions there. Often, however you may want to allow the training method to perform the actual allocation of structure, rather than fixing it a priori. To do this, you would simply put appropriate constraints on the parameters to only allow them to move in the region where more accuracy is needed.

In any of these three cases, the constraints can be captured by requiring that

$$\theta(j) \in \Theta(j)$$

for all $j \geq 0$ where $\Theta(j)$ is the (known) parameter "constraint set" at iteration j. Often, we know that $\Theta(j) = \Theta$; that is, that the constraint set is the same at each iteration. For convenience, assume this in the discussion below since the case for a time-varying constraint set is similar.

There are effective ways to incorporate into the algorithm constraints on parameter values that are known a priori.

How do we ensure that $\theta(j) \in \Theta$ for all $j \geq 0$? First, we initialize so that $\theta(0) \in \Theta$ so that all we need to concern ourselves with is the case for $j > 0$. In general, the normal approach is to assume that Θ is a convex set. Then, if for some update the gradient method places $\theta(j+1)$ outside Θ, you simply require $\theta(j+1)$ to stay on the boundary of Θ. If it is on the boundary, and the gradient update says to put it outside the boundary, leave it on the boundary. But if it says to leave it on the boundary, or place it within Θ, you let it do that. While it is not difficult to characterize this (and "this" is called a gradient method with "projection" since we project the updates back into the convex feasible parameter set) precisely in mathematical terms, in practice the implementation is often easy and we will simply explain this.

The most common case in practice is when we know scalars θ_i^{min} and θ_i^{max}, $i = 1, 2, \ldots, p$, such that we want

$$\theta_i^{min} \leq \theta_i(j) \leq \theta_i^{max} \tag{11.11}$$

for all $j \geq 0$ (this specifies a convex set $\Theta = \{\theta : \theta_i^{min} \leq \theta_i \leq \theta_i^{max}, i = 1, 2, \ldots, p\}$). Then, each time you use a gradient update formula to generate $\theta(j+1)$, you test Equation (11.11) for each $i = 1, 2, \ldots, p$ and if it has chosen

$$\theta_i(j+1) > \theta_i^{max}$$

you let

$$\theta_i(j+1) = \theta_i^{max}$$

and if it has chosen

$$\theta_i(j+1) < \theta_i^{min}$$

you let

$$\theta_i(j+1) = \theta_i^{min}$$

If any generated $\theta_i(j+1)$ is within the range specified by Equation (11.11), then you accept the update $\theta_i(j+1)$ with no modification.

In this way, for all $j \geq 1$ we will never update the parameter vector $\theta(j)$ to lie outside Θ. You can see that this "projection method" is very easy to implement in practice, and hence, it is often easy to avoid the problems outlined at the beginning of this subsection.

Parameter Update Termination

In an offline function approximation problem, there is a need to specify a termination criterion so that when this criterion is met, the algorithm is stopped and the final computed value of the parameters is taken to be the best solution (which below, we will call $\theta^\star$, since the algorithm may not have found the optimal solution, i.e., it may be that $\theta^\star \neq \theta^*$). It is also possible in an online method, where you decide to do multiple iterations of the gradient method from one sampling instant to the next (an issue that is discussed more in the next section), to use a type of termination criterion as we will discuss below.

It is best to use "scale-free" termination criteria such as the following:

Choice of termination criteria is governed by the desire to terminate with the best possible approximation.

1. *Terminate if Parameter Change Rate is Low:* Terminate if

$$(\theta(j+1)-\theta(j))^\top(\theta(j+1)-\theta(j)) \leq \epsilon\theta(j)^\top\theta(j)$$

for some $\epsilon > 0$ and let $\theta^\star = \theta(j+1)$. This requires the relative amount of change in the parameter values to decrease enough before termination. In this case, it terminates since the parameters are not changing much anyway, so the algorithm is not making much progress. Other times, tests that check that the parameters have not changed much over the last fixed number of iterations are used.

2. *Terminate if the Gradient is Small:* Terminate if

$$\nabla J(\theta(j),G)^\top \nabla J(\theta(j),G) \leq \epsilon \nabla J(\theta(0),G)^\top \nabla J(\theta(0),G)$$

for some $\epsilon > 0$ and let $\theta^\star = \theta(j)$. In this case, when the size of the gradient is small relative to its size in the beginning, then you terminate since the algorithm is operating slowly at this point and will probably not make many further improvements.

While such methods are often used, in practice they are often augmented (i.e., used in conjunction with) other criteria such as the following:

1. *Terminate Based on a "Validation Set":* For many problems you have an idea of how much you would like to reduce the cost function and when you get to this value, you terminate. Along these lines, in the function approximation problem one common approach is to pick a "validation set"

$$V = \{(x(i), y(i)) : i = 1, 2, \ldots, M_v\}$$

on which the cost function value will be evaluated. While some training data from G may be used in V, it is best if V contains a significant amount of data that are not in G so that $J(\theta(j), V)$ is a measure of $J(\theta(j), \Gamma)$ (recall that Γ is the "test set") and hence also quantifies how well the function approximator "generalizes" (i.e., interpolates between training data points) and achieves its overall function approximation task. Now using V, terminate when $J(\theta(j), V)$ starts increasing since this will stop the algorithm before it starts generalizing poorly. Other times, you may terminate when $J(\theta(j), V) < \epsilon$ (i.e., when you have reduced a measure of the function approximation error to less than some value) or

$$J(\theta(j), V) \leq \epsilon J(\theta(0), V)$$

(i.e., you have reduced a measure of the function approximation error to some percentage of its initial value) and let $\theta^\star = \theta(j)$. Both of these approaches could make sense for particular applications.

2. *Terminate After a Maximum Number of Iterations:* In practical problems, you simply have to set a maximum number of iterations that you will allow. Otherwise, the above criteria may either never be met or take too long to reach. If N_{max} is the maximum number of iterations that you will allow, upon termination you will let $\theta^\star = \theta(N_{max})$.

Regardless of which termination method(s) you choose, it is important to view them as something that may also need to be tuned (i.e., modified for each application to get the best or at least an adequate approach).

11.1.7 Offline and Online, Serial and Parallel Data Processing

In this section we discuss how data can be processed by gradient algorithms. We focus on issues of order of data processing and parallel versus serial processing. We will, however, discuss such issues in the context of whether we are doing offline or online (i.e., real time) processing of training data.

The manner in which you process the given training data can significantly affect the performance of the approximator.

Offline Processing

In this case we know G a priori and hence its size M is fixed. The gradient methods discussed up till now (except some in the linear in the parameters case) process the data set G in "parallel" by repeated application of the gradient update formula to the entire data set. There are, however, ways to process the data in G serially and this can, in some cases, lead to savings in computational complexity, and improved convergence properties.

For instance, sometimes a sequential fixed-order processing of single data pairs from G can improve the rate of convergence over the case where the data in G is used in a batch fashion. In this case, you simply order the data in G and process it in that order many times (cycle through the finite data set G repeatedly). For each data pair you could execute one (or more) iterations of the gradient update formula. Normally, in this case you would use the parameter set last generated by the algorithm to initialize the algorithm when you start processing another piece of data.

Other times, when G is known a priori, you can process the data from G one at a time in a random order (perhaps with more than one iteration of the gradient update formula for each piece of data), and sometimes this has been found to provide better approximation accuracy.

In addition, whether you use fixed-order cycling through the data or a random order, you could process a *subset* of the data in G (i.e., more than one piece of training data) and perhaps execute more than one iteration for each subset (again, when you start the processing for one subset you normally would initialize the gradient update formula with the parameter set found at the last iteration for the last subset considered). Or, you could process subsets of different sizes at different times. For example, you could start by processing the data pairs one at a time and then increase the size of the subset processed at each subsequent step (the rate of the increase in subset size would be a design parameter) until all the data pairs in G were processed in a batch fashion. The algorithm could then continue in the normal batch processing mode until some termination criterion was met.

No matter which type of processing you choose for your application, it is important to keep in mind that step size choice and data processing choices are interrelated (e.g., sometimes you want to diminish your step size if you grow the size of the batch of data that you process at each step).

Online Processing

The data processing issues are different for the online case since we do not know G a priori since $M \to \infty$ (of course, the number of data considered is never really infinite, it is just convenient to think of it that way). The most common case in online (real time) processing is to use one training data pair per time step and take one iteration of the gradient formula; this aligns time steps, data acquisitions, and iterations of the gradient update formula.

But, of course, you could gather multiple pieces of data, and iterate the gradient update formula multiple times for each of these data sets (in this case, we acquire data at a higher rate than we initiate processing of the gradient update formula). Again, when you start the processing for one subset, you normally would initialize the gradient update formula with the parameter set, found at the last iteration for the last subset considered. Just as with the offline approach, you could process varying sizes of subsets of data at each step, and you will have to pay attention to the effects of data processing choice on selection of the step size.

11.1.8 Stochastic Gradient Optimization Basics

Suppose that we seek to minimize $J(\theta) \geq 0$ where $\nabla J(\theta(j))$ is Lipschitz in θ. Suppose that we use the gradient method

$$\theta(j+1) = \theta(j) + \lambda_j d(j)$$

with

$$d(j) = -\left(\nabla J(\theta(j)) + n(j)\right) \tag{11.12}$$

where

$$\sum_{j=0}^{\infty} \lambda_j = \infty, \sum_{j=0}^{\infty} \lambda_j^2 < \infty \tag{11.13}$$

(so the step size results in a persistent parameter update). Also, $n(j) \in \Re^p$ is a vector noise term with

$$E[n(j)|\mathcal{H}_j] = 0 \tag{11.14}$$

where $\mathcal{H}_j = \{\theta(i), d(i), \lambda_i : i = 0, 1, 2, \ldots, j\}$ holds the past information from the algorithm and $E[\cdot]$ denotes the expected value, and

$$E[n(j)^\top n(j)|\mathcal{H}_j] = c_1 + c_2 \nabla J(\theta(j))^\top \nabla J(\theta(j)) \tag{11.15}$$

where c_1 and c_2 are two positive constants. Under these conditions the sequence $J(\theta(j))$ converges, $\lim_{j\to\infty} \nabla J(\theta(j)) = 0$, and every limit point of $\theta(j)$ is a stationary point of J.

Equation (11.12) is a standard deterministic steepest descent approach modified at each step by perturbing the steepest descent direction with the direction $n(j)$. How could this help with the performance of the optimization process? On average, due to Equation (11.14), the algorithm will move in the steepest descent direction. Due to Equation (11.15), the noise perturbations will not be so large that they will destroy convergence properties. On a surface $J(\theta)$ that has many local minima, it may be that $n(j)$ will move the updates so as to avoid local minima (i.e., it could help "jump" out of local minima).

11.2 Levenberg-Marquardt and Conjugate Gradient Methods

In this section we introduce Newton's, conjugate gradient, and Levenberg-Marquardt methods, the latter two of which have been found to be quite effective in solving practical function approximation problems.

11.2.1 Newton's Method

For the nonlinear in the parameter approximator, the cost function $J(\theta, G)$ is not a quadratic function of θ as it is in the linear in the parameter case. For this reason, $J(\theta, G)$ can take on very complex shapes (with many local minima) with high slope regions in some areas of the parameter space and very low slope regions in other areas. In the low slope regions, the gradient is (very) small so if you use a constant step size, the changes to $\theta(j)$ will generally be small and convergence will generally be (very) slow. In the high slope regions, if a constant step size is used, the changes to $\theta(j)$ can be too large so that convergence is not achieved. While there are a variety of approaches to try to solve these problems with the steepest descent approach (e.g., adaptive step size methods, modifications to the descent direction, such as the "momentum term" approach), in practice such modifications are often found to be lacking. For example, the traditional "backpropagation" algorithm is a (stochastic) steepest descent method whose direct application has often been found to lead to slow convergence for practical problems. It is for these reasons that more sophisticated methods have been employed for tuning θ that rely on the more sophisticated use of information from $F(x, \theta)$. The methods we are referring to are the Newton, conjugate gradient and quasi-Newton, Gauss-Newton, and Levenberg-Marquardt methods.

Newton's Parameter Update Formula

Let

$$\nabla^2 J(\theta(j), G) = \left[\frac{\partial^2 J(\theta, G)}{\partial \theta_i \theta_j} \right] \Bigg|_{\theta=\theta(j)}$$

be the (symmetric) $p \times p$ "Hessian matrix," whose elements are the second partials of $J(\theta, G)$ at $\theta = \theta(j)$. In Newton's method we make a quadratic approximation of $J(\theta, G)$ at $\theta(j)$ at each iteration j and minimize this to generate $\theta(j+1)$. Let the quadratic approximation at $\theta(j)$ be the second order Taylor series expansion of $J(\theta, G)$ at $\theta(j)$ (i.e., a Taylor series expansion truncated after the second term), which we will denote by

Newton's method is based on producing a quadratic approximation to the cost function at the current parameter values and then choosing the next parameters to be those that minimize that quadratic cost.

$$\begin{aligned} J_q(\theta, G) &= J(\theta(j), G) + \nabla J(\theta(j), G)^\top (\theta - \theta(j)) + \\ & \frac{1}{2}(\theta - \theta(j))^\top \nabla^2 J(\theta(j), G)(\theta - \theta(j)) \end{aligned}$$

Since this is quadratic in θ, if we take the derivative with respect to θ and set it equal to zero and solve for θ, its value will be $\theta(j+1)$, the parameter vector that minimizes $J_q(\theta(j), G)$. In particular,

$$\nabla J_q(\theta, G) = \nabla J(\theta(j), G) + \nabla^2 J(\theta(j), G)(\theta - \theta(j))$$

and if we let $\nabla J_q(\theta, G) = 0$, we get $\theta = \theta(j+1)$ and Newton's update formula is

$$\theta(j+1) = \theta(j) - \lambda_j \left(\nabla^2 J(\theta(j), G)\right)^{-1} \nabla J(\theta(j), G) \qquad (11.16)$$

where we have added the step size $\lambda_j > 0$. In the case where $\lambda_j = 1$ for all j, the method is called the "pure form" of Newton's method. In this case, the Newton direction (which may not be a descent direction since it may be that $\nabla^2 J(\theta(j), G)$ is not positive definite, so it may be that J is not decreased at each iteration) is

Newton's method is not generally practical since it depends on computation of the inverse of the Hessian of the cost function.

$$d(j) = -\left(\nabla^2 J(\theta(j), G)\right)^{-1} \nabla J(\theta(j), G) \qquad (11.17)$$

Notice here that we have to assume that $\nabla^2 J(\theta(j), G)$ is invertible (e.g., if it is positive definite, then it is invertible). Note also that this pure form for Newton's method can be attracted by both local minima and maxima. There are many methods that have been developed to try to solve these problems with Newton's method.

The Linear in the Parameter Case

For a function $J(\theta, G)$ that is quadratic in θ (the linear in the parameter case), Newton's method provides convergence in one step. To see this, first consider our simple scalar quadratic example in Figure 11.3, where $J(\theta, G) = \theta^2$. Recall that we had

$$\nabla J(\theta, G)|_{\theta=\theta(j)} = 2\theta(j)$$

so that

$$\nabla^2 J(\theta, G)\big|_{\theta=\theta(j)} = 2$$

Hence, Newton's method for this simple example is given by

$$\theta(j+1) = \theta(j) - \left(\frac{1}{2}\right)(2\theta(j))$$

if we pick $\lambda_j = \lambda = 1$ for all j (i.e., a "pure" Newton update). Clearly, if we guess any value of $\theta(0)$, we get

$$\theta(1) = 0$$

so we get convergence in one step (i.e., Newton's method modifies the descent direction so that it converges very fast). Of course, this one step convergence (to the least squares value) only works for quadratic functions and it works because the quadratic approximation $J_q(\theta, G)$ is exact.

In tuning only parameters that enter linearly, the quadratic approximation used in Newton's method is exact so its first update is a batch least squares solution and convergence occurs in one step.

In the general linear in the parameter case, using derivatives from the last section,

$$\nabla J(\theta, G)|_{\theta=\theta(j)} = -\Phi^\top E\big|_{\theta=\theta(j)}$$

and

$$\nabla^2 J(\theta, G)\big|_{\theta=\theta(j)} = \Phi^\top \Phi$$

so that the pure Newton method is

$$\begin{aligned}\theta(j+1) &= \theta(j) + (\Phi^\top\Phi)^{-1}\Phi^\top E\big|_{\theta=\theta(j)} \\ &= \theta(j) + (\Phi^\top\Phi)^{-1}\Phi^\top(Y - \Phi\theta(j)) \\ &= (\Phi^\top\Phi)^{-1}\Phi^\top Y\end{aligned}$$

and we see that we get one-step convergence no matter what the initial guess $\theta(0)$ is. Note that this simply shows that for the linear in the parameter case, Newton's method is equivalent to a batch least squares approach and hence it relies on the existence of the inverse shown in Equation (11.16). We pay the price for fast convergence by assuming the existence of the inverse and computing it.

11.2.2 Conjugate Gradient and Quasi-Newton Methods

In the nonlinear in the parameter case, it is very difficult to guarantee the existence of the inverse in the Newton update formula, and difficult to compute it. Hence, Newton's method is rarely used in practice for the tuning of function approximators. Newton's method does, however, set up a goal for us in terms of convergence rate and hence, many methods try to approximate it but still avoid the problems with the computation of the inverse. The Levenberg-Marquardt method is one approach to avoid the computations necessary for the Newton method but still try to achieve its nice rate of convergence properties. It is discussed later.

In this section we briefly study other methods to try to speed up the steepest descent method without the extra computations associated with Newton's method.

Conjugate Gradient Methods

Conjugate gradient methods were originally developed for quadratic optimization problems to try to keep the descent directions properly aligned to speed the convergence of the steepest descent approach. In fact, for a quadratic minimization problem with p variables, they can be shown to converge in p steps. For nonlinear optimization problems they cannot in general be shown to provide this fast convergence property; however, for many problems they do provide good convergence and rate of convergence properties. They achieve this without using the Hessian or any matrix inversion; hence, they have sometimes been found to be useful for problems with a large value of p.

There are many variations on the parameter update formula for conjugate gradient methods (many of these are equivalent for the quadratic case but different for the nonlinear case). Here, we pick just the most common one that is given by

$$\theta(j+1) = \theta(j) + \lambda_j d(j)$$

where λ_j is generated by a line minimization so that

$$J(\theta(j) + \lambda_j d(j), G) = \min_{\lambda \in [0, \lambda^0]} J(\theta(j) + \lambda d(j), G)$$

The accuracy of the line minimization can affect the performance of the algorithm and often you need to experiment with the choice of parameters of the line minimization method to get good performance. Here, we will assume that the Armijo step size rule is used to specify λ_j. The descent direction $d(j)$ is given by

$$d(j) = -\nabla J(\theta(j), G) + \beta(j) d(j-1) \tag{11.18}$$

where

$$\beta(j) = \frac{\nabla J(\theta(j), G)^\top \left(\nabla J(\theta(j), G) - \nabla J(\theta(j-1), G)\right)}{\nabla J(\theta(j-1), G)^\top \nabla J(\theta(j-1), G)}$$

is called the "Polak-Ribiere" formula.

In the practical application of the method, it has been found to be useful to make certain modifications to the method to improve its convergence properties. There are many ways to modify the algorithm. For instance, the method could be modified by using a steepest descent direction at the first step. Then, every N_{cg} steps, the algorithm is "restarted" by using a steepest descent update direction.

Quasi-Newton Methods

In "quasi-Newton methods" you try to avoid problems with existence and computation of the inverse in Equation (11.17) by choosing

$$d(j) = -\Lambda(j)\nabla J(\theta(j), G)$$

where $\Lambda(j)$ is a positive definite $p \times p$ matrix for all j and that is chosen to approximate $\left(\nabla^2 J(\theta(j), G)\right)^{-1}$. In this way, $d(j)$ may approximate a Newton direction and we may get the associated fast convergence without all the extra computations.

For example, in some cases the approximation is performed by letting $\Lambda(j)$ be a diagonal matrix with its elements set to the corresponding diagonal elements of $\left(\nabla^2 J(\theta(j), G)\right)^{-1}$ and in this case, the method is often referred to as the "diagonally scaled steepest descent method." In some practical applications this method can be quite effective.

Generally, if $\Lambda(j)$ is chosen properly, for some applications much of the convergence speed of Newton's method can be achieved. In other more sophisticated approaches, $\Lambda(j)$ is chosen to form an approximation to the inverse of the Hessian. Here, we outline just one, the "Broyden-Fletcher-Goldfarb-Shanno" (BFGS) method. For this, we have

$$\theta(j+1) = \theta(j) + \lambda_j d(j)$$

where λ_j is generated by a line minimization (e.g., the Armijo step size rule), and

$$d(j) = -\Lambda(j)\nabla J(\theta(j), G)$$

We define

$$c(j) = \theta(j+1) - \theta(j)$$

and

$$g(j) = \nabla J(\theta(j+1), G) - \nabla J(\theta(j), G)$$

Then, we let $\Lambda(0)$ be an arbitrary positive definite matrix, and

$$\begin{aligned} \Lambda(j+1) &= \Lambda(j) + \frac{c(j)c(j)^\top}{c(j)^\top g(j)} - \frac{\Lambda(j)g(j)g(j)^\top \Lambda(j)}{g(j)^\top \Lambda(j) g(j)} \\ &\quad + g(j)^\top \Lambda(j) g(j) h(j) h(j)^\top \end{aligned}$$

where

$$h(j) = \frac{c(j)}{c(j)^\top g(j)} - \frac{\Lambda(j)g(j)}{g(j)^\top \Lambda(j) g(j)}$$

Even though storage of $\Lambda(j)$ and other computational requirements can be heavy for the BFGS method, there are situations where the BFGS method may be preferred to the conjugate gradient method. The BFGS method can provide fast convergence when it is near a solution, and generally seems to be less sensitive to line minimization accuracy. Depending on the complexity of computing $J(\theta(j), G)$ and its gradient, you may, however find one method preferred over the other.

11.2.3 Gauss-Newton and Levenberg-Marquardt Methods

Next, we consider the Gauss-Newton method that is used to solve a (nonlinear) least squares problem, such as finding θ to minimize $J(\theta, G)$ in Equation (11.1) when we do not use a linear in the parameter approximator. To develop the Gauss-Newton method, we have

$$J(\theta, G) = \frac{1}{2}\sum_{i=1}^{M} |y(i) - F(x(i), \theta)|^2$$

and let the $\bar{N} \times 1$ vectors

$$\epsilon(i) = y(i) - F(x(i), \theta)$$

(these are the function approximation errors arising at each piece of training data) and define the $\bar{N}M \times 1$ vector

$$\begin{aligned}\epsilon(\theta, G) &= [\epsilon(1)^\top, \epsilon(2)^\top, \ldots, \epsilon(M)^\top]^\top \\ &= [\epsilon_1, \epsilon_2, \ldots, \epsilon_{\bar{N}M}]^\top\end{aligned}$$

(where ϵ_j, $j = 1, 2, \ldots, \bar{N}M$, are scalars) to be a vector containing all of the approximation errors. Note that

$$J(\theta, G) = \frac{1}{2}\sum_{i=1}^{M} \epsilon(i)^\top \epsilon(i) = \frac{1}{2}\epsilon(\theta, G)^\top \epsilon(\theta, G)$$

In the Gauss-Newton method, the approximation error is linearized about the current parameter values and then least squares is used to minimize the linearized error value to provide the next parameter update.

For functions $J(\theta, G)$ that are quadratic in θ (scalar or vector case), Newton's method gave very fast convergence (in one step). For the function approximation problem, to get $J(\theta, G)$ quadratic in θ, we use a linear in the parameter approximator $F(x, \theta) = \theta^\top \phi(x)$ so that the approximation errors $\epsilon(i)$ and hence $\epsilon(\theta, G)$ are linear (affine) with respect to the parameters θ, and then $J(\theta, G)$ is quadratic in θ. If $F(x, \theta)$ is nonlinear in the parameters, then so are $\epsilon(i)$ and $\epsilon(\theta, G)$.

Gauss-Newton Parameter Update Formula

To tune nonlinear in the parameter approximators in the Gauss-Newton approach, at each iteration j we proceed according to the following steps:

1. Linearize the error $\epsilon(\theta, G)$ about the current value of $\theta(j)$.
2. Solve a least squares problem to minimize the linearized error value and provide the next guess at the parameter, $\theta(j+1)$.

Compared to Newton's method, in the Gauss-Newton method you create a quadratic approximation to the function you want to minimize at each iteration, but now it is done via linearization, rather than using second derivative information. We discuss these two steps in more detail next.

First, linearize $\epsilon(\theta, G)$ around $\theta(j)$ using a truncated Taylor series expansion to get

$$\hat{\epsilon}(\theta, \theta(j), G) = \epsilon(\theta(j), G) + \nabla\epsilon(\theta, G)^\top\big|_{\theta=\theta(j)} (\theta - \theta(j))$$

where $\hat{\epsilon}(\theta, \theta(j), G)$ is an approximation of $\epsilon(\theta, G)$ since we omitted the higher order terms (second order and higher) in the Taylor series expansion. We use the notation $\hat{\epsilon}(\theta, \theta(j), G)$ to emphasize the dependence on both θ and $\theta(j)$. Here,

$$\nabla\epsilon(\theta, G) = \begin{bmatrix} \frac{\partial \epsilon_1}{\partial \theta_1} & \cdots & \frac{\partial \epsilon_{\bar{N}M}}{\partial \theta_1} \\ & \ddots & \\ \vdots & & \vdots \\ & \ddots & \\ \frac{\partial \epsilon_1}{\partial \theta_p} & \cdots & \frac{\partial \epsilon_{\bar{N}M}}{\partial \theta_p} \end{bmatrix} = [\nabla\epsilon_1, \nabla\epsilon_2, \ldots, \nabla\epsilon_{\bar{N}M}] \quad (11.19)$$

is a $p \times \bar{N}M$ matrix and $\nabla\epsilon(\theta, G)^\top$ is the "Jacobian."

Second, minimize the (scaled) squared norm,

$$J_q(\theta, G) = \frac{1}{2}\hat{\epsilon}(\theta, \theta(j), G)^\top \hat{\epsilon}(\theta, \theta(j), G)$$

which is a quadratic approximation to $J(\theta, G)$ (at $\theta(j)$), which is nonlinear in θ, and different from the one used in Newton's method. Let

$$\begin{aligned} \theta(j+1) &= \arg\min_\theta J_q(\theta, G) \\ &= \arg\min_\theta \frac{1}{2}\hat{\epsilon}(\theta, \theta(j), G)^\top \hat{\epsilon}(\theta, \theta(j), G) \end{aligned}$$

(here, "$\arg\min_\theta$" is simply mathematical notation for the value of θ that minimizes the norm—it is the "argument" that provides the value that achieves the minimization).

We know how to solve this problem. It is the same as the batch least squares problem for the linear in the parameters case. To see this, note that

$$J_q(\theta, G) = \frac{1}{2}E^\top E$$

with $E = \hat{\epsilon}(\theta, \theta(j), G)$. Recall that we had

$$E = Y - \Phi\theta$$

Here, we have

$$\hat{\epsilon}(\theta, \theta(j), G) = \left(\epsilon(\theta(j), G) - \nabla\epsilon(\theta, G)^\top\big|_{\theta=\theta(j)} \theta(j)\right) + \nabla\epsilon(\theta, G)^\top\big|_{\theta=\theta(j)} \theta$$

so if we let

$$Y = \left(\epsilon(\theta(j), G) - \nabla\epsilon(\theta, G)^\top\big|_{\theta=\theta(j)} \theta(j)\right)$$

and

$$\Phi = - \left. \nabla\epsilon(\theta, G)^\top \right|_{\theta=\theta(j)}$$

our least squares solution (the value of the parameter at the next iteration) is given by Equation (10.2) as

$$\theta(j+1) = (\Phi^\top \Phi)^{-1} \Phi^\top Y$$

so $\theta(j+1)$ is

$$-\left(\nabla\epsilon(\theta(j), G)\nabla\epsilon(\theta(j), G)^\top\right)^{-1} \nabla\epsilon(\theta(j), G)\left(\epsilon(\theta(j), G) - \nabla\epsilon(\theta(j), G)^\top \theta(j)\right)$$

where

$$\nabla\epsilon(\theta(j), G)^\top = \left. \nabla\epsilon(\theta, G)^\top \right|_{\theta=\theta(j)}$$

so that the resulting Gauss-Newton update formula is

$$\theta(j+1) = \theta(j) - \left(\nabla\epsilon(\theta(j), G)\nabla\epsilon(\theta(j), G)^\top\right)^{-1} \nabla\epsilon(\theta(j), G)\epsilon(\theta(j), G) \quad (11.20)$$

(If we had included a step size parameter, then the method is sometimes referred to as a "damped" Gauss-Newton approach.) Notice that compared with Newton's method, we do not need the Hessian, only the Jacobian. Essentially, a Gauss-Newton iteration is an approximation to a Newton iteration (in the sense that the quadratic approximation at each iteration tries to approximate the one in Newton's method that uses second derivative information in its quadratic approximation) so it can typically provide for faster convergence than, for instance, steepest descent, but generally not as fast as a pure Newton method. It is also interesting to note that the Gauss-Newton method is the same as the "extended Kalman filter" (EKF) except where the linearizations are performed. (In the EKF, where we process one data pair at a time, we perform the linearizations at each point; for the Gauss-Newton method where batch processing is used, we perform the linearizations for each batch.) To make the methods the same simply involves changing how the data are processed.

The Gauss-Newton method is equivalent to the extended Kalman filter if the data are processed in the same way.

Levenberg-Marquardt Parameter Update Formula

To avoid problems with computing the inverse in Equation (11.20), the method is often implemented as

$$\theta(j+1) = \theta(j) - \left(\nabla\epsilon(\theta(j), G)\nabla\epsilon(\theta(j), G)^\top + \Lambda(j)\right)^{-1} \nabla\epsilon(\theta(j), G)\epsilon(\theta(j), G) \quad (11.21)$$

where $\Lambda(j)$ is a $p \times p$ diagonal matrix such that

$$\nabla\epsilon(\theta(j), G)\nabla\epsilon(\theta(j), G)^\top + \Lambda(j)$$

is positive definite so that it is invertible. Sometimes, a "Cholesky factorization" is used to specify $\Lambda(j)$ at each iteration. In the Levenberg-Marquardt method, you choose $\Lambda(j) = \lambda I$ where $\lambda > 0$ and I is the $p \times p$ identity matrix. The parameter λ serves a role similar to the step size. When $\lambda = 0$, we get the standard Gauss-Newton method and as you increase λ, the descent direction moves towards the gradient. Hence, generally thinking of λ as a step size, we expect that for a small value of λ, we will get fast convergence; for a larger value, we should get slower convergence.

The Linear in the Parameter Case

As a simple example, notice that in the case where $J(\theta, G) = \theta^2$, $\epsilon(\theta, G) = \theta$, and if we pick $\lambda_j = \lambda = 1$, $\Lambda(j) = \Lambda = I$ for all j, then $\nabla\epsilon(\theta, G) = 1$ so the Gauss-Newton method is

$$\theta(j+1) = \theta(j) - \theta(j) = 0$$

(and the Levenberg-Marquardt method would be the same if you pick $\lambda_j = \lambda = 2$). Hence, no matter what the choice is for the initial guess $\theta(0)$, $\theta(1) = 0$, and we get convergence in one step (similar to Newton's method for this example). Generally, however, this only occurs in the case where $J(\theta, G)$ is quadratic and we have a linear in the parameters approximator.

11.3 Matlab for Training Neural Networks

There exist many software packages for solving optimization problems with gradient methods (you may want to search the Web to find some public-domain ones), and one that is particularly well-suited for the gradient training of neural networks is the Matlab Neural Networks Toolbox.

11.3.1 Motivation to Use Software Packages

This toolbox provides a variety of tools that facilitate the construction of neural network structures and the training of the parameters in these structures. The training methods include steepest descent, conjugate gradient methods, Levenberg-Marquardt, and others. Moreover, many numerical issues for these algorithms have been tested to help ensure their robustness and numerical accuracy for ease of use.

Excellent software packages exist for optimization and its application to training neural networks and fuzzy systems.

It is recommended that if you want to construct complex multilayer neural networks (e.g., a perceptron with two or more hidden layers) or regularly work with sophisticated practical applications, that you use some software package. In this part we have shown the basic concepts, but there are more issues to deal with when the networks get more complex, in addition to how sophisticated the gradient method is. For instance, for more complex multilayer neural networks, the recognition of the repeated calculations that are necessary in gradient update formulas (particularly steepest descent) led to the "backpropagation" method, which is a method for saving computations in the application of the gradient method (even though, most often, the term "backpropagation" is used to refer to the scheme for saving computations, *and* the gradient method employed).

The Matlab toolbox exploits the repeated calculations using a backpropagation method and frees you from these somewhat tedious details so that you can focus on the fundamental issues in training that are discussed here.

It is also important to point out that many software packages, including Matlab, provide functions for general nonlinear least squares minimization (e.g., using the Levenberg-Marquardt method) and all you have to do is find the gradients and provide the proper information to the software and it will provide a solution. Hence, with such packages it is not only possible to train neural networks but also Takagi-Sugeno fuzzy systems or other approximator structures.

An exercise at the end of the chapter asks you to solve a simple function approximation problem with software such as the Matlab toolbox. Next, we show how to train a multilayer perceptron with the Matlab Neural Networks Toolbox.

11.3.2 Example: Matlab Neural Networks Toolbox

In this section, we will show how to use the Matlab Neural Networks Toolbox to tune a multilayer perceptron to match the training data shown in Figure 9.10 (this defines G and in our case, we have $M = 121$). In particular, we will train a two layer multilayer perceptron with $n_1 = 11$ hidden layer neurons that have logistic activation functions and a linear activation function in the output layer (as shown in Figure 9.13). We will test different training methods, and will use 500 training "epochs" in each case. The code used is given at the Web site for the book listed in the Preface.

Gradient Descent Training

In this case we use the training option `traingd` which indicates that we want to use a gradient descent approach (this is the classical "backpropagation" approach). When you execute the program, it displays data indicating how the algorithm is performing (e.g., the mean squared error and size of the gradient) and a plot of the mean square error versus the epoch number as shown in Figure 11.4. Notice that as training progresses, the mean squared error decreases. The quality of the approximation for this case is shown in Figure 11.5, where we can see by inspection that a reasonably good approximation was achieved. Note that if you run the code at the Web site you will almost surely get a different plot, since the data are presented in a random order for the training.

Conjugate Gradient Training

In this case we use the training option `traincgp`, which indicates that we want to use a conjugate gradient approach (actually the Polak-Ribiere method). When you execute the program, it displays data indicating how the algorithm is performing (e.g., the mean squared error and size of the gradient) and a plot of the mean square error versus the epoch number as shown in Figure 11.6. Notice that as training progresses, the mean squared error decreases at a faster rate

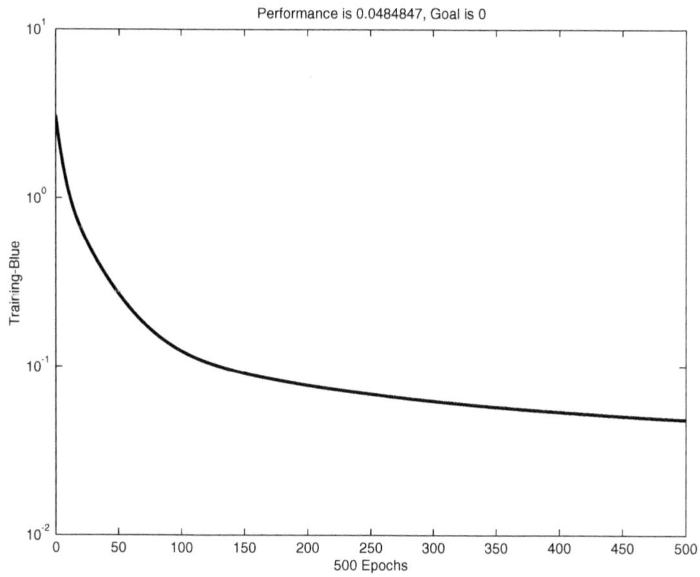

Figure 11.4: Mean squared error vs. epoch number for backpropagation training.

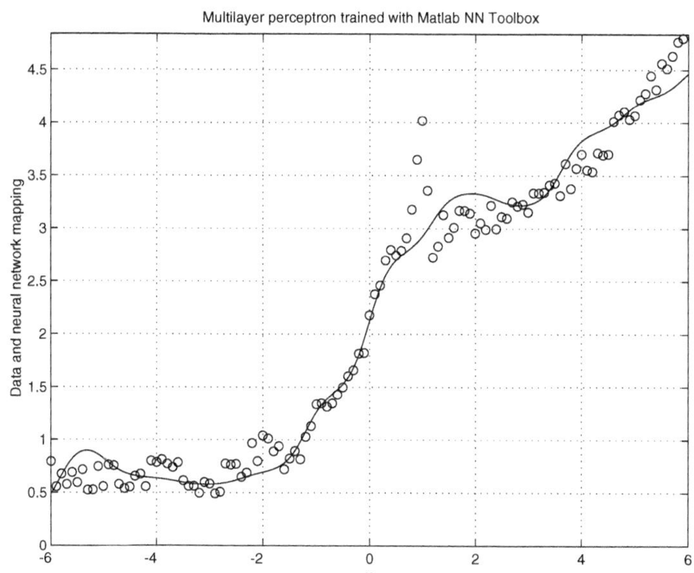

Figure 11.5: Approximator mapping and data for training with backpropagation.

and achieves a lower value than in the standard backpropagation method shown in Figure 11.4.

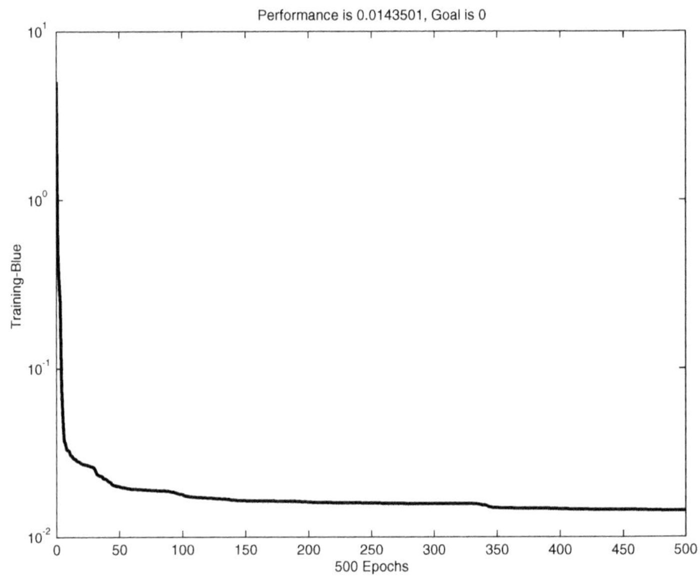

Figure 11.6: Mean squared error vs. epoch number for conjugate gradient training.

The quality of the approximation for this case is shown in Figure 11.7, where we can see by inspection that better approximation was achieved (for this number of training epochs and specific training run) than was achieved for backpropagation in Figure 11.5. It is typical to find slow training times for standard backpropagation and improvements on convergence rates and approximation errors if you compare to a conjugate gradient method.

Levenberg-Marquardt Training

In this case we use the training option `trainlm`, which indicates that we want to use a Levenberg-Marquardt training approach. When you execute the program, it displays data indicating how the algorithm is performing (e.g., the mean squared error and size of the gradient) and a plot of the mean square error versus the epoch number as shown in Figure 11.8. Notice that as training progresses, the mean squared error decreases at a relatively fast rate (even faster than what was obtained in the above training run for the conjugate gradient method in Figure 11.6), then levels off at about the same value as that which was obtained with the conjugate gradient method.

The quality of the approximation for this case is shown in Figure 11.9, where we can see by inspection that better approximation was achieved (for this num-

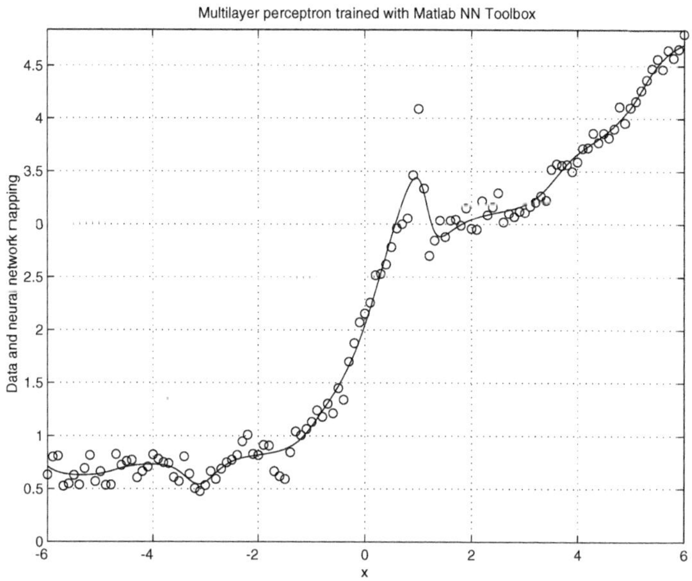

Figure 11.7: Approximator mapping and data for training with a conjugate gradient method.

ber of training epochs and specific training run) than was achieved for backpropagation in Figure 11.5. The result is, however, different from the one obtained with the conjugate gradient method in Figure 11.7, in that it does not adjust the map to the high frequency peak even though it achieves similar accuracy (of course, this is just for this training run; you should not reach any general conclusions by this).

To understand the "canned" software packages, it is useful to build a "homemade" optimization algorithm for approximator tuning.

11.4 Example: Levenberg-Marquardt Training of a Fuzzy System

In this section we study the use of the Levenberg-Marquardt method for training a Takagi-Sugeno fuzzy system with $R = 11$ rules. We will tune all 44 parameters of the approximator. Here, we consider offline batch processing of a data set $G = \{(x(i), y(i)) : i = 1, 2, \ldots, M\}$ from Figure 9.10 (where in this case $n = 1$).

In this case, our Takagi-Sugeno fuzzy system is given by

$$y = F_{ts}(x, \theta) = \frac{\sum_{i=1}^{R} g_i(x)\mu_i(x)}{\sum_{i=1}^{R} \mu_i(x)}$$

where $g_i(x) = a_{i,0} + a_{i,1}x_1$ and the $a_{i,j}$, $i = 1, 2, \ldots, R$, $j = 0, 1$ are constants.

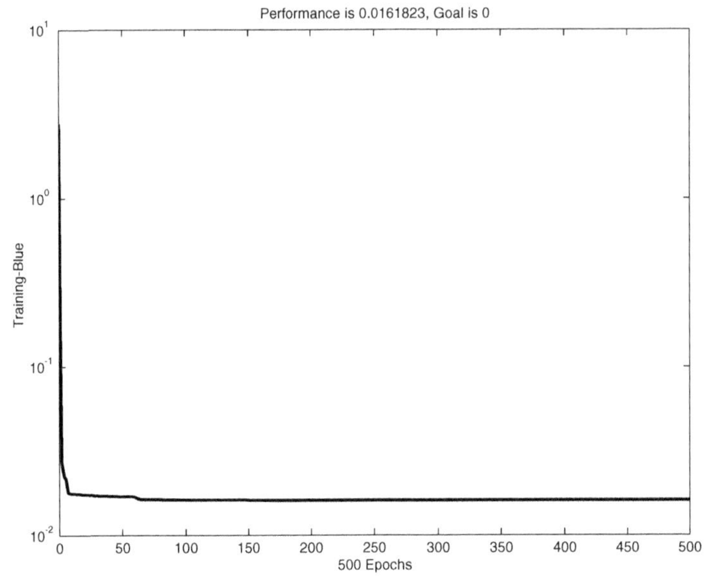

Figure 11.8: Mean squared error vs. epoch number for Levenberg-Marquardt training.

Also,

$$\mu_i(x) = \prod_{j=1}^{n} \exp\left(-\frac{1}{2}\left(\frac{x_j - c_j^i}{\sigma_j^i}\right)^2\right) = \exp\left(-\frac{1}{2}\left(\frac{x_1 - c_1^i}{\sigma_1^i}\right)^2\right)$$

where c_j^i is the point in the jth input universe of discourse where the membership function for the ith rule achieves a maximum, and $\sigma_j^i > 0$ is the relative width of the membership function for the jth input and the ith rule (since $n = 1$, the premise membership functions are the same as the input membership functions). Recall that we had defined

$$\xi_j = \frac{\mu_j(x)}{\sum_{i=1}^{R} \mu_i(x)}$$

$j = 1, 2, \ldots, R$. For our case, we have

$$\begin{aligned} \theta \quad = \quad & [c_1^1, \ldots, c_1^R, \sigma_1^1, \ldots, \sigma_1^R, \\ & a_{1,0}, a_{2,0}, \ldots, a_{R,0}, a_{1,1}, a_{2,1}, \ldots, a_{R,1}]^\top \end{aligned}$$

for a total of $p = 4R = 44$ parameters to tune.

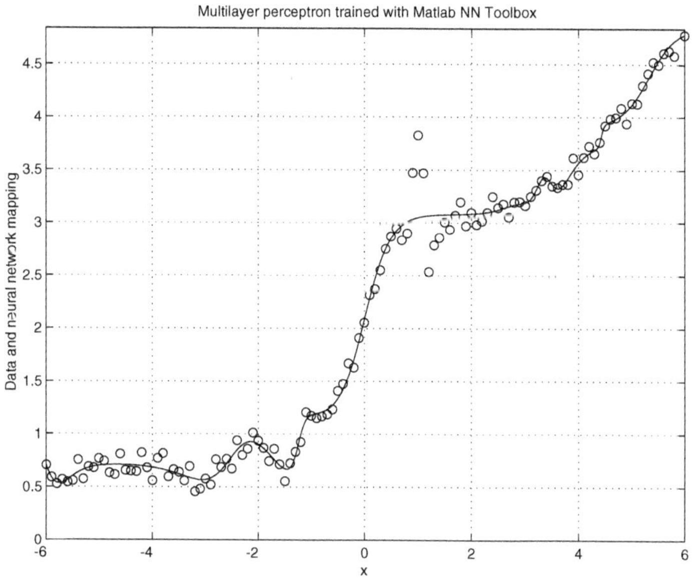

Figure 11.9: Approximator mapping and data for training with the Levenberg-Marquardt method.

11.4.1 Update Formulas

The update formula, given in Equation (11.21), is

$$\theta(j+1) = \theta(j) - \left(\nabla\epsilon(\theta(j), G)\nabla\epsilon(\theta(j), G)^\top + \Lambda(j)\right)^{-1} \nabla\epsilon(\theta(j), G)\epsilon(\theta(j), G) \tag{11.22}$$

where $\Lambda(j) = \lambda I$ where $\lambda > 0$ is a tuning parameter (where if λ is small, we can generally expect faster convergence, but we may need it to be larger to ensure the existence of the inverse) and I is the $p \times p$ identity matrix.

To make the computations for the update formula we need, for $\bar{N} = 1$, the $p \times M$ matrix $\nabla\epsilon(\theta(j), G)$ and the $M \times 1$ vector $\epsilon(\theta(j), G)$. With $\bar{N} = 1$, the scalars

$$\epsilon_i = \epsilon(i) = y(i) - F_{ts}(x(i), \theta)$$

for $i = 1, 2, \ldots, M$, and so $\epsilon(\theta, G) = [\epsilon_1, \epsilon_2, \ldots, \epsilon_M]^\top$. Here,

$$\nabla\epsilon(\theta, G) = \begin{bmatrix} \frac{\partial\epsilon_1}{\partial\theta_1} & \cdots & \frac{\partial\epsilon_M}{\partial\theta_1} \\ & \ddots & \\ \vdots & & \vdots \\ & \ddots & \\ \frac{\partial\epsilon_1}{\partial\theta_p} & \cdots & \frac{\partial\epsilon_M}{\partial\theta_p} \end{bmatrix}$$

Now, notice that for $i = 1, 2, \ldots, M$, $j = 1, 2, \ldots, p$,

$$\begin{aligned}\frac{\partial \epsilon_i}{\partial \theta_j} &= \frac{\partial}{\partial \theta_j}\left(y(i) - F_{ts}(x(i), \theta)\right) \\ &= -\frac{\partial}{\partial \theta_j} F_{ts}(x(i), \theta)\end{aligned}$$

It is convenient to compute this partial by considering various components of the vector in sequence (not forgetting about the minus sign in front of the partials).

Developing the update formulas simply requires the use of the partial of the approximator with respect to the parameters.

First, consider the update formulas for the centers of the premise membership functions. We will use indices i^* and j^* to help avoid confusion with the indices i and j. We find, for $j^* = 1, 2, \ldots, R$,

$$\frac{\partial}{\partial c_1^{j^*}} F_{ts}(x(i^*), \theta) = \frac{\partial}{\partial c_1^{j^*}} \left(\frac{\sum_{i=1}^R g_i(x(i^*)) \mu_i(x(i^*))}{\sum_{i=1}^R \mu_i(x(i^*))} \right)$$

where

$$\mu_i(x(i^*)) = \exp\left(-\frac{1}{2} \left(\frac{x(i^*) - c_1^i}{\sigma_1^i} \right)^2 \right)$$

(we replaced x_1 with x, since they are the same) and

$$\xi_{j^*}(x(i^*)) = \frac{\mu_{j^*}(x(i^*))}{\sum_{i=1}^R \mu_i(x(i^*))}$$

Hence, we have

$$\begin{aligned}\frac{\partial}{\partial c_1^{j^*}} F_{ts}(x(i^*), \theta) &= \frac{\left(\sum_{i=1}^R \mu_i(x(i^*))\right)\left(g_{j^*}(x(i^*)) \frac{\partial}{\partial c_1^{j^*}} \mu_{j^*}(x(i^*))\right)}{\left(\sum_{i=1}^R \mu_i(x(i^*))\right)^2} \\ &\quad - \frac{\left(\sum_{i=1}^R g_i(x(i^*))\mu_i(x(i^*))\right)\left(\frac{\partial}{\partial c_1^{j^*}} \mu_{j^*}(x(i^*))\right)}{\left(\sum_{i=1}^R \mu_i(x(i^*))\right)^2} \\ &= \left(\frac{g_{j^*}(x(i^*)) - F_{ts}(x(i^*), \theta)}{\sum_{i=1}^R \mu_i(x(i^*))} \right) \frac{\partial}{\partial c_1^{j^*}} \mu_{j^*}(x(i^*))\end{aligned}$$

For this, let

$$\bar{x}^{j^*} = -\frac{1}{2} \left(\frac{x(i^*) - c_1^{j^*}}{\sigma_1^{j^*}} \right)^2$$

so that using the chain rule from calculus

$$\frac{\partial}{\partial c_1^{j^*}} \mu_{j^*}(x(i^*)) = \frac{\partial \mu_{j^*}(x(i^*))}{\partial \bar{x}^{j^*}} \frac{\partial \bar{x}^{j^*}}{\partial c_1^{j^*}}$$

We have

$$\frac{\partial \mu_{j^*}(x(i^*))}{\partial \bar{x}^{j^*}} = \mu_{j^*}(x(i^*))$$

and

$$\frac{\partial \bar{x}^{j^*}}{\partial c_1^{j^*}} = \frac{x(i^*) - c_1^{j^*}}{\left(\sigma_1^{j^*}\right)^2}$$

so

$$\frac{\partial}{\partial c_1^{j^*}} F_{ts}(x(i^*), \theta) = \left(\frac{g_{j^*}(x(i^*)) - F_{ts}(x(i^*), \theta)}{\sum_{i=1}^{R} \mu_i(x(i^*))}\right) \mu_{j^*}(x(i^*)) \frac{\left(x(i^*) - c_1^{j^*}\right)}{\left(\sigma_1^{j^*}\right)^2}$$

Next, for the spreads on the premise membership functions, we use the same development above to find

$$\frac{\partial}{\partial \sigma_1^{j^*}} F_{ts}(x(i^*), \theta) = \left(\frac{g_{j^*}(x(i^*)) - F_{ts}(x(i^*), \theta)}{\sum_{i=1}^{R} \mu_i(x(i^*))}\right) \mu_{j^*}(x(i^*)) \frac{\left(x(i^*) - c_1^{j^*}\right)^2}{\left(\sigma_1^{j^*}\right)^3}$$

since

$$\frac{\partial \bar{x}^{j^*}}{\partial \sigma_1^{j^*}} = \frac{\left(x(i^*) - c_1^{j^*}\right)^2}{\left(\sigma_1^{j^*}\right)^3}$$

Next, for the parameters of the consequent functions, notice that

$$\frac{\partial}{\partial a_{j^*,0}} F_{ts}(x(i^*), \theta) = \frac{\partial}{\partial a_{j^*,0}} \left(g_{j^*}(x(i^*)) \xi_{j^*}(x(i^*))\right) = \xi_{j^*}(x(i^*))$$

and

$$\frac{\partial}{\partial a_{j^*,1}} F_{ts}(x(i^*), \theta) = x_1(i^*) \xi_{j^*}(x(i^*))$$

This gives us all the elements for the $\nabla \epsilon(\theta, G)$ matrix, and hence, we can implement the Levenberg-Marquardt update formula.

11.4.2 Parameter Constraint Set and Initialization

The chosen parameter constraint set simply forces the centers to lie between -6 and $+6$ (hence, we assume that we know the maximum variation on the input domain a priori) and spreads to all between 0.1 and 1 and uses projection to maintain this for each iteration. We place the constraints on the spreads for two reasons. First, we must keep the values of the spreads above some fixed value to ensure that we do not have a divide-by-zero error in computing the partials needed for the update formula. Second, it seems reasonable not to have spreads cover too much of the input domain, since then its corresponding consequent (a line) would have to produce an approximation over that large portion of the domain. We put no constraints on the parameters of the consequent functions.

The centers are initialized to be on a uniform grid across the input space, a reasonable choice if you do not know where high frequency behavior occurs; however, if you know that there is a region with higher frequency oscillations, then it may be advantageous to put more centers in that region. In particular, we choose $c_1^1 = -5$, $c_1^2 = -4$, up to $c_1^{11} = 5$. The spreads are all initialized to be 0.5 so that there is a reasonable amount of separation between them when one consequent function of one rule turns on and the other turns off. We will, however, experiment with the effects of the size of the initial spreads on the performance of the method. The parameters of the consequent functions are simply initialized to be all zero. It must be emphasized that while these choices make sense for this problem, and as you will see, work reasonably well for this problem, other initializations may work better (and others, much worse).

11.4.3 Approximator Tuning Results: Effects on the Nonlinear Part

Here, we first consider the $M = 121$ case for the function shown in Figure 9.10. We will simply show the mapping shape at various iterations and hence, will not implement a termination criterion. We choose $\lambda = 0.5$ (you can easily tune this parameter where, if you make it smaller, it tends to make bigger updates). Figure 11.10 shows the mapping after just one iteration. Clearly, even after one iteration, even though it has not tuned the centers and spreads much, the method has chosen reasonable values for the consequent functions and this is not surprising considering the performance of the batch least squares method for this approach and the similarities to that method.

Next, we will focus on how the method tunes the nonlinear part of the approximator (i.e., the μ_i, and hence ξ_i functions) but we must keep in mind that the linear part is also being tuned at the same time. Figure 11.11 shows that by the second iteration, there is already significant and successful tuning of the nonlinear part so that approximation errors are reduced, particularly in the region around $x = -2$.

As the algorithm continues, it continues to tune the nonlinear part of the approximator. In particular, consider Figure 11.12 at $j = 5$, and we see that at this point, the training method has done quite a good job at shaping the nonlinear part to obtain good accuracy around $x = -2$. Note that here it is exploiting the parameters that enter in a nonlinear fashion to achieve interesting shapes for the nonlinearity (you could think of this as illustrating the inherent tuning flexibility associated with approximators, where we tune both the parameters that enter linearly and the ones that enter in a nonlinear fashion).

Adjustments to the parameters that enter nonlinearly provide significant tuning flexibility for the shape of the mapping.

As the algorithm continues, it still continues to tune the nonlinear part of the approximator, both in the region around $x = -2$ and in the high frequency region around $x = 1$. In particular, consider Figure 11.13 at $j = 12$, and we see that while it has tuned the parameters some, it is not much different in the $x = -2$ region. It is, however, having difficulties in the $x = 1$ region due to

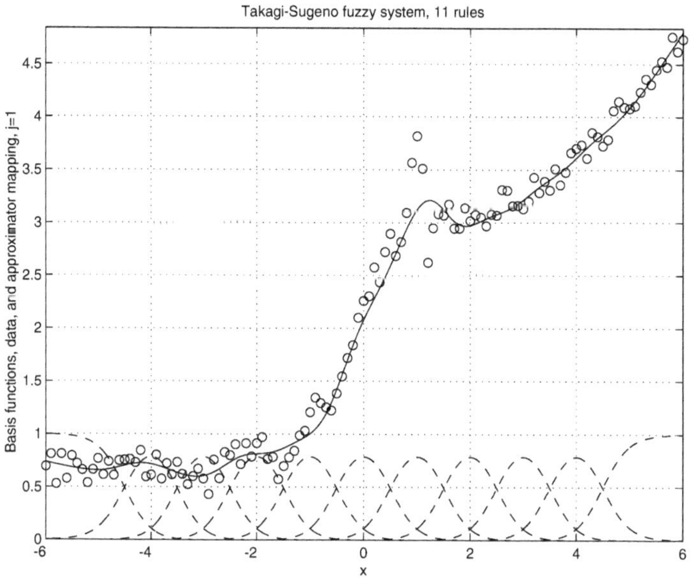

Figure 11.10: Levenberg-Marquardt training of a Takagi-Sugeno fuzzy system, mapping shape at iteration $j = 1$.

the high frequency behavior. It seems that for this example, for higher numbers of iterations, it tends to leave the approximator structure near $x = -2$ pretty much as it is and it tries to "fix" the part near $x = 1$. Consider the mapping at iteration $j = 15$, which is shown in Figure 11.14. Notice that in the $x = 1$ region, there is a significant change in the nonlinear shape. It tends to keep moving this shape around near $x = 1$ to try to improve accuracy.

Now, this is where the issue of termination arises. Do you terminate at $j = 12$ and declare success? Do you try to run the algorithm for many more iterations to see if it can "allocate" more approximator structure to the $x = 1$ high frequency region to improve the accuracy further? If you use more iterations will the overall approximation accuracy improve? Or, will it get even worse that it is here? These are all important issues, but they tend to be very application dependent. It is best if you are simply aware of all these issues and experiment with the particular application at hand to try to get the best possible results (where the definition of "best" certainly depends on the constraints of the particular application).

11.4.4 Approximator Tuning Results: Effects of Initialization

Next, consider the same initial parameters as above except let the spreads all be 0.2 instead of 0.5. Figure 11.15 shows the mapping shape at $j = 1$ and we see

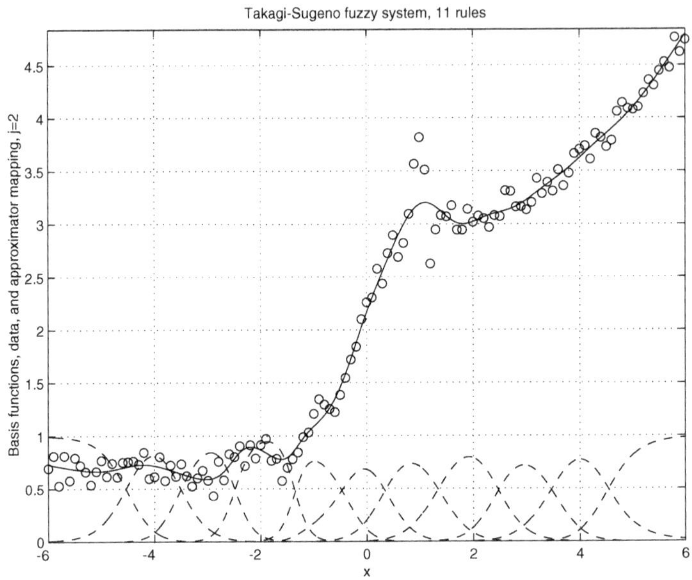

Figure 11.11: Levenberg-Marquardt training of a Takagi-Sugeno fuzzy system, mapping shape at iteration $j = 2$.

that the approximator is not performing too well. The small spread results in sharp transitions between the rules so that there is a sharp transition between the lines that are used in the consequents.

Figure 11.16 shows the mapping shape at $j = 2$. We see that the centers are updated to values that were similar to the 0.5 initialization case; the algorithm quickly recovers from what appeared to be a poor initialization (and then the behavior is qualitatively similar to the case where the spreads were initialized with 0.5 after $j = 2$).

Next, we will use the same initial parameters as above, except let the spreads all be 1 instead of 0.5. Figure 11.17 shows the mapping shape at $j = 1$ and this shows that as we smooth out the membership functions, we tend to get a smoothed out function. This time, however, the method does not recover from this initialization as fast as when the spreads were initialized at 0.2. For instance, notice that by $j = 15$ the mapping shape, which is shown in Figure 11.18, is not much better in the region around $x = -2$; it has, however, done something interesting: up to this point, the algorithm has focused on trying to allocate approximator structure to the high frequency region to try to improve approximation accuracy there.

Overall, these simulations show that the performance of the algorithm clearly depends on the initialization. We would like to start with the best possible initialization; however, for practical problems it can be particularly difficult to get a good one for a particular application without having significant insights

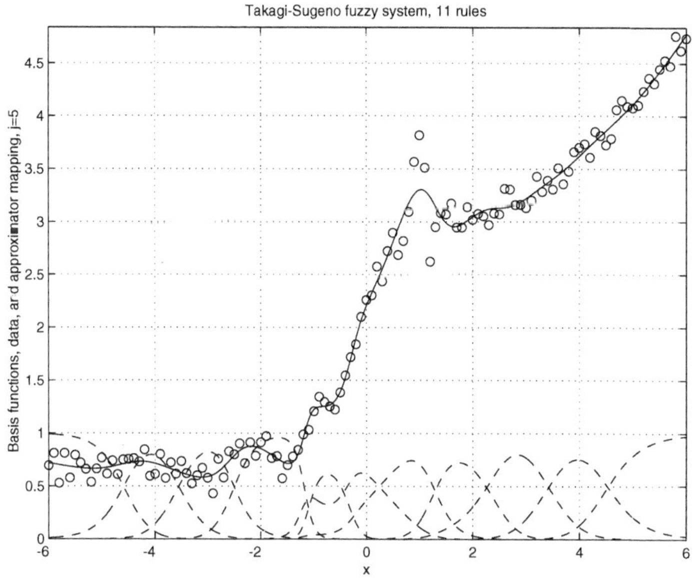

Figure 11.12: Levenberg-Marquardt training of a Takagi-Sugeno fuzzy system, mapping shape at iteration $j = 5$.

into the physics of the problem or by performing analysis on the data before training. Hence, even in practical problems, you may want to use the same basic approaches that we use here for this simple problem.

11.4.5 Overtraining, Overfitting, and Generalization

Next, we consider the case where $M = 13$, which is a much smaller data set than used above. We still use $R = 11$ rules and tune 44 parameters, so our number of parameters is greater than the number of data points. We use our earlier choice of initial parameters as $c_1^1 = -5$, $c_1^2 = -4$, up to $c_1^{11} = 5$ with all the spreads as 0.5. Also, we use $\lambda = 0.5$ as earlier. In Figure 11.19, we show the mapping shape at $j = 1$, and we see that it picks a reasonable shape considering how little information it has been given. There is, however, a problem when we train with so few data and so many parameters, that becomes even clearer if we allow a few more iterations to occur. In particular, consider Figure 11.20, where the mapping is shown at $j = 12$. We see that the algorithm, in one sense, does a very good job. It matches the training data almost exactly at every point. However, this causes a problem since at points outside the training data, the matching to the unknown function is poor (consider, e.g., the large peak near $x = 1$, where even though the mapping goes through one point in that region, we know its shape is not appropriate for the problem at hand). This is called poor "generalization." If the approximator generalizes well, then it will produce

Poor generalization, which is bad interpolation between training data, can occur if the approximator is too complex relative to the amount of information in the training data. You want your approximator simple to help avoid poor generalization, yet complex enough to provide flexibility to match the unknown function.

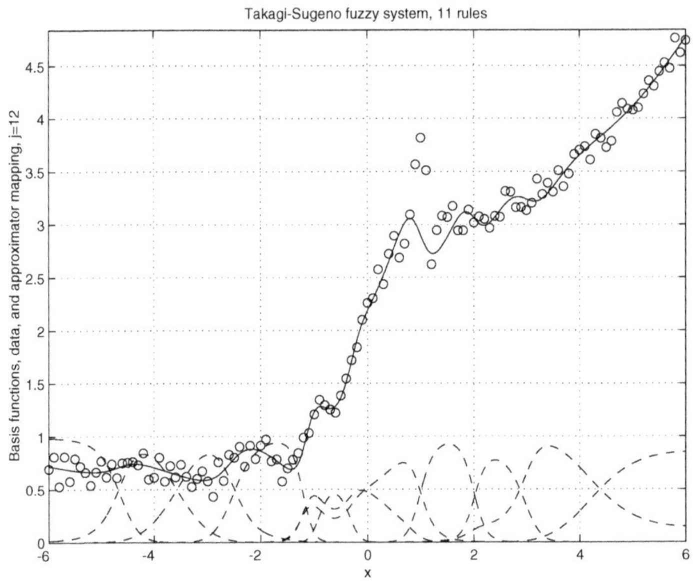

Figure 11.13: Levenberg-Marquardt training of a Takagi-Sugeno fuzzy system, mapping shape at iteration $j = 12$.

a good interpolation between the training data, not one that provides large oscillations between the data. Moreover, if you study Figure 11.20 carefully (and compare it to Figure 9.9 when noise is not added to the function), as we have seen in the least squares case, the approximator is failing also in the sense that it is trying to match the noise in the function (i.e., it is exhibiting overfitting).

How do we avoid these problems? First, you would normally never pick $p > M$; that is, you will normally have fewer parameters than training data pairs. Next, in some applications you need to make sure that you do not "overtrain;" that is, use too many iterations of the gradient update method. Sometimes this can result in forcing the approximator to match exactly at the data pairs at the expense of performing poor generalization (i.e., poor interpolation between the training data). Sometimes, the use of a "validation set" can help to detect when poor generalization is occurring and the updating can be terminated.

11.4.6 Approximator Reparameterization for Flexibility and Complexity Reduction

Sometimes an approximator has too much flexibility, in the sense that there are many ways to tune the parameters to get good approximation accuracy. One way to reduce this flexibility, and thereby simplify the parameter update method, is to make some of the parameters of the approximator that enter

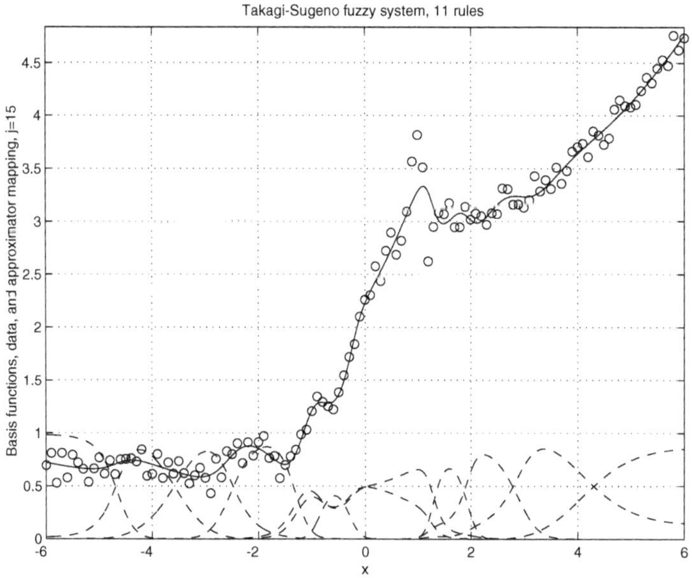

Figure 11.14: Levenberg-Marquardt training of a Takagi-Sugeno fuzzy system, mapping shape at iteration $j = 15$.

in a nonlinear fashion a function of some of the other parameters. In this way we reduce tuning flexibility, but not so much as to reduce it to the case where we only tune the parameters that enter in a linear fashion. For example, one way to do this for the Takagi-Sugeno fuzzy system is to simply make the spreads a function of the centers. One way to do this is to pick the spreads so that neighboring premise membership functions always cross over each other at 0.5. This way, when many centers are allocated to a region to try to improve approximation accuracy, the choice of the spreads will allow for the "turning on" and "off" of the appropriate consequent functions for a high density of membership functions. This approach could be good for some applications, but it should be emphasized that it does *reduce* approximator flexibility and so for some applications, it may not be a good choice. Moreover, the exact methods to specify the function specifying how the spreads change based on the centers will depend on the particular application.

11.4.7 Approximation Error Measures: Using a Test Set

To focus on other issues, we have been glossing over the issues of the use of a "test set" Γ for evaluating the approximation quality of our approximators. Instead we have been relying on visual inspection of the plots to comment on approximation accuracy. Generally, for more complex multidimensional applications, this is not a good approach and you will want to use some type of

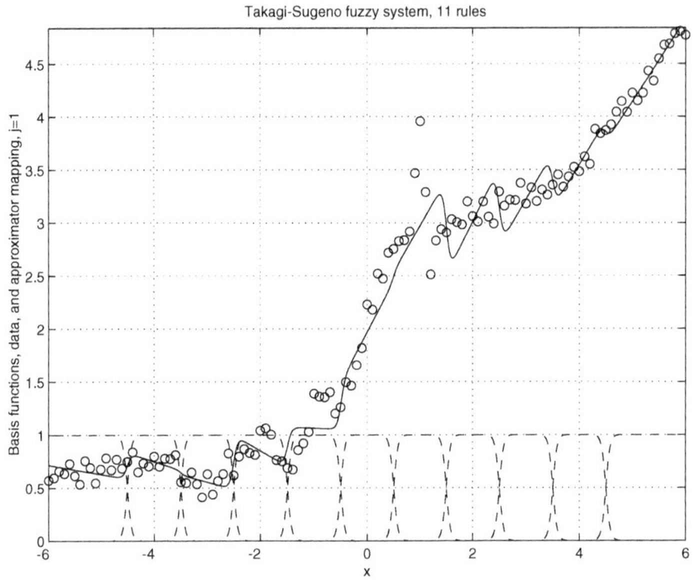

Figure 11.15: Levenberg-Marquardt training of a Takagi-Sugeno fuzzy system, mapping shape at iteration $j = 1$, different initialization.

numerical measure of approximation accuracy where you measure the accuracy *both* at the training data points and at points in between these. In addition, you will often want to evaluate the approximator for points where it is "extrapolating" from the data (e.g., at the end points of the input domains).

Such approximation error measures, based on, for example, a sum of squares or the maximum error over the domain, provide a way to quantify accuracy, and hence to compare different training methods and approximation structures.

11.5 Example: Online Steepest Descent Training of a Neural Network

For online function approximation, we must choose how we will process the data that we gather online. Here, we simply use $G_k = \{(x(k), y(k))\}$ so that we acquire and process one data pair at each time step. We will assume that $\bar{N} = 1$ so that there is only one output and hence $y(k)$ is a scalar (the development is similar for many outputs). Once again we will train the neural network to match the function in Figure 9.10.

We will use a single hidden layer neural network. Recall that ϕ_j, $j = 1, 2, \ldots, n_1$ denotes the output of the jth neuron in the hidden layer, and b_j is its bias. We defined

$$w^j = [w_{1,j}, w_{2,j}, \ldots, w_{n,j}]^\top$$

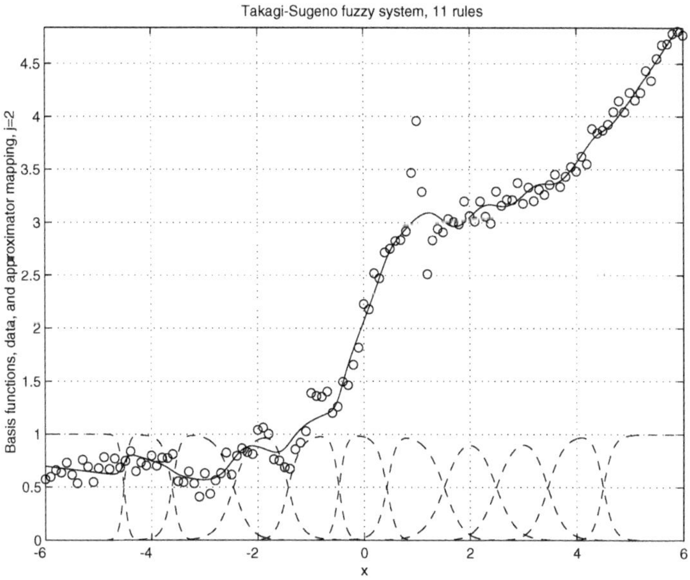

Figure 11.16: Levenberg-Marquardt training of a Takagi-Sugeno fuzzy system, mapping shape at iteration $j = 2$, different initialization.

so $\phi_j = f(b_j + (w^j)^\top x)$. Here, for every neuron in the hidden layer, we use the activation function

$$f(\bar{x}) = \frac{1}{1 + \exp(-\bar{x})}$$

Recall that w_j, $j = 1, 2, \ldots, n_1$ denotes a weight in the output layer and b is the bias for the output layer neuron. We have $w = [w_1, w_2, \ldots, w_{n_1}]^\top$. With a linear activation function in the output layer, the approximator is

$$y = F_{mlp}(x, \theta) = b + \sum_{j=1}^{n_1} w_j \left(f(b_j + (w^j)^\top x) \right)$$

If we tune all the parameters of this approximator, both the ones that enter linearly and in a nonlinear fashion, we let

$$\theta = [(w^1)^\top, b_1, (w^2)^\top, b_2, \ldots, (w^{n_1})^\top, b_{n_1}, w^\top, b]^\top$$

In this case, if n is the number of inputs to the approximator, the number of parameters to be tuned is $p = nn_1 + n_1 + n_1 + 1 = n_1(n+2) + 1$.

Here, we use the steepest descent training method to update the parameter vector $\theta = [\theta_1, \theta_2, \ldots, \theta_p]^\top$ and use a constant step size. We will only execute one iteration of the gradient update formula for each piece of data gathered; hence, we are aligning gradient iterations with time steps. In particular, for our

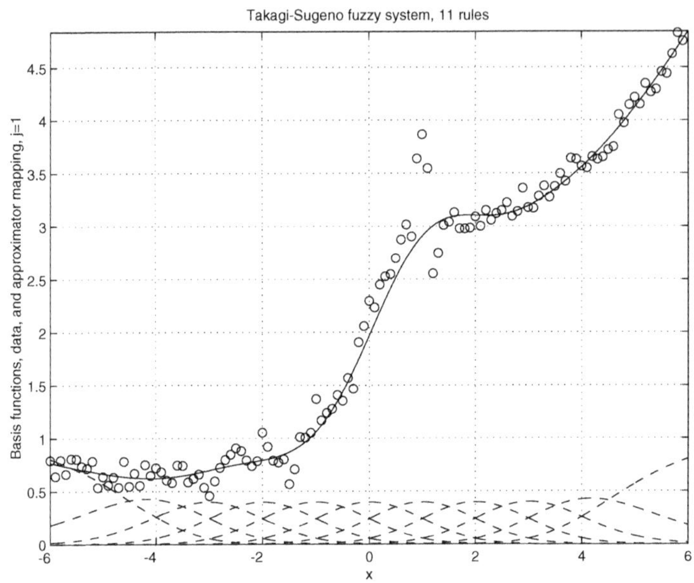

Figure 11.17: Levenberg-Marquardt training of a Takagi-Sugeno fuzzy system, mapping shape at iteration $j = 1$, different initialization.

online case, our update formula is given in Equation (11.6), which we repeat here as

$$\theta(k+1) = \theta(k) - \lambda \left. \frac{\partial J(\theta, G_k)}{\partial \theta} \right|_{\theta=\theta(k)}$$

where $\lambda > 0$ is the constant step size. Recall that we have a cost function given by Equation (11.1), which in our case is

$$J(\theta, G_k) = \frac{1}{2} \left(y(k) - F_{mlp}(x(k), \theta)\right)^2$$

From this, in order to fully specify the parameter update law, it is clear that we must provide

$$\frac{\partial J(\theta, G_k)}{\partial \theta}$$

for this case. This is what we do next.

11.5.1 Update Formulas

Clearly, we have

$$\begin{aligned} \frac{\partial J(\theta, G_k)}{\partial \theta} &= \frac{1}{2} \frac{\partial}{\partial \theta} \left(y(k) - F_{mlp}(x(k), \theta)\right)^2 \\ &= -\epsilon \frac{\partial F_{mlp}(x(k), \theta)}{\partial \theta} \end{aligned}$$

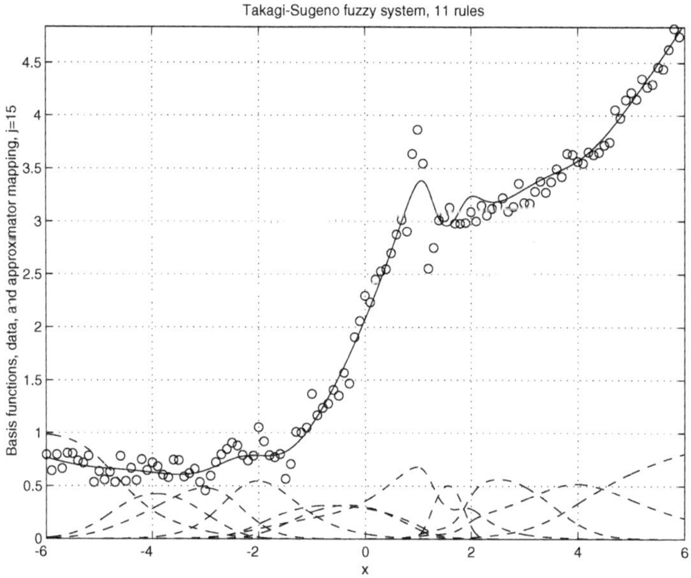

Figure 11.18: Levenberg-Marquardt training of a Takagi-Sugeno fuzzy system, mapping shape at iteration $j = 15$, different initialization.

where we let the scalar $\epsilon(k) = y(k) - F_{mlp}(x(k), \theta)$. Now, using the definition of the approximator structure

$$\frac{\partial F_{mlp}(x(k), \theta)}{\partial \theta} = \frac{\partial}{\partial \theta}\left(b + \sum_{j=1}^{n_1} w_j f\left(b_j + (w^j)^\top x\right)\right)$$

At this point, it is convenient to develop the update formula for different components of the θ vector individually, since there will be special cancellations in each case. First, we derive the case for the weights of the hidden layer, then its biases. Then we will proceed to the case for the parameters that enter linearly, the weights and bias of the output layer.

To help avoid confusion with the use of the indices, we will use j^* and i^* to denote the particular parameter value that we seek to derive the update formula for. Hence, we seek to find, for $j^* = 1, 2, \ldots, n_1$, and $i^* = 1, 2, \ldots, n$,

Development of update formulas simply requires the chain rule from calculus and some algebra.

$$\frac{\partial F_{mlp}(x(k), \theta)}{\partial w_{i^*,j^*}} = \frac{\partial}{\partial w_{i^*,j^*}}\left(b + \sum_{j=1}^{n_1} w_j f\left(b_j + (w^j)^\top x\right)\right)$$

Now, taking the partial we find, using the chain rule from calculus,

$$\frac{\partial F_{mlp}(x(k), \theta)}{\partial w_{i^*,j^*}} = w_{j^*} \frac{\partial}{\partial w_{i^*,j^*}} f\left(b_{j^*} + (w^{j^*})^\top x\right)$$

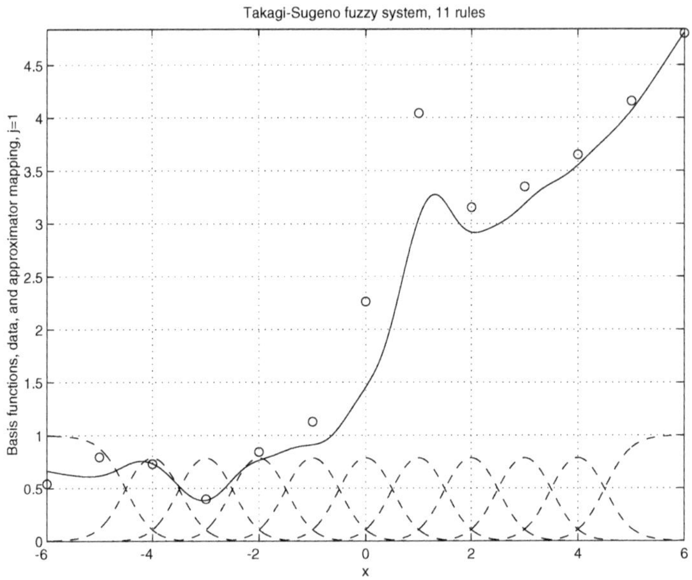

Figure 11.19: Levenberg-Marquardt training of a Takagi-Sugeno fuzzy system, mapping shape at iteration $j = 1$, $M = 13$.

$$= \quad w_{j^*} \frac{\partial f}{\partial \bar{x}_{j^*}} \frac{\partial \bar{x}_{j^*}}{\partial w_{i^*,j^*}}$$

Here, $\bar{x}_{j^*} = b_{j^*} + (w^{j^*})^\top x$ and note that using simple rules from calculus, with the above definition for the logistic function,

$$\frac{\partial f}{\partial \bar{x}_{j^*}} = f(\bar{x}_{j^*})(1 - f(\bar{x}_{j^*}))$$

If we had used the hyperbolic tangent for the activation functions f, then

$$\frac{\partial f}{\partial \bar{x}_{j^*}} = 1 - (f(\bar{x}_{j^*}))^2$$

Returning to the logistic function case, notice that

$$\frac{\partial \bar{x}_{j^*}}{\partial w_{i^*,j^*}} = x_{i^*}$$

so

$$\frac{\partial F_{mlp}(x(k), \theta)}{\partial w_{i^*,j^*}} = w_{j^*} f(\bar{x}_{j^*})(1 - f(\bar{x}_{j^*}))x_{i^*}$$

Hence, the update formula for the weights in the hidden layer is

$$\begin{aligned} w_{i,j}(k+1) &= w_{i,j}(k) + \\ \lambda\epsilon(k) w_j f&\left(b_j(k) + (w^j)^\top(k)x(k)\right)\left(1 - f\left(b_j(k) + (w^j)^\top(k)x(k)\right)\right) x_i(k) \end{aligned} \tag{11.23}$$

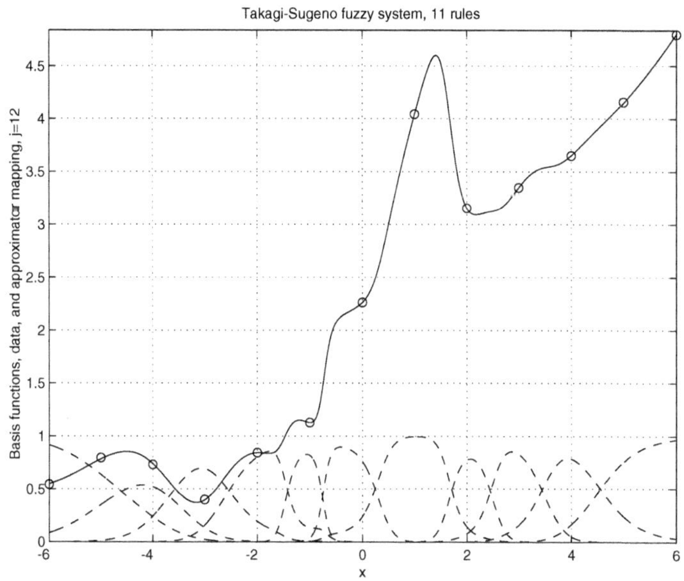

Figure 11.20: Levenberg-Marquardt training of a Takagi-Sugeno fuzzy system, mapping shape at iteration $j = 12$, $M = 13$.

for $j = 1, 2, \ldots, n_1$, and $i = 1, 2, \ldots, n$, where $\epsilon(k) = y(k) - F_{mlp}(x(k), \theta(k))$.

Next, we will derive the update formula for the biases that enter the n_1 neurons in the hidden layer. For this, for $j^* = 1, 2, \ldots, n_1$,

$$\begin{aligned}\frac{\partial F_{mlp}(x(k),\theta)}{\partial b_{j^*}} &= w_{j^*}\frac{\partial}{\partial b_{j^*}} f\left(b_{j^*} + (w^{j^*})^\top x\right) \\ &= w_{j^*}\frac{\partial f}{\partial \bar{x}_{j^*}}\frac{\partial \bar{x}_{j^*}}{\partial b_{j^*}} \\ &= w_{j^*} f(\bar{x}_{j^*})(1 - f(\bar{x}_{j^*}))\end{aligned}$$

Note that

$$\frac{\partial \bar{x}_{j^*}}{\partial b_{j^*}} = 1$$

Hence, we get the update formula

$$\begin{aligned}&b_j(k+1) = b_j(k) + \\ &\lambda\epsilon(k) w_j(k) f\left(b_j(k) + (w^j)^\top(k)x(k)\right)\left(1 - f\left(b_j(k) + (w^j)^\top(k)x(k)\right)\right)\end{aligned} \tag{11.24}$$

for $j = 1, 2, \ldots, n_1$, where $\epsilon(k) = y(k) - F_{mlp}(x(k), \theta(k))$.

Next, we derive the update formula for the n_1 weights in the output layer. For this, for $j^* = 1, 2, \ldots, n_1$,

$$\frac{\partial F_{mlp}(x(k),\theta)}{\partial w_{j^*}} = f\left(b_{j^*} + (w^{j^*})^\top x\right)$$

Hence, we get the update formula

$$w_j(k+1) = w_j(k) + \lambda\epsilon(k) f\left(b_j(k) + (w^j)^\top(k)x(k)\right) \tag{11.25}$$

for $j = 1, 2, \ldots, n_1$, where $\epsilon(k) = y(k) - F_{mlp}(x(k), \theta(k))$. Notice that this is the update formula for parameters that enter linearly, and you will generally find such a relationship for this case.

Finally, we derive the scalar update formula for the bias b in the output layer. For this

$$\frac{\partial F_{mlp}(x(k), \theta)}{\partial b} = 1$$

Hence, we get the update formula

$$b(k+1) = b(k) + \lambda\epsilon(k) \tag{11.26}$$

where $\epsilon(k) = y(k) - F_{mlp}(x(k), \theta(k))$.

To summarize, the update formulas for $\theta(k)$ are given by Equations (11.23), (11.24), (11.25), and (11.26). Clearly, while we use only one constant step size, you could use different ones for the different update formulas.

Notice that, as a practical computational issue, there are many shared calculations that are used in the update formulas. It is for this reason that it is probably best to first update the output layer bias, the output layer weights, the hidden layer biases, then finally the hidden layer weights (and then each update can use some of the calculations needed for the previous update).

11.5.2 Parameter Constraints and Initialization

Notice that for the update formulas we derived, there are no particular values of parameters that will cause, for instance, the functions on the right side of the update formulas to be undefined (which could cause, e.g., a divide-by-zero error). Hence, we will not have to constrain the parameters to avoid such situations. Moreover, in this simple example, we will not assume that, due to implementation concerns, the parameters must lie in certain bounded regions. For this reason, we will not put any constraints on the parameter update laws from a parameter constraint set. We emphasize, however, that generally the more information you have about the underlying function, the more you tend to know about how to initialize the approximator. Here, for the sake of illustration, we assume that we know nothing useful for the initialization (even though we could certainly analyze the data to learn some useful ideas for initialization, just as we have done in Section 10.5).

How do we initialize the algorithm? That is, how do we specify $\theta(0)$? There are many ways to choose this, but often in practice the parameters are simply chosen to be random small values (here in one case we choose $\theta_i(0)$ to be uniformly distributed on $[-0.1, 0.1]$). This often tends to be a good choice for several reasons. First, we get some initial random distribution of the biases that place the sigmoid functions across at least some small region of the space (sometimes, if you know the range of possible values on some input space a priori, then you can spread the sigmoids randomly across this range). Next, by choosing the weights to be small but random, we start with "steps" that are going both up and down with small slopes and this tends to make sure that the gradient is not too small initially. Finally, the small values for the output layer provide something close to picking the values at zero, which as we saw in the recursive least squares case, can be a good choice.

There are many cases for practical applications where you can determine what may be a better initialization than simply using small random values. For instance, in Section 10.5, we chose initial values for the parameters of the approximator that enter in a nonlinear fashion (i.e., the hidden layer weights and biases) in a way that when it was tuned with the recursive least squares method, it determined a good approximation to the function after 300 iterations. It did this whether we chose the initial values for the parameters that enter linearly (the output layer weights and bias) as all zero, or if we used values perturbed off the ones that batch least squares finds. Such initialization by some educated guessing at the nonlinear part and using batch least squares to specify the linear part is generally a good approach and one that we will study here. We must keep in mind, however, that in practical applications you are sometimes limited by how many data are available a priori so that initialization with batch least squares is not always possible. It is for this reason that we will also study the case where we simply pick the parameters that enter linearly to be zero.

11.5.3 Approximator Tuning Results: Effects of Step Size

For our example, we pick $n_1 = 25$, the same as we have studied in the recursive least squares case in Section 10.5. Notice that now, however, we will tune all 76 parameters of the neural network. Tuning this many parameters is probably not necessary for this problem to get a reasonable level of accuracy (e.g., consider the similar effects on the shape of the nonlinearity for the weights in the hidden and output layers), but we will use this example simply for illustration. Without much tuning, we picked $\lambda = 0.1$ or $\lambda = 0.01$ to illustrate the differences in the algorithm's behavior (and we note that if you pick it too much larger, the algorithm will diverge in some cases, as it did for the simple scalar quadratic example considered earlier).

When you use the batch least squares initialization, with $\lambda = 0.1$, you get the results shown in Figure 11.21 for the first 10 steps, in Figure 11.22 for the last 10 steps, and in Figure 11.23 at $k = 1000$ steps. Notice that the algorithm quickly tunes the shape to be a reasonable approximation, but that it does not ultimately achieve the kind of approximation accuracy that was achieved with the recursive least squares method in Section 10.5, even though it has what is most likely a better initialization (the actual values found from batch least squares, rather than the perturbed ones used there).

For a fixed step size, under very general conditions, asymptotically the map will "oscillate" by persistently trying to match the most recent data. Smaller step sizes result in smaller asymptotic oscillations, but slower convergence.

If you examine the shapes in Figure 11.22 for the last 10 steps, you find that the accuracy found at $k = 1000$ is also found at earlier steps and the shape changes at each iteration try to accommodate the new piece of training data

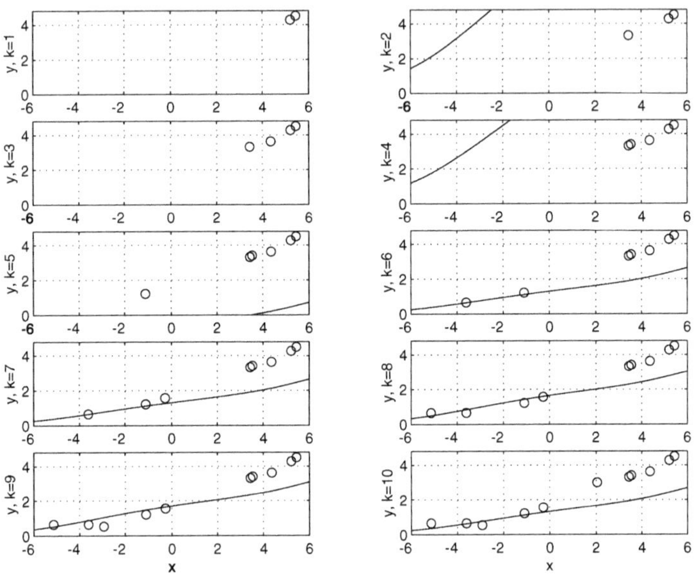

Figure 11.21: Steepest descent training of a neural network, mapping shapes for first 10 steps, batch least squares initialization, step size $\lambda = 0.1$.

(hence, the result in Figure 11.23 should only be taken as representative of the shapes found). The shape is still moving around at $k = 1000$ (and will for higher numbers of iterations also); it is not fixed at that point.

Next, if you use the batch least squares initialization, with $\lambda = 0.01$, you get the results shown in Figure 11.24 for the first 10 steps, in Figure 11.25 for the last 10 steps, and in Figure 11.26 after 1000 steps. Notice that with a smaller value for λ, the shape initially changes slowly and also near the end. Basically, the algorithm is less aggressive in trying to match each new piece of training data. This may be a desirable characteristic of an algorithm for online operation in some applications. Generally, larger step sizes will tend to force the method to pay more significant attention to each new piece of data, while smaller ones allow for it to partially ignore new data. There is generally a good choice that will allow the algorithm to slowly shape the nonlinear mapping as new information is gathered, allowing new information to partially reshape the nonlinearity, but not too much so that the information encountered in the past is not forgotten (some think of the algorithm as being "greedy" in seeking to achieve the minimization, which in this case means that it tries to approximate the information provided by the new piece of training data, with the amount of greed proportional to the step size). Sometimes, to keep the shape from moving around too much at each step, you have to use a very small step size, and then generally you need more steps in the algorithm to get convergence.

Next, recall that we are presenting data to the algorithm where x is uniformly

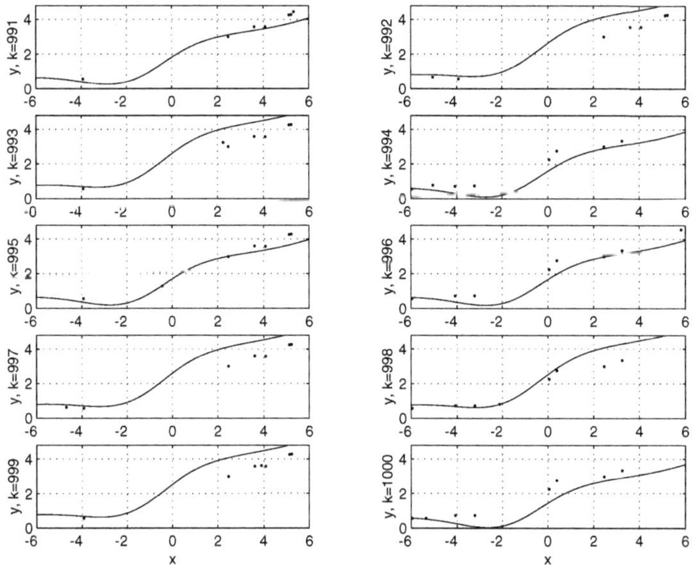

Figure 11.22: Steepest descent training of a neural network, mapping shapes for last 10 steps, batch least squares initialization, step size $\lambda = 0.1$.

distributed on $[-6, 6]$. Now, if we are unlucky and we only get data in one region of the x domain over several initial steps, then we generally will not get the kind of initial accuracy that you see in Figure 11.21. Clearly if it does not have data in certain regions, then it generally will do poor approximation in that region (of course you may get lucky and it may do a good extrapolation). Generally, the performance and convergence properties of the algorithm depend on the order of presentation of the training data. Finally, note that while we have run the algorithm for many iterations and the parameters did not diverge, we must emphasize that this does not *prove* that they will not; it could be that after only a few more iterations they will diverge. Generally, you must be very careful to ensure boundedness for parameters that you adjust online and one way to do this is to use a parameter constraint set (which we did not do here just to keep things simple).

11.5.4 Approximator Tuning Results: Effects of Initialization

In this subsection, we will assume that $\lambda = 0.01$. First, we initialize the parameters that enter linearly to be all zero, as we did for the recursive least squares method in Section 10.5. Using this initialization, we get the approximator mapping shown in Figure 11.27 after 1000 iterations (the plots for the first and last ten steps are omitted as they are similar to the case above where we initialized

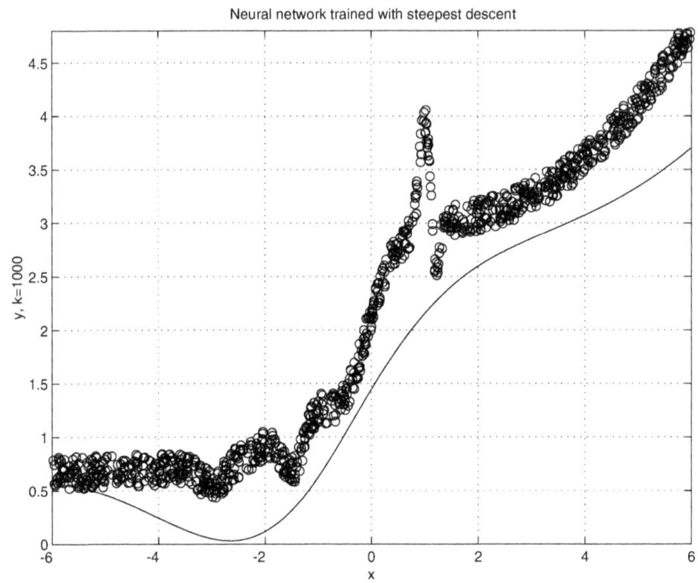

Figure 11.23: Steepest descent training of a neural network, mapping shape after 1000 steps, batch least squares initialization, step size $\lambda = 0.1$.

with batch least squares). At 1000 iterations, this approximator shape provides an approximation accuracy that is clearly close to that shown in Figure 11.23. It seems that in this case for this choice of training data (which is random), the steepest descent method ultimately picked the parameters just as well as when it had (what was probably) a better initialization. We can say that it seemed to overcome the poor initialization in this case (of course, we cannot always expect this).

When we initialize with all small random values, the results are shown in Figure 11.28 after 1000 steps. The mapping shapes for the first 10 iterations are not shown, but basically, it is as you would guess: little progress is seen in coming up with a good approximation since λ is small and the initialization is not very good. The mapping shapes for the last 10 iterations are close to the one shown in $k = 1000$ in Figure 11.28, showing that it appears that the mapping shape has converged. Hence, it seems that we have found that this initialization, which is often used when you know nothing better about how to initialize the mapping, results in poorer approximation accuracy as compared to the others.

11.5.5 Can We Improve Approximation Accuracy?

Well, there are many things that you can try, but the choices depend on the particular application. For the simple example we have been studying, there is

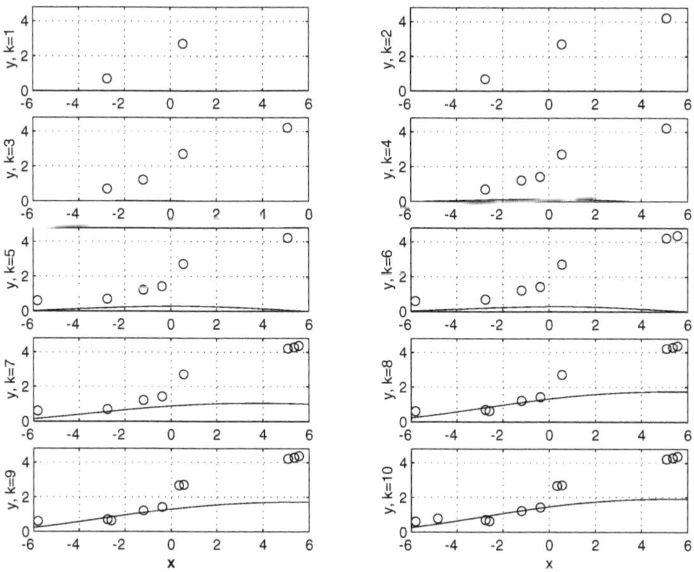

Figure 11.24: Steepest descent training of a neural network, mapping shapes for first 10 steps, batch least squares initialization, step size $\lambda = 0.01$.

clearly room for improvement of approximation accuracy, and there are several options that become apparent after completion of the above investigations.

First, you could try to use a diminishing step size rule, such as the one that starts with a certain step size and then decreases it to some minimum value. For some applications, this can ensure that the initial data are quickly used to tune the approximator to get a reasonable accuracy, but then the step size decreases so that the oscillations in the shape of the mapping do not occur at later iterations after it has learned more about the shape.

Second, you could try processing more than one data pair per step, for instance, by "windowing" the data. Then, at each iteration, you would execute several iterations of the gradient method to try to get the approximator to match the function (often you would simply terminate the iterations after some fixed number, since you will often be constrained by processor resources; however, other times you could use a termination criterion at each step). This can help alleviate the problems with the algorithm being too aggressive in seeking to match the data pair just encountered. In such an approach, you could weight the old data as being less important than the new data, just as we did in the least squares approach with a forgetting factor. To do this, you would need to add weighting factors to the cost function you are trying to minimize. Overall, such an approach can offer improved accuracy but you are certainly paying for it in computational complexity.

Third, you could try to use a different gradient method such as the Levenberg-

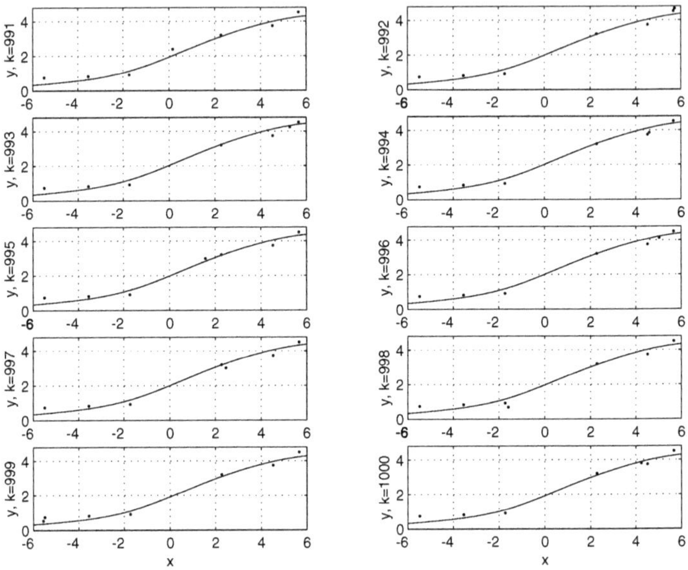

Figure 11.25: Steepest descent training of a neural network, mapping shapes for last 10 steps, batch least squares initialization, step size $\lambda = 0.01$.

Marquardt method, and again you may want to consider serially processing batches of data as we discussed above (with the computational complexity generally increasing with an increase in the batch size). Why might this have a chance at improving approximation accuracy? First, it should try to approximate a Newton method so that it should get fast convergence, but even with tuning, you may get the type of behavior seen with the steepest descent method where the mapping shape oscillates. Second, experience has shown that the Levenberg-Marquardt approach is generally better than the steepest descent algorithm for offline training, so we might find the same or similar benefits for online training. At the same time, using a more sophisticated method can raise other problems, such as ensuring that the inverse for the Levenberg-Marquardt update formulas can be computed.

11.5.6 Local Vs. Global Tuning/Learning

It is interesting to consider how the mapping shape changes over time as we have done in the recursive least squares case. To do this, we will return to the first case where we had initialized with the batch least squares and had $\lambda = 0.1$ (see Figures 11.21, 11.22, and 11.23), since this will most dramatically illustrate the ideas here. Figure 11.21 shows the approximator nonlinearity for the first 10 steps, and notice that for the first 5 steps, the approximator does not have much data and hence the quality of approximation is quite poor. Next,

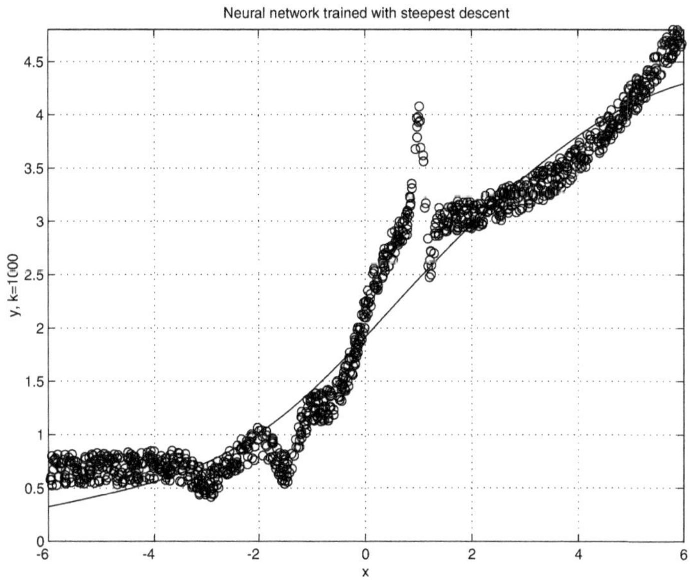

Figure 11.26: Steepest descent training of a neural network, mapping shape after 1000 steps, batch least squares initialization, step size $\lambda = 0.01$.

however, notice that at $k = 6$ a data pair is obtained, and the approximator is tuned to provide a reasonable approximation to the given data (although it does not match the data near $x = 5$ very well). At times $k = 7, 8, 9$, data are obtained on the left side and the approximator shape changes very little. Now, at $k = 10$, a data point is obtained on the right, but it does not modify the approximator shape much to try to improve the accuracy, because the step size is relatively small.

What would we have liked to see in this initial sequence? Well, by $k = 4$, we had data on the right side that the approximator did not match very well, and we would have liked to see it do better. Then, when at steps $k = 5, 6, 7, 8, 9$, it got data on the left side we would have liked to see it let the approximator pass through the data gathered earlier, but also force the approximator to pass through these new data. Then, when the data pair is gathered at $k = 10$, we would like to have seen it adjust the approximator on the right, without disturbing (forgetting) what it had already done on the left. In summary, we would have liked it to have made "local" adjustments to the approximator nonlinearity, depending on where it gathered data, so that it incrementally learns the proper shape.

For some approximator structures trained with some methods, learning in the present can destroy what has been learned in the past (the stability-plasticity dilemma).

Such problems arise for a variety of reasons, such as step size choice, the choice of using a gradient method, and only processing one data point at each iteration; however, one other significant contributing factor can be the choice of the approximator structure. For neural networks with sigmoid nonlineari-

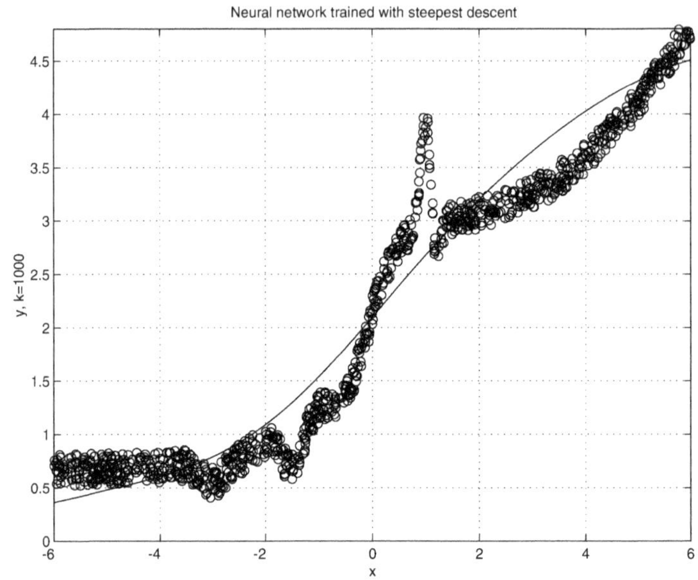

Figure 11.27: Steepest descent training of a neural network, mapping shape after 1000 steps, parameters that enter linearly initialized to zero.

ties, and other approximators, a change in one parameter can change the whole mapping shape (like the bias on the output layer, which shifts the whole plot vertically up and down) so that when it should only be shaping the mapping locally, where it got the training data it does so "globally." At times, this is not a problem as the method can sometimes be designed so that it shapes the nonlinearity appropriately, or since the neural network is a universal approximator, it can provide for local learning too if it picks the parameters properly. Sometimes, however, it is difficult to achieve this with the neural network or with other approximator structures. At times, it can be beneficial to *force* a type of local learning to help overcome this problem by picking the nonlinear part of the approximator to have functions that approximately have "local support" (i.e., they are only positive in a certain domain of the input space) so that only local adjustments are made. Radial basis function neural networks can achieve local support as well as the Takagi-Sugeno fuzzy system with Gaussian input membership functions.

11.6 Clustering for Classifiers and Approximators

It is important to realize that gradient methods are very general and applicable to many optimization problems you can encounter in engineering. In particular,

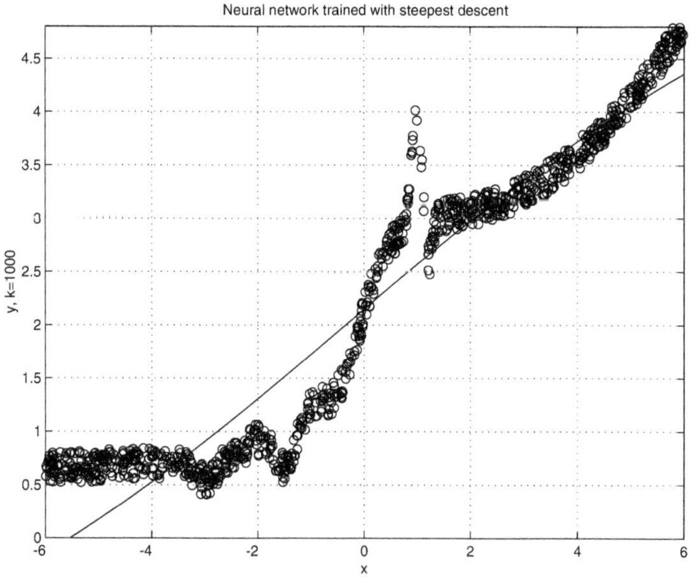

Figure 11.28: Steepest descent training of a neural network, mapping shape after 1000 steps, parameters initialized to small random values.

they have been found to be useful in many roles in the development of intelligent control systems. One area that they can be particularly useful is in the tuning of nonlinearities to partition data vectors into different "classes" (the data vectors can be numeric representations of the parameters of many different kinds of objects, from computer vision and image processing data, to speech signals and plant input-output data). These can then be used in "classifiers" that can take a given input vector and indicate which of a finite set of classes that input corresponds to. There is a wide variety of "pattern recognition" problems that can use classification methods. In some approaches, the data vectors are grouped into "clusters" that partition the data. Then, when an input is given, the membership in each cluster is determined and the one it best matches is declared to be the cluster that the data vector belongs to. In this way, even if we get an input vector that is somewhat different from the center of the cluster (e.g., due to noise), it can still correctly classify the object to the proper class.

There are also times when it is useful to use clustering to tune a portion of an approximator to solve a function approximation problem. For instance, using the "input portion" x of the training data set G, we can form clusters around similar x vectors. Then we can use these clusters in the nonlinear portion of the approximator (i.e., in the ϕ function) and train the remaining linear portion of the approximator to solve a function approximation problem. Indeed, the classification problem discussed above can be thought of as a type of function approximation problem where the output portion of the training data y simply indicates which class x belongs to, where $(x, y) \in G$. It is for this reason that many different approximators and training methods of the previous sections can be used for the classification task.

In this section, after we explain how to use approximators as classifiers, we show how to form clusters around data. The resulting methods will be shown to provide either classifiers or function approximators. You can think of the methods of this chapter as a different approach to tune nonlinear in the parameter approximators. Here, you first use a cost function that characterizes quality of clustering to get the clusters, and hence the nonlinear portion of the approximator. Then you can use a linear least squares criterion that characterizes approximation accuracy to specify a least squares method to find the parameters that enter linearly.

Classifiers indicate which group of data (cluster) an input vector belongs to (is associated with).

11.6.1 Using Approximators to Solve Classification Problems

We must emphasize that *any* of the methods developed in the previous sections can be used as function approximators to solve a classification problem. To explain how this is done in a bit more detail, note that the key to formulating the classification problem as a function approximation problem is to start by picking the data set and this will suggest whether you use a single- or multiple-output approximator.

Single-Output Classifiers

One way is to assume that we have n_c different classes that objects can belong to. Here, our objects are characterized by (parameterized by) a vector of n numbers. Suppose that these classes are simply labeled with numbers $1, 2, \ldots, n_c$. Suppose that we have M examples that pair objects with their classes, such as $(x(i), y(i))$ where $x(i)$ is a specific data vector and $y(i) \in \{1, 2, \ldots, n_c\}$ is its class. Clearly, we can use these data to specify the training data set G and the resulting approximator (which has n inputs and one output) can be trained to classify the data. In such an approach, the output will be a scalar and you will have to specify a method to determine which integer $1, 2, \ldots, n_c$ the output is closest to in order to classify it into one of the finite number of possible classes.

Note that the issue of approximator structure choice can be very important in the design of a classifier. For instance, suppose that $n_c = n = 2$ and that the input space is simply split by a line where vectors on one side of the line belong to the first class and the ones on the other side belong to the second class. In this case, it may be good to use a neural network with a logistic function since it can then be tuned to provide for the splitting of the space along the line mentioned above. If the two classes were defined by being in or out of a circular region, then a different nonlinearity might work better. Which one? Usually the choice is very application-dependent and requires significant insight into the problem; however, in this case you may consider a normalized Gaussian function (like ξ_i that we had used for the fuzzy systems) since it can then provide a function that naturally comes on in a circular region, and a function that comes on everywhere but in a circular region. (Develop and sketch one to convince yourself of this.)

Similar issues in structure choice arise in the multi-output classifiers that we discuss next.

Multiple-Output Classifiers

Another perhaps more common way to formulate the classification problem as a function approximation problem is to construct a multi-output approximator with n_c outputs. View this multi-output system as n_c multi-input single-output systems. Consider how to train the jth output to classify whether the input vector x is a member of class j where $j \in \{1, 2, \ldots, n_c\}$. Suppose that we have M examples that pair objects with their classes but in a different way than in the last subsection. Here, suppose that we have M_j data pairs $(x(i), y(i))$ where $x(i)$ is a specific data vector and $y(i) \in \{0, 1\}$ where if $x(i)$ is in class j, then $y(i) = 1$ and if it is not in class j, then $y(i) = 0$. The entire data set for training the classifier is simply the union of the data sets used to train each classifier (then $M = \sum_{j=1}^{n_c} M_j$ for the data set G). Now, suppose that we have trained the n_c approximators with these data sets.

How does the classification process work? Suppose that we consider only the jth classifier that tries to decide if the input vector is in the jth class. Suppose that we call the approximator that was trained for this task $F_j(x, \theta)$ (of course, θ is the parameter vector that resulted from the training process). For a given x, we could test if $F_j(x, \theta) \geq 0.5$ and if it is, then we could indicate that x has class j (and if it is not, then it is not of class j). There are several possible problems with such an approach. First, for a given x there may be more than one output that is greater than 0.5 so that a single vector could be classified as being in two different classes (and often you would not want this). Second, it is possible that there is no j such that the value of $F_j(x, \theta) \geq 0.5$ and in this case, it does not know how to classify.

Hence, the common approach is to pick the output, say j^*, that has a maximum value and then indicate that x has class j^*. Mathematically, we say that we decide that the given input x is of class j^* where

$$j^* = \arg \max_{j=1,2,\ldots,n_c} \{F_j(x, \theta)\}$$

(if there is more than one value that has the maximum value, then you simply arbitrarily pick one). Note that with this approach, we will always have a unique classification. But, of course, if all the values of $F_j(x, \theta)$ are close to zero, we may not be very confident in the classification. In fact, in some applications it may make sense to use the output of the classifier to indicate the confidence in the classification.

11.6.2 Clustering Methods: Gradient Approaches

In this section we take a different approach to the classification problem from in the last subsection. Here, we specify functions that are designed to partition data in certain ways and try to adjust the parameters of these functions so that they group the data into clusters. We do not explicitly focus on a function approximation problem; however, we note that these methods can be used with other methods to construct approximators (e.g., see the next section, where we couple a clustering method with a least squares approach to form a function approximator). The clustering methods of this section could be used in a similar role.

Cluster Functions

First, we give some examples of how to specify what we will call "cluster functions" that are nonlinear functions designed to partition data. There are a wide variety of possibilities for such functions and we only consider two here (the first one will be studied in more detail in the next section).

Polynomial-Based Function: Let

$$v^j = [v_1^j, v_2^j, \ldots, v_n^j]^\top$$

denote the jth "cluster center" where $j = 1, 2, \ldots, R$. Let

$$p_j(x) = \left[\sum_{k=1}^{R} \left(\frac{|x - v^j|^2}{|x - v^k|^2} \right)^{\frac{1}{m-1}} \right]^{-1} \tag{11.27}$$

$j = 1, 2, \ldots, R$ be the "polynomial-based" cluster functions. Here, we must have $m > 1$. Note that m controls the "width" of all the clusters.

As an example, consider the case where $n = 1$ and $R = 3$. Let $m = 2$. A plot of p_j, $j = 1, 2, 3$ for the case where $v^1 = -5$, $v^2 = 0$, and $v^3 = 5$ is shown in Figure 11.29 (see top plot). We use a solid line for p_1, a dashed line for p_2, and a dotted line for p_3. Notice that the clusters provide "soft" partitions for the x domain. When one function is near one, the others are near zero. At the outer edges of the domain of x (i.e., for large $|x|$ values), the cluster function values all approach the same value.

A plot of p_j, $j = 1, 2, 3$ for the case where $v^1 = -3$, $v^2 = 6$, and $v^3 = 1$ is shown in Figure 11.30 (see top plot). In this case, notice that it also achieves a good partitioning of the x axis.

Gaussian-Based Function: For $j = 1, 2, \ldots, R$, let

$$\mu_j(x) = \prod_{i=1}^{n} \exp\left(-\frac{1}{2} \left(\frac{x_i - c_i^j}{\sigma_i^j} \right)^2 \right)$$

where c_i^j is the point in the ith input x_i where the function achieves a maximum, and $\sigma_i^j > 0$ is the "width" of the function for the ith input. (This is simply the Gaussian premise membership function used earlier for fuzzy systems.) We will

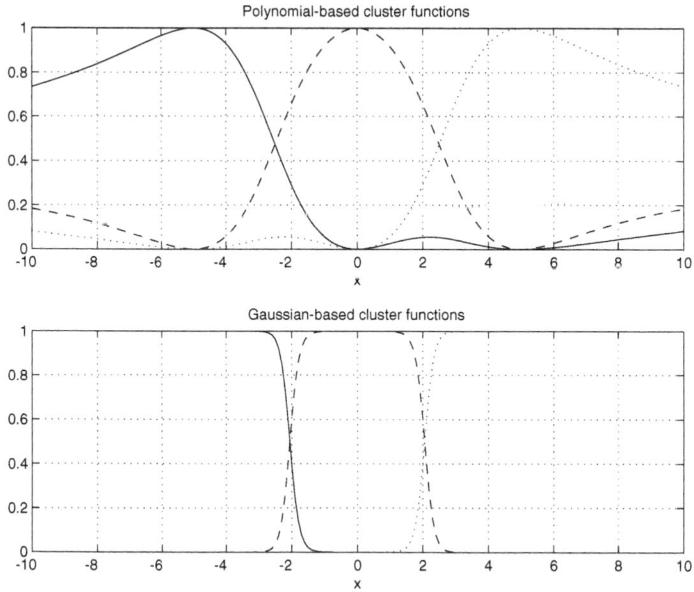

Figure 11.29: Polynomial and Gaussian-based cluster functions.

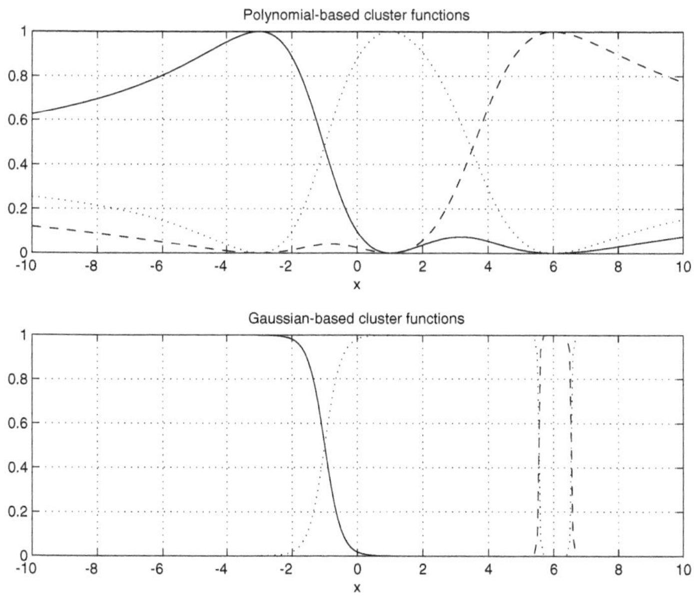

Figure 11.30: Polynomial and Gaussian-based cluster functions.

use this function to construct another type of cluster function. In particular, we "normalize the Gaussian functions" by letting

$$\xi_j = \frac{\mu_j(x)}{\sum_{i=1}^{R} \mu_i(x)}$$

$j = 1, 2, \ldots, R$. We can use these functions as cluster functions. Note that in Takagi-Sugeno fuzzy systems, we used them to turn on and off different consequent functions.

As an example, consider the case where $n = 1$ and $R = 3$. A plot of ξ_j, $j = 1, 2, 3$, for the case where $c_1^1 = -5$, $c_1^2 = 0$, and $c_1^3 = 5$, with $\sigma_1^1 = 1$, $\sigma_1^2 = 0.7$, and $\sigma_1^3 = 1$, is shown in Figure 11.29 (see bottom plot). We use a solid line for ξ_1, a dashed line for ξ_2, and a dotted line for ξ_3. Notice that these functions provide a different type of partitioning of the domain from the polynomial-based function. Besides the shapes being different, at the outer edges of the domain, all the points would be grouped together into the outermost cluster. Notice also that the widths for all the functions can be different, while in the polynomial-based case, the widths are all controlled by one parameter. This added flexibility may or may not be useful (and of course, the polynomial-based function can be modified to provide this characteristic).

It is also possible that a cluster of this type exhibits other shapes that may be useful. For instance, it can be the case that one cluster can come on (i.e., achieve a value near one) in more than one region of the space. However, note that due to the normalization (i.e., the division by the sum of the μ_i), the sum of the cluster function values at any one x point must be one. This ensures that as one cluster function increases, the others must decrease so that any input is in a cluster in varying amounts and never completely in more than one cluster.

As an example, a plot of ξ_j, $j = 1, 2, 3$, for the case where $c_1^1 = -3$, $c_1^2 = 6$, and $c_1^3 = 1$, with $\sigma_1^1 = 1$, $\sigma_1^2 = 0.1$, and $\sigma_1^3 = 1$, is shown in Figure 11.30 (see bottom plot). Compare this to the result from the polynomial-based function and notice that the result is quite different. Notice that here ξ_3 is very near one (i.e., it is on) for $x \in [0, 5]$ *and* $x \geq 7$. This characteristic of this cluster function could be useful in some applications (but may be bad for others) and provides for some interesting cluster shapes (not just the standard ones). For instance, in the case where $n = 2$, it is possible to have what you may call circular concentric clusters with a circle in the middle and doughnut-shaped clusters centered around it.

Clustering Cost Functions

In this section we provide cost functions that we will seek to minimize to make the cluster functions partition the data.

Clustering via a gradient method involves adjusting functions representing clusters to minimize a measure of how the data are grouped and separated.

Cost for Polynomial-Based Function: Consider the function

$$J(\theta) = \sum_{i=1}^{M} \sum_{j=1}^{R} (\mu_{ij})^m |x(i) - v^j|^2 \tag{11.28}$$

where $m > 1$, v^j are the cluster centers, typically $M >> R$, and the μ_{ij} are scalars. Here, the parameter vector θ holds both the cluster centers and the μ_{ij} scalars. Intuitively, the μ_{ij} for $i = 1, ..., M$ and $j = 1, ..., R$ are the grades of membership of $x(i)$ in the jth cluster. Typically, you would require that for each $i = 1, 2, \ldots, M$, $\sum_{j=1}^{R} \mu_{ij} = 1$ so that the centers are placed so that no more than one will have a value of 1 at any point on the input domain (this forces us to solve a nonlinear optimization problem with constraints, and this will be discussed below). The terms $|x(i) - v^j|^2$ are included to try to get the clusters to be in the middle of the data. The $(\mu_{ij})^2$ values weight the terms $|x(i) - v^j|^2$ and they are adjusted so that the cluster centers will separate to find different groups of data.

Cost for Gaussian-Based Function: Recall that c_i^j is the point in the ith input where the jth cluster center reaches a maximum. Let

$$c^j = [c_1^j, c_2^j, \ldots, c_n^j]^\top$$

and think of this as a cluster center. Consider the function

$$J(\theta) = \sum_{i=1}^{M} \sum_{j=1}^{R} \xi_j(x(i)) |x(i) - c^j|^2 \tag{11.29}$$

Here, θ can hold both the c^j and σ_i^j values. Sometimes, however, it may be convenient to use the same value for all the σ_i^j and you may only want to adjust that single value. Alternatively, you may simply want to fix the values of the spreads, for instance, to be all the same value (this can simplify the optimization problem). Conceptually, the cost function is closely related to the one used for the polynomial-based function. Notice, however, that the clustering problem for the Gaussian-based function is a nonlinear optimization problem *without* constraints.

Cluster Adjustment Methods

For the cost function for the Gaussian-based function defined in the last section, it is possible to define a gradient update formula and use it to iteratively update the parameters of the cluster functions. The gradient $\nabla J(\theta)$ can be found and used with the methods of the last section. For instance, you may want to use a steepest descent or Levenberg-Marquardt method to solve the minimization problem. Clearly, standard initialization and termination issues for gradient algorithms are relevant. Also, we must emphasize that there are no convergence guarantees, so we will not know if we have found a global minimum of the cost function.

Specification of gradient update formulas for cluster functions requires the same general approach as for approximators.

The cost function for the polynomial-based function can also be minimized but in doing so, we must guarantee that the method ensures that for each $i = 1, 2, \ldots, M$, $\sum_{j=1}^{R} \mu_{ij} = 1$ (this is a constrained minimization problem). In the next section we will show one method to do this.

11.6.3 Fuzzy C-Means Clustering and Function Approximation

We can use clustering methods to tune the parameters that enter nonlinearly, and linear least squares to tune the parameters that enter linearly. This is just one of many possible "hybrid" training methods.

As indicated above, "clustering" is the partitioning of data into subsets or groups based on similarities between the data. Here, we will introduce a method to perform fuzzy clustering, where we seek to use fuzzy sets to define soft boundaries to separate data into groups. The methods here are related to conventional ones that have been developed in the field of pattern recognition. In the c-means approach, we continue in the spirit of the previous methods in that we use optimization to pick the clusters and, hence, the premise membership function parameters. The consequent parameters are chosen using the weighted least squares approach developed earlier. In this way, we show one way to use a clustering method in the construction of function approximators. The combined least squares-clustering method has been called "clustering with optimal output predefuzzification."

Clustering for Specifying Rule Premises

Fuzzy clustering is the partitioning of a collection of data into fuzzy subsets or "clusters" based on similarities between the data, and can be implemented using an algorithm called fuzzy c-means.

C-Means Cost Function: Fuzzy c-means is an iterative algorithm used to find grades of membership μ_{ij} (scalars) and cluster centers v^j (vectors of dimension $n \times 1$) to minimize the cost function

$$J(\theta) = \sum_{i=1}^{M} \sum_{j=1}^{R} (\mu_{ij})^m |x(i) - v^j|^2 \tag{11.30}$$

where $m > 1$ is a design parameter. Here, M is the number of input-output data pairs in the training data set G, R is the number of clusters (number of rules) we wish to calculate, $x(i)$ for $i = 1, ..., M$ is the input portion of the input-output training data pairs, $v^j = [v_1^j, v_2^j, \ldots, v_n^j]^\top$ for $j = 1, ..., R$ are the cluster centers, μ_{ij} for $i = 1, ..., M$, and $j = 1, ..., R$ is the grade of membership of $x(i)$ in the jth cluster. Also, $|x| = \sqrt{x^\top x}$ where x is a vector. Intuitively, minimization of J results in cluster centers being placed to represent groups (clusters) of data.

The Premises and Fuzzy System to be Constructed: Fuzzy clustering will be used to form the premise portion of the If-Then rules in the fuzzy system we wish to construct. The process of "optimal output predefuzzification" (least squares training for consequent parameters) is used to form the consequent portion of the rules. We will combine fuzzy clustering and optimal output predefuzzification to construct multi-input single-output fuzzy systems. Extension of our discussion to multi-input multi-output systems can be done by repeating the process for each of the outputs.

In this section we utilize a Takagi-Sugeno fuzzy system in which the consequent portion of the rule-base is a function of the crisp inputs such that

$$\textbf{If } H^j \textbf{ Then } g_j(x) = a_{j,0} + a_{j,1}x_1 + \cdots + a_{j,n}x_n \tag{11.31}$$

where n is the number of inputs and H^j is an input fuzzy set given by

$$H^j = \{(x, \mu_{H^j}(x)) : x \in \mathcal{X}_1 \times \cdots \times \mathcal{X}_n\} \tag{11.32}$$

where $\mathcal{X}_i$ is the ith universe of discourse, and $\mu_{H^j}(x)$ is the membership function associated with H^j that represents the premise certainty for rule j; and $g_j(x) = \underline{a}_j^\top \hat{x}$ where $\underline{a}_j = [a_{j,0}, a_{j,1} \ldots, a_{j,n}]^\top$ and $\hat{x} = [1, x^\top]^\top$ where $j = 1, \ldots, R$. The resulting fuzzy system is a weighted average of the output $g_j(x)$ for $j = 1, ..., R$ and is given by

$$F_{ts}(x, \theta) = \frac{\sum_{j=1}^{R} g_j(x)\mu_{H^j}(x)}{\sum_{j=1}^{R} \mu_{H^j}(x)} \tag{11.33}$$

where R is the number of rules in the rule-base. Next, we will use the Takagi-Sugeno fuzzy model, fuzzy clustering, and optimal output defuzzification to determine the parameters $\underline{a}_j$ and $\mu_{H^j}(x)$, which define the fuzzy system. We will do this via a simple example.

Clustering Algorithm

We first discuss the choice of some of the parameters and initialization. Then we will provide a method to iteratively update the cluster centers and μ_{ij}. To do this, we will use a simple example with the training data set

$$G = \left\{ \left(\begin{bmatrix} 0 \\ 2 \end{bmatrix}, 1 \right), \left(\begin{bmatrix} 2 \\ 4 \end{bmatrix}, 5 \right), \left(\begin{bmatrix} 3 \\ 6 \end{bmatrix}, 6 \right) \right\} \tag{11.34}$$

as shown in Figure 11.31. For the clustering method, we will only use the input portion of the training data; however, when we seek to form our approximator, we will also use the output data.

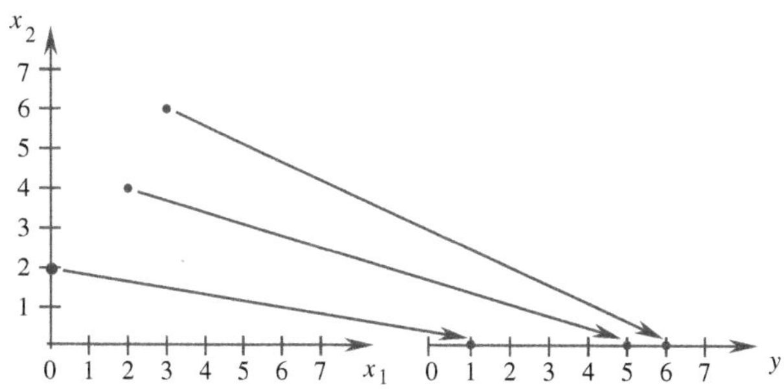

Figure 11.31: A simple training data set G.

Initialization: To specify the clustering algorithm, we first specify a "fuzziness factor" $m > 1$, which is a parameter that determines the amount of overlap of the clusters. If $m > 1$ is large, then points with less membership in the jth cluster have less influence on the determination of the new cluster centers. Next, we specify the number of clusters R we wish to calculate. The number of clusters R equals the number of rules in the rule-base and must be less than or equal to the number of data pairs in the training data set G (i.e., $R \leq M$). We also specify the error tolerance $\epsilon_c > 0$, which is the amount of error allowed in calculating the cluster centers. We initialize the cluster centers v_0^j via a random number generator so that each component of v_0^j is no larger (smaller) than the largest (smallest) corresponding component of the input portion of the training data. The selection of v_0^j, although somewhat arbitrary, may affect the final solution.

For our simple example, we choose $m = 2$ (a typical choice) and $R = 2$, and let $\epsilon_c = 0.001$. Our initial cluster centers were randomly chosen to be

$$v_0^1 = \begin{bmatrix} 1.89 \\ 3.76 \end{bmatrix}$$

and

$$v_0^2 = \begin{bmatrix} 2.47 \\ 4.76 \end{bmatrix}$$

so that each component lies in between $x_1(i)$ and $x_2(i)$ for $i = 1, 2, 3$ (see the definition of G in Equation (11.34)).

Cluster Center Calculations: Next, we compute the new cluster centers v^j based on the previous cluster centers to try to minimize the cost function in Equation (11.30). The necessary conditions for minimizing J are given by using Lagrange multiplier theory as

$$v_{new}^j = \frac{\sum_{i=1}^M x(i)(\mu_{ij}^{new})^m}{\sum_{i=1}^M (\mu_{ij}^{new})^m} \tag{11.35}$$

where

$$\mu_{ij}^{new} = \left[\sum_{k=1}^R \left(\frac{|x(i) - v_{old}^j|^2}{|x(i) - v_{old}^k|^2} \right)^{\frac{1}{m-1}} \right]^{-1} \tag{11.36}$$

for each $i = 1, \ldots, M$ and for each $j = 1, 2, \ldots, R$ such that $\sum_{j=1}^R \mu_{ij}^{new} = 1$ (and $|x|^2 = x^\top x$). In Equation (11.36), we see that it is possible that there exists an $i = 1, 2, \ldots, M$ such that $|x(i) - v_{old}^j|^2 = 0$ for some $j = 1, 2, \ldots, R$. In this case, the μ_{ij}^{new} is undefined. To fix this problem, let μ_{ij} for all i be any nonnegative numbers such that $\sum_{j=1}^R \mu_{ij} = 1$ and $\mu_{ij} = 0$, if $|x(i) - v_{old}^j|^2 \neq 0$.

Using Equation (11.36) for our example with $v_{old}^j = v_0^j$, $j = 1, 2$, we find that $\mu_{11}^{new} = 0.6729, \mu_{12}^{new} = 0.3271, \mu_{21}^{new} = 0.9197, \mu_{22}^{new} = 0.0803, \mu_{31}^{new} = 0.2254,$

and $\mu_{32}^{new} = 0.7746$. We use these μ_{ij}^{new} from Equation (11.36) to calculate the new cluster centers

$$v_{new}^1 = \begin{bmatrix} 1.366 \\ 3.4043 \end{bmatrix}$$

and

$$v_{new}^2 = \begin{bmatrix} 2.5410 \\ 5.3820 \end{bmatrix}$$

using Equation (11.35).

Testing for Termination: Next, we compare the distances between the current cluster centers v_{new}^j and the previous cluster centers v_{old}^j (which for the first step is v_0^j). If $|v_{new}^j - v_{old}^j| < \epsilon_c$ for all $j = 1, 2, \ldots, R$, then the cluster centers v_{new}^j accurately represent the input data, the fuzzy clustering algorithm is terminated, and we proceed to the optimal output defuzzification algorithm (see below) where we use a least squares method. Otherwise, we continue to iteratively use Equations (11.35) and (11.36) until we find cluster centers v_{new}^j that satisfy $|v_{new}^j - v_{old}^j| < \epsilon_c$ for all $j = 1, 2, \ldots, R$. For our example, $v_{old}^j = v_0^j$, and we see that $|v_{new}^j - v_{old}^j| = 0.6328$ for $j = 1$ and 0.6260 for $j = 2$. Both of these values are greater than ϵ_c, so we continue to update the cluster centers.

Proceeding to the next iteration, let $v_{old}^j = v_{new}^j$, $j = 1, 2, \ldots, R$ from the last iteration, and apply Equations (11.35) and (11.36) to find $\mu_{11}^{new} = 0.8233, \mu_{12}^{new} = 0.1767, \mu_{21}^{new} = 0.7445, \mu_{22}^{new} = 0.2555, \mu_{31}^{new} = 0.0593$, and $\mu_{32}^{new} = 0.9407$ using the cluster centers calculated above, yielding the new cluster centers

$$v_{new}^1 = \begin{bmatrix} 0.9056 \\ 2.9084 \end{bmatrix}$$

and

$$v_{new}^2 = \begin{bmatrix} 2.8381 \\ 5.7397 \end{bmatrix}$$

Computing the distances between these cluster centers and the previous ones, we find that $|v_{new}^j - v_{old}^j| > \epsilon_c$, so the algorithm continues. It takes 14 iterations before the algorithm terminates (i.e., before we have $|v_{new}^j - v_{old}^j| \le \epsilon_c = 0.001$ for all $j = 1, 2, \ldots, R$). When it does terminate, name the final membership grade values μ_{ij} and cluster centers v^j, $i = 1, 2, \ldots, M$, $j = 1, 2, \ldots, R$.

Finding the Final Cluster Center Values: For our example, after 14 iterations the algorithm finds $\mu_{11} = 0.9994$, $\mu_{12} = 0.0006$, $\mu_{21} = 0.1875$, $\mu_{22} = 0.8125$, $\mu_{31} = 0.0345$, $\mu_{32} = 0.9655$,

$$v^1 = \begin{bmatrix} 0.0714 \\ 2.0725 \end{bmatrix}$$

and

$$v^2 = \begin{bmatrix} 2.5854 \\ 5.1707 \end{bmatrix}$$

Notice that the clusters have converged so that v^1 is near $x(1) = [0, 2]^\top$ and v^2 lies in between $x(2) = [2, 4]^\top$ and $x(3) = [3, 6]^\top$.

Specifying the Premise Membership Function: The final values of v^j, $j = 1, 2, \ldots, R$, are used to specify the premise membership functions for the ith rule. In particular, we specify the premise membership functions as

$$\mu_{H^j}(x) = \left[\sum_{k=1}^{R} \left(\frac{|x - v^j|^2}{|x - v^k|^2} \right)^{\frac{1}{m-1}} \right]^{-1} \tag{11.37}$$

$j = 1, 2, \ldots, R$ where v^j, $j = 1, 2, \ldots, R$ are the cluster centers from the last iteration that uses Equations (11.35) and (11.36). It is interesting to note that for large values of m, we get smoother (less distinctive) membership functions. This is the primary guideline to use in selecting the value of m; however, often a good first choice is $m = 2$. Next, note that $\mu_{H^j}(x)$ is a premise membership function that is different from any that we have considered. With the premises of the rules defined, we next specify the consequent portion.

Least Squares for Specifying Rule Consequents

We apply "optimal output predefuzzification" to the training data to calculate the function $g_j(x) = \underline{a}_j^\top \hat{x}$, $j = 1, 2, \ldots, R$ for each rule (i.e., each cluster center), by determining the parameters $\underline{a}_j$. There are two methods you can use to find the $\underline{a}_j$.

Approach 1: For each cluster center v^j, in this approach we wish to minimize the squared error between the function $g_j(x)$ and the output portion of the training data pairs. Let $\hat{x}(i) = [1, (x(i))^\top]^\top$ where $(x(i), y(i)) \in G$. We wish to minimize the cost function J_j given by

We may use the μ_{ij} from the clusters to weight the batch least squares calculation so that the linear approximations pertain to each cluster.

$$J_j = \sum_{i=1}^{M} (\mu_{ij})^2 \left(y(i) - (\hat{x}(i))^\top \underline{a}_j \right)^2 \tag{11.38}$$

for each $j = 1, 2, \ldots, R$ where μ_{ij} is the grade of membership of the input portion of the ith data pair for the jth cluster that resulted from the clustering algorithm after it converged, $y(i)$ is the output portion of the ith data pair from G, $(x(i), y(i))$, and the multiplication of $(\hat{x}(i))^\top$ and $\underline{a}_j$ defines the output $g_j(x)$ associated with the jth rule for the ith training data point.

Looking at Equation (11.38), we see that the minimization of J_j via the choice of the $\underline{a}_j$ is a weighted least squares problem. From Equation (10.2) on page 424, the solution $\underline{a}_j$ for $j = 1, 2, \ldots, R$ to the weighted least squares problem is given by

$$\underline{a}_j = (\hat{X}^\top D_j^2 \hat{X})^{-1} \hat{X}^\top D_j^2 Y \tag{11.39}$$

where

$$\hat{X} = \begin{bmatrix} 1 & \dots & 1 \\ x(1) & \dots & x(M) \end{bmatrix}^\top$$
$$Y = [y(1), \dots, y(M)]^\top,$$
$$D_j^2 = (\text{diag}([\mu_{1j}, \dots, \mu_{Mj}]))^2$$

For our example, the parameters that satisfy the linear function $g_j(x) = \underline{a}_j^\top \hat{\underline{x}}(i)$ for $j = 1, 2$ such that J_j in Equation (11.38) is minimized, were found to be $\underline{a}_1 = [3, 2.999, -1]^\top$ and $\underline{a}_2 = [3, 3, -1]^\top$, which are very close to each other.

Approach 2: As an alternative approach, rather than solving R least squares problems, one for each rule, we can use the least squares methods to specify the consequent parameters of the Takagi-Sugeno fuzzy system. To do this, we simply parameterize the Takagi-Sugeno fuzzy system in Equation (11.33) in a form so that it is linear in the consequent parameters; then we can use batch or recursive least squares methods to find the parameters. Unless we indicate otherwise, we will always use approach 1 in this book.

Testing the Approximator

Suppose that we use approach 1 to specify the rule consequents. To test how accurately the constructed fuzzy system represents the training data set G in Figure 11.31 on page 537, suppose that we choose the test point x' such that $(x', y') \notin G$. Specifically, we choose

$$x' = \begin{bmatrix} 1 \\ 2 \end{bmatrix}$$

We would expect from Figure 11.31 that the output of the fuzzy system would lie somewhere between 1 and 5. The output is 3.9999, so we see that the trained Takagi-Sugeno fuzzy system seems to interpolate adequately. Notice also that if we let $x = x(i)$, $i = 1, 2, 3$ where $(x(i), y(i)) \in G$, we get values very close to the $y(i)$, $i = 1, 2, 3$, respectively. That is, for this example, the fuzzy system nearly perfectly maps the training data pairs. We also note that if the input to the fuzzy system is $x = [2.5, 5]^\top$, the output is 5.5, so the fuzzy system seems to perform good interpolation near the training data points.

Finally, we note that the $\underline{a}_j$ will clearly not always be as close to each other as for this example. For instance, if we add the data pair $([4, 5]^\top, 5.5)$ to G (i.e., make $M = 4$), then the cluster centers converge after 13 iterations (using the same parameters m and ϵ_c as we did earlier). Using approach 1 to find the consequent parameters, we get

$$\underline{a}_1 = [-1.458, 0.7307, 1.2307]^\top$$

and

$$\underline{a}_2 = [2.999, 0.00004, 0.5]^\top$$

For the resulting fuzzy system, if we let $x = [1, 2]^\top$ in Equation (11.33), we get an output value of 1.8378, so we see that it performs differently from the case for $M = 3$ but still provides a reasonable interpolated value.

11.7 Neural or Fuzzy: Which is Better? Bad Question!

If you are asking this question, it shows that you do not understand the fundamental concepts!

- You should be concerned about whether your training data carries the proper information to perform good approximation. Is the training data set large enough? Is the test set large enough? Does your measure of approximation accuracy properly reflect your approximation goals? Did you start with a simple linear (or affine) approximator and a linear least squares method? For many applications this can be sufficient; you only need all the capabilities of neural or fuzzy system approximators if you have a nonlinear approximation problem.

- You should realize that for practical applications, the choice of which is the best approximator structure is very difficult and you cannot quickly conclude that one is better than another.

- You should be concerned with approximator complexity and approximator tunability for your application. For example, via experience have you found that a certain type of structure works well? Or, based on physical insights, can you use nonlinear functions of input data as inputs to your approximator?

- You should ask whether to use a local or globally supported basis function (i.e., one that only has a local influence on the approximator mapping, or one that, if it is changed, can change the shape over the whole region of the mapping).

- You should ask whether you have too many inputs (i.e., too large a value for n) so that it is not possible to use a grid if you are using a locally supported basis function (i.e., you should be concerned with the impact of how many inputs you have on the computational complexity in approximator structure choice).

- You should ask whether you should use a linear or nonlinear in the parameter approximator, since this affects tuning flexibility and training algorithm performance.

The names "neural" or "fuzzy" are largely attached simply for historical purposes due to the fields that they came from. Really you need to think of the basics, not this terminology. Focus on generalization, overfitting, complexity, and composition of the data set. Generally, structure choice is quite difficult so it needs attention; however, methods to *automate* the construction of the structure are discussed in the "For Further Study" section at the end of this part.

11.8 Exercises and Design Problems

Exercise 11.1 (Matlab for Neural Network Training):

(a) Suppose you use a multilayer perceptron with two layers, the first layer has $n_1 = 11$ logistic function neurons, and the output layer has a single linear neuron. Use a software package (e.g., the Matlab Neural Networks toolbox) to match the training data shown in Figure 9.10 (this defines G and here use $M = 121$). Train with the Levenberg-Marquardt method. While you train with 121 data pairs, test with about 10 times that many. Plot the approximator mapping and data on the same plot to evaluate the accuracy of the interpolation.

(b) Train with a conjugate gradient method and compare to the result in (a).

(c) Train with steepest descent and compare to the results in (a) and (b).

Exercise 11.2 (Levenberg-Marquardt Update Formulas for Neural Networks):

(a) Derive the Levenberg-Marquardt parameter update formulas for a two-layer multilayer perceptron with a linear output layer and hyperbolic tangent activation functions in the hidden layer. Assume that you update all weights and biases in the network (i.e., both the parameters that enter linearly and those that enter in a nonlinear fashion).

(b) Repeat (a), but for a radial basis function neural network where the output is computed as a sum of Gaussian receptive field units. Assume that you update all parameters in the network (i.e., both the parameters that enter linearly and those that enter in a nonlinear fashion).

(c) For both (a) and (b), solve the function approximation theme problem given in this chapter. Clearly explain all your choices for the approximator structure and training method. Illustrate generalization properties of the approximators after they are trained.

Design Problem 11.1 (Fuzzy C-Means and Least Squares for Approximator Tuning):

(a) Use fuzzy c-means and least squares for tuning the special type of Takagi-Sugeno fuzzy system given in the chapter to solve the function approximation theme problem studied throughout this chapter.

(b) Illustrate its generalization capabilities and that it makes reasonable choices for cluster placement (plot the final clusters on the same plot as the function you are trying to approximate).

(c) Compare the results to what you obtained in Exercise 11.2 where neural networks and Levenberg-Marquardt training were used.

Design Problem 11.2 (Structural Plasticity and Approximators)*: The human brain and the brains of many other animals learn not only via parameter adjustment (the adjustment of strengths between connections in the biological neural network), but also by growing new neuronal connections and destroying others. In this problem you will learn methods for constructing the structure of approximators. For example, in the case of multilayer perceptrons, some methods automatically pick the number of neurons used in each layer, and some of the methods use biomimicry concepts based on biological neural networks. Alternatively, with our unified view of approximators, we can consider how to construct the structure of a fuzzy system. For instance, we may study how to automatically pick the number of rules or membership functions.

(a) First, you must conduct some background research. Read the papers [295, 431] and see the book [412] for ideas on how to construct the number of rules in a fuzzy system.

(b) Explain in detail how the neural network methods can be used for fuzzy systems. To do this, pick a standard fuzzy system and define the algorithms for its construction. Are there methods developed in the area of neural networks that do not seem to apply to any fuzzy system?

(c) Choose a method from one of the above references, specify a structure construction/destruction algorithm, develop code to implement it, and test it for the theme problem that was studied in this chapter. For many methods this will involve specifying how structure is adjusted, and the use of a standard training method (e.g, gradient or least squares) as found in the chapter. Be sure to use appropriate training and test sets, and clearly illustrate the performance of the method. If possible, compare it to the results in the chapter where only the parameters were tuned, not the structure.

(d) Invent a method for tuning structure of an approximator. You choose the type of approximator you want to study. Hint: Suppose that you have a low-dimensional function to approximate (e.g., one output and two inputs). Suppose that your training data set is G and test set is Γ. Suppose that you grid the input space, calling each subregion a "cell," and label these c_i, $i = 1, 2, \ldots, N_g$ where N_g is the number of cells created by the number of partitions on the jth input space x_j (we assume that the number of divisions on each input dimension is the same, but clearly this is just for convenience). Suppose that you use a test set Γ_a and a cost function J_a that is defined to be the approximation error in each cell c_i between the approximator and the actual function for Γ_a. In particular, if c_i is a cell (e.g., a square if the dimension of the input space is 2), then $J_a(c_i)$ could be defined to be the average mean squared error over points in the test set Γ_a that lie in c_i (clearly, then, to make this a reasonable definition you would want Γ_a to have points in each cell created by the input space gridding). Suppose that there are no common points in Γ_a, G, and Γ. Suppose that $|G| < |\Gamma|$ (with the difference in size large enough so that you can achieve good function approximation for a fixed size approximator, and for any value that you adjust p to be in your structure adjustment method). Also, suppose that $|\Gamma_a|$ is large enough to be representative of the approximation error, no matter how you adjust the structure (e.g., it could be that $|\Gamma_a| = |\Gamma|$).

Now, view the approximator construction problem as a two-level optimization problem. In particular, we will view it as a type of multilevel reinforcement learning approach, and hence, it is a gradient-type method. For any fixed structure (i.e., fixed p), we will tune with a standard gradient method (e.g., Levenberg-Marquardt). This tuning will occur interleaved with structure adjustments; there will be a structure adjustment, then multiple steps of the standard gradient method will be executed (e.g., until some termination criterion is satisfied) before the next structure adjustment. How do we then make structure adjustments? There are many ways. One way is to adjust the structure of the approximator to try to achieve the minimization of J_a. Choose some threshold $\varepsilon > 0$ that represents what you consider to be an acceptable level of approximation error in any cell c_i. Suppose that we try to adjust an approximator structure that is based on gridding the input space with basis functions (e.g., the radial basis function neural network or several types of fuzzy systems). To be more concrete, suppose that we adjust radial basis function neural networks with their radial basis functions defined to be Gaussian functions with centers that our structure adjustment method will place (for simplicity, let the parameters that enter linearly be adjusted only after structure adjustments are made in the step where we use gradient training). Adjust structure as follows:

1. Compute $J_a(c_i)$, $i = 1, 2, \ldots, N_g$, over the test set Γ_a.
2. If for some i, $J_a(c_i) > \varepsilon$, then randomly place $\lambda_{add} int(|J_a(c_i) - \varepsilon|)$ ($int(\cdot)$ is the integer part of its argument) new radial basis functions in the region c_i, where $\lambda_{add} > 0$ can be thought of as a step size for the structure adjustment algorithm in the case where structure is added.
3. If for some i, $J_a(c_i) \leq \varepsilon$, then randomly remove $\lambda_{sub} int(|J_a(c_i) - \varepsilon|)$ radial basis functions from the region c_i, where $\lambda_{sub} > 0$ can be thought of as a step size for the structure adjustment

algorithm in the case where structure is deleted.

4. Go to standard gradient method for parameter adjustments.

The goal of the method is to try to achieve an error of ε in each cell. Why not just try for zero approximation error in each cell? This will in general require an infinite number of radial basis functions; we pay for accuracy with complexity. The addition of more radial basis functions allows for more accurate function approximation in regions where they are added. Removal of radial basis functions can result in lower approximation accuracy where they are removed. The algorithm will tend to redistribute the centers so as to allocate them where more accuracy is needed.

Fully test this algorithm for both $n = 1$ and $n = 2$, showing how it reshapes the approximator mapping (show plots) and reallocates the radial basis function centers. Explain why you can view the above approach as a reinforcement-based learning method for structure, and in particular, write down the update equation that clearly shows it is a gradient-type method. What is the reinforcement function? Next, can you achieve a simpler approximator structure with this approach than with the one you would construct manually? Does the gridding approach that this method is based on make it impossible to apply to high-dimensional function approximation problems? If not, explain. If so, which method from the literature would do better? Next, explain how you could redesign the algorithm so that it can be used for online function construction.

Chapter 12

Adaptive Control

Chapter Contents

To illustrate how function approximation and online function approximation can be used as the basis for learning in control, in this chapter we will show how they can be used in adaptive control. We first consider online approximation based control where neural networks or fuzzy systems are used as the approximator structure for nonlinear discrete-time systems. It should be clear that we could have just as easily used other approximators (e.g., polynomials or wavelets). We consider the discrete-time case since we developed our parameter update formulas for approximators (e.g., least squares and gradient) in this part for this case and they are easy to understand. For the discrete-time case, we will not prove that the resulting closed-loop systems are stable, and indeed we will at times violate certain assumptions that are typically needed to ensure stability. Hence, strictly speaking, like Section 9.4, the first part of this chapter focuses on the *heuristic* construction of adaptive controllers. In contrast to Section 9.4 where we use simple bio-inspired adaptation heuristics, here the focus is on how to use optimization methods to construct adaptive controllers. Of course, both cases can be considered to be methods to use heuristics to construct adaptive systems; it is just that for the case where we consider discrete-time systems, the optimization methods form a clear basis for our heuristics.

In the last section of this chapter, we will consider the continuous time case and focus on showing how to achieve stable adaptive control when either neural or fuzzy systems are used as online approximators. This section is, however, brief, and the interested reader is referred to the "For Further Study" section at the end of this part for more detailed treatments.

12.1 Strategies for Adaptive Control

First, we briefly overview the two strategies for adaptive control that are our primary focus in this chapter. A variety of other ways to incorporate learning into control systems are discussed in Section 9.4.5. Keep in mind that all of these methods bear very close relationships to the work in conventional adaptive control (see "For Further Study" for references).

12.1.1 Indirect Adaptive Control

There are (at least) two general approaches to adaptive control, and in the first one, which is depicted in Figure 12.1, we use an online identification method to estimate the plant input-output mapping (by estimating the parameters of an "identifier model") and a "controller designer" module to subsequently specify the parameters of the controller. Generally, you can think of indirect adaptive control as automating the model-building and control design process that we use for fixed controllers (e.g., in Part II).

Indirect adaptive control entails estimating a model of the plant and using it to specify the control input.

If the plant input-output mapping changes, the identifier will provide estimates of these changes and the controller designer will subsequently tune the controller. It is inherently assumed that we are certain that the estimated plant mapping is equivalent to the actual one at all times (this is called the "certainty

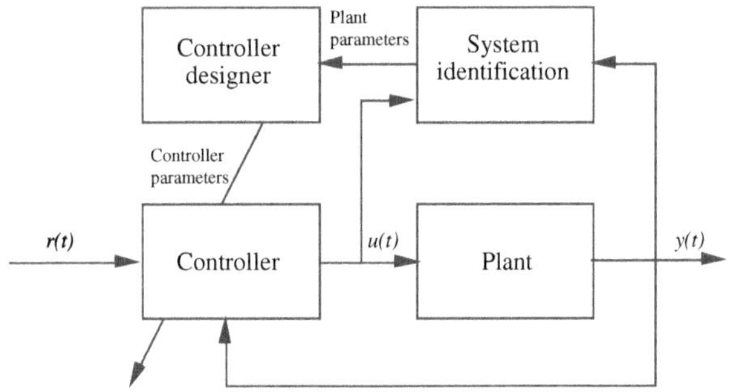

Figure 12.1: Indirect adaptive control.

equivalence principle"). Then if the controller designer can specify a controller for each set of plant parameter estimates, it will succeed in controlling the plant. The overall approach is called "indirect adaptive control" since we tune the controller indirectly by first estimating the plant parameters (as opposed to direct adaptive control, which is discussed below, where the controller parameters are estimated directly without first identifying the plant parameters).

The structure used for the identifier model could be linear with adjustable coefficients. Alternatively, it could be a neural or fuzzy system with tunable parameters that enter in a linear or nonlinear fashion (e.g., membership function parameters or weights and biases). In this case, the model that is being tuned is a nonlinear function. Since the plant is assumed to be unknown, the nonlinear mapping it implements is unknown. In adjusting the nonlinear mapping implemented by the neural or fuzzy system to match the unknown nonlinear mapping of the plant, we are solving an online function approximation problem. Normally, gradient or least squares methods are used to tune neural or fuzzy systems for indirect adaptive control (although sometimes problem-dependent heuristics have been found to be useful for practical applications).

Alternatively, you could use an optimization method based on biomimicry of an individual foraging animal (see Part V); in this case, we think of parameter adjustments as foraging for model information. Other times, a genetic algorithm has been employed for such online model tuning, or more generally for the evolution of the entire indirect adaptive control strategy. In one genetic adaptive control strategy, a set (population) of identifier models is evolved online, and the best one is used at each time instant in a certainty equivalence control law to specify the control input. This is studied in Section 16.5.

12.1.2 Direct Adaptive Control

In the second general approach to adaptive control we study in this chapter, which is shown in Figure 12.2, the "adaptation mechanism" observes the signals from the control system and adapts the parameters of the controller to maintain performance even if there are changes in the plant. Sometimes, in either the direct or indirect adaptive controllers, the desired performance is characterized with a "reference model," and the controller then seeks to make the closed-loop system behave as the reference model would, even if the plant changes. This is called "model reference adaptive control" (MRAC).

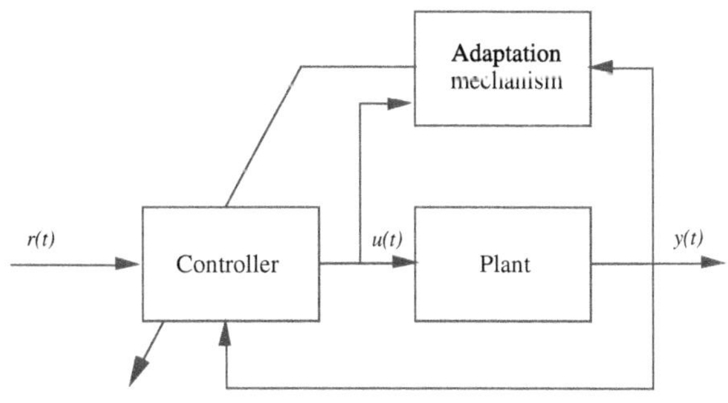

Figure 12.2: Direct adaptive control.

Direct adaptive control entails tuning the controller to improve a performance measure that leads to improved closed-loop performance.

In neural control or adaptive fuzzy control, the controller is implemented with a neural or fuzzy system, respectively. Normally, gradient or least squares methods are used to tune the controller. Also, many heuristic direct adaptive control methods have been developed, for instance, based on reinforcement learning control. Two such methods were studied in Section 9.4.

Alternatively, you could use an optimization method based on biomimicry of foraging; in this case, we think of controller parameter adjustments as foraging for information on how the controller should behave (see Part V). Clearly, since the genetic algorithm is also an optimization method, it can be used to tune neural or fuzzy system mappings when they are used as controllers also. The key to making such a controller work is to provide a way to define a fitness function for evaluating the quality of a population of controllers (in one approach, a model of the plant is used to predict into the future how each controller in the population will perform). Then, the most fit controller in the population is used at each step to control the plant. This is studied in Section 16.5.

12.2 Classes of Nonlinear Discrete-Time Systems

In this section we will outline some classes of plants that we will consider in this chapter, and we provide some simple examples to highlight some key characteristics of these.

12.2.1 Nonlinear Discrete-Time Systems

Consider the control of plants that can be described with

$$y(k+d) = f(x(k), u(k)) \tag{12.1}$$

where $f(x(k), u(k))$ is a smooth (but unknown) function of its arguments, $u(k)$ is the measurable (scalar) input, $y(k)$ is the measurable (scalar) output, $d \geq 1$ is the delay between the input and output, and $x(k)$ is the state vector where

$$x(k) = [y(k), y(k-1), \ldots, y(k-n), u(k-1), u(k-2), \ldots, u(k-m)]^\top \quad (12.2)$$

for some $n \geq 0$ and $m \geq 0$. Clearly, this class of plants is quite general. It allows for unknown nonlinear dependencies on plant parameters and the past values of the inputs and outputs. Note, however, that it is assumed that f depends on $x(k)$ and $u(k)$, not k (i.e., we do not consider $f(x(k), u(k), k)$). Hence, we consider time-invariant systems; the mapping shape is fixed but unknown. Also, it is assumed that d, m, and n are constant and known (we do not allow for a time-varying delay between the input and output). In some practical problems you may not know them, so you may have to try different choices, perhaps with guidance from some experiments designed to help find their values (e.g., you may be able to apply some inputs and observe the outputs to see how many time steps it takes for the effect of the input to reach the output, and this can be useful to estimate d). Also, note that for practical problems it may be that some of the elements of the $x(k)$ vector may be missing, yet we will typically include them unless we have some good a priori knowledge about the plant. We do this simply because, without knowing any better, it seems best to include all past inputs and outputs, up to the one that last affected $y(k+d)$ (note that the amount of "memory" in the system affects the size of n and m). For the direct adaptive control case, we will consider certain subclasses of this type of nonlinear system.

In the indirect adaptive control case, we also consider a special subclass of plants that can be represented by

$$y(k+d) = \alpha(x(k)) + \beta(x(k))u(k) \quad (12.3)$$

where $\alpha(x(k))$ and $\beta(x(k))$ are unknown smooth functions of the state $x(k)$ that is defined in Equation (12.2). Notice that while in this case we cannot consider plants where $y(k+d)$ is a nonlinear function of $u(k)$, it can be a nonlinear function of past values of u. We will also make certain assumptions about $\beta(x(k))$, such as requiring it to be bounded away from zero, or bounded from above by a constant. This class of plants is one that will also work for the direct adaptive control case we consider.

12.2.2 Example: Linear Systems with Unknown Constant Coefficients

Clearly, the above classes of systems include linear discrete-time systems with constant but unknown coefficients. In particular, let $d = 1$ and

$$\begin{aligned} y(k+1) &= a_1 y(k) + a_2 y(k-1) + \cdots + a_{n+1} y(k-n) \\ &\quad + b_1 u(k) + b_2 u(k-1) + \cdots + b_{m+1} u(k-m) \end{aligned}$$

where the a_i, $i = 1, 2, \ldots, n+1$, and b_i, $i = 1, 2, \ldots, m+1$, are unknown but constant scalars. If we let

$$\alpha(x(k)) = a_1 y(k) + a_2 y(k-1) + \cdots + a_{n+1} y(k-n) + b_2 u(k-1) + \cdots + b_{m+1} u(k-m)$$

and

$$\beta(x(k)) = b_1$$

then

$$y(k+1) = \alpha(x(k)) + \beta(x(k))u(k)$$

so that the model clearly fits the form in Equation (12.3).

12.3 Indirect Adaptive Neural/Fuzzy Control

Here, we introduce an indirect adaptive control scheme based on the use of online approximation with neural networks or fuzzy systems. We will focus on developing the controller to track a reference input $r(k)$ where $|r(k)|$ is bounded by a finite constant. We will define the tracking error as $e(k) = r(k) - y(k)$.

12.3.1 Estimators and Certainty Equivalence Controller

As indicated above, we will consider the discrete-time system of the form

$$\begin{aligned} y(k+d) &= \alpha(x(k)) + \beta(x(k))u(k) & (12.4)\\ &= \alpha_u(x(k)) + \alpha_k(k) + (\beta_u(x(k)) + \beta_k(k))u(k) & (12.5) \end{aligned}$$

The functions $\alpha_u(x(k))$ and $\beta_u(x(k))$, defined above, represent the unknown nonlinear dynamics of the plant (of course, we assume these functions are smooth). It is these functions that we want to estimate so that we can specify a controller. Notice also that we have defined $\alpha_k(k)$ and $\beta_k(k)$ to be *known* parts of the plant dynamics. While these can be set to zero and the approach to be developed will still work, there are times in practical applications where you know portions of the nonlinear dynamics and hence, it makes sense to include this sort of knowledge in the adaptation scheme.

We assume $\beta(x(k))$ to satisfy

$$0 < \beta_0 \leq \beta(x(k)) \tag{12.6}$$

for some known $\beta_0 > 0$ for all $x(k)$. This places a restriction on the class of plants that we can consider. Intuitively, we require that the gain on $u(k)$ be bounded from below due to how an estimate of β will be used to specify the control. It is also possible to develop a scheme where $\beta(x(k))$ is known to be negative and bounded from above by a constant that is less than zero.

Estimating an Unknown Ideal Controller

If we assume $r(k+d)$ is known, which is reasonable for many applications since this is specified by the user, we know that there exists an "ideal" controller

$$u^*(k) = \frac{-\alpha(x(k)) + r(k+d)}{\beta(x(k))}, \tag{12.7}$$

that linearizes the dynamics of Equation (12.4) such that $y(k) \to r(k)$. To see this, substitute $u(k) = u^*(k)$ in Equation (12.3) to obtain

$$y(k+d) = r(k+d)$$

so that we achieve tracking of the reference input within d steps. The problem is that $u^*(k)$ above is not known because we do not know $\alpha(x(k))$ or $\beta(x(k))$. Here, we will develop an estimator for these plant nonlinearities and use them to form an approximation to $u^*(k)$.

In particular, we estimate the unknown $\alpha(x(k))$ and $\beta(x(k))$ mappings with two different linear in the parameter approximators. Let

$$\alpha_u(x(k)) = \theta_\alpha^{*\top} \phi_\alpha(x(k)) + w_\alpha(k+d) \tag{12.8}$$

$$\beta_u(x(k)) = \theta_\beta^{*\top} \phi_\beta(x(k)) + w_\beta(k+d) \tag{12.9}$$

Here, w_α and w_β are the representation errors that arise due to the use of a finite size approximator. (Recall that from the universal approximation property, if we let p, the number of parameters of the approximator, be arbitrarily large, we could make the representation errors arbitrarily small on a closed and bounded set.) For design, we will often have to experiment with different values of p to find an approximator that is simple, but still flexible enough, to be tuned to find the proper shape of the unknown nonlinearity.

Note that above, we assume that θ_α^* and θ_β^* are the "ideal parameters" for $\alpha_u(x(k))$ and $\beta_u(x(k))$, In particular, these are

$$\begin{aligned}
\theta_\alpha^* &= \arg\min \theta_\alpha \in \Omega_\alpha \left(\sup_{x \in S_x} \left| \theta_\alpha^\top \phi_\alpha(x) - \alpha_u(x) \right| \right) \\
\theta_\beta^* &= \arg\min \theta_\beta \in \Omega_\beta \left(\sup_{x \in S_x} \left| \theta_\beta^\top \phi_\beta(x) - \beta_u(x) \right| \right)
\end{aligned} \tag{12.10}$$

Here, $\Omega_\alpha \subset \Re^{p_\alpha}$ and $\Omega_\beta \subset \Re^{p_\beta}$ are the convex and closed bounded sets containing the feasible parameter vectors θ_α and θ_β, respectively. Here, we will simply assume that these are defined similar to those in Section 11.1.6; in particular, we simply assume that we know feasible ranges of values for each component of the indicated vectors. Note that we assume that θ_α^* and θ_β^* are unknown constants. We do not need to know them to specify the controller, but will develop methods to estimate them to try to specify the control. We will see that even if they cannot be approximated very well, it may still be possible to achieve good tracking.

Next, note that $S_x \subset \Re^{n+m+1}$ is assumed to be a closed and bounded set and we assume that we know an a priori bound on its size; we will think of it as the range of validity of the approximators. In particular, we will define our approximators to have good accuracy over a range of their domain and then let S_x be this region of the domain. Basically, you cannot expect to do good control in regions of the domain where you have not allocated enough approximator structure; otherwise, you are expecting it to perform good extrapolation which is often not possible. For physical plants it is often clear how to specify the approximator to cover an appropriate S_x.

Certainty Equivalence Controller

Using a "certainty equivalence approach," the control input is defined as

$$u(k) = \frac{-\hat{\alpha}(x(k)) + r(k+d)}{\hat{\beta}(x(k))}, \tag{12.11}$$

The certainty equivalence approach entails specifying a control input using an estimate of the plant model that would cancel appropriate plant dynamics and achieve good tracking if the estimate was accurate.

where $\hat{\alpha}(x(k))$ and $\hat{\beta}(x(k))$ are estimates of $\alpha(x(k))$ and $\beta(x(k))$, respectively, defined as

$$\hat{\alpha}(x(k)) = \theta_\alpha^\top(k)\phi_\alpha(x(k)) + \alpha_k(k) \tag{12.12}$$

$$\hat{\beta}(x(k)) = \theta_\beta^\top(k)\phi_\beta(x(k)) + \beta_k(k). \tag{12.13}$$

We will use optimization methods (e.g., gradient or least squares) to pick $\theta_\alpha(k)$ and $\theta_\beta(k)$ to try to minimize the approximation error. We include $\alpha_k(k)$ and $\beta_k(k)$ to provide knowledge that we may have about the nonlinear form of the plant dynamics (if you have good information, then the approximator will simply try to estimate the deviation from the known dynamics and sometimes this is easier and performance can be improved).

Projection algorithms can be used to ensure that $\theta_\alpha(k) \in \Omega_\alpha$ and $\theta_\beta(k) \in \Omega_\beta$ for all k. Projection algorithms may also be used to ensure that $\hat{\beta}(x(k)) \geq \beta_0$ so that the control signal is well defined. Note that, for instance, if we know that each element of the ϕ_β vector is always positive (which it is for a radial basis function neural network with receptive field units that are Gaussian functions and for certain fuzzy systems), then to ensure that $\hat{\beta}(x(k)) \geq \beta_0$, we can simply use a projection method to keep *each* component of $\theta_\beta(k)$ greater than or equal to β_0.

12.3.2 Error Equations and Representation Error Bounds

Next, we derive an expression for the tracking error that results from the above definitions. Also, we quantify the "combined" approximation error that results from using the two linear in the parameter approximators.

Linear Error Equations

The tracking error $e(k) = r(k) - y(k)$ and if we advance time by d steps

$$e(k+d) = r(k+d) - y(k+d)$$

$$
\begin{aligned}
&= r(k+d) - \alpha(x(k)) - \beta(x(k))u(k) \\
&= (\hat{\alpha}(x(k)) - \alpha(x(k))) + \left(\hat{\beta}(x(k)) - \beta(x(k))\right) u(k)
\end{aligned}
$$

where we used the value of $r(k+d)$ obtained from Equation (12.11). It is interesting to note that if you define

$$\hat{y}(k+d) = \hat{\alpha}(x(k)) + \hat{\beta}(x(k))u(k)$$

then with this certainty equivalence control law, the output tracking error

$$e(k+d) = \hat{y}(k+d) - y(k+d)$$

which can be viewed as the "identification error" (i.e., it is a measure of the quality of the model that we are tuning to represent the plant).

The parameter errors for the indirect adaptive controller are defined as

$$\tilde{\theta}_\alpha(k) = \theta_\alpha(k) - \theta_\alpha^*$$

and

$$\tilde{\theta}_\beta(k) = \theta_\beta(k) - \theta_\beta^*$$

With these definitions, the tracking error becomes

$$e(k+d) = \tilde{\theta}_\alpha^\top(k)\phi_\alpha(x(k)) + \tilde{\theta}_\beta^\top(k)\phi_\beta(x(k))u(k) - w_\alpha(k+d) - w_\beta(k+d)u(k) \tag{12.14}$$

With this we see that the representation errors affect the tracking error. Notice that even though $e(k+d)$ is measurable (since we assume that $y(k+d)$ and $r(k+d)$ are measurable), $\tilde{\theta}_\alpha(k)$ and $\tilde{\theta}_\beta(k)$ are not known because we assume that we do not know the ideal parameters.

Next, we seek to find a linear error equation; this error equation is actually the quantity that the optimization method seeks to minimize. First, let

$$\tilde{\theta} = \left[\tilde{\theta}_\alpha^\top, \tilde{\theta}_\beta^\top\right]^\top \tag{12.15}$$

and

$$\theta(k) = \left[\theta_\alpha^\top(k), \theta_\beta^\top(k)\right]^\top$$

and

$$\phi(x(k), u(k)) = \left[\phi_\alpha^\top(x(k)), \phi_\beta^\top(x(k))u(k)\right]^\top$$

Here, notice that $\phi(x(k), u(k))$ is a function of $u(k)$ due to how we multiply $u(k)$ by $\phi_\beta^\top(x(k))$, and we do this to ensure that we get a linear error equation below.

Next, let

$$w(k) = w_\alpha(k) + w_\beta(k)u(k-d)$$

and we will think of this as an equation for the "combined" representation error that arises from approximating both $\alpha(x(k))$ and $\beta(x(k))$.

With this, Equation (12.14) becomes the linear error equation (with an offset defined by the representation error)

$$e(k+d) = \tilde{\theta}^\top(k)\phi(x(k), u(k)) - w(k+d) \tag{12.16}$$

Due to the current state of the theory, we generally need this type of linear error equation so that we can apply gradient and recursive least squares methods and be assured we can get stable adaptive systems.

Bounds on the Approximation Errors

Here, we assume that we know W_α and W_β a priori such that

$$0 \leq |w_\alpha(k)| < W_\alpha$$

and

$$0 \leq |w_\beta(k)| < W_\beta$$

Notice that with these, we have

$$|w(k)| = |w_\alpha(k) + w_\beta(k)u(k-d)| < W_\alpha + W_\beta|u(k-d)|$$

as a time-varying bound on the combined representation error.

It is important to emphasize that we assume that we know bounds W_α and W_β on the *ideal* approximation errors (i.e., the errors that result from the best possible way to tune the given approximator, using any adjustment method, not from the current representation error that arises from the current parameter estimates). Note that for practical applications, you will not typically know these bounds a priori, because it is difficult to know how good a given approximator can be tuned to match the unknown nonlinearity. There are, however, ways to estimate these bounds (or sometimes experience gives insights into their choice). You could do some experiments with the plant before applying a control technique and compute estimates of the bounds for finite size data sets. Often, however, you may want to simply view W_α and W_β as tuning parameters for the algorithm.

12.3.3 Adaptation Methods

Here, we introduce two methods, normalized gradient and recursive least squares, for the training of $\theta(k)$ to try to approximate $\alpha(x(k))$ and $\beta(x(k))$.

Normalized Gradient Method

Here, we will use the normalized gradient method that was discussed in Chapter 11. Recall that this is a steepest descent approach with a special choice for the step size (see the section on step size choice). If we consider a cost function

$$J(\theta) = \frac{1}{2}e^2(k)$$

(i.e., only one data pair is used in the optimization), which is a measure of the size of the tracking error (and identification error) that we want to minimize, then the update formula would be

$$\theta(k) = \theta(k-d) + \lambda_k d(k)$$

where λ_k is the step size and $d(k)$ is the descent direction (d here is the constant scalar delay, and $d(k)$ is the notation used in Chapter 11 for a $p \times 1$ vector descent direction). Note that the delay d appears in $\theta(k-d)$ due to the fact that $e(k)$ depends on $\theta(k-d)$. Suppose that we choose a normalized gradient approach and hence,

$$\begin{aligned}
d(k) &= -\frac{\partial J(\theta)}{\partial \theta} \\
&= -e(k)\frac{\partial e(k)}{\partial \theta} \\
&= -e(k)\frac{\partial}{\partial \theta}\left(\theta^\top(k-d)\phi(x(k-d), u(k-d))\right) \\
&= -e(k)\phi(x(k-d), u(k-d))
\end{aligned}$$

Achieving a linear relationship between model error and tracking error ensures that if our online optimization approach seeks to minimize tracking error, it will also try to minimize model error.

Now, if we pick λ_k as discussed in Chapter 11 (see selection of step size, normalized gradient approach), use the notation $\phi(k-d) = \phi(x(k-d), u(k-d))$, and use a "robustification" approach to modify the tracking error and produce $e_\epsilon(k)$, the adaptation law becomes

$$\theta(k) = \theta(k-d) + \frac{\kappa_1 \eta \phi(k-d)}{1 + \gamma|\phi(k-d)|^2} e_\epsilon(k), \tag{12.17}$$

where $\kappa_1 = -1$, γ is a design parameter (see Chapter 11 for a discussion on how to choose it), and we assume that the constant "adaptation gain" η is such that

$$0 < \eta < \frac{2\gamma}{\kappa_2}$$

where for the indirect adaptive control case $\kappa_2 = 1$ (the κ_i parameters are different for the direct adaptive control case that is treated below). This constraint on the adaptation gain is specified to help ensure stability. You should think of it as a step size parameter where we need $\eta > 0$ to ensure that we use the negative gradient and we need $\eta < 2\gamma$ to ensure that the step size is small enough to maintain stability (consider the example in Chapter 11, where we showed a case where if the step size is too large, the parameters can diverge).

We define $e_\epsilon(k)$ to be a "dead zone modification" of the tracking error $e(k)$. In a standard normalized gradient approach, you would use $e(k)$ in place of $e_\epsilon(k)$, but due to our finite size approximators, there will be a representation error that necessitates the use of $e_\epsilon(k)$ to try to achieve stability (this is what is sometimes referred to as a "robustification method"). In particular, we define

$$\epsilon(k) = W_\alpha + W_\beta|u(k-d)|$$

to be a bound on the combined representation error. Next, we define

$$e_\epsilon(k) = \begin{cases} e(k) - \epsilon(k) & \text{if } e(k) > \epsilon(k) \\ 0 & \text{if } |e(k)| \leq \epsilon(k) \\ e(k) + \epsilon(k) & \text{if } e(k) < -\epsilon(k) \end{cases} \tag{12.18}$$

which is shown in Figure 12.3(a) for the case where $\epsilon(k) = 1$ for all k. As $\epsilon(k)$ varies the size of the portion of the plot around zero where "$e(k)$ is in the dead zone" and hence $e_\epsilon(k) = 0$ varies. Basically, if we have large representation errors, then we are doing a poor job at approximating the plant dynamics. If the tracking error is smaller than the combined representation error, we turn off the parameter update (so $\theta(k) = \theta(k-d)$) since in this case, updates are futile since the approximator cannot do much better anyway (and updating in this case when it cannot do better can cause an accumulation of errors that results in parameter divergence). Clearly we would like to make $\epsilon(k)$ as small as possible and to do this, we must do a good job in designing our approximators (e.g., we may need many parameters and a good choice for the nonlinear part of the structure). Finally, note that unfortunately in the indirect approach, $\epsilon(k)$ depends on $|u(k-d)|$ (due to the need to obtain a linear relationship between tracking error and identification error) and hence, this can adversely affect the size of the dead zone and hence, quality of tracking.

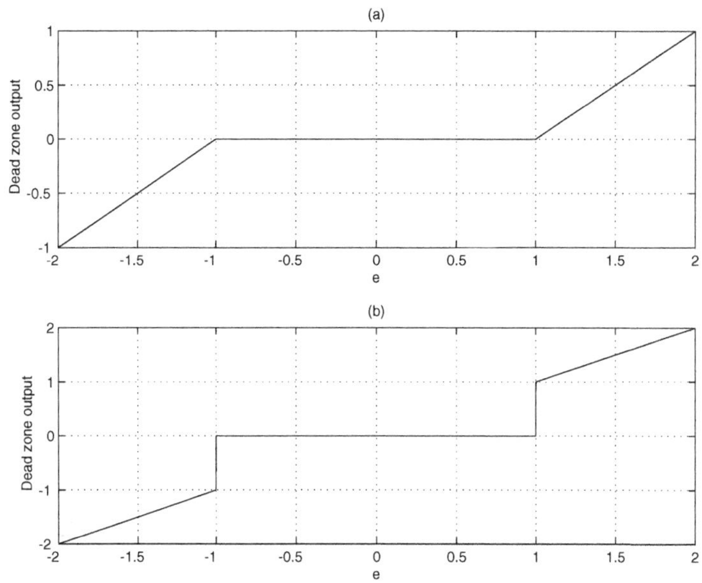

Figure 12.3: Dead zone nonlinearities used for adaptation laws ((a) used for normalized gradient, (b) used for recursive least squares).

Recursive Least Squares

The recursive least squares adaptation law (with forgetting factor $\lambda = 1$) is

$$\theta(k) = \theta(k-d) - \frac{P(k-2d)\phi(k-d)\bar{e}_\epsilon(k)}{1 + a(k)\phi^\top(k-d)P(k-2d)\phi(k-d)} \quad (12.19)$$

$$P(k-d) = P(k-2d) - \frac{a(k)P(k-2d)\phi(k-d)\phi^\top(k-d)P(k-2d)}{1 + a(k)\phi^\top(k-d)P(k-2d)\phi(k-d)} \quad (12.20)$$

Here, $\theta(k)$ and $\phi(k)$ are $p \times 1$ vectors and $P(k)$ is a $p \times p$ matrix. The indices $k-2d$ and $k-d$ result from the delay d. Notice that to compute $\theta(0)$ we need $\theta(-d)$ and in fact, to compute $\theta(k)$ for $k = 0, 1, 2, \ldots, d-1$, we need $\theta(-d), \theta(1-d), \ldots, \theta(-1)$. Hence, to initialize the algorithm you need to specify d initial parameter values. One way to do this is to just use all the same values, either a good guess at their values or zero vectors for each case. Next, note that to compute $P(k-d)$ when $k = 0$, we need $P(-2d)$ and in fact, to compute $P(k-d)$ for $k = 0, 1, 2, \ldots, d-1$, we need $P(-2d), P(1-2d), \ldots, P(d-1-2d)$. Hence, to initialize the algorithm, you need d initial matrices. One way to pick these is to simply use a $p \times p$ diagonal matrix αI where $\alpha > 0$ is some suitably large value, for all the initial values. Finally, note that due to the delay, there is a need to store several past values to implement the update law.

For the above adaptive law,

$$\bar{e}_\epsilon(k) = \begin{cases} e(k) & \text{if } e(k) > \epsilon_d(k) \\ 0 & \text{if } |e(k)| \le \epsilon_d(k) \\ e(k) & \text{if } e(k) < -\epsilon_d(k) \end{cases} \quad (12.21)$$

where

$$\epsilon_d(k) = \sqrt{2}\epsilon(k)\sqrt{1 + \phi^\top(k-d)P(k-2d)\phi(k-d)} \quad (12.22)$$

and $\epsilon(k)$ is the same as for the indirect adaptive control case. This dead zone is shown in Figure 12.3(b) for the case where $\epsilon_d(k) = 1$ for all k (note that in general, the dead zone size is time-varying). Note that $P(k)$ is positive definite for all k so that $\phi^\top(k-d)P(k-2d)\phi(k-d) \ge 0$, which will be shown below. This ensures that $\epsilon_d(k)$ is properly defined (i.e., so that the term in the square root will not be negative). Next, we define $a(k)$ as

$$a(k) = \begin{cases} 0 & \text{if } |e(k)| \le \epsilon_d(k) \\ 1 & \text{otherwise} \end{cases} \quad (12.23)$$

and we think of it as a variable that indicates whether or not the tracking error is in the dead zone.

When $|e(k)| \le \epsilon_d(k)$, the tracking error is in the dead zone where the error dynamics are possibly driven not by the parameter error, but by the ideal approximation error. Therefore, if we update the parameters, there is the possibility that they will not be updated in the proper direction, so $\bar{e}_\epsilon(k) = 0$, and hence, $\theta(k) = \theta(k-d)$ and we do not update the parameter vector. The turning on and off of the updates for $P(k)$ via $a(k)$ is more complex. First, note that to compute $\theta(k)$, we only need $P(k-2d)$. Now, if at time k, $a(k) = 0$ so the tracking error is in the dead zone, we could get an update for $P(k-2d)$. However, if $a(k-d) = 0$ (i.e., d steps ago the tracking error was in the dead zone), then the $P(k-2d)$ we generated was such that $P(k-2d) = P(k-3d)$ and we see that the update to the covariance matrix is turned off. Notice that $a(k)$ also affects the parameter updates.

Finally, we note that while the transient responses from the gradient and recursive least squares methods may be different, the asymptotic properties for the parameters and tracking error are basically the same as those we discussed for the gradient case in the last subsection (in particular, $\bar{e}_\epsilon(k) \to 0$ as $k \to \infty$).

Application to Plants with Known Linear in the Parameter Nonlinearities

Suppose that the underlying system is linear with constant but unknown coefficients and that n and m are known and $d = 1$. Suppose that we use linear in the parameter approximators with the same structure as those in the plant (i.e., with regression vectors with the same elements as the underlying unknown linear functions that we are trying to estimate). Notice that in this case the ideal approximation errors are zero, so $W_\alpha = W_\beta = 0$, the size of the dead zone is zero, and the normalized gradient or RLS method will force the tracking error (not the dead zone modification of it, with a finite size dead zone) to zero, so that asymptotically we get perfect tracking.

It is interesting, and sometimes practically useful, to note that if there are nonlinearities in the plant that are known and parameterized linearly in unknown coefficients, we can get the same type of result. If there are various types of uncertainty in the plant (e.g., an unknown bias term), the case reverts to the situation where we get convergence of the tracking error to a neighborhood of zero, where the size of the neighborhood generally depends on the "size" of the uncertainty in the plant (similar to the case presented earlier).

12.3.4 Discussion: Multiple Model Adaptation Strategies

Another approach to indirect adaptive control is to use multiple identifier models, each with its own identification strategy. Then, you can specify an online measure of identification error accuracy (e.g., a sum of squares of past identification errors) and use this to pick which model has been adjusted the best, and hence, should be chosen for use in a certainty equivalence strategy. Sometimes you may want to adapt several models, other times you may want to use a set of guesses at the plant model that are simulated online, and one adaptive mechanism that attempts to estimate the plant model. This way, if one of the fixed models is close to the actual model, it will be chosen for use in the certainty equivalence strategy. However, if none of the fixed models is accurate, then the adaptive one can be relied on to estimate the plant characteristics. There have been a variety of studies of such "multiple model adaptive control" approaches, some of which have focused on derivation of stability properties. See the "For Further Study" section for more details.

We will not discuss the use of gradient and least squares methods for multiple model adaptive control where the underlying approximators are neural networks or fuzzy systems. Instead we will study in Section 16.5 a variety of approaches, including direct adaptive methods (to achieve adaptive model predictive control), which rely on multiple models (or controllers) and employ evolutionary algorithms for approximator adjustment (to estimate models, controllers, or both). Moreover, in Design Problem 18.4, we ask you to use an optimization/search algorithm that is based on biomimicry of foraging to implement the same types of adaptive control schemes. Both the evolutionary and foraging methods offer approaches to tune not only parameters, but structure of approximators. Also, they show how to use the entire set of model estimates to help update the estimates for each model (i.e., they show how to share good information about estimation among the whole set of estimators). If you study the evolutionary and foraging approaches to multiple model adaptive control, it will be clear how to use least squares or gradient approaches.

12.4 Design Example: Indirect Neural Control for a Process Control Problem

In this section we apply the indirect neural controller to the simple process control problem studied in Section 6.4.1 to illustrate some characteristics of the behavior of the closed-loop system. For the model there, we choose $A(h(t)) = |\bar{a}h(t) + \bar{b}|$ with $\bar{a} = 0.01$ and $\bar{b} = 0.2$. Also, we use $\bar{c} = 1$ and $\bar{d} = 1$. Assume that you know the reference trajectory a priori and assume that $r(t) \in [0.1, 8]$ and that we will not have $h(t) > 10$. Assume that $h(0) = 1$.

12.4.1 The Neural Controller Development

Putting the Plant into the Proper Form

To approximate the tank dynamics, we will ignore the saturation at the actuator input and the fact that the liquid level never goes negative, and view the dynamics as

$$h(k+1) = \left(h(k) + T\frac{-\bar{d}\sqrt{19.6h(k)}}{|\bar{a}h(k) + \bar{b}|} \right) + \left(\frac{\bar{c}T}{|\bar{a}h(k) + \bar{b}|} \right) u(k)$$

and using our notation from the previous section, we let

$$\alpha(h(k)) = \left(h(k) + T\frac{-\bar{d}\sqrt{19.6h(k)}}{|\bar{a}h(k) + \bar{b}|} \right)$$

and

$$\beta(h(k)) = \left(\frac{\bar{c}T}{|\bar{a}h(k)+\bar{b}|}\right)$$

so we have

$$h(k+1) = \alpha(h(k)) + \beta(h(k))u(k)$$

Hence, we have $n = m = 0$ and $d = 1$. Also, we assume that $\alpha_k = \beta_k = 0$ so that we are not assuming that we have any a priori knowledge of the plant dynamics.

We assume that we know that $\bar{a}$, $\bar{b}$, and $\bar{c}$ only vary in ways so that

$$\beta_0 = \frac{1}{4} \leq \beta(h(k)) \leq \frac{1}{2}$$

For instance, using nominal values for $\bar{a}$, $\bar{b}$, and $\bar{c}$, and the above definitions, together with the constraint that $h(k) \in [0.001, 10]$, we know

$$\beta_0 < \frac{1}{3} \leq \beta(h(k)) \leq \frac{1}{2}$$

(recall that $T = 0.1$).

Specifying Estimators for the Plant Nonlinearities

Suppose that we use two linear in the parameter approximators

$$\hat{\alpha}(h(k)) = \theta_\alpha^\top \phi_\alpha(h(k))$$

and

$$\hat{\beta}(h(k)) = \theta_\beta^\top \phi_\beta(h(k))$$

where θ_α and θ_β, the parameter vectors, and ϕ_α and ϕ_β, are defined so that these represent a radial basis function neural network that we will define below. In particular, to estimate $h(k)$, we use $\hat{h}(k)$ where

$$\hat{h}(k+1) = \hat{\alpha}(h(k)) + \hat{\beta}(h(k))u(k)$$

or

$$\hat{h}(k+1) = \theta^\top \phi(h(k), u(k))$$

where

$$\theta = \left[\theta_\alpha^\top, \theta_\beta^\top\right]^\top$$

and

$$\phi = \left[\phi_\alpha^\top, u\phi_\beta^\top\right]^\top$$

However, to save computations, we let $\phi_\alpha = \phi_\beta$ and we define this as ϕ_h. Note that this way, we use the same nonlinear part for the approximators for $\alpha(h(k))$ and $\beta(h(k))$, but different parameters that enter linearly so that different nonlinearities $\alpha(h(k))$ and $\beta(h(k))$ can be represented. Clearly, however, in some applications you will want to use different ϕ_α and ϕ_β nonlinearities, especially if you know that one nonlinear function has many more oscillating nonlinearities than the other.

Specifying the Approximator Structures

For the radial basis function neural network, we will use

$$F_{rbf}(x, \theta_{rbf}) = \sum_{i=1}^{n_R} b_i R_i(x)$$

where θ_{rbf} will be either θ_α or θ_β and

$$R_i(x) = \exp\left(-\frac{|x - c^i|^2}{(\sigma^i)^2}\right)$$

and

$$c^i = [c_1^i, c_2^i, \ldots, c_n^i]^\top$$

and σ^i is a scalar (and if z is a vector then $|z| = \sqrt{z^\top z}$). The c^i and σ^i are parameters that enter in a nonlinear fashion. Note that we can use initialization methods (i.e., gridding on the input space) for the centers and spreads that are similar to those that we discussed earlier for the Takagi-Sugeno fuzzy system.

If we let

$$\theta_{rbf} = [b_1, b_2, \ldots, b_{n_R}]^\top$$

and

$$\phi_{rbf} = [R_1, R_2, \ldots, R_{n_R}]^\top$$

then

$$F_{rbf}(x, \theta_{rbf}) = \theta_{rbf}^\top \phi_{rbf}$$

is the form of the linear in the parameter approximator that we will use. While for the Takagi-Sugeno fuzzy system we thought of the basis functions as turning different lines on and off in different regions, here the R_i functions turn different constants b_i on and off in different regions (i.e., they interpolate between different constants), where the position and size of region i is fixed by c^i and σ^i.

Specifying the Control and Projection

Next, note that we will use the estimates $\hat{\alpha}(h(k))$ and $\hat{\beta}(h(k))$ in the certainty equivalence control law

$$u(k) = \frac{-\hat{\alpha}(h(k)) + r(k+1)}{\hat{\beta}(h(k))}$$

and either a gradient or recursive least squares update law. Notice that with this choice for the control law, as indicated earlier, we have to ensure that

$$\hat{\beta}(h(k)) = \theta_\beta^\top \phi_h(h(k)) \geq \beta_0 > 0$$

for all $h(k)$ to keep the update algorithm from picking θ_β so that $\hat{\beta}(h(k)) = 0$, which would make the control $u(k)$ above undefined. To do this, we simply use a projection method to keep each element of θ_β greater than or equal to $\beta_0 = 0.25$.

12.4.2 Indirect Neural Control Results

In this section we discuss how to choose the design parameters for the controller, provide the results of the neural control scheme, and illustrate several key characteristics of the closed-loop system behavior.

Approximator and Adaptation Law Design Parameter Choices

First, we chose $n_R = 50$ so that we are using 50 receptive field units for ϕ_h. To initialize the parameters of θ_β we let all the elements take on the same value, the one in the middle of the range $[\beta_0, \beta_1]$ (or 0.375). For simplicity we chose the values of the elements of θ_α to all be zero; hence, the initialization for $\hat{\beta}$ is better than the one for $\hat{\alpha}$. We chose all the spreads $\sigma^i = 0.2$ for the receptive field units. We let $c_1^1 = 0$, $c_1^i = c_1^{i-1} + 0.2$ for $i = 2, 3, \ldots, n_R$ (for a uniform spacing).

We chose the gradient update law, and $\eta = 1.25$ and $\gamma = 1$ after a bit of tuning. Basically, keeping γ constant, if you lower the adaptation gain η, the adaptation slows so that tracking suffers. We chose $W_\alpha = W_\beta = 0.01$ simply by trying out several values (i.e., we view them as tuning parameters). We must emphasize that we are not guaranteed with this choice to have the respective approximation errors less than these values and hence, we are not guaranteed stability (but of course, we are also not guaranteed stability due to the saturation constraints on the input and the liquid level).

Convergence of Tracking Error to a Dead Zone

In this first set of results, we show how the control system responds to a square wave input for the desired liquid level height and focus, in particular, on how it tries to force the tracking error to converge to a dead zone. Figure 12.4 shows that for this case, the tracking performance improves over time, but that both at the lower and upper value, there is a tracking error even after the value of the liquid height settles out. This error results from the use of the dead zone and the finite size approximator. To remove this error you need to use a better approximator structure. Note that in this case, the radial basis function does not seem to be a particularly good choice as it is easy to see that significant errors can arise unless very many receptive field units are used.

Estimator Performance: Nonlinear Mapping Shape Mismatches

Next, we compare the estimates and actual values of h, α, and β. Figure 12.5 shows that, for this case, the estimate of h improves over time, but that the steady state error results from improper estimates of α and especially inaccurate ones of β.

Next, we note that Figure 12.6 and Figure 12.7 show that for this case, the change in the parameter error is decreasing as we would like, and that the error is slowly approaching the dead zone (which is depicted there with a black bar—study the beginning of the plot where the black region is small, and then later

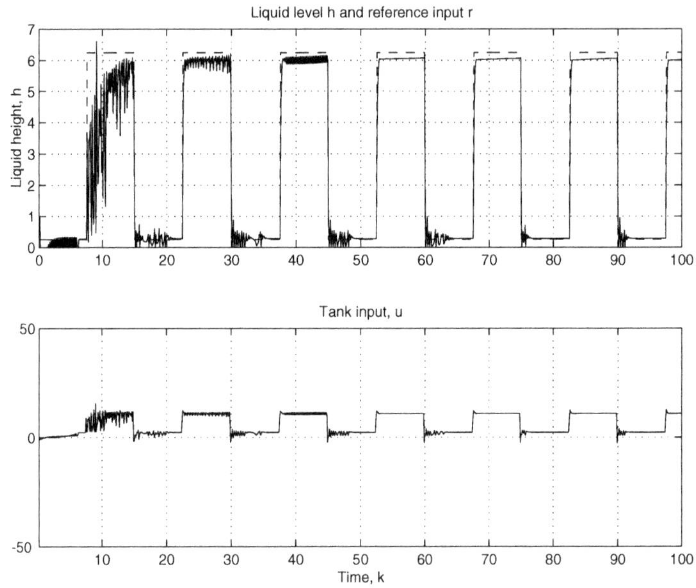

Figure 12.4: Indirect neural controller, reference input, plant output, and plant input.

the error decreases toward this). Again, however, we must emphasize that this is not guaranteed for this case. Figure 12.6 nicely illustrates that the parameter vector seems to be converging. We must emphasize however, that for a different reference input, it may behave differently.

Accurate plant model estimation is not necessarily needed to achieve good tracking performance in indirect adaptive control.

It is important, however, to note that the mapping shapes at the last iteration (1000 iterations total) are not very close to the actual shapes of the unknown α and β as shown in Figure 12.8. Hence, it is clear that the quality of the estimate produced by the output of the estimator is not necessarily a good indication of the quality of the matching between the estimator nonlinear map that is being trained and the unknown nonlinear map (even though in our discussion for the development of the controller, you may have thought of it that way).

Persistency of Excitation to Get Better Matching of Nonlinearities

Accurate modeling can sometimes be achieved if the reference input is "persistently exciting."

So, how do we get the estimator nonlinear mapping to more closely match the plant nonlinearity that it is trying to match? The reference input signal has to be "persistently exciting" in a proper way to ensure that the whole map gets updated properly. Here, rather than a square wave, we put in a noise sequence that is uniformly distributed on the same range of values used for the square wave. In this case we get the results in Figure 12.9, where, comparing to Figure 12.8, we see that we get quite good matching of the estimator and unknown

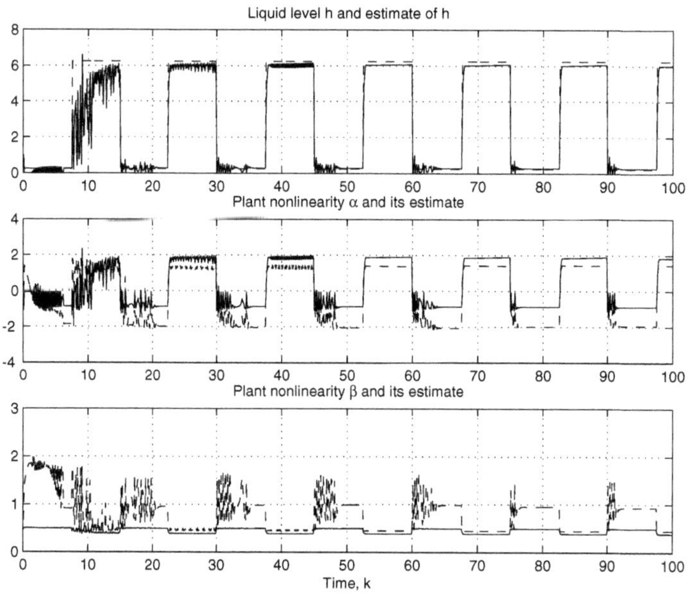

Figure 12.5: Indirect neural controller, estimates and actual values of h, α, and β.

plant nonlinearities in the region where the plant output visited frequently (i.e., roughly in the range $[0.001, 6]$). You should be reminded of the discussions on input space coverage and persistency of excitation, and their impact on approximator accuracy from Chapter 11.

12.5 Direct Adaptive Neural/Fuzzy Control

In this section we will discuss direct adaptive neural/fuzzy control where we do not explicitly estimate the plant dynamics. Instead, we simply design an optimization algorithm to search for a controller that will stabilize the plant and try to force a low tracking error.

12.5.1 Development of Controller Estimator and Error Equations

The Class of Plants

Here, we can consider more general classes of discrete-time nonlinear systems than in the indirect adaptive control case. As before, we assume that $r(k+d)$ is known. Rather than providing an explicit form of nonlinear system that we can treat, we will instead require that there exists some $u^*(x(k), r(k+d))$

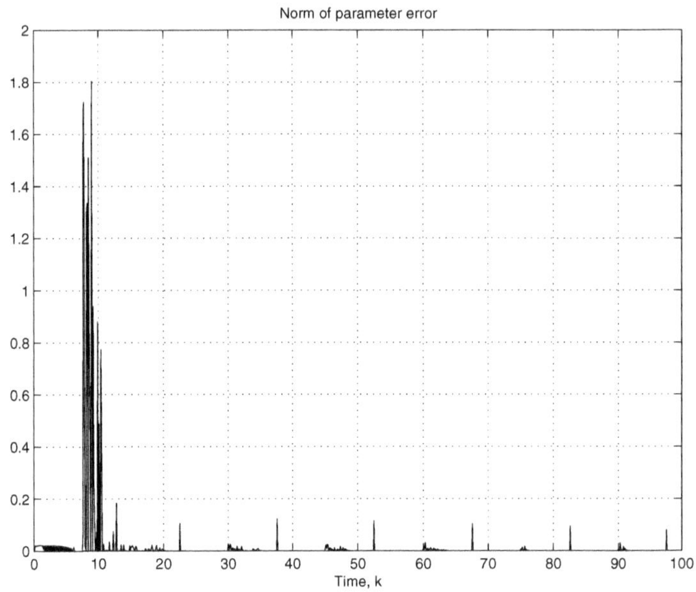

Figure 12.6: Indirect neural controller, parameter error.

(that is continuous in its arguments) such that the error dynamics $e(k+d) = r(k+d) - y(k+d)$ may be expressed as

$$e(k+d) = -\psi(x(k))(u(k) - u^*(k)) + \nu(k) \tag{12.24}$$

where $\psi(x(k))$ is such that

$$0 < \psi_0 \leq \psi(x(k)) \leq \psi_1$$

where ψ_0 and ψ_1 are known constants related to the plant dynamics, and

$$\sup_k |\nu(k)| \leq \mathcal{V}$$

for some known $\mathcal{V}$. Hence, we provide the error dynamics and any plant that can be transformed to have its error dynamics in this form will fall into the class of plants that the direct adaptive controller will apply to.

Next, we show that control laws u^* exist, which transform the error dynamics into the proper form. We must emphasize, however, that we will show general ways to transform the dynamics, and in particular, show that the class of plants that can be considered in the indirect adaptive case can also be considered for the direct adaptive case. You could use similar approaches to transform the nonlinear dynamics into the form above for particular applications.

Suppose that $x(k) = [y(k), y(k-1), \ldots, y(k-n), u(k-1), u(k-2), \ldots, u(k-m)]^\top$ and consider a nonlinear discrete-time system of the form

$$y(k+d) = \alpha(x(k)) + \beta(x(k))u(k) \tag{12.25}$$

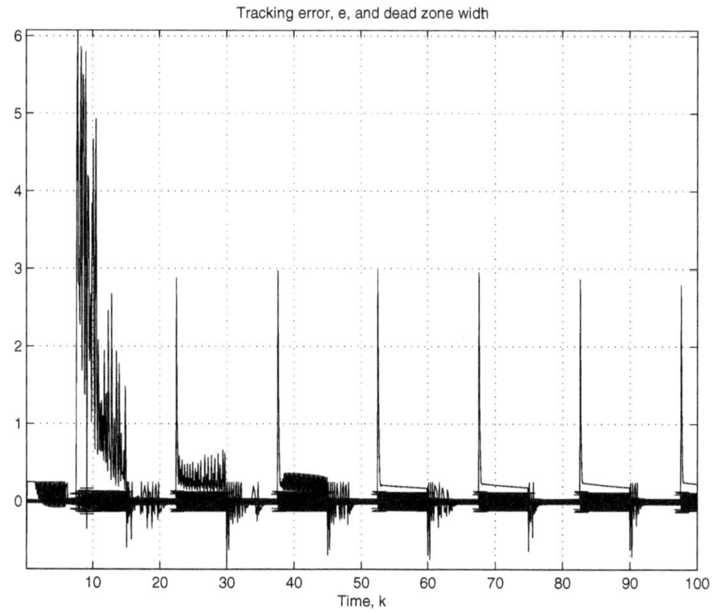

Figure 12.7: Indirect neural controller, tracking error and dead zone (shaded black).

which is "linear in the control input" in the sense that $u(k)$ is multiplied by what we think of as a gain $\beta(x(k))$, and that gain is not a function of the input $u(k)$. We assume $\beta(x(k))$ to satisfy

$$0 < \beta_0 \leq \beta(x(k)) \leq \beta_1 < \infty \tag{12.26}$$

We use $\beta_0 \leq \beta(x(k))$ to ensure existence of a control law below, and the parameter β_1 in the specification of constraints on the adaptation gain for a gradient update law.

Note that for this class of plants, the control law

In direct adaptive control, we view the ideal controller as one that the adaptation mechanism is searching for.

$$u^*(x(k), r(k+d)) = \frac{-\alpha(x(k)) + r(k+d)}{\beta(x(k))}$$

exists if $\beta(x(k))$ is bounded away from zero. With this choice of u^*, we find that

$$\begin{aligned} e(k+d) &= r(k+d) - y(k+d) \\ &= r(k+d) - \alpha(x(k)) - \beta(x(k))u(k) \\ &= \beta(x(k))u^*(k) - \beta(x(k))u(k) \\ &= -\beta(x(k))(u(k) - u^*(k)) \end{aligned}$$

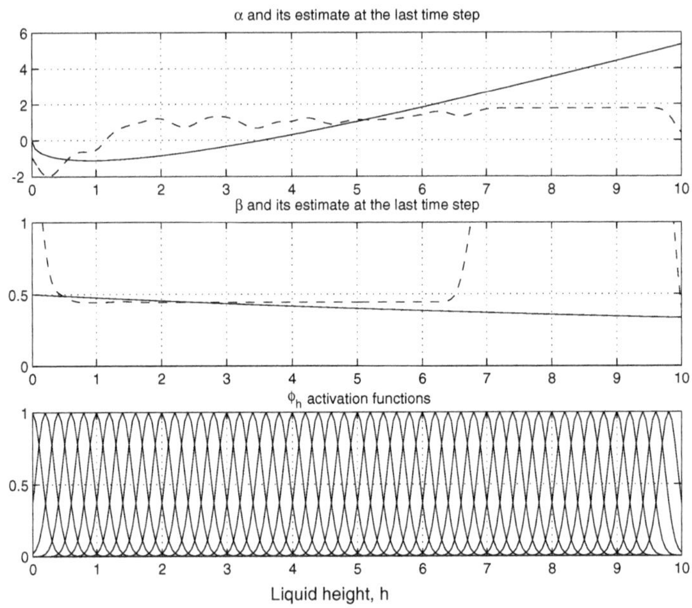

Figure 12.8: Indirect neural controller, final estimator mapping shapes compared to the actual nonlinearities they are trying to estimate (solid lines for top two plots are actual nonlinearities, dashed lines are the estimates).

From this last equation, considering Equation (12.24), $\psi(x(k)) = \beta(x(k))$, $\psi_0 = \beta_0$, $\psi_1 = \beta_1$, and $\nu(k) = 0$; hence, this class of plants can be considered in the direct adaptive control case. In this case we seek to tune the neural network or fuzzy system so that it finds a feedback linearizing controller where an important assumption in the existence of the feedback linearizing controller is that the control enters linearly. However, we do not necessarily require this; we only require that a plant is used so that the error equation (Equation (12.24)) results and more general classes of plants can satisfy this (see the "For Further Study" section at the end of this part for more details).

The linear relationship between controller estimate error and tracking error allows us to adjust the controller to minimize tracking error and thereby also try to minimize the error between the controller estimate and the unknown ideal controller.

We must emphasize that just like the indirect case, we will not be able to ensure that we will get parameter convergence and hence, while we may think of the estimator to be developed below as searching for a feedback linearizing controller, it may never find it, and it may still stabilize the plant. It may find a different controller that performs well, or its continual search may proceed in such a way that it will stabilize the plant. Finally, it is interesting to note that $e(k + d)$ is a measure of the (instantaneous) tracking error and how close the current control is to some current ideal one (notice the analogy with the indirect case where the tracking error is also a measure of the instantaneous identification error).

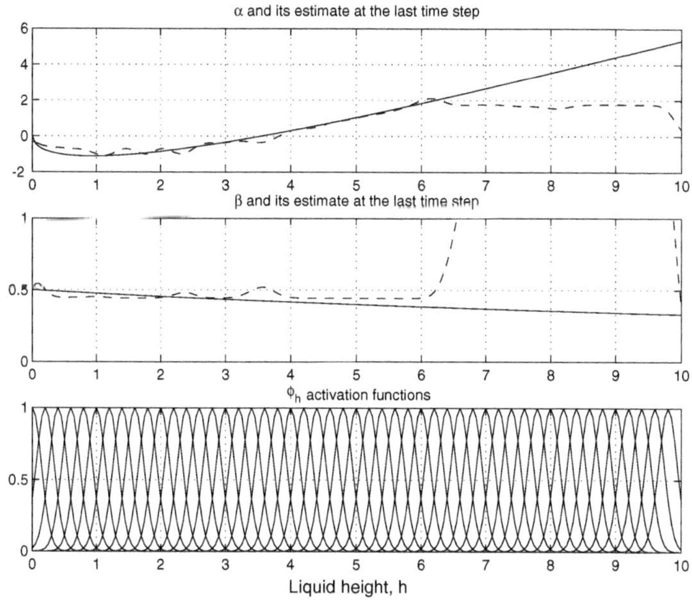

Figure 12.9: Indirect neural controller, noise for reference input (solid lines for top two plots are actual nonlinearities, dashed lines are the estimates).

Error Dynamics and Controller Estimator

To develop the error dynamics for the direct adaptive control case, first express u^* as

$$u^*(x(k), r(k+d)) = u_u(x(k), r(k+d)) + u_k(k) \tag{12.27}$$

where $u_u(x(k), r(k+d))$ and $u_k(k)$ represent the unknown and known portions of the ideal control law, respectively (notice the analogy with the indirect adaptive case where we had unknown and known parts of the dynamics that we were trying to estimate). In an analogous manner to the indirect adaptive case, the signal $u_k(k)$ is not required for implementation and we may assign $u_k(k) = 0$, for all $k \geq 0$. Sometimes, however, it can be convenient to include u_k. If you know a controller that works reasonably well for nominal operation of the plant, then it can be included as u_k and the neural network or fuzzy controller that is tuned will then tend to correct for its deficiencies and subsequent performance improvements can be obtained in some cases.

Using the online approximation-based approach, we may represent $u^*(k)$ as

$$u^*(k) = \theta_u^{*\top} \phi_u(x(k), r(k+d)) + u_k(k) + w_u(k) \tag{12.28}$$

where

$$\theta_u^* = \arg\min \theta_u \in \Omega_u \left(\sup_{x \in S_x, r \in S_r} |\theta_u^\top \phi_u(x(k), r(k+d)) - u_u(x,r)| \right) \tag{12.29}$$

$\Omega_u \subset \Re^{p_u}$ is the convex compact set of allowable controller parameters, $S_x \subset \Re^{n+m+1}$ is the known compact set through which the state may travel, and S_r is the space through which the reference input may travel (we assume that $r(k)$ is bounded as we did in the indirect adaptive case so naturally, S_r is bounded). The current estimate of the ideal controller is given by

$$u(k) = \theta_u^\top(k)\phi_u(x(k), r(k+d)) + u_k(k), \tag{12.30}$$

where the current parameter vector $\theta_u(k)$ will be updated online and a projection algorithm can be used to make sure that $\theta_u(k) \in \Omega_u$, if that is needed. It is important to highlight the fact, however, that sometimes we may have no constraints on the controller parameters so then we will not need a projection method for the direct adaptive control case. Note that for plants of the form given in Equation (12.25), while we needed to assume that $\beta(x(k))$ was bounded away from zero to ensure the existence of u^*, we are not trying to estimate $\beta(x(k))$, so there is no need for projection to ensure that an estimate of it does not take on a value of zero.

Next, defining the parameter error as

$$\tilde{\theta}_u(k) = \theta_u(k) - \theta_u^*$$

the output error dynamics $e(k+d) = r(k+d) - y(k+d)$ may be expressed as

$$e(k+d) = -\psi(x(k))\tilde{\theta}_u^\top(k)\phi_u(x(k), r(k+d)) + \psi(x(k))w_u(k) + \nu(k) \tag{12.31}$$

where $0 \leq |w_u(k)| < W_u$. We will assume that we can define W_u, such that if $|\tilde{\theta}_u|^2$ is bounded, then $0 \leq |w_u(k)| < W_u$ for all k. It can be shown that $|\tilde{\theta}_u|^2$ is bounded so as long as you pick W_u to overbound the ideal representation error.

12.5.2 Adaptation Method

Here, we will only discuss the use of the gradient method, which can be derived in a similar manner to the indirect adaptive control case by tuning the parameter vector to minimize the (instantaneous) tracking error and hence, the instantaneous error between the current control and some ideal one. Doing this for the direct adaptive control case, with attention given to specifying the adaptive laws to achieve a stable closed-loop system, results in the adaptation law in Equation (12.17) with $\kappa_1 = 1$ (the minus sign difference comes from the minus sign in Equation (12.31)), $\kappa_2 = \psi_1$, and

$$\epsilon(k) = \psi_1 W_u + \mathcal{V}$$

for all $k \geq 0$, where for the class of plants we discussed above, we had $\psi_1 = \beta_1$ and $\mathcal{V} = 0$. Hence, for the direct adaptive control case, we get a fixed size dead zone, not one that is time-varying as in the indirect adaptive control case. Note that in the direct adaptive control case, we must have a bound on the nonlinear gain on the plant input and this affects the possible choices for the adaptation gain.

Considering that we do not necessarily need a projection method to implement the adaptation law, and the possibility that a more general class of plants can be used, sometimes makes the direct adaptive approach more desirable than the indirect approach. However, in some applications, the indirect approach may show some advantages such as the fact that, unlike the direct case, the approximators do not need $r(k+d)$ as an input. So, in general, they may need fewer inputs and this can affect the complexity of the approximators (on the other hand, we need two approximators for the indirect case and only one for the direct case).

12.6 Design Example: Direct Neural Control for a Process Control Problem

Here, we use the same plant as for the indirect adaptive control example, with all the same parameters. For the approximator for the controller, we use a radial basis function neural network with $n = 2$ inputs, $h(k)$ and $r(k+1)$, and $n_R = 100$, so we will adjust 100 strengths that are loaded into the vector θ_u and we initialize them all to zero. We let $u_k = 0$. We use the same spread $\sigma^i = 1$ for all the $R_i(h(k), r(k+1))$ and create a uniform grid for the c^i centers, $i = 1, 2, \ldots, n_R$. In particular, recall that $r(k) \in [0.1, 8]$ and $h(k) \in [0.001, 10]$. For convenience, we simply create a uniform grid with its four outer corners at $(0,0)$, $(9,0)$, $(0,9)$, and $(9,9)$ (hence it has $n_R = 100$ centers). For the adaptation, we use $\eta = 1.25$ and $\gamma = 1$. We chose $W_u = 0.01$.

12.6.1 Direct Neural Controller Results

The results for the direct adaptive controller are shown in Figure 12.10, where we see that due to the poor initialization of the controller, we get poor performance early in the simulation; however, the adaptive controller quickly recovers from this and provides better tracking as time progresses. The bottom plot shows the control input generated by the online approximator and the "ideal" feedback linearizing control input. It is interesting to note that in this case, the gradient method seems to have adjusted the controller to provide the ideal control. (Note, however, that this does not imply that the entire mapping produced by the approximator will match the mapping produced by the ideal controller; see the discussion below.)

To gain more insight into the behavior of the closed-loop system, and the type of convergence properties we expect, see Figures 12.11 and 12.12, where we show how the change in the parameter error and tracking error behave. Note that while this shows a nice parameter convergence and the tracking error is decreasing, we must emphasize that this is not guaranteed in general, due to the added constraints on the plant and due to the choice of W_u.

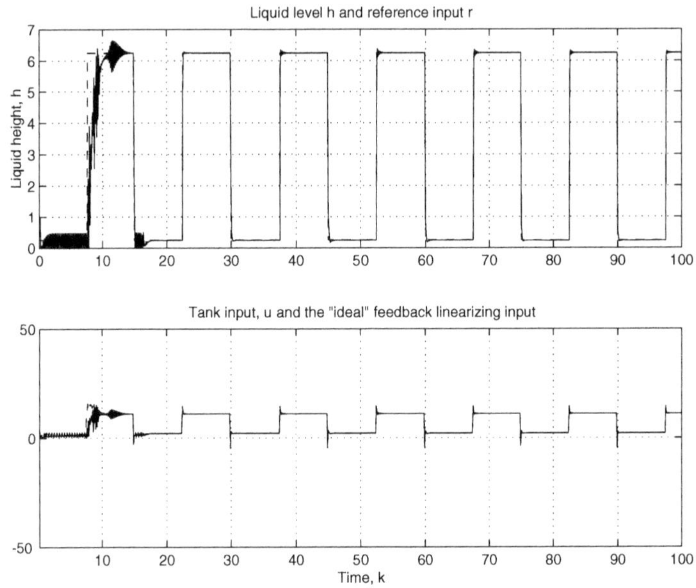

Figure 12.10: Direct neural control, closed-loop behavior.

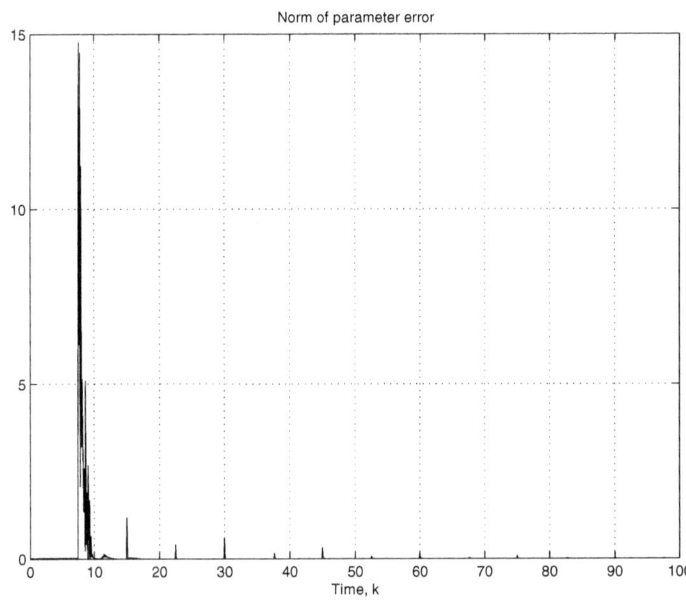

Figure 12.11: Direct neural controller, parameter error.

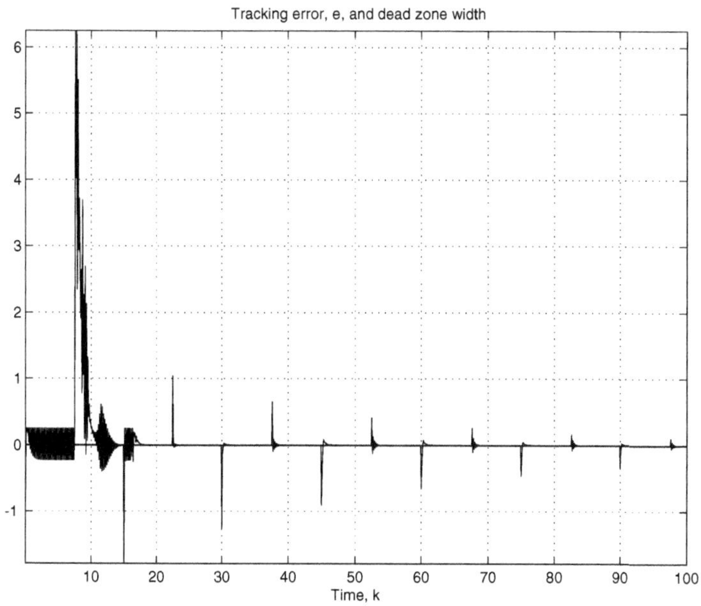

Figure 12.12: Direct neural controller, tracking error and dead zone (shaded black).

12.6.2 Tuned and Ideal Controller Mapping Shapes

To highlight some of the issues with parameter convergence, we provide plots of the neural controller mapping at the last step in the above simulation in Figure 12.13, and the corresponding "ideal" controller mapping in Figure 12.14. Notice that while the approximator has grossly approximated part of the mapping, it clearly has not learned the shape of the mapping in every region. Note that this is consistent with the result shown in Figure 12.11, since that plot simply shows that the mapping has largely stopped changing its shape by the end of the simulation. One reason it has not converged to the exact nonlinear control mapping, is that the input has not driven the system into every region frequently enough so that it has not been able to learn properly in those regions. It is possible that, if you use a reference input that more consistently drives the $h(k)$ and $r(k)$ into different regions, that the mapping shape will become closer to the ideal one. (Just as in the indirect adaptive control case, this is an issue of persistency of excitation.) For instance, if you use noise as the reference input as we have in the indirect adaptive control case, the mapping is adjusted to become a bit closer to the ideal one.

An accurate estimate of the ideal controller is not necessarily required to obtain good tracking error.

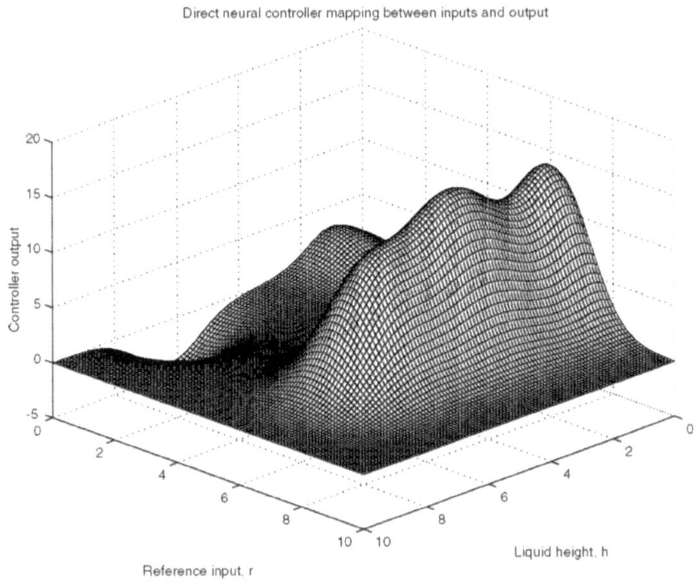

Figure 12.13: Direct neural controller mapping shape at the last iteration.

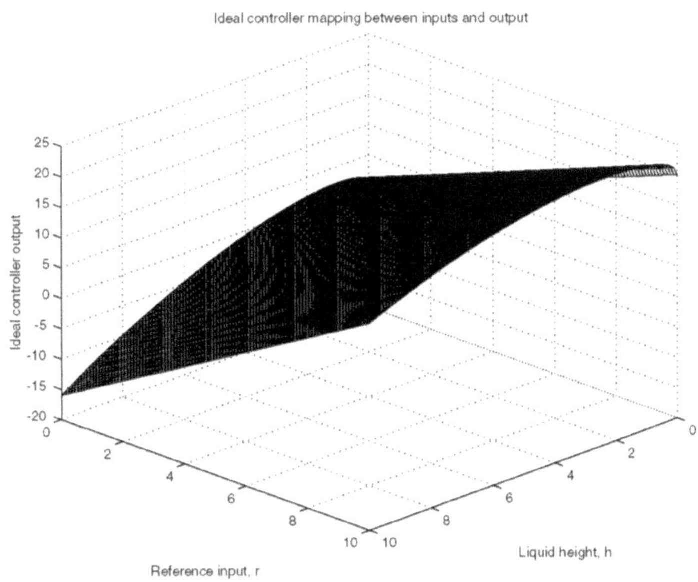

Figure 12.14: Ideal controller mapping shape.

12.7 Stable Adaptive Fuzzy/Neural Control

In this section we will develop stable adaptive fuzzy/neural controllers for a class of continuous time nonlinear systems. We will introduce both indirect and direct adaptive control approaches and show that under certain conditions, each provides asymptotic tracking of a reference signal and boundedness of all signals. Several of the concepts in this chapter build on those of the previous section; however, to develop conditions for stability, we will only consider gradient update laws for the indirect and direct methods.

For stable adaptive control, the focus is on methods that can be proven to possess closed-loop properties.

12.7.1 Class of Nonlinear Systems

In this section we will provide a description of the class of systems that we consider in this chapter, and provide an example physical plant that falls in this class.

Feedback Linearizable Continuous Time Nonlinear Systems

Consider the plant

$$\dot{x} = f(x) + g(x)u \tag{12.32}$$
$$y = h(x) \tag{12.33}$$

where $x = [x_1, \ldots, x_n]^\top$ is the state vector, u is the (scalar) input, y is the (scalar) output of the plant, and functions $f(x)$, $g(x)$, and $h(x)$ are smooth.

Let $L_g^d h(x)$ be the dth Lie derivative of $h(x)$ with respect to g. In particular, define

$$L_g h(x) = \left(\frac{\partial h}{\partial x}\right)^\top g(x)$$

and, for example,

$$L_g^2 h(x) = L_g(L_g h(x))$$

A system is said to have "strong relative degree" d if

$$L_g h(x) = L_g L_f h(x) =, \cdots, = L_g L_f^{d-2} h(x) = 0$$

and $L_g L_f^{d-1} h(x)$ is bounded away from zero for all x. If the system has strong relative degree d, then

$$\begin{aligned}
\dot{z}_1 &= z_2 = L_f h(x) \\
&\vdots \\
\dot{z}_{d-1} &= z_d = L_f^{d-1} h(x) \\
\dot{z}_d &= L_f^d h(x) + L_g L_f^{d-1} h(x) u
\end{aligned} \tag{12.34}$$

with $z_1 = y$, which, if we let $y^{(d)}$ denote the dth derivative of y, may be rewritten as

$$y^{(d)} = (\alpha_k(t) + \alpha(x)) + (\beta_k(t) + \beta(x))u \tag{12.35}$$

Here, we assume that $y = h(x) = x_1$. We will assume that $d = n$ here, since it simplifies the stability analysis. (For a treatment of the more general case, see the "For Further Study" section at the end of the part.) We will assume that if the z_i, $i = 1, 2, \ldots, d = n$ (i.e., $y, \dot{y}, \ldots, y^{(d)}$), are bounded, then so are the x_i, $i = 1, 2, \ldots, d = n$.

Furthermore, it is assumed that for some $\beta_0 > 0$, we have

$$|\beta_k(t) + \beta(x)| \geq \beta_0$$

so that it is bounded away from zero (for convenience, we assume that $\beta_k(t) + \beta(x) > 0$, however, the following analysis may easily be modified for systems which are defined with $\beta_k(t) + \beta(x) < 0$). We will assume that $\alpha_k(t)$ and $\beta_k(t)$ are known components of the dynamics of the plant (that may depend on the state) or known exogenous time-dependent signals and that $\alpha(x)$ and $\beta(x)$ represent nonlinear dynamics of the plant that are unknown. It is assumed that if x is a bounded state vector, then $\alpha(x)$, $\beta(x)$, $\alpha_k(t)$, and $\beta_k(t)$ are bounded signals. Throughout the analysis to follow, both $\alpha_k(t)$ and $\beta_k(t)$ may be set to zero for all $t \geq 0$.

Example: Ball on a Beam Experiment

We have found via experimentation in our laboratory (using physical modeling and some system identification techniques) that the ball on a beam experiment shown in Figure 12.15 can be accurately represented with

$$\begin{aligned} \dot{x}_1(t) &= x_2(t) \\ \dot{x}_2(t) &= \bar{a}\tan^{-1}(\bar{b}x_2(t))\left(\exp(-\bar{c}x_2^2(t)) - 1\right) - \bar{d}u(t) \end{aligned} \tag{12.36}$$

where $x_1(t)$ is the distance from the center of the ball to one end of the beam, $u(t)$ is the angle the beam makes with the horizontal that is controlled with a motor, and $\bar{a} = 9.84$, $\bar{b} = 100$, $\bar{c} = 10^4$, and $\bar{d} = 514.96$. While $\bar{d}$ is unknown, we assume due to modeling considerations and physical constraints $\bar{d} \in [500, 525]$.

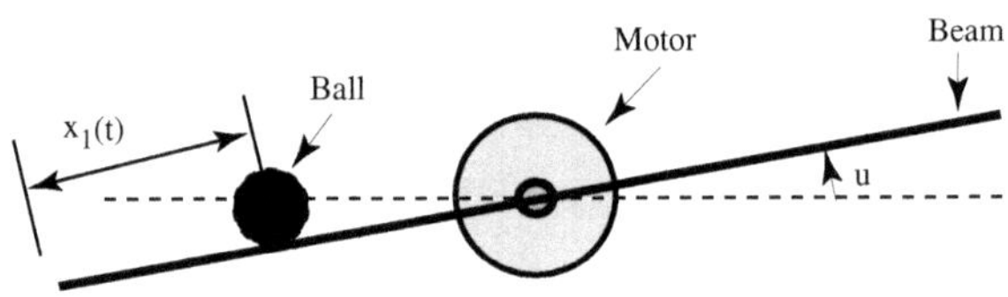

Figure 12.15: Ball on a beam experiment.

We have $y = x_1$ and

$$\dot{x} = f(x) + g(x)u$$

where

$$f(x) = \begin{bmatrix} x_2 \\ \bar{a}\tan^{-1}(\bar{b}x_2(t))\left(\exp(-\bar{c}x_2^2(t)) - 1\right) \end{bmatrix}$$

and

$$g(x) = \begin{bmatrix} 0 \\ -\bar{d} \end{bmatrix}$$

Notice that f and g are smooth. First, we will determine the relative degree of the plant. To do this, we simply take derivatives of the output until the input appears. In particular,

$$\begin{aligned}\dot{y} &= \dot{h}(x) = \left(\frac{\partial h}{\partial x}\right)^\top \dot{x} \\ &= \left(\frac{\partial h}{\partial x}\right)^\top f(x) + \left(\frac{\partial h}{\partial x}\right)^\top g(x)u \\ &= L_f h(x) + L_g h(x) u\end{aligned}$$

and for our case,

$$\frac{\partial h}{\partial x} = [1, 0]^\top$$

so

$$\dot{y} = L_f h(x) = x_2$$

(which is easy to see from the definition of the plant). Since $L_g h(x) = 0$, we need to find the second derivative of the plant output. Doing this, we find (defining $\alpha_k = \beta_k = 0$

$$\begin{aligned}\ddot{y} &= L_f^2 h(x) + L_g L_f h(x) u \\ &= \dot{x}_2 \\ &= \bar{a}\tan^{-1}(\bar{b}x_2(t))\left(\exp(-\bar{c}x_2^2(t)) - 1\right) - \bar{d}u(t) \\ &= \alpha(x) + \beta(x)u\end{aligned}$$

and

$$L_g L_f h(x) = \left(\frac{\partial x_2}{\partial x}\right)^\top \begin{bmatrix} 0 \\ -\bar{d} \end{bmatrix} = [0, 1] \begin{bmatrix} 0 \\ -\bar{d} \end{bmatrix} = -\bar{d} \neq 0$$

so that the relative degree is $d = n = 2$. Also, notice that if $y = z_1$ and $\dot{y} = z_2$ are bounded, then x_1 and x_2 are bounded. Notice, however, that there does not exist a $\beta_0 > 0$ such that $\beta(x) \geq \beta_0$ since $\beta(x)$ is known to lie in a fixed interval of *negative* numbers.

12.7.2 Indirect Adaptive Control

As with the discrete-time case we will seek to approximate the underlying plant nonlinearities and we will use these in a certainty equivalence controller.

Reference Model, Trajectory to Be Tracked

We want the output $y(t)$ and its derivatives $\dot{y}(t), \ldots, y^{(d)}(t)$ to track a "reference trajectory" $y_m(t)$ and its derivatives $\dot{y}_m(t), \ldots, y_m^{(d)}(t)$, respectively. We will assume that $y_m(t)$ and its derivatives $\dot{y}_m(t), \ldots, y_m^{(d)}(t)$, are bounded. A convenient way to specify the reference trajectory signals is to use a "reference model." While such trajectories can be generated by a nonlinear system, we will explain how to generate them with a linear system, since this is often how it is done in practice. In particular, if we have a reference input $r(t)$, with Laplace transform $R(s)$, and $Y_m(s)$ is the Laplace transform of $y_m(t)$,

$$\frac{Y_m(s)}{R(s)} = \frac{q(s)}{p(s)} = \frac{q_0}{s^d + p_{d-1}s^{d-1} + \cdots + p_0}$$

is a reference model where $p(s)$ is the pole polynomial with stable roots and q_0 is a constant.

As an example, suppose that $r(t) = 0$, $t \geq 0$, so we want $y(t) \to 0$ as $t \to \infty$. For this, we could simply choose

$$y_m(t) = \dot{y}_m(t) = \cdots = y_m^{(d)}(t) = 0$$

and this would represent a (perhaps challenging) request to immediately have the output and its derivatives track zero. To provide a request that the output go to zero "more gently," or according to some dynamics, we could use $r(t) = 0$, $t \geq 0$, so $R(s) = 0$ and

$$p(s)Y_m(s) = 0$$

or

$$(s^d + p_{d-1}s^{d-1} + \cdots + p_0)Y_m(s) = 0$$

or

$$y_m^{(d)}(t) + p_{d-1}y_m^{(d-1)}(t) + \cdots + p_0 y_m(t) = 0$$

The parameters $p_{d-1}, \ldots, p_0$ specify the dynamics of how $y_m(t)$ evolves over time and hence, specifies how we would like $y(t)$ and its derivatives to evolve over time.

Online Approximators for Plant Nonlinearities

We will approximate the functions $\alpha(x)$ and $\beta(x)$ with

$$\theta_\alpha^\top \phi_\alpha(x)$$

and

$$\theta_\beta^\top \phi_\beta(x)$$

by adjusting the θ_α and θ_β. The parameter vectors, θ_α and θ_β, are assumed to be defined within the compact parameter sets Ω_α and Ω_β, respectively. In addition, we define the subspace $S_x \subseteq \Re^n$ as the space through which the state trajectory may travel under closed-loop control (a known compact set). Notice that

$$\alpha(x) = \theta_\alpha^{*\top} \phi_\alpha(x) + w_\alpha(x) \tag{12.37}$$
$$\beta(x) = \theta_\beta^{*\top} \phi_\beta(x) + w_\beta(x) \tag{12.38}$$

where

$$\theta_\alpha^* = \arg\min_{\theta_\alpha \in \Omega_\alpha} \left(\sup_{x \in S_x} |\theta_\alpha^\top \phi_\alpha(x) - \alpha(x)| \right) \tag{12.39}$$

$$\theta_\beta^* = \arg\min_{\theta_\beta \in \Omega_\beta} \left(\sup_{x \in S_x} |\theta_\beta^\top \phi_\beta(x) - \beta(x)| \right) \tag{12.40}$$

so that $w_\alpha(x)$ and $w_\beta(x)$ are approximation errors, which arise when $\alpha(x)$ and $\beta(x)$ are represented by finite size approximators. We assume that

$$W_\alpha(x) \geq |w_\alpha(x)|$$

and

$$W_\beta(x) \geq |w_\beta(x)|$$

where $W_\alpha(x)$ and $W_\beta(x)$ are known state dependent bounds on the error in representing the actual system with approximators. Since we will use universal approximators, both $|w_\alpha(x)|$ and $|w_\beta(x)|$ may be made arbitrarily small by a proper choice of the approximator, since $\alpha(x)$ and $\beta(x)$ are smooth (of course, this may require an arbitrarily large number of parameters p). It is important to keep in mind that $W_\alpha(x)$ and $W_\beta(x)$ represent the magnitude of error between the actual nonlinear functions describing the system dynamics and the approximators when the "best" parameters are used, and we do not need to know these best parameters.

The approximations of $\alpha(x)$ and $\beta(x)$ of the actual system are

$$\hat{\alpha}(x) = \theta_\alpha^\top(t)\phi_\alpha(x) \tag{12.41}$$

$$\hat{\beta}(x) = \theta_\beta^\top(t)\phi_\beta(x) \tag{12.42}$$

where the vectors $\theta_\alpha(t)$ and $\theta_\beta(t)$ are updated online. The parameter errors are

$$\tilde{\theta}_\alpha(t) = \theta_\alpha(t) - \theta_\alpha^* \tag{12.43}$$

$$\tilde{\theta}_\beta(t) = \theta_\beta(t) - \theta_\beta^* \tag{12.44}$$

Consider the indirect adaptive control law

$$u = u_{ce} + u_{si} \tag{12.45}$$

The control law is comprised of a "certainty equivalence" control term u_{ce} and a "sliding mode" term u_{si}. We will introduce each of these next.

Certainty Equivalence Control Term

Let the tracking error be

$$e(t) = y_m(t) - y(t)$$

Let

$$K = [k_0, k_1, \ldots, k_{d-2}, 1]^\top$$

be a vector of design parameters (whose choice we will discuss below) and

$$e_s(t) = e^{(d-1)}(t) + k_{d-2}e^{(d-2)}(t) + \cdots + k_1\dot{e}(t) + k_0 e(t)$$

Also, for convenience, below we let

$$\bar{e}_s(t) = k_{d-2}e^{(d-1)}(t) + \cdots + k_0\dot{e}(t)$$

so that

$$\bar{e}_s(t) = \dot{e}_s(t) - e^{(d)}(t)$$

Let

$$L(s) = s^{d-1} + k_{d-2}s^{d-2} + \cdots + k_1 s + k_0$$

and assume that the design parameters in K are chosen so that $L(s)$ has its roots in the (open) left half plane.

Our goal is to drive $e_s(t) \to 0$ as $t \to \infty$. Notice that $e_s(t)$ is a measure of the tracking error. As an example, consider the case where $d = 2$ so $K = [k_0, 1]^\top$ and

$$e_s(t) = \dot{e}(t) + k_0 e(t)$$

For $L(s)$ to have its roots in the left half plane, we have $k_0 > 0$. Suppose that we have $e_s(t) = 0$. Then,

$$\dot{e}(t) + k_0 e(t) = 0$$

so that

$$\dot{e}(t) = -k_0 e(t)$$

so that $e(t) \to 0$ as $t \to \infty$ and hence, $y(t) \to y_m(t)$ as $t \to \infty$. The shape of the error dynamics is dictated by the choice of k_0. A large k_0 represents that we would like $e(t)$ to go to zero fast, while a small value of k_0 represents that we can accept that $y(t)$ may not achieve good tracking of $y_m(t)$ as fast. Note that you do not always want to choose k_0 large because, if you make an unreasonable request in the speed of the response, the controller may try to use too much control energy to achieve it.

The certainty equivalence control term is defined as

$$u_{ce} = \frac{1}{\beta_k(t) + \hat{\beta}(x)} \left(-\left(\alpha_k(t) + \hat{\alpha}(x)\right) + \nu(t)\right) \tag{12.46}$$

where

$$\nu(t) = y_m^{(d)} + \gamma e_s + \bar{e}_s$$

and $\gamma > 0$ is a design parameter whose choice we will discuss below. As in the discrete-time case, we will use projection to ensure that $\beta_k(t) + \hat{\beta}(x)$ is bounded away from zero so that u_{ce} is well-defined.

The dth derivative of the output error is $e^{(d)} = y_m^{(d)} - y^{(d)}$ so

$$e^{(d)} = y_m^{(d)} - \left(\alpha_k(t) + \alpha(x)\right) - \left(\beta_k(t) + \beta(x)\right) u(t)$$

and since $u = u_{ce} + u_{si}$

$$\begin{aligned} e^{(d)} &= y_m^{(d)} - (\alpha_k(t) + \alpha(x)) - \\ &\quad \frac{\beta_k(t) + \beta(x)}{\beta_k(t) + \hat{\beta}(x)} \left(- (\alpha_k(t) + \hat{\alpha}(x)) + \nu(t)\right) - (\beta_k(t) + \beta(x))u_{si} \end{aligned} \tag{12.47}$$

Note that the first two terms

$$\begin{aligned} y_m^{(d)} - (\alpha_k(t) + \alpha(x)) &= y_m^{(d)} - (\alpha_k(t) + \hat{\alpha}(x)) - \alpha(x) + \hat{\alpha}(x) \\ &= (- (\alpha_k(t) + \hat{\alpha}(x)) + \nu(t)) - \alpha(x) + \hat{\alpha}(x) - \nu(t) + y_m^{(d)} \\ &= (- (\alpha_k(t) + \hat{\alpha}(x)) + \nu(t)) - \alpha(x) + \hat{\alpha}(x) - \gamma e_s - \bar{e}_s \end{aligned}$$

Substituting this into Equation (12.47), we get

$$\begin{aligned} e^{(d)} &= \left(1 - \frac{\beta_k(t) + \beta(x)}{\beta_k(t) + \hat{\beta}(x)}\right) (-(\alpha_k(t) + \hat{\alpha}(x)) + \nu(t)) - \alpha(x) + \hat{\alpha}(x) \\ &\quad -\gamma e_s - \bar{e}_s - (\beta_k(t) + \beta(x))u_{si} \\ &= (\hat{\alpha}(x) - \alpha(x)) + \left(\hat{\beta}(x) - \beta(x)\right) u_{ce} \\ &\quad -\gamma e_s - \bar{e}_s - (\beta_k(t) + \beta(x))u_{si} \end{aligned} \tag{12.48}$$

Since $\bar{e}_s = \dot{e}_s - e^{(d)}$

$$\dot{e}_s + \gamma e_s = (\hat{\alpha}(x) - \alpha(x)) + \left(\hat{\beta}(x) - \beta(x)\right) u_{ce} - (\beta_k(t) + \beta(x))u_{si} \tag{12.49}$$

We get a type of linear relationship between a tracking error measure and model error.

Parameter Update Laws

Consider the following Lyapunov function candidate

$$V_i = \frac{1}{2}e_s^2 + \frac{1}{2\eta_\alpha}\tilde{\theta}_\alpha^\top \tilde{\theta}_\alpha + \frac{1}{2\eta_\beta}\tilde{\theta}_\beta^\top \tilde{\theta}_\beta \tag{12.50}$$

where $\eta_\alpha > 0$ and $\eta_\beta > 0$ are design parameters whose choice we will discuss below. This Lyapunov function quantifies both the error in tracking and in the parameter estimates. Using vector derivatives, the time derivative of Equation (12.50) is

$$\dot{V}_i = e_s \dot{e}_s + \frac{1}{\eta_\alpha}\tilde{\theta}_\alpha^\top \dot{\tilde{\theta}}_\alpha + \frac{1}{\eta_\beta}\tilde{\theta}_\beta^\top \dot{\tilde{\theta}}_\beta \tag{12.51}$$

Substituting in the derivative of the tracking error, $\dot{e}_s$ from Equation (12.49), yields

$$\begin{aligned} \dot{V}_i &= e_s \left(-\gamma e_s + (\hat{\alpha}(x) - \alpha(x)) + (\hat{\beta}(x) - \beta(x))u_{ce} - (\beta_k(t) + \beta(x))u_{si}\right) \\ &\quad + \frac{1}{\eta_\alpha}\tilde{\theta}_\alpha^\top \dot{\tilde{\theta}}_\alpha + \frac{1}{\eta_\beta}\tilde{\theta}_\beta^\top \dot{\tilde{\theta}}_\beta \end{aligned} \tag{12.52}$$

Notice that

$$\hat{\alpha}(x) - \alpha(x) = \theta_\alpha^\top \phi_\alpha(x) - \theta_\alpha^{*\top} \phi_\alpha(x) - w_\alpha(x) = \tilde{\theta}_\alpha^\top \phi_\alpha(x) - w_\alpha(x)$$

and similarly for $\hat{\beta}(x) - \beta(x)$. Hence,

$$\begin{aligned}\dot{V}_i &= -\gamma e_s^2 \quad (12.53)\\ &+ \left(\tilde{\theta}_\alpha^\top \phi_\alpha(x) - w_\alpha(x) + \tilde{\theta}_\beta^\top \phi_\beta(x) u_{ce} - w_\beta(x) u_{ce} - (\beta_k(t) + \beta(x)) u_{si}\right) e_s \\ &+ \frac{1}{\eta_\alpha} \tilde{\theta}_\alpha^\top \dot{\tilde{\theta}}_\alpha + \frac{1}{\eta_\beta} \tilde{\theta}_\beta^\top \dot{\tilde{\theta}}_\beta \quad (12.54)\end{aligned}$$

The update laws implement a continuous time version of the gradient optimization update method.

Consider the following update laws

$$\dot{\theta}_\alpha(t) = -\eta_\alpha \phi_\alpha(x) e_s \quad (12.55)$$
$$\dot{\theta}_\beta(t) = -\eta_\beta \phi_\beta(x) e_s u_{ce} \quad (12.56)$$

We see that $\eta_\alpha > 0$ and $\eta_\beta > 0$ are adaptation gains. Picking these gains to be larger will indicate that you want a faster adaptation.

Note that since we assume that the ideal parameters are constant $\dot{\tilde{\theta}}_\alpha = \dot{\theta}_\alpha$ and $\dot{\tilde{\theta}}_\beta = \dot{\theta}_\beta$. Now with this, notice that

$$\frac{1}{\eta_\alpha} \tilde{\theta}_\alpha^\top \dot{\tilde{\theta}}_\alpha = -\tilde{\theta}_\alpha^\top \phi_\alpha(x) e_s$$

and

$$\frac{1}{\eta_\beta} \tilde{\theta}_\beta^\top \dot{\tilde{\theta}}_\beta = -\tilde{\theta}_\beta^\top \phi_\beta(x) e_s u_{ce}$$

so

$$\begin{aligned}\dot{V}_i &= -\gamma e_s^2 + \left(\tilde{\theta}_\alpha^\top \phi_\alpha(x) - w_\alpha(x) + \tilde{\theta}_\beta^\top \phi_\beta(x) u_{ce} - w_\beta(x) u_{ce}\right) e_s \quad (12.57)\\ &\quad -(\beta_k(t) + \beta(x)) u_{si} e_s - \tilde{\theta}_\alpha^\top \phi_\alpha(x) e_s - \tilde{\theta}_\beta^\top \phi_\beta(x) e_s u_{ce}\end{aligned}$$

and

$$\dot{V}_i = -\gamma e_s^2 - (w_\alpha(x) + w_\beta(x) u_{ce}) e_s - (\beta_k(t) + \beta(x)) u_{si} e_s \quad (12.58)$$

Projection Modification to Parameter Update Laws

The above adaptation laws in Equations (12.55) and (12.56) will not guarantee that $\theta_\alpha \in \Omega_\alpha$ and $\theta_\beta \in \Omega_\beta$, so we will use projection to ensure this (e.g., to make sure that $(\beta_k(t) + \hat{\beta}(x)) \geq \beta_0$). Suppose in particular that we know that the ith component of θ_α^* (θ_β^*) is in the (known) interval

$$\theta_{\alpha_i}^* \in [\theta_{\alpha_i}^{min}, \theta_{\alpha_i}^{max}]$$

and

$$\theta_{\beta_i}^* \in [\theta_{\beta_i}^{min}, \theta_{\beta_i}^{max}]$$

Suppose we place the initial values of the parameters in these ranges. Also, if $\theta_{\alpha_i}(t)$ and $\theta_{\beta_i}(t)$ are strictly within these ranges, then you use the update given by the update formulas in Equations (12.55) and (12.56). If, however, $\theta_{\alpha_i}(t)$ or $\theta_{\beta_i}(t)$ is on the boundary of its interval and the update formula indicates that it should be moved outside the interval, then you leave it on the boundary of the interval. However, if it is on the boundary and the update law indicates that it should be moved on the boundary or to within the interval, then the update from Equations (12.55) and (12.56) is allowed.

To show this in more detail, consider how to implement projection for the case for θ_β when it is a scalar (the general vector case follows easily, and clearly the approach is similar for θ_α). In this case, suppose that we know

$$\theta_\beta^* \in [\theta_\beta^{min}, \theta_\beta^{max}]$$

(we remove the i index since we consider the scalar case) and so we want

$$\theta_\beta(t) \in [\theta_\beta^{min}, \theta_\beta^{max}]$$

Define

$$\theta_\beta^m \in [\theta_\beta^{min}, \theta_\beta^{max}]$$

to be a point in the acceptable range (actually, any such point will work). Let

$$\theta_\beta^{ud}(t) = -\eta_\beta \phi_\beta e_s u_{ce}$$

to be the update that would result if we did not use projection. Projection is implemented by the following tests:

- If $\theta_\beta(t) \leq \theta_\beta^{min}$ and
  - If $\theta_\beta^{ud}(t) > 0$, use $\dot{\theta}_\beta(t) = \theta_\beta^{ud}(t)$
  - If $\theta_\beta^{ud}(t) \leq 0$, use $\dot{\theta}_\beta(t) = 0$ (since the update would move it outside the interval)
- If $\theta_\beta(t) \geq \theta_\beta^{max}$ and
  - If $\theta_\beta^{ud}(t) \geq 0$, use $\dot{\theta}_\beta(t) = 0$ (since the update would move it outside the interval)
  - If $\theta_\beta^{ud}(t) < 0$, use $\dot{\theta}_\beta(t) = \theta_\beta^{ud}(t)$

Or, we can summarize these conditions more concisely by using the following rule: If $\theta_\beta(t) \notin \left(\theta_\beta^{min}, \theta_\beta^{max}\right)$ and $\theta_\beta^{ud}(t)(\theta_\beta(t) - \theta_\beta^m) \geq 0$ let $\dot{\theta}_\beta(t) = 0$ and otherwise let $\dot{\theta}_\beta(t) = \theta_\beta^{ud}(t)$.

Returning to the stability analysis, clearly since $\theta_{\alpha_i}^*$ and $\theta_{\beta_i}^*$ are within the allowable ranges, this projection modification to the update laws will always result in a parameter estimation error that will decrease V_i at least as much as if the projection were not used; hence, the right-hand side of Equation (12.58) will overbound the $\dot{V}_i$ that would result if projection is used. For this reason, we conclude that

$$\dot{V}_i \leq -\gamma e_s^2 - (w_\alpha(x) + w_\beta(x) u_{ce}) e_s - (\beta_k(t) + \beta(x)) u_{si} e_s \tag{12.59}$$

Sliding Mode Control Term

To ensure that Equation (12.59) is less than or equal to zero, we choose[1]

$$u_{si} = \frac{(W_\alpha(x) + W_\beta(x)|u_{ce}|)}{\beta_0}\text{sgn}(e_s) \tag{12.60}$$

where

$$\text{sgn}(e_s) = \begin{cases} 1 & e_s > 0 \\ -1 & e_s < 0 \end{cases} \tag{12.61}$$

Note that

$$-(w_\alpha(x) + w_\beta(x)u_{ce})e_s \leq (|w_\alpha(x)| + |w_\beta(x)u_{ce}|)\,|e_s|$$

Hence,

$$\begin{aligned} \dot{V}_i \;\leq\; & -\gamma e_s^2 + (|w_\alpha(x)| + |w_\beta(x)u_{ce}|)|e_s| \\ & -e_s(\beta_k(t) + \beta(x))\left(\frac{(W_\alpha(x) + W_\beta(x)|u_{ce}|)}{\beta_0}\text{sgn}(e_s)\right) \end{aligned} \tag{12.62}$$

Now, considering the last term in this equation and noting that

$$\frac{(\beta_k(t) + \beta(x))}{\beta_0} \geq 1$$

we have

$$\dot{V}_i \leq -\gamma e_s^2 + |w_\alpha(x)||e_s| + |w_\beta(x)u_{ce}||e_s| - e_s\text{sgn}(e_s)W_\alpha(x) - e_s\text{sgn}(e_s)W_\beta(x)|u_{ce}| \tag{12.63}$$

Notice that $|e_s| = e_s\text{sgn}(e_s)$ (except at $e_s = 0$) and recall that $|w_\alpha(x)| \leq W_\alpha(x)$ and $|w_\beta(x)| \leq W_\beta(x)$ so

$$|w_\alpha(x)||e_s| - e_s\text{sgn}(e_s)W_\alpha(x) = |e_s|(|w_\alpha(x)| - W_\alpha(x)) \leq 0$$

and

$$|w_\beta(x)u_{ce}||e_s| - e_s\text{sgn}(e_s)W_\beta(x)|u_{ce}| = |e_s|(|w_\beta(x)u_{ce}| - W_\beta(x)|u_{ce}|) \leq 0$$

so

$$\dot{V}_i \leq -\gamma e_s^2 \tag{12.64}$$

Since $\gamma e_s^2 \geq 0$ this shows that V_i, which is a measure of the tracking error and parameter estimation error, is a nonincreasing function of time. Notice that $\gamma > 0$ has an influence on how fast $V_i \to 0$. By picking γ larger you will often get faster convergence of the tracking error.

[1] Note that we are introducing a discontinuity here so strictly speaking, we are not guaranteed that solutions to the differential equation representing the closed-loop system exist and are unique. This issue has received significant attention in the literature, particularly, the sliding mode control literature. Here, we simply highlight the issue and note that if you want to avoid the problem, you can use the "smoothed control law" defined later in this section so that you are guaranteed existence and uniqueness of solutions.

Asymptotic Convergence of the Tracking Error and Boundedness of Signals

Given the plant assumptions (that the reference signals are bounded, x is measurable, $d = n$, and that projection ensures that the u_{ce} term is well-defined), the following hold:

Asymptotic tracking of reference inputs with bounded signals can be achieved via indirect adaptive control.

- The plant output is such that $y, \dot{y}, \ldots, y^{(d-1)}$ are bounded.
- The input signals u, u_{ce}, and u_{si} are bounded.
- The parameters $\theta_\alpha(t)$ and $\theta_\beta(t)$ are bounded.
- We get asymptotic tracking, that is,

$$\lim_{t\to\infty} e(t) = 0$$

To see this, first note that since V_i is a positive function and

$$\dot{V}_i \le -\gamma e_s^2 \tag{12.65}$$

we know that e_s, θ_α, and θ_β are bounded. Since e_s is bounded and y_m and its derivatives (i.e., $y_m, \dot{y}_m, \ldots, y_m^{(d-1)}$) are bounded, we know that $y, \dot{y}, \ldots, y^{(d-1)}$ are bounded. Hence, by assumption we have that z and hence, x are bounded. Hence, $\alpha(x)$, $\hat{\alpha}(x)$, $\alpha_k(t)$, $\beta(x)$, $\hat{\beta}(x)$, and $\beta_k(t)$ are bounded. Since x is bounded and $(\beta_k(t) + \hat{\beta}(x)) \ge \beta_0$, u_{ce} and u_{si} and hence, u are bounded. Next, note that

$$\int_0^\infty \gamma e_s^2 dt \quad \le \quad -\int_0^\infty \dot{V}_i dt = V_i(0) - V_i(\infty) \tag{12.66}$$

This establishes that $e_s \in \mathcal{L}_2$ ($\mathcal{L}_2 = \{z(t) : \int_0^\infty z^2(t)dt < \infty\}$) since $V_i(0)$ and $V_i(\infty)$ are bounded. Note that via Equation (12.49), $\dot{e}_s$ is bounded. Hence, since e_s and $\dot{e}_s$ are bounded and $e_s \in \mathcal{L}_2$, we have that $\lim_{t\to\infty} e_s(t) = 0$ (this is what is called Barbalat's lemma). It should be clear then, via the definition of $e_s(t)$, that $\lim_{t\to\infty} e(t) = 0$.

Smoothed Control Law

It is possible to augment the above control law with a "bounding control term" that will ensure that the states stay bounded within some region. This can be useful in defining the approximator structures to provide good approximation properties in the region where the states will be. The sliding mode control term in effect compensates for the approximation errors that arise, since we are using finite sized approximators. If the approximators are not defined so that it is possible to make them accurate in the region where the system will operate, then $W_\alpha(x)$ and $W_\beta(x)$ will have to be large and significant actions are then taken by the sliding mode control term (and these generally result in a "high gain effect" that can cause undesirable oscillations). It is possible to reduce the high frequency signals that can result from the sliding mode control term by using a "smoothed version" of this signal (i.e., one that has a smooth transition from negative to positive values, not the sgn(e_s) term). In this case, however, you only get convergence to an ϵ-neighborhood of $e_s = 0$ (it seems that our ability to use the high gain effect from the sliding mode term in continuous time systems allows us to get asymptotic tracking, not just to a neighborhood of zero as we had in the discrete-time case). We will discuss the details of such an approach next.

Define the error

$$e_\epsilon = e_s - \epsilon \mathrm{sat}\left(e_s/\epsilon\right) \tag{12.67}$$

where $\epsilon > 0$ and

$$\mathrm{sat}(x) = \begin{cases} 1 & \text{if } 1 \leq x \\ x & \text{if } -1 < x < 1 \\ -1 & \text{if } x \leq -1 \end{cases} \tag{12.68}$$

Hence, e_ϵ measures the distance between e_s and the desired "boundary layer," so that $e_\epsilon = 0$ when e_s is within the boundary layer.

The certainty equivalence controller is now defined to be

$$u_{ce} = \frac{1}{\beta_k(t) + \hat{\beta}(x)} \left(-\left(\alpha_k(t) + \hat{\alpha}(x)\right) + \nu_\epsilon(t)\right) \tag{12.69}$$

where

$$\nu_\epsilon(t) = y_m^{(d)} + \gamma e_\epsilon + \bar{e}_s$$

with $\bar{e}_s$ as defined before. With these changes, Equation (12.49) becomes

$$\dot{e}_s + \gamma e_\epsilon = \left(\hat{\alpha}(x) - \alpha(x)\right) + \left(\hat{\beta}(x) - \beta(x)\right) u_{ce} - \left(\beta_k(t) + \beta(x)\right) u_{si} \tag{12.70}$$

Now, consider Equation (12.50) with the e_ϵ as the tracking error measurement

$$V_i = \frac{1}{2} e_\epsilon^2 + \frac{1}{2\eta_\alpha} \tilde{\theta}_\alpha^\top \tilde{\theta}_\alpha + \frac{1}{2\eta_\beta} \tilde{\theta}_\beta^\top \tilde{\theta}_\beta \tag{12.71}$$

Consider the following update laws

$$\dot{\theta}_\alpha(t) = -\eta_\alpha \phi_\alpha(x) e_\epsilon \tag{12.72}$$

$$\dot{\theta}_\beta(t) = -\eta_\beta \phi_\beta(x) e_\epsilon u_{ce} \tag{12.73}$$

where $\eta_\alpha > 0$ and $\eta_\beta > 0$ are adaptation gains. Use an appropriate projection algorithm.

With this, Equation (12.62) is expressed as

$$\dot{V}_i \leq -\gamma e_\epsilon^2 + \left(|w_\alpha(x)| + |w_\beta(x) u_{ce}|\right)|e_\epsilon| - \left(\beta_k(t) + \beta(x)\right) e_\epsilon u_{si} \tag{12.74}$$

We now redefine the control term

$$u_{si} = \frac{W_\alpha(x) + W_\beta(x)|u_{ce}|}{\beta_0} \mathrm{sat}(e_s/\epsilon) \tag{12.75}$$

so that we now have smooth control action. Notice that

$$e_\epsilon \text{sat}(e_s/\epsilon) = |e_\epsilon|$$

to see that

$$\dot{V}_i \leq -\eta e_\epsilon^2$$

which ensures asymptotic stability of e_ϵ using Barbalat's lemma as before. This implies that e_s will converge asymptotically to an ϵ-neighborhood of $e_s = 0$, and also e will converge to a neighborhood of $e = 0$ (and the size of that neighborhood is proportional to ϵ). Hence, by choosing ϵ to be small, we get a higher gain control, and better convergence results (i.e., to a smaller neighborhood). Clearly, as $\epsilon \to 0$, we get the same results as for the earlier case when we used the high gain term.

The smoothed control law results in convergence of the tracking error to a neighborhood whose size can be set a priori.

12.7.3 Direct Adaptive Control

In addition to the assumptions we made in the indirect adaptive control case, we require

$$\beta_k(t) = \alpha_k(t) = 0$$

for all $t \geq 0$, and that there exists positive constants β_0 and β_1 such that

$$0 < \beta_0 \leq \beta(x) \leq \beta_1$$

(which often holds in practical applications). Also, we assume that we can specify some function $B(x) \geq 0$ such that

$$|\dot{\beta}(x)| = \left| \left(\frac{\partial \beta}{\partial x} \right)^\top \dot{x} \right| \leq B(x)$$

for all $x \in S_x$. This requirement is often met in practice since, if we think of $\dot{\beta}(x)$ as the rate of change of the "gain" on the input term, it is often the case that this will be bounded. For example, notice that if $\beta(x)$ is a constant that we know lies in a fixed positive interval, then all these conditions are satisfied with $B(x) = 0$, for all x.

Controller Approximator

We know that there exists some ideal controller

$$u^* = \frac{1}{\beta(x)}(-\alpha(x) + \nu(t)) \tag{12.76}$$

where $\nu(t)$ is defined the same as in the indirect adaptive control case. Let

$$u^* = \theta_u^{*\top} \phi_u(x, \nu) + u_k(t) + w_u(x, \nu) \tag{12.77}$$

where u_k is a known part of the controller (e.g., one that was designed for the nominal plant) and

$$\theta_u^* = \arg\min_{\theta_u \in \Omega_u} \left(\sup_{x \in S_x, \nu \in S_m} |\theta_u^\top \phi_u(x,\nu) - (u^* - u_k)| \right) \tag{12.78}$$

so that $w_u(x,\nu)$ is the approximation error. We assume that $W_u(x,\nu) \geq |w_u(x,\nu)|$, where $W_u(x,\nu)$ is a known bound on the error in representing the ideal controller. The approximation is

$$\hat{u} = \theta_u^\top(t)\phi_u(x,\nu) + u_k \tag{12.79}$$

where the matrix $\theta_u(t)$ is updated online. The parameter error is

$$\tilde{\theta}_u(t) = \theta_u(t) - \theta_u^* \tag{12.80}$$

Consider the control law

$$u = \hat{u} + u_{sd} \tag{12.81}$$

which is the sum of an approximation to an ideal control law, and a sliding mode control term. With this, the dth derivative of the tracking error becomes

$$e^{(d)} = y_m^{(d)} - \alpha(x) - \beta(x)\left(\hat{u} + u_{sd}\right) \tag{12.82}$$

Adding and subtracting $\beta(x)u^*$ and then using the definition of u^*, we get

$$e^{(d)} = y_m^{(d)} - \alpha(x) - \beta(x)u^* - \beta(x)\left(\hat{u} - u^*\right) - \beta(x)u_{sd} \tag{12.83}$$

$$= -\gamma e_s - \bar{e}_s - \beta(x)\left(\hat{u} - u^*\right) - \beta(x)u_{sd}. \tag{12.84}$$

or in a manner analogous to the indirect case,

$$\dot{e}_s + \gamma e_s = -\beta(x)\left(\hat{u} - u^*\right) - \beta(x)u_{sd}. \tag{12.85}$$

Controller Parameter Updates

Consider the following Lyapunov function candidate

$$V_d = \frac{1}{2\beta(x)}e_s^2 + \frac{1}{2\eta_u}\tilde{\theta}_u^\top\tilde{\theta}_u \tag{12.86}$$

where $\eta_u > 0$. Since $0 < \beta_0 \leq \beta(x) \leq \beta_1$, V_d is radially unbounded. The Lyapunov candidate, V_d, is used to measure both the error in tracking and the error between the desired controller and current controller. Taking the time derivative of Equation (12.86) yields

$$\dot{V}_d = \frac{e_s}{\beta(x)}\dot{e}_s - \frac{\dot{\beta}(x)e_s^2}{2\beta^2(x)} + \frac{1}{\eta_u}\tilde{\theta}_u^\top\dot{\tilde{\theta}}_u \tag{12.87}$$

(the second term arises since $\beta(x)$ depends on time). Substituting $\dot{e}_s$, as defined in Equation (12.85), we find

$$\dot{V}_d = \frac{e_s}{\beta(x)}\left(-\gamma e_s - \beta(x)\left(\hat{u} - u^*\right) - \beta(x)u_{sd}\right) - \frac{\dot{\beta}(x)e_s^2}{2\beta^2(x)} + \frac{1}{\eta_u}\tilde{\theta}_u^\top\dot{\tilde{\theta}}_u \tag{12.88}$$

Use the update law

$$\dot{\theta}_u(t) = \eta_u \phi_u(x, \nu) e_s(t) \tag{12.89}$$

so $\eta_u > 0$ is an adaptation gain. Since $\dot{\tilde{\theta}}_u = \dot{\theta}_u$,

$$\dot{V}_d = -\frac{\gamma}{\beta(x)} e_s^2 - \left(\tilde{\theta}_u^\top \phi_u(x, \nu) - w_u(x, \nu) + u_{sd} \right) e_s - \frac{\dot{\beta}(x) e_s^2}{2\beta^2(x)} + \tilde{\theta}_u^\top \phi_u(x, \nu) e_s \tag{12.90}$$

and so

$$\dot{V}_d = -\frac{\gamma}{\beta(x)} e_s^2 - \left(\frac{\dot{\beta}(x) c_s}{2\beta^2(x)} - w_u(x, \nu) \right) e_s - e_s u_{sd} \tag{12.91}$$

We use a projection method to ensure that $\theta_u \in \Omega_u$ so in an analogous manner to the indirect case,

$$\dot{V}_d \leq -\frac{\gamma}{\beta(x)} e_s^2 - \left(\frac{\dot{\beta}(x) e_s}{2\beta^2(x)} - w_u(x, \nu) \right) e_s - e_s u_{sd} \tag{12.92}$$

Sliding Mode Control Term and Stability Properties

We once again use a sliding mode control term to compensate for the approximation error in modeling u^* by a finite size approximator. Notice that

$$\dot{V}_d \leq -\frac{\gamma}{\beta_1} e_s^2 + \left(\frac{|\dot{\beta}(x)||e_s|}{2\beta^2(x)} + |w_u(x, \nu)| \right) |e_s| - e_s u_{sd} \tag{12.93}$$

(12.94)

We now define the sliding mode control term for the direct adaptive controller as

$$u_{sd} = \left(\frac{B(x)|e_s|}{2\beta_0^2} + W_u(x, \nu) \right) \operatorname{sgn}(e_s) \tag{12.95}$$

which ensures that

$$\dot{V}_d \leq -\gamma e_s^2 / \beta_1$$

so that V_d is a nonincreasing function of time.

This gives us the same type of results that we obtained in the indirect case. In particular, all the signals are bounded and $e(t) \to 0$ as $t \to \infty$, so we get asymptotic tracking. There are practical applications where some u_k can be designed so that the resulting transient performance can then sometimes be improved. In an analogous manner to the indirect case, it is possible to define a smoothed version of the sliding mode control term that will only result in e_s reducing to a neighborhood of zero.

The same type of asymptotic tracking result can be obtained for the direct adaptive control case.

12.7.4 Design Example: Aircraft Wing Rock Regulation

In this section we provide an example of how to design continuous time indirect and direct adaptive controllers for an aircraft wing rock regulation problem.

The Aircraft Wing Rock Regulation Problem

Aircraft wing rock is a limit cycling oscillation in the aircraft roll angle ϕ and roll rate $\dot{\phi}$. If δ_A is the actuator output, a model of this phenomenon is given by

$$\ddot{\phi} = a_1\phi + a_2\dot{\phi} + a_3\dot{\phi}^3 + a_4\phi^2\dot{\phi} + a_5\phi\dot{\phi}^2 + b\delta_A$$

where a_i, $i = 1, 2, 3, 4, 5$, and b, are constant but unknown. We assume that we know the sign of b. Choose the state vector $x = [x_1, x_2, x_3]^\top$ with $x_1 = \phi$, $x_2 = p = \dot{\phi}$, and $x_3 = \delta_A$. Suppose that we use a first order model to represent the actuator dynamics of the aileron (the control surface at the outer part of the wing). Then we have

$$\begin{aligned}
\dot{x}_1 &= x_2 \\
\dot{x}_2 &= a_1x_1 + a_2x_2 + a_3x_2^3 + a_4x_1^2x_2 + a_5x_1x_2^2 + bx_3 \\
\dot{x}_3 &= -\frac{1}{\tau}x_3 + \frac{1}{\tau}u \\
y &= x_1
\end{aligned}$$

where u is the control input to the actuator and τ is the aileron time constant. For an angle of attack of 21.5 degrees, $a_1 = -0.0148927$, $a_2 = 0.0415424$, $a_3 = 0.01668756$, $a_4 = -0.06578382$, $a_5 = 0.08578836$. Also, $b = 1.5$ and $\tau = \frac{1}{15}$. Take these as constant nominal values that you do not know. Suppose, however, that you know that $b \in [1, 2]$ and $\tau \in [\frac{1}{20}, \frac{1}{10}]$. Also, assume that you know that there is a constant but unknown gain that multiplies the input u (i.e., assume that you know it is not a nonlinear function of x); however, suppose that you do not know that the particular plant nonlinearities are of the form indicated above, or that the parameters appear as they do (i.e., do not use the fact that they enter linearly).

Suppose that you want the output $y(t)$ to track the reference signal $y_m(t)$ that is zero, and has all its derivatives identical to zero, for all time. Assume that you have sensors to measure y, $\dot{y}$, and $\ddot{y}$ (for the simulation, we will use the above model to simulate the sensing of all these values, and in particular, use $\ddot{y} = a_1\phi + a_2\dot{\phi} + a_3\dot{\phi}^3 + a_4\phi^2\dot{\phi} + a_5\phi\dot{\phi}^2 + b\delta_A$ to simulate the sensing of $\ddot{y}$). We will use $x(0) = [0.4, 0, 0]^\top$ in all our simulations.

Indirect Adaptive Controller Development and Results

The relative degree is $d = n = 3$. Assume that $\alpha_k = \beta_k = 0$ so that we assume that we have no special information about the form of the underlying nonlinearities. Next, we need to find a $\beta_0 > 0$ and later for the direct adaptive controller a $\beta_1 \geq \beta_0$ such that $\beta_0 \leq \beta(x) \leq \beta_1$. We use $\beta_0 = 10$ and $\beta_1 = 40$. After a bit of tuning, we chose $k_0 = 100$, $k_1 = 20$, and $\gamma = 2$. Also, we tuned the adaptation gains to get fast enough adaptation to meet the objectives. In particular, we used $\eta_\alpha = \eta_\beta = 2$.

To design the approximators that we need, we first note that since we assume that we know that b is an unknown constant, we can simply use a constant to estimate it (i.e., an affine approximator with only the constant term). To keep things simple, this is what we will do. We use projection to make sure that the estimate of b stays in the proper range. To estimate the α term, we will use a Takagi-Sugeno fuzzy system but as an input to the premise terms, we will only use x_1 and x_2 (key system variables where nonlinearities enter), while we will use all three state variables as inputs to the consequent (we are trying to avoid problems with computational complexity). We placed the centers of the membership functions on the two universes of discourse at -2, 0, and 2, with the spread values all equal to 2, and used all possible combinations of rules so we get $R = 9$ rules. This means that we will tune 36 parameters for our approximator. We chose $W_\alpha = 0.01$ (simply a guess) and $W_\beta = 0$ (since we know that ideally our approximator can succeed).

The results showing the quality of the tracking are shown in Figure 12.16. There, we also show the actuator output. Notice that we get fast tracking. Next, see Figures 12.17 and 12.18, where we show the time history of the parameters that are used in the approximators. Notice that the parameters estimates move significantly in the beginning, but reach a steady state.

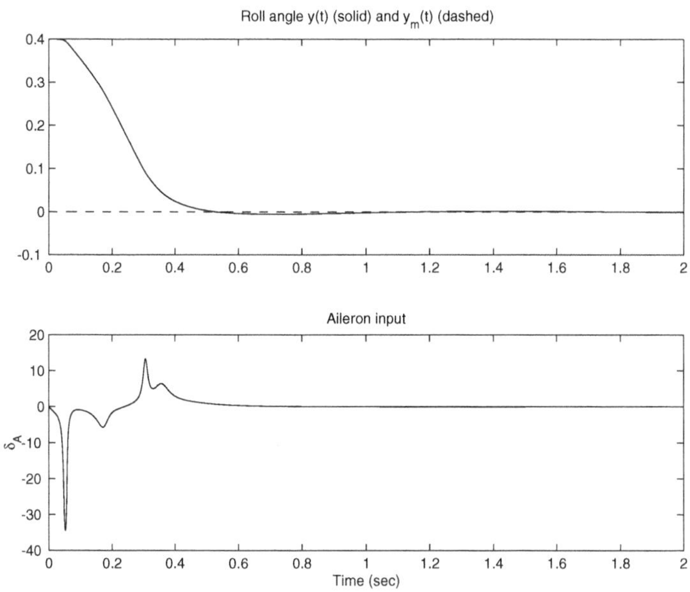

Figure 12.16: Wing rock controller results, roll angle.

Direct Adaptive Controller Development and Results

For the direct adaptive controller, we use all the same parameters as in the indirect case but choose $W_u = 0.01$ and $\eta_u = 2$. For convenience, the controller approximator structure is implemented using the same approximator structure

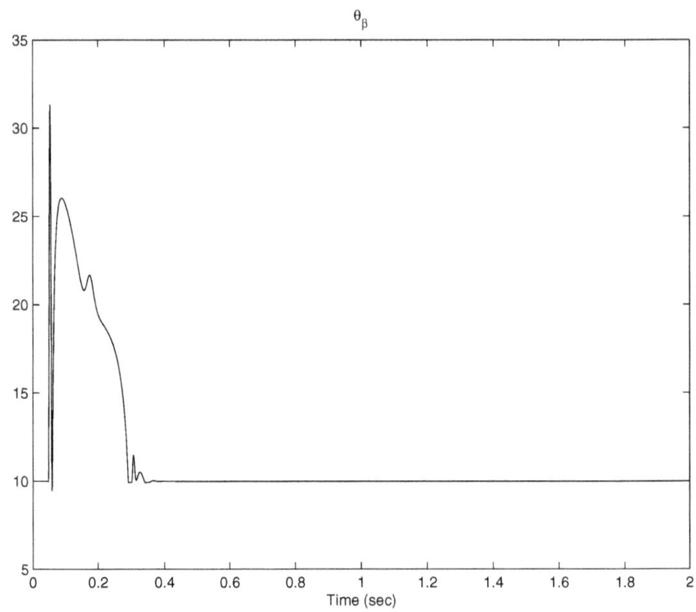

Figure 12.17: Wing rock controller results, θ_β.

as we used in the indirect adaptive controller for the α term. Hence, it has inputs of x_1 and x_2 for the premise, but all the state variables as inputs to the consequents. We do not use ν as an input to the controller (even though this may help further improve performance).

The results showing the quality of the tracking are shown in Figure 12.19. There, we also show the actuator output. Notice that we get fast tracking as we did in the indirect case. Next, see Figure 12.20, where we show the time history of the parameters that are used in the approximator. Notice that the parameters estimates reach a steady state after a brief transient period.

12.8 Discussion: Tuning Structure and Nonlinear in the Parameter Approximators

Tuning approximator structure or nonlinear in the parameter approximators may lead to a better exploitation of approximator flexibility.

There are several ways to tune nonlinear in the parameter approximators that are used in either direct or indirect adaptive control schemes. For instance, you could linearize the approximator structure, assume that the errors from the linearization are bounded, and then the error equations for the adaptation laws will remain linear so that the same normalized gradient and recursive least squares methods used in the adaptive neural/fuzzy control approaches in the previous subsections can be used to tune the approximators. Of course in this case, the error that results from the linearization may adversely affect the performance

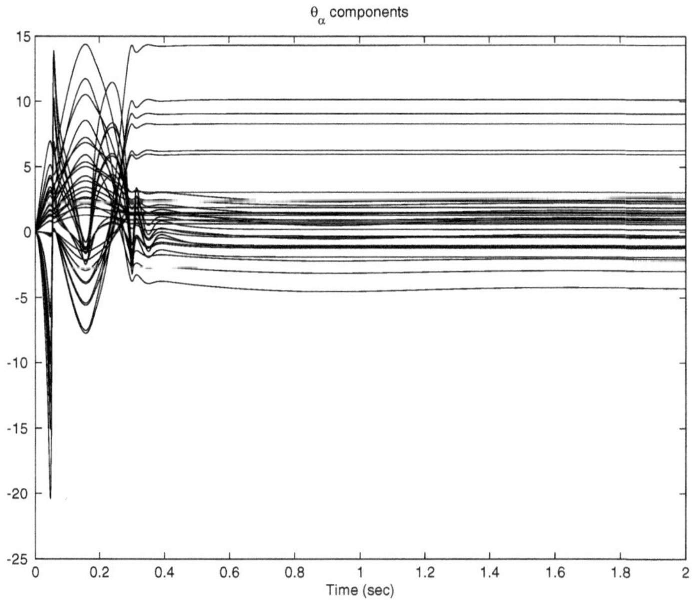

Figure 12.18: Wing rock controller results, components of θ_α.

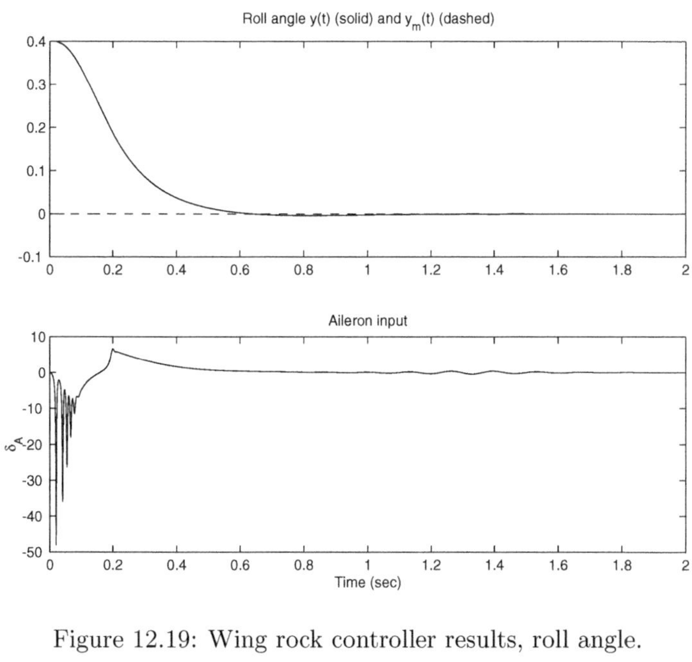

Figure 12.19: Wing rock controller results, roll angle.

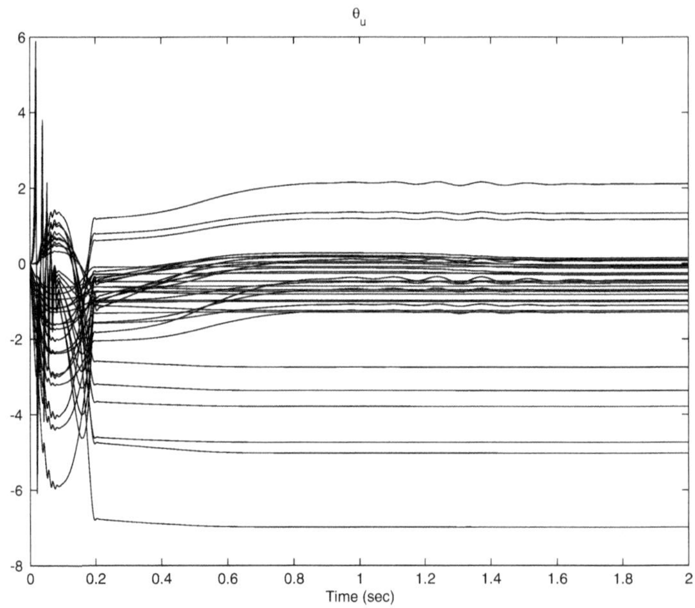

Figure 12.20: Wing rock controller results, θ_u.

of the methods. See the "For Further Study" section at the end of this part for more details.

Another approach to tune nonlinear in the parameter approximators is to simply apply the gradient methods of Chapter 11 directly to the tuning of all the parameters of the approximator. Clearly, this will be possible so long as we can define the gradient of the approximator with respect to its parameters (and for the approximators we defined in this book, we can always do this). We can proceed in this case to tune parameters based on the acquisition of only one data pair (as we did in the last two sections) or many data pairs.

In Section 9.4.5 we overviewed several heuristic methods to use attentive systems to adjust the parameters of approximators that enter in a nonlinear fashion. Basically, these methods seek to allocate approximator structure to the region of interest (e.g., where the system is being regulated to) by tuning the parameters that enter in a nonlinear fashion. Approaches based on adaptively tuning approximator structures that are typically used in clustering applications can also be used.

Finally, note that ideas on approximator structure tuning could be integrated into the adaptive control methods. We leave such an endeavor to the interested reader. Nongradient methods (e.g., foraging or genetic) could be used to tune structure, or perhaps ideas from Section 9.4.6 and Design Problem 11.2. Principles of tuning structure of approximators are not very well established, or frequently used in applications, even though they may be quite useful.

12.9 Exercises and Design Problems

Exercise 12.1 (Optimization-Based Adaptive Control Strategies): Define other adaptive control strategies that use the online optimization viewpoint of learning models or controllers. Include block diagrams and detailed explanations of how the controllers could operate. Include in the discussion methods for hierarchical adaptive control (e.g., via supervisory methods), adaptive model predictive control, and multiple model adaptive control.

Exercise 12.2 (Stable Adaptive Fuzzy/Neural Control: Derivations): In this problem you will derive stable adaptive laws under different assumptions from the ones considered in the chapter.

(a) Suppose that we know that

$$(\beta_k(t) + \beta(x)) \leq \bar{\beta}_0$$

for some $\bar{\beta}_0 < 0$. Specify an indirect adaptive controller that provides the same stability and boundedness properties as the case where we used $\beta_0 > 0$ (assuming all the other assumptions hold).

(b) Repeat (a) but for the direct adaptive control case assuming that

$$\bar{\beta}_1 \leq \beta(x) \leq \bar{\beta}_0$$

for some $\bar{\beta}_1$ and $\bar{\beta}_0$ such that $\bar{\beta}_1 \leq \bar{\beta}_0 < 0$.

(c) Suppose that for the continuous time indirect adaptive control case, you used adaptation laws

$$\dot{\theta}_\alpha(t) = -Q_\alpha^{-1}\phi_\alpha(x)e_s \tag{12.96}$$
$$\dot{\theta}_\beta(t) = -Q_\beta^{-1}\phi_\beta(x)e_s u_{ce} \tag{12.97}$$

where Q_α and Q_β are positive definite and diagonal matrices (hence, there is a different adaptation gain for each component of the parameter vector). Show that in this case we can also obtain a stable indirect adaptive controller. Hint: Begin with a different choice for the Lyapunov function.

(d) Suppose that for the continuous time direct adaptive control case, you used adaptation law

$$\dot{\theta}_u(t) = Q_u^{-1}\phi_u(x,\nu)e_s(t) \tag{12.98}$$

where Q_u is a positive definite and diagonal matrix (hence, there is a different adaptation gain for each component of the parameter vector). Show that in this case we can also obtain a stable direct adaptive controller. Hint: Begin with a different choice for the Lyapunov function.

Design Problem 12.1 (Optimization-Based Adaptive Fuzzy/Neural Control): In this problem you will study the development of indirect and direct neural controllers for the process control problem studied in Sections 12.4 and 12.6.

(a) Develop an indirect neural controller for the process control problem. You may use the same controller developed in the chapter, or you may use a different approximator structure (e.g., the multilayer perceptron). Regardless of which approach you use, verify the operation of your controller in the same manner as was done in Section 12.4. Study the effect of the choice of the reference input on the ability of the approximator mappings to match the underlying unknown nonlinearities. Provide plots to illustrate the quality of the matching as was done in the chapter.

(b) Develop a direct neural controller for the process control problem. You may use the same controller developed in the chapter, or you may use a different approximator structure (e.g., the multilayer perceptron). Regardless of which approach you use, verify the operation of your controller in the same manner as was done in Section 12.6. Study the effect of the choice of the reference input on the ability of the approximator mapping to match the "ideal" controller nonlinearity discussed in the chapter. Provide plots to illustrate the quality of the matching as was done in the chapter.

(c) Compare the performance of the indirect and direct methods and discuss.

(d) Optional: For fun, repeat (a)-(c), but for the case where a Takagi-Sugeno fuzzy system is used for the approximator structures.

(e) Develop and evaluate a multiple model adaptive control strategy for this problem.

Design Problem 12.2 (Stable Adaptive Fuzzy/Neural Control for Balancing a Ball on a Beam): In this problem you will develop stable (continuous time) indirect and direct adaptive controllers for the ball on a beam problem described in Section 12.7.1. Note that you need a result from Exercise 12.2 to solve this problem. Use $x(0) = [1, 0]^\top$ (one unit corresponds to 0.75 inches on the beam), and $y_m(t) = 3$ and $\dot{y}_m(t) = 0$ for all $t \geq 0$. Take the parameters $\bar{a}$, $\bar{b}$, $\bar{c}$, and $\bar{d}$ as constant nominal values that you do not know. Also, assume that you know that there is a constant but unknown gain that multiplies the input u (i.e., assume that you know it is not a nonlinear function of x); however, suppose that you do not know that the particular nonlinearities are of the form indicated in Section 12.7.1, or that the parameters appear as they do (explain, however, in the development of the controllers below what you do need to assume about the plant).

(a) Find a $\beta_0 < 0$ and $\beta_1 \leq \beta_0$ such that $\beta_0 \leq \beta(x) \leq \beta_1$. Define the signals e_s, $\bar{e}_s$, and ν. What are the sliding mode terms u_{si} and u_{sd}? With the above assumptions, what is a reasonable choice for W_β (noting that you typically want to choose its value as small as possible) for the indirect adaptive controller? For the direct adaptive controller, what is a good choice for $B(x)$?

(b) What signals need to be measured so that they can be used as inputs to the direct and indirect adaptive controllers? Is it the same set of inputs to both controllers?

(c) Develop a stable indirect adaptive controller for the ball on a beam problem and illustrate its performance in simulation. Plot the plant output, tracking error, and plant input u to illustrate the performance of the system. Illustrate the effects of the choice of the adaptation gains and other design parameters, and the choice of the approximator structures.

(d) Develop a stable direct adaptive controller for the ball on a beam problem and illustrate its performance in simulation. Plot the plant output, tracking error, and plant input u to illustrate the performance of the system. Illustrate the effects of the choice of the adaptation gains and other design parameters, and the choice of the approximator structure.

Design Problem 12.3 (Stable Adaptive Control for Aircraft Wing Rock): Aircraft wing rock is a limit cycling oscillation in the aircraft roll angle ϕ and roll rate $\dot{\phi}$. A model of this phenomenon is given by

$$\ddot{\phi} = a_1 + a_2\phi + a_3\dot{\phi} + a_4|\phi|\dot{\phi} + a_5|\dot{\phi}|\dot{\phi} + b\delta_A$$

where the constants a_i, $i = 1, 2, 3, 4, 5$ are constant but unknown. If we model first-order actuator dynamics of the aileron (the control surface at the outer part of the wing), the model is

$$\begin{aligned}
\dot{\phi} &= p \\
\dot{p} &= a_1 + a_2\phi + a_3 p + a_4|\phi|p + a_5|p|p + b\delta_A \\
\dot{\delta}_A &= -\frac{1}{\tau}\delta_A + \frac{1}{\tau}u \\
y &= \phi
\end{aligned}$$

where δ_A is the aileron deflection angle, u is the control input, τ is the aileron time constant, and b is an unknown but constant parameter (that we know the sign of). For an angle of attack of 30 degrees, $a_1 = 0$, $a_2 = -26.67$, $a_3 = 0.76485$, $a_4 = -2.9225$, $a_5 = 0$, and $b = 1.5$, and $\tau = \frac{1}{15}$. Take these as constant nominal values that you do not know. Suppose, however, that you know that $b \in [1, 2]$ and $\tau \in [\frac{1}{20}, \frac{1}{10}]$. Also, assume that you know that there is a constant but unknown gain that multiplies the input u (i.e., assume that you know it is not a nonlinear function of x); however, suppose that you do not know that the particular nonlinearities are of the form indicated above, or that the parameters appear as they do (explain, however, what you do need to assume about the plant dynamics in the development of the controllers below). Suppose that you want the output $y(t)$ to track the reference signal $y_m(t)$ where

$$(s+10)(s^2+4s+24.25)Y_m(s)=0$$

Assume that you have sensors to measure y, $\dot{y}$, and $\ddot{y}$ (for your simulation, you can use the above model to simulate the sensing of all these values, and in particular, use $\ddot{y}=\dot{p}=a_1+a_2\phi+a_3p+a_4|\phi|p+a_5|p|p+b\delta_A$ to simulate the sensing of $\ddot{y}$). Use $x(0)=[0.4,0,0]^\top$, and $y_m(0)=0.5$, with the higher order derivatives of y_m initialized at zero.

(a) Choose the state vector $x=[x_1,x_2,x_3]^\top$ with $x_1=\phi$, $x_2=p=\dot{\phi}$, and $x_3=\delta_A$. Define f and g for Equation (12.32). Show that the relative degree $d=n=3$. Assume that $\alpha_k=\beta_k=0$ and find the form of the plant in Equation (12.35). Find a $\beta_0>0$ and $\beta_1\geq\beta_0$ such that $\beta_0\leq\beta(x)\leq\beta_1$.

(b) What signals need to be measured so that they can be used as inputs to the direct and indirect adaptive controllers? Is it the same set of inputs to both controllers?

(c) Develop a stable (continuous time) indirect adaptive controller for the wing rock problem and illustrate its performance in simulation. Plot the plant output, reference signal y_m, tracking error, and plant input u to illustrate the performance of the system. Illustrate the effects of the choice of the adaptation gains and other design parameters, and the choice of the approximator structures.

(d) Develop a stable (continuous time) direct adaptive controller for the wing rock problem and illustrate its performance in simulation. Plot the plant output, reference signal y_m, tracking error, and plant input u to illustrate the performance of the system. Illustrate the effects of the choice of the adaptation gains and other design parameters, and the choice of the approximator structure.

Chapter 13

For Further Study

To deepen your understanding of the methods of this part, first you can study optimization theory, as this forms the basis of all the methods. Two good books on optimization that, among others, have influenced the development here are [68], and the earlier book [337]. To learn more about neural networks and their training, see [130, 238]. The reader wishing to strengthen her or his background in conventional adaptive control should consult [254] (or the earlier books [448, 376, 219, 30]). For an in-depth treatment of stable adaptive estimation and control using fuzzy and neural systems, see [484].

Cognitive Neuroscience of Learning: The descriptions of classical and operant conditioning were based on [152, 223, 268, 269]. The description of the conditioned learning mechanisms in the *Aplysia* was taken from [267, 223, 269]; for more relevant literature in this area, see [130]. It is interesting to note that habituation can occur in microorganisms (e.g., *Vorticella* and nematodes), and it seems that learning of simple behavioral rules can occur in flatworms [161]. The discussion here on Hebbian learning is based on [206, 130]. For more details on modeling and analysis of learning processes from the field of theoretical neuroscience, see [130] and the references therein. Of particular relevance to this book is their coverage of the modified Rescorla-Wagner model used at the neural level for representing classical conditioning, the discussions on modeling of Hebbian learning and its connections with both deterministic and stochastic gradient methods, and the "tuning curves" (e.g., see pp. 14–17) and their connection to function approximation (see pp. 316–321) by viewing them as basis functions. The Rescorla-Wagner model studied in psychology and related mathematical and computer representations of the learning process are studied in [152] (pp. 109–119) and the references therein.

An early study growing from the field of cybernetics is given in [26] where the author seeks to explain the origins of adaptive behavior (learning). While research has often focused on organisms with a neural network when studying learning, there have been studies of microorganisms that can demonstrate behavioral plasticity via training [208].

Function Approximation and System Identification: It is helpful if you study the theory of system identification and for this, it is recommended that you see [331]. It could be helpful to study approximation and regularization theory, and one window into the mathematical literature you may want to consider is [422]. The section on approximation theory, and in particular, the section on whether to use linear or nonlinear in the parameter approximators, used the ideas in [45]. An introduction to the topical area of this part is given in [469, 265], where the authors also cover wavelets, and other approximators and properties in some detail. Wavelet methods for nonlinear identification are studied in [175].

Neural Control: A method that has been popular in the control of robots is the cerebellar model articulation controller (CMAC), which was first introduced in [9], and later applied in a different form in [362, 286]. An early paper on neural networks for control is given in [378]. A very nice introduction to learning control is given in [176]. Although developed independently, the FMRLC approach discussed in Section 9.4, is related to the neural control method in [286]. A nice overview of neural control methods is given in [171] and in [252, 361]. For a method that also adapts the structure of the neural network, see [322]. A related topic is that of "neural dynamic optimization for control," where optimal control laws are approximated [460].

Adaptive Fuzzy Control: The FMRLC was introduced in [300, 301] and uses ideas from the linguistic self-organizing controller (SOC) presented in [429] (with applications in [451, 507, 255, 121, 120, 119, 547]) and ideas in conventional model reference adaptive control. The ship steering application was developed from the work in [30, 376, 301, 412] and other applications of the FMRLC are studied in [555, 297, 370, 302, 300, 560, 303]. Other methods and relevant work are contained in [155, 266, 80, 530, 46, 505, 466, 435, 222, 221, 117, 118, 49, 80, 528, 80, 320], but note that there are many other methods that have been developed and reported in the literature.

Expert, Planning, and Attentive Systems in Adaptive Control: In addition to [193], the authors in [521, 395, 132, 516, 373] study fuzzy supervisory controllers that tune conventional controllers, especially ones that tune PID controllers (there are many conventional PID auto-tuning methods [28, 311]). Conventional gain scheduling has been studied extensively in the literature, especially for a wide range of practical applications. See [461, 443, 462] for some theoretical studies of gain scheduling. The connections between fuzzy supervision and gain scheduling have been highlighted by several researchers. A more detailed mathematical study of the connections is provided in [399]. The idea of using a supervisor for conventional adaptive controllers was studied earlier in [27, 21]. A case study for supervisory control of a two-link flexible robot was presented in [371]. The approach to supervision there bears some similarity to the one in [319]. A case study for a fault-tolerant aircraft control problem, where a rule based system supervises an adaptation mechanism to achieve performance adaptive control, is given in [297]. General issues in hierarchical fuzzy control are discussed in [133].

A survey of model predictive control is given in [192]. Adaptive fuzzy model predictive control is studied for a process control problem in [389] and there have been other similar studies for fuzzy model predictive control on which some of this work was based [237, 236]. The section on dynamically focused learning, an attentive mechanism for adaptive fuzzy systems, is based on [296].

Next, note that multisensor integration [339] and a variety of applications [11] utilize a concept called "world modeling" where a model of the environment is built while the system operates and information from the model is used in decision-making. While there are some relationships between systems that exploit a world model, and those in adaptive model predictive control, general world modeling is an important topic in its own right as it represents a very general philosophy on model building.

Finally, note that the area of learning automata is relevant to the topics studied in this part (e.g., in modeling learning systems and analysis of stochastic learning systems). For an introduction to that area, see [380]. The area of learning Bayesian networks from data is covered in [383].

Linear Least Squares: There are many methods to train neural networks and fuzzy systems. There are many books on neural networks (see, e.g., [238, 262]). For other methods to train fuzzy systems, consider [412], [262], [530, 531], or [242, 37].

The idea of using least squares to train fuzzy systems was first introduced in [504] and later studied in [532] and other places. Numerical issues for least squares methods are discussed in [331, 103, 332] and model validity is studied in [70]. The controller construction problem where process operator data is used was taken from [498], as was the CO_2 estimation problem for the gas furnace studied in an exercise at the end of Chapter 10. Issues in how to determine which inputs to use for an estimator are discussed in [331, 104, 260, 498, 262].

Gradient Methods: If you are interested in connections between gradient methods and learning in neuroscience, the first area to study is modeling Hebbian learning [241], specifically when it is modeled as a gradient method. For this, you can study [130, 238] and the references therein.

For more details on gradient methods, see [337, 68]. The brief discussion on the stochastic gradient method is based on [69]. For more background on stochastic optimization, see [439], where the "stochastic approximation" method was introduced, and [293] (the classical backpropagation method is a stochastic gradient approach, since it uses a steepest descent gradient approach and random presentation of data from the training data set).

The hybrid methods (e.g., methods that may use one optimization method for the nonlinear part of the approximator structure and another for the linear part) have been used by a variety of researchers; a particularly nice set of applications were studied in [262, 259, 261]. Some clustering methods are overviewed in [43]. A method that combines an online clustering and least squares method is given in [102]. A variety of clustering methods are discussed in [147], where the authors also focus on construction of local models that are useful for the development of control systems.

Some other methods related to the topics in this part are given in [251, 48, 545, 306, 1, 253, 378, 80].

Stable Adaptive Fuzzy/Neural Control: Here, the treatment was only meant to introduce the topic of stable adaptive fuzzy/neural control. A recent text that covers the full details of many stable adaptive neural/fuzzy methods is [484]. There, more general direct and indirect adaptive control methods are introduced, the output feedback and multivariable cases are discussed in detail, many examples and applications/implementations are provided, and discrete-time and decentralized adaptive control are covered. It seems that the field of stable neural control started with [424] and has been significantly affected by the work in [423, 424, 441, 172, 174, 167, 321, 552, 316, 379, 447, 101, 546, 445, 425, 105], where the authors make use of neural networks as approximators of nonlinear functions. In [497, 248, 529, 304, 99, 530, 486], the authors use fuzzy systems for the same purpose and [441, 379] use dynamical neural networks. The neural and fuzzy approaches are most of the time equivalent, differing between each other mainly in the structure of the approximator chosen. Indeed, to try to bridge the gap between the neural and fuzzy approaches, several researchers (e.g., in [486]) introduce adaptive schemes using a class of parameterized functions that include both neural networks and fuzzy systems. Linear in the parameter approximators are used in much of the above-referenced work, and for example, [497, 248, 423, 424, 92, 172, 167, 99, 447, 445, 486] and nonlinear in, for example, [321, 552, 316, 379, 101, 546, 425]. Note that most of the papers deal with indirect adaptive control, whereas very few authors use the direct adaptive control approach (see, however, [486, 442]). Research on decentralized adaptive neural/fuzzy control is given in [487] and for the MIMO case (both direct and indirect) in [396]. Persistency of excitation issues are studied in [173, 172]. An interesting study on issues related to the use of local (finite support) approximators in adaptive control can be found in [170]. Implementation studies for adaptive neural fuzzy control are given in [397]. For more information on multiple model adaptive control, see [31, 347, 333, 377, 375] and the references therein.

The aircraft wing rock model used in the design problem in Section 12.7 was taken from [382, 165], which is based on wind tunnel data as studied in [309]. The aircraft wing rock design problem at the end of Section 12.7 was taken from [289] and is based on [247].

Approximator Structure Learning: For an overview of methods for automatically constructing or pruning neural networks, see [295, 431] and for discussion on some methods to adjust structure of fuzzy systems, see [412]. The issue of structure choice is treated in [103] for linear in the parameter approximators where you want to choose the number of basis functions. There, least squares methods are used to eliminate basis functions that do not significantly contribute to making the approximation accurate. Such methods could be employed in another method to initialize the nonlinear portion of the approximator [331] where you simply pick a very large approximator structure, then eliminate the parts that do not contribute in a significant way. Some work in the direction of trying to tune structure of an approximator in an adaptive control system is contained in [198] (but also see the references there).

Immune Networks: While not discussed in this chapter, there has been recent interest in biomimicry of immune systems and subsequent engineering applications (e.g., in pattern recognition and control) [124]. For more recent work, see [125]. Some in fact think of the immune system as a second type of "brain" in the human body, with the ability to learn (e.g., it encounters a type of pathogen, then "learns" how to more easily recognize it the next time) and make decisions (e.g., how and when to attack foreign invaders). Immune networks are models of immune systems and some such networks are "connectionist" in similar ways to how neural networks are [168], and some types of immune networks have underlying mechanisms that are sometimes thought of as being similar to evolutionary optimization (e.g., the genetic algorithm) [182], since their learning strategy can be viewed as a type of nongradient stochastic optimization process. There have been several studies of underlying mechanisms for learning in immune networks [169, 417, 66, 63, 67, 249, 250, 65, 471] and the application of these ideas to control [63, 64] has been considered.

A recent study [146] focuses on tuning approximator structure and parameters and these ideas are more firmly connected with the ideas and methods of this part.

Temporal Difference Learning and Neuro-Dynamic Programming: An introduction to the area of reinforcement learning, and in particular "temporal difference learning" is given in [499], and connections to neuroscience are discussed. Related methods and analysis of temporal difference learning are studied in the area of "neuro-dynamic programming" [69], where the authors also study the application to a number of multi-stage decision-making system problems. In neuro-dynamic programming, an approximator, such as a neural network, is used to approximate the "optimal cost to go" in the dynamic programming methodology, and then it is used to make choices of decision variables. Many other references are available in this general area, so you should search the current literature.

Part IV

EVOLUTION

Part Contents

Sequence of Essential Concepts

- Evolution is a type of search that continually and incrementally redesigns the structure and parameters of organisms to maximize organism fitness for survival in an uncertain environment. To do this, it tends to optimize the design of the organism for typical characteristics of its environment (i.e., some type of average environment it encounters) and thereby produce an organism that is "robust" for survival in its habitat. (Extinction processes are due to insufficient adaptation rate or "traps" in the search space due to coupled constraints like physiology and environment.) Engineering design is analogous to evolution, but for technological products and systems.

- Genetic algorithms simulate evolution and hence, can serve as a general tool for parallel stochastic nongradient based optimization. There are, however, many closely related deterministic and stochastic conventional approaches (i.e., not biologically motivated), including response surface, pattern search, simplex, and stochastic optimization methods. These provide insights into the operation of biologically motivated optimization methods, such as the genetic algorithm (or the foraging algorithms of Part V). Moreover, they provide practical approaches for solving engineering problems that involve robust optimal design.

- Evolution is best viewed as a type of global optimization process ("global" in time and population space) that can act on all aspects of the organism, including its ability to perform learning, which can be viewed as a "local" adaptive search. The environment is the fundamental driver of this optimization process. Learning is "local" in time, since it applies to a single generation and local in space, since it occurs in a single individual (but of course, culture has more global influences on learning in groups of organisms). Learning can accelerate evolution (the "Baldwin effect") and evolution can shape learning (it can design every aspect of the learning system). Evolution may produce an optimal balance between instincts and learning capabilities that is dependent on characteristics of the environment (e.g., the stochastic nature of the environment) and organism. These ideas provide some principles in the design of robust optimal complex decision-making systems.

- Genetic algorithms are optimization processes that can be employed to tune approximators in closed-loop systems and hence, can achieve real-time adaptive control. Direct and indirect adaptive controllers and adaptive model predictive controllers can be designed using genetic algorithms.

Chapter 14

The Genetic Algorithm

Chapter Contents

Darwin pioneered the idea that biological organisms develop and adapt over long periods of time via "descent with modification." For example, organism "parents," each with their own genetic makeup, mate, and their children's genetic makeup is a mixture of their parents so that often in appearance, you see characteristics of both parents (ideas originally studied by Mendel for varieties of garden peas). Sometimes, there are molecular "mutations" where an abnormal gene arises, which then affects the formation of the child. Both the mating and the mutations result in children growing up to be more or less fit to survive and mate in the environment that they live in. Children who are more fit tend to have more offspring, while children who are less fit often do not get the chance to mate, or have fewer offspring. There is then a "natural selection" that proceeds gradually over time so that populations evolve to be more fit for their environment.

The genetic algorithm (GA) is a computer simulation that incorporates ideas from Darwin's theory on natural selection, and Mendel's work in genetics on inheritance, and it tries to simulate natural evolution of biological systems. From an engineering perspective, the genetic algorithm is an optimization technique that evaluates more than one area of the search space and can discover more than one solution to a problem. (Some would call it a type of stochastic direct search method.) In particular, it provides a stochastic optimization method where, if it "gets stuck" at a local optimum, it tries, via multiple search points, to simultaneously find other parts of the search space and "jump out" of the local optimum to a global one that represents the highest fitness individuals.

Evolution is the theory and mechanism that is ubiquitous and fundamental to all of biology (bacteria, plants, and animals are all subject to the mechanisms of evolution). One would expect it to have a similar pervasive role in all of intelligent control. As discussed in this part, it applies to evolution of neural, fuzzy, expert, planning, attentional, and learning systems.

14.1 Biological Evolution

The basic process of biological evolution was explained in Section 2.5 in Part I on page 80 and it would be good to review that material before proceeding. Here, however, we also give a brief overview of evolution in biological systems so that you can easily form appropriate analogies to biological systems as you learn about genetic algorithms. At the basis of evolution lies selection, mating, and mutation, and each of these is outlined next.

"Survival of the fittest" refers to fitness in terms of reproductive success. Natural selection is the process where organisms with higher reproductive success generate offspring, and hence, propagate their DNA through time. Less fit individuals do not have offspring (or have fewer of them) and hence can become extinct over time. For instance, consider the cormorant (large fish-eating seabirds that catch their prey under water). There are more than one species of this bird. The one found on the coast of South America has sufficient wingspan to fly, but the only member of its family that does not fly is the "flightless" cormorant that is found on the remote Galapagos Islands (part of Ecuador), which are quite far from mainland South America in the Pacific Ocean. There are no natural predators on the islands and a plentiful supply of fish immediately offshore. Its loss of flight does not seem to have harmed them, and in fact can be viewed as beneficial since then it does not have to use the extremely energy-expensive activity for obtaining food. Apparently, there was a selective pressure on wing length that drove the evolutionary history of the flightless cormorant. This example illustrates the powerful force that selection provides in "designing" organisms.

It is often useful to think of the environment as the designer of the organism.

Mating and reproduction drive evolution. Mating is a process of mixing (combining) chromosomes of the parents, and it tends to "homogenize" the gene pool of the population. Different parts of chromosomes from each parent are combined in a child. It is via this swapping of genetic material on the chromosomes that a child inherits some characteristics of each parent (and thereby, as you would imagine, characteristics of each grandparent). On *average* we may think of each individual as being composed of half of one parent and half of the other, one fourth of each grandparent, and so on. This swapping of genetic material, and hence blending of outward characteristics, is apparent in many living organisms that mate. For example, for some types of corn, pollination from a dark colored corn stalk to a corn stalk with a "white" gene bears an ear of corn that is half dark and half white.

Mutations result in variations in the offspring that result from mating. The mutation rate is dictated by the probability of error in gene replication in biological systems (and certain "mutagens," whose source can lie in the environment, can affect this rate). While we typically think of mutations as something undesirable in biological systems (e.g., some may think of "mutants" in science-fiction movies), they can also lead to more fit individuals (i.e., the mutation may represent a jump to a region of the space where the reproductive fitness increases significantly).

14.2 Representing the Population of Individuals

The "fitness function" measures the fitness of an individual to survive, mate, and produce offspring in a population of individuals for a given environment. The genetic algorithm will seek to *maximize* the fitness function $\bar{J}(\theta)$ by selecting the individuals that we represent with θ (note that we place the bar over the cost function to emphasize that we seek to maximize this function, where we always sought to minimize "J" in the studies on optimization for approximation in Part III).

14.2.1 Strings, Chromosomes, Genes, Alleles, and Encodings

To represent the genetic algorithm in a computer, we make θ a string. A string represents a chromosome in a biological system and one is shown in Figure 14.1.

A chromosome is a string of "genes" that can take on different "alleles." In a computer, we often use number systems to encode alleles. Here, we adopt the convention that a gene is a "digit location" that can take on different values from a number system (i.e., different types of alleles).

We encode the parameters of the optimization problem on the chromosome, which can simply be a sequence of base-10 numbers.

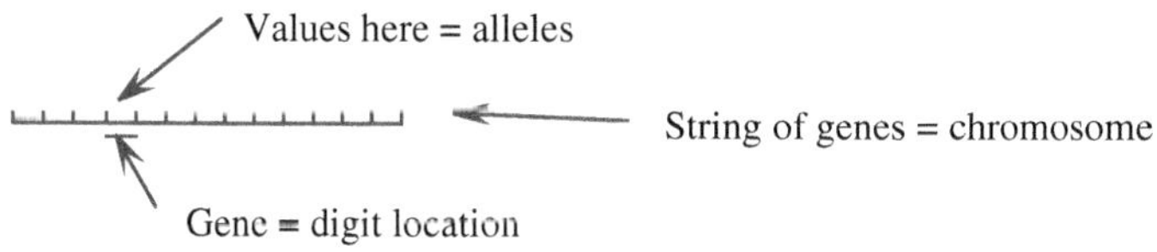

Figure 14.1: String for representing an individual.

In a base-2 number system, alleles come from the set $\{0, 1\}$, while in a base-10 number system, alleles come from the set $\{0, 1, 2, 3, 4, 5, 6, 7, 8, 9\}$. Hence, a binary chromosome has zeros or ones in its gene locations. As an example, consider the binary chromosome

$$1011110001010$$

which is a binary string of length 13. If we are seeking to optimize parameters of a system that come in a base-10 number system, then we will need to encode the numbers into the binary number system (using the standard method for the conversion of base-10 numbers to base-2 numbers). We will also need to decode the binary strings into base-10 numbers to use them. Here, we will develop the genetic algorithm for base-2 or base-10 number systems, but we will favor the use of the base-10 representation in our examples, since encoding and decoding is simple in that case (and it can be computationally expensive for online, real-time applications if the proper representation is not used).

As an example of a base-10 chromosome, consider

$$8219345127066$$

which has 13 gene positions. For such chromosomes we add a gene for the sign of the number (either "+" or "−") and fix a position for the decimal point. For instance, for the above chromosome we could have

$$+821934.5127066$$

where there is no need to carry along the decimal point; the computer will just have to remember its position. Note that we could also use a floating point representation where we could code numbers in a fixed-length string plus the number in the exponent. The ideas developed here work just as readily for this number representation system as for standard base-2 or base-10.

14.2.2 Encoding Examples

It is possible to encode many different problems so that artificial evolution can be applied. We discuss a few representations here, using a base-10 number system, that are particularly relevant to the field of intelligent control.

Proportional-Integral-Derivative Controllers

We can encode a wide variety of structures (e.g., controllers) into a form that the genetic algorithm can operate on; hence, the genetic algorithm is a very general optimization tool (and this creates potential for its misuse).

Suppose that you want to evolve a proportional-integral-derivative (PID) controller (e.g., using a fitness function that quantifies closed-loop performance and is evaluated by repeated simulations). In this case, suppose that we have three gains, K_p, K_i, and K_d, and that at some time we have

$$K_p = +5.12,\ K_i = 0.1,\ K_d = -2.137$$

then we would represent this in a chromosome as

$$+051200 + 001000 - 021370$$

which is a concatenation of the digits, where we assume that there are six digits for the representation of each parameter (two before the decimal point and four after it) plus the sign digit (this is why you see the extra padding of zeros). The computer will have to keep track of where the decimal points are. We see that each chromosome will have a certain structure (its "genotype" in biological terms, and the entire genetic structure is referred to as the "genome"). Each chromosome represents a point in the search space of the genetic algorithm. Here, we will use the term "phenotype" from biology to refer to the whole structure of the controller that is to be evolved; hence, in this case the phenotype is

$$K_p e(t) + K_i \int_0^t e(\tau)d\tau + K_d \frac{de(t)}{dt}$$

where $e = r - y$ is the error input to the PID controller, r is the reference input, and y is the output of the plant.

Clearly, a similar approach can be used to encode lead-lag compensators, state feedback controllers, nonlinear controllers, adaptive controllers, and so on.

Elements of Decision Making

In this case, you simply concatenate the parameters of the fuzzy or neural system that you would like to evolve. For instance, for a Takagi-Sugeno fuzzy system you may simply encode the consequent parameters, or you may want to also encode the membership function parameters (centers and spreads). Similarly, for a neural network, you could encode all the weights and biases. Hence, in this case we have θ defined similar to the case where we adjust it with least squares or gradient methods (except here, θ is often treated as a string of concatenated parameters, rather than a vector of parameters).

Another interesting, and useful, possibility to consider is the case when you encode the "structure" of the fuzzy or neural system. For instance, you may encode the number of rules, different forms of consequent functions (e.g., different nonlinearities), different membership functions (e.g., both Gaussian and triangular), different inference methods (e.g., use of product or minimum in premise quantification with fuzzy logic), or different defuzzification methods (e.g., center-of-gravity and center-average). In a neural network you could encode the number of layers, number of neurons in each layer, existence of connections, methods to combine the inputs to the activation function (e.g., linear, polynomial functions of the inputs, other nonlinear functions), and activation function types (e.g., logistic, hyperbolic tangent, linear). Sometimes you may want to encode certain characteristics of both fuzzy and neural systems (e.g., the number of rules and neurons), and then allow them to coexist within a population.

In these cases, you can evolve the structure of the fuzzy or neural system. The genetic algorithm should be viewed as a very general optimization tool that can be used to adapt both parameters and structure of intelligent systems (of course, structure can be quantified with parameters and this is often done in practice). Indeed, many think of the genetic algorithm as having its most natural role in the adjustment (or design) of structure of systems rather than parameter tuning, which is often associated with learning over a lifetime. Next, we discuss some very general ways to encode higher-level cognitive functions.

Just like the fuzzy system, the number of rules used by an expert controller can be encoded. Moreover, all the other functional components can be encoded, such as the conflict resolution strategies. For planning systems, we could encode the look-ahead horizon length, the models used for projecting ahead, and so on. For attentional systems, we can make similar general encodings (e.g., on response times of refocusing, or intensity levels needed to evoke a response).

14.2.3 The Population of Individuals

A population is a set of candidate solutions (chromosomes).

Next, we develop a notation for representing a whole set of individuals (i.e., a population). Let k denote the generation number. Let $\theta_i^j(k)$ be a single parameter at time k (a fixed-length string with sign digit). Here, we number the chromosomes and the superscript j refers to the jth chromosome. Also, we number the "traits" on each chromosome. (Note that strictly speaking, a trait in a biological system is most often thought of as a property of the phenotype, like hair or eye color, while in our systems, as in the PID controller example above, genes, which are digits of parameters, are "expressed" as traits of the phenotype; hence, we think of strings of genes as being expressed as traits in the phenotype.) With this, the i subscript on $\theta_i^j(k)$ refers to the ith trait on the jth chromosome. Suppose that chromosome j is composed of p of these parameters (traits).

Let

$$\theta^j(k) = \left[\theta_1^j(k), \theta_2^j(k), \ldots, \theta_p^j(k)\right]^\top$$

be the jth chromosome. Note that earlier we had concatenated elements in a string, while here we simply take the concatenated elements and form a vector from them. We do this simply because this is probably the way that you will want to code the algorithm in the computer (e.g., in Matlab). We will at times, however, still let θ^j be a concatenated string when it is convenient to do so. It will be clear from the context which form of representation we are using.

The population of individuals at time k is given by

$$P(k) = \left\{\theta^j(k) : j = 1, 2, \ldots, S\right\} \tag{14.1}$$

(not to be confused with the covariance matrix used in recursive least squares) and the number of individuals in the population is given by S. We want to pick S to be big enough so that the population elements can cover the search space. However, we do not want S to be too big, since this increases the number of computations we have to perform. In an optimization example below, we will discuss some issues in the choice of the size of S.

Note that while we will not do so here, you could allow the size of the population to vary with time (more like it does in nature) and the size of the population could depend on resources in the environment, competition between individuals (e.g., as measured by fitness), and physical constraints. Population size plays a fundamental role in nature. Nature often exploits the use of very high numbers of individuals and allows for many individuals to be killed so that population size may vary significantly (at least in some regions). Our simulations of evolution often do not exploit this due to a lack of computational resources.

14.3 Genetic Operations

The population $P(k)$ at time k is often referred to as the "generation" of individuals at time k. Basically, according to Darwin, the most qualified individuals survive to mate and produce offspring. We quantify "most qualified" via an individual's fitness $\bar{J}(\theta^j(k))$ at time k. For selection, we create a "mating pool" at time k, something every individual would like to get into, which we denote by

$$M(k) = \left\{m^j(k) : j = 1, 2, \ldots, S\right\} \tag{14.2}$$

The mating pool is the set of chromosomes that are selected for mating. Here, we perform selection to decide who gets in the mating pool, mate the individuals via crossover, then induce mutations. After mutation we get a modified mating pool at time k, $M(k)$. Below, we will outline the operations involved in creating the mating pool, performing mating for individuals in the mating pool, and subsequent mutations. This will explain how the $m^j(k)$ in $M(k)$ above are created and modified.

To form the next generation for the population, we let

$$P(k+1) = M(k)$$

Evolution occurs as we go from a generation at time k to the next generation at time $k+1$. Hence, in this artificial environment mating is done in parallel and is synchronized with a clock, which is far different from how it typically occurs in nature. Parallel asynchronous versions of the algorithm can, however, also be developed.

14.3.1 Selection

There are many ways to perform selection, but by far the most common one used in practice is fitness-proportionate selection.

Fitness-Proportionate Selection

In this case, we select an individual (the ith chromosome) for mating by letting each $m^j(k)$ be equal to $\theta^i(k) \in P(k)$ with probability

$$p_i = \frac{\bar{J}(\theta^i(k))}{\sum_{j=1}^{S} \bar{J}(\theta^j(k))} \tag{14.3}$$

To clarify the meaning of this formula and hence the selection strategy, you can use the analogy of spinning a unit circumference roulette wheel where the wheel is divided like a pie into S regions where the ith region is associated with the ith individual of $P(k)$. Each pie-shaped region has a portion of the circumference that is given by p_i in Equation (14.3). You spin the wheel, and if the pointer points at region i when the wheel stops, then you place θ^i into the mating pool $M(k)$. You spin the wheel S times so that S elements end up in the mating pool and the population size stays constant.

Selection dictates that the best points in the search space should be given preferential treatment in specifying the composition of the population at the next step.

Clearly, individuals who are more fit will end up with more copies in the mating pool; hence, chromosomes with larger-than-average fitness will embody a greater portion of the next generation. At the same time, due to the probabilistic nature of the selection process, it is possible that some relatively unfit individuals may end up in the mating pool $M(k)$.

Other Selection Strategies

There are many other options that have been considered for selection besides the fitness-proportionate approach above. For instance, sometimes the individuals in the population are ranked by order of fitness and a fixed number of the least fit ones are "killed" and only the ones in the remaining set are used in the selection process, perhaps with a fitness-proportionate method (this eliminates the possibility of very unfit individuals from mating). Other times, some subset of very fit individuals are allowed to get into the mating pool without spins of the roulette wheel. Such strategies are often called "elitist" strategies since the individuals who are most fit (the elite ones) are assured to be able to get into the mating pool. Sometimes there is only one elite individual that is allowed, and at other times you could allow more than one. Often, when such elitist strategies are used, the elite individual(s) are allowed to proceed directly to the next generation, without modification via the crossover and mutation operations that are discussed next. (In this sense, we can think of the elite individuals as having an ability to clone themselves so that their "offspring" are exact copies of themselves.)

Using Gradient Information Before or After Selection

There are many relationships between genetic algorithms and other optimization methods. For instance, there are many stochastic methods for optimization of nonlinear and nonconvex functions that bear similarities to the genetic algorithm (see the next chapter). The advantages and disadvantages of these methods relative to the genetic algorithm tend to be very application dependent, so we will not comment on relative merits of the methods and which to pick in a particular situation. There are, however, ways to use ideas from conventional gradient optimization methods in genetic algorithms and we briefly discuss this here.

Traditional gradient methods can be integrated into genetic algorithms.

Traditionally, it has been said that one of the key advantages of the genetic algorithm is that it does not rely on the existence and use of gradient information (as the gradient methods do). In some contexts, such as for estimation and control problems, however, it could be that there is useful gradient information available that you may not want to ignore. For instance, we know that in adjusting a nonlinear in the parameter approximator (that serves as an estimator or controller), it can be very difficult for a gradient method to find the global extremum value since it may get stuck at a local extremum. In such cases, it is possible that a genetic algorithm can help find the way out of such local extrema to find the global ones. Now, in such cases, for optimization algorithm design, you could start with a gradient method and add certain features from the genetic algorithm (e.g., evolving a population, use of random excursions by random re-initialization at various steps, etc.). In addition, for your algorithm design, you could think of the genetic algorithm as the main vehicle for optimization and interleave gradient updates. To accomplish this, you could, for instance, perform one or more gradient-based parameter update steps for every individual before (or after) selection is used to place individuals in the mating pool. In this way, it is hoped that we gain the benefits of using the directional information used in the gradient updates, and the benefits of parallel search and random excursions given by the genetic algorithm. See Section 15.5 at the end of the next chapter for more discussion on such "interleaved" and hybrid methods. See the "For Further Study" section at the end of the part for a reference that studies other approaches.

14.3.2 Reproduction Phase, Crossover

We think of crossover as mating in biological terms, which at a fundamental biological level involves the process of combining (mixing) chromosomes.

Crossover adds a mechanism for both local and global search, but near fit individuals.

For the computer simulation of evolution, the crossover operation operates on the mating pool $M(k)$ by "mating" different individuals there. First, you specify the "crossover probability" p_c (usually chosen to be near 1 since, when mating occurs in biological systems, genetic material is certainly swapped between the parents). There are many types of crossover (i.e., ways to swap genetic material on chromosomes), but the simplest one is "single-point" crossover.

Single-Point Crossover

The procedure for single-point crossover consists of the following steps:

1. Randomly pair off the individuals in the mating pool $M(k)$. There are many ways to do this. For instance, you could simply pick each individual from the mating pool and then randomly select a different individual for it to mate with. Or, you could just have all the individuals mate with the ones that are right next to each other (where "right next to each other" is defined by how you label the individuals with a number system). In this approach, if there are an odd number of individuals in $M(k)$, then, for instance, you could simply take the last individual and pair it off with another individual who has already been paired off (or you could pair it off with the individual with the highest fitness).

2. Consider the chromosome pair θ^j, θ^i that was formed in step 1. Generate a random number $r \in [0, 1]$.

   (a) If $r < p_c$, then cross over θ^j and θ^i. To cross over these chromosomes, select a "cross site" at random and exchange all the digits to the right of the cross site of one string with those of the other. This process is pictured in Figure 14.2. In this example, the cross site is position 5 on the string (be careful in how you count positions), and hence, we swap the last eight digits between the two strings. Clearly, the cross site is a random number that is greater than or equal to 1, and less than or equal to the number of digits in the string minus 1.

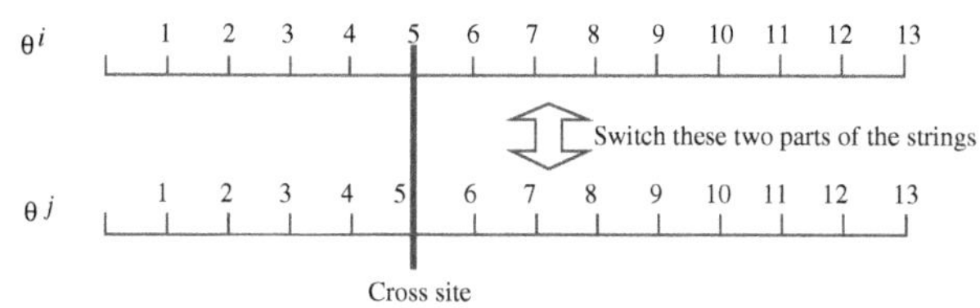

Figure 14.2: Crossover operation example.

   (b) If $r > p_c$, then we will not cross over; hence, we do not modify the strings, and we go to the mutation operation below.

3. Repeat step 2 for each pair of strings in the mating pool $M(k)$.

As an example, suppose that $S = 10$ and that in step 1 above, we randomly pair off the chromosomes. Suppose that θ^5 and θ^9 ($j = 5, i = 9$) are paired off where

$$\theta^5 = +2.9845$$

and

$$\theta^9 = +1.9322$$

Suppose that $p_c = 0.9$ and that when we randomly generate r, we get $r = 0.34$. Hence, by step 2 we will cross over θ^5 and θ^9. According to step 2, we randomly pick the cross site. Suppose that it is chosen to be position 3 on the string (include the sign as a position). In this case, the strings that are produced by crossover are

$$\theta^5 = +2.9322$$

and

$$\theta^9 = +1.9845$$

Besides the fact that crossover helps to model the mating part of the evolution process, why should the genetic algorithm perform crossover? Basically, the crossover operation perturbs the parameters near good positions to try to find better solutions to the optimization problem. It tends to help perform a localized search around the more fit individuals (since on average the individuals in the mating pool $M(k)$ at time k should be more fit than the ones in the population $P(k)$ at time k) that could be near each other on the fitness landscape. However, on a complex landscape, two relatively well-fit individuals may be on very different parts of the landscape, so that an offspring may lie "between" them (or extrapolated along a line connecting the two, but not too far away) at points that represent poor fitness. In this case, you would *not* think of crossover as producing local search. Indeed, in this situation it results in a global type of search.

Other Crossover Methods

There are many other crossover methods that have been studied. For instance, you could use a two-point crossover, where you pick two crossover points on each chromosome and swap the elements in between the two points. Or, more generally, you could have a multi-point crossover where you have one or more crossovers per trait. Generally, when elitism is used, the elite individuals would not undergo any type of crossover.

Another option is to make the crossover rate change with time. For instance, you could start with $p_c = 1$ and then reduce it as the overall fitness of the population (as measured by, for example, the average of the fitness values of all the individuals) increases. This way, there will be fewer explorations into close-by regions when we are likely to be near a local maxima. Sometimes, however, this can cause "premature convergence" where the algorithm locks on to some values and does not properly explore other parts of the space to find the global maximum. Sometimes, for online applications, especially when $\bar{J}$ is time-varying, you want to keep the crossover rate at $p_c = 1$ for the entire time, since this will ensure good exploration of the space. Note that you can think of the crossover probability as being under genetic control so that its value could be adapted also.

In other methods, similarity measures between individuals are developed and only similar individuals are allowed to mate and hence, cross over (you can then think of the population as having multiple species, with mating only within species). This may be a way to cope with having multiple types of structures (e.g., fuzzy and neural systems) within a population. Or, you could just allow individuals of similar fitness to mate, or you could pick the most fit individual and have everyone else mate with that one.

Sometimes, you may want to "spatially" restrict mating so that only those individuals who are "close" (e.g., with close defined in terms of the Euclidean distance between two individuals) are allowed to mate and swap genetic material (otherwise it is possible that two very different individuals mate). This would then restrict crossover to a local type search.

14.3.3 Reproduction Phase, Mutation

Like crossover, mutation modifies the mating pool (i.e., after selection has taken place). The operation of mutation is normally performed on the elements in the mating pool after crossover has been performed. The biological analog of our mutation operation is the random mutation of genetic material. Again, there are many ways to perform mutations. Below, we will discuss the most common methods.

Mutation provides a mechanism to jump out of local maxima and to randomly explore local and wide areas of the search space.

Gene Mutations

To perform mutation in the computer, first choose a mutation probability p_m. With probability p_m, change (mutate) each gene location on each chromosome randomly to a member of the number system being used. For instance, in a base-2 genetic algorithm, we could mutate

$$1010111$$

to

$$1011111$$

where the fourth bit was mutated to one. For a base-10 number system, you would simply pick a number at random to replace a digit, if you are going to mutate a digit location (normally we do not consider a replacement to be valid if we replace a digit with the same value).

Besides the fact that this helps to model mutation in a biological system, why should the genetic algorithm perform mutation? Basically, it provides random excursions into new parts of the search space. It is possible that we will get lucky and mutate to a good solution. It is the main mechanism (crossover can also help) that tries to make sure that we do not get stuck at a local maxima and that we seek to explore other areas of the search space to help find a global maximum for $\bar{J}(\theta)$. Usually, the mutation probability is chosen to be quite small (e.g., less than 0.01), since this will help guarantee that all the individuals in the mating pool are not mutated, so that any search progress that was made is lost (i.e., we keep it relatively low to avoid degradation to exhaustive search via a random walk in the search space).

Keep in mind that we can think of mutation as providing both a local and global search component to the genetic algorithm. If, for instance, the mutation of a gene at a particular location on the chromosome represents a small (large) magnitude change, then a random local (global, respectively) search behavior is exhibited.

Other Mutation Methods

Sometimes, you could mutate an entire trait (i.e., a set of genes). Other times, you may want to restrict mutation by only allowing certain genes, traits, or individuals to be mutated. There could be reasons to vary the mutation rate; indeed, in biological systems, the mutation rate is under genetic control (i.e., it could evolve to an optimal level to make sure that the population can properly adapt to its environment). As an example, in some applications you may want to start with a relatively high mutation rate and then decrease the mutation rate as the overall fitness of the population increases. Other times you may simply want to code the mutation rate in a chromosome and try to evolve it.

In applications where the fitness function is fixed in time, you often do not want to have a great reliance on mutation in generating new solutions, as it is by simple luck that mutation succeeds. However, when the fitness function changes with time (e.g., in biological "coevolution"), you may want a higher mutation rate to ensure that many options are considered. For instance, in online applications where the fitness function is time-varying, there is sometimes a need for an exceptionally high mutation rate to ensure that you do not at any point get stuck in a local maxima (since the actual maxima points can be changing) and that you actively pursue many different solution options so as to ensure active adaptation. You have to be careful, however, not to have the mutation rate too high or any search progress made by the algorithm at earlier stages can be destroyed. Also, typically in such online approaches, an "elitism" strategy is used to ensure that at least one good solution is available at all times (i.e., the elite individual is not subjected to any mutations). Even in situations where the fitness function is not time-varying, elitism has been used very effectively as a way to keep the best solution available while searching for others (this will be illustrated in Section 14.5 for an optimization problem).

14.4 Programming the Genetic Algorithm

In this section we briefly discuss how to code the genetic algorithm, issues you encounter in choosing the method to code it, memory and computation time requirements, and termination and initialization issues.

14.4.1 Pseudocode for a Simple Genetic Algorithm

To summarize the operations of the genetic algorithm, and provide some guidance on how to implement the algorithm in a computer, we provide some high-level pseudocode that could be useful in programming the genetic algorithm in any computer language. Here, we assume for convenience that we use fitness-proportionate selection, single-point crossover with probability p_c (below, `pc`), gene mutation with probability p_m (below, `pm`), and we terminate after some fixed number of iterations, N_{ga} (below, `Nga`).

1. Define the GA parameters (e.g., crossover and mutation probability, population size, termination parameters).
2. Define the initial population $P(0)$.
3. For $k = 1$ to N_{ga} (main loop for producing generations).

   Compute the fitness function for each individual.

   Selection: From $P(k)$, form $M(k)$, the mating pool at iteration k, using fitness-proportionate selection.

   Reproduction: For each individual in $M(k)$, select another individual in $M(k)$, mate the two via crossover with probability p_c, mutate each gene position with probability p_m. Take the S children produced by this process, and put the children in $P(k+1)$, the next generation.

4. Next k (i.e., return to step 3).
5. Provide results.

Clearly this only provides a high-level view of the operation of the genetic algorithm. The details of the various steps depend on how you design your particular genetic algorithm and on the programming language you use.

14.4.2 Alternative Sequencing of Operations

The above pseudocode shows one common way to implement the GA. There are many possible variations on this approach. First, you could use any of the options for fitness, crossover, and mutation listed earlier. Moreover, the very way that the steps are sequenced is sometimes different from shown above.

For instance, note that one common way to implement the GA is to use selection to choose two individuals, then when the two parents are used in mating, they are allowed to form *two* children via crossover, and these children are both subjected to mutation and kept in the next generation (if the population size S is odd, then one child is randomly removed). In this way, the genetic material of the two parents is not lost. Above, when a parent mates, it produces one child who will, in general, only have part of each parent's genetic material while the other part is lost (this is what generally happens in biological systems).

Next, note that with the above approach, there are two ways that we could end up with identical individuals in the mating pool at any iteration:

1. Due to the way that fitness-proportionate selection works, the same individual (e.g., the most fit one) can have more than one copy of itself in the mating pool.

2. Two individuals could have different ancestors, but just happen to end up with the same genetic makeup due to crossover and mutation operations (e.g., by random chance two individuals' genetic makeup could be the same).

With normal choices for representations and parameters, case 1 would be much more common than case 2. Using a biological system analogy, it seems satisfactory to mate two individuals with different ancestry (ignoring that in most species it takes a male and female to mate); however, if they were simply due to the multiple copies of one individual getting in the mating pool due to selection, it seems inappropriate. In either case, note that if the individual happens to mate with itself, then crossover has no effect, although mutation still does. Some researchers like to avoid this situation altogether and one way to do this, is to allow only different individuals to mate (and if they are all the same, then simply terminate). In this case, the above pseudocode would have to be modified to represent the addition of the operations to ensure no identical individuals mate. Note that if you use the approach discussed above where you do not generate the whole mating pool at one time, and just two parents, you could generate one parent, then generate the other one with selection by repeating the process until a different individual is generated. Of course, in this case that would not be a pure fitness proportionate selection approach.

Finally, note that you may want to choose the method based on your understanding of the manner in which evolution occurs in biological systems in nature. However, we would emphasize that any of the variations described here are likely only very rough approximations of what is actually happening in nature; hence, in this book we will focus more on the view of the GA as a stochastic optimization method. If one way of defining and sequencing operations works better for some application, we will accept it whether or not it models some biological system (i.e., ours is an engineering focus, not one where we are focusing on the science of computer modeling of biological systems).

14.4.3 Representations, Complexity, Termination, and Initialization

For programming the genetic algorithm, one issue that you will encounter is whether to use special "string operations" in a computer language (some have better features than others in this regard). Such operations allow, for instance, conversion of numbers to strings and strings to numbers (which, depending on how you code the GA, you may need, since we need the numbers in string format to cross over and mutate, and in numeric format for fitness evaluation to perform selection, and of course, to plot certain results). They may allow for concatenation, swapping, and other features that could be useful in a string-based approach to implementing genetic algorithms. Another approach to implementing the genetic algorithm is to only use standard numeric representations but ones that allow for us to perform crossover and mutation (in this case, often integer representations are used).

Next, in choosing which number system to use (e.g., base-2 or base-10), be careful in considering that the parameters will have to be encoded to, and decoded from, this representation. In some computer languages, there are special functions that perform these functions and these can be quite useful in programming the genetic algorithm.

Finally, note that there is often a need to include parameter constraints and one way to do this is to use the "projection method" used for the gradient methods.

In implementation, you are typically concerned with memory and computation time and there are certain choices that can affect these significantly. First, due to the structure of the algorithm, it is clear that increasing the size of the population S will increase both memory and computation time requirements. In online applications, you typically only execute a fixed number (often one) of iterations of the genetic algorithm (an iteration in a genetic algorithm is the act of producing the next generation from the current one using the genetic operations) per sampling period. Clearly, just as in the case of the gradient methods, if you are solving, for instance, a function approximation problem, then you need to carefully consider issues involved in how big a batch of data to process at each step, and how many iterations to perform per sampling period because these can significantly affect memory and computation time requirements. Finally, the choice of the parameters of the genetic algorithm can significantly affect memory and computation time requirements, simply by how they affect performance of the algorithm and hence, how fast it finds a solution.

The discussion in the previous section showed how to produce successive generations and thereby simulate evolution. While the biological evolutionary process continues, perhaps indefinitely, there are many times when we would like to terminate our artificial one and find the following:

- The individual of the population—say, $\theta^\star(k)$—that best maximizes the fitness function (note that we do not use the notation θ^* as we reserve this for a global maximum point if it exists (they exist)). Notice that to determine this, we also need to know the generation number k where the most fit individual existed (it is not necessarily in the last generation). You may want to design the computer code that implements the genetic algorithm to always keep track of the highest $\bar{J}$ value, and the generation number and individual that achieved this value of $\bar{J}$.

- The value of the fitness function $\bar{J}(\theta^\star(k))$. While for some applications this value may not be important, for others it may be useful (e.g., in many function optimization problems).

- The average of the fitness values in the population.

- Information about the way that the population has evolved, which areas of the search space were visited, and how the fitness function has evolved over time. You may want to design the code that implements the genetic algorithm to provide plots or printouts of all the relevant genetic algorithm data.

There is then the question of how to terminate the genetic algorithm. There are many ways to terminate a genetic algorithm, many of them similar to termination conditions used for conventional (gradient) optimization algorithms. To introduce a few of these, let $\epsilon > 0$ be a small number and $M_1 > 0$ and $M_2 > 0$ be integers. Consider the following options for terminating the genetic algorithm:

- Stop the algorithm after generating generation $P(M_1)$—that is, after M_1 generations.
- Stop the algorithm after at least M_1 generations have occurred and at least M_2 steps have occurred where the maximum (or average) value of $\bar{J}$ for all population members has increased by no more than ϵ.
- Stop the algorithm once $\bar{J}$ takes on a value above some fixed value.

Of course, there are other possibilities for termination conditions (see the discussion in Section 11.1.6). The above ones are easy to implement on a computer but sometimes you may want to watch the parameters evolve and decide yourself when to manually stop the algorithm.

By "initialization" of the genetic algorithm, we mean, for instance, how to select the initial population. (Of course, to start a genetic algorithm, you need to specify other parameters.) Sometimes, the initial population is simply set to be random values. Other times, domain-specific information can be useful in establishing the initial population. Similar to the gradient methods, we generally expect better algorithm performance if we start with a better initialization. You should note, however, that unlike gradient methods, we get to initialize the parameters to S *different* values if we want. So we can think of the initial population as a set of best guesses at the solution. If even one of these is close to the global maximum of $\bar{J}$, then it is possible that the performance of the genetic algorithm will be improved.

14.5 Example: Solving an Optimization Problem

In engineering design problems, there are many times when it is useful to solve some sort of optimization problem, since we often try to produce the "best" designs within a wide range of constraints (which include, e.g., cost). In practical engineering problems, such optimization problems can be very difficult and at times it can be useful to turn to the genetic algorithm. In this section, to illustrate the operation of the genetic algorithm, how to tune its parameters, and how to program the genetic algorithm, we study its application to a relatively simple optimization problem.

Suppose, in particular, that we want to find the maximum of the function shown in Figure 18.10 using a genetic algorithm. Such a surface is sometimes called a "fitness landscape" by analogy with mountain climbing. Notice that it has many hills and valleys that could confuse, for instance, a gradient optimization algorithm. We will act as though we do not know an analytical expression for the underlying function. We assume, however, that we can provide candidate solutions to the function, and it will return (in finite time) the fitness of these candidate individuals. This is a necessary feature for implementing any genetic algorithm.

14.5.1 Genetic Algorithm Design

The genetic algorithm used to solve this problem was coded in Matlab using a base-10 encoding. We use two digits before the decimal point and four after it for a total number of six digits. (Clearly, this constrains the accuracy that we can achieve in the solution to the optimization problem.)

We will either initialize with a random population (with values uniformly distributed on the known optimization variable domains) or with all the parameters initially at zero. We assume we know the size of the domain that we want to optimize over (it is $[0, 30]$ for each dimension) and use "projection" to keep the parameters in this range.

Note that since the function goes below zero in Figure 18.10, we will shift the whole plot up by a constant (a value of 5 in this case). This will not change where the extrema occur on the landscape but it will shift the fitness values computed and displayed. Why perform this "shift"? For our selection method, we require that we have all positive fitness values since a negative one can result in a negative probability of being placed in the mating pool.

We use fitness-proportionate selection, single-point crossover, and to pair off individuals for mating, we pick each one in the mating pool and randomly select a mate for it. We use gene mutation. We will explore the use of different values for the population size S, the crossover probability p_c, and mutation probability p_m. We will also study the effects of using elitism, with a single elite member.

For a termination criterion, we allow no more than a fixed maximum number of iterations (here, $M_1 = 1000$). However, we also add another termination criterion that may stop the algorithm before this maximum number of iterations is achieved. In particular, we terminate the program if the best fitness in the population has not changed more than $\epsilon = 0.01$ over the last $M_2 = 100$ generations.

14.5.2 Algorithm Performance and Tuning

In this section, we run the genetic algorithm program under a variety of conditions to provide insights into its operation, and to provide ideas on how to tune a genetic algorithm's parameters.

Random Initial Population

To illustrate the operation of the genetic algorithm, we begin with $p_c = 0.8$ and $p_m = 0.05$ and a population size of $S = 20$. A random initial population is chosen, so that each parameter is uniformly distributed on [0, 30]. To better view the results of the optimization process, we will plot the contour map of the function in Figure 18.10 and place points on this plot that represent individuals *at some iteration.* Figures 14.3 and 14.4 illustrate the operation of the genetic algorithm. We see that for these choices, the algorithm performs well, and it does find the best individual, but then later loses it. Note that if you run the algorithm again, it may not do as well, since it may be unlucky in its random initial choices. (This shows why you may want a big population size; if it is big, it is more likely that it will make at least one good initial choice.) The scatter pattern shown in Figure 14.3, where there are horizontal/vertical groupings (bands), is the result of our genetic operator choices (e.g., the group of points above the global maximum point results from crossover and gene mutation in *one* dimension).

Initialization can significantly affect genetic algorithm performance.

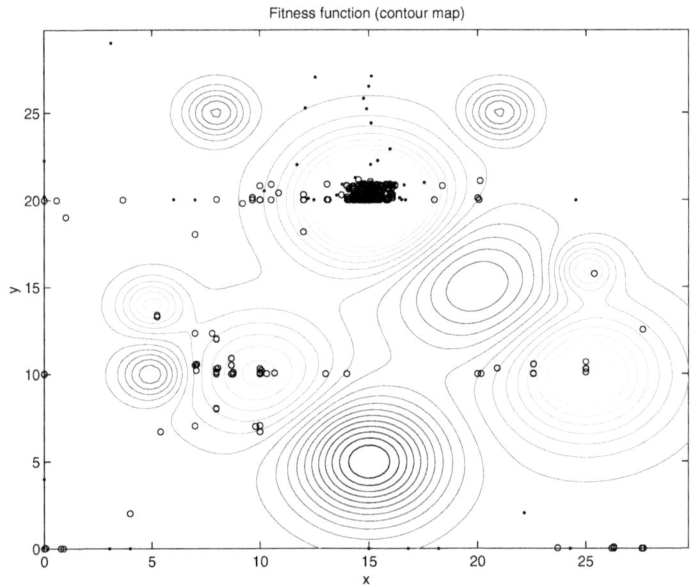

Figure 14.3: Contour plot of surface in Figure 18.10.

Initial Population of all Zeros

Here, with all the other parameters the same, we choose an initial population with all zeros for the parameter values. With this poor initialization, it fails to find the optimum point (see Figure 14.5) by the time the algorithm terminates. If you choose a different termination criterion that allows the algorithm to evolve more generations, it may find a good solution. Also, note that if you run the algorithm again, it may be the case that it will succeed, since the algorithm may make some lucky mutations or crossovers that result in better guesses.

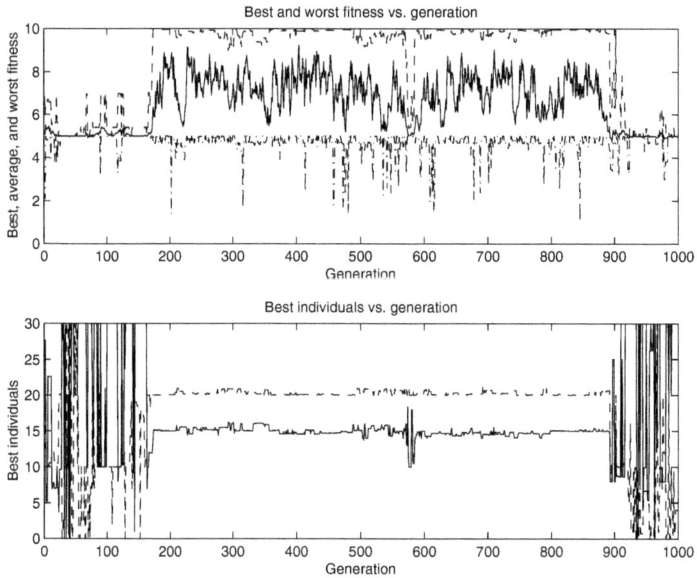

Figure 14.4: Fitness and optimization parameter evolution.

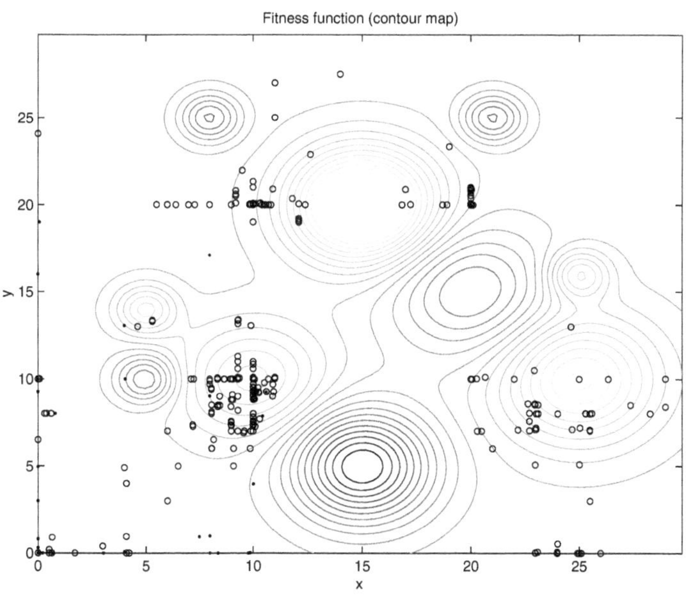

Figure 14.5: Contour plot of surface in Figure 18.10.

Increased Mutation Probability

Large mutation probabilities can lead to random and exhaustive search.

Next, let $p_c = 0.8$ and $p_m = 0.1$ (a larger value than above) and consider a population size of $S = 20$. As you can see in Figures 14.6 and 14.7, with the higher value for the mutation probability, it fails to lock on to the global maximum point. Basically, this happens since mutation is destroying good solutions (i.e., it destroys the progress of the method). From this, it should be clear that if you pick the mutation probability too high, the algorithm executes what can be considered a "random walk" in the parameter space so, while it may find a good solution at some point, it may take a long time to do so and we would basically attribute its success to "dumb luck."

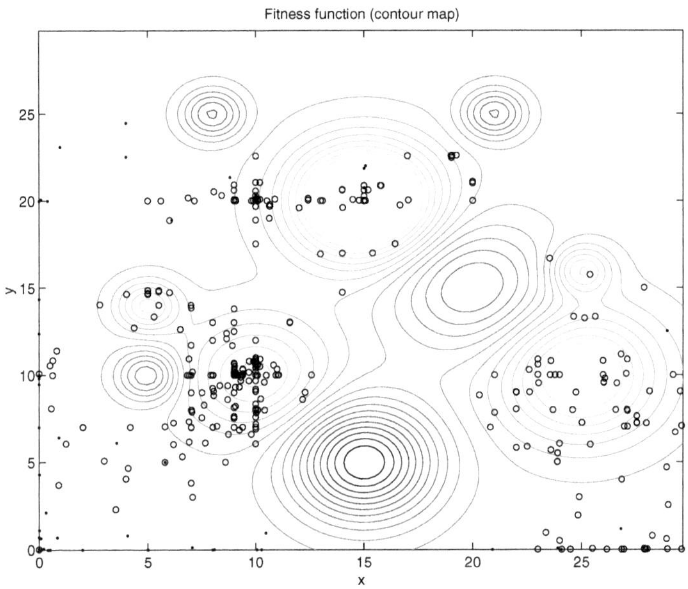

Figure 14.6: Contour plot of surface in Figure 18.10 with random initial population and higher mutation probability.

Decreased Crossover Probability

Next, let $p_c = 0.5$ (a smaller value than above) and $p_m = 0.05$ (i.e., return it to its earlier value) and consider a population size of $S = 20$. With the lower value for the crossover probability, it does less local search between good solutions (see Figures 14.8 and 14.9). Note that if you make $p_c = 0.1$, it fails to find a local optimum (at least for one time the algorithm was run). In this case, it is passing too many individuals through the mating process without mixing genetic material; hence, it stagnates.

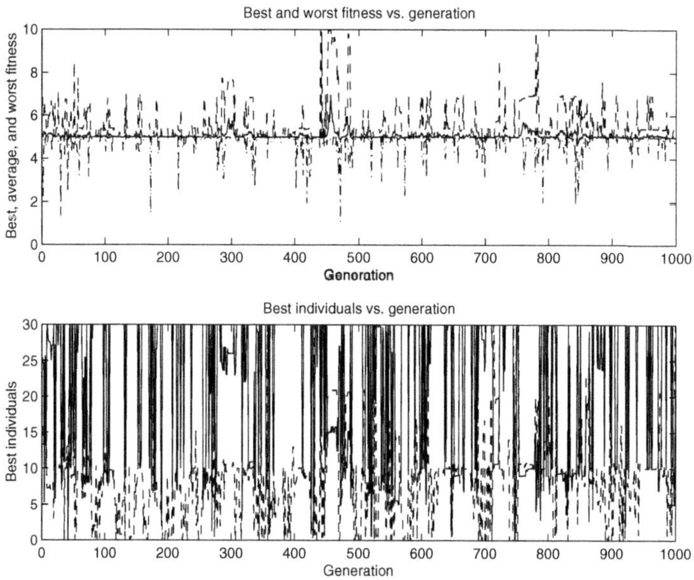

Figure 14.7: Fitness and optimization parameter evolution, with random initial population and higher mutation probability.

Increased Population Size

Next, let $p_c = 0.8$ and $p_m = 0.05$ and consider a population size of $S = 40$ (i.e., twice as big as earlier). With a bigger population size, we still get convergence, but this shows that increasing its size is not necessarily good (see Figures 14.10 and 14.11). Of course, we have to qualify this statement by saying "for this run of the program, with these termination criteria, etc." This simulation was produced simply to make the point that bigger is not always better (even though for the population, for some applications, this may generally be true).

Effects of Elitism

Next, let $p_c = 0.8$ and $p_m = 0.05$ and a population size of $S = 20$. Now, however, we use elitism with a single elite member. See Figures 14.12 and 14.13. With elitism we get much quicker convergence (notice that the early termination criterion was invoked) since crossover and mutation do not alter the best individual. Elitism has, in fact, been found to provide qualitatively similar results for a variety of applications. Basically, elitism ensures that there is a highly fit individual that survives in each generation. The other individuals are allowed to mate with the elite individual, so less fit individuals that do mate with this very fit individual will tend to have more fit children. This tends to accelerate convergence, while avoiding "premature convergence," since all the other individuals are allowed to explore the search space (provided that the other pa-

In practice, elitism has often been found to be useful, especially in real-time control where you cannot afford to use anything but the best-known controller.

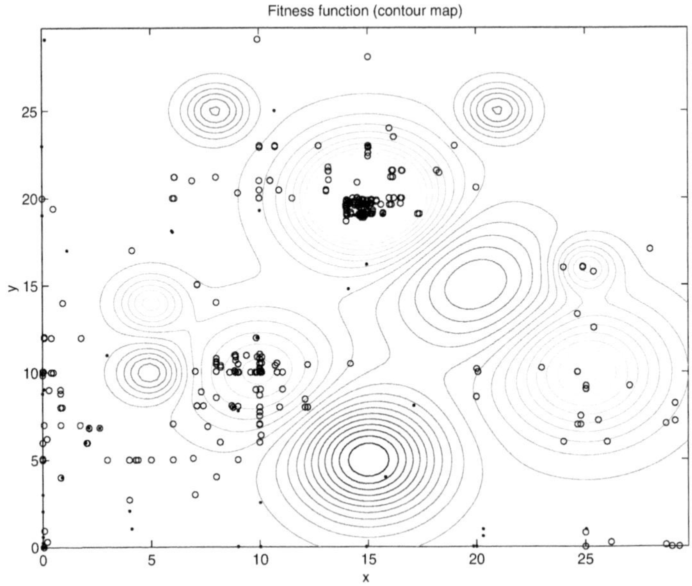

Figure 14.8: Contour plot of surface in Figure 18.10 with random initial population and lower crossover probability.

rameters, particularly the mutation rate, are set properly). It tends, for some applications, to provide a nice trade-off between focusing and exploration.

14.6 Approximations to Reduce Algorithm Complexity

When you start programming genetic algorithms, you often get ideas on how to modify the algorithm, either to better model how evolution works in nature, or to improve the performance of the algorithm (which may result in an algorithm that is successful, but quite unlike anything in nature). In this section, we briefly study some ways to make approximations to the genetic operations so that computational complexity of the algorithm can be reduced.

14.6.1 Reducing Algorithm Complexity

First, note that the encoding and decoding, even with a base-10 encoding, causes extra computations because you must convert the base-10 numbers to strings of integers and back. The only reason that we needed to do this was because we needed to perform crossover and mutation. We can, however, approximate these two operations and get reductions in complexity simply because no conversions will be necessary. Moreover, when we remove the conversions to strings, we get

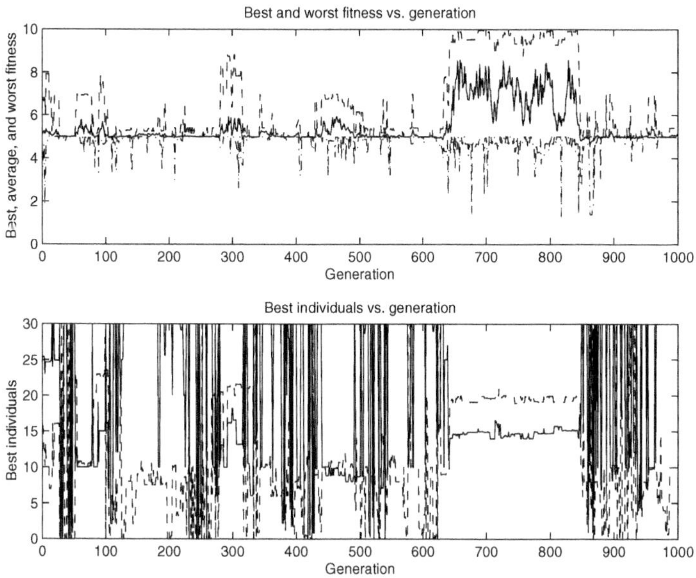

Figure 14.9: Fitness and optimization parameter evolution, with random initial population and lower crossover probability.

all standard operations on vectors so we will then be simply encoding traits with the natural representation in the computer (e.g., in Matlab, a chromosome will be a vector with elements that are traits, and the traits are simply the parameters of the problem).

For mutation we will perform a simple "trait mutation" where, with probability p_m, we mutate each trait in each individual. If we mutate, we simply switch the trait to be any number on the known domain of that trait (e.g., for the optimization problem in the last section, between 0 and 30). With this approach we will know that mutation cannot generate an out-of-range value so we will not need to use "projection" to fix it. For other problems, you could consider simply adding on a random value to the trait, but then this may place the value out of range, but it can be fixed with projection.

There are simple ways to modify the genetic algorithm so that it still grossly emulates evolution, but computational complexity is reduced.

Next, for crossover there are many possibilities. Here, we will consider two and to illustrate how they work, we will apply them to the function optimization problem studied in the last section. In both cases, we use $p_c = 0.8$ and $p_m = 0.05$ and a population size of $S = 20$. A random initial population is chosen, one so that each parameter is uniformly distributed on $[0, 30]$.

14.6.2 Crossover Option 1

First, we generate a method that approximates what happens when we cross over strings. In particular, consider an approach where we cross over at a random

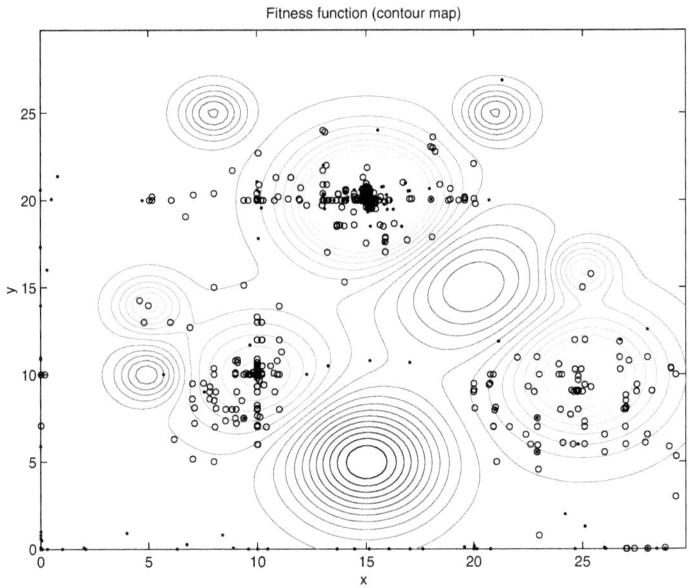

Figure 14.10: Contour plot of surface in Figure 18.10 with random initial population and increased population size.

"trait site." We use the standard crossover probability p_c and suppose that we choose to cross over the vectors θ^i and θ^j where $i \neq j$ and $i, j \in \{1, 2, \ldots, S\}$. To do this, we generate a random trait site number, say i^*, where $1 \leq i^* \leq p$ (where p is the number of traits, i.e., parameters), a random number $\alpha \in (0, 1)$, and let the child θ have

$$\theta_m = \theta^i_m$$

for $m = 1, 2, \ldots, i^* - 1$,

$$\theta_{i^*} = \alpha\theta^i_{i^*} + (1 - \alpha)\theta^j_{i^*}$$

(we think of this as "splitting" the trait at the trait site split point) and

$$\theta_m = \theta^j_m$$

for $m = i^*, \ldots, p$. When this approach is used, we get the results shown in Figures 14.14 and 14.15. Notice that the approach finds the maximum point.

14.6.3 Crossover Option 2

Next, we do everything the same as in "crossover option 1" except we perform crossover differently. Suppose we choose to cross over the vectors θ^i and θ^j where $i \neq j$ and $i, j \in \{1, 2, \ldots, S\}$. To do this, we generate a random number $\alpha \in (0, 1)$, and let the child θ be

$$\theta = \alpha\theta^i + (1 - \alpha)\theta^j$$

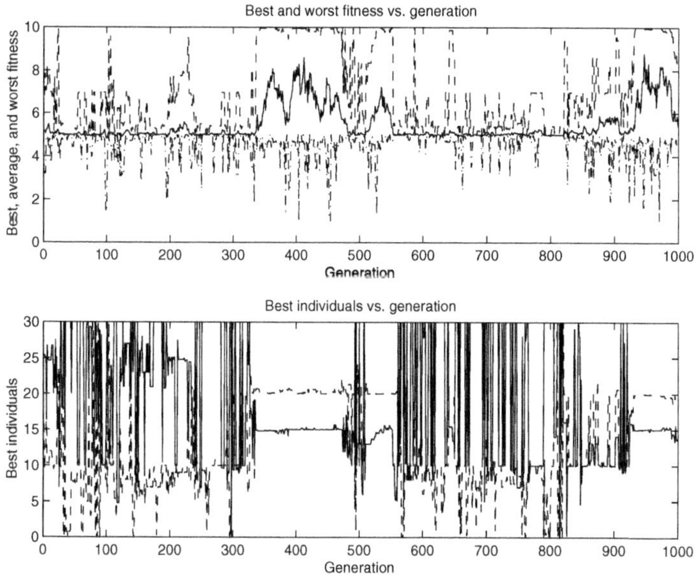

Figure 14.11: Fitness and optimization parameter evolution, with random initial population and increased population size.

This θ is a point on the line joining θ^i and θ^j. It is a simple type of interpolation between θ^i and θ^j. We think of this as "search in the neighborhood" of the two individuals, which is effectively what crossover tries to do. When this approach is used, we get the results shown in Figures 14.16 and 14.17. Notice that the approach finds the maximum point.

14.7 Exercises and Design Problems

Exercise 14.1 (Genetic Algorithms for Optimization): In this problem you will use the genetic algorithm to solve some simple optimization problems. You may use the code that is given at the Web site listed in the Preface.

(a) Suppose that you are given the function

$$f(x) = x \sin(10\pi x) + 1$$

which is taken from [352]. Design and implement on a computer a genetic algorithm for finding the maximum of this function over the range $x \in [-0.5, 1]$. Plot the best individual, best fitness, and average fitness against the generation. Plot the function to verify the results.

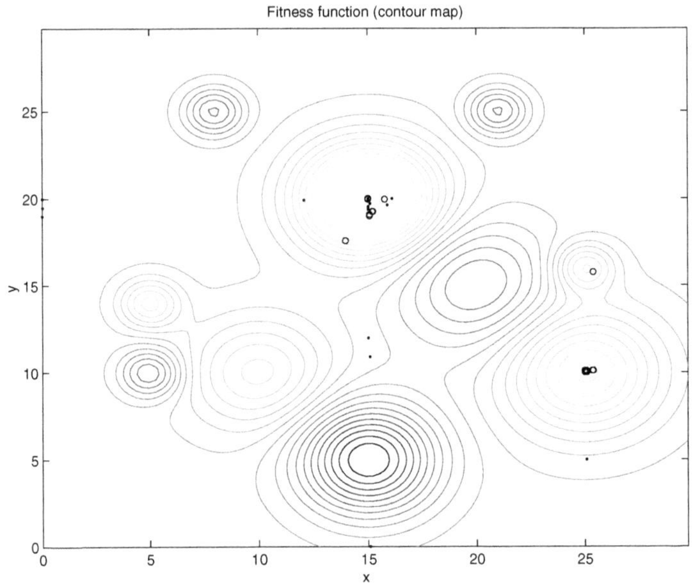

Figure 14.12: Contour plot of surface in Figure 18.10 with random initial population and elitism.

(b) Suppose that you are given the function

$$f(x) = \mathrm{sinc}(x+2) = \frac{\sin(\pi(x+2))}{\pi(x+2)}$$

Design and implement on a computer a genetic algorithm for finding the maximum of this function over the range $x \in [-10, 10]$. Plot the best individual, best fitness, and average fitness against the generation. Plot the function to verify the results.

(c) Suppose that you are given the function

$$\begin{aligned} z &= 0.8x \exp\left(-x^2 - (y+1.3)^2\right) + x \exp\left(-x^2 - (y-1)^2\right) \\ &\quad + 1.15x \exp\left(-x^2 - (y+3.25)^2\right) \end{aligned}$$

Design and implement on a computer a genetic algorithm for finding the maximum of this function over the range $x \in [-5, 2]$, $y \in [-2, 2]$. Plot the best individual, best fitness, and average fitness against the generation. Plot the function to verify the results.

(d) Suppose that you are given the function

$$z = 1.5\mathrm{sinc}(x) + 2\mathrm{sinc}(y) + 3\mathrm{sinc}(x+8) + \mathrm{sinc}(y+8) + 2$$

Design and implement on a computer a genetic algorithm for finding the maximum of this function over the range $x \in [-12, 12]$, $y \in$

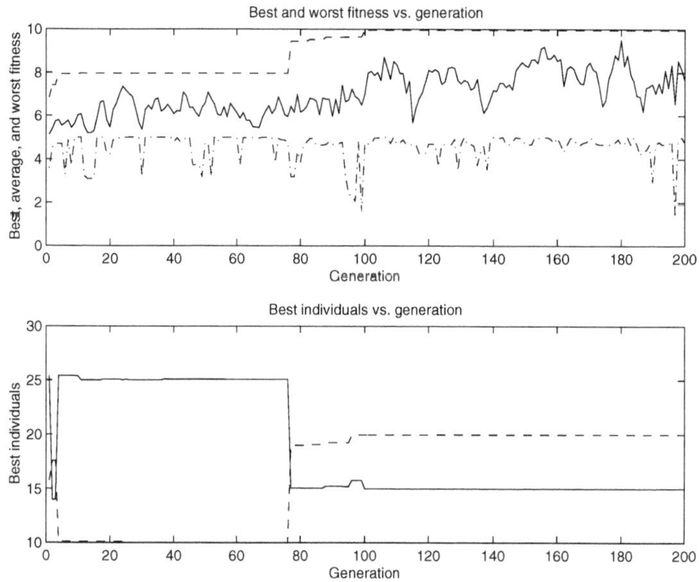

Figure 14.13: Fitness and optimization parameter evolution, with random initial population and elitism.

$[-12, 12]$. Plot the best individual, best fitness, and average fitness against the generation. Plot the function to verify the results.

Exercise 14.2 (Approximations of Genetic Algorithms):

(a) Repeat Exercise 14.1(c), but where you use one of the approximations discussed in Section 14.6. Code the algorithm and evaluate its performance (both complexity and ability to find the global minimum) in simulation.

(b) Repeat (a) but for Exercise 14.1(d).

Design Problem 14.1 (Design of Genetic Operators for Genetic Algorithms):

(a) Repeat Exercise 14.1(c), but where you use a genetic algorithm with different genetic operators. You choose which operators to use, but make the selection, crossover, and mutation operators different from those coded into the program given at the Web site. Code the algorithm and evaluate its performance (both complexity and ability to find the global minimum) in simulation.

(b) Repeat (c) but for Exercise 14.1(d).

Design Problem 14.2 (Optimization for Finding Mountain Peaks and Coffee-Growing Regions in Topographical Data for Colombia):

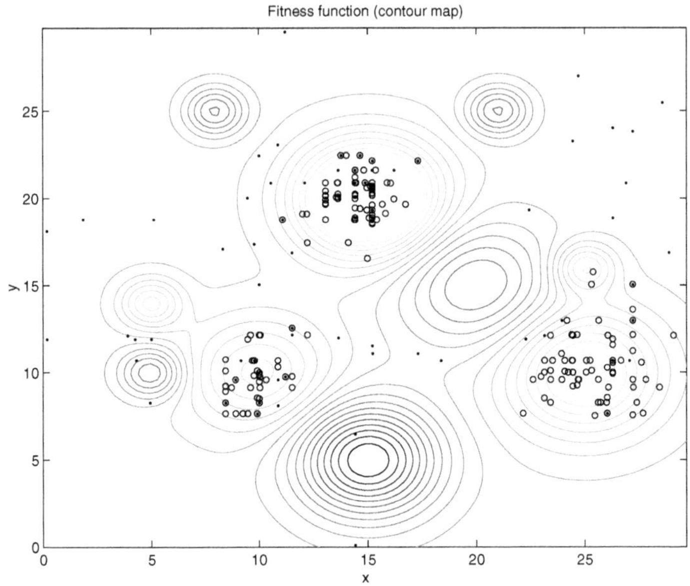

Figure 14.14: Contour plot of surface in Figure 18.10 with approximations to genetic algorithm, crossover option 1.

In this problem, you will study how to use genetic algorithms to search for the highest mountain peak in a region of the earth. To do this, you will need to go to the Web site for the book and download a topographical data set and a program that shows you how to work with the data (the topographical data were obtained from the US Dept. of Commerce, National Inst. of Geophysical Data). The topographical map for the region around Colombia is given in Figure 14.18. Notice that the data includes underwater data as negative elevations, and a black line was added at zero elevation to show roughly where the shorelines are with the Carribean Sea and Pacific Ocean.

After you download the data set and associated program, study the code to understand how to work with the data, plot it, and how to interpolate the data so that you can estimate the elevation for points that are not given in the data set. Clearly, you do not have analytical gradient information for this problem; however, you could go to the library or world atlas and find the solution to the problem for any fixed region on the earth. Hence, you should simply view the topographical data as providing an interesting cost function to search over.

(a) Design a genetic algorithm and simulate its operation on the topographical map of Colombia. As in the chapter, study choices of the population size and other genetic algorithm parameters.

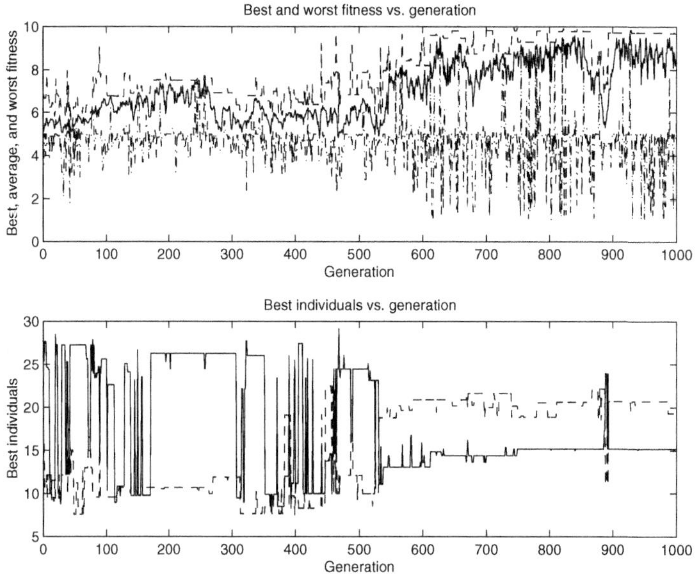

Figure 14.15: Fitness and optimization parameter evolution, with approximations to genetic algorithm, crossover option 1.

(b) What is the highest point on the map? Does this correspond to what a world atlas (or other source) tells you about the highest mountain in this region? What is the name of that mountain?

(c) In Colombia, coffee grows best at altitudes between 1000 and 2100 meters. Define a cost function that indicates where it is best to grow coffee on the region defined by the topographical map of Colombia. Formulate and solve a problem that evolves where coffee growers should live in Colombia (assuming they live near their farm). Illustrate the performance of the algorithm. Do the points where the population evolves to correspond to where coffee is actually grown in Colombia (e.g., *"la zona cafetera"*)?

Design Problem 14.3 (Genetic Algorithms for Approximator Structure Construction)*: Read Design Problem 11.2, where we give ideas on how to construct the structure of an approximator using gradient-type algorithms. In this problem you want to develop a genetic algorithm that can evolve the structure of an approximator for a particular function approximation problem.

(a) First, you must conduct some background research. Search the literature, evaluate existing methods, and summarize these.

(b) Using ideas from the literature, and perhaps your own ideas, design a genetic algorithm that can construct the structure of a function

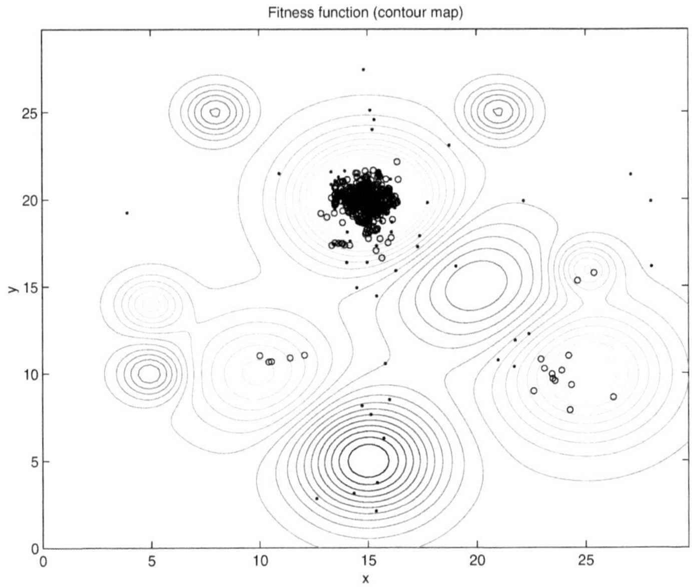

Figure 14.16: Contour plot of surface in Figure 18.10 with approximations to genetic algorithm, crossover option 2.

approximator. You must decide whether to use a genetic algorithm in conjunction with a least squares or gradient method, and all the issues associated with representation of the approximator in the genetic algorithm. Test the performance of the algorithm. Hint: Use the function approximation problem that was used throughout Part III and a fitness function that quantifies the inverse of the approximation error as measured by some test set.

Design Problem 14.4 (Artificial Immune Systems and Evolutionary Algorithms)*: First, see the discussion in the "For Further Study" section of this part. Second, study [125, 124] on artificial immune systems.

(a) Choose an artificial immune system and simulate it. Choose one that provides the capability for either learning or optimization, or both.

(b) Explain in detail the relationships between the algorithm you implement in (a) and the standard genetic algorithm. Be sure to identify the fitness function, selection, crossover, and mutation analogies.

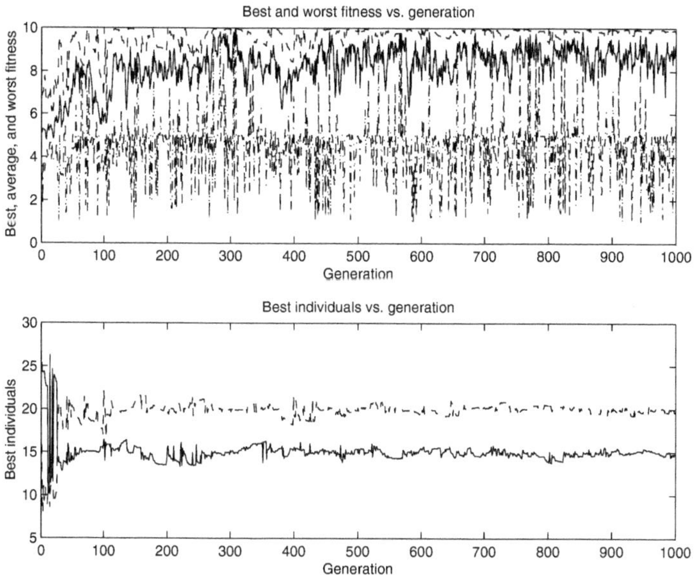

Figure 14.17: Fitness and optimization parameter evolution, with approximations to genetic algorithm, crossover option 2.

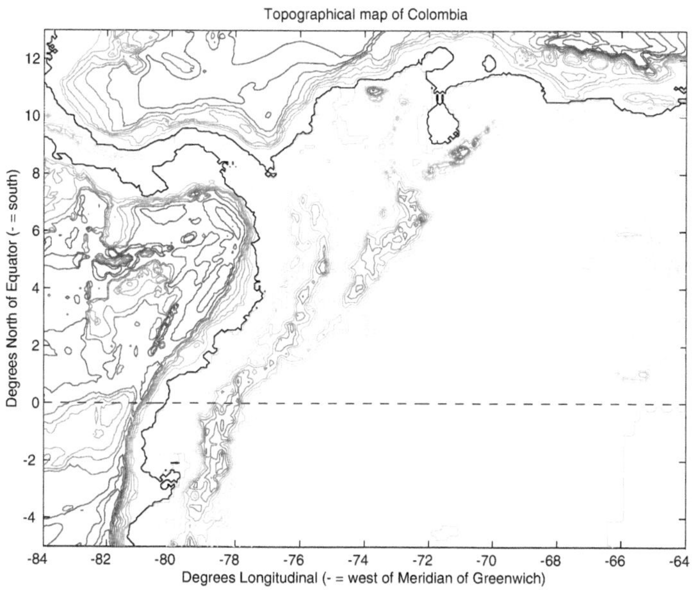

Figure 14.18: Topographical map of region around Colombia, South America.

Chapter 15

Stochastic and Nongradient Optimization for Design

Chapter Contents

Evolution is *the* fundamental theory that unites biology. Here, we will view evolution as a design process for constructing life forms and use that metaphor as a basis for our view of stochastic and nongradient optimization algorithms for designing control and automation systems. As briefly discussed in the last chapter, we can encode both the parameters and structure of neural networks, fuzzy systems, expert systems, planning systems, attentional systems, and learning systems, and use a genetic algorithm to evolve (optimize) these systems according to some fitness function. Here, rather than exploring the many possible approaches to genetic algorithms for design, we seek to examine fundamental issues you encounter when taking a stochastic or nongradient optimization approach to design (e.g., robustness trade-offs). We will do this via general stochastic and nongradient optimization algorithms, but throughout and at the end of the chapter, we will return to the genetic algorithm and discuss its use in design.

We will overview some relevant theory about how organisms evolve to be robust in their environment (to achieve "robust yet fragile" operation via "highly optimized tolerance"). Next, we discuss the use of stochastic optimization algorithms for computer-aided control system design (CACSD) for a tanker ship heading regulation problem. We introduce the "response surface methodology" (RSM) that has been used in computer-aided engineering, but our focus will be on control system design. We will discuss certain robustness trade-offs in the designs via response surfaces. We will introduce a variety of nongradient optimization algorithms, each of which involves iteratively making multiple simultaneous cost function evaluations to try to find a minimum point. We introduce the stochastic and nongradient "simultaneous perturbation stochastic approximation" (SPSA) algorithm and show how it can be used in control design. We discuss "set-based" stochastic optimization algorithms, of which the genetic algorithm is a special case, and show how they can be used in control design. Finally, while the focus here is on the use of simulations to perform the design, we will also briefly discuss a hardware-based "Darwinian design" methodology for control systems.

15.1 Design of Robust Organisms and Systems

Evolution is the design of optimized robust organisms. Design follows the environment.

Evolution is a design process for organisms. It produces organisms that are "robust" in the sense that they are able to survive and reproduce in the face of a variety of adverse influences from their uncertain environment. The environment tests the fitness of the organism for a wide variety of situations, and organisms best able to cope will survive to reproduce. The species may become optimized for the "typical" types of events it encounters in its environment; it may become "robust" to the environment it lives in. In the evolutionary process, the organism's complexity may increase to enable it to cope (e.g., instincts, planning, or learning capabilities may evolve) or fill a new ecological niche. Other times, simpler designs are maintained, since they fit the ecological niche and lead to highest reproduction rates (e.g., for bacteria). In optimizing reproduction rates, evolution chooses the appropriate level of complexity to achieve robustness for an ecological niche. The characteristics of the environment an organism lives in have fundamental influences on its design, since the environment (which includes adversaries) helps define the cost function that is optimized by evolution to produce an organism's design.

Engineering design focuses on construction of optimized robust systems.

Engineering design has produced a wide variety of robust and successful systems, from the cruise controller on an automobile to the variety of autonomous vehicles available today. The design constraints presented by the problem at hand constrain the complexity of the design (e.g., you may only be given an 8-bit microprocessor for the cruise controller) and the design is optimized to cope with typically encountered characteristics of the environment it is to operate in (e.g., in cruise control, hills and wind). Included in this "optimization" are constraints such as the time allowed for design, the intelligence of the people doing the design, and so on. Increased complexity in the cruise controller design (e.g., via a faster 16-bit processor, a vision system with sophisticated algorithms) can allow the design to cope with its environment better (e.g., via anticipating hills using a camera), but the sales market also defines part of its environment and frequency of use (its "propagation" in the marketplace as partly dictated by the cost of the product). The engineering design process generally leads to simple designs that are "good enough" and that can adequately cope with the most critical aspects of the environment it will operate in. Sometimes the simplest designs are best, but to solve some problems, it has been necessary to design increasingly complex systems (e.g., autonomous vehicles, advanced avionics, or high levels of automation in manufacturing). For these complex systems, there is typically a type of progression of design complexity used to achieve higher levels of automation (e.g., see the discussion in Part I on cruise and temperature control). However, even for these complex systems, the main focus is on trying to optimize the system for typically encountered events, and for operation in an uncertain environment.

Optimal robust designs are for a certain set of conditions, and these designs are typically sensitive to conditions they were not optimized for.

It is impossible to ignore the analogy between biological evolution and engineering design that is highlighted in the previous two paragraphs, and the fundamental role of optimization in achieving robust operation for both organisms and engineered automation systems. The goal of engineering is to produce the "best" design, so even if we do not explicitly think of design as an optimization process, it usually underlies most engineering design activities. In fact, recently these relationships have been quantified via the introduction of the concept of "highly optimized tolerance" (HOT) (see the "For Further Study" section at the end of this part). This term refers to systems that have been designed, via engineering methodology or evolution, to have optimized performance in an uncertain environment. The studies of systems with HOT focuses, however, not only on the successful aspects of the optimized designs, but also on the problems that arise due to trying to optimize designs. That is, the main focus is on trade-offs inherent in design processes.

To identify these trade-offs, note that when organisms are highly optimized (highly evolved) to tolerate adverse influences from their environment, roughly speaking, they are only designed to cope with the types of events (situations) that have been encountered over the species' existence (they are only optimized to cope with what the species has encountered in evolutionary time in their environment). If some event occurs that has never occurred before, even though it may seem rather harmless, the organism might not survive it, even though it has incredible complexity to be able to cope with a wide array of other adverse influences. (The organism is said to be "robust yet fragile.") Such an event may occur during the lifetime of a single organism of a species or may be species-wide via some relatively fast environmental change (e.g., a meteor hitting near México). Hence, HOT explains deaths and extinctions by considering how the design process operates to optimize performance for only certain situations, and therefore, it naturally becomes sensitive to other types of failures. There is a type of robustness trade-off. Due to limited complexity of the organism, and the focus on achieving a good design for typically encountered events, the optimization process of evolution can be thought of as shifting robustness across the spectrum of possible events, with a focus on getting the most robust performance for the events encountered most often. Evolution makes organisms tolerant to some events at the *expense* of being sensitive to others (hence, you may view the trade-offs as characteristics of a type of "conservation principle"). If the environment drives the design of a very complex system, and this system is highly optimized, it can be especially fragile. Highly optimized complex systems can be especially intolerant to small unexpected adverse events.

Robustness principles state that it is impossible to obtain arbitrarily good performance for all possible situations; it is necessary to allocate good behavior across typically encountered situations.

The control engineer familiar with linear robust control will likely see the connection to one of the central issues in linear robust control design. The "sensitivity function" shows that if you get highly robust control in some frequency range, it is at the expense of having a sensitive design for other frequencies (see any book on robust control, such as the ones listed in the "For Further Study" section of Part I). This same concept seems to hold for robust organism design via evolution, and design of complex automation systems via the enterprise of engineering or stochastic optimization tools used for design.

The optimization theorist is likely to think of the "no free lunch theorems" while reading this section (see the "For Further Study" section at the end of this part for references). For that theory, the problem of how to pick an optimization algorithm for a particular problem is discussed, and a nice perspective on "black box" optimization approaches is given. For instance, it has been shown that for a wide class of optimization problems, the average performance of any pair of algorithms across all possible problems is identical; hence, if one algorithm is superior to some other algorithm over some set of problems, then it must be inferior to it for another set of problems (clearly this is related to the "robust yet fragile" idea if we think of "problem" as "event"). Also, ideas from the no free lunch theorems explain how average performance of an algorithm is determined by how it is "aligned" with the underlying probability distribution of the optimization problems over which it is run; again, this is relevant to the HOT viewpoint. Clearly, it is also connected to an evolutionary perspective if we view the optimization algorithm itself as the organism (which is reasonable for many types of activities that organisms are engaged in, such as foraging, which is discussed in the next part).

15.2 Response Surface Methodology for Design

While the use of stochastic optimization methods (e.g., the genetic algorithm) for design has several appealing features, especially since it provides a way to model environmental uncertainties and account for these in design optimization, it also has the following drawbacks:

1. *Algorithm convergence:* For typical design problems, there are multiple local optima; while stochastic optimization methods typically have mechanisms for avoiding these, there are no guarantees. In practical problems there are typically *many* optimization variables. This can make the optimization problem much more difficult (e.g., to sufficiently explore the "design space"), and it may be more difficult to determine if the optimization algorithm has converged to a design that is optimal (or close to optimal).

2. *Computational complexity:* Simulation of all possible conditions is impossible in practical problems, especially when there are many optimization variables and conditions to test and high uncertainty in the "environment of the design." Is your problem such that you can simulate enough *representative situations* so that the set of simulations represents typically encountered problems so you are convinced that you have sufficiently explored the design space? Are you sure that you are not missing some likely situations?

3. *The "don't think, compute" mentality:* Do not get trapped into a mentality of "don't think, compute" when using stochastic optimization methods for design. "Design insights" are very important. You should not ignore the underlying physics and models when they can be quite useful in design. For instance, if you have gradient information, this can often be quite useful for an optimization algorithm and hence, it should not be ignored. Or, if you have some constraints given by the physics of the problem, these should be incorporated into the optimization process.

In this section we will introduce the "response surface methodology" (RSM) that addresses some of these issues and provides useful insights into design. RSM tries to address item 1 via visualization and optimization over the values of the cost function in order to pick the best design. It tries to address item 2 by using strategies to pick which design points should be tried, and not by just trying all possible designs. It addresses item 3 via an established methodology for process optimization. Here, we only give the essentials of RSM as it is a rich field composed of a large set of ideas and techniques; the interested reader should see the "For Further Study" section at the end of this part for more information.

15.2.1 Controller Design for a Tanker Ship

In this section we will explain response surface methodology by designing a proportional-derivative (PD) controller for the tanker ship heading regulation problem that was first introduced in Section 4.3. We will discuss issues typically encountered in RSM, and seek to highlight robustness trade-offs in control design.

Recall that the ship heading is ψ and the rudder input is δ. The reference input is ψ_r. We use RSM to design (choose) the K_p and K_d gains for a PD controller of the form

$$\delta = K_p e + K_d c$$

where $e = \psi_r - \psi$ and c denotes an Euler approximation to the derivative (the "change-in-error"). Hence, we keep the design problem as simple as possible so that we can focus only on certain principles, and can show two- or three-dimensional plots to provide insights. When we want to refer to the ith design, we will use K_p^i and K_d^i.

We need to set a goal, or performance objective, in order to take the optimization approach to design. Here, to specify the performance objective, we will use a linear first-order reference model

$$G(s) = \frac{\frac{1}{150}}{s + \frac{1}{150}}$$

We discretize it with a bilinear (Tustin) transformation for simulation purposes, use ψ_r as the input to this system, and name its output ψ_m. Next, suppose that we conduct, for a specified reference input sequence and the ith controller design, a simulation that produces N_s values (we choose this value via knowledge of typical settling times that are possible for the tanker ship heading). Here, we simulate the ship for 1200 s, to obtain 1201 values of the variables. We quantify the quality of the performance over this entire simulation via

$$J_{cl}(K_p^i, K_d^i, \psi_r) = w_1 \sum_{j=0}^{N_s} (\psi(j) - \psi_m(j))^2 + w_2 \sum_{j=0}^{N_s} (\delta(j))^2 \tag{15.1}$$

so that we seek a design which minimizes the deviation in the response from that of a reference model, and the amount of control energy used to achieve that response. Here, we choose $w_1 = 1$ and $w_2 = 0.01$.

We will begin by considering "nominal" conditions where we have "ballast" conditions, a speed of $u = 5$ meters/sec., no wind, and no heading sensor noise. Moreover, we will restrict our attention to a specific reference input sequence that involves wanting to hold the heading constant at zero for a period, and then change the heading to 45 deg.

Control system design can be formulated as an optimization problem. Even if a cost function is not made explicit, one is often kept in mind as a controller is tuned.

15.2.2 Response Surface and "Optimal Gains"

Suppose that we grid the (K_p, K_d) "design space" with each grid point defining a pair of proportional and derivative gains for the tanker ship. Due to our experience in PD controller design we consider a range of gains, such that

$$K_p \in [-5, -0.5]$$

and

$$K_d \in [-500, -100]$$

Suppose that we compute $J_{cl}(K_p^i, K_d^i, \psi_r)$ in Equation (15.1) for nominal conditions, the same reference input sequence ψ_r, w_1, and w_2. The plot of J_{cl} for this case is shown in Figure 15.1. Each intersection point on the mesh showing the response surface shows a design point (K_p, K_d) that we computed J_{cl} for. This plot shows very clearly how to think about tuning the PD gains in terms of this performance measure, and nicely highlights that tuning is an optimization process.

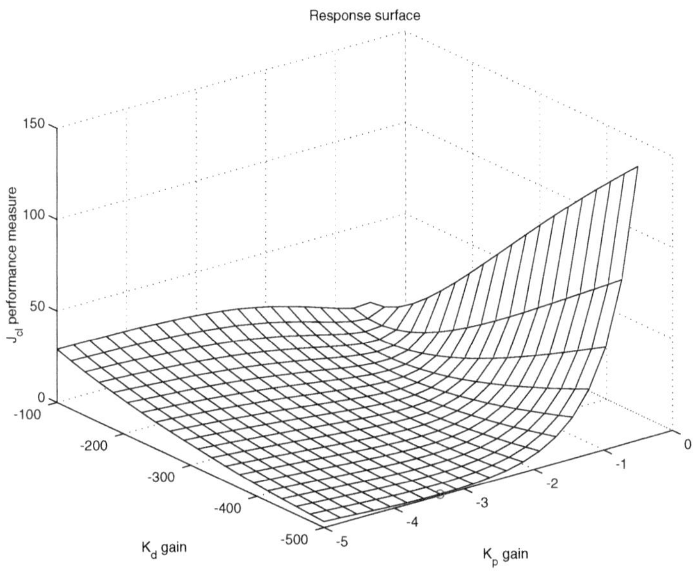

Figure 15.1: Response surface of J_{cl} for PD controller designs.

The best gains (the ones that result in the lowest point on the surface in Figure 15.1 that is designated there with a circle) are $K_p = -3.3421$ and $K_d = -500$, and the closed-loop response for the tanker ship for this choice is shown in Figure 15.2.

15.2.3 Response Surface and "Optimal Gains" for Off-Nominal Conditions

Next, we simply repeat what we did in the last subsection, but for the case where we have "full" rather than "ballast" conditions. For this case, we get the

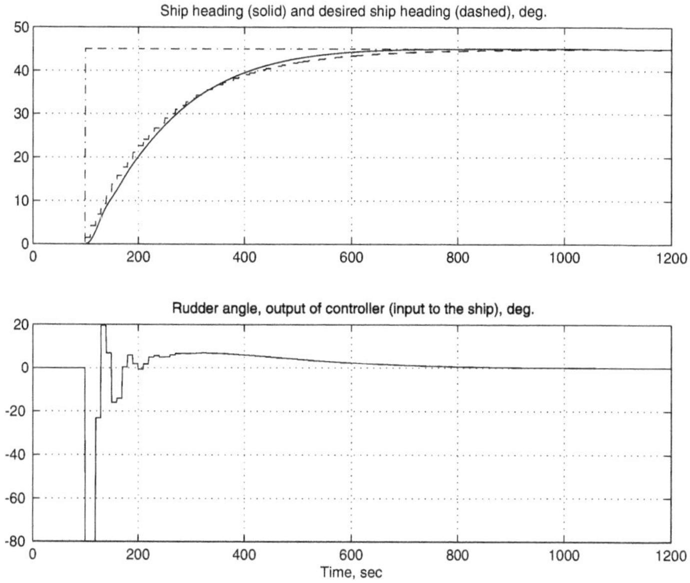

Figure 15.2: Closed-loop tanker response for best (K_p, K_d) gains as indicated by the response surface in Figure 15.1.

response surface in Figure 15.3. Notice that the shape of the surface changes, since the underlying plant that is simulated to generate each point on the surface is different from the last section.

The best gains (the ones that result in the lowest point on the surface in Figure 15.3) are $K_p = -2.8684$ and $K_d = -500$, and the closed-loop response for the tanker ship for this choice is shown in Figure 15.4. Notice that the best performance that can be achieved is not as good as in the last section, and the best proportional gain is smaller for the "full" case than for the ballast case (which makes sense, since slower changes in the input are needed for a ship that does not weigh as much).

15.2.4 Design Optimization Over Multiple Response Surfaces: Robustness Trade-Offs

You could construct response surfaces for different ψ_r, noise, wind, speeds, ship weights, etc. This would give a theoretically infinite number of response surfaces, each one corresponding to the infinite number of possible conditions that the plant (environment) can present. Theoretically, you could then combine the response surfaces to obtain a single performance measure and then find the minimum point on the resulting surface and call it the "best design" for all the conditions tested. Such an optimization could "weight" doing better for some conditions compared to others.

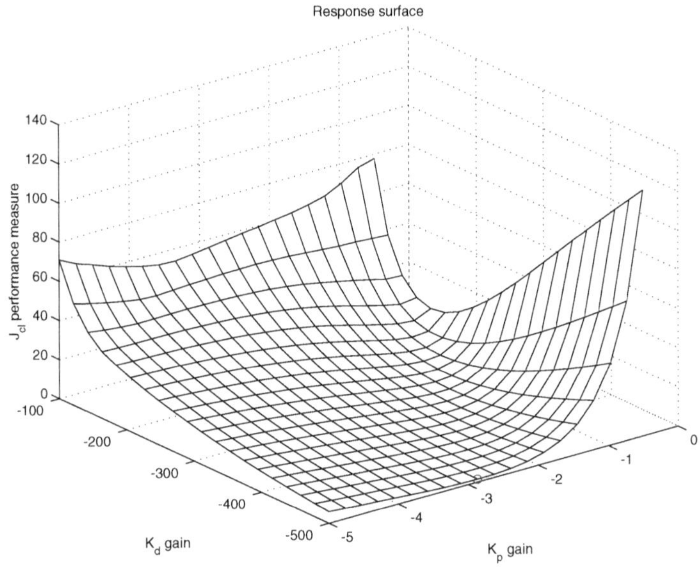

Figure 15.3: Response surface of J_{cl} for PD controller designs, "full" tanker ship.

As an example, suppose that we sum the response surfaces for the full and ballast conditions to get a "combined" response surface that perhaps indicates which PD controller is best, if we will encounter both situations with equal probability. The resulting combined response surface is shown in Figure 15.5.

The best gains (the ones that result in the lowest point on the surface in Figure 15.5) are $K_p = -3.1053$ and $K_d = -500$. Notice that this value of K_p is *in between* the two values that were found for the "ballast" and "full" cases earlier. The best gains are the ones that try to satisfy both conditions. The best gains to cope with both situations are *not* as good as the ones specially designed for each individual situation; there is a performance trade-off. Scaling of the sum of the response surfaces could result in a redistribution of the performance, showing favor to achieving good performance in one situation versus the other.

This illustrates, *in a very simple way,* fundamental trade-offs in design. Note, however, that in practical situations, there are an infinite number of response surfaces. For instance, note that all the above results depended on the length of the simulation used for the evaluation of the closed-loop performance. Moreover, there are clearly an infinite number of possible reference trajectories, and hence, corresponding response surfaces. (This shows the fundamental challenge in design of a control system for good tracking of any signal that is presented to the control system versus regulation of the output to a fixed trajectory.) Next, note that in any physical plant, there are stochastic effects so that actually you may need to run many simulations for each point on the response surface (e.g.,

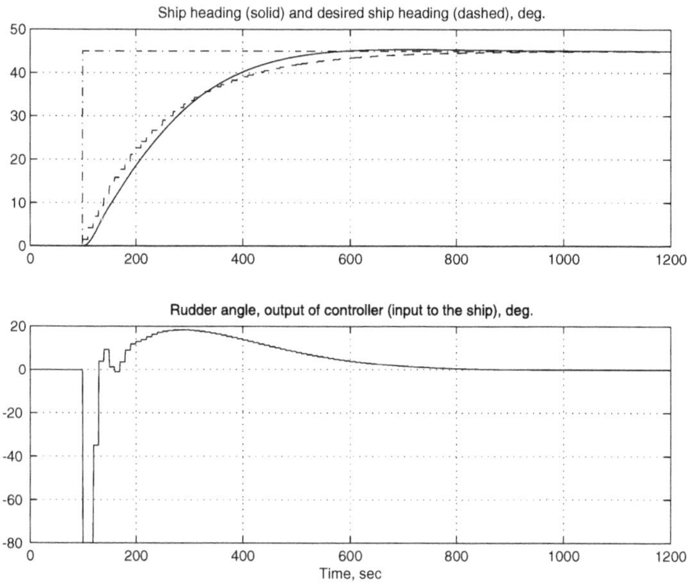

Figure 15.4: Closed-loop tanker response for best (K_p, K_d) gains as indicated by the response surface in Figure 15.3.

you could then plot the average closed-loop performance). For example, for the tanker ship, there can be noisy heading sensor measurements. Finally, there are certainly other possible plant variations that may need to be considered in simulating the plant. For instance, for the tanker ship, you may need to consider a range of ship speeds, ship weights, and wind characteristics. Clearly, the construction of response surfaces can require significant computational resources for practical problems (or experiments if you gather data from an actual experiment for the design).

Now, it should be clear that if your response surface is constructed by simulating stochastic effects and uncertainty in the environment for a practical problem, the probability that a response surface representing effects from only a few uncertain characteristics would *perfectly match* a response surface when a different set of characteristics is considered is essentially zero. Similarly, the probability that the set of "optimal gains" indicated by one surface is highly unlikely to perfectly correspond to the "optimal gains" suggested by another surface. The implication is that, while you may be able to optimize a design under a narrow range of conditions, performance will certainly suffer under conditions other than what the design is for. This clearly shows fundamental trade-offs in robustness.

Performance and hence, optimal design choices are different for different plant conditions; hence, finding the "best" single design entails allocating good performance across different situations.

To illustrate the need to perform multiple simulations to characterize performance when there is uncertainty, in Section 16.2, we will study robust approximator design and in this case, we will need to run many simulations to get a

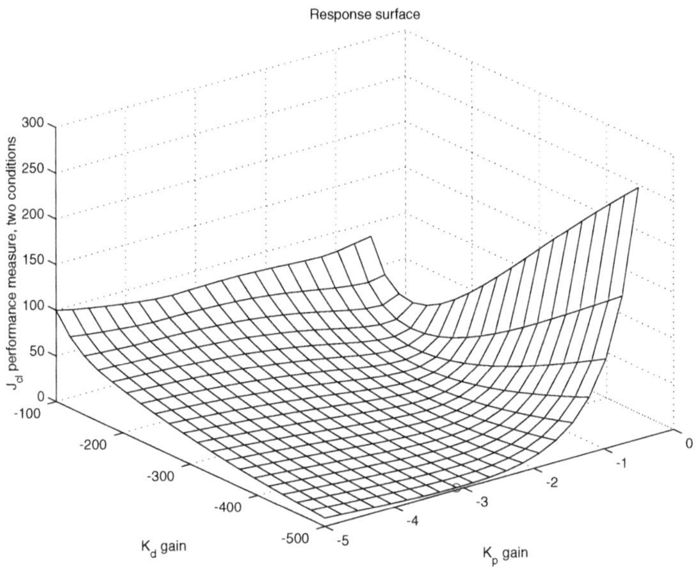

Figure 15.5: Response surface resulting from the sum of the response surfaces for PD controller designs for the "ballast" and "full" conditions.

single value on the response surface that will represent the average performance for a certain design point.

15.2.5 Choosing Design Points, Approximating Response Surfaces, and Optimizing Designs

We close this section with a discussion of three central topics in response surface methodology: how to pick "design points," how to approximate a complex response surface, and how to use optimization methods to find an optimal design. It is possible to make the discussion quite brief, since many of the concepts are discussed in Part III (and other related ones are discussed in this part).

Choosing Design Points: "Design of Experiments"

Above, we have ignored issues of complexity that arise due to the possibility that there may be many "independent" design variables. For the tanker ship we only consider PD controller design and hence, the number of design variables was $p = 2$. In this case, it was simple to grid on each of the two dimensions, and plot the resulting three-dimensional response surface. What if p is much larger (e.g., all the parameters of an instinctual neural or fuzzy controller)? Certainly visualization becomes much more difficult, essentially requiring you to fix all but two (or three) variables and plot the response surface for the other ones.

The other problem that arises is due to the "curse of dimensionality" if you grid the design variable space with a uniform grid. For instance, suppose that there are n_G points (grid partitions) on each dimension, with a total of p design variables. Then, using uniform gridding, there will be

$$(n_G)^p$$

"design points." For example, for the tanker ship we had $p = 2$ and $n_G = 20$ for a total of 400 design points. What if we had decided to tune a PID controller (i.e., one that also had an integral gain)? Then, $p = 3$ and the number of design points is 8000 if $n_G = 20$. Next, suppose that we tuned an instinctual neural or fuzzy controller that had 121 parameters (not unrealistic). Then if $n_G = 20$, there would be $20^{121} = 2.6585 \times 10^{157}$ design points and this would likely present a serious computational challenge, no matter what type of computer you own!

In practical problems, there are often many possible design points; hence, you must try to choose a few representative ones in searching for the best design.

The common solution to such a problem is to be very careful about which design points are considered. We were *not* careful for the tanker ship, since the problem is not very complex, and even there it should be clear that we "wasted" some computations (e.g., notice that in flat regions of the response surface we could have computed far fewer points). The typical approach in response surface methodology to choose design points is to use concepts from "design of experiments" (DOE). For instance, suppose that there are only two values for each design variable on each dimension. In this case, $n_G = 2$ so that the number of design points becomes 2^p. This choice of design points is called the "2^p factorial design" and is a common choice for practical response surface methodology (at least for initial "screening" to determine which variables are key factors influencing response surface shapes). As an example, notice that for the tanker ship, this choice would correspond to picking the four corners of the design space considered earlier. How would this have worked for the tanker ship? By studying the earlier response surfaces, you will notice that this would have resulted in a choice of $K_p = -5$ and $K_d = -500$, which is certainly not as good as when we used more design points (but it may be acceptable). Note that the 2^p factorial design considers simple slopes in each dimension and hence can, in general, provide a rough approximation to where the optimal design is. (Of course, however, nonlinearities in the parameters can result in the approximation being quite inaccurate.) However, this may be necessary when you work with practical problems when there are many dimensions (design variables).

Notice, however, that with the 2^p factorial design approach, the curse of dimensionality still holds, due to the exponent. In practical problems, for large p, it then sometimes becomes necessary to use a "fractional factorial design," where a fraction of the set of 2^p design points is used (e.g., only one point on some dimensions). Clearly, for the tanker ship such an approach would generally result in an even worse design than the 2^p factorial design would provide. Generally, then, you see that the ability to test more design points, *provided that they are chosen judiciously*, will generally result in better designs. We pay for getting a good design by additional computations, *and* by the need for good insights into how to choose representative design points.

Response Surface Construction is Function Approximation

While in this section we simply plotted the surface using actual data from the simulations, it is clearly also possible to approximate the data using the function approximation methods discussed in Part III. To do this, you simply choose an approximator structure and optimization method for constructing the approximator by tuning its parameters. In traditional RSM, linear or polynomial models are often used ("first-" or "second-order" models), and least squares is used to fit the models to the data. Clearly, the methods of Part III suggest that it may be productive to use neural or fuzzy systems as approximators, and least squares or gradient methods to tune them. DOE provides a method to choose the training data set and choice of the data set is one of the key issues in RSM, just as it is in approximator design.

The pairing of design points to resulting performance is a function approximation problem; approximator construction corresponds to constructing an interpolator that is called the "response surface."

It is important to note that the fundamental principles of learning as function approximation still hold here. For example, the size of the approximator should be bounded by the number of design points, otherwise poor generalization can occur. In this case, the surface could suggest an optimal design point that is far different from the actual one. Clearly, if you only have a few design points, you will only be able to use a simple (e.g., linear) approximator structure. Moreover, it is important to keep in mind that if you have a finite amount of data, it may be just as good to simply find the minimum computed value of the response surface (just as we did for the tanker ship). Clearly, however, it could be possible that an interpolation of the data could suggest a better design.

Design Via Optimization Over A Response Surface Without Considering Additional Design Points

As shown earlier, to obtain an "optimal design," you find the minimum point on the response surface (or combination of such surfaces). If you do not use a simple brute force approach where you find the minimum of all computed points on the response surface, you can perform an optimization over the response surface via some gradient method. In fact, the most common approach in traditional RSM is to use steepest descent gradient methods. Clearly, however, it may be better to use some other optimization method (e.g., conjugate gradient or Levenberg-Marquardt).

Finally, note that it is best to think of the approximation of the surface as providing a way to consider design points without actually having to compute them; if you have a good interpolator, then working with the response surface is almost the same as working with the simulator (or experiment) used to generate response surface points. Hence, practitioners often think of the use of the approximator and optimization over that surface as showing a path of design points that leads to an optimal design, that *you do not have to compute*, since they are similar to points that you have already computed the performance for. Hence, the use of the approximator for the actual surface is thought of as a way to help cope with computational complexity (or experimental complexity).

15.3 Nongradient Optimization

In this section, we introduce a variety of deterministic and stochastic methods to perform nongradient optimization (including methods that actually formed the foundation for the *introduction* of the genetic algorithm model of evolution treated in the last chapter). You can think of these as methods to optimize over response surfaces, and in the stochastic methods, as optimizing over a whole set of such response surfaces.

15.3.1 Pattern and Coordinate Search

In this subsection, we discuss relatively simple *deterministic* nongradient optimization methods in order to establish some basic concepts. In the last two subsections, we will consider more sophisticated deterministic nongradient methods. We begin by building on the methods of Part III on optimization for learning by discussing how to use approximations to the gradient in optimization algorithms.

Approximations of the Gradient

In some problems, it may be possible to approximate the gradient and then use this in any of the standard gradient update formulas in place of the analytically determined gradients. In particular, we can use a "central difference formula" to provide an approximation to the gradient with respect to each parameter (that we denote with $g_i(j)$, the approximation of the gradient with respect to the ith parameter at the jth iteration, where $\theta = [\theta_1, \ldots, \theta_p]^\top$).

We define this as

$$\frac{\partial J(\theta(j))}{\partial \theta_i} \approx \frac{1}{2c}\left(J(\theta(j) + ce_i) - J(\theta(j) - ce_i)\right) = g_i(j) \tag{15.2}$$

If you cannot find an analytical expression for the gradient at a point, it is possible to specify approximations of it via computations of the cost in a region about the point.

where c is a positive scalar and e_i is the ith column of the $p \times p$ identity matrix (i.e., it is the ith "unit vector"). To gain insight into the approximation, it is useful to consider Figure 15.6, where the $p = 2$ case is depicted for an iteration j where $\theta(j) = [2, 2]^\top$. Here, we show the pattern of points defined by the unit vectors e_1, e_2, $-e_1$, and $-e_2$ that are used in Equation (15.2) as perturbations to $\theta(j)$ in order to calculate an approximation to the gradient. The overall direction of the update to $\theta(j)$ would normally be modified by a step size and could point in *any* direction from $\theta(j)$, depending on the computed values of the cost. What direction will the update point if $J = \theta^\top\theta$? Note that to approximate the gradient with this approach, we need $2p$ calculations of the cost function. We avoided the need for analytical gradient information but we now need these calculations of the value of the cost to approximate the gradient.

The quality of the approximation specified in Equation (15.2) can significantly affect the performance of the optimization algorithm. Sometimes you can use the same c for all p of the θ_i values in Equation (15.2), while for some

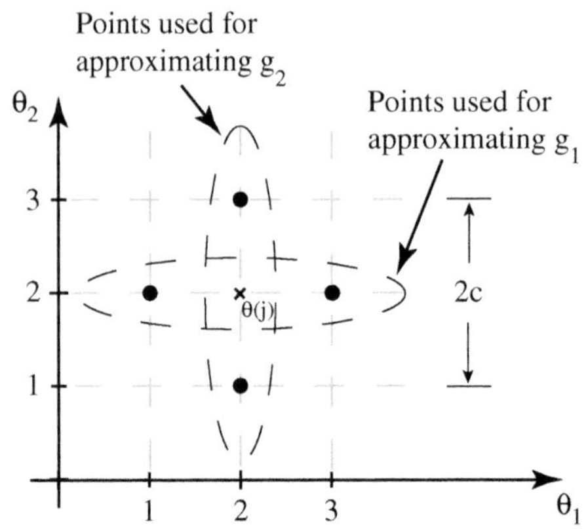

Figure 15.6: Pattern of points for approximation to the gradient, $p = 2$ case.

problems you may need different choices to get good approximations. Sometimes the "forward difference" approximation is used in place of the "central difference formula" in Equation (15.2), since it takes fewer computations; however, it is generally not as accurate. If you use the Hessian in the update formula (e.g., in Newton's method), then you could specify an approximation to the second derivatives; however, approximations for second-order methods will not be discussed here (see the "For Further Study" chapter at the end of this part). It should be clear, however, that you could use the above approximation in *any* gradient method that only used first-order information (i.e., first derivatives).

Simple Pattern Search Methods

In this and the next subsection, we will introduce the idea of pattern search by outlining two simple approaches. Methods discussed later in this chapter (e.g., the multidirectional search method) are also pattern search methods. The basic idea in pattern search is to compute the cost at each point in a pattern of "exploratory" points around the current estimate $\theta(j)$, and then decide how to move the estimate and pattern of points at the next iteration so as to reduce the cost. In some approaches, you can think of the computations of the cost as being used for types of *approximations to the gradient in a region* and the subsequent movements of the pattern as being driven by these approximations. This idea is discussed a bit more in Section 15.3.4, where the multidirectional search method is discussed.

In pattern search, you compute the cost at a pattern of points near the current estimate and iteratively update the pattern so that it moves toward a minimum point.

At iteration j, we start with a pattern

$$P = \{\theta^0, \theta^1, \theta^2, \ldots\}$$

and the method generates another set of points for iteration $j+1$, $j = 0, 1, 2, \ldots$. We will suppose that θ^0 is always $\theta(j)$ (and in the implementation of the algorithm, we will maintain this). We consider $\theta(j)$ to be the current estimate of an optimum point. The number of points in the pattern P depends on the pattern choice, and we will specify some different options for the pattern below. You should think of the points in $P(j)$ as being candidate solutions to the optimization problem. Hence, we think of the method as searching in "parallel" in the sense that it considers $|P|$ candidate solutions at each step, rather than just one as we did for the gradient methods of Part III.

Pattern search methods are considered "parallel" in the sense that they evaluate more than one region of the search space at each iteration.

To define the pattern of points, let C denote a matrix whose columns specify perturbations to the current estimate $\theta(j)$ to specify the pattern P; hence, C has p rows and a number of columns equal to the number of exploratory points in the pattern. Let $\theta_s^i(j)$ denote the ith exploratory perturbation from $\theta(j)$, $i = 1, 2, \ldots, |C|$. Next, we provide two methods to specify the pattern of points.

Evolutionary Operation Using Factorial Designs: In this method, the columns of the matrix C are chosen to have elements that are all possible combinations of $\{-1, 1\}$ and one column of zeros (if the scaling for the problem is such that some components of θ are much larger than others, you can choose B as a diagonal matrix of positive values and use BC in place of C). This corresponds to choosing the 2^p corners of a hypercube centered at $\theta(j)$, plus a column of zeros, which represent the center point $\theta(j) = \theta^0(j)$. (What is the relationship to the design of experiments choice for response surface methodology?) With this, $|C| = 2^p + 1 = |P|$. Let c^i denote the ith column of C.

As an example, if $p = 2$, then

$$C = \begin{bmatrix} 1 & 1 & -1 & -1 & 0 \\ 1 & -1 & -1 & 1 & 0 \end{bmatrix}$$

Notice that the columns of C specify a pattern P that for the $p = 2$ case are large black dots in Figure 15.7(a). Notice that in general, we will compute the cost at each of the 2^p points of the hypercube around $\theta(j)$ and the cost at $\theta(j)$.

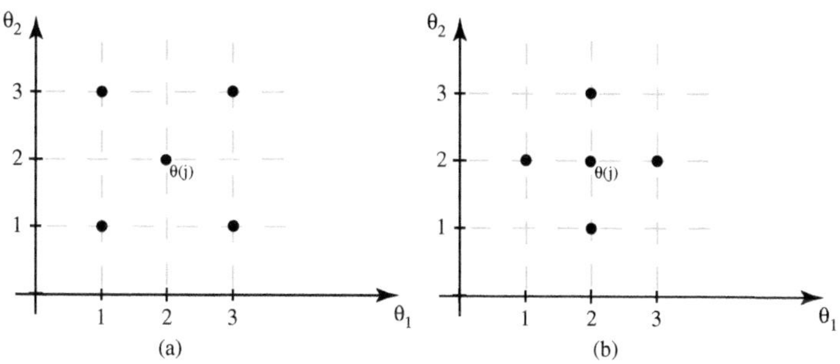

Figure 15.7: The pattern of points for the $p = 2$ case in evolutionary operation using factorial designs is shown in (a), and in (b) for simple coordinate search.

Simple Coordinate Search: In what we will call "simple" coordinate search (more sophisticated coordinate search methods can be found in the literature referenced in the "For Further Study" section of this part), we choose a different pattern of points than the one given above. In particular, we use

$$C = [I \ -I \ 0]$$

where I is the $p \times p$ identity matrix, so that C is a $p \times (2p+1)$ matrix. For $p = 2$, this choice corresponds to the pattern of dots in Figure 15.7(b). Hence, C simply specifies that we should search along each coordinate. As in the last section, if some components of θ are much larger than others, you may specify some matrix B with positive diagonal scaling elements and use BC in place of C above. With this, perturbations in some directions will have a larger magnitude than others (of course, you would typically want larger perturbations along coordinates that have higher magnitudes) so that in the $p = 2$ case, the dots would lie on a rectangle, not a box.

Pattern Search Algorithm: Suppose that we initialize the algorithm with $\theta(0)$. Note that we *do not* have to specify the entire initial pattern $P(0)$, since this is specified via the choice of C once the single point $\theta(0) = \theta^0(0)$ is chosen. We use a parameter γ_c to specify how the pattern should contract at the next iteration if a lower cost exploratory point was not found in the current pattern (the usual choice for this parameter is $\gamma_c = \frac{1}{2}$). If a lower cost point was found on the current pattern, then the pattern is not contracted and the parameter λ_j, for example with $\lambda_0 = 1$, will be used to specify the actual contraction that the algorithm takes at the next step. Let $\theta_s(j)$ denote the perturbation from $\theta(j)$ that is chosen at step j as the *best* point in the pattern of exploratory points (if $\theta_s(j) = 0$ at some step, this represents that no better point was found, so $\theta(j+1) = \theta(j)$, and the pattern is contracted). At each iteration, let J_{min} denote a scalar that is the lowest cost point found so far in computing the cost at each point in the pattern.

The algorithm for general pattern search is the following:

1. For $j = 0, 1, 2, \ldots$:
2. Compute $J(\theta(j))$.
3. Exploratory moves:
   (a) Let $\theta_s(j) = 0$, $\rho_j = 0$, and $J_{min} = J(\theta(j))$.
   (b) For $i = 1, 2, \ldots, |C| - 1$:
      - Compute $J(\theta^i(j))$ where

$$\theta_s^i(j) = \lambda_j c^i$$

and

$$\theta^i(j) = \theta(j) + \theta_s^i(j)$$

- If $J(\theta^i(j)) < J_{min}$ let $\rho_j = J(\theta(j)) - J(\theta^i(j))$, $J_{min} = J(\theta^i(j))$, and choose $\theta_s(j) = \theta_s^i(j)$.
- Next i.

4. Update/contract:
   - If $\rho_j > 0$, then a better point was found on the pattern, so let $\theta^0(j+1) = \theta(j+1)$ where
   $$\theta(j+1) = \theta(j) + \theta_s(j)$$
   and let
   $$\lambda_{j+1} = \lambda_j$$
   so that we do not contract, since this pattern size seems to be making good progress.
   - If $\rho_j \leq 0$, a better point was not found on the pattern, so let
   $$\theta(j+1) = \theta(j)$$
   and contract the pattern by letting
   $$\lambda_{j+1} = \gamma_c \lambda_j$$
5. Next j.

Clearly, there is a need to add some type of stopping criterion (e.g., based on changes in the parameter or cost values, or on how close λ_j is to zero). This could, for instance, be placed after step 4.

Algorithm Complexity and Convergence: Notice that at each iteration, there is the need to compute $|C|$ cost values so that the algorithm complexity is dictated partly by the choice of the size of the pattern used. Most pattern search methods do not guarantee a cost decrease for every change in the pattern, but under mild restrictions, some guarantee a decrease for a certain sequence of patterns and this leads to convergence properties. In particular, for both the evolutionary operation using factorial designs method and the simple coordinate search method under mild restrictions, it can be shown [513] that

Under mild restrictions, these pattern search algorithms can be shown to possess certain convergence properties.

$$\lim_{j \to \infty} ||\nabla J(\theta(j))|| = 0$$

Finally, note that if a hypercube is used to specify bounds on the allowable parameters, then the simple coordinate search method can be modified so as to still provide convergence properties [317].

Coordinate Descent Using Gradient Approximations and Line Search

In "coordinate descent" approaches (originally developed for the case when gradient information is available), optimization proceeds by only searching along one dimension at a time, often in some cyclic order. For instance, in the $p = 2$ case, it could search alternatively along the *north-south* direction, then along the *east-west* direction. If there is no analytical gradient information available, you can simply perform a "line minimization" ("line search") at each step for the dimension that is considered. There are several methods for solving such line minimization. For instance, you can compute the cost function at several points and fit a curve to it (e.g., a quadratic or cubic polynomial), then find the minimum point on this curve and use it as an approximation to the minimum along a dimension. Another approach is the "golden section" method. See the "For Further Study" chapter at the end of the part for references for such approaches.

Another approach to coordinate descent is to simply use the discrete approximation to the gradient discussed in the last section. Then, for instance, for a steepest descent-based method you simply pick a dimension, compute an approximation to the gradient in this dimension using the central difference formula, and then update the parameter in that dimension using the standard update formula, but only for that one dimension. That is, you may cycle through the dimensions, and when you update the ith dimension (i.e., the ith parameter), you use

$$\theta_i(j+1) = \theta_i(j) - \lambda_j g_i(j)$$

where $g_i(j)$ was defined in the last subsection to be an approximation to the gradient. Other gradient methods can be modified in an analogous way.

Here, we will not investigate line search methods as it should be clear via the earlier discussions how to specify the algorithms. Moreover, the multidirectional pattern search method that we will study in Section 15.3.4 below is a type of multidirectional line search method, and it is a pattern search method. It is interesting to note that convergence of line search methods has been studied extensively, and recent results on convergence and integration of pattern and line search methods are given in [336].

Example: Solving an Optimization Problem with Simple Coordinate Search

Here, we solve a low dimensional ($p = 2$) optimization problem to illustrate the operation of the simple coordinate search method. We will, in particular, use the simple coordinate search method to find the minimum of the function in Figure 18.10 (note that the point $[15, 5]^\top$ is the global minimum point and $[20, 15]^\top$ is a local minimum).

As an example, if we let $\theta(0) = [20, 2]^\top$, then we get the results in Figures 15.8 and 15.9, so that, since it started near the global minimum, it found the global minimum. On the other hand, if $\theta(0) = [16, 21]^\top$, then we get the results in Figures 15.10 and 15.11, where we see that it successfully climbed down from the high peak, but then got trapped by a local minimum. Notice that in either case, the movements are along the two coordinates of the optimization problem. Also, we use a stopping criterion based on how close ρ_j is to zero with a maximum possible number of steps as 200; hence, in both cases, the stopping criterion was invoked.

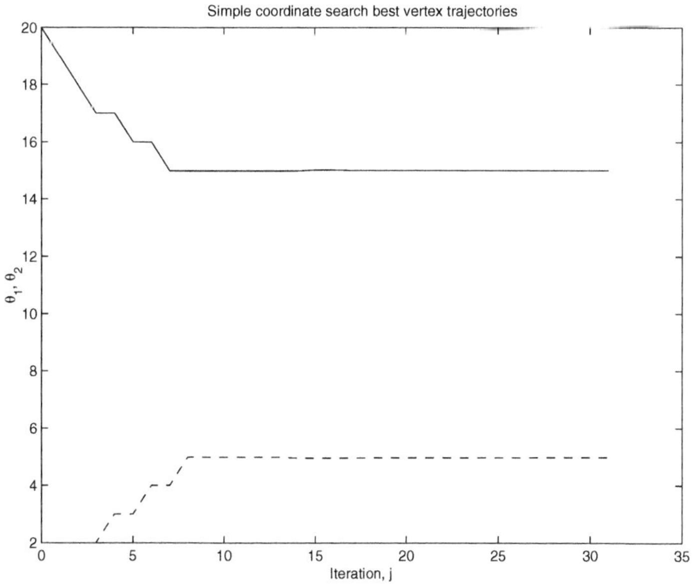

Figure 15.8: Parameter trajectories for simple coordinate search optimization problem.

Finally, note that if we had used the evolutionary operation using factorial designs method, we would get similar results, qualitatively speaking. For that method, in this problem, we would have the same number of cost evaluations per step, but that is just since $2p = 2^p$ for $p = 2$. For larger values of p, the simple coordinate search method will use far fewer cost evaluations per step. So, is the simple coordinate search method preferred? It is not appropriate to jump to such a conclusion with only the analysis presented here. Note that for some cost functions, the evolutionary operation using factorial designs method may find better directions to move along, and hence, the cost calculations at each step may be worthwhile. For other cost functions, it may be useful to only use the $2p + 1$ points on the simple coordinate search algorithm, since it may find good directions using only these. Finally, note that we do not modify the simple coordinate search method so that once it finds one pattern point that is better than the middle one, it uses that one as the new center point.

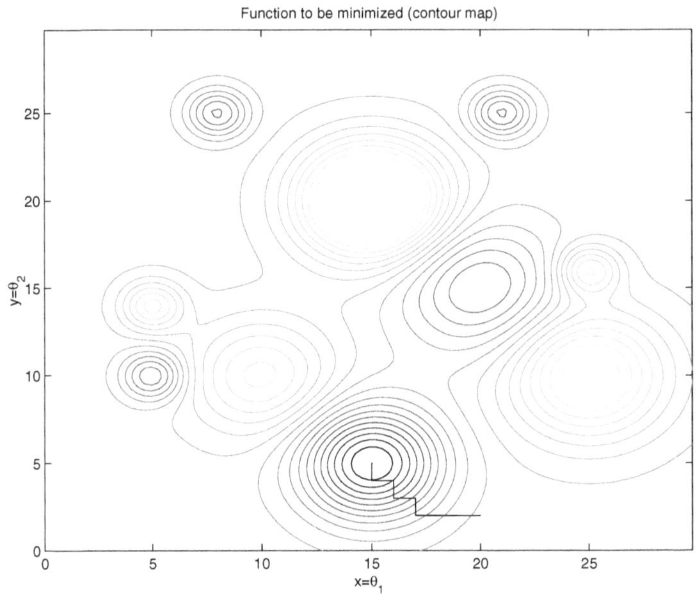

Figure 15.9: Parameter trajectories for simple coordinate search optimization problem, plotted on the contour plot of the cost function.

15.3.2 Simultaneous Perturbation Stochastic Approximation Algorithm

In this section, we introduce the simultaneous perturbation stochastic approximation (SPSA) algorithm, compare it with the use of gradient approximation approaches (e.g., as in Section 15.3.1), and then give an example of how to use the approach for a simple optimization problem. In the next section, we will show how it can be used to design a PD controller for the tanker ship. You should think of SPSA as performing optimization over a set of cost functions (response surfaces) and hence, as a good candidate for optimal design of robust systems.

SPSA Algorithm

Consider minimizing $J(\theta)$ by adjusting $\theta \in \Re^p$. We assume that the gradient $\nabla J(\theta)$ is not known analytically and that we cannot measure or compute values of $\nabla J(\theta)$ for any $\theta \in \Re^p$. Assume that given any θ, we can compute or measure $J(\theta)$ to obtain

SPSA is designed for the case when the cost function has noise.

$$J_n(\theta) = J(\theta) + w$$

where w is noise, so that we can obtain noisy measurements of $J(\theta)$ at θ (e.g., it could be that the expected value $E[w] = 0$ and variance $E[w^2]$ is finite). As an example, note that in the function approximation problem of Part III,

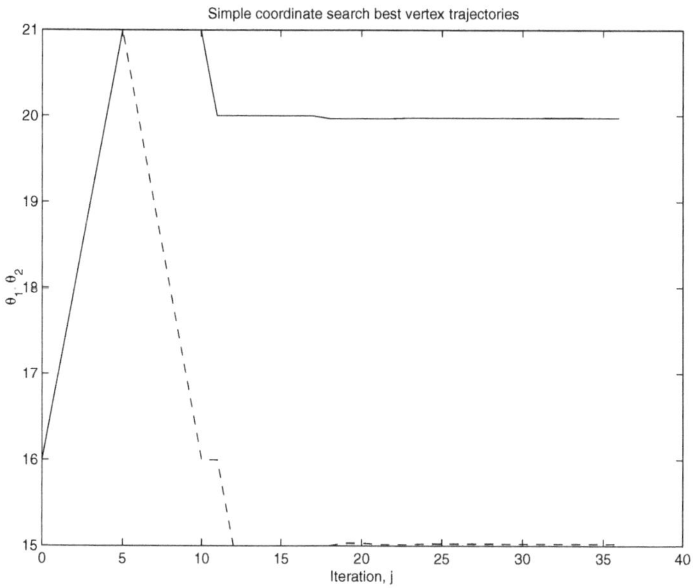

Figure 15.10: Parameter trajectories for simple coordinate search optimization problem.

we use the unknown function $G(x, z)$ where z could be noise; this makes our standard measures of approximation accuracy that we use as cost functions stochastic (even though, unlike in the last section, we often ignored this fact in the development of our algorithms there). We basically think of noise w leading to multiple cost functions.

Consider the parameter update formula

$$\theta(j+1) = \theta(j) - \lambda_j g(\theta(j), j) \tag{15.3}$$

where $g(\theta(j), j) \in \Re^p$ is an estimate of $\nabla J(\theta(j))$ at $\theta(j)$ and $\lambda_j > 0$ is a scalar step size. (A typical choice for λ_j is one where its value decreases in size as the number of iterations increases.) The dependence on j is included in $g(\theta(j), j)$, since at two different iterations with the same $\theta(j)$, we may use different approximations to the gradient (more details will be given on this below). Note that a standard projection method (as defined in Part III) can be used to keep the parameters within a known bounded (convex) region.

An approximation to the gradient is used at each step that is computed using cost evaluations at two points specified via a simultaneous perturbation from the current estimate.

Here, a "simultaneous perturbation" approximation is used for $g(\theta(j), j)$. In particular, each component of the approximation, $i = 1, 2, \ldots, p$, to the gradient is chosen as

$$g_i(\theta(j), j) = \frac{J_n(\theta(j) + c_j\Delta(j)) - J_n(\theta(j) - c_j\Delta(j))}{2c_j\Delta_i(j)} \tag{15.4}$$

where $c_j > 0$ for all j (a typical choice is to use a sequence of c_j whose values

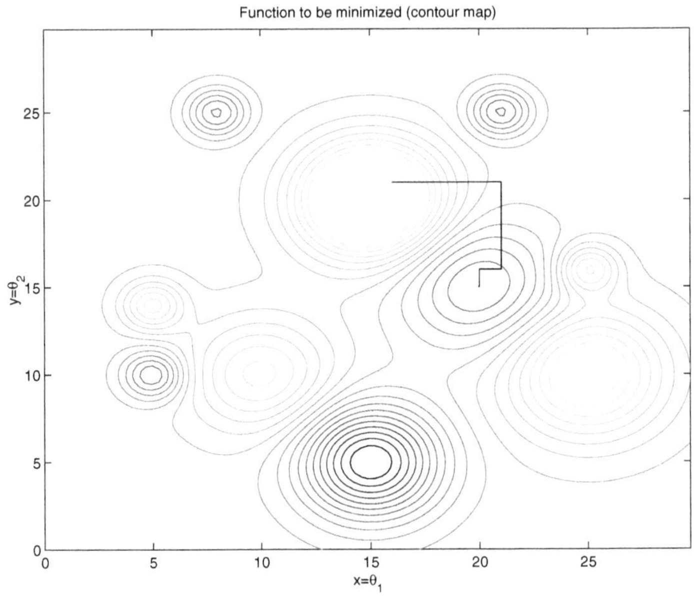

Figure 15.11: Parameter trajectories for simple coordinate search optimization problem, plotted on the contour plot of the cost function.

decrease in size as the number of iterations increases) and

$$\Delta(j) = \begin{bmatrix} \Delta_1(j) \\ \vdots \\ \Delta_p(j) \end{bmatrix}$$

is a random perturbation vector. The components of the vector $\Delta(j)$ should be independently generated from a zero mean probability distribution and one theoretically valid choice is to use a Bernoulli ± 1 distribution for each ± 1 outcome. In this way, the $\theta(j) \pm c_j \Delta(j)$ lie on corners of a hypercube centered at $\theta(j)$; it is at these values that J_n is computed in Equation (15.4). Note that projection can be used to keep the generated parameters in a known (convex) bounded region.

Note that if $p = 2$, then the $\Delta(j)$ are the corners of a unit square (i.e., one with unit magnitude for each edge) so for each j

$$\Delta(j) \in \left\{ \begin{bmatrix} 1 \\ 1 \end{bmatrix}, \begin{bmatrix} 1 \\ -1 \end{bmatrix}, \begin{bmatrix} -1 \\ 1 \end{bmatrix}, \begin{bmatrix} -1 \\ -1 \end{bmatrix} \right\}$$

In general, there are 2^p possible $\Delta(j)$ values. (How does this compare to the 2^p-factorial design used for RSM?) If $\theta(j) = [2, 2]^\top$ and $c_j = 1$, then the four corners of the square centered at $\theta(j)$ where we *might* make calculations for values of J_n in Equation (15.4) are shown in Figure 15.12. For this example, if $\Delta(j) = [1, 1]^\top$, then

$$\begin{aligned} \theta^+(j) &= \theta(j) + c_j\Delta(j) = \begin{bmatrix} 3 \\ 3 \end{bmatrix} \\ \theta^-(j) &= \theta(j) - c_j\Delta(j) = \begin{bmatrix} 1 \\ 1 \end{bmatrix} \end{aligned}$$

which are the upper right and lower left corners (denoted with large black dots) of the square centered at $\theta(j)$ in Figure 15.12 (and these values are swapped if $\Delta(j) = [-1, -1]^\top$). The points on the other diagonal of the square are chosen if $\Delta(j) = [1, -1]^\top$ or $\Delta(j) = [-1, 1]^\top$. For the $\Delta(j) = [1, 1]^\top$ case, Equation (15.4) gives, for this example,

SPSA uses the end points of a randomly selected diagonal of the hypercube centered at $\theta(j)$ to compute the estimate to the gradient.

$$g_1(\theta(j), j) = g_2(\theta(j), j) = \frac{J_n(\theta^+(j)) - J_n(\theta^-(j))}{2c_j}$$

so the approximation to the gradient is computed via a type of "central difference" approximation (see Section 15.3.1). If $\Delta(j) = [-1, -1]^\top$, then you get *exactly* the same approximation for the gradient as for when $\Delta(j) = [1, 1]^\top$ (why?). Also, the points $\Delta(j) = [1, -1]^\top$ or $\Delta(j) = [-1, 1]^\top$ both lead to the same approximations for the gradient, but generally a different one from that obtained when $\Delta(j) = [-1, -1]^\top$. (Why?)

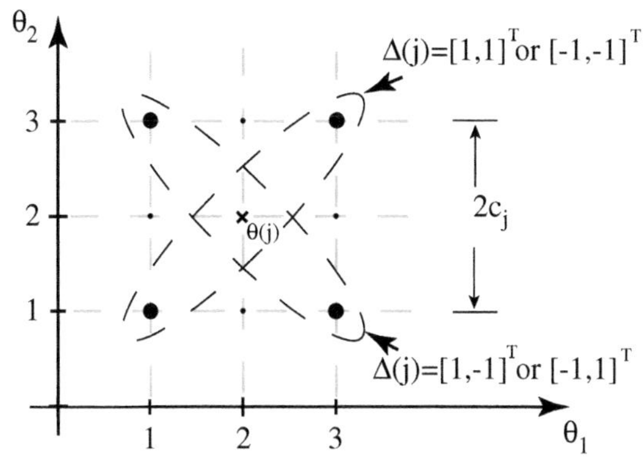

Figure 15.12: Illustration of what values are used in computing an approximation to the gradient in the SPSA method via Bernoulli perturbations.

It is useful to envision, via Figure 15.12, how the SPSA algorithm operates. Intuitively, at each iteration it generates at random two points on a hypercube centered at $\theta(j)$ that are on a diagonal. Then it uses these to compute the approximation to the gradient that is used to specify the direction of the update in Equation (15.3). For instance, consider the case for the above example where $J_n(\theta) = \theta^\top\theta$ with no noise (so $\theta = 0$ is the global minimum) and note that for one approximation to the gradient it can choose, it will be that $J_n(\theta^+(j)) - J_n(\theta^-(j)) = 0$ (so there will be no update), and for the other choice, $J_n(\theta^+(j)) - J_n(\theta^-(j)) > 0$ so that the parameters will be updated to move toward the optimum point (and if the step size is of appropriate size, it will not overshoot the origin). Also note that if you had a no-noise case with $J_n(\theta) = (\theta - \theta^*)^\top(\theta - \theta^*)$ so θ^* is the global minimum, then clearly there could be a choice of θ^* such that for the above case, one gradient approximation would lead to a parameter update that would decrease the cost, while for the other one, it could *increase* the cost. Next, note that since the c_j and λ_j sequences of values decrease at each iteration, the size of the hypercube and the size of changes to the parameter values at each iteration do also; hence, for iterations at the beginning, the algorithm will generally make larger updates and thereby explore larger regions of the optimization space. As time progresses, the size of the hypercube and parameter updates decreases and hence (under reasonably general conditions), the algorithm will converge.

Algorithm Complexity and Convergence

In general, for SPSA if you use the Bernoulli ± 1 distribution for Δ, there are 2^p possible $\Delta(j)$ vectors and $2^{(p-1)}$ possible diagonals on the hypercube centered at $\theta(j)$. Hence, if there is no noise (i.e., $w(j) = 0$ for all j), then using the SPSA approach there are in general $2^{(p-1)}$ possible approximations to the gradient at each iteration, and hence, one of $2^{(p-1)}$ possible update directions is chosen at each iteration. (Note that it is not 2^p possible directions since, for the no-noise case, specific values of the cost are computed and hence, the sign of their difference is fixed once the points at which the costs are computed are fixed.) Contrast this with the stochastic gradient method that was discussed briefly in Section 11.1.8, and where analytical gradient information is used, no noise was used for the measurement of $J(\theta)$, and one of an *infinite* number of possible update directions is chosen. For the stochastic gradient method we had constraints on the perturbations to the gradient; here, the use of points on the hypercube centered at $\theta(j)$ constrains the size and directions of the update. Moreover, choosing c_j and λ_j as decreasing sequences for SPSA, is conceptually consistent with the constraints needed for convergence for the stochastic gradient method.

Notice that for the SPSA, there are only *two* perturbations taken and these are used to compute all components of $g(\theta(j), j)$. The approach in Section 15.3.1, where the central-difference approximation is used for the gradient, has been called the "Keifer-Wolfowitz finite difference stochastic approximation" (FDSA) algorithm when it is applied to minimization of J_n. Notice that with FDSA, if there is no noise in measuring $J(\theta)$, then there is only *one* possible update direction at each iteration. Notice that for FDSA, p scaled unit vectors are used, one for computing each component of $g(\theta(j), j)$. To contrast FDSA and SPSA, use the above example and note that for FDSA, the four points that are used to compute $g(\theta(j), j)$, are the small black dots shown in Figure 15.12. Note that FDSA needs all four of these points; the large black dots in Figure 15.12

FDSA computes the cost at 2p points (two per dimension) and uses two of these per dimension to approximate the gradient.

represent points that SPSA *might* need.

Hence, the above example illustrates that while there are $2p$ calculations of J_n needed by FDSA at each iteration for the computation of the approximation to the gradient, only two such calculations are needed by SPSA (where, if $p = 1$, the methods are equivalent). In fact, it has been shown in [476] that under reasonably general conditions, SPSA and FDSA achieve the same level of statistical accuracy for a given number of iterations and

$$\frac{\text{Number of measurements of } J_n(\theta) \text{ in SPSA}}{\text{Number of measurements of } J_n(\theta) \text{ in FDSA}} \to \frac{1}{p}$$

as the number of measurements gets large. This is an important property, since it shows that SPSA could be more efficient for large scale optimization problems, and that this is certainly the case when comparing to the FDSA method. The types of conditions for convergence that are needed for the SPSA method include certain conditions on c_j and λ_j (that are guaranteed with the choices to be outlined in the next section), the variance on the noise on the cost function must satisfy certain bounds, the size of $\theta(k)$ must "almost surely" be bounded for all k, a stationary point must be "attractive" in a certain way, and the estimate must visit a certain region near the stationary point infinitely often [476].

Guidelines for Choosing SPSA Parameters

There are several parameters that must be specified for the SPSA algorithm and here we outline some of the guidelines in [477] for their choice. First, choose

$$\lambda_j = \frac{\lambda}{(\lambda_0 + j)^{\alpha_1}}$$

where $\lambda > 0$, $\lambda_0 > 0$, and $\alpha_1 > 0$, and

$$c_j = \frac{c}{j^{\alpha_2}}$$

where $c > 0$ and $\alpha_2 > 0$. However, if the θ_i have very different magnitudes, you may want to use a different λ_j for each of the p dimensions. This can be difficult at times in practice, however, so another approach is to scale the parameter values themselves.

Some actual values that have been found useful in applications are

$$\alpha_1 = 0.602$$

and

$$\alpha_2 = 0.101$$

which are effectively the lowest allowable ones that satisfy theoretical conditions. However, values $\alpha_1 \in [0.602, 1]$ and $\alpha_2 \in [0.101, \frac{1}{6}]$ may also work. In fact, $\alpha_1 = 1$ and $\alpha_2 = \frac{1}{6}$ are the "asymptotically optimal" values so, if the algorithm runs for a long time, it may be beneficial to switch to these values. With this choice, if the noise is significant, you may need to choose λ smaller and c larger than in a low-noise case.

With the Bernoulli ± 1 choice for the components of $\Delta(j)$, set c to a level that is approximately the standard deviation of the noise $w(j)$ to keep the components of $g(\theta(j), j)$ from being too large in magnitude. If there is no noise term $w(j)$, then you should choose some small value $c > 0$. You can choose λ_0 to be approximately 10% of the maximum number of iterations and λ to try to achieve a certain amount of change in the cost function values at each iteration.

Example: Solving an Optimization Problem with SPSA

Here, we solve a low dimensional ($p = 2$) optimization problem to illustrate the operation of the SPSA algorithms. Hence, our goal here is not to illustrate the full advantages of the approach, which are realized for *high* dimensional problems, but simply to introduce the method. We will, in particular, use the SPSA method to find the minimum of the noise-free function in Figure 18.10 (note that the point $[15, 5]^\top$ is the global minimum point and $[20, 15]^\top$ is a local minimum).

Convergence to Local Minimum: Using the guidelines above, and a bit of tuning, we selected $\lambda = \lambda_0 = 1$, $c = 0.01$, $\alpha_1 = 0.602$, and $\alpha_2 = 0.101$. We use projection to keep both parameters in $[0, 30]$. We use 100 iterations of the algorithm and for $\theta(0) = [15, 15]^\top$, we get the results shown in Figures 15.13 and 15.14. In this case, due to the initialization, the SPSA converges to a local minimum. Clearly, if you set the initial condition closer to the global optimum value, the algorithm will converge to it.

Convergence to the Global Minimum: Next, we use all the same parameters for the SPSA except we use $\lambda = 20$ and run the algorithm for 200 iterations. We choose this value of λ simply to make the step sizes larger so that the algorithm will make larger steps, and, hopefully, avoid the local minimum that it got stuck in for the last case. We ran the algorithm for twice as many steps so that the step size will adequately decrease and we will get convergence if it finds a local (or global) minimum.

In this case, the results are shown in Figures 15.15 and 15.16. In this case, the algorithm overcomes the bad initialization and finds the global minimum since, in the early steps of the optimization process, it aggressively investigates the region around where it was initialized and is lucky to find itself near the global minimum. Note, however, that to get this result, the algorithm had to be run several times. Sometimes the algorithm seemed to get stuck at the boundary of the optimization region. Other times it converged to a local minimum.

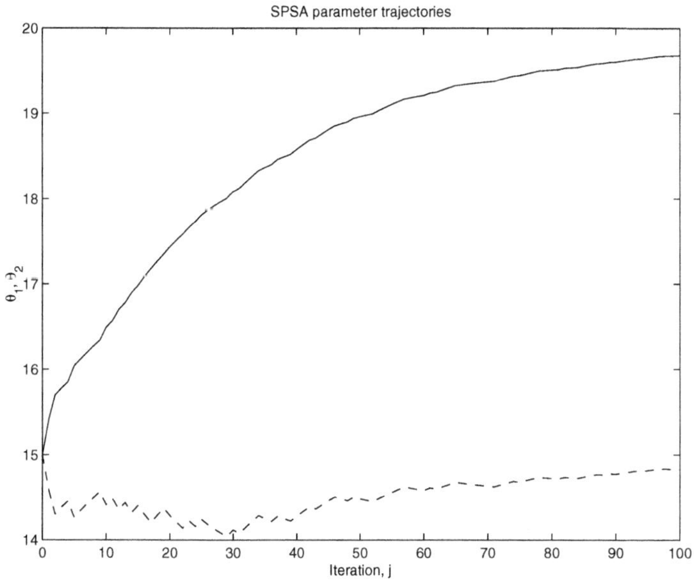

Figure 15.13: Parameter trajectories for SPSA optimization example.

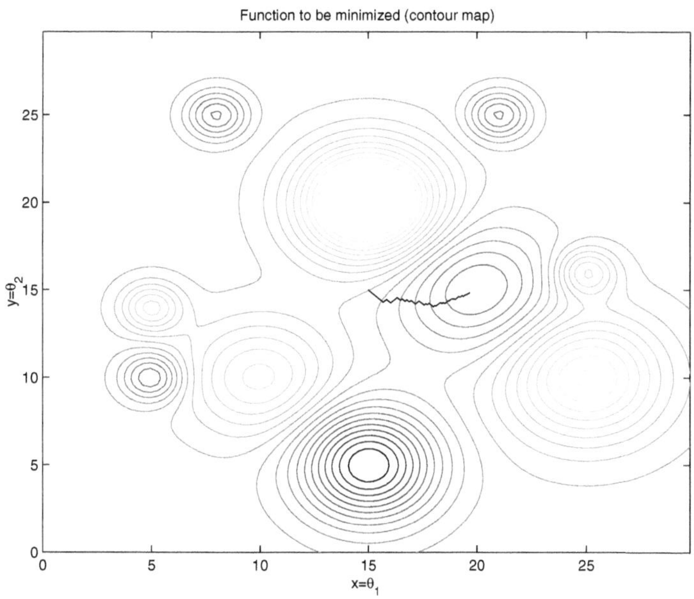

Figure 15.14: Parameter trajectories for SPSA optimization example, plotted on contour plot of cost function.

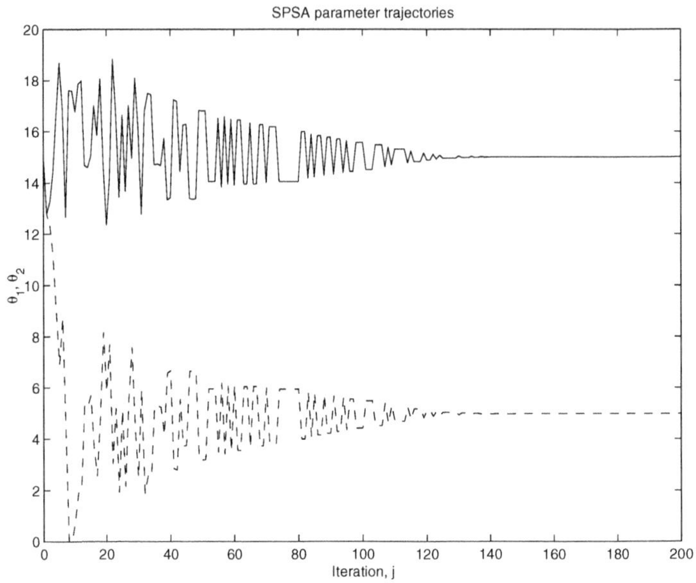

Figure 15.15: Parameter trajectories for SPSA optimization example.

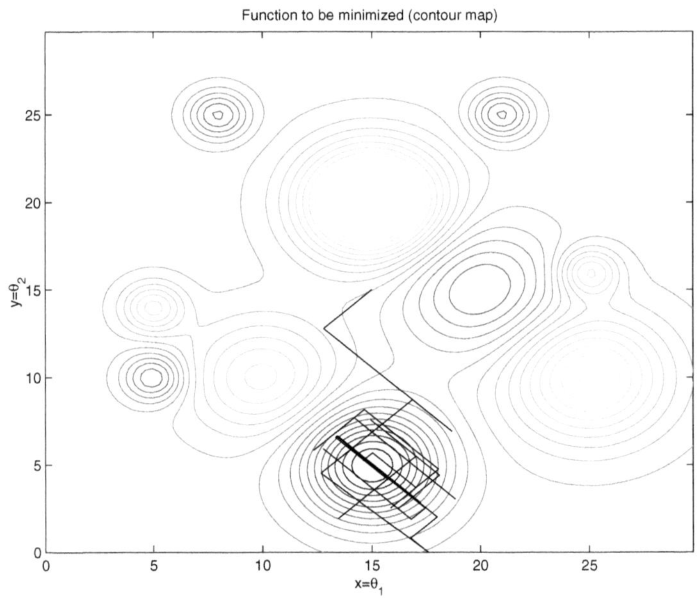

Figure 15.16: Parameter trajectories for SPSA optimization example, plotted on contour plot of cost function.

15.3.3 Nelder-Mead Simplex Method

One nongradient method that has enjoyed success in several practical optimization problems is the Nelder-Mead "simplex method" (not to be confused with the simplex methods often used for solving optimization problems with linear cost functions and linear inequality constraints).

Candidate Solutions as Vertices of a Simplex

For this method, at iteration j, we start with a "simplex" (a "convex hull") of a set of $p+1$ points (vertices)

$$P = \{\theta^0, \theta^1, \ldots, \theta^p\} \subset \Re^p$$

and the method generates another simplex for iteration $j+1$, $j = 0, 1, 2, \ldots$. Note that each vertex of the simplex is a $p \times 1$ parameter vector that we will think of as a candidate solution to the optimization problem.

Using cost information at each candidate vertex, the simplex shape is adjusted by moving the vertices in directions that appear to be promising for reducing the cost function.

First, we define a convex hull. A "convex combination" of elements of P is a vector of the form

$$\sum_{i=0}^{p} \alpha_i \theta^i$$

where $\theta^i \in P$, $i = 0, 1, \ldots, p$, and $\alpha_0, \ldots, \alpha_p$ are scalars such that $\alpha_i \geq 0$ with

$$\sum_{i=0}^{p} \alpha_i = 1$$

The convex hull of P, which is denoted by $conv(P)$, is the set of all convex combinations of elements of P.

The set $conv(P)$ is a "convex set." Also, $conv(P)$ is the intersection of all convex sets containing P. A set $C \subset \Re^p$ is convex if $\alpha\theta^1 + (1-\alpha)\theta^2 \in C$, for all $\theta^1, \theta^2 \in C$, and all $0 < \alpha < 1$ (i.e., it is convex if for any two points in C, the line segment between these two points is in C). There are several useful properties of convex sets. First, there are set operations that preserve convexity, in the sense that, if you start with a convex set and perform these set operations, the resulting set that is produced is also a convex set. If C is a convex set, and b is a real number, the set $\{\theta : \theta = b\theta_c, \theta_c \in C\}$ is a convex set (i.e., we can scale the set and still get a convex set). If C and D are convex sets, then $C + D = \{\theta : \theta = \theta_c + \theta_d, \theta_c \in C, \theta_d \in D\}$ is convex (i.e., if we add all possible combinations of elements from C and D, then the resulting set is convex). Also, the intersection of any collection of convex sets is a convex set.

Initialization and Algorithm Overview

Next, given one simplex at iteration j with the set of vertices

$$P(j) = \{\theta^0(j), \theta^1(j), \ldots, \theta^p(j)\}$$

we explain how to generate the next set of vertices $P(j+1)$ for the next simplex. We will need to initialize the algorithm with a set of vertices $P(0)$ and one way to do this is to let the elements of $P(0)$ be

$$\theta^i = \theta^0 + \beta_i e_i$$

where $i = 1, 2, \ldots, p$ where θ^0 is a guess at where the solution is, the β_i are scalars, and e_i is the ith column of the $p \times p$ identity matrix. Although not necessary, one approach that is sometimes used is to try to initialize so that the minimum point of the cost function is contained within the convex hull of $P(0)$. This idea provides one set of guidelines on how to choose θ^0 and the β_i for the above initialization approach (e.g., pick the β_i large enough so that the solution is likely to be in the initial convex hull). If you do not, however, use this approach, the method can often still reorient and shift the simplex so that the optimal solution can still be found. (Here, all we seek to do is provide one reasonable method to initialize the algorithm; this is certainly not the best method in all cases.)

To proceed, first define $\theta^{min}(j)$ and $\theta^{max}(j)$ to be the best and worst vertices of the set $P(j)$ so

$$\begin{aligned} J(\theta^{min}(j)) &= \min_{i=0,1,\ldots,p} J(\theta^i(j)) \\ J(\theta^{max}(j)) &= \max_{i=0,1,\ldots,p} J(\theta^i(j)) \end{aligned}$$

Also, let $\theta^{cent}(j)$ be the centroid of the face of the simplex formed by all the vertices in $P(j)$ *except* $\theta^{max}(j)$. Hence,

$$\theta^{cent}(j) = \frac{1}{p}\left(-\theta^{max}(j) + \sum_{i=0}^{p} \theta^i(j)\right)$$

The algorithm seeks to replace $\theta^{max}(j)$ by a point with a lower cost (which we will call $\theta^{new}(j)$ below). To explain how it does this, consider the case where $p = 2$ so we have 3 vertices and each simplex is a triangle. Follow the discussion below by continually referring to the associated figures, where, for instance, in Figure 15.17, the shaded triangles outlined with solid lines represent the (current) simplex at iteration j, triangles outlined with dashed lines represent simplices that could be generated at iteration $j+1$, and the ovals represent a contour map of the surface $J(\theta)$ that we are trying to find a minimum point on (the smaller ovals indicate that the cost function $J(\theta)$ has decreased).

To begin with, a "reflection point" $\theta^{ref}(j)$ is generated. (We will think of this as a test point that will help us decide how to reorient the simplex.) To do this by hand, draw an imaginary line that passes through $\theta^{max}(j)$ and bisects the line between $\theta^{min}(j)$ and $\theta^i(j)$ (the vertex with intermediate cost) at $\theta^{cent}(j)$ (see Figure 15.17, where this imaginary line is the dotted one). Place $\theta^{ref}(j)$ at that point on that line that would result by reflecting $\theta^{max}(j)$ about the $\theta^{cent}(j)$ point. See Figure 15.17. Next, we evaluate the cost $J(\theta^{ref}(j))$ and then decide what to do. There are three possibilities to consider.

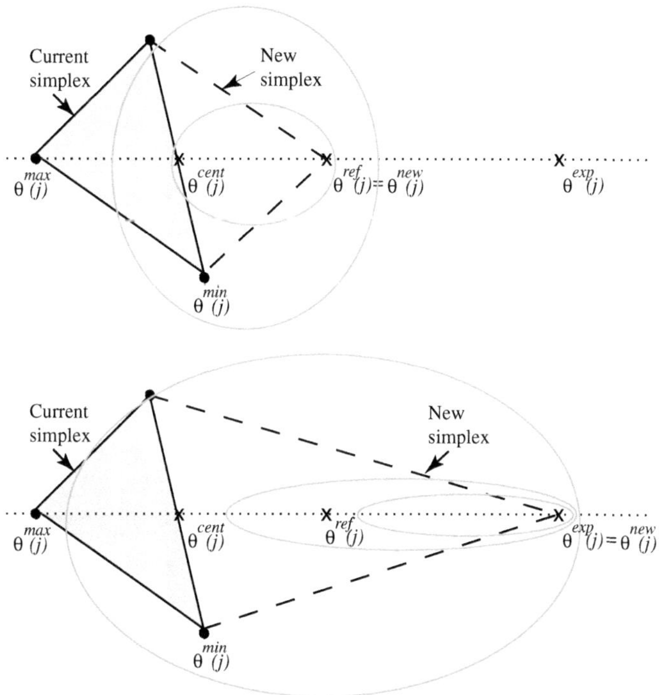

Figure 15.17: Direct search sample steps, for case when $\theta^{ref}(j)$ is better than any vertex on the current simplex (top sketch, expansion is not used; bottom sketch, expansion is used).

First, it could be that $\theta^{ref}(j)$ could have the lowest cost of any of the vertices. This is the case in Figure 15.17. In this case, the direction along the line that bisects $\theta^{min}(j)$ and $\theta^{i}(j)$ appears to be a good one, so we create an "expansion point" $\theta^{exp}(j)$ that is the point that would result if we reflected $\theta^{max}(j)$ about $\theta^{ref}(j)$. If this $\theta^{exp}(j)$ point is even better than $\theta^{ref}(j)$, then we will replace the worst point in the existing simplex with $\theta^{exp}(j)$ to form the new simplex (see the bottom sketch in Figure 15.17). However, if $\theta^{exp}(j)$ is worse than $\theta^{ref}(j)$, we simply use $\theta^{ref}(j)$ to replace $\theta^{max}(j)$ to form the new vertex (see the top sketch in Figure 15.17). In Figure 15.17, the new simplices for each case are shown in dashed lines. Carefully study the relations between the choice of the new simplex and the quality of the vertices and other points.

Second, it could be that $\theta^{ref}(j)$ has an intermediate cost relative to the vertices, besides the worst one (i.e., it is better than the second to the worst vertex, but not as good as the best one). This is depicted in Figure 15.18. In this case, we simply form the new simplex by replacing $\theta^{max}(j)$ by $\theta^{ref}(j)$, since the resulting simplex is better in the sense that we will have reduced the cost of the worst vertex.

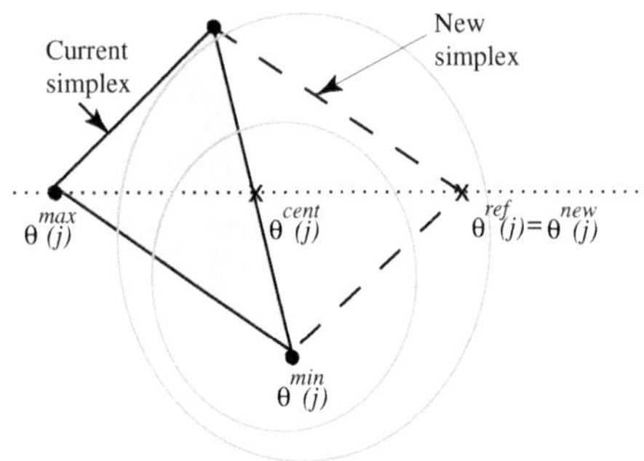

Figure 15.18: Direct search sample step, for case when $\theta^{ref}(j)$ has an intermediate cost.

Third, it could be that $\theta^{ref}(j)$ has a cost that is worse than all the vertices, possibly even $\theta^{max}(j)$. In this case, we perform a "contraction" of the simplex, since stretching it in the direction of $\theta^{ref}(j)$ did not seem to be a good approach. To contract, we simply let the new vertex be an interpolation between $\theta^{max}(j)$ and $\theta^{cent}(j)$ if $\theta^{max}(j)$ was better than $\theta^{ref}(j)$ (see the bottom sketch in Figure 15.19). On the other hand, we let the new vertex be an interpolation between $\theta^{ref}(j)$ and $\theta^{cent}(j)$ if $\theta^{max}(j)$ was worse than $\theta^{ref}(j)$ (see the top sketch in Figure 15.19). Hence, this tries to place $\theta^{new}(j)$ on the line through $\theta^{max}(j)$ and $\theta^{cent}(j)$, nearer to $\theta^{max}(j)$ if it is a better point, and nearer to $\theta^{ref}(j)$ if it was better than $\theta^{max}(j)$.

Finally, note that it is possible that in the contraction step (either case), the resulting $\theta^{new}(j)$ is worse than $\theta^{max}(j)$. For instance, if we use the contraction step and $\theta^{max}(j)$ is worse than $\theta^{ref}(j)$, then we can have the situation in Figure 15.20, where $\theta^{new}(j)$ is worse than $\theta^{max}(j)$. Clearly, it does not make sense to use this new vertex so instead, the new simplex is formed by simply "shrinking" the current simplex towards the best vertex $\theta^{min}(j)$ as shown in Figure 15.20. A simple redrawing of the contours can show the case where $\theta^{max}(j)$ is better than $\theta^{ref}(j)$, and in the contraction step, the algorithm generates a $\theta^{new}(j)$ that is between $\theta^{max}(j)$ and $\theta^{cent}(j)$, but with $\theta^{new}(j)$ worse than $\theta^{max}(j)$. In this case, the same new simplex shown in Figure 15.20 is used.

Generally, the method reorients the simplex to the local landscape of the cost function. It elongates down long inclined planes (e.g., via the expansion step), changes direction when encountering a valley at an angle, and contracts in the neighborhood of a minimum (e.g., via shrinking).

The Steps of the Algorithm

We are now prepared to specify the algorithm, which is given by the following steps:

1. Given $P(0)$, let $j = 0$. Find $\theta^{min}(j)$, $\theta^{max}(j)$, and $\theta^{cent}(j)$.

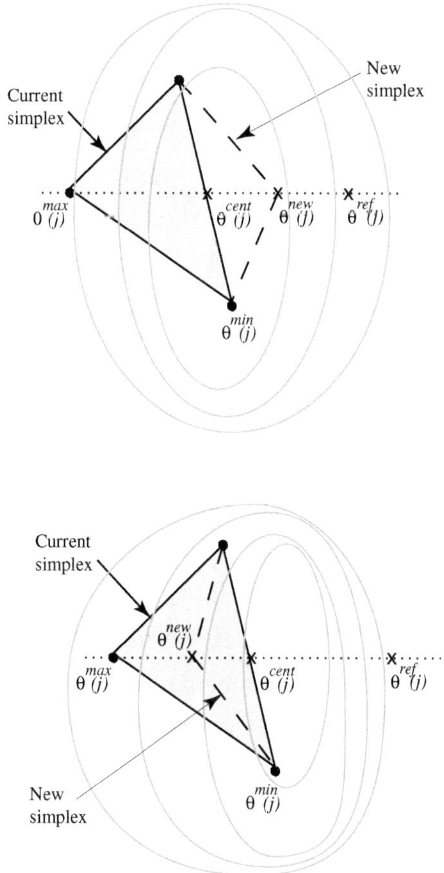

Figure 15.19: Direct search sample steps, for case when $\theta^{ref}(j)$ has a bad (high) cost relative to other vertices.

2. Compute and evaluate the cost of the reflection point. Find

$$\theta^{ref}(j) = \theta^{cent}(j) + \beta\left(\theta^{cent}(j) - \theta^{max}(j)\right)$$

   where one common choice for the "reflection coefficient" is $\beta = 1$. Now, depending on how good $\theta^{ref}(j)$ is, we take three different actions:

   - If $J(\theta^{min}(j)) > J(\theta^{ref}(j))$, then go to step 3 (in this case, $\theta^{ref}(j)$ has the minimum cost so it is a good value).
   - If

$$\max\left\{J(\theta^i(j)) : \theta^i(j) \neq \theta^{max}(j)\right\} > J(\theta^{ref}(j)) \geq J(\theta^{min}(j))$$

     then go to step 4 (in this case, $\theta^{ref}(j)$ is not the best or worst vertex, it has an intermediate cost).

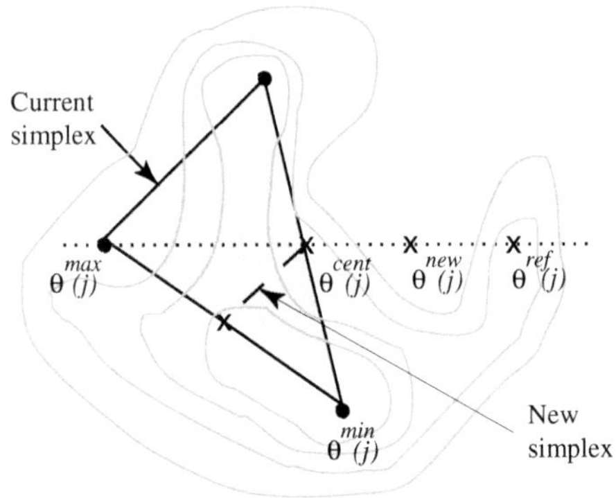

Figure 15.20: Direct search sample steps, for case when $\theta^{new}(j)$ is worse than $\theta^{max}(j)$, so we shrink the simplex toward the best point $\theta^{min}(j)$.

- If
$$J(\theta^{ref}(j)) \geq \max\left\{J(\theta^i(j)) : \theta^i(j) \neq \theta^{max}(j)\right\}$$
then go to step 5 (in this case, $\theta^{ref}(j)$ has the worst cost, except possibly that of $\theta^{max}(j)$, so it is not a good value).

3. Reflection point is the new minimum point so attempt expansion (see Figure 15.17). Find
$$\theta^{exp}(j) = \theta^{ref}(j) + \gamma\left(\theta^{ref}(j) - \theta^{cent}(j)\right) \tag{15.5}$$
(think of the "expansion point" $\theta^{exp}(j)$ as an attempt to move further in the direction where it has found cost improvement by finding $\theta^{ref}(j)$) and let
$$\theta^{new}(j) = \begin{cases} \theta^{exp}(j) & \text{if } J(\theta^{exp}(j)) < J(\theta^{ref}(j)) \\ \theta^{ref}(j) & \text{otherwise} \end{cases} \tag{15.6}$$
so that if the expansion point was even better, we take it as the new point in our simplex (so the simplex tries to stretch toward the minimum). The size of the expansion is governed by the expansion coefficient γ where one common choice is $\gamma = 1$. Next, go to step 7.

4. Reflection point has an intermediate cost, so replace worst vertex with it (see Figure 15.18). Define
$$\theta^{new}(j) = \theta^{ref}(j)$$
In this case, we use $\theta^{ref}(j)$ as a new vertex, since it is better than $\theta^{max}(j)$, the one that will be replaced. Next, go to step 7.

5. Reflection point is not a good one so "contract" the simplex (see Figure 15.19). Define

$$\theta^{new}(j) = \begin{cases} \alpha\theta^{max}(j) + (1-\alpha)\theta^{cent}(j) & \text{if} J(\theta^{max}(j)) \leq J(\theta^{ref}(j)) \\ \alpha\theta^{ref}(j) + (1-\alpha)\theta^{cent}(j) & \text{otherwise} \end{cases} \tag{15.7}$$

where α is the "contraction coefficient" $(0 < \alpha < 1)$. In this case, $\theta^{ref}(j)$ was not a good point in step 2 (it was worse than all those, except possibly $\theta^{max}(j)$). Now, if $\theta^{ref}(j)$ is worse than $\theta^{max}(j)$, then we create $\theta^{new}(j)$ by interpolating between $\theta^{max}(j)$ and $\theta^{cent}(j)$ with α parameterizing the interpolation. If, on the other hand, $\theta^{ref}(j)$ is better than $\theta^{max}(j)$, then we create $\theta^{new}(j)$ by interpolating between $\theta^{ref}(j)$ and $\theta^{cent}(j)$. Next, go to step 6.

6. If the cost at the new vertex is worse than any existing one, then shrink the simplex towards the best vertex to form the new simplex (see Figure 15.20). If

$$J(\theta^{new}(j)) > J(\theta^{max}(j))$$

then it does not make sense to replace $\theta^{max}(j)$ with $\theta^{new}(j)$. In this case, replace all vertices in $P(j)$ by letting

$$P(j+1) = \left\{ \frac{1}{2}\left(\theta^{i}(j) + \theta^{min}(j)\right) : i = 0, 1, 2, \ldots, p \right\}$$

This "shrinks" the old $P(j)$ simplex towards the best vertex $\theta^{min}(j)$. Go to step 8.

7. The new vertex is better than the worst existing one, so replace the worst one with the new one to form the new simplex. In this case, $\theta^{new}(j)$ is good (at least better than $\theta^{max}(j)$) so we form a new simplex using

$$P(j+1) = P(j) - \{\theta^{max}(j)\} + \{\theta^{new}(j)\}$$

(i.e., remove the worst vertex and replace it with the new one, so the mathematical addition and subtraction operations in the previous equation are set operations). Go to step 8.

8. Let $j = j + 1$ and go to step 2.

This completes the specification of the Nelder-Mead simplex method. There are no convergence guarantees for this particular direct search algorithm.

Algorithm Complexity

There are many issues that affect the complexity of the algorithm. First, note that for high dimensional optimization problems, p is large and there are $p+1$ vertices that will require memory and computations. Next, note that the algorithm requires computing $J(\theta)$ for a variety of values of θ, vector addition, and scaling of vectors, in addition to computations of maximum values of sets of values of the cost function, and so on.

For many applications, the computations of the cost $J(\theta)$ are the most computationally intensive; hence, we consider how many such calculations are needed of the cost function. First, suppose that initially you compute the costs of all the vertices; this would entail $p+1$ calculations of the cost initially. Next, *at each iteration*, the algorithm requires the following computations for various paths that the computations can take in the algorithm:

- Steps 2, 4, 7, 8: $J(\theta^{ref})$ (one cost calculation)
- Steps 2, 3, 7, 8: $J(\theta^{ref})$, $J(\theta^{exp})$ (two cost calculations)
- Steps 2, 5, (6), 8: $J(\theta^{ref})$ (one cost calculation, not including step 6). Note, however, that Step 6 is optional and if you put in this optional "shrink" step, then you must also compute $J(\theta^{new}(j))$ and if $J(\theta^{new}(j)) > J(\theta^{max}(j))$, then to perform the shrink operation, you need to compute costs *for all* the vertices of the simplex except the one for $\theta^{min}(j)$. Hence, the total possible calculations if step 6 is included in the algorithm is $p+1$.

In summary, if the shrink step is not included, then each iteration only requires two calculations of the cost at most. If the shrink step is included, then there could be as many as $p+2$ calculations of the cost. Clearly, then you may want to evaluate the utility of adding the shrink step; however, use of the algorithm in practice has shown that in many applications, it is only used rarely, so the overhead in calculations may be worthwhile. Keep in mind that in all these discussions, there is no guarantee that there will be a reduction in cost from one iteration to the next, and this must be taken into consideration when comparing the complexity of this method to others. Indeed, there are counter-examples that show that the Nelder-Mead simplex method can get stuck at a nonstationary point and this also must be taken into consideration when evaluating computational complexity.

There are counterexamples that show that the Nelder-Mead simplex method can get stuck at a nonstationary point.

Example: Solving an Optimization Problem with Direct Search

In this section, we apply the Nelder-Mead simplex method to finding the minimum point of the function shown in Figure 18.10 (note that the point $[15,5]^\top$ is the global minimum point and $[20,15]^\top$ is a local minimum).

Starting with a Good Guess for the Initial Simplex: Here, $p=2$, so we have three vertices. We use $\beta=\gamma=1$ and $\alpha=0.5$. We start the initial set of vertices as

$$P(0)=\left\{\begin{bmatrix}20\\5\end{bmatrix},\begin{bmatrix}15\\0\end{bmatrix},\begin{bmatrix}15\\15\end{bmatrix}\right\}$$

To see how the algorithm operates in this case, see Figure 15.21, where we show the simplices on top of a contour plot of Figure 18.10. Here, the vertices of $P(0)$ are shown with circles, and the simplex of $P(0)$ is outlined with solid lines, in the upper left plot. As the algorithm operates, it reorients the simplex. For instance, from iteration $j = 2$ to iteration $j = 3$, it accepts the reflection point that was generated as a new vertex, so it essentially rotates the simplex to find the new one. Figure 15.22 shows that the simplex is properly reoriented so that it "falls" into the portion of the surface where the global minimum is located.

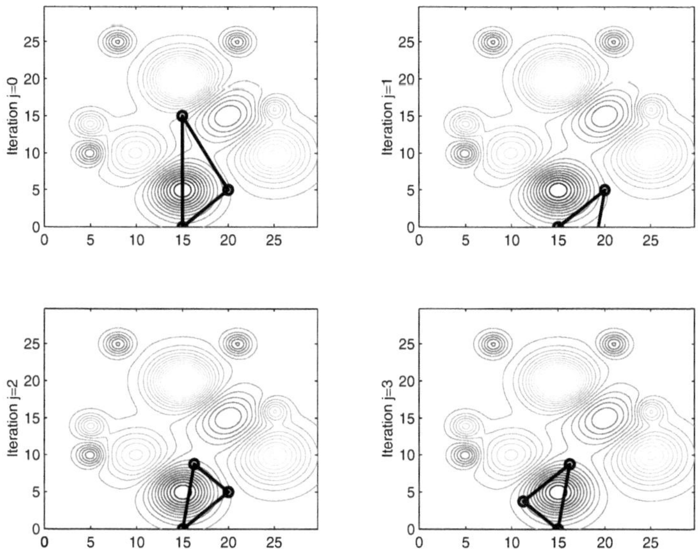

Figure 15.21: First four simplices of Nelder-Mead simplex algorithm on top of contour plots of Figure 18.10, where the global minimum is at $(15, 5)$.

To get an idea of how the method operates for later iterations, see Figure 15.23, where we show the best vertex of simplex at each iteration (notice that it converges) and the cost of the best vertex at each iteration.

Other Choices for the Initial Simplex: First, consider what happens if we simply place the simplex so that the points are on the boundaries of the search region. If you choose

$$P(0) = \left\{ \begin{bmatrix} 0 \\ 0 \end{bmatrix}, \begin{bmatrix} 15 \\ 30 \end{bmatrix}, \begin{bmatrix} 30 \\ 0 \end{bmatrix} \right\}$$

or

$$P(0) = \left\{ \begin{bmatrix} 0 \\ 15 \end{bmatrix}, \begin{bmatrix} 30 \\ 0 \end{bmatrix}, \begin{bmatrix} 30 \\ 30 \end{bmatrix} \right\}$$

or

$$P(0) = \left\{ \begin{bmatrix} 0 \\ 0 \end{bmatrix}, \begin{bmatrix} 0 \\ 30 \end{bmatrix}, \begin{bmatrix} 30 \\ 15 \end{bmatrix} \right\}$$

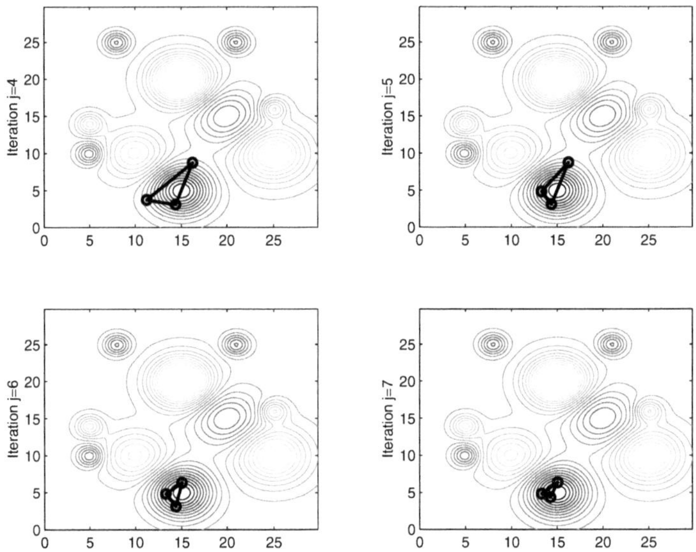

Figure 15.22: Simplices for iterations $j = 4$ to $j = 7$ of Nelder-Mead simplex algorithm on top of contour plots of Figure 18.10, where the global minimum is at $(15, 5)$.

the algorithm will find the global minimum at $(15, 5)$. However, if you start it at

$$P(0) = \left\{ \begin{bmatrix} 0 \\ 30 \end{bmatrix}, \begin{bmatrix} 15 \\ 0 \end{bmatrix}, \begin{bmatrix} 30 \\ 30 \end{bmatrix} \right\}$$

the algorithm finds the local minimum at $(20, 15)$. It will also find this local minimum if the simplex starts close to it. This shows that a good initialization can be important for convergence (not surprising).

15.3.4 The Multidirectional Search Method

In this section, we introduce another direct search method. It is one that, compared to the Nelder-Mead simplex method, has guaranteed convergence properties and is a type of pattern search method (there is an underlying pattern of points that is investigated).

The Multidirectional Search Algorithm

Again, consider minimizing $J(\theta)$, $\theta \in \Re^p$, and assume that J is continuous in θ and that $\nabla J(\theta)$ *exists* (but we neither assume that we have an analytical description of $\nabla J(\theta)$ nor that we can obtain measurements of the gradient).

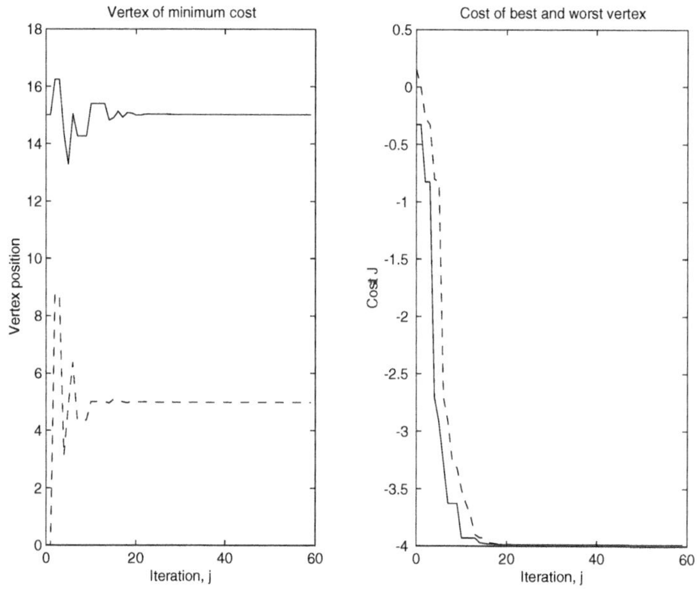

Figure 15.23: Best vertex of the simplex at each iteration (where the global minimum is at (15, 5)) and the cost of the best vertex at each iteration (where the cost at (15, 5) is $J = -4$).

Similar to the Nelder-Mead simplex method, multidirectional search iterates on a simplex of $p+1$ candidate solutions

$$P = \left\{\theta^0, \theta^1, \ldots, \theta^p\right\} \subset \Re^p$$

Next, we provide some intuition behind the operation of the multidirectional search algorithm. First, suppose that *at each iteration* $\theta^0(j)$ is the best vertex of $P(j)$ so

$$J(\theta^0(j)) \leq J(\theta^i(j)),\ i = 0, 1, \ldots, p$$

The goal at each iteration is to find another candidate solution with a cost that is strictly less than $J(\theta^0(j))$. To do this, it searches along lines passing through $\theta^0(j)$ and its p adjacent vertices. To see how it does this, consider the $p = 2$ case in Figure 15.24, where the current simplex $P(j)$ has its vertices connected by black lines. First, there is a "rotation" step (analogous to the reflection step in the Nelder-Mead method), where $\theta^1(j)$ and $\theta^2(j)$ are reflected about $\theta^0(j)$ on the gray dashed lines to obtain $\theta^1_{rot}(j)$ and $\theta^2_{rot}(j)$, respectively (hence the simplex $P(j)$ is *rotated* about $\theta^0(j)$). If the cost at $\theta^i_{rot}(j)$, $i = 1, 2$, is better than the one at $\theta^0(j)$, then an "expansion" is computed by reflecting about $\theta^0(j)$, but more than twice as far as in the rotation step, to produce $\theta^i_{exp}(j)$, $i = 1, 2$. If the cost at any of these two new vertices, $\theta^i_{exp}(j)$, $i = 1, 2$, is better than the cost at $\theta^i_{rot}(j)$, $i = 1, 2$, then accept the minimum cost vertex from the expansion as the new minimum cost vertex ($\theta^0(j+1)$) and it, with $\theta^i_{exp}(j)$ (i so that $\theta^i_{exp}(j) \neq \theta^0(j+1)$) and $\theta^0(j)$ define the new simplex $P(j+1)$. If expansion does not result in points with lower cost than the costs for the new vertices from the rotation step, then you accept the minimum cost vertex from the rotation step as $\theta^0(j+1)$ and it, together with $\theta^0(j)$ and the other vertex from the rotation step, form the new simplex. Now, if $\theta^i_{rot}(j)$, $i = 1, 2$, did not result in a lower cost than the one at $\theta^0(j)$ (so expansion was not used), then you take a "contraction" step where you move the adjacent vertices $\theta^1(j)$ and $\theta^2(j)$ toward $\theta^0(j)$ along the gray dashed lines to the points $\theta^i_{cont}(j)$, $i = 1, 2$. The process of rotation, expansion, and contraction repeats until a stopping criterion is satisfied. As an example, consider the $p = 1$ case and if you sketch a few iterations of the algorithm, you will see a similarity to "line search methods."

Multidirectional search can be thought of as a type of multidimensional line search where it expands in directions that appear promising, and contracts if promising points were not found.

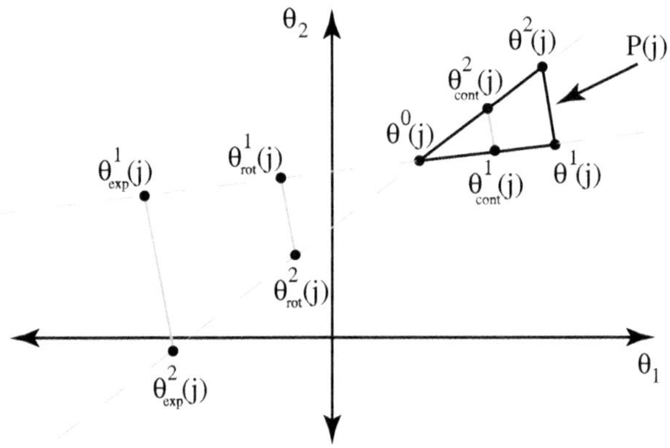

Figure 15.24: Illustration of search directions and possible next simplices for multidirectional search for $p = 2$.

Simplex methods try to ignore "noise" perturbations on the cost by using a region-based approximation to the gradient of the underlying smooth function that dictates how the simplex moves.

Finally, note that while the explicit analytical gradient information is not available and we do not assume we can get measurements of the gradient, there is a type of approximation to the gradient that is being used in this simplex method. The method explores points and uses these to set the direction of updates of the simplex. Since it does this *over a region*, we can hope that it will not get caught in local minima, but try to follow the gross characteristics of the function to find lower cost values. In fact, a mathematical definition of the "simplex gradient" is given in [275], where it is explained that simplex methods can often be effectively used to perform optimization for functions like the one in Figure 15.25. Note that standard gradient methods will get trapped in one of the many local minima, depending on how the algorithm is initialized. The simplex methods sample the cost function in a way to ignore the "noise" perturbations and try to approximate the gradient of the underlying smooth function and hence, try to move in directions to minimize its value.

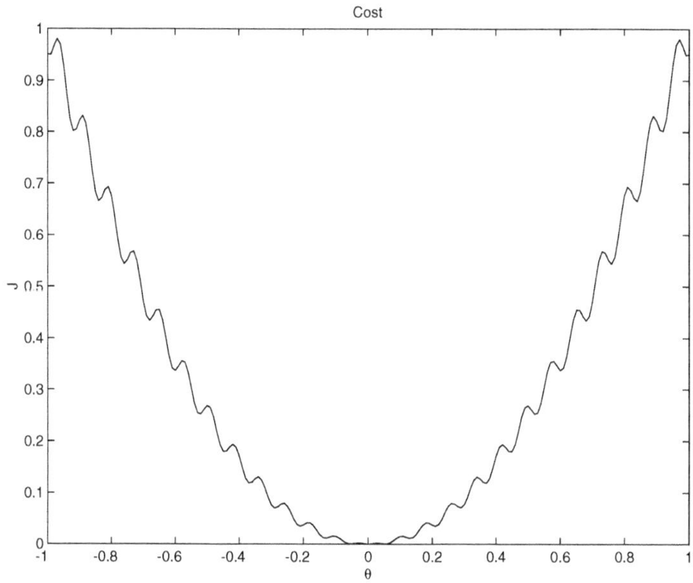

Figure 15.25: An example cost function with multiple local minima.

Steps of the Algorithm

The multidirectional search algorithm uses two parameters: an expansion factor γ_e that is a rational number (i.e., there exist two integers p_1 and $p_2 \neq 0$ such that $p_1/p_2 = \gamma_e$) with

$$\gamma_e \in (1, \infty),$$

and a contraction factor

$$\gamma_c = \frac{1}{\gamma_e}$$

For example, you may use $\gamma_e = 2$ and $\gamma_c = \frac{1}{2}$ as we did in the example in Figure 15.24. To initialize the algorithm, compute $J(\theta^i(0))$, $i = 0, 1, \ldots, p$, and let $j = 0$. The algorithm then proceeds according to the following steps until a stopping criterion is satisfied:

1. Find the best new vertex of $P(j)$. Let

$$i^* = \arg\min\{J(\theta^i(j)) : i = 0, 1, \ldots, p\}$$

   and swap $\theta^0(j)$ and $\theta^{i^*}(j)$ (hence, we always keep $\theta^0(j)$ as the current best vertex). Check the stopping criterion.

2. Rotation step. Compute for $i = 1, 2, \ldots, p$

$$\theta^i_{rot}(j) = \theta^0(j) - (\theta^i(j) - \theta^0(j))$$

   and $J(\theta^i_{rot}(j))$. Go to step 3.

3. Expansion step. If

$$\min\{J(\theta^i_{rot}(j)) : i = 1, \ldots, p\} < J(\theta^0(j))$$

compute for $i = 1, 2, \ldots, p$

$$\theta^i_{exp}(j) = \theta^0(j) - \gamma_e(\theta^i(j) - \theta^0(j))$$

and $J(\theta^i_{exp}(j))$, then decide whether to form the new simplex with the expansion or rotation and form $P(j+1)$. In particular, if

$$\min\{J(\theta^i_{exp}(j)) : i = 1, \ldots, p\} < \min\{J(\theta^i_{rot}(j)) : i = 1, \ldots, p\} \quad (15.8)$$

then accept the expansion by letting $\theta^0(j+1) = \theta^0(j)$ and

$$\theta^i(j+1) = \theta^i_{exp}(j), \; i = 1, 2, \ldots, p$$

but if Equation (15.8) is not true, then accept the rotation by letting $\theta^0(j+1) = \theta^0(j)$ and

$$\theta^i(j+1) = \theta^i_{rot}(j), \; i = 1, 2, \ldots, p$$

Go to step 1.

4. Contraction step. If

$$\min\{J(\theta^i_{rot}(j)) : i = 1, \ldots, p\} \geq J(\theta^0(j))$$

then compute for $i = 1, 2, \ldots, p$

$$\theta^i_{cont}(j) = \theta^0(j) + \gamma_c(\theta^i(j) - \theta^0(j))$$

and $J(\theta^i_{cont}(j))$. If

$$\min\{J(\theta^i_{cont}(j)) : i = 1, \ldots, p\} < J(\theta^0(j))$$

form the new simplex $P(j+1)$ by letting $\theta^0(j+1) = \theta^0(j)$ and

$$\theta^i(j+1) = \theta^i_{cont}(j), \; i = 1, 2, \ldots, p$$

and then go to step 1. However, if

$$\min\{J(\theta^i_{cont}(j)) : i = 1, \ldots, p\} \geq J(\theta^0(j))$$

replace $\theta^i(j)$ with $\theta^i_{cont}(j)$, $i = 1, 2, \ldots, p$ and then go to step 2 (in this case we accept the contraction, but do not go to the next iteration $j+1$ since no better vertices were found in rotation or contraction).

Note that the algorithm will not go to the next iteration until it has found a lower cost vertex at the current iteration (for "iterations" in j).

Algorithm Complexity and Convergence

In practical applications it is often the case that computing $J(\theta)$ is the most computationally intensive operation; hence, we will focus on it here as we did for the Nelder-Mead method. First, note that we need $p+1$ cost evaluations for initialization. Then for rotation, expansion, and contraction, we need p cost evaluations each. If rotation is successful, then expansion is used and if rotation is not successful, then contraction is used; hence, in either case, $2p$ cost evaluations are needed. Note, however, that if contraction fails to find a better cost, then steps 2, 3, and 4 can occur repeatedly and hence, you can incur $2p$ cost evaluations in each sequence of these steps *before* the next iteration; hence, for N such sequences, we would need $2Np$ cost evaluations.

How does this compare to the Nelder-Mead method? It is difficult to compare the two, since the Nelder-Mead method is not guaranteed to provide any cost reduction at an iteration even if the shrink step is used. The multidirectional search method, however, guarantees that under mild restrictions there will be cost reductions, and convergence results exist [512] of the type stated for the simple pattern search methods.

Under mild restrictions, the multidirectional search method possesses convergence properties.

Example: Solving an Optimization Problem with Multidirectional Search

In this section, we apply the multidirectional search algorithm to finding the minimum point of the function shown in Figure 18.10 (note that the point $[15,5]^\top$ is the global minimum point and $[20,15]^\top$ is a local minimum). We do not use projection (to do that, you would need to manage how the vertices along each dimension are generated).

Here, $p=2$, so we have three vertices. We use $\gamma_e = 2$ and $\gamma_c = \frac{1}{2}$ and a stopping criterion that is based on how much the parameters change, with the maximum number of iterations set at 200. If we start the initial set of vertices as

$$P(0) = \left\{ \begin{bmatrix} 15.2 \\ 20.6 \end{bmatrix}, \begin{bmatrix} 14.9 \\ 20.1 \end{bmatrix}, \begin{bmatrix} 16 \\ 19.3 \end{bmatrix} \right\}$$

we get the results in Figures 15.26 and 15.27, and we see that the termination criterion was satisfied so the maximum number of iterations was not used. Here, we see that the algorithm finds a local minimum. If you start the algorithm on the other side of the peak from what the above initialization does, then it can drop into the local minimum in the upper left-hand corner of Figure 15.27.

As another example, if we start the initial set of vertices as

$$P(0) = \left\{ \begin{bmatrix} 30 \\ 0 \end{bmatrix}, \begin{bmatrix} 0 \\ 0 \end{bmatrix}, \begin{bmatrix} 15 \\ 30 \end{bmatrix} \right\}$$

(vertices on outer edges of the domain we are interested in), we get the results in Figures 15.28 and 15.29, where we obtain convergence to the global minimum. Clearly, the convergence properties of the algorithm will depend on the initialization of the simplex.

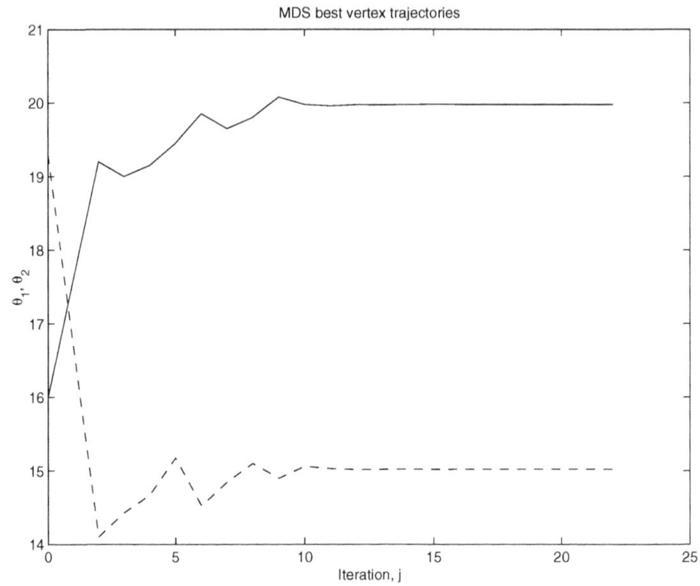

Figure 15.26: Multidirectional search example, parameter trajectories.

15.4 SPSA for Decision-Making System Design: Examples

In this section we study the use of the SPSA algorithm for the design of two types of decision-making systems, a controller for the tanker ship heading regulation problem, and an attentional strategy for predator/prey problem.

Nongradient optimization methods can be used for a wide variety of design problems.

15.4.1 Design Example: SPSA for Tanker Ship PD Controller Design

Here, we will investigate the use of the SPSA algorithm for the design of a PD controller for the tanker ship, a problem that was formulated and solved with an RSM method in Section 15.2. The only difference in the design problem is that we will add sensor noise to the heading sensor measurement; hence, the performance measure for a closed-loop response is a random variable. While here we study a low-dimensional design problem ($p = 2$), and the shapes of the response surfaces in Section 15.2 should give you confidence that the method will find a good design (provided the noise effects are not too significant, and the algorithm is designed properly), it should be clear that the SPSA could also be a valuable tool for higher-dimensional design problems.

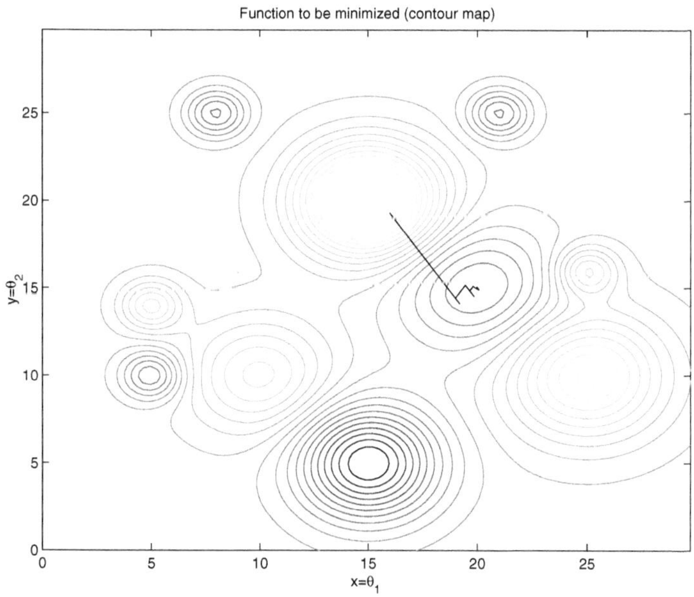

Figure 15.27: Multidirectional search example, parameter trajectory on the contour plot of the cost function.

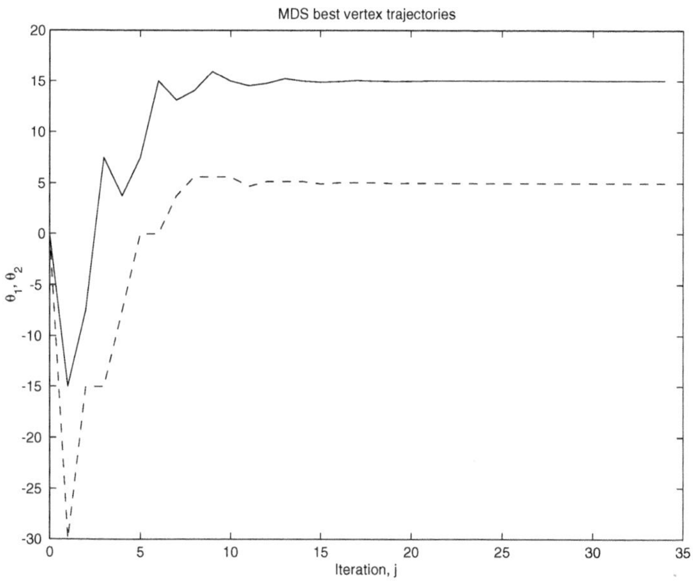

Figure 15.28: Multidirectional search example, parameter trajectories.

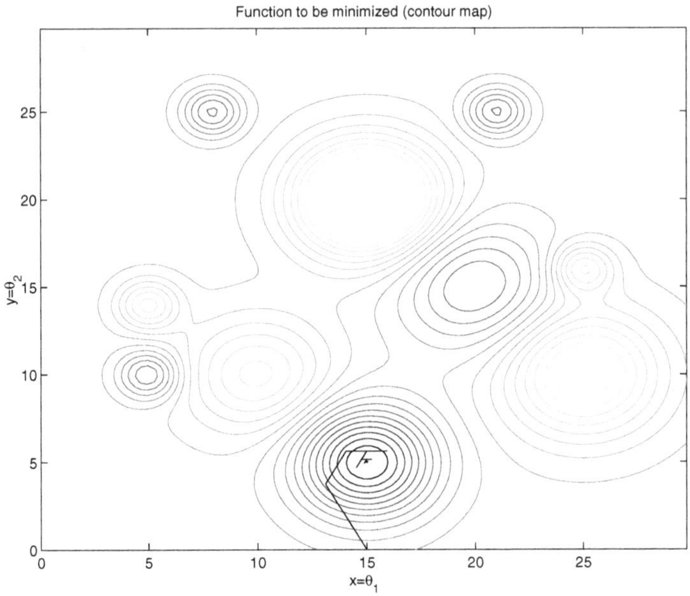

Figure 15.29: Multidirectional search example, parameter trajectory on the contour plot of the cost function.

Design Problem and Algorithm Design

We consider optimization over a domain with $K_p \in [-1.5, -0.5]$ and $K_d \in [-500, 100]$ as we did in the RSM. We seek to minimize the performance index $J_{cl}(K_p^i, K_d^i, \psi_r)$ in Equation 15.1 when there is sensor noise, and use the same reference input sequence ψ_r as with RSM, $w_1 = 1$ and $w_2 = 0.01$, and other variables that we used earlier.

We use 50 iterations of the SPSA algorithm. Due to the scale differences on the K_p and K_d domains, we introduce $\lambda_p = 1$ and $\lambda_d = 500$ (in place of λ) so that the step sizes on each dimension are different, but use the same $\lambda_0 = 1$ for both dimensions. Moreover, we use $c_p = 0.5$ and $c_d = 50$ (in place of c) so that the pattern of points considered is scaled different on each dimension and hence, the gradient is computed accordingly. We use the same $\alpha_1 = 0.602$, and $\alpha_2 = 0.101$ for both dimensions. To illustrate the operation of the algorithm we choose $\theta(0) = [-0.5, -300]^\top$.

The values above were arrived at after a bit of tuning. Note that if the c_p and c_d values are not large enough, the algorithm will not explore in a wide enough region, and if λ_p and λ_d are too small, then convergence can be very slow. Also, if you pick larger values for α_1 and α_2, you can get a type of "premature convergence," where the algorithm locks onto values that may be far from the best ones. Based on our experience with the RSM method earlier, we certainly could have guessed better values for the initial gains, but we use these to illustrate the performance improvements that can be obtained if you guess at poor values.

SPSA Design Results

A typical run for the SPSA algorithm is illustrated by showing the values of the cost function $J_{cl}(K_p^i, K_d^i, \psi_r)$ in Equation 15.1 for $\theta^+(j)$ and $\theta^-(j)$ in Figure 15.30, and the corresponding values for $\theta(j)$ in Figure 15.31. Notice that as the algorithm converges, the values of $J_{cl}(K_p^i, K_d^i, \psi_r)$ at $\theta^+(j)$ and $\theta^-(j)$ become nearly the same. Figure 15.31 shows the gains that are considered in the design space. The gains found in the last iteration are $K_p = -2.7442$ and $K_d = -407.4084$, which for convenience we take as the best gains. (These seem reasonable, if the noise effects are not too significant, considering the RSM results in Figure 15.1.)

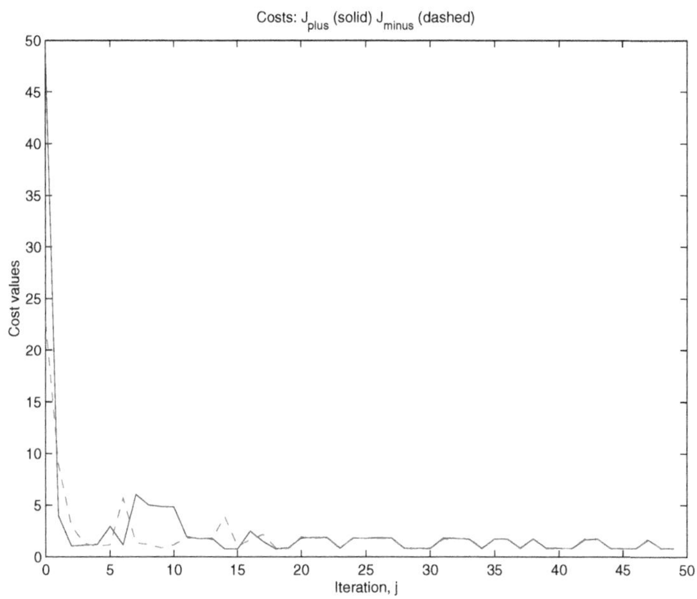

Figure 15.30: Cost function $J_{cl}(K_p^i, K_d^i, \psi_r)$ for $\theta^+(j)$ and $\theta^-(j)$.

Next, we plot the closed-loop response using the gains provided by the SPSA in Figures 15.32 and 15.33, where we see that reasonably good performance was achieved, in spite of the sensor noise that is apparent in Figure 15.33.

Note that we cannot be confident that the gains that the SPSA found are the "best" ones. Why? Each time you run the algorithm it will provide different values. Was there premature convergence for the SPSA? What if we initialized the algorithm differently; would it find a different "optimum" design point? You should not view the SPSA as an algorithm that will guarantee an optimal design, just one that will help to make design improvements. Clearly, you would

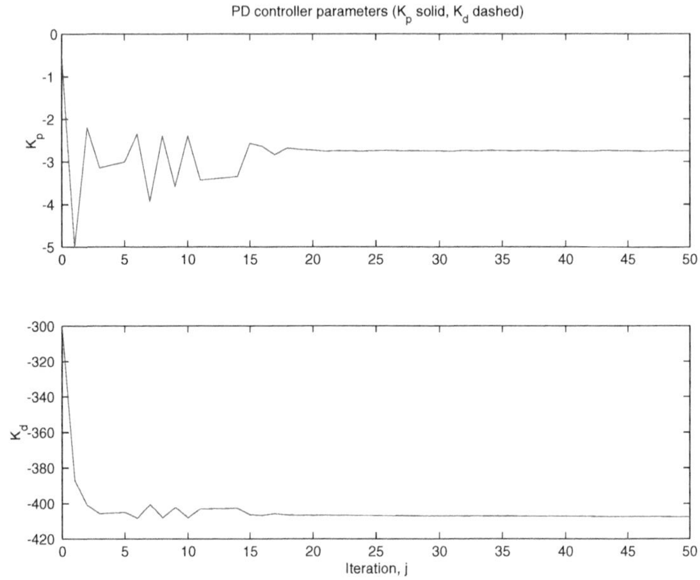

Figure 15.31: K_p and K_d gains generated by SPSA.

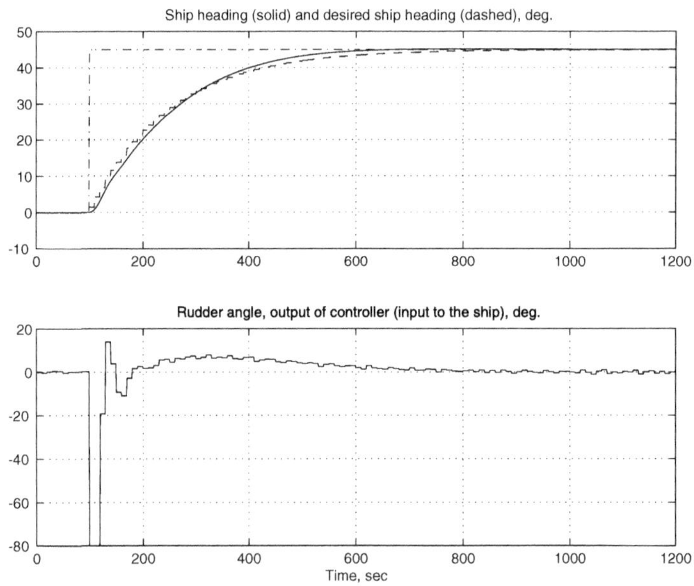

Figure 15.32: Closed-loop tanker response for SPSA gains.

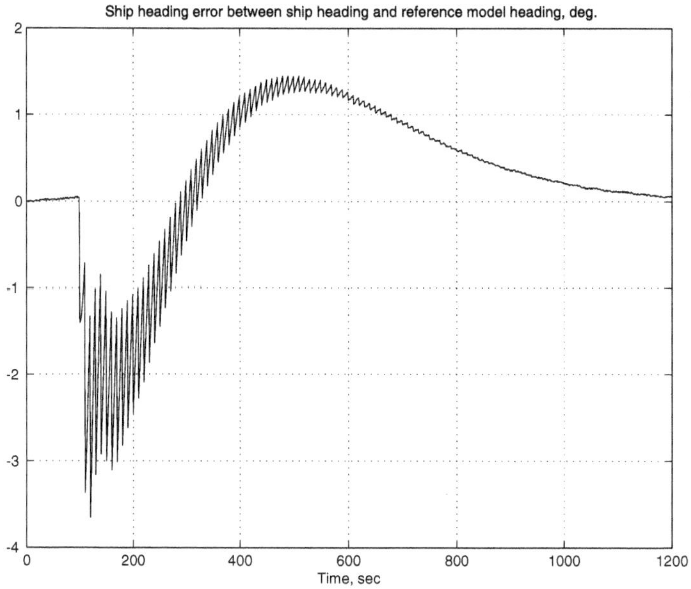

Figure 15.33: Tanker ship heading error (shows measurement noise).

generally have to try multiple initial conditions and trial runs to increase your confidence in the quality of the design. If you think of SPSA as operating over an infinite number of response surfaces (using the RSM view), then it should be clear that achieving an "optimal design" is a difficult problem, and difficult to verify.

15.4.2 Design Example: SPSA for Attentional Strategy Design

Next, we show how to use SPSA for the design of attentional strategies, by continuing the study in Section 7.4.4. There, we tuned the w_i parameters of the attentional strategy in Equation (7.8). Here, we will use the SPSA to automate the tuning of the parameters of the attentional strategy.

Design Problem and Algorithm Design

Our goal is to obtain performance that is better than that which we obtained via manual tuning, where we obtained a time average of the average values of the lengths of times waited of 3.2755 for $w_1 = 4$, $w_2 = 2$, $w_3 = 1$, and $w_4 = 4$. One approach would be to start with these values to initialize the parameter vector. Here, we do not do this in order to consider the case where we had done no a priori tuning. Here, we start with $w_i = 1$, $i = 1, 2, 3, 4$.

Recall that we had $N = 4$ and

$$\delta^1 = 1.05,\ \delta^2 = 1.15,\ \delta^3 = 1.25,\ \delta^4 = 1.35$$

These are bounds for appearance periods given in Figure 7.7. We have $\delta_s = 0.03$. We have

$$a_1 = 0.1,\ a_2 = 0.2,\ a_3 = 0.3,\ a_4 = 0.1$$

and this gives $\sum_{i=1}^{4} a_i = 0.7$.

For the SPSA algorithm, we define the cost function to be the time average of the average amount of time each predator/prey is ignored. We bound the parameter values to be between 0.1 and 10, $\alpha_1 = 0.602$, $\alpha_2 = 0.101$, $\lambda = 0.5$, $\lambda_0 = 10$, and $c = 0.25$. We allow the SPSA algorithm to run for 100 iterations.

SPSA Design Results

The results of running the algorithm are shown via a plot of the cost function in Figure 15.34, and a plot of the parameters values explored during the search in Figure 15.35. We see that the cost decreases from the initial value, then the algorithm searches, but does not improve the performance.

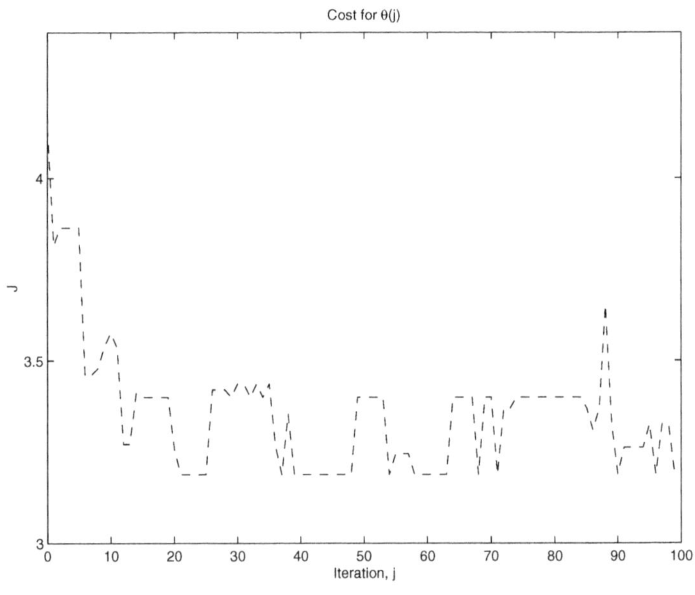

Figure 15.34: Values of cost function for SPSA for each iteration.

At iteration 22 the cost is 3.1871 and the parameter values are

$$w_1 = 1.9184,\ w_2 = 0.7888,\ w_3 = 0.4585,\ w_4 = 1.4590$$

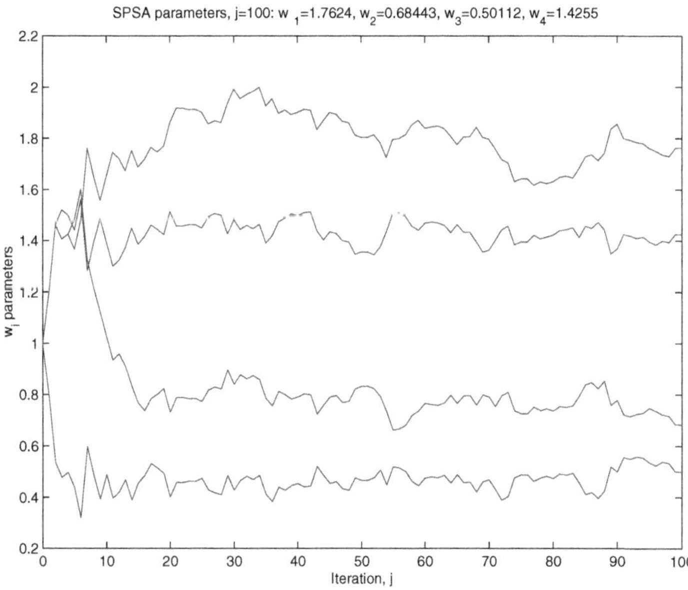

Figure 15.35: Parameter values for attentional strategy explored by SPSA for each iteration.

Compare this with the values that we obtained via manual tuning. First, the SPSA did better in terms of reducing the time average of the average amount of time that predators/prey are ignored (i.e., better than 3.2755). Note, however, that the SPSA found a similar *relative* weighting to what we had found manually, in the sense that it found that the strength of weighting, from highest to lowest, should be w_1, w_4, w_2, w_3. Recall that this ordering was arrived at in the manual tuning case by iterative tuning, where we raised a weight if a certain predator/prey was not getting enough attention. SPSA seems to have tried to automatically achieve a sort of balancing of focusing in order to improve performance.

15.5 Parallel, Interleaved, and Hierarchical Nongradient Methods

"Parallel" methods ("set-based" techniques) are implemented by executing N copies of, for instance, a gradient method from Part III at the same time, with different initial conditions; note that we use the word "parallel" in a sense that is similar to how it is thought of in parallel computing. It should be clear that in this case each of the N algorithms could, for instance, converge to a different local minimum. Then, at the end of the optimization process, the best one could be chosen. This provides for an approach to explore more than one region of

Parallel operation involves executing N copies of an algorithm simultaneously.

the search space; it is equivalent to simply restarting the same algorithm with different initial conditions. You could also use such a "parallelization" approach for the SPSA, line, pattern, and simplex methods so that explicit values of the gradient are not needed. When N parallel SPSA algorithms are used, we obtain a stochastic parallel search method (with no communication between the methods). Of course, the genetic algorithm could also be classified as a stochastic nongradient parallel search method; however, it shares information across different individuals in the population, whereas the parallel SPSA would not (at least in the form we are thinking of at this point). For Part V, you may think of N foragers that do not communicate as implementing N parallel algorithms.

Interleaving involves alternating between the use of different algorithms as the optimization process proceeds.

"Interleaving" the operation of optimization approaches means that you devise a strategy to switch from one optimization method to another, perhaps many times, as the algorithm progresses (you think of interleaving in time). Such methods have been used for many years (e.g., in gradient methods where you switch from one method to another when you are near a stationary point to try to speed convergence). Many interleaving strategies are possible, some where a fixed schedule of alternation between different algorithms is used (you could think of coordinate search via line search along each successive direction as one approach), others where various conditions dictate changing algorithms. Information gathered during the process may allow for modifications to be made to the methods during or in between applications of a particular method, or to the schedule of the alternation between the algorithms. Moreover, there are different methods to pass information from the current method to the one that it switches to (e.g., you may initialize the algorithm that you switch to with the parameter values that were found at the last iteration of the algorithm that was just used). Some such interleaving strategies were discussed in Section 11.1.7 on page 489, when we discussed offline and online processing of data for function approximation. Relative to the next part on foraging, it seems possible that foragers use different strategies depending on cues that they obtain from their environment.

Hierarchical methods employ one algorithm to supervise the operation of several algorithms.

"Hierarchical" methods may combine the concepts of parallel and interleaved methods by using one type of optimization strategy to periodically manage the operation of a parallel set of N algorithms. While multiple levels are possible, the basic concepts can be illustrated with a two-level hierarchy, where one type of optimization algorithm (which we call the "high-level" algorithm) manages N other "low-level" algorithms that operate in parallel. To perform management of the low-level algorithms, the high-level algorithm may use information from each of the N low-level algorithms (e.g., how much progress it has made at reducing cost in the region in which it is operating). Also, it may "tune" each of the low-level algorithms (e.g., it may try to optimize the operation of the low-level algorithms by adjusting the step sizes that they use). Relative to the next part on foraging, you may think of the two-level hierarchy as a type of social foraging where there is a "leader" (the animal at the top level); clearly, however, you can think of social foraging as an optimization process that does not use or need a hierarchy.

In the next two sections, we briefly discuss two hierarchical methods, one where pattern search is used to manage N algorithms, and another where an evolutionary strategy is used to manage the algorithms. Unlike the other methods in this chapter, we do not specify the full details of such algorithms, as their specification should be clear from the discussions in this and the last chapter. A design problem will, however, be given at the end of the chapter, where you are asked to design your own hierarchical method and apply it to the optimization problem studied earlier in this chapter.

15.5.1 Pattern Search

Suppose that in this approach you use pattern search with pattern $P_h(j)$ at the jth iteration for the higher-level algorithm that will manage the N algorithms. Suppose that there are $|P_h|$ elements in the pattern and that this number is fixed. Suppose that for the high-level manager, we use any of the pattern search or simplex methods discussed in this chapter. For instance, if we used the simple coordinate search method, we would have $|P_h| = p+1$, so we would need to have $N = |P_h| = p+1$ algorithms running in parallel.

Next, suppose that each of the N algorithms that are managed are also pattern search or simplex methods (possibly different from the high-level algorithm). Suppose that we use $P_i(j)$ at the jth iteration to denote the set of parameter values that are used for the ith pattern. If, for instance, simple coordinate search is used for each of the N algorithms, then $|P_i| = 2p+1$ for $i = 1, 2, \ldots, N$. Note that standard gradient methods and the SPSA here can be thought of as having $|P_i| = 1$ or $|P_i| = 2$, respectively, so that only one (two) parameter vector(s) is (are) updated at each iteration; using this fact, it should be clear how to manage the operation of these methods once you understand how to manage a set of pattern search algorithms.

How does the high-level algorithm manage the N algorithms? While there are many approaches, explaining one should serve to illustrate the basic idea. Suppose that we specify a way to *abstract* the information from the N patterns of the low-level algorithms. For instance, note that if simple coordinate search is used for each of the N low-level algorithms, each pattern P_i would have $2p+1$ elements (hence, there are $N(2p+1)$ cost evaluations at each iteration of *all* the low-level algorithms). One approach would be to use "representative points" for each of the N patterns. For instance, suppose that we use the centroid (see definition in Section 15.3.3) of pattern P_i to represent it (in simple coordinate search, the centroid is simply the center). Alternatively, we could use the best (in terms of least cost) point in P_i to represent it (which for simple coordinate search would also be the center). By "represent" we mean that it should somehow represent the progress of the entire pattern for that algorithm. Suppose that we define these N representative points as θ_h^i and use these to specify the pattern P_h.

Pattern search can be used to manage a set of algorithms, if you view each point in the pattern as representing the current operation of an algorithm.

The operation of the method proceeds, for instance, in the following manner, if we use simple coordinate search for both the high-level and N low-level algorithms:

1. Execute a fixed number of steps of the N low-level algorithms.
2. Form the N representative points θ_h^i (which, hopefully, will not result in degeneration of directions in the high-level algorithm), and via these, form P_h.
3. Take a single iteration of the high-level pattern search method. Use the N vectors that result from this iteration to define the centers of the patterns P_i to restart the low-level algorithms.
4. Go to step 1.

In this way, the N low-level algorithms are busy searching in various regions of the space and there is a periodic intervention from the high-level algorithm to try to *redistribute the patterns*, so that they are more effective in finding a global minimum (it should be clear how more than two layers could be used in the hierarchy of algorithms). The choice of the frequency at which the higher-level algorithm intervenes to reinitialize the N algorithms is an important design parameter. Note that for some approaches, such as when we use simple coordinate search at both the higher and lower levels, we must use $N(2p+1)$ cost evaluations at each iteration of the low-level algorithms (this assumes that they take steps all at the same time), and then $2p+1$ cost evaluations each time the higher-level algorithm executes. With the evolutionary strategies in the next section there may be no constraints on the number of algorithms that run in parallel, or the number of points each of the N algorithms works with.

15.5.2 Evolutionary Strategies

The overall approach to managing N algorithms that operate in parallel should be clear at this point; hence, all that we will comment on here is how to use evolutionary strategies for the management of N parallel pattern search methods (from this, strategies for other algorithms should be clear). Suppose, for convenience, that we use the simple coordinate search method for each of the N low-level algorithms and a genetic algorithm for the high-level algorithm. Of course, you could use N genetic algorithms or N SPSA algorithms for the low-level algorithms. Moreover, it should be clear that other non-evolutionary strategies could be designed to manage the set of low-level algorithms, ones that are not based on pattern search methods. For example, suppose that you use a set of N SPSA algorithms where N is not chosen according to standard pattern search ideas, but according to how many regions of the search space you think need to be investigated. In this case, N could be much larger than the one used in pattern search. How do we manage the N algorithms? You could simply take the $N_1 < N$ algorithms that have done the best and shift their starting points at the next iteration towards the centroid of the points that represent their current progress. Many other similar strategies are possible.

Returning to the issue of how to use evolutionary strategies for the management of N algorithms, we will use the management algorithm explained in the last section; however, we replace step 3 with an evolutionary strategy. Also, suppose that we use the centroid of each pattern (the center, for the simple coordinate search method) in selection, crossover, and mutation to define the reinitialization for the next sequence of iterations of the low-level algorithms. (Some other representative point could be used also, or it may be possible to base crossover on patterns rather than points.) It is important that you be able to envision how the resulting method will operate. The evolutionary strategy uses a set of N patterns, each of which is searching in a region of the search space, and when evolution takes place, it will move these search methods around based on the basic principles of evolution. For instance, note the following:

- Via selection, the genetic algorithm will tend to focus *more algorithms* in regions where the most optimization progress has been made.
- Via crossover, the genetic algorithm will generally take two relatively successful algorithms and generate *"offspring" algorithms* that search in a region near their parents' region, since that was working well.
- Via mutation, the genetic algorithm will randomly perturb the *initial conditions* of a low-level algorithm to make sure that all regions of the space get explored.

The use of elitism would correspond to allowing, for instance, the algorithm that has found the lowest cost to proceed without modification. Other strategies from genetic algorithms can be applied also. Moreover, genetic algorithms could be designed to try to optimize the operation of the N lower-level algorithms by tuning their parameters (e.g., step sizes).

The methods of this section illustrate the significant flexibility that exists in the development and implementation of optimization algorithms. While convergence properties for some of the proposed methods may follow from the analysis of the algorithms on which the methods are based, it should be clear that guarantees of convergence to a global minimum cannot generally be made. With the hierarchical methods in this section, we pay a price in computational complexity for the *possibility* that a global minimum may be found.

15.6 Set-Based Stochastic Optimization for Design

At each iteration SPSA considers two design points. The evaluation of the cost at each of these two points could be computed independently and this provides an opportunity to parallelize the algorithm on a computer with multiple processors. In fact, you could run N_p SPSA algorithms and compute $2N_p$ cost evaluations on a parallel computer. This provides a way to consider multiple initial conditions and SPSA algorithm parameters, and, hopefully, obtain a better design.

Suppose that we consider a "set-based" optimization algorithm to be one that explores multiple parameter vectors at each step. SPSA is set-based with two parameter vectors considered in the set, but a parallel version of SPSA could contain many more values in the set. Are there other types of set-based optimization methods? Genetic algorithms are one class of set-based stochastic optimization methods with the size of the set given by the number of individuals in the population. We discuss another such algorithm next; this algorithm will borrow certain characteristics from the genetic algorithm.

Basic characteristics of evolutionary algorithms can be captured and effectively employed by stochastic optimization methods, without the "overhead" of trying to emulate biological evolution.

15.6.1 A Set-Based Stochastic Optimization Method

The optimization method introduced here is based on an evolutionary approach, but only in certain principles of its operation. We specify the algorithm here to be applied to the PD controller design problem for the tanker ship, but it should be clear that it is a general method applicable to other problems.

We use a set (population) of controllers,

$$P(k) = \left\{\theta^i(k) | i = 1, 2, \ldots, S\right\}$$

with S members. Each member of the population is

$$\theta^i = \left[\begin{array}{c} K_p^i \\ K_d^i \end{array} \right]$$

where the ith controller is given by

$$\delta = K_p^i e + K_d^i c$$

We constrain the simulation to search for gains such that

$$K_p \in [-5, 0]$$

and

$$K_d \in [-500, 0]$$

(a wider range than we used earlier). We initialize the population by choosing $P(0)$ with θ^i such that the K_p^i and K_d^i are uniformly distributed (randomly) on this range.

Now, given the initial population of controllers, how do we produce subsequent estimates (generations)? For each controller, $i = 1, 2, \ldots, S$, we simulate the closed-loop system and compute the $J_{cl}(K_p^i, K_d^i, \psi)$ performance index given above. This gives us a ranking of the quality of all the controllers. Next, using this performance measure, we select the "best" controller and name it i^*. We then create the next generation of controllers by letting

$$\theta^{i^*}(k+1) = \theta^{i^*}(k)$$

(hence, we use what could be called "elitism," since we keep the best controller) and for $i \neq i^*$,

$$\theta^i(k+1) = \theta^{i^*}(k+1) + \left[\begin{array}{c} \beta_1 r_1 \\ \beta_2 r_2 \end{array} \right]$$

where r_1 and r_2 are random numbers drawn from a normal distribution with zero mean and unit variance. The parameters β_1 and β_2 scale the variance. In this way, intuitively, to produce the next generation, we generate a "cloud" of design points centered at the best design point found in the current generation (and the cost at each of these points could be computed in parallel). Clearly, if the β_i are chosen too small, then a sufficient region around the best point may not be explored. Also, if they are too large, the cloud may result in poor focusing in the search. A normal distribution is used to allow for the possibility of generating points at some distance from the mean. Here, after a few simulations, we used $\beta_1 = 0.5$ and $\beta_2 = 50$ (one-tenth the size of the search region for each gain); however, it was found that other values of the same magnitude worked just as well. Next, if the generation of gains resulted in values outside the above ranges, projection was used to place them at the boundaries of the ranges in the usual way. Note that the above approach tends to select for good controller characteristics and explore designs in the region of good designs (analogous to the operations of selection and crossover in genetic algorithms).

Finally, to try to ensure that the method would not get "stuck," a mutation-type mechanism was added. For this, we simply selected one $i \neq i^*$ and for this i generated the design point in the next generation by taking the gains randomly from a uniform distribution on the range of allowable values defined above.

15.6.2 Design Example: Tanker Controller Design

We use a population size of $S = 4$ and 40 optimization steps. For one typical simulation run, the values of J_{cl} for the "best" controller for each generation is shown in Figure 15.36. Also, in Figures 15.37 and 15.38, we show the trajectories of the PD controller gains for the best controller for each generation. Here, we see that the stochastic optimization approach seems to have been able to improve on the design as the optimization progressed. Note that while this is a typical plot provided by the program, it can be different. For instance, if via the initialization you get lucky and it guesses a good design, there may be little or no improvement over that design. Alternatively, if it guessed all bad designs, it may show more drastic performance improvements. Moreover, the algorithm can get stuck in oscillations (how?). To more fully test the stochastic optimization method, you may want to run many such simulations and take averages.

Next, to get an idea of how good the design is, we take the best controller from the last generation and in Figure 15.39, show the closed-loop performance. Note that relatively good performance is achieved, for this reference input, and the nominal conditions. How does this design work for other cases? Suppose that we consider how it performs for a ship that is "full." The particular gains that it evolved are $K_p = -2.9345$ and $K_d = -436.6634$. The performance for this case is shown in Figure 15.40. The design that was "optimized" for the nominal condition does not perform nearly as well for the perturbed condition as we would expect.

Optimizing the design for one condition typically results in poorer performance for a condition that was not designed for.

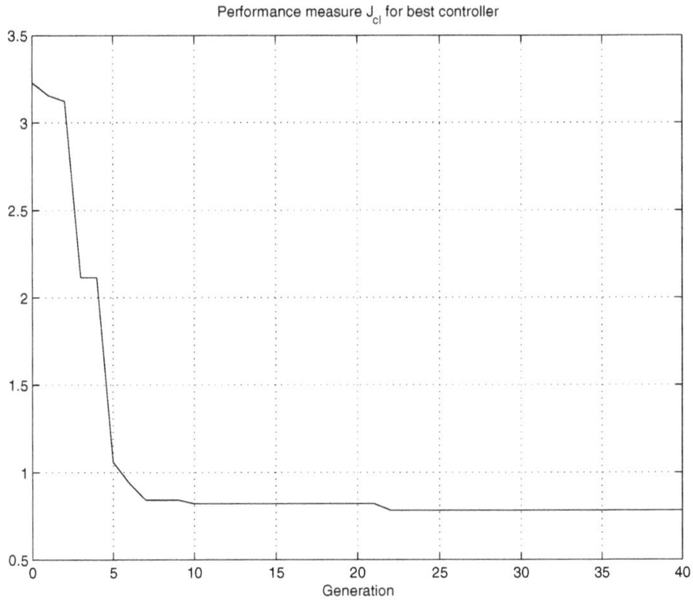

Figure 15.36: Closed-loop tanker control system performance for best controller for each generation.

15.7 Discussion: Evolutionary Control System Design

In this section, we briefly discuss the use of evolutionary algorithms for the design of control systems, both in simulation and in an experimental setting.

15.7.1 Genetic Algorithms for Computer-Aided Control System Design

The genetic algorithm can be used in the (offline) computer-aided design of control systems, since it can artificially evolve an appropriate controller that meets the performance specifications to the greatest extent possible. To do this, the genetic algorithm maintains a population of strings that each represent a different controller (digits on the strings characterize parameters of the controller), and it uses a fitness measure that characterizes the closed-loop specifications.

Suppose, for instance, that the closed-loop specifications indicate that you want, for a step input, a (stable) response with a rise-time of t_r^*, a percent overshoot of M_p^*, and a settling time of t_s^*. We need to define the fitness function so that it measures how close each individual in the population at time k (i.e., each controller candidate) is to meeting these specifications. Suppose that we let t_r, M_p, and t_s denote the rise-time, overshoot, and settling time, respectively,

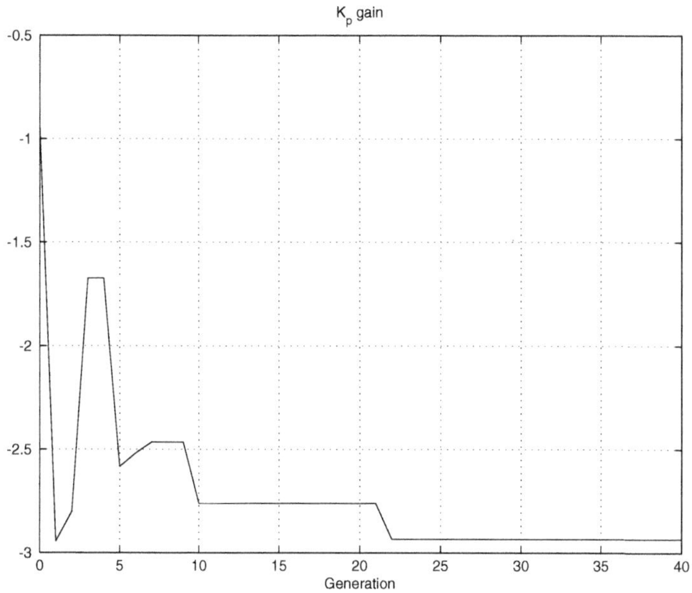

Figure 15.37: Gain K_p for the best controller for each generation.

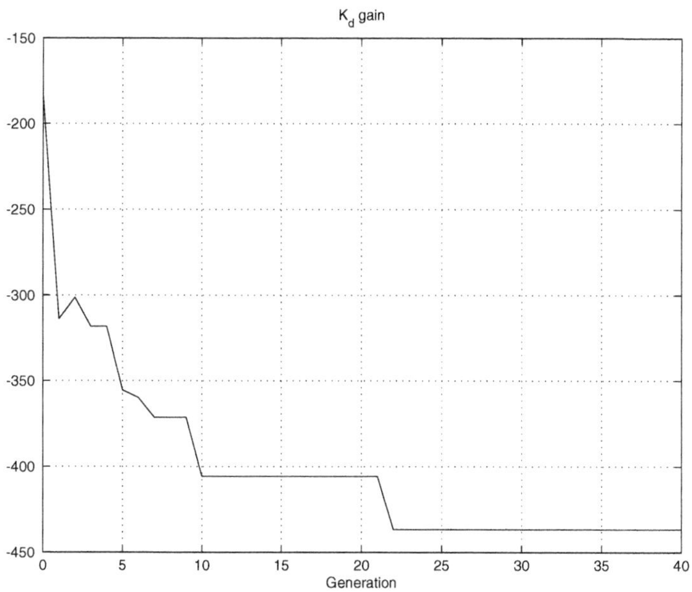

Figure 15.38: Gain K_d for the best controller for each generation.

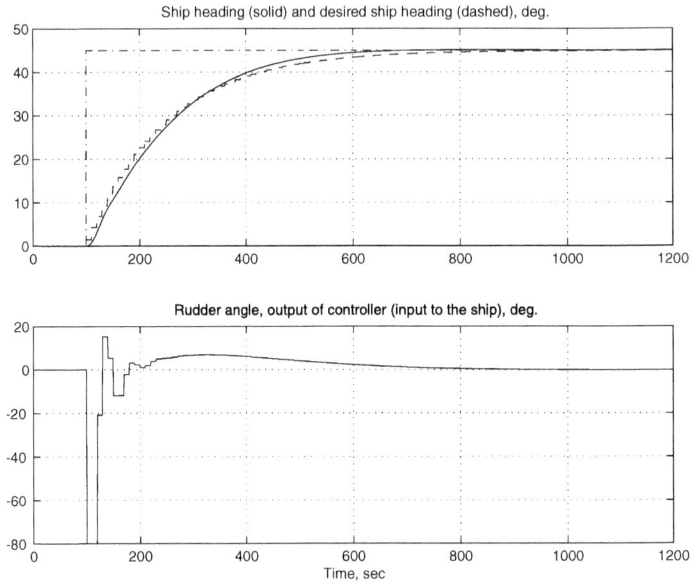

Figure 15.39: Closed-loop performance of the best controller for the last evolution step.

for a given individual (we compute these for an individual in the population by performing a simulation of the closed-loop system with the candidate controller and a model of the plant; of course, we can only perform the simulation for a finite amount of time). Given these values, we let (for each individual and every time step k)

$$J = w_1(t_r - t_r^*)^2 + w_2(M_p - M_p^*)^2 + w_3(t_s - t_s^*)^2$$

where $w_i > 0$, $i = 1, 2, 3$, are positive weighting factors. The function J characterizes how well the candidate controller meets the closed-loop specifications where, if $J = 0$, it meets the specifications perfectly. The weighting factors can be used to prioritize the importance of meeting the various specifications (e.g., a high value of w_2 relative to the others indicates that the percent overshoot specification is more important to meet than the others).

Now, we would like to minimize J, but the genetic algorithm is a maximization routine for $\bar{J}$. To minimize J with the genetic algorithm, we can choose the fitness function

$$\bar{J} = \frac{1}{J + \epsilon}$$

where $\epsilon > 0$ is a small positive number. Maximization of $\bar{J}$ can only be achieved by minimization of J, so the desired effect is achieved. Another way to define

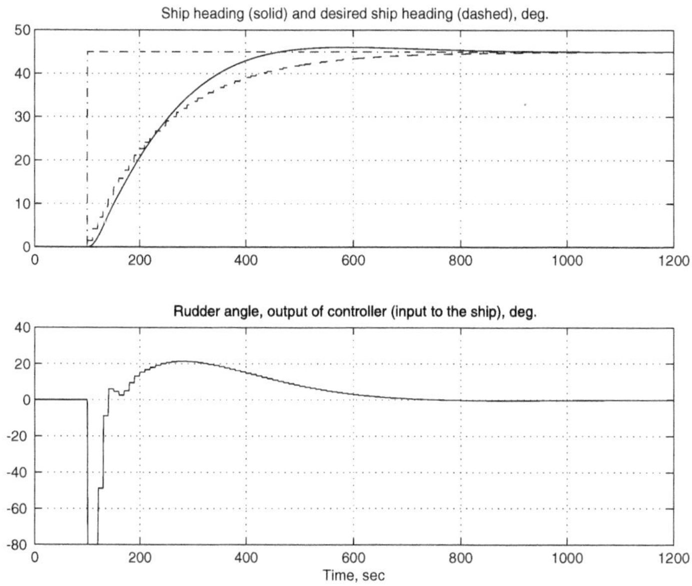

Figure 15.40: Closed-loop performance of the best controller for the last evolution step (design for nominal condition, test for "full" condition).

the fitness function is to let

$$\bar{J}(\theta(k)) = -J(\theta(k)) + \max_{\theta(k)}\{J(\theta(k))\}$$

The minus sign in front of the $J(\theta(k))$ term turns the minimization problem into a maximization problem (to see this, consider $J(\theta) = (\theta)^2$, where θ is a scalar, as an example). The $\max_{\theta(k)}\{J(\theta(k))\}$ term is needed to shift the function up so that $\bar{J}(\theta(k))$ is always positive. We need it positive since, in selection, Equation (14.3) defines a probability that must always be positive and between one and zero.

This completes the definition of how to use a genetic algorithm as a set-based stochastic optimization algorithm for computer-aided control system design, what you might call "evolutionary control system design." Note that the above approach depends in no way on whether the controller that is evolved is a conventional controller (e.g., a PID controller), neural network, a fuzzy system, expert system, planning system, or learning system, all possibly augmented with an attentional system; hence, we can use this approach to design any of these types of controllers. Typically, however, the problems with this approach are that you need to define an appropriate fitness function that avoids degradation to exhaustive search and the connected problems of computational complexity.

Clearly, the above approach can also be used in approximator construction, just as we used gradient optimization for these problems. One possible advantage that the GA approach could offer over, for example, a gradient method is that it may be able to better avoid local optima and hence, find the global optimum. One possible disadvantage is that it may be difficult to design an effective evolutionary algorithm, especially one that is computationally efficient (e.g., one that will not take an extraordinary number of generations).

15.7.2 Darwinian Design for Physical Control Systems

We close this chapter with a brief discussion on the use of evolutionary methods for the design of "physical" control systems. This represents either a design aid or alternative to conventional design methods.

Evolution in the Marketplace

Suppose that your company manufactures a product that has a control system in it, such as an automobile with a cruise controller. Suppose that you engineer a cruise controller using any design methodology that you like. Next, we will put a whole fleet of vehicles into operation with this controller. Suppose that we have enough memory and computing power on the vehicle to monitor its operation and compute and store a performance measure that quantifies the quality of the behavior of the control system (e.g., in terms of the rise-time, overshoot, ability to track). Suppose that minimization of the performance measure results in meeting the design objectives. Next, suppose that we define the ability of each controller to survive and reproduce to be inversely proportional to the performance measure. (A controller that performs well has a low value for its performance index and is likely to survive and reproduce and hence, have more copies in the next generation; hence, its reproductive "fitness" is *inversely* proportional to the performance index.) Now, suppose that every time that a vehicle comes in for maintenance and is connected to a computer diagnostic system, we obtain the overall measure of performance that was stored (of course, wireless communications may make it unnecessary to physically bring the vehicle in). We do this for the whole fleet of vehicles. Also, while in maintenance, we install in the vehicle a "next generation" controller that is an immediate descendant of the "population" controllers that are currently used (ones that we have gathered performance measures for). This controller is one that was more likely to be a good one for operation on the fleet of vehicles. After several iterations (generations), the controllers that are in use will tend to adapt to the overall conditions that they are used in. We have then evolved the control system for this fleet of vehicles, and it may even be *robust* if implemented on new vehicles of the same type (why?).

Darwinian design methodology for physical systems provides a way to design control systems that is analogous to how evolution designs organisms.

In addition to the fact that the standard "robust yet fragile" concept discussed earlier holds, there could be problems with the above approach. First, besides typical difficulties in defining a fitness function, if the initial population of controllers is not properly tested, then early versions of the product may not perform so well (you may be reminded of actual products you buy, and how later versions are improved). Second, the approach requires processor power and memory to compute and store the fitness measure. Third, you must properly constrain how the controller can evolve or it may be possible that mutations and perhaps crossovers will produce very substandard controllers (so that after maintenance where the new generation controller is installed, the product actually performs worse; clearly, there would be unhappy customers). You want to explore the design space to find what is both good and bad, but it will cost you to find where the designs are bad. Fourth, it may perform very badly for certain unforeseen situations that may arise due to inherent robustness trade offs. Fifth, it may be difficult or impossible to guarantee that the process will converge to an optimal robust design.

Darwinian Design Via Parallel Networked Experiments

A solution to some of these problems would be to set up an artificial environment to evolve the control system before it is released to the public. Then on release you could set it up so that it continues to evolve, or you may simply fix the design. The type of experiment needed to evolve physical control systems is shown in Figure 15.41. At the lower part of the figure, suppose that we have S computers (S is the population size), interfaced via a data acquisition system to S copies of the plant to be controlled. Keep in mind that while we say "copies," clearly they will not be exact duplicates; component and manufacturing differences will result in at least slightly different plants. Now, suppose that you design a first guess at a control system (or alternatively, you may have up to S different members in your control systems group who each design a controller for the plant), each of the same structure (e.g., PID controllers). Suppose that we use these initial design(s) to initialize the controllers for the S plants. Also, place these same controllers in the population of controllers in a separate "central" computer as shown in the figure. Place a genetic algorithm with appropriately chosen genetic operations in the central computer (or some other set-based stochastic optimization algorithm). Also, connect this central computer to each of the computers used to control each of the plants via a computer network (e.g., the Internet).

Now we are prepared to exercise the system to simulate the evolution of a population of control systems for this plant. First, suppose that we have a standard set of experiments that the computers perform on each of the plants, all at the same time (i.e., in parallel). Typical choices for this could be a step input, a square wave input, or a reference input that is likely to be encountered in actual operation of the plant. While the experiments operate, suppose that we collect data and compute performance measures that can be represented in a fitness function. Once the experiments are complete, they will provide a fitness value that quantifies the performance of each of the S controllers. These S fitness values are passed to the central computer. Using these, the central computer makes the calculations necessary to produce the next generation using the genetic operations (and as indicated above, there may be a need to use constraints on what values the controller parameters can take on). Then, it passes the controllers back to the S computers, and the process repeats.

After several generations it is hoped that the population converges to a single

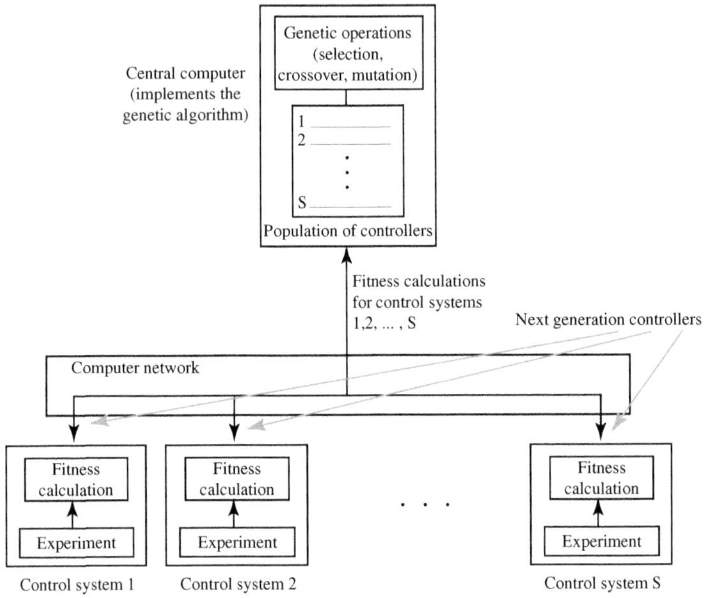

Figure 15.41: Experiment to perform Darwinian design for physical control systems.

controller (or at least a population of very similar controllers). This controller is then implemented on the actual system and the product is distributed. Clearly, there could be problems with this approach in some situations. First, it could be that it takes prohibitively long to complete the process, so that too many experiments have to be performed, and hence, it takes too long to come up with a design. On the other hand, the method has great potential for some applications where it is easy to produce many copies, and set up the experiment described above. It may produce a robust controller (at least it may end up being robust to the types of behavior that the S plants exhibit), and at the end of the design process, you have some confidence in how the controller will perform on the actual system, since it has already been exercised on it many times (i.e., there is a type of verification process embedded in the design process).

15.8 Exercises and Design Problems

Exercise 15.1 (Design Strategies that Exploit Robustness Trade-Offs: Application to Farming): Read the paper [79]. Explain how the "design strategy," where scientists recommend planting nongenetically modified crops near genetically modified ones (e.g., ones that are resistant to some insects) can be explained in the context of robustness trade-offs in design. Note that the planting strategies of farmers that try to maximize yield are actually "redesigning" insects in the "arms-race" between farmers and pests.

Exercise 15.2 (Pattern Search Methods):

(a) For a cost function, suppose that you use "Rosenbrock's function" $y = f(x)$, $x = [x_1, x_2]^\top$, which is a parabolic valley

$$y = 100(x_2 - x_1^2)^2 + (1 - x_1)^2 \tag{15.9}$$

with a minimum at $(1, 1)$. Program the simple coordinate search method and use it to search for the minimum. Experiment with different choices for the initial condition. Plot the trajectories for the estimates of the minimum as a function of optimization algorithm step. Be sure to pick your scale for your plot appropriately so that you can illustrate the behavior of the algorithm in the plot.

(b) Repeat (a) but for the evolutionary operation via factorial designs method.

Exercise 15.3 (Nelder-Mead Simplex Method):

(a) Redraw the contours to illustrate the step where $\theta^{max}(j)$ is better than $\theta^{ref}(j)$, in the contraction step the algorithm generates a $\theta^{new}(j)$ that is in between $\theta^{max}(j)$ and $\theta^{cent}(j)$, but with $\theta^{new}(j)$ worse than $\theta^{max}(j)$. This should be a sketch like the one in Figure 15.20 that illustrates how the simplex is reoriented in this case.

(b) Suppose that you use Rosenbrock's function in Equation (15.9) as a cost function. Program the Nelder-Mead simplex method and use it to search for the minimum. Experiment with different choices for the initial simplex.

Exercise 15.4 (Multidirectional Search Method):

(a) For $p = 2$, draw an example simplex on the function $J = \theta_1^2 + \theta_2^2$. Develop a definition for the "simplex gradient" and explain, using geometric arguments, why the update direction suggested by it makes sense.

(b) Suppose that you use Rosenbrock's function in Equation (15.9) as a cost function. Program the multidirectional search method and use it to search for the minimum. Experiment with different choices for the initial simplex.

Exercise 15.5 (Simultaneous Perturbation Stochastic Approximation Method):

(a) Suppose that you use Rosenbrock's function $y = f(x)$ given in Equation (15.9), and which has a minimum at $(1, 1)$. This will be the "noise-free" cost. Define a noisy cost function to be one that has an additive white Gaussian noise with zero mean and a variance of 0.1. Program the SPSA method and test its performance for different initial conditions and the noise-free and noise cases.

(b) Program the FDSA method and test its performance for different initial conditions and the noise-free and noise cases. Compare its performance to the results in (a) for SPSA. Compare the complexity of the two algorithms.

(c) (This part of this problem was provided by J. Spall.) Consider the skewed-quartic cost function,

$$J(\theta) = \theta^\top B^\top B\theta + 0.1 \sum_{i=1} p(B\theta)_i^3 + 0.01 \sum_{i=1} p(B\theta)_i^4$$

where $J_n(\theta) = J(\theta) + w$ and $(\cdot)_i$ represents the ith component of the argument vector and B is such that pB is an upper triangular matrix of ones. The minimum occurs at $\theta^* = 0$ with $J(\theta) = 0$. Let us compare SPSA with a valid perturbation distribution and two invalid perturbation distributions. In the valid case, let the perturbations ($\Delta(j)$ components) be independent and identically distributed (i.i.d.) Bernoulli ± 1 distributed and in the invalid cases, let the perturbations be i.i.d. uniformly distributed over $[-\sqrt{3}, \sqrt{3}]$ and i.i.d. normally distributed with zero mean and unit variance. Note that the valid and invalid perturbation distributions all have zero mean and a variance of one. Let $p = 10$ and the gains be in the form of $a_k = a/(k+1+A)^\alpha$, $c_k = c/(k+1)^\gamma$. Suppose the noise in the measurements (i.e., the cost function) w is i.i.d. normal with zero mean and a variance of 0.1^2. Furthermore, let 2000 cost function measurements be used in the search process based on an initial condition of the parameter vector of $[1, 1, \ldots, 1]^\top$ and let $a = 0.5$, $A = 50$, $c = 0.1$, $\alpha = 0.602$, and $\gamma = 0.101$. With no constraints imposed on the movement of the parameter vector, compare one replication of 2000 measurements for the valid and invalid implementations and comment on the relative performance of the implementations.

Exercise 15.6 (SPSA for Tanker Controller Design): In this problem you will study characteristics of SPSA design approach for the tanker controller in Section 15.4.1.

(a) Explain, via simulations, the effects of changing λ_p and λ_d on algorithm performance.

(b) Explain, via simulations, the effects of changing c_p and c_d on algorithm performance.

(c) Explain, via simulations, the effects of changing α_1 and α_2 on algorithm performance.

Design Problem 15.1 (Response Surface Methodology for Design): In this problem you will expand on the response surface design approach discussed in Section 15.2 for the tanker ship.

(a) Repeat the same general steps of the design process, but where you use a ship in "ballast" conditions, and consider the nominal case to be one where the speed is $u = 5$ and the "off-nominal" case to be where $u = 3$ (or some other appropriate value). What are the optimal gains for each case, and the performance of these gains for the case that they were not designed for? Support your answers with response surfaces and appropriate analysis.

(b) Consider nominal conditions for the ship. Show that the response surface can be different for different reference input sequences. Provide at least one response surface that is different from the one provided in the chapter for nominal conditions and the reference input trajectory provided there. What is the fundamental reason why the response surface changes when there are different reference input sequences?

(c) Use response surface methodology to design a fuzzy controller for the tanker ship. Explain all of your choices and compare to the manual designs that were conducted in Chapter 5. Hint: The simplest approach is to simply tune the three normalizing gains. Another approach is to use RSM to tune the centers of the output membership functions, the parameters of the input membership functions, or the number of rules.

(d) Use RSM to design the parameters of an adaptive controller for a tanker ship.

Design Problem 15.2 (Nongradient Optimization for Tanker Controller Design): In this problem you will develop deterministic nongradient optimization algorithms for the design of the tanker controller in Section 15.4.1.

(a) Design a simple coordinate search method that can design a tanker ship controller. Illustrate its performance in simulation and explain your choice of the algorithm parameters.

(b) Repeat (a), but design a Nelder-Mead simplex algorithm.

(c) Repeat (a), but design a multidirectional search algorithm.

(d) Compare and contrast the performance of the algorithms in (a), (b), and (c) according to their computational complexity and convergence properties for this problem.

Design Problem 15.3 (Deterministic Nongradient Optimization for Approximator Construction): For the single input function approximation (theme) problem of Part III that is studied, for instance, in Section 10.2, design deterministic nongradient optimization algorithms to tune the approximator. Evaluate the resulting approximation accuracy, and explain your choices for the algorithm parameters.

(a) Use the simple coordinate search method.

(b) Use the Nelder-Mead method.

(c) Use the multidirectional search method.

Design Problem 15.4 (Stochastic Pattern Search)*: Suppose that you are given the cost function with $p = 2$ in Figure 18.10, but that you do not have gradient information. The SPSA method chooses only two points and makes cost evaluations for these, then decides how to update the parameter vector. Design what you might call a stochastic pattern search method that uses only a few (e.g., two) of the points on a pattern at iteration j to compute how to update the parameter estimate. You should consider modifying the simple coordinate search method, or the multidirectional search method.

Design Problem 15.5 (Set-Based Stochastic Optimization for Tanker Ship Controller Design): Your objective in this problem is to expand on the analysis of the set-based stochastic optimization algorithm that was introduced and briefly evaluated in Section 15.6.

(a) Explain the effects of different cloud sizes S, and number of optimization steps. Clearly, with other choices you may obtain different performance from what was found in the chapter. Illustrate via simulation results.

(b) Explain the effect of w_1 and w_2 and support your explanation with simulation results.

(c) How would you perform convergence analysis via simulations for this approach? Optional: Illustrate your approach with a simulation-based analysis.

Design Problem 15.6 (Design of Hierarchical Methods for Function Optimization)*: Suppose that you are given the cost function in Figure 18.10, but that you do not have gradient information. Design a hierarchical parallel interleaved method for finding the global minimum. Fully specify the pseudocode for the algorithm in an analogous way to how it was done in the chapter for the other methods, and write a computer program to simulate the algorithm. Evaluate its performance by trying different initial conditions. Develop a useful way to graphically illustrate the operation of the algorithm and explain how to choose the parameters that specify the method (e.g., step sizes). You may consider using either pattern search or an evolutionary strategy for managing the N parallel algorithms. Use the same type of algorithm for each of the N parallel algorithms; however, you may choose the specific type of algorithm to use (e.g., you may use SPSA or simple coordinate search)

Design Problem 15.7 (Nongradient Optimization for Approximator Structure Construction)*: Design and test a deterministic nongradient algorithm to solve Design Problem 11.2.

Design Problem 15.8 (Design Problem Formulation and Solution)*: We have studied *design* of instinctual neural controllers (mapping shape), fuzzy controllers (mapping shape), planning systems for control (prediction model, optimization method, planning horizon), attentional systems (attentional strategy parameters), learning systems (approximator size, instinct-learning balance, optimization strategy design, structure construction), and adaptive controllers (e.g., adaptation gain, initial model or controller). Choose a design problem considered earlier in the book. Show how to formulate it as an optimization problem. Design a stochastic optimization method and illustrate its performance in design. Discuss algorithm performance (e.g., convergence, rate of convergence, quality of solution) and computational complexity. Also, discuss robustness tradeoffs.

Chapter 16

Evolution and Learning: Synergistic Effects

Chapter Contents

Evolution is a design process for constructing life forms, including ones that have learning capabilities. We can encode both the parameters and structure of learning systems, and use a genetic algorithm to evolve (optimize) these systems according to some fitness function to operate in a robust fashion. In the last chapter we discussed how organisms evolve to be robust in their environment to achieve "robust yet fragile" operation via "highly optimized tolerance." Here, we focus in particular on the interactions between the two adaptive processes of learning and evolution. We discuss how organisms evolve learning, and how a balance between learning and instincts may be achieved and maintained, based on characteristics of the environment that the organism lives in.

We show how the response surface methodology (RSM) of the last chapter can be used to design learning systems, by showing how to use response surface methods to find the best approximator size p for a function approximation problem (a robust learning system design problem from Part III). Moreover, we show how to use RSM and the set-based stochastic optimization methods of the last chapter to study instinct-learning balance. We discuss how the instinct-learning balance concepts help to quantify and clarify certain fundamental design trade-offs in adaptive controller design. Finally, we show how genetic algorithms can be used in adaptive control in cases where learning methods were traditionally used (e.g., the ones in Part III).

16.1 Relevant Theories of Biological Evolution

Learning provides one approach to achieve robust behavior.

One way for organisms to achieve robustness to changes in their environment is to use learning. It seems that learning has evolved in many species since it provides a selective advantage; animals that can learn may have greater reproductive success in their environments. In Chapter 2, Figure 2.8, on page 85, we described an experiment where learning capability (based on operant conditioning) seemed to evolve in a small population of rats. Even though drawing significant conclusions from such studies is problematic, it seems logical that many aspects of learning could evolve. Indeed it seems likely that sophisticated learning strategies may evolve from simple ones. For instance, it seems like simple control and learning functions implemented by networks of neurons, such as those discussed in Section 4.1.2 on page 109, could incrementally evolve to form more complex organisms. For example, did the *Aplysia*, which can exhibit classical conditioning as discussed in Section 9.1.3 on page 328, evolve from simple instinctual neural networks by "inventing" (in the evolutionary sense) a simple learning mechanism? Could the neural networks that implement simple learning functions, such as classical conditioning, be built upon to achieve the types of operant conditioning that are possible for pigeons, mice, and rats (see, e.g., Section 9.1.4 on page 335)?

In this section, we explain how evolution affects learning (e.g., via evolving learning capabilities) and how learning affects evolution (e.g., via the Baldwin effect). Also, we discuss the evolution of the balance between instincts and learning capabilities.

16.1.1 Genetics of Learning

We begin by providing an example of how genetic influences on learning have been quantified. This highlights the fact that there is a strong connection between genes and higher-level cognitive capabilities, and hence, clearly highlights that learning can be influenced by evolution.

It is possible to show how *single gene* mutations ("gene knockout") can affect learning in mice, and this provides direct evidence for a link between learning and genetics [268]. In particular, new methods have made it possible to study the adverse effects of gene knockout (e.g., for the *fyn* gene) on long-term potentiation and spatial learning in the intact animal. To begin with, a group of mice with the *fyn* gene knocked out are grown. Then another group of "wild type" mice (without gene knockout) are used. All the mice are trained (for seven days) to escape from an underwater maze by swimming to find a platform located beneath the surface of the water. (You can then think of the platform as a reward.) Then, the platform was removed and the mice swam for 60 sec. The wild type mouse's pattern of swimming (see [268]) clearly indicates that it remembered where the platform was. The *fyn*-deficient mouse, however, clearly never remembered how to find the platform (i.e., it has a deficiency in spatial memory).

16.1.2 The Evolution of Learning

The characteristics of the environment that an organism lives in affect its design over long periods of time. One particular characteristic that is affected is learning. To see how, suppose that some environment and organism satisfy the following assumptions:

Learning is invented and shaped by evolution.

- *Environment:* Suppose that the environment in which the organism lives has some features that do not change over very long periods of time (e.g., gravity), and others that may change quite frequently in *somewhat* unpredictable ways (e.g., the wind direction, how dangerous some predator is, how fast some prey runs). Hence, there are certain features that can be thought of as environmental "constants" and other aspects that have some uncertainty associated with them, but which are not completely unpredictable (e.g., any two predators of the same species have variability and hence, different performance characteristics, but within certain bounds). Certain events can occur to change the "static" parts of the environment (e.g., global temperature change), or change the probability distributions on other characteristics (e.g., evolution of a predator may shift the probability density function describing the maximum running speed of the predator to have a higher mean value).
- *Organism:* Suppose that the organism is of lesser complexity than its environment. We will think of the environment of a particular organism as being composed of everything that is not the organism; hence, the environment for an organism will typically contain other members of its

species and predators/prey. Suppose that the organism can in some way sense only certain aspects of its environment. It can store some information about its environment, but since it is simpler than its environment, it cannot store a perfect representation of its environment. Suppose that the storage and use of information about the environment "costs" the organism in a physiological and evolutionary sense. (For example, it has been argued that we lose our ability to learn language as we get older due to selective pressures [420]; hence, if an organism does not need them, there will be a selective pressure to abandon information storage and learning capabilities.)

These constraints imply that the organism will operate in the presence of uncertainty. It will not, however, operate in complete uncertainty, since we suppose that it has reliable ways to gather information and store it. Also, since there is not complete uncertainty, it should be able to make effective "decisions" that lead to the organism meeting its goals (e.g., a survival goal).

Next, suppose that the organism is such that it is highly evolved in its environment (but it seems unlikely that we could ever be certain that this is the case). First, will the static part of the environment be, in a sense, encoded in the organism's genes and passed from generation to generation? It seems that this will at times be the case, at least for some characteristics, for some organisms where it is worth the physiological cost and is physically possible. For instance, humans seem to be born with a sense of the influences of gravity (and hence, knowledge of up and down), and as far as tests can tell, "object permanence" (e.g., for very young babies, if you cover an object up that they can see, they *do* know that it is still there, and has not magically disappeared—babies instinctively know some physics). Second, what influence will the nonstatic but predictable part of the environment have on the organism's evolution? It is what leads to the evolution of learning. For uncertain but predictable environments, if the organism has learning capabilities, it can store information about its experiences, try to predict what will happen when similar situations occur, and will tend to be more successful than organisms that do not have such capabilities (or ones that have less effective learning capabilities). Hence, the uncertain but predictable characteristics that are encountered during the lifetime of each organism in an environment lead to a selective pressure over many generations for organisms in the species to be able to learn about the environment. Does this mean that every organism has a capability to learn, or will evolve one? No. It depends on the ecological niche, physiological costs of maintaining information storage and learning, and the random nature of evolution. Moreover, it depends on characteristics of the convergence of the evolutionary process, or the lack thereof.

Characteristics of the environment drive the evolution of specific types of learning capabilities.

What are the implications of the robustness concepts discussed in the last chapter on the evolution of learning? Clearly, they apply in the same basic manner. For instance, due to the finite complexity of an organism, there is no way that it can know how to cope with all situations, and in particular, it is likely to be optimized to learn about certain types of uncertain environmental characteristics, ones that can be quite different from the low-probability events that can lead to the demise of the organism or species. Learning is a way to achieve robustness, and organisms typically have good learning capabilities for aspects of the environment that are most important to their survival. Learning is the result of an optimization process that minimizes complexity of an organism operating in a structured uncertain environment with correlated events. Learning is simply an approach to achieve robustness, but it does not allow the organism to beat the fundamental robustness trade-offs in the design of complex systems.

16.1.3 The Baldwin Effect: Learning Can Accelerate Evolution of Instincts

To explain the Baldwin effect, we again think of the environment as driving the design of characteristics of learning capabilities of an organism. Suppose that after a long time, some organism has evolved some learning capabilities so that over its lifetime it can effectively cope with some uncertain but predictable part of its environment (i.e., it is robust to a certain extent). Now, suppose that some portion of that predictable part of the environment becomes quite consistent in its characteristics (i.e., that for some reason, one part of the environment becomes quite predictable). Further, suppose that each organism then basically learns the same predictable part of its environment during its lifetime, and this occurs over many generations. Suppose that learning this characteristic quickly provides a selective advantage. For example, suppose that many organisms are killed if they do not know about this characteristic of their environment, but organisms that encountered this characteristic and survived it will generally also survive a second encounter with it. Now, suppose that at some point there is a type of mutation in some organism that represents an encoding of the information about the environment that allows the organism to instinctively know how to predict the characteristic of the environment and hence, know how to cope with it to survive, *even on the first encounter* (the organism does not have to engage in the dangerous activity of learning about how to cope with the adverse event). This mutation would be a successful mutation that would propagate through the species, since the organisms that needed to learn about the characteristic would generally be less successful in survival.

Learning can accelerate evolution by speeding the incorporation of instincts.

This "genetic encoding" of information about some aspect of the environment can actually be thought of as accelerating evolution; hence, we can think of learning as providing a method to accelerate evolution (this is the "Baldwin effect"). How? Suppose for a moment that there was no learning capability in the species. Suppose that there is a beneficial mutation for one organism that allows it to survive. Unfortunately, it is not likely that another organism in that species will have the same successful mutation, and even less likely that those two will mate so that their offspring will also have the characteristic. Hence, genetically encoding the useful information for survival from the environment is highly improbable. On the other hand, with learning as a general characteristic of the species, the possibility of *many* organisms surviving the characteristic improves drastically, and this makes it more likely that *several* of their offspring may have successful mutations (or at least mutations that are "partially successful" in the sense that they lead to the encoding of part of the information that gives a slight advantage), and hence, the likelihood that two of the offspring mate and produce offspring with the useful mutation is increased. In this way, the learning results in an additional selective pressure to encode consistently useful information in the environment. In this way we see a shifting of learning capabilities to instincts via evolutionary pressure, and a faster acquisition of instincts when learning is present.

Does this imply that all learning will eventually evolve into instincts? No. Again, the physiological costs for learning and storage of information can be significant. Hence, in some ecological niches, learning capabilities may stay constant. Moreover, a fundamental characteristic of the stochastic dynamics of evolution of the organism may make it so that there is no convergence on a best design (e.g., there may be an oscillation in the designs over evolutionary time).

16.1.4 Evolving an Instinct-Learning Balance

Uncertain but predictable portions of our environment lead to selective pressures for evolution of learning capabilities, and static portions can lead to genetic encoding of instincts. The Baldwin effect explains one way that an organism with learning capabilities has a selective pressure for genetic encoding of portions of the environment that are found over long time epochs to be static (completely predictable). Both instincts and learning capabilities carry a cost for the organism (e.g., a maintenance cost for the underlying physiology) so there is a tendency to be conservative in increasing either of these, and for different organisms, the costs associated with instincts may be more or less expensive than those associated with learning capabilities.

Characteristics of the environment drive the construction of an optimal balance between instincts and learning.

There is then a type of balance between how much instinct an organism has and how many learning capabilities it has, and this balance can change if the environment changes. For example, via the Baldwin effect, there may be a shift where an organism gains more instincts. If the environment changes to be more static, the organism may lose some learning capabilities, whereas if the environment acquires certain characteristics that are uncertain but predictable (e.g., a new predator), it may gain more learning capabilities. Is this partition between instinct and learning "stable" (e.g., if there is some type of environmental change, will it settle to an new equilibrium)? Suppose that we wanted to experimentally verify these concepts for some organism. Where is the equilibrium for a particular organism? Could we change the environment in a way that will increase or decrease instinctual or learning capabilities of an organism? Could we use it to predict how many learning capabilities the organism has? This would amount to experiments in how to redesign cognitive capabilities of organisms (not just learning, but redesigning the brain to function differently). Could we perform analogous experiments or theoretical investigations for control and automation?

Is the point of equilibrium in the balance between instincts and learning a HOT point? Can we show how the HOT point shifts based on changing characteristics of the environment? For example, can we show that the HOT point moves towards more learning capabilities if some part of the environment becomes uncertain but predictable? Will the HOT point move towards more instincts if some part of the environment becomes static?

What are the implications for the design of adaptive control systems? Could we evolve in real time an optimal level of learning capabilities? Could we evolve the best model complexity for adaptive model predictive control? Could we evolve (optimize) the complexity of a controller?

16.1.5 Cultural Influences on Learning and Evolution

Jean-Baptiste Lamarck had a theory that organisms could inherit acquired characteristics of their parents (e.g., he proposed that offspring could take advantage of information acquired during the lifetime of the parent). Our modern understanding of genetics and reproduction shows that such information is not directly carried in DNA (e.g., advances in farming methodology learned by a parent are not somehow encoded in the genes and transferred to children). But, can what the parent learns help the child survive? Yes, perhaps, but in an indirect way. Since children often can learn from parents via the parents directly teaching the children, they can gain survival advantages via education. Of course, education can be provided not only by parents, but from the society in general via schools and conversation with friends and relatives. The information is stored and transmitted via word-of-mouth, books, libraries, and the Internet in many cultures. Hence, for example, libraries provide a type of species-wide memory that if properly exploited may help us to survive. For example, construction methodology is relatively well-understood, and is well-documented. Hence, you can go to the library and get a book that will tell you how to build a house that will withstand the elements and help you to survive. Moreover, you can go to the library to learn about effective farming methods that could help you and your family survive. Clearly, you can find similar information via the Internet; it provides an electronic library.

It should be clear then that learning can be accelerated via cultural influences (like libraries and conversation). Moreover, it should be clear that cultural practices can influence the survival of a group of people. In some situations, cultural influences may provide the necessary information to allow individuals in the group to be more likely to survive and reproduce. Is the environment we are living in changing in ways that influences the balance between our instincts and learning capabilities? Is the human species becoming more robust to certain situations and more sensitive to others? Is there a Baldwin effect at work in human evolution (i.e., can cultural influences accelerate learning and thereby accelerate evolution)? What is the human fitness function? How robust are humans? Do these concepts teach us anything about how to ensure the survival of the human species? What are the costs of ensuring such survival?

16.2 Robust Approximator Size Design

In this section, we study the approximator design problem studied in Section 10.2, but with a focus on how to design a "robust" approximator. To do this, we discuss robust optimal approximator design, the RSM approach to solving this problem, then show the results for a particular RSM.

16.2.1 Approximator Size Design Problem

What is meant by a "robust" approximator? Note that the function in Figure 9.10 is actually just one realization for one set of probability draws that generate the data (i.e., one possible z, for the function $G(x, z)$). A robust approximator is the single best approximator, in the sense that it will seek to optimize its performance for all possible training data sets of a certain size. Note, however, that here the "optimality" is with respect to the approximator structure choice and tuning method. Here, our only objective will be to find the size p of the best approximator for the function in Figure 9.10, for a fixed data set size M, for a Takagai-Sugeno fuzzy system (a nonlinear interpolator between lines), trained by a batch least squares method.

Response surface methodology can be used to make basic design choices for approximators.

First, note that due to the problem of generalization (overfitting) that can arise if you use more parameters than training data pairs, we will assume in our RSM that $p \leq M$. Second, we note that via our studies in Part III, we found that $M = 121$ generally led to reasonably good approximation of the underlying function. Hence, in order to reduce computational complexity, we choose $M = 100$ here (and we make it uniform across the domain $[-6, 6]$). Third, note that for the studies of the function in Figure 9.10, we know that you need at least $p = 4$ to get reasonable approximation accuracy (see Figures 9.16 and 9.18), but that with $p = 40$, we should be able to get very good approximation accuracy. This defines a range of values to consider. Note that such insights should not be ignored, as they can lead to significant computational savings in constructing the response surface.

Another principle that we learned in Part III, was that you need a larger test set than a training set. Hence, we choose $M_\Gamma = 2M = 200$ and place the input portion of the data on a uniform grid across the domain $[-6, 6]$. This will then test for approximation accuracy and will be likely to uncover problems with generalization if there are any.

Next, since there is noise on the unknown function, we will need to make multiple designs for each fixed set of training data (with fixed $M = 100$, each time you generate training data it will be different). Hence, for each value of p we consider, we will generate $N_t = 100$ trials, where we construct an approximator and compute the mean squared error relative to the test set.

For a fixed p, each time we construct an approximator, we will need to pick the centers and spreads of the membership functions. Based on the insights we gathered in Part III, here we simply place the centers on a uniform grid in the range $[-6, 6]$ and set all the spreads to be half the distance between two centers.

For the response surface, we first compute the mean squared error (MSE) for the N_t approximator designs for each p. Note that in this case, the MSE is a random variable whose characteristics are driven by the noise on the unknown function, the value of M, the approximator structure, and the training method (here, BLS). Clearly, then, it is difficult to know the probability density function of the MSE. Here, as a simple measure of MSE for a design point, we simply take an average of all the MSE values at each design point. We chose N_t to try to obtain reasonable computational demands, and yet large enough so that the average has a low variance.

16.2.2 Average Mean-Squared Error Response Surface for Approximator Size

Using the design choices in the last section, for illustrative purposes, we construct the response surface for the MSE for $N_t = 1$ approximator of size p for a range of p values, but we do this five times so we obtain five response surfaces. The response surfaces are shown in Figure 16.1. Each time we run the program, we get a response surface with a different shape. Why? It is due to the noise that is on the function that we are trying to approximate. In order to get a reliable assessment of the quality of the approximation, we will next run several trials for each design point (i.e., each p value) and then compute the average of those values.

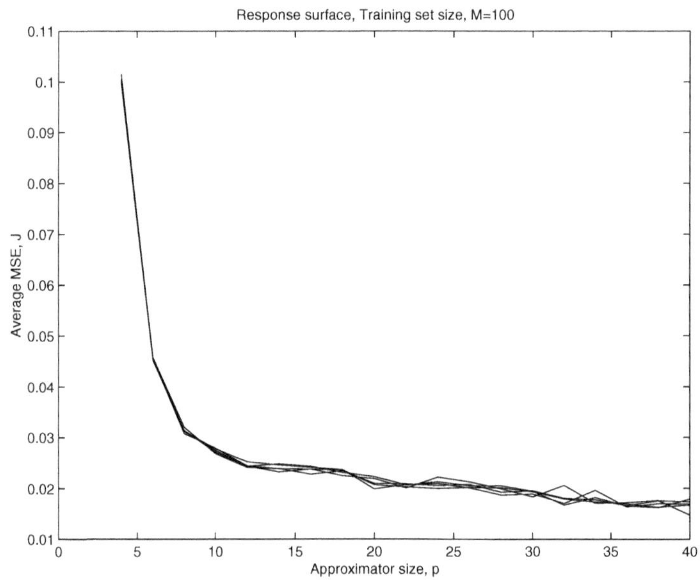

Figure 16.1: MSE response surface for approximator size, $N_t = 1$.

Next, we construct the response surface for the average of the MSE for $N_t = 100$ approximators of size p over a range of p values. The response surface is shown in Figure 16.2. How do you know that the surface will not change if you increase the number of trials N_t? One simple way, but one which generally takes lots of computations, is to simply run the program that constructs the response surface several times to see if nearly the same surface is generated for increasing values of N_t. Here, suppose that we run the program five times to generate five surfaces. In this case, we obtain Figure 16.3, where we have changed the axes to try to highlight the slight differences in the five cases. Notice, however, that there is very little difference between each of the five different trials. This gives us confidence that the plot shown in Figure 16.2 can be used to guide design decisions.

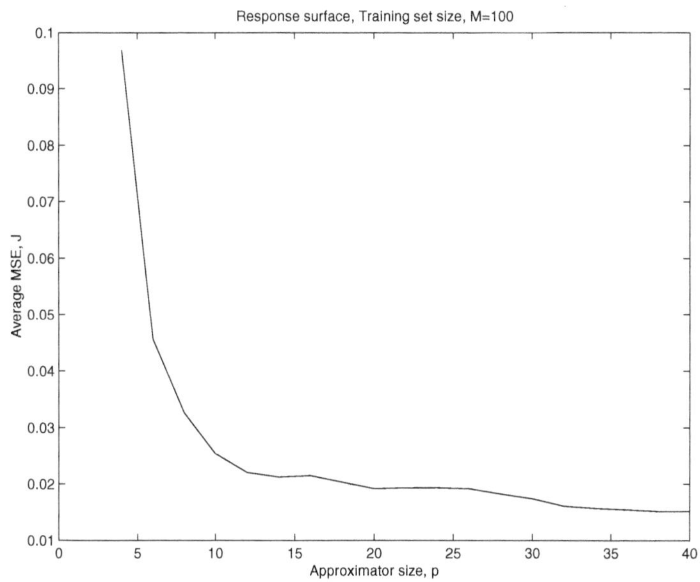

Figure 16.2: Average MSE response surface for approximator size, $N_t = 100$.

In addition to finding that the surface in Figure 16.2 is (essentially) the same shape on each trial, we also find that its basic shape makes sense. For this range of p, the estimation error generally decreases for increasing p. We find a typical "knee" in the curve that shows that it may not pay to arbitrarily increase the approximator size for a function approximation curve (for increasingly large p, little is gained in approximation accuracy). For different problems, you will find that the location of the knee in the curve will shift, but generally you will find a similar shape. What will the shape of the curve be for larger values of p? Generally, at some point (e.g., when $p > M$), the average approximation error will *increase* due to problems with overfitting and generalization.

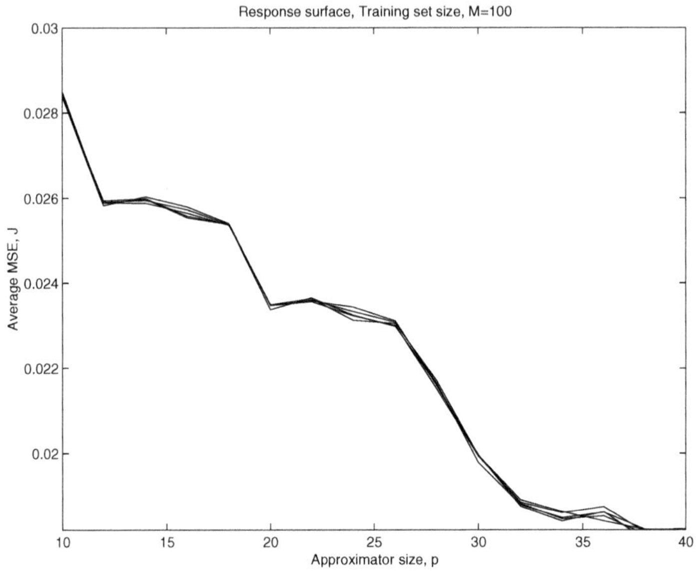

Figure 16.3: Average MSE response surface for approximator size, $N_t = 100$, 5 surfaces.

16.3 Instinct-Learning Balance in an Uncertain Environment

In this section, we will use a simple design problem for a linear (time-varying) estimator that is set up to have analogies with instincts and learning in an uncertain environment. This example is used to study characteristics that influence the "optimal balance" between the use of instincts and learning. In other words, we will set up a simple model of the trade-offs and processes involved in the evolution of a balance between how many instincts (or the quality of the instincts) and the amount (or quality) of learning. Clearly, as discussed earlier in this chapter, characteristics of the uncertainty present in the environment will affect where the optimal balance is. For instance, with more uncertainty it may be necessary to have more learning capabilities, since fixed instincts will not predict characteristics of the environment very well. However, if there is not much uncertainty, then it may be best to simply use instincts and no learning, and hence little or no memory is needed. The objective here is not to show how the instinct-learning balance idea is useful in engineering design; indeed, the estimator design problem considered here is already very well studied so that there is no contribution to engineering per se. The objective here is more modest. It is simply to show that the concept of instinct-learning balance is already used in engineering.

Instinct-learning balance concepts can be useful in engineering design (e.g., to pick an optimal trade-off between amount and quality of a priori information vs. information gathered online).

We will be discussing this problem as if we are going to use an evolutionary algorithm, but we do not actually do so in this section. We will construct the "fitness function" (actually a response surface that represents average performance values). We will show the shape of this surface. We will show how parameters affect the shape of the surface and hence, the optimum values for quality of instinct and level of learning capability. Later, in Section 16.3.3, we will study the use of an evolutionary algorithm for evolution of the optimal parameters; the development here will provide intuitive insights into why such an algorithm should work.

16.3.1 Estimator Design Problem and Analogies with Instincts and Learning

Recall from Part III that one definition of learning is: "any process through which experience at one time can alter an individual's behavior at a future time" [223] (i.e., any system with memory has the potential to be a learning system). We will use this definition here. In particular, we will define *learning ability* to be the number n of values that can be remembered by an organism in performing some activity during a lifetime. The value of n affects how the organism performs the activity (e.g., in helping it to perform the activity better). The value of n is how much memory the organism has, a key component of learning. We avoid more sophisticated learning, where the values remembered are used to change the manner in which the activities are performed. Hence, we study a *primitive* form of learning, where there is an action taken that depends on values that are remembered and the actions taken affect the performance in executing some task in the environment of the organism.

We assume that the organism can sense some aspect of its environment and store this in a (scalar) variable $y(k, \ell)$ at time k, $k = 1, 2, \ldots, N_L$, for "generation" ℓ, $\ell = 1, 2, \ldots, N_g$. Hence, N_L is a measure of the length of the lifetime of the organism and N_g is the number of generations of evolution (here, generations simply correspond to lifetimes of a single individual organism).

We use a very simple model of the "environment" that the organism performs the task in, where

$$y(k, \ell) = x(k, \ell) + z(k, \ell)$$

Here, we will assume that $x(k, \ell) = \bar{x} = 2$ is a constant that the organism wants to estimate in order to be successful in performing some activity (which enhances its reproductive success), and $z(k, \ell)$ is noise representing uncertainty in the environment (and perhaps measurement noise). The particular value $\bar{x} = 2$ is not important, and here is simply chosen for illustrative purposes. We suppose that $z(k, \ell)$ is drawn from a normal distribution with zero mean and a variance

$$\sigma_z^2(k, \ell)$$

that could change over the lifetime, or over generations. Here, at first, we will consider the case where $\sigma_z^2 = 0.5$ for all k and ℓ.

We assume that our organism can sense and store (remember) n values of $y(k, \ell)$ at each time k and that it estimates $x(k, \ell)$ via

$$\hat{x}(k, \ell) = \frac{1}{n} \sum_{j=k-n+1}^{k} y(j, \ell)$$

Hence, it simply uses a "sliding window" of n values that it computes the mean of as an estimate of the x value (note that the sample mean is a least squares estimator, assuming that the scalar to be estimated is a constant corrupted by additive white Gaussian noise). Hence, the task of the organism is to estimate x, and if it does a good job at this, we assume it has good reproductive success. It is recognized that in a computer you do not need n memory locations to store the n values, since you can simply add each new sensed value to one location. Here, we consider a type of system (organism) where all n values have to be stored (or at least where the size of n is proportional to the learning capabilities of the organism, and also to the cost of physiological maintenance of the estimator).

We model instincts as the initial conditions of the estimator. We denote the initial condition value (instinct value) as $\hat{x}_0$; this is a value that we will want to evolve. Notice that at $k = 1$, we need initial values at $k-1, k-2, \ldots, k-n+1$. Here, we assume that *all* past initial values needed before $k = 1$ are equal to $\hat{x}_0$. Hence, for example, if $n = 3$ and $k = \ell = 1$, we will use the sensed value $y(1, 1)$, and two values of $\hat{x}_0$, for a total of three values in the computation of the mean.

For evolution we need to define what we mean by the "fittest." Here, we quantify this with

$$J(\hat{x}_0, n, \ell) = w_1 n + w_2 \frac{1}{N_L} \sum_{j=1}^{N_L} (x(j, \ell) - \hat{x}(j, \ell))^2 + w_3 \exp\left(-\frac{(x(1, \ell) - \hat{x}_0)^2}{\sigma^2}\right) \tag{16.1}$$

with $w_i \geq 0$ as weighting factors and $\sigma = 0.1$. First, note that compared to our study of genetic algorithms, we consider the organism "fit" if the above cost value is minimized, not maximized. The three above terms in J can be explained as follows:

- The first term, $w_1 n$, quantifies the cost of remembering more sensed values. It may represent a cost of maintaining physiology of the mechanisms needed by the organism to sense and store the sensed values (or neural hardware to sum n values). If w_1 is big, this represents that there is a high cost, in terms of survival ability, to sensing and storing (adding) values.

- The second term is simply the mean-squared estimation error that measures the quality of estimation for a given organism over its entire lifetime (one with particular instincts $\hat{x}_0$ and learning abilities n). Notice that generally speaking, better instincts should lead to better estimation performance, and using more values to compute the mean (higher n) should also improve performance. We think of performance improvements as leading to better reproductive success.

- The third term models the cost of accurate instincts (e.g., physiological costs) via a Gaussian function centered at the actual value of the variable to be estimated. The parameter σ^2 specifies how quick the cost component decreases as the instinct quality moves away from the accurate value. Hence, accurate instincts cost, and the value of w_3 can be used to specify the magnitude of that cost relative to the other two components.

Clearly, other choices can make sense for the components of the cost function. To summarize, the cost function J quantifies that good instincts may help estimation quality, but they cost. More learning capability generally helps estimation quality, but it costs also. Intuitively, there should be an optimum balance between quality of instincts and learning capability in order to get the lowest value of J. This point where the minimum is achieved is the optimum balance point for trade-offs between learning and instinct.

16.3.2 Response Surfaces for Optimal Instinct-Learning Balance

Here, in order to gain insights, we are going to take an RSM approach, and hence, remove the mechanisms of evolution. We will show the surface (fitness landscape) that evolution will operate over in trying to find an optimum instinct-learning balance for a given environment. From this, it should be easy to envision how evolution will operate over this landscape to find the minimum over the response surface and hence, evolve the optimum instinct-learning balance.

The fitness landscape is defined via the average of J for $N_g = 100$ runs for each $\hat{x}_0$ and n considered. We consider a range of values of $\hat{x}_0$, where

$$\hat{x}_0 \in [0, 4]$$

so that we consider when the instinct value is too low and too high, and there is no difference in cost for having the instinct value deviate by the same amount above or below the $\bar{x}$ value. We consider a range of n values, $n = 1, 2, \ldots, 20$.

For $w_1 = 0.01$, $w_2 = 1$, and $w_3 = 0.05$, we get the response surface in Figure 16.4. In this case, the lowest (optimal) point on this surface, shown via the dot, is

$$\hat{x}_0^* = 1.6410, \; n^* = 5$$

so that, since very good instincts are costly and learning abilities are relatively inexpensive (low w_1), it is best to have a somewhat accurate instinct, coupled with an ability to sense and remember several values from the environment.

Next, suppose that you keep all the other parameters the same, except you let $w_1 = 0.1$, which is ten times the above value. This indicates that it costs a lot to sense and store values. How will the response surface change, and what will the new resulting optimum point be? What do you expect? Consider Figure 16.5, where we show the resulting response surface. In this case

$$\hat{x}_0^* = 1.5385, \; n^* = 2$$

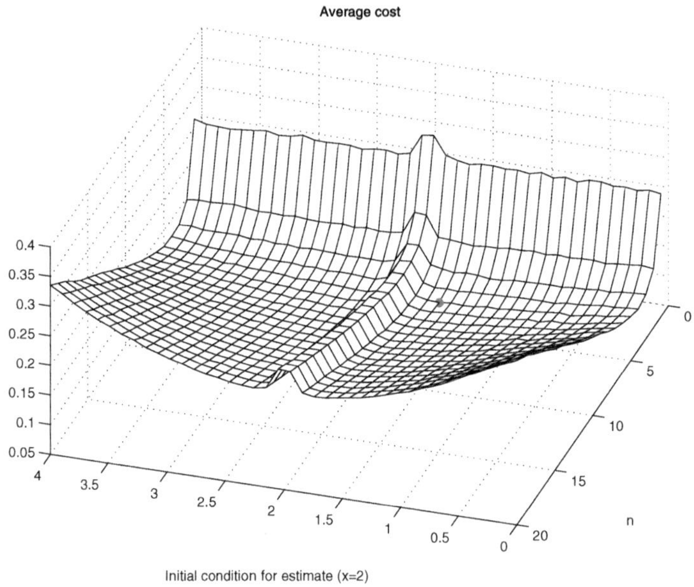

Figure 16.4: Response surface (fitness landscape) for instinct-learning balance problem, $\sigma_z^2 = 0.5$, optimum point shown with a dot.

Does this make sense? Yes. It is now better to use fewer stored values since storage costs more. Notice that the ridge due to the cost of instincts is still present so that it is best not to maintain perfect instincts.

Next, to study the effects of changes in the environment, we let the variance be $\sigma_z^2 = 0.75$, which is higher than above. This increases the uncertainty in the environment and hence, we expect that this should shift where the optimum balance is between instinct and learning capabilities. Also, for the sake of illustration, we use $w_1 = 0.01$ and $w_2 = 1$ (same as above), but $w_3 = 0.005$, which is one tenth of the above. This indicates that it is less expensive to have accurate instincts. How should the optimum point shift? We expect it to be such that it gets better instincts (closer to two) and uses more learning capabilities since with the higher variance, it will need more values in the computation of the mean in order to get good estimation accuracy. If you run the program with the values, you get the response surface shown in Figure 16.6, and from this

$$\hat{x}_0^* = 2.2564, \; n^* = 8$$

which validates what we would expect. Notice that due to the lower value of w_3, the ridge in Figure 16.4 is significantly attenuated (but is still there if you study the surface carefully). It uses a higher value of n (more learning capability) and more accurate instincts (within 0.2564, compared with $0.359 = 2 - 1.6410$). What would happen if we lower the variance to $\sigma_z^2 = 0.25$, with the same values for everything else? What is the effect of σ if we had made it bigger?

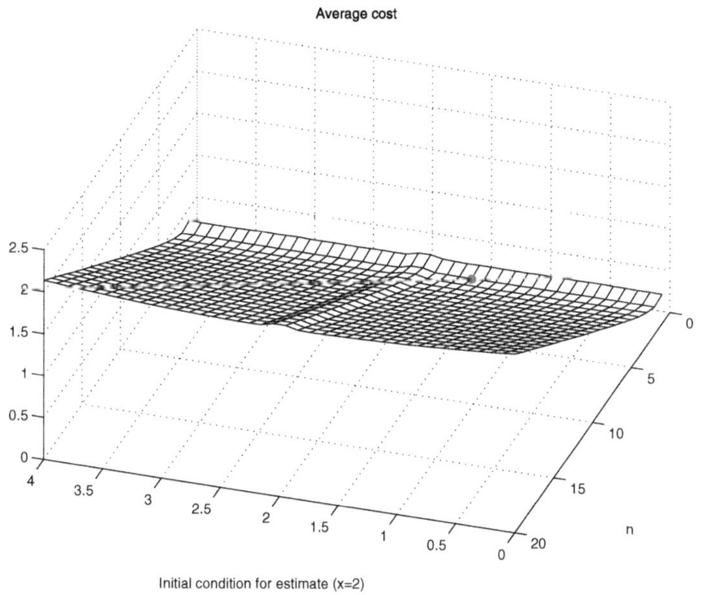

Figure 16.5: Response surface (fitness landscape) for instinct-learning balance problem, $\sigma_z^2 = 0.5$, $w_1 = 0.1$, optimum point shown with a dot.

In summary, we have developed an optimization model of achieving instinct-learning balance. Clearly, the various parameters of the cost function J (including N_L) model various aspects of the organism and the environment, and hence, different optimum instinct-learning balances are achieved for different J. It is best, then, to think of the optimum point as being time-varying, since the response surface is time-varying if the parameters of the problem change. The above plot illustrated that the optimum point shifts if the environment becomes more uncertain. What if we changed the variance slowly? We would get a slow shift of the optimum point, and next we will show how evolution can adapt to (track) such changes in the environment.

16.3.3 Evolving Instinct-Learning Balance Via Set-Based Optimization

Would a set-based stochastic optimization algorithm obtain results similar to those in the previous subsection? Yes. Here, we will show how. Moreover, we will show that if the variance of the signal that the organism is trying to estimate changes, since the fitness function shifts (changes shape), the algorithm will shift its estimates of the optimum point.

A set-based stochastic optimization algorithm would have to operate over the positive integers for the n dimension, and at the same time over a continuous range of values for the $\hat{x}_0$ dimension. Here, we use $\beta = 0.01$ for the generation

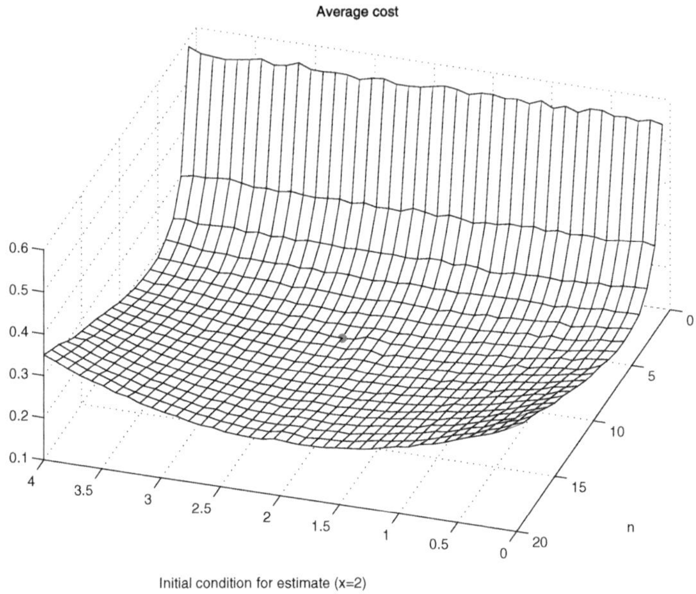

Figure 16.6: Response surface (fitness landscape) for instinct-learning balance problem, $\sigma_z^2 = 0.75$, $w_3 = 0.005$, optimum point shown with a dot.

of the "cloud" of points for the $\hat{x}_0$ dimension, and simply randomly generate n values about the best one at the current generation (e.g., here, if at some generation $n = 6$ for the best individual, we randomly place $S - 1$ points at values of either $n = 5$, $n = 6$, or $n = 7$). We use weights $w_1 = 0.01$, $w_2 = 1$, $w_3 = 0.05$, and a mutation probability $p_m = 0.1$. When mutation occurs, we place $\hat{x}_0$ and n randomly on their allowable domain. We use projection to keep $\hat{x}_0 \in [0, 4]$ and $n \in [1, 2, \ldots, 20]$ (their allowable domains). We use $S = 100$ individuals and generate $N_g = 200$ generations. In the beginning we use $\sigma_z^2 = 0.5$, but for generations ℓ, such that $\ell \geq 100$, we change the variance to $\sigma_z^2 = 0.75$. We initialize with all $S = 100$ organisms with $n = 1$ (insignificant learning capability) and place the instincts of all the organisms at $\hat{x}_0 = 1$ (representing poor quality instincts). Recall that the actual value of the signal we are trying to estimate is $\bar{x} = 2$.

The results are shown in Figures 16.7, 16.8, and 16.9. Notice that before $\ell = 100$ generations, the algorithm found that n should be about 4, which is consistent with our findings with the response surface methodology. When the standard deviation changed, it increased the amount of learning capability to around 7, which is close to what we had found via the response surface methodology (note that for that case, we had also adjusted another weight). Note that since we did not change the weight on the initial condition cost term in the cost function, there is little change in the best value of the initial condition. During the first 100 generations, the average of the values in Figure 16.8 is

1.9911, while for the last 100 generations, it is 1.9811.

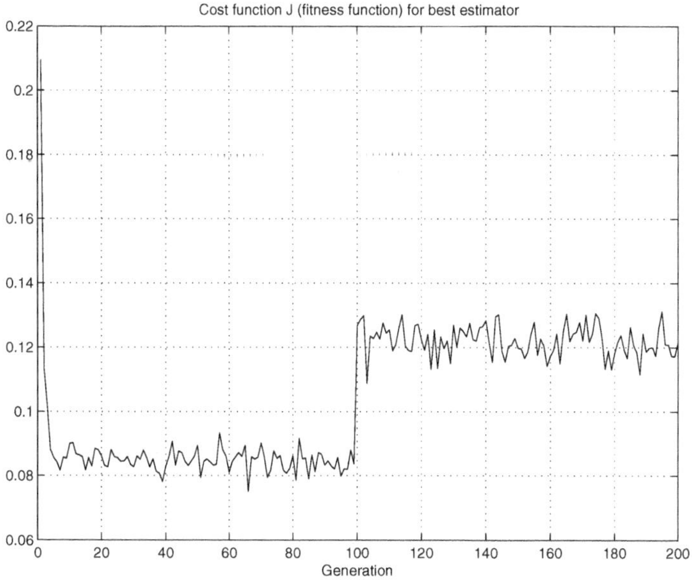

Figure 16.7: Cost J (fitness function) for set-based stochastic optimization for instinct-learning balance.

This clearly shows how time-varying features of the environment shift the fitness landscape, and hence, bring about changes in the design of an organism. You will obtain qualitatively similar results if you use a genetic algorithm to perform the evolution.

16.4 Discussion: Instinct-Learning Balance for Adaptive Control

It should be clear that there are interesting relationships between evolution and online learning for adaptive control as it was studied in the last part. Evolutionary algorithms could be used to gain insight into (i) challenging adaptive controller design problems, such as the choice of the size of the approximator(s) in direct and indirect adaptive control, (ii) choice and needed quality of initial conditions for estimators, and (iii) interactions between (i) and (ii), such as whether a poor initialization can be overcome by a more sophisticated approximator or whether with a good initialization and little plant uncertainty, how you can get away with less tuning flexibility (i.e., a smaller size approximator). Such issues arise when you consider the balance between instinct and learning in adaptive controller design.

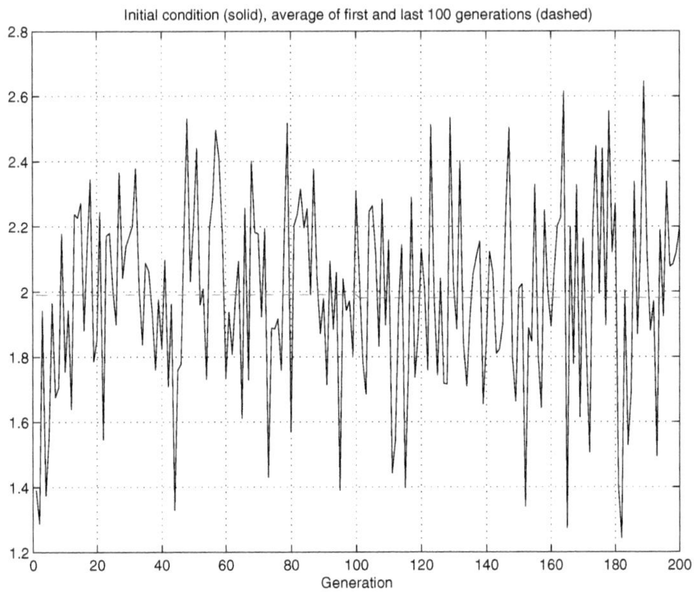

Figure 16.8: Initial condition $\hat{x}_0$ for set-based stochastic optimization for instinct-learning balance.

How would you study instinct-learning balance for adaptive controller design? For a given plant with a quantified level of uncertainty, you could try to evolve the size of the approximator structure; if you penalize size, there should be an optimal value to reach a certain performance level. It may be that if the uncertainty is increased, then a better initial condition is needed, or a larger sized approximator. Also, it should be clear that if there is too much uncertainty, increasing the size of the approximator may have no benefit. It would be nice to know how to quantify relationships among uncertainty, quality of initial condition, approximator capability, learning capability, and evolution, as this will offer the potential to better understand some fundamental and challenging problems in the design of adaptive controllers.

16.5 Genetic Adaptive Control

Suppose that each organism in a population of organisms is engaged throughout its lifetime in controlling some system. Suppose that we measure its fitness via a measure of the quality of output tracking that it achieves, and use the usual selection, crossover, and mutation mechanisms. If designed properly, the population should evolve ("learn" via genetic encoding?) an ability to achieve high performance control. This approach was discussed earlier, but there we largely took the view that we were performing offline a priori design of the control

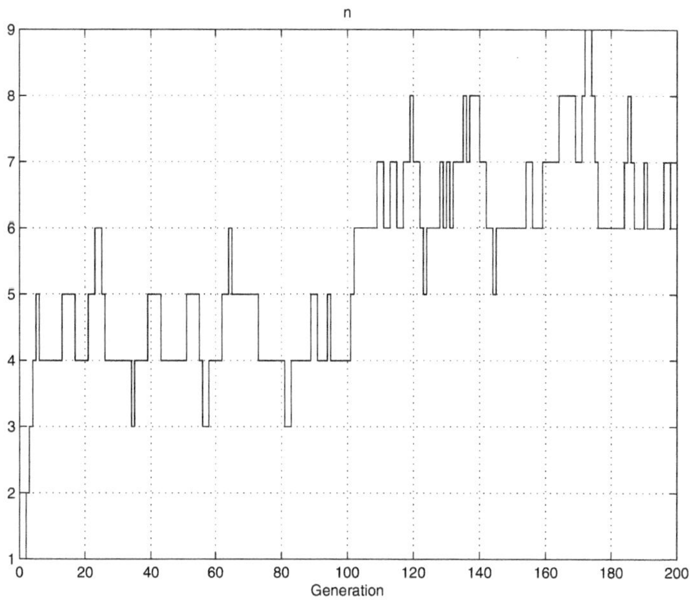

Figure 16.9: Learning capability n for set-based stochastic optimization for instinct-learning balance.

system(s). Here, we will take the view that we will use an online evolutionary process. Now, one approach that you may consider to perform online evolutionary control is to extend the concepts from the last chapter to the online case. For instance, you could set up a laboratory experiment,[1] where a controller is applied to a plant for a period of time, and a performance measure (fitness function) is computed to measure tracking performance for a typical reference input sequence. Then, the process is repeated for each controller in a set (population) of controllers by repeating the experiment. (By "repeating," we simply mean that the same sequence of reference inputs is provided; of course, this could be done by concatenating the reference sequences.) This provides values of the fitness function for each controller that are then used by standard selection, crossover, and mutation methods to generate a new population of controllers. Each controller in this population is then tested in sequence, and the process repeats. The evolved set of controllers is designed for the plant in an online fashion. Clearly, such an approach may not work for some fast processes; however, it may be quite satisfactory for others.

In this section, we take a different approach to the online evolution of controllers, one that borrows ideas from conventional control, and the adaptive

[1] A successful implementation of such an experiment was demonstrated to me by Prof. Manuel Betancur and his students at Universidad Pontificia Bolivariana, in Medellín, Colombia, in Aug. 1997.

control approaches studied in Part III. Here, we show how to use a genetic algorithm as an online optimization technique in the estimators that are used for indirect and direct adaptive control. How do these methods in this section differ from the ones we already studied in Part III? They are similar, in the sense that we also take an online optimization approach to specify adaptation mechanisms. They are different in that they utilize significantly different optimization methods. In particular, the methods here are based on sets (populations) of estimators for models of the plant or controllers and hence, are related to "multiple model" adaptive control methods (see "For Further Study" section at the end of this part). They are not, however, equivalent to multiple model methods; in addition to using different optimization methods, they show how to combine information from different model (controller) estimates in the set of estimates. Also, we show how to use the ideas in a type of multiple model direct adaptive control technique, whereas traditional multiple model methods have typically been achieved via indirect adaptive control. Moreover, the gradient methods we studied earlier, depend on explicit knowledge of the gradient of the approximators with respect to the tunable parameters; hence, those methods do not allow for the possibility of adjusting the structure of the approximators (since the gradient with respect to parameter vector *dimension changes* and basis function-type changes is not well defined). Methods like genetic algorithms (and foraging algorithms and pattern search, as you will see in Part V) do not need gradient information and hence, they are all candidates for nongradient adaptation strategies. Here, we will not fully explore this wide range of possibilities of using genetic algorithms for adaptive control. We will focus on the basic principles by establishing one class of generic adaptation schemes that are designed to be able to incorporate a wide range of nongradient optimization methods. For instance, in addition to the potential for constructing and adjusting the structure of neural and fuzzy approximator structures, the methods considered here could also be used for tuning of expert controllers, planning system-based controllers, and attentional mechanisms when gradient information is not available.

16.5.1 Indirect Genetic Adaptive Control

Like other indirect schemes, here we tune a plant model in order to specify the controller parameters. Here, however, we tune a *set* (population) of models (approximators), and the optimization method we use to tune the set is a nongradient method such as an online version of a genetic algorithm.

Population of Estimators

We assume that we use the class of plants in Equation (12.3) with the restriction that $\beta(x(k)) \geq \beta_0$ for some known $\beta_0 > 0$, and for convenience we assume that $d = 1$ and $\alpha_k = \beta_k = 0$ (if we have $d > 1$, then we need to store a set of d populations). The method to be developed will, however, work for more general classes of plants. The main reason for focusing on this class of plants is so that we have a way to specify, via a certainty equivalence approach (i.e., Equation (12.11)), the control from the estimate of the parameters.

Here, we use a *set* of approximators to estimate the unknown $\alpha(x(k))$ and $\beta(x(k))$. Following the development for the neural/fuzzy case, let

$$\begin{aligned} \alpha(x(k)) &= F_\alpha(x(k), \theta_\alpha^*) + w_\alpha \\ \beta(x(k)) &= F_\beta(x(k), \theta_\beta^*) + w_\beta \end{aligned}$$

where F_α and F_β are possibly nonlinear in the parameter approximators and w_α and w_β are once again the resulting approximation errors. Here, θ_α^* and θ_β^* are defined as in Equation (12.10), with $\theta_\alpha^\top \phi_\alpha(x)$ and $\theta_\beta^\top \phi_\beta(x)$ replaced by $F_\alpha(x, \theta_\alpha)$ and $F_\beta(x, \theta_\beta)$, respectively. Of course, if the dimension p is specified to change by the optimization method, this would have to be incorporated into the definitions.

A set (population) of models is estimated and the best one is used at each step in a certainty equivalence controller.

The certainty equivalence controller is given in Equation (12.11) with the estimates

$$\hat{\alpha}(x(k)) = F_\alpha(x(k), \theta_\alpha(k))$$

and

$$\hat{\beta}(x(k)) = F_\beta(x(k), \theta_\beta(k))$$

and we may need to use projection methods to keep $\theta_\alpha(k)$ and $\theta_\beta(k)$ values within appropriate bounds.

Due to the fact that, in general, we use a nonlinear in the parameter approximator, the error $e(k) = r(k) - y(k)$ is not a linear function of the parameters (or parameter error). We do, however, still know that

$$e(k) = \hat{y}(k) - y(k)$$

Suppose that we have a set of S approximators for α and β where the ith ones are denoted by

$$F_\alpha(x, \theta_\alpha^i)$$

and

$$F_\beta(x, \theta_\beta^i)$$

for $i = 1, 2, \ldots, S$. Let the ith estimate of the output and identification error be

$$\hat{y}^i(k+1) = F_\alpha(x(k), \theta_\alpha^i(k)) + F_\beta(x(k), \theta_\beta^i(k))u(k)$$

and

$$e^i(k) = \hat{y}^i(k) - y(k)$$

for $i = 1, 2, \ldots, S$. We choose the ith individual (e.g., chromosome) at time k to be given by

$$\theta^i(k) = \left[\theta_\alpha^{i\top}(k), \theta_\beta^{i\top}(k)\right]^\top$$

$i = 1, 2, \ldots, S$. From an evolutionary perspective, we view θ^i as the chromosome of the ith individual of the population. This completes the definition of the population of estimators.

Searching for Estimator Parameters (Models)

First, consider

$$\begin{aligned} J(\theta^i(k-1)) &= \left(e^i(k)\right)^2 \\ &= \left(\hat{y}^i(k) - y(k)\right)^2 \\ &= \big(F_\alpha(x(k-1), \theta^i_\alpha(k-1)) + \\ &\quad F_\beta(x(k-1), \theta^i_\beta(k-1))u(k-1) - y(k)\big)^2 \end{aligned}$$

which measures the size of the (instantaneous) estimation error for the ith estimate. Notice that $J(\theta^i(k-1))$ is computable, since we know $x(k-1)$, $u(k-1)$, and $y(k)$ at time k. We want to minimize $J(\theta^i(k-1))$.

The optimization achieved by evolutionary algorithms can be put to use in seeking to reduce model identification error and hence, tracking error.

In a genetic algorithm, we will evolve parameters to maximize the fitness function

$$\bar{J}(\theta^i(k-1)) = \frac{1}{\gamma + J(\theta^i(k-1))}$$

where $\gamma > 0$ is a design parameter (normally you would choose it to have a small value, since it is included simply to ensure that we do not divide by zero). With this, we will have $\bar{J}(\theta^i(k-1))$ for $i = 1, 2, \ldots, S$ and we pick $\theta^i(k)$, the population at the next generation (time step), for $i = 1, 2, \ldots, S$ using genetic operations (e.g., selection, crossover, and mutation, perhaps with elitism and other modifications). In an evolutionary strategy, we view the fitness function as defining how good each individual is doing in modeling aspects of the plant. Note that there is a type of information sharing between different estimators in that the fitness function compares the quality of estimation using each model (each with, in general, its own parameter values) and then tends to propagate the good ones over the generations (time steps) and kill ones that are not performing well.

No matter which parameter adjustment (optimization) method is used for parameter adjustments, to pick $\theta(k)$, the parameter estimates at time k that are to be used for producing the estimates $F_\alpha(x(k), \theta_\alpha(k))$ and $F_\beta(x(k), \theta_\beta(k))$ that are used in the certainty equivalence control law, we use

$$\theta(k) = \arg\min \left\{ J(\theta^i(k-1)) : i = 1, 2, \ldots, S \right\} \tag{16.2}$$

which can be viewed as the parameters for the best model of the plant in the population of estimators at the last time step. If a genetic algorithm with elitism is used, as it often is in online genetic schemes, then we are picking the elite element to define the model parameters, since the parameters that minimize the above expression are the same ones that maximize the fitness function.

The above $\theta(k)$ is determined by considering only the instantaneous value of the estimation error. Notice, however, that it is possible to define the cost function so that it quantifies the estimation error that would have resulted if we had used the model specified via $\theta(k-1)$ over *several* past time steps. In particular, let

$$\begin{aligned} J_s(\theta^i(k-1), N) &= \sum_{j=k}^{k-N} \left(e_s^i(j)\right)^2 \\ &= \sum_{j=k}^{k-N} \left(\hat{y}_s^i(j) - y(j)\right)^2 \end{aligned} \tag{16.3}$$

where

$$\hat{y}_s^i(j) = F_\alpha(x(j-1), \theta_\alpha^i(k-1)) + F_\beta(x(j-1), \theta_\beta^i(k-1))u(j-1)$$

This $J_s(\theta^i(k-1), N)$ measures the size of the approximation error over the last N steps (at $N+1$ values) when $\theta^i(k-1)$ is used for the parameters for the estimator (notice that we obtained $\hat{y}_s^i(j)$ from $\hat{y}^i(j)$ by replacing $\theta^i(j-1)$ by $\theta^i(k-1)$). Hence, $J_s(\theta^i(k-1), N)$ does not consider the actual parameter estimates that were used to specify the controls over the past time steps; it tests what the approximation errors would have been if $\theta^i(k-1)$ had been used.

Notice that if at time k we have some $\theta^i(k-1)$, we can explicitly compute $J_s(\theta^i(k-1), N)$, since it depends only on known data from the plant. At time k we know

$$y(k-N), \ldots, y(k)$$

To compute $\hat{y}_s^i(j)$ for $j = (k-N), \ldots, k$, we need

$$u(k-N-1), \ldots, u(k-1)$$

and note that since

$$x(j-1) = [y(j-1), \ldots, y(j-n-1), u(j-2), \ldots, u(j-m-1)]^\top$$

we need

$$x(k-N-1), \ldots, x(k-1)$$

In summary, we need the values

$$u(k-N-m-1), \ldots, u(k-1)$$

and

$$y(k-N-n-1), \ldots, y(k)$$

to be able to compute $J_s(\theta^i(k-1), N)$. We have all these values, so $J_s(\theta^i(k-1), N)$ is computable at time k using past data.

To specify $\theta(k)$, we can use the same approach as above in Equation (16.2), where we use the parameter adjustment method to pick the $\theta^i(k)$, for $i = 1, 2, \ldots, S$, then choose the best estimator of the last generation (where best is measured by $J_s(\theta^i(k-1), N)$) to pick the $\theta(k)$ to use in the certainty equivalence control law.

Fixed Models (Controllers) in the Population

If you know good guesses at the approximators (e.g., via application-specific heuristics or training with a gradient method on plant data), then these guesses can be placed in the population of models and could be kept fixed. With an evolutionary approach, fixed members can be viewed as "seeds" in the mating pool. In particular, if their fitness is good enough, they could be mated with other individuals to form children. You could think of this as having static "immortal" individuals who, when their fitness is good, will tend to mate with others in the population to improve the overall fitness of the population at the next generation.

Essentially, this shows one way to incorporate a priori information into the estimation process. We can think of the fixed population members as helping to guide the estimation process by the continual infusion of new ideas about how to estimate. If the plant operating conditions change, then one fixed member may contribute more than one that had a significant impact in the past when the operating condition was different.

Using Gradient Methods to Iteratively Pick a Model

The above scheme shows just one way to use genetic algorithms in an indirect adaptive scheme. Another approach is to, at each step, take several iterations of the genetic algorithm to try to make a better choice for the estimator at each step. Alternatively, you could try to use a gradient method for these inter-sample iterative optimization procedures (if gradient information is available). Regardless, such an approach may provide certain performance benefits, but clearly you pay for it in computational complexity.

16.5.2 Direct Genetic Adaptive Control

For simplicity, we will consider the same class of plants as we did in the indirect genetic adaptive control case, except we also assume that $\beta(x(k))$ is bounded from above by a known constant. For the direct case we seek to evolve a set of controllers without explicitly estimating the plant dynamics. We will use a controller of the form

$$u(k) = F_u(x(k), r(k+1), \theta(k))$$

At each iteration, we will use the best controller in the population at that time to choose the control input to the plant, as we explain next.

In a direct method, a set of controllers is evolved that seeks to minimize tracking error.

Searching for Controllers

Recall that for the direct adaptive control case, we had

$$e(k+1) = r(k+1) - y(k+1) = -\beta(x(k))(u(k) - u^*(k))$$

so that the (instantaneous) tracking error can be viewed as proportional to the (instantaneous) error between the actual and ideal control. This motivates the choice

$$e^i(k) = r(k) - y^i(k)$$

where

$$y^i(k) = \alpha(x(k-1)) + \beta(x(k-1))u^i(k-1)$$

and

$$u^i(k-1) = F_u(x(k-1), r(k), \theta^i(k-1))$$

Choose

$$\begin{aligned} J(\theta^i(k-1)) &= \left(e^i(k)\right)^2 = \left(r(k) - y^i(k)\right)^2 \\ &= \left(r(k) - \alpha(x(k-1)) - \right. \\ &\quad \left. \beta(x(k-1))F_u(x(k-1), r(k), \theta^i(k-1))\right)^2 \end{aligned}$$

where you may want to choose $\theta(k)$ to be the best $\theta^i(k-1)$. Notice, however, that this is not possible, since we cannot compute $J(\theta^i(k-1))$ due to its dependence on $\alpha(x(k-1))$ and $\beta(x(k-1))$, which are unknown. Hence, we must take a different approach.

Basically, we cannot proceed as we had in the indirect case, since making evaluations of different controllers in a population based on past data does not make sense, since alternative sequences of *past* control inputs cannot be considered, since we have already applied one sequence of inputs to the plant. (In the case where you have S physical plants of the same type with an ability to communicate between the controllers implemented for these plants, it is possible to develop an online direct adaptive scheme along similar lines, as we discussed in Chapter 14). One way to overcome this problem is to *assume the existence of a model of the plant* and to use it to predict how each controller in the population will perform (note the similarity to a planning strategy).

In particular, suppose that our model is given by

$$y_m(j+1) = f_m(x_m(j), u^i(j))$$

for $j = 0, 1, 2, \ldots, N-1$. Notice that this model can be quite general if needed; however, in practice, sometimes a linear model is all that is available and this may be sufficient. Let $y^i_m(k, j)$ denote the jth value generated at time k using $\theta^i(k)$; similarly for $u^i(k, j)$ and $x_m(k, j)$. Suppose that at each time k, you compute for $j = 0, 1, 2, \ldots, N-1$,

In a direct approach, you need a model to predict how each controller will behave so that fitness can be computed.

$$y^i_m(k, j+1) = f_m(x_m(k, j), u^i(k, j))$$

where

$$u^i(k, j) = F_u(x_m(k, j), r(k+1+j), \theta^i(k))$$

At time k to simulate ahead in time, for $j = 0$ you initialize with $x_m(k, 0) = x(k)$. Then, generate $y_m(k, j+1)$, $j = 1, 2, \ldots, N-1$, using the model (note that you will need to appropriately shift values in x_m at each step) and computed values of $u^i(k,j)$, $j = 1, 2, \ldots, N-1$.

There are many possible ways to define the cost function. One cost function that we could use would be

$$J(\theta^i(k), N, k) = w_1 \sum_{j=1}^{N} \left(r(k+j) - y_m^i(k,j)\right)^2 + w_2 \sum_{j=0}^{N-1} \left(u^i(k,j)\right)^2$$

where $w_1 > 0$ and $w_2 > 0$ are scaling factors that are used to weight the importance of achieving the tracking error closely (first term) or minimizing the use of control energy (second term). Other cost functions could use the output of a reference model to obtain a model reference adaptive scheme (see "For Further Study" for more details), an error measure on the other past values of the inputs and outputs, or an error measure on some other system variable.

To specify the set of controllers at the next time step, you use genetic operations to specify $\theta^i(k+1)$. If you use a genetic algorithm, you will probably want to use elitism to ensure that at least one good controller exists in the population (and when elitism is used, then the mutation rate can generally be increased to try to force the search method to explore many regions of the space to find a better controller). To specify the control at time k, you simply take the best $\theta^i(k)$, as measured by $J(\theta^i(k), N, k)$, and call it $\theta(k)$, and generate the control using $u(k) = F_u(x(k), r(k+1), \theta(k))$.

Note that this specifies a variety of method to achieve what is called "adaptive model predictive control" in conventional control theory. Clearly, different optimization methods will lead to different closed-loop system performance characteristics. It can be difficult to know which optimization method to choose for a particular application.

Direct/Indirect Method, Fixed Controllers, Alternative Schemes

It is interesting to note that it is possible to develop a type of "direct/indirect" scheme, where the model that is generated by the indirect scheme is used to predict in the direct scheme. In this way, a genetic algorithm is essentially used to adjust the cost function for the direct adaptive method. It creates a model of its environment and uses this model to evaluate how to create controllers that operate in that environment. Of course, you could use a hybrid approach where a gradient-based method is used for either the direct or indirect scheme, along with a genetic algorithm for the other.

The prediction model could be adaptive to achieve adaptive model predictive control.

In an analogous way to the indirect case, we could seed the population with some fixed controllers that we consider to be good guesses at how to control the plant and these could be preserved in the population as immortal individuals. These could help guide the search for a suitable controller. For instance, let the fixed controllers be a set of controllers where each one is designed to perform well at a different operating condition. Then, as plant conditions change, these different controllers may have their cost function decrease and hence, be more likely to influence subsequent parameter adjustments.

In addition, we could, at each iteration, run several iterations of an optimization algorithm (e.g., the genetic algorithm or a gradient method) to try to find an optimal controller to specify the control sequence. Moreover, we could tune the structure of the controller in addition to its parameters.

16.5.3 Design Example: Process Control Problem

For the surge tank problem, we have $n = m = 0$ so $x(k) = h(k)$. Next, we pick the approximator structures, design a genetic algorithm and adaptation mechanism, and provide results showing how the approach works for the tank example.

Approximator Choice

First, we must pick the approximator structures to be used for the two approximators for the plant nonlinearities. To keep the approach very simple, we use

$$F_\alpha(h(k), \theta_\alpha(k)) = \theta_\alpha(k)h(k)$$

and

$$F_\beta(h(k), \theta_\beta(k)) = \theta_\beta(k)$$

where $\theta_\alpha(k)$ and $\theta_\beta(k)$ are scalars (hence, we try to use a linear approximator for a nonlinear system). Note that in this case, we clearly know the partial of the approximator with respect to the tunable parameters so clearly, we can use a gradient method for tuning and we are not then fully illustrating the potential of the nongradient-based methods of this section. Here, we only seek to illustrate the operation of the method, and do not explore the interesting issues involved in tuning approximator structure where we do not have gradient information. We assume that we know bounds on the underlying nonlinearities so that we know $-2 \leq \theta_\alpha(k) \leq 6$ and $0.25 \leq \theta_\beta(k) \leq 0.5$, and we will use projection to ensure that these inequalities hold at each time step.

It is our hope that the approximators will be able to readjust $\theta_\alpha(k)$ and $\theta_\beta(k)$ in a way so that it continually retunes their values to maintain an accurate enough approximation to achieve good control. This will be necessary, since the underlying functions we are trying to approximate are nonlinear functions of the liquid height. Clearly, other choices for the approximator structure may produce an algorithm that performs better. For instance, you may use a fuzzy or neural approximator structure, and perhaps tune the parameters that enter in a nonlinear fashion, or the approximator structure. The algorithm is then, however, more complex, since you are tuning more parameters.

Indirect Genetic Adaptive Control

We use a representation where we simply have θ_α as one trait that is concatenated with θ_β. Hence, the ith population element is $\theta^i = [\theta_\alpha^i, \theta_\beta^i]^\top$, $i = 1, 2, \ldots, S$. We choose $S = 10$ as the population size. We use a base-10 genetic algorithm, where we simply produce one generation per time step (i.e., in this case, every $T = 0.1$ sec.). We fix the decimal position so that there is one digit to its left. We use five digits total, so there are four to the right of the decimal point. We have an additional gene for the sign of each of the two traits. We initialize the population of estimators to $\theta^i_\alpha(0) = 2$ and $\theta^i_\beta(0) = 0.5$, $i = 1, 2, \ldots, S$, and pick the initial estimates as $\theta_\alpha(0) = 2$ and $\theta_\beta(0) = 0.5$. We use the fitness evaluation procedure with $J_s(\theta^i(k-1), N)$ in Equation (16.3) with $N = 100$. We use fitness-proportionate selection. We choose $\gamma = 0.001$ to be the parameter used to form the fitness function $\bar{J}(\theta^i(k-1))$ from $J_s(\theta^i(k-1), N)$. We choose the crossover probability $p_c = 0.9$ (for single-point crossover) and the mutation probability $p_m = 0.05$ (for gene mutation). We use elitism and base the choice of the best individual on J_s. The best individual is used in the certainty equivalence control law at each time step.

The performance of the closed-loop system is illustrated in Figure 16.10, where we see that, after an initial transient period that results in part due to the poor initialization of estimators, we get reasonably good tracking of the reference input. Next, Figure 16.11 shows that the estimate of the tank liquid level is quite good, even though at times the individual estimates of the nonlinearities are not.

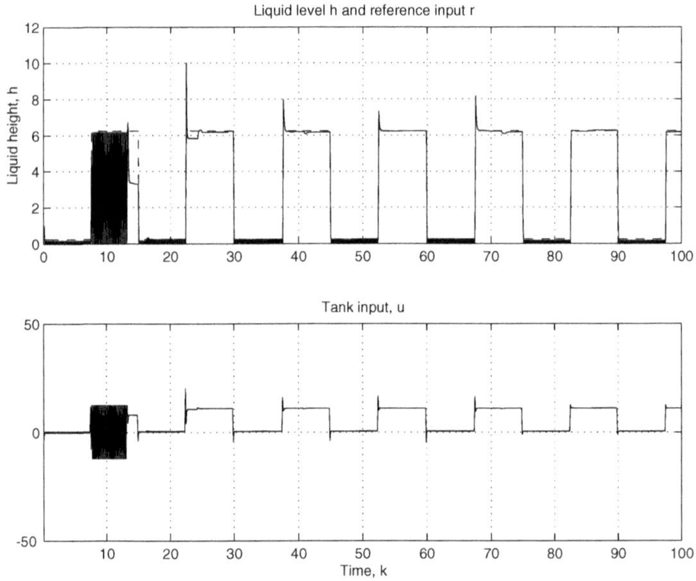

Figure 16.10: Indirect genetic adaptive controller closed-loop response.

To further illustrate some properties of the genetic adaptive controller, see Figure 16.12, where we plot the average fitness of the population (the average is $\frac{1}{S}\sum_{i=1}^{S}\bar{J}(\theta^i)$) and the index i of the best individual in the population for every time step. First, note that early in the simulation, the average fitness is

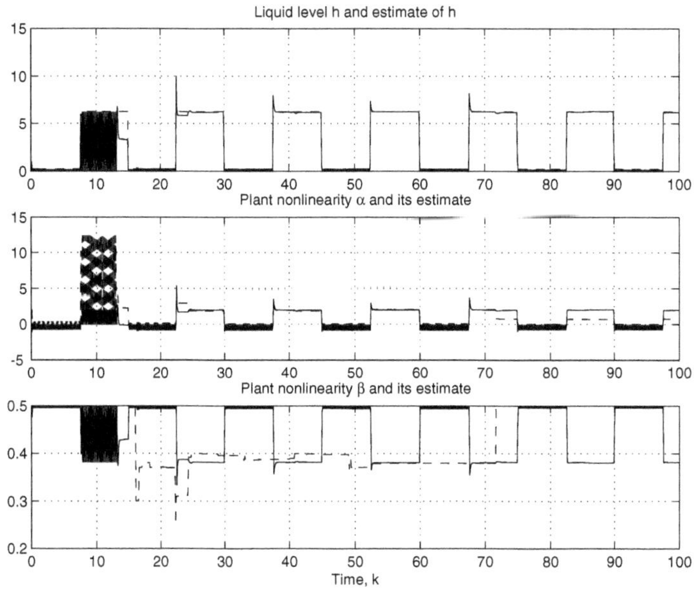

Figure 16.11: Indirect genetic adaptive controller, estimates of liquid level and nonlinearities.

very low because of the poor choice for the initial population (remember, we are trying to maximize the fitness so low values are bad). After some time, however, the genetic operations are somewhat successful at adjusting the population members so that the average estimation error decreases and hence, the average fitness increases. Note, however, that the fitness does not simply increase over time. It can also decrease and one cause of this can be the change in the reference input. Next, note that in the bottom plot, we show the index of the best individual in the population at each step. Notice that there are some relatively long stretches of time where the best individual does not change (and elitism helps to ensure this); however, there is a significant amount of switching between different members of the population that provide better estimates at different times. For some choices of genetic algorithm parameters and some applications, it could be that all members of the population become the same early in the operation of the control system (i.e., we could get "premature convergence"); however, the choice of a reasonably large mutation rate can ensure that new model parameters are continually tested and elitism tries to ensure that these random explorations will not adversely affect the estimate that is actually used in the certainty equivalence controller. Hence, we think of using mutation (and crossover to some extent) to encourage the search for better models and selection, and elitism to try to keep a good model available so that a good estimate of the plant dynamics is available for use in the certainty equivalence controller.

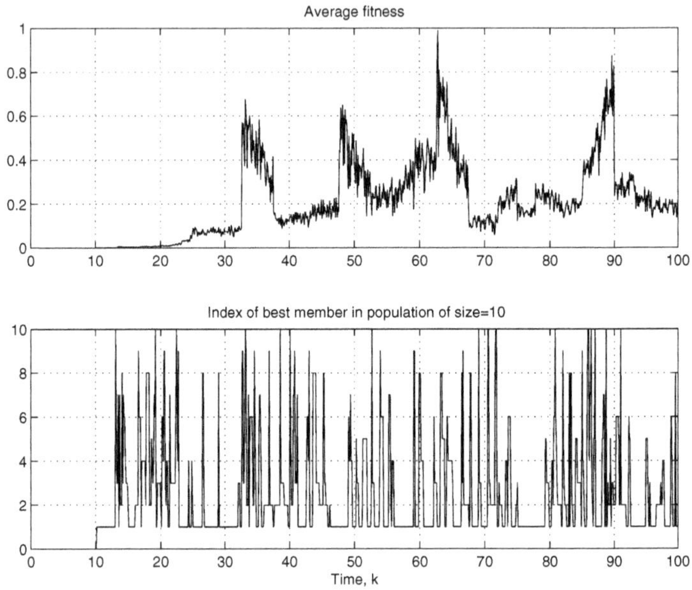

Figure 16.12: Indirect genetic adaptive controller, average fitness and index of best individual in the population.

16.6 Exercises and Design Problems

Exercise 16.1 (Indirect Genetic Adaptive Control for the Tank): In this problem, you will study the indirect adaptive controller for the surge tank that is developed in the chapter in Section 16.5.3.

(a) Run the code given at the Web site and verify all the results provided in the chapter.

(b) Tune the algorithm to obtain better performance. To do this, you may want to consider changing S, N, and the parameters of the genetic operators. Be sure to quantify performance properly (considering the stochastic nature of the algorithm).

Exercise 16.2 (Alternative Indirect Genetic Adaptive Controller for the Tank): In this problem, you will study the tank problem in Section 16.5.3; however, here, compared with Exercise 16.1, you will use a different optimization algorithm. In particular, use one of the approximations in Section 14.6 in place of the genetic algorithm and evaluate performance relative to that obtained in Exercise 16.1.

Design Problem 16.1 (Response Surface Methodology for Choice of Approximator Structure Size and Type): In this problem you will investigate the use of response surface methodology for the choice of approximator structure size and type. Be careful not to overgeneralize the conclusions you draw from this problem: your analysis used to solve the problems below is only for one particular function approximation problem.

(a) Repeat the approach of Section 16.2, but for the design of a multilayer perceptron. Clearly explain all of your choices and illustrate your results via simulations.

(b) Repeat the approach of Section 16.2, but for the design of a radial basis function neural network. Clearly explain all of your choices and illustrate your results via simulations.

(c) This problem combines the work in the chapter with the results of (a) and (b). Suppose that you decide that your criterion for choice of which type of approximator structure to use is based solely on approximator size as measured by the number of tuned parameters p. Create a fair comparative study of approximator complexity for the multilayer perceptron, radial basis function neural network, and Takagi-Sugeno fuzzy system. Draw appropriate conclusions and support them with analysis and response surfaces.

(d) Repeat (a)–(c), but for the case where the Levenberg-Marquardt method is used for training all the parameters of the approximators. Is there any fundamental difference compared to (a)–(c)?

Design Problem 16.2 (Stochastic Nongradient Optimization for Approximator Construction):

(a) For the function approximation (theme) problem of Part III that is studied, for instance, in Section 10.2, design an SPSA algorithm to tune the approximator. Evaluate the resulting approximation accuracy, and explain your choices for the algorithm parameters.

(b) Repeat (a) but for the set-based stochastic optimization method.

Design Problem 16.3 (Evolving Instinct-Learning Balance): In Section 16.3.3, we studied the use of a set-based stochastic optimization method for evolving instinct-learning balance. Repeat the study there, but instead use a genetic algorithm as the evolutionary mechanism. You should be able to show that you get results that are qualitatively similar to the set-based stochastic optimization method. Next, study the effect of changing the variance to 0.25 from 0.5. What happens? Illustrate with simulation results. Also, study the effects of changing the constant value $\bar{x}$ to 3. What are the effects on the optimal balance? Illustrate with simulation results.

Design Problem 16.4 (Indirect/Direct Genetic Adaptive Control): In this problem, you will study the development of indirect and direct genetic adaptive controllers for the process control problem studied in Sections 12.4 and 12.6, and in Design Problem 18.4. Here, however, you will use a genetic algorithm as the basis for specifying the parameter update laws.

(a) Repeat (a)–(c) in Design Problem 12.1, but design genetic operations for the parameter adjustments for both the direct and indirect strategies. Investigate the relationship between the use of elitism and the choice of the mutation rate. Illustrate the utility of using fixed models (indirect case) or controllers (direct case) in the adaptive controllers.

(b) Repeat (a) in Design Problem 18.4, but develop an "indirect/direct" strategy where you use a genetic algorithm to tune the model that is used in the cost evaluation for the direct genetic adaptive controller.

Design Problem 16.5 (Response Surface Methodology for Process Controller Design)*: Use RSM to design controllers of the following types for the process control problem in Section 6.4.1:

(a) Instinctual neural controller. In this case, focus on the design of the number of neurons used in the hidden layer of a two-layer network.

(b) Fuzzy controller. In this case, focus on the design of the number of rules.

(c) A controller based on a planning methodology. In this case, focus on the design of the planning horizon length.

(d) An adaptive controller. In this case, focus on choice of the size of the approximator, or the initialization of the approximator mapping(s).

Design Problem 16.6 (Approximator Structure Construction)*: Design and test an SPSA algorithm to solve Design Problem 11.2.

Design Problem 16.7 (Instinct-Learning Balance in Control)*: In Section 16.4, we discussed the relevance of the instinct-learning balance to adaptive control. Develop an application and simulation that will illustrate how the instinct-learning balance concept is useful in adaptive controller design.

Design Problem 16.8 (Evolution of Stable Adaptive Systems)*: This problem is a continuation of Exercise 2.3 on page 93.

(a) Read [273, 272], focusing on the parts where stability and "adaptation to the edge of chaos" are discussed. Key issues to focus on are self-organization in complex systems, chaos, and Kauffman's idea that evolution keeps complex biological systems at the edge of chaos.

(b) Explain how Kauffman might view parameter adaptation in an adaptive control system (e.g., in light of his views in Chapter 5 of [273]).

(c) Persistency of excitation for system identification makes sure that the plant input signal is rich enough, so that the input-output data will contain a sufficient amount of information, so that good system identification can be performed. In indirect adaptive control, the control input to the plant is the signal that excites the plant dynamics; the type of input determines the quality of the online identification of the plant dynamics. The problem is that the types of control inputs that are generated to achieve good tracking are not necessarily ones that lead to good plant identification; a well-designed adaptive control system must then achieve a type of balance in choice of the control input to the plant. It must be rich enough so that it can figure out the plant dynamics, yet still try to achieve tracking. Suppose that you think of a chaotic signal as being persistently exciting for some adaptive control system. Kauffman says that complex systems adapt to the "edge of chaos" and this is the best place for the system to be in order to adapt. Is there a relationship between his ideas and the well-understood properties of indirect adaptive controllers discussed above? What is the relationship? Explain. Can you support your claims with simulation results for a particular plant?

Chapter 17

For Further Study

A nice introduction to genetic algorithms is given in [363], and another treatment where many related optimization methods are treated (e.g., evolutionary programming) is given in [355]. Mathematical models of evolution are studied in [245, 534]. Response surface methodology is covered in [351]. An overview of stochastic optimization and simulation is given in [480].

Genetic Algorithms and Evolutionary Programming: For more details on genetic algorithms, see the books [363, 355, 216, 353, 527, 179, 344, 39] or article [488]. For a study on how to integrate gradient methods and evolutionary programming, see [474]. For more details on "genetic hill-climbing algorithms," see [3]. For a mathematical treatment of genetic and related algorithms, see [527, 39], and parts of some of the other books above.

Genetic algorithms presented in this part are special cases of the more general "evolutionary programming" methods, where the same sorts of Darwinian/Mendelian principles are applied to, for example, general data structures or to evolve the code needed to perform a task (e.g., to evolve the lines of C code needed to implement a nonlinear controller). Such general evolutionary programming methods are discussed in, for example, [284, 285] (perhaps of particular interest is the fact that there the author also considers evolving code for nonlinear controllers) and in [353, 42]. The topic of how to evaluate the general success of one genetic optimization strategy over another is a deep and important problem (see some recent results in [354]).

Robustness, Optimization, and Evolutionary Background: To learn more about highly optimized tolerance and robustness trade-offs in the design of complex systems, see [93, 94]. "No free lunch theorems" for optimization are covered in [543]. An accessible easy-to-read overview of a biologist's perspective on the evolution of complexity is given in [438]. The Baldwin effect is discussed in [243, 53, 363] and related earlier studies are given in [26]. A computational perspective on the influence of cultural factors on evolution is given in [53]. Lamarckism is discussed in [139] and a computational perspective on Lamarckian influences on evolution for optimization is given in [5, 4]. To learn more about neuroscience, learning, and evolution of language, see [420]. The study of genetic influences on learning and memory in mice was taken from [268], and more recent research is reported in [515, 269]. An area relevant to the topics of this part is evolutionary game theory [472, 473] (for more comments on this area, see the "For Further Study" section at the end of the next part). Evolution/optimization of social insect colony composition is discussed in [539].

Nongradient Optimization: A good introduction to deterministic nongradient methods for optimization is given in [68, 337]. Evolutionary operation via factorial designs was introduced in [74]. The "simple" coordinate search method was based on [512], where more sophisticated coordinate search methods and the Hooke-Jeeves method are also explained. Line search methods are discussed in most optimization books (e.g., see [68]), and relations between and convergence results for combined pattern and line search methods are given in [336]. The Nelder-Mead direct search method was introduced in [385]. It is shown that it can converge to a nonstationary point in [349], and its convergence properties in low dimensional problems is studied in [298]. The section on multidirectional search is based on [512], where its convergence properties are studied; the implementation of such methods on parallel machines is studied in [140]. Convergence properties of general pattern search methods (e.g., coordinate search and multidirectional search) are studied in [513]. Convergence properties of pattern search methods for constrained minimization are studied in [317] (where the "simple" coordinate search method studied here is also studied). In [275], the author discusses the "simplex gradient" and direct search methods and provides Matlab code for these.

It is common in several areas of optimization theory to use "surrogate models." For instance, in nongradient optimization, when it is expensive to compute values of the cost function, sometimes a surrogate model of the cost function is constructed from the samples taken. Often this model is simply an approximator structure and the methods discussed in the last part for approximator tuning can be used. Then, the surrogate model can be used to decide where to take the next sample of the cost function (sometimes this can be done by executing an optimization method over the surrogate model where computation at points on it are typically much less expensive than for the actual cost function). The surrogate model method is described in [514], and other such approaches are given in [108] and the references therein. Surrogate model methods are related to "response surface methods" [351].

Stochastic Nongradient Optimization: Response surface methodology is covered in detail in [351]. There, you can also learn more about "Taguchi's parameter design," which can be useful in the design of robust processes in some cases (e.g., for variance reduction). Moreover, you will find that the approach in [351] of "continuous process improvement with evolutionary operation" bears similarity to several concepts discussed in this part, but especially to the use of online evolutionary algorithms for adaptive control.

An introduction to stochastic search and optimization for estimation, simulation, and control is given in [480]. The SPSA is also treated there. The SPSA section was developed using [476, 477, 478, 479]. The SPSA method was introduced in [476] and some theory is given there. An overview of the method is given in [478], where Matlab code is included. Guidelines for choosing parameters of the SPSA algorithm are given in [477]. Monte Carlo optimization methods are relevant to the methods studied in this book, as are "simulated annealing" techniques that are motivated by physical inorganic phenomena rather than biological ones. For an overview of the simulated annealing optimization approach, and its relation to the SPSA algorithm, see [480, 479]. For some theoretical comparisons of stochastic optimization methods, including some evolutionary methods, see [480, 481].

Set-based nongradient deterministic and stochastic optimization methods are discussed in more detail in Part V. The set-based stochastic optimization method in Section 15.6 grew out of ideas in Part V, the SPSA method, and [345] (and discussions with the author of that book).

Evolutionary Design: Computer-aided design of fuzzy controllers via genetic algorithms has been studied in a variety of places, including [271, 256, 402, 523, 305]. An example of how to use a genetic algorithm to design direct, adaptive, and supervisory fuzzy controllers for a robot is given in [81]. Design of robust control systems with genetic algorithms is studied in [452, 98]. For more information on using genetic algorithms for constructing neural networks, see [549]. Darwinian design has been used in several areas, including circuit design [509] (the authors of this paper also study evolution of controllers implemented with electronics). There is a growing interest in evolvable hardware, and in connection to this topic and others, you may want to study the area of "evolutionary robotics" [388].

Evolutionary Adaptive Estimation and Control: The use of genetic algorithms in system identification is studied in [180, 288], and several researchers have applied genetic algorithm methods to system identification for various problems. Genetic adaptive estimation is studied in [548, 386, 506, 224, 226]. Genetic adaptive observers and state estimation is studied in [427, 225, 227].

For more information on multiple model adaptive control, see [31, 347, 333, 377, 375] and the references therein. The study of evolving neural controllers is related to the work in [5], where evolutionary reinforcement learning is studied in an artificial life simulation. Indirect genetic adaptive control was first introduced in [288]. The genetic model predictive control method was first introduced in [426, 428] (there, it was called the "genetic model reference adaptive controller" (GMRAC), since it is an MRAC scheme that adapts by evolving population of controllers). Other genetic adaptive strategies (e.g., the direct/indirect scheme) are studied in [307] and compared to conventional adaptive control methods for a specific application. Genetic adaptive control methods were implemented and compared to other nonlinear and adaptive fuzzy control methods in [366, 367].

Evolutionary Models of Physiological/Behavioral Phenomena: Evolution creates optimization processes: this has been a central theme of this book (e.g., it created the underlying optimization processes in planning, attention, and learning systems). As evolution itself can be viewed as an optimization process, it is natural that we can find physiological processes that seem to demand the use of an evolutionary model (some have even suggested that the basic cognitive processes in the human brain can be profitably thought of as an evolutionary process). Now, there exists great potential for confusion on this point, or wonderful opportunities to gain a deeper understanding of systems level biological processes. The confusion can arise in having persons intimately familiar with evolutionary models (which might not be all that accurate) come to view every stochastic optimization process that operates in nature as being a type of evolutionary algorithm, perhaps even without any way to verify that (e.g., experimentally). At the same time, it seems unwise to ignore the possibility that evolution can create other physical evolutionary processes in living organisms.

So, where are evolutionary models being used in physiological systems? First, note that there are many connections between evolutionary algorithms and "artificial immune system" (AIS) models; see [125, 124] and the additional references sited on this topic in the "For Further Study" section at the end of the last part. Also, see Design Problem 14.4. AIS models show a way to model one aspect of a single organism as an evolutionary process. Second, we will see in Part V that with some types of social foraging of bacteria (e.g., *E. coli*), evolutionary processes play an important role, and the optimization algorithm that the foraging is represented by, bears many similarities to evolutionary algorithms. Third, one part of the physiology of social foraging of honey bees can be modeled as a type of evolutionary process (and in this case, even the experimentalists that study social foraging use the evolution metaphor to explain how the overall process operates). Connections between behavior and evolution are also studied in the field of "evolutionary game theory;" for references on that topic, see the "For Further Study" section at the end of the next part.

Part V

FORAGING

Part Contents

Sequence of Essential Concepts

- Due to fine-tuning by evolution, the activity of animal foraging can be modeled as a constrained optimization (search) process. For example, some bacteria engage in nutrient hill-climbing, and this is an optimization process, if we view the cost function as modeling the concentration of nutrients.

- Some animals communicate with others of their species in order to help each other forage. This is called cooperative or social foraging and it too can be modeled as an optimization process. Social foraging of groups of animals can exhibit the same basic decision-making elements as in a single intelligent animal. For instance, bacteria use simple rules to govern their foraging, and ants lay pheromones to *learn* about their environment and can later use the pheromone trails to decide where to go. Honey bees use a type of distributed resource allocation of foragers that can be thought of as a type of attentional process (with more attention given to more profitable sites).

- Biomimicry of the optimization processes used by animals in (social) foraging provides new nongradient, stochastic, and parallel optimization methods for solving control, automation, and other engineering problems. Good examples of application areas in automation for the methods of this part are in the areas of mobile robots or cooperative robotics.

- Cohesiveness in swarms can be characterized as a stability property, and conventional Lyapunov stability analysis methods can be used to study the qualitative behavior of swarms.

- Animal competitive and cooperative foraging behaviors can be modeled as games. Social foraging can be modeled as a cooperative game, and competitive foraging as an adversarial game. Game-theoretic strategies, such as the Nash and Pareto solutions, take on specific and useful meanings in the context of foraging games.

- Higher animals use learning, planning, attention, and team cooperation and competition in foraging and fighting. The design of intelligent algorithms to mimic social foraging animals presents significant and interesting challenges.

Chapter 18

Cooperative Foraging and Search

Chapter Contents

For many organisms, the survival-critical activity of foraging involves trying to find and consume nutrients in a manner that maximizes energy obtained from nutrient sources per unit time spent foraging, while at the same time minimizing exposure to risks from predators. If the organism has a decision-making mechanism (e.g., a brain), then we can view this mechanism as the controller and the remainder of the organism and environment as the "plant" (process to be controlled). "Decisions" involve where and when to move, what to eat, and so on. Decision heuristics, planning, attention, and learning can all play a role in foraging (if the organism is endowed with such capabilities). Moreover, evolution can be viewed as a process that optimizes the foraging strategy; it redesigns (tunes) all supporting physiology and the foraging decision-making strategy.

In this chapter, we outline the foundations of foraging and basic concepts related to how animals work together to "socially" forage. We do this by modeling the foraging process of a species of bacteria as optimization processes. You will see that they use elements of gradient-type search (e.g., approximations of gradients) without relying on explicit gradient information. Hence, foraging methods naturally build on the gradient optimization methods of Part III. However, they also incorporate evolutionary aspects, and are nongradient methods, so they provide a natural bridge between the optimization methods of Part III and Part IV, by giving ideas for how the various types of optimization algorithms can be merged (e.g., viewing learning as gradient-based and occurring over a lifetime, and evolution as nongradient, population-based, and occurring over long time epochs). You will also see relationships between the foraging algorithms and the pattern search and SPSA methods of Part IV. Later in the chapter, we show how to model swarms, characterize their cohesiveness as a stability property, and provide conditions under which they will converge to maintain a cohesive group. In simulation, we will study how stable swarms forage. Moreover, we show how a cooperative robotics problem of guiding a group of robots around some obstacles to a goal position in a factory can be formulated and solved via the ideas from foraging swarms.

Broadly speaking, you could simply view this chapter and the next as building on all the earlier chapters in the sense that the focus here is on the development of control strategies for guidance of an organism (e.g., the last section of this chapter will show a firm connection to Chapter 6, where we study path planning for obstacle avoidance by an autonomous vehicle) or a group of organisms. Here, however we add details on how specific animals forage (i.e., how they solve the control problem of guiding themselves successfully through their environment) and the relevant connections to optimization theory. We study the case of cooperative foraging, where animals work together in this chapter, and in the next, we study the case where they compete. Moreover, the final section in the next chapter serves to challenge the reader to develop an intelligent foraging team, and to do this, many of the concepts developed in this and the next chapter will be quite useful, not to mention the rest of the book.

18.1 Foraging Theory

In this section we outline the basic principles of foraging theory and highlight aspects of social foraging where animals cooperate in foraging activities.

18.1.1 Elements of Foraging Theory

Foraging is an optimization process created by evolution.

A large portion of foraging theory is based on the assumption that animals search for and obtain nutrients in a way that maximizes their energy intake E per unit time T spent foraging. Hence, some animals try to maximize a function like

$$\frac{E}{T}$$

Maximization of such a function gives them the nutrient sources to survive and additional time for other important activities (e.g., fighting, fleeing, mating, reproducing, sleeping, or shelter-building). Other possible currencies may be used in foraging such as "energetic efficiency," which is the net energy gained, divided by the energy invested to get it. Some animals seem to switch between optimizing the net rate of energy gain E/T to optimizing energetic efficiency. Sometimes variance in energetic gain is the primary variable that drives foraging behavior (e.g., when there is a need to meet a daily energetic intake requirement before nightfall), while other times, predation is a significant factor. Regardless, most biologists argue that animals seek to optimize some variable that is correlated with fitness.

Clearly, foraging characteristics can be very different for different species. Herbivores generally find food easily, but must eat a lot. Carnivores generally find it difficult to find food, but do not have to eat as much, since their food is of high energy value. Some other activities are related to foraging. For instance, seeking favorable environments and avoiding harmful ones (e.g., finding shelter from the weather), or searching for a suitable mate, are both related to foraging.

The "environment" of the animal establishes the pattern of nutrients that are available (e.g., via what other organisms and nutrients are available, constraints such as rivers and mountains, and weather patterns), and it places constraints on obtaining that food (e.g., small portions of food may be separated by large distances). It also affects the availability of resources (e.g., weather). Some foragers have a search rhythm (e.g., daily, nightly, etc.), but others forage opportunistically and, depending on their needs, independent of such a rhythm. During foraging there can be risks due to predators, the prey may be mobile so it must be chased, and the physiological characteristics of the forager constrain its capabilities and ultimate success. Often, in biology, researchers think of evolution as having optimized the foraging behavior of a species for an ecological niche and within its physiological constraints (which may change due to evolution also).

For some animals, there are multiple prey types that could be chosen, and the choice may depend on its diet, the abundances of the prey types, and how easy they are to find. For other animals, nutrients are distributed in "patches" (e.g., a lake, a meadow, a bush with berries, a group of trees with fruit). Foraging involves finding such patches, deciding whether to enter a patch and search for food, and whether to continue searching for food in the current patch or to find another patch that, hopefully, has a higher quality and quantity of nutrients than the current patch. Patches or prey are generally encountered sequentially and sometimes great effort and risk is needed to travel from one patch or prey to another. Some patches of food or prey naturally appear and disappear, and appearance can "trigger" certain foraging behavior (as can hunger). Generally, if an animal encounters a nutrient-poor patch or undesirable prey, but based on past experience it expects that there should be a better patch or prey, then it will consider risks and efforts to find another patch or prey and, if it finds them acceptable, it will seek another patch or prey. Also, if an animal has been in a patch for some time, it can begin to deplete its resources, so there should be an optimal time to leave the patch and venture out to try to find a richer one. It does not want to waste resources that are readily available, but it also does not want to waste time in the face of diminishing energy returns.

Decision-making in foraging is a control strategy for organism guidance.

Optimal foraging theory formulates the foraging problem as an optimization problem and via computational or analytical methods, can provide an optimal foraging "policy" that specifies how foraging decisions are made. There are quantifications of what foraging decisions must be made, measures of "currency" (the opposite of cost), and constraints on the parameters of the optimization. For instance, researchers have studied how to maximize long-term average rate of energy intake, where only certain decisions and constraints are allowed. Constraints due to incomplete information (e.g., due to limited sensing capabilities), predation, and risks (e.g., due to predators) have been considered.

Here, the interesting fact is that the foraging problem can be formulated as an optimization problem that results in an optimal decision policy, if the optimization problem can be solved (traditionally, dynamic programming formulations have been used [490]). Essentially, these optimization approaches seek to construct an optimal controller (policy) for making foraging decisions. Some biologists have questioned the validity of such an approach, arguing that no animal can make optimal decisions. However, the optimal foraging formulation is only meant to be a model that explains what optimal behavior would be like. Biologists have shown that foraging decision heuristics are used very effectively by animals to approximate optimal policies, given the physiological (and other) constraints that are imposed on the animal. Such an approach is quite rational, even in engineering applications. In the construction of a computer decision-making system, dynamic programming is sometimes found to be impractical due to computational complexity issues. Heuristics and approximations are then sometimes used to try to provide a suboptimal solution, but one that is as good as possible, given the available computing resources.

18.1.2 Behavioral/Sensory Ecology of Search

Foraging has been studied for many years from both experimental and theoretical perspectives. One aspect of foraging that is particularly relevant here is the search strategy (an optimization process) that is employed in finding food. Here, we outline some relevant theory pertaining to search strategies for foraging. Later, we briefly discuss how nongradient optimization methods might be useful in modeling some search strategies in foraging.

Cruise, Saltatory, and Ambush Search

In one approach to the study of foraging search strategies [390], predation is broken into components that are similar for many animals. First, predators must search for and locate prey. Next, they pursue and attack the prey. Finally, they handle and ingest the prey. The importance of various components of the foraging behavior depends on the relationship between the predator and the prey. If the prey is larger than the predator, then the pursuit, attack, and handling can be most important. The prey may be easy to find, but the prey's size gives it an advantage. If the prey are smaller than the predator, then generally the search component of foraging is most important. Small size can be an advantage for the prey. Since prey are often smaller than predators, for many animals they must be consumed often and in large numbers; this makes the search time limit other components of the predation cycle. Here, suppose we consider cases where the searching behavior is the dominant factor in foraging. This is the case for foragers, such as many birds, fish, lizards, and insects.

Animals use a variety of search strategies in foraging that are each optimization methods.

Some animals are "cruise" or "ambush" searchers. For the cruise approach to searching, the forager moves continuously through the environment constantly searching for prey at the boundary of the area being searched (tuna fish and hawks are cruise searchers). In ambush search, the forager sits and waits for prey to cross into strike range (rattlesnakes are ambushers). The search strategies of many species are actually in between the cruise and ambush extremes. In particular, in "saltatory search strategies," an animal will intermittently cruise and sit and wait, possibly changing direction at various times when it stops and possibly while it moves. To envision this strategy, consider Figure 18.1, where distance traveled while searching is plotted versus time. In cruise search, distance increases at a constant rate dictated by how fast the animal moves in search. At the other extreme, the ambusher sits and waits for a long time, then makes a move to try to obtain a prey. In between, there are many possible saltatory search strategies that are based on an alternating sequence of cruising and waiting. Many animals' foraging strategies seem to lie somewhere on the continuum between ambush and cruise, and hence, are saltatory search strategies.

While cruisers tend to search at the boundary of the search space and ambushers stay in one place, saltatory searchers generally move throughout the search space. Saltatory search can be adjusted to suit the environment by changing rates of movement during cruises, and the lengths of cruises and waits. For instance, some fish are known to pause more briefly and swim farther and faster during repositioning when searching for large prey compared to small ones. This is consistent with foraging theory in that the fish is willing to spend more effort to obtain more food (energy).

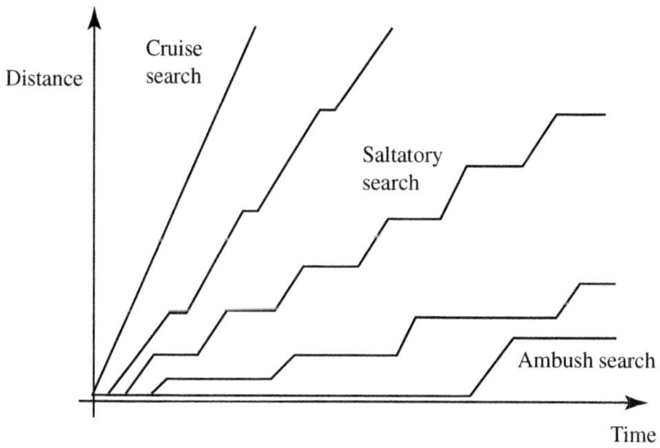

Figure 18.1: Illustration of the range of search strategies for foraging animals (figure adapted from [390], © Sigma Xi, The Scientific Research Society, and used with permission).

Scanning and Repositioning Relationships

Studies of foraging for some fish have shown that they do not search for prey while they are moving, but only during the stationary pause between repositioning moves. They stop and look around. Hence, the repositioning moves serve only to move the fish into regions where they have not looked before. Generally, if the animal searches only during pauses, then the repositioning moves (length and direction) depend on the sensing capabilities of the animal. For instance, consider Figure 18.2, where, at the top, the animal is imagined to be at the center of the circle which represents a local scan range. Suppose that for this particular animal, it can search in the pie-shaped region that is shaded. (Other animals have different shaped regions, some that almost fill the entire circle.) How large should the repositioning move be? As illustrated, if the move is too short, then a significant portion of the search space is searched again after the move and this is generally a waste of resources. If the move is too large, then there is no overlap and there can be some part of the space that is not searched, representing possible missed opportunities. For some intermediate length moves, there will be some overlap but not too much search space ignored. As illustrated on the bottom, changes in direction are dependent on the shape of the scan area also. Finally, note that as the geometry of the shaded region changes (e.g., the pie-shaped region becomes a larger piece of the pie), then identical intermediate-sized moves result in larger overlap of the search space (draw a figure and convince yourself of this). It has been shown that some fish choose repositioning moves so as to maximize net energy gain, consistent with basic ideas in foraging theory.

In many species the pauses are used for orienting the animal toward prey. That is, they stop and change their direction based on their scan information.

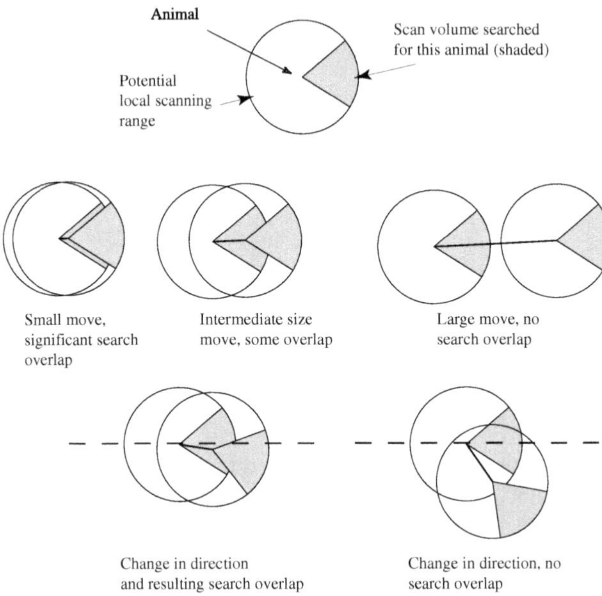

Figure 18.2: Illustration of trade-offs between repositioning and scanning; trying to scan everywhere and not rescan already visited areas (figure adapted from [390], © Sigma Xi, The Scientific Research Society, and used with permission).

Then, for instance, if there is not an abundance of prey, in some species of fish, fewer pursuits follow pauses. Also, in some fish, the length of the stop and wait generally decreases when they are looking for large, easily located prey. Often as the difficulty of the search increases, the pauses get longer. In environments where there are few prey, the fish persistently search.

Finally, note that there are also effects of likelihood and frequency of encounter of prey that would influence repositioning and direction-changing behavior. Effects of risk of moving (e.g., from some predator) should also be taken into consideration. All these aspects can result in the animal dynamically adjusting its saltatory search strategy.

18.1.3 Cooperative/Social Foraging

The foraging concepts discussed above were for individual animals. Foragers, however, live in environments with both a biotic and abiotic part, so a more complete formulation includes the other foragers in the environment. There can be advantages to group cooperative (or "social") foraging. Some method of sharing information is necessary for cooperative foraging. The shared information could come in many forms. In humans, this could include language. In other animals, it might be certain movements, noises, or "trail-laying" mechanisms. Such information is in the form of cues or signals. Shared information could also arise via possession of shared genetic material.

The advantages of group foraging include:

Group foraging requires group search strategies and hence, distributed cooperation within the group.

- The per capita rate of energetic gain for each animal may be higher if the animal is in a group. This can be the case even if the gains are not split evenly in the group.
- In some cases, the net rate of energetic gain of each individual may go down if it joins a group. However, it may still be a good strategy to join a group if the variance in the rate of energetic gain is lower. Sometimes, when there are more animals searching for nutrients, the likelihood of finding nutrients may increase. When one animal finds some nutrients, it can tell others in the group where the nutrients are. You may think of joining a group as gaining access to an "information center" for helping with survival.
- There may be increased capabilities to cope with larger prey. The group can "gang up" on a large prey and kill and ingest it, while a single small predator may not be able to do this.
- There may be protection from predators that can be provided by members of the group (e.g., in some species of birds, the members in the middle of the group are protected by the ones at the edges).
- In some cases, phenotypic diversity is profitably exploited by groups to produce highly efficient coordinated group behavior (e.g., for some ants and bees).

Sometimes it is useful to think of a group of animals as a *single* living creature, where via grouping, each individual essentially gains additional physiological capabilities that help it to succeed in foraging (and the gains may offset the possibility of food-competition problems in groups). Some call this the "super-organism" viewpoint.

For group foraging, you may think of how a pack of wolves hunts, or a flock of birds, colony of ants, or school of fish behave. Connections between optimization, engineering applications, and foraging behavior of colonies of ants have been studied [73]. There, it is explained how colonies of ants can solve shortest path problems, minimum spanning tree problems, and traveling salesperson problems (all combinatorial optimization problems) among other engineering applications. (The resulting computer algorithms are called, for instance, "ant colony optimization" algorithms.) These ants use "indirect" communications called "stigmergy," where one ant can modify its environment and later, another ant can change its behavior due to that modification. For instance, if an ant goes out foraging, it may search far and wide in a relatively random pattern; however, once it finds a food source, it goes back to the ant hill, laying a trail of "pheromone" (that can evaporate, but normally stays in place, possibly up to several months). Then, when other ants go out foraging, they tend to follow the pheromone trail and find food more easily. You can then think of the first ant as having "recruited" additional foragers and the trails as a type of memory for the whole ant colony (i.e., using communications and working together, they gain the important physiological capability of learning). Viewed as a superorganism, you may even detect elements of planning. Communications, memory, and learning, result in more efficient foraging for the group. Other social insects use other communication methods. For instance, after successful foraging, a bee will come back to the hive and communicate the profitability and location of the food source via a "dance." These dances then proportionally recruit foragers based on forage site profitability, and the result is a dynamic proportioning of foragers across a wide area.

Social behavior enables what can be thought of as higher-level cognitive functions of the group.

To be more concrete about the connections between foraging and optimization, consider Figure 18.3. There, initially boxes 1 and 2 hold two colonies of the Argentine ant *Iridomyrmex humilis*. These colonies interact via ants traveling between the two colonies. Since trail densities on the shorter path tend to grow faster, more and more ants tend to choose that path and thereby avoid the longer path (the figure shows ants traveling after they have, as a group, found the shortest path). A similar behavior is found when one box holds a colony of ants and the other holds a food source [25] and this illustrates that in foraging, the ants work together to find the shortest path to food sources.

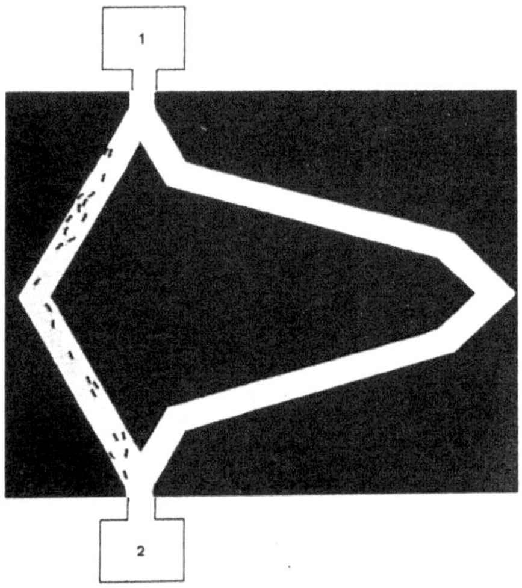

Figure 18.3: Experiment showing that ants will select a shortest path between two colonies (figure taken from [25], © Springer-Verlag GmbH and Co., and used with permission).

Finally, while we discuss "intelligent foraging" in more detail in Section 19.6, we briefly note that the individual characteristics of the animal can significantly affect its success in foraging. If an individual forager can pay *attention* to the critical parts of the environment and *learn* about the environment (e.g., by developing and storing a "cognitive map") and characteristics of its prey, then it can probably increase its foraging success. If, based on such learned information it can *plan* its foraging, then it may gain further increases in efficiency. Furthermore, if groups of foragers can learn and plan their activities together, it is possible that even greater success might be obtained. Indeed, it seems that humans often act as group foragers that can collectively learn and plan. We can think of many individuals working together to achieve something that is unachievable by any individual.

In the next section, we will consider individual and group foraging in bacteria, organisms that are much more simple than an ant or human, yet which can still work together for the benefit of the group. First, however, we explain some connections to nongradient optimization.

18.1.4 Nongradient Optimization Models

Before we turn to specific foraging models, it is important to point out that there are close connections to some of the optimization methods considered earlier in this book, both gradient and nongradient methods. First, note that it is impossible for most animals (e.g., bacteria) to know an analytical expression for the gradient of a nutrient concentration profile (i.e., a mathematical expression for how the nutrient concentration will change as the bacterium makes small changes in its position). This is both because it does not have the memory to store it, and also due to the high level of uncertainty about the environment it lives in (e.g., time-varying and stochastic effects). Moreover, even with sufficient physiology for remembering an analytical gradient, in general, it is impossible for any animal (besides perhaps a human) to know an analytical form for the gradient of the surface being searched (e.g., one that represented food locations, predators, risk, etc.) with respect to its location in the search domain. Animals sequentially decide where to explore, and in doing so they encounter new parts of a search domain, and the environment has significant random effects. In foraging, animals conduct an optimization process without use of an analytical expression for the gradient and hence, we say that they perform nongradient optimization or "search."

Nongradient optimization models can form a set of tools for modeling social foraging.

Motivated by the earlier sections, and studies on search strategies for foraging, in this section, we discuss several "conventional" (i.e., not biologically motivated) deterministic and stochastic approaches to perform optimization without the use of analytical gradient information or measures of the gradient. As you read about the methods, it will be useful to draw analogies with the basic search mechanisms of the bacteria discussed later in this chapter. For instance, in one way or another, most nongradient methods use measurements of the cost function and form approximations to the gradient to decide which direction to move. (Some of these are what might be called "regional" approximations, since they use a pattern of points over possibly a large region to provide a gross approximation to the gradient.) In the context of foraging, you can then think of the process of obtaining measurements and deciding where to move (i.e., the steps of the algorithms we cover in this section) in one of three following ways:

1. You can view the taking of a measurement of the cost function as a single forager going to the location in the search domain and taking a single measurement by using, for example, its vision to assess what food is at that location (or the likelihood that it is there). The forager may make several such local "exploratory" moves from its current position before it tries to move in what it considers to be generally a good direction to find food. (Of course, it may get lucky and get something worth stopping and eating in this sampling process.)

2. You can think of the forager as being at a single location, taking several local measurements at locations in the search domain, processing these, and then deciding which direction to move. We think of the forager as having sensors that it can focus on different nearby regions of the optimization domain, and only moving after it has scanned its environment. Some lower life forms (e.g., *E. coli*) cannot sense at a distance; they must go to a position to find out if there is food there. Higher life forms can generally sense at a distance and this saves them the energy needed to travel to every position to find out if food is there.

3. You can view the algorithm as modeling a social foraging process, where there are a finite number of foragers who each make measurements of the cost function and then via communications decide how to move the entire group of foragers. For example, in the "pattern search methods" of the last part, we may think of each point in the pattern of (local) cost function measurements as representing a location of a forager. We think of the group as having a capability to communicate how well they are doing to all the other members of the group, and come to an agreement on which is the best way to move the group to ensure foraging success.

In a sense, you may ask yourself if biology mimics any of the conventional strategies; is there an animal that forages according to a particular pattern search method? If so, why did evolution create this search strategy? What features of the environment drove the creation of the strategy?

18.2 Bacterial Foraging: *E. coli*

The *E. coli* bacterium is shown in Figure 18.4. It has a plasma membrane, cell wall, and capsule that contain, for instance, the cytoplasm and nucleoid. The pili (singular, pilus) are used for a type of gene transfer to other *E. coli* bacteria, and flagella (singular, flagellum) are used for locomotion. (Only one is shown, but in the actual cell there are as many as six.) The cell is about $1\mu m$ in diameter, and $2\mu m$ in length. The *E. coli* cell only weighs about 1 picogram, and is composed of about 70% water. *Salmonella typhimurium* is a similar type of bacterium.

Think of the E. coli *as a small underwater vehicle, a nanotechnology.*

The *E. coli* bacterium is probably the best understood microorganism. Its entire genome has been sequenced; it contains 4,639,221 of the A, C, G, and

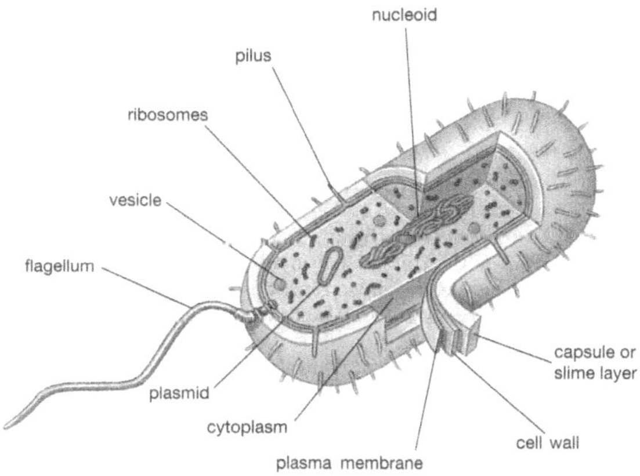

Figure 18.4: *E. coli* bacterium (figure taken from [34], © Pearson Education Inc., used with permission).

T "letters"—adenosine, cytosine, guanine, and thymine—arranged into a total of 4,288 genes. When *E. coli* grows, it gets longer, then divides in the middle into two "daughters." Given sufficient food and held at the temperature of the human gut (one place where they live) of 37 deg. C, *E. coli* can synthesize and replicate everything it needs to make a copy of itself in about 20 min.; hence, growth of a population of bacteria is exponential with a relatively short "time to double" the population size. For instance, following [61], if at noon today you start with one cell and sufficient food, by noon tomorrow there will be $2^{72} = 4.7 \times 10^{21}$ cells, which is enough to pack a cube 17 meters on one side. (It should be clear that with enough food, at this reproduction rate, they could quickly cover the entire earth with a knee-deep layer!)

The *E. coli* bacterium has a control system that enables it to search for food and try to avoid noxious substances (the resulting motions are called "taxes"). For instance, it swims away from alkaline and acidic environments, and towards more neutral ones. To explain the motile behavior of *E. coli* bacteria, we will explain its actuator (the flagella), "decision-making," sensors, and closed-loop behavior (i.e., how it moves in various environments—its "motile behavior"). You will see that *E. coli* perform a type of "saltatory search," a concept that is discussed in Section 18.1.2.

18.2.1 Swimming and Tumbling via Flagella

Locomotion is achieved via a set of relatively rigid flagella that enable it to "swim" via each of them rotating in the same direction at about 100 – 200 revolutions per second (in control systems terms, we think of the flagella as providing for actuation). Each flagellum is a left-handed helix configured so that as the base of the flagellum (i.e., where it is connected to the cell) rotates counterclockwise, as viewed from the free end of the flagellum looking towards the cell, it produces a force against the bacterium so it pushes the cell. You may think of each flagellum as a type of propeller. If a flagellum rotates clockwise, then it will pull at the cell. From an engineering perspective, the rotating shaft at the base of the flagellum is quite an interesting contraption that seems to use what biologists call a "universal joint" (so the rigid flagellum can "point" in different directions, relative to the cell). In addition, the mechanism that creates the rotational forces to spin the flagellum in either direction is described by biologists as being a biological "motor" (a relatively rare contraption in biology even though several types of bacteria use it) as shown in Figure 18.5. The motor is quite efficient in that it rotates a complete revolution using only about 1000 protons and thereby *E. coli* spends less than 1% of its energy budget for motility.

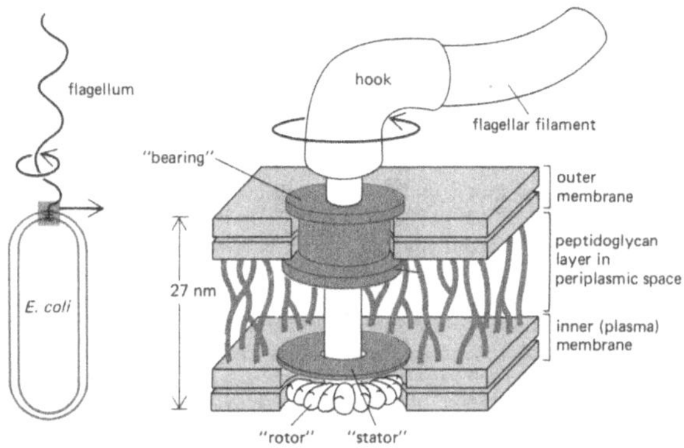

Figure 18.5: *E. coli* bacterium, flagellar connection, and biological "motor" (figure taken from [8], © Garland Science/Taylor and Francis Books, Inc., used with permission).

An *E. coli* bacterium can move in two different ways: it can "run" (swim for a period of time) or it can "tumble," and it alternates between these two modes of operation its entire lifetime (i.e., it is rare that the flagella will stop rotating). First, we explain each of these two modes of operation. Following that, we will explain how it decides how long to swim before it tumbles.

If the flagella rotate clockwise, each flagellum pulls on the cell and the net effect is that each flagellum operates relatively independent of the others and so the bacterium "tumbles" about (i.e., the bacterium does not have a set direction of movement and there is little displacement). See Figure 18.6(a). To tumble after a run, the cell slows down or stops first; since bacteria are so small they experience almost no inertia, only viscosity, so that when a bacterium stops swimming, it stops within the diameter of a proton. Call the time interval during which a tumble occurs a "tumble interval." Under certain experimental conditions (an isotropic, homogeneous medium—one with no nutrient or noxious substance gradients) for a "wild type" cell (one found in nature), the mean tumble interval is about 0.14 ± 0.19 sec. (mean $\pm$ standard deviation, and it is exponentially distributed) [61, 62]. After a tumble, the cell will generally be pointed in a random direction, but there is a slight bias toward being placed in a direction it was traveling before the tumble.

There are two modes of operation of the actuator: one leading to swimming and the other to tumbling.

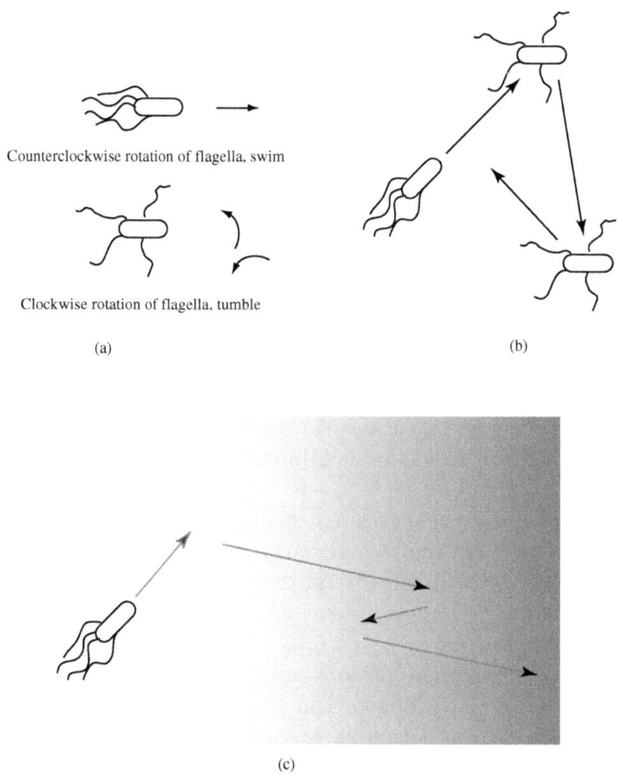

Figure 18.6: Bundling phenomenon of flagella shown in (a), swimming and tumbling behavior of the *E. coli* bacterium is shown in (b) in a neutral medium and in (c) where there is a nutrient concentration gradient, with darker shades indicating higher concentrations of the nutrient. (Note: Relative sizes of the bacteria and lengths of runs are not to scale.)

If the flagella move counterclockwise, their effects accumulate by forming a "bundle" (it is thought that the bundle is formed due to the viscous drag of the medium) and hence, they essentially make a "composite propeller" and push the bacterium so that it runs (swims) in one direction (see Figure 18.6(a)). On a run, bacteria swim at a rate of about $10 - 20$ μmeters/sec., or about 10 body lengths per second (assuming the faster speed and an *E. coli* that is 2 μ meters long, a typical length), but in a rich medium they can swim even faster

[335]. This is a relatively fast rate for a living organism to travel; consider how fast you could move through water if you could swim at 10 of your body lengths per second. (You would certainly win a gold medal in the Olympics!) Call the time interval during which a run occurs the "run interval." Under certain experimental conditions (an isotropic, homogeneous medium—the same as the one mentioned above) for a wild type cell, the mean run interval is about 0.86 ± 1.18 sec. (and it is exponentially distributed) [61, 62]. Also, under these conditions, the mean speed is 14.2 ± 3.4 $\mu m/sec$. Runs are not perfectly straight since the cell is subject to Brownian movement that causes it to wander off course by about 30 deg. in 1 sec. in one type of medium, so this is how much it typically can deviate on a run. In a certain medium, after about 10 sec. it drifts off course more than 90 deg. and hence, essentially forgets the direction it was moving [61]. Finally, note that in many bacteria, the motion of the flagella can induce other motions, e.g., rotating the bacteria about an axis.

18.2.2 Bacterial Motile Behavior: Climbing Nutrient Gradients

The motion patterns (called "taxes") that the bacteria will generate in the presence of chemical attractants and repellents are called "chemotaxes." For *E. coli*, encounters with serine or aspartate result in attractant responses, while repellent responses result from the metal ions Ni and Co, changes in pH, amino acids like leucine, and organic acids like acetate. What is the resulting emergent pattern of behavior for a whole group of *E. coli* bacteria? Generally, as a group they will try to find food and avoid harmful phenomena, and when viewed under a microscope, you will get a sense that a type of intelligent behavior has emerged, since they will seem to intentionally move as a group (analogous to how a swarm of bees moves).

Simple control rules are used to decide whether to swim or tumble.

To explain how chemotaxis motions are generated, we simply must explain how the *E. coli* decides how long to run since, from the above discussion, we know what happens during a tumble or run. First, note that if an *E. coli* is in some substance that is neutral, in the sense that it does not have food or noxious substances, and if it is in this medium for a long period of time (e.g., more than one minute), then the flagella will simultaneously alternate between moving clockwise and counterclockwise so that the bacterium will alternately tumble and run. This alternation between the two modes will move the bacterium, but in random directions, and this enables it to "search" for nutrients (see Figure 18.6(b)). For instance, in the isotropic homogeneous environment described above, the bacteria alternately tumble and run with the mean tumble and run lengths given above, and at the speed that was given. If the bacteria are placed in a homogeneous concentration of serine (i.e., one with a nutrient but no gradients), then a variety of changes occur in the characteristics of their motile behavior. For instance, mean run length and mean speed increase and mean tumble time decreases. They do, however, still produce a basic type of searching behavior; even though it has some food, it persistently searches for more. As an example of tumbles and runs in the isotropic homogeneous medium described above, in one trial motility experiment lasting 29.5 sec., there were 26 runs, the maximum run length was 3.6 sec., and the mean speed was about 21 $\mu m/sec.$ [61, 62].

Next, suppose that the bacterium happens to encounter a nutrient gradient (e.g., serine) as shown in Figure 18.6(c). The *change* in the concentration of the nutrient triggers a reaction such that the bacterium will spend more time swimming and less time tumbling. As long as it travels on a positive concentration gradient (i.e., so that it moves towards increasing nutrient concentrations) it will tend to lengthen the time it spends swimming (i.e., it runs farther). The directions of movement are "biased" towards increasing nutrient gradients. The cell does not change its *direction* on a run due to changes in the gradient—the tumbles basically determine the direction of the run, aside from the Brownian influences mentioned above.

The bacterial foraging behavior results in a type of hill-climbing optimization algorithm.

On the other hand, typically if the bacterium happens to swim down a concentration gradient (or into a positive gradient of noxious substances), it will return to its baseline behavior so that essentially it tries to search for a way to climb back up the gradient (or down the noxious substance gradient). For instance, under certain conditions, for a wild-type cell swimming up serine gradients, the mean run length is 2.19 ± 3.43 sec., but if it swims down a serine gradient, mean run length is 1.40 ± 1.88 sec. [62]. Hence, when it moves up the gradient, it lengthens its runs. The mean run length for swimming down the gradient is the one that is expected, considering that the bacteria are in this particular type of medium; they act basically the same as in a homogeneous medium so that they are engaging their search/avoidance behavior to try to climb back up the gradient.

Finally, suppose that the concentration of the nutrient is constant for the region it is in, after it has been on a positive gradient for some time. In this case, after a period of time (not immediately), the bacterium will return to the same proportion of swimming and tumbling as when it was in the neutral substance so that it returns to its standard searching behavior. It is never satisfied with the amount of surrounding food; it always seeks higher concentrations. Actually, under certain experimental conditions, the cell will compare the concentration observed over the past 1 sec. with the concentration observed over the 3 sec. before that and it responds to the difference [61]. Hence, it uses the past 4 sec. of nutrient concentration data to decide how long to run [459]. Considering the deviations in direction due to Brownian movement discussed above, the bacterium basically uses as much time as it can in making decisions about climbing gradients [60]. In effect, the run length results from how much climbing it has done recently. If it has made lots of progress and hence, has just had a long run, then even if for a little while it is observing a homogeneous medium (without gradients), it will take a longer run. After a certain time period, it will recover and return to its standard behavior in a homogeneous medium.

Basically, the bacterium is trying to swim from places with low concentrations of nutrients to places with high concentrations. An opposite type of behavior is used when it encounters noxious substances. If the various concentrations move with time, then the bacteria will try to "chase" after the more favorable environments and run from harmful ones. Clearly, nutrient and noxious substance diffusion and motion will affect the motion patterns of a group of bacteria in complex ways.

18.2.3 Underlying Sensing and Decision-Making Mechanisms

The decision-making is implemented by chemical reactions driven by sensing of chemicals in the environment.

Consider Figure 18.7, where a cross-section of one corner of the *E. coli* bacterium is shown. The sensors are the receptor proteins, which are signaled directly by external substances (e.g., in the case for the pictured amino acids) or via the "periplasmic substrate-binding proteins." The "sensor" is very sensitive, in some cases requiring less than 10 molecules of attractant to trigger a reaction, and attractants can trigger a swimming reaction in less than 200 ms. You can then think of the bacterium as having a "high gain" with a small attractant detection threshold (detection of only a small number of molecules can trigger a doubling or tripling of the run length). On the other hand, the corresponding threshold for encountering a homogeneous medium after being in a nutrient rich one is larger. Also, there is a type of time-averaging that is occurring in the sensing process. The receptor proteins then affect signaling molecules inside the bacterium. Also, there is in effect an "adding machine" and an ability to compare values and to arrive at an overall decision about which mode the flagella should operate in; essentially, the different sensors add and subtract their effects, and the more active or numerous have a greater influence on the final decision. Even though the sensory and decision-making system in *E. coli* is probably the best understood one in biology, we are ignoring the underlying chemistry that is needed for a full explanation (the interested reader can see the "For Further Study" chapter at the end of this part to find references that explain it in detail).

It is interesting to note that the "decision-making system" in the *E. coli* bacterium must have some ability to sense a *derivative*, and hence, it has a type of memory! At first glance it may seem possible that the bacterium senses concentrations at both ends of the cell and finds a simple difference to recognize a concentration gradient (a spatial derivative); however, this is not the case. Experiments have shown that it performs a type of sampling, and roughly speaking, it remembers the concentration a moment ago, compares it with a current one, and makes decisions based on the difference (i.e., it computes something like an Euler approximation to a time derivative). Actually, in [553] the authors show how internal bacterial decision-making processes involve some type of integral feedback control mechanism.

In summary, we see that with memory, a type of addition mechanism, an ability to make comparisons, a few simple internal "control rules," and its chemical sensing and locomotion capabilities, the bacterium is able to achieve a complex type of searching and avoidance behavior. Evolution has designed this control system. It is robust and clearly very successful at meeting its goals of survival when viewed from a population perspective.

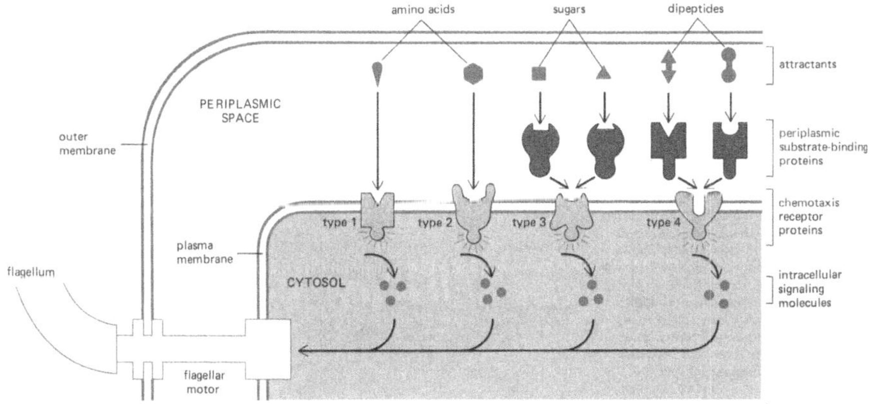

Figure 18.7: Sensing and internal mechanisms for control in the *E. coli* bacterium (figure taken from [8], © Garland Science/Taylor and Francis Books, Inc., used with permission).

18.2.4 Elimination and Dispersal Events

It is possible that the local environment where a population of bacteria lives changes either gradually (e.g., via consumption of nutrients) or suddenly due to some other influence. There can be events such that all the bacteria in a region are killed or a group is dispersed into a new part of the environment. For example, local significant increases in heat can kill a population of bacteria that are currently in a region with a high concentration of nutrients (you can think of heat as a type of noxious influence). Or, it may be that water or some animal will move populations of bacteria from one place to another in the environment. Over long periods of time, such events have spread various types of bacteria into virtually every part of our environment, from our intestines, to hot springs and underground environments, and so on.

What is the effect of elimination and dispersal events on chemotaxis? It has the effect of possibly destroying chemotactic progress, but it also has the effect of assisting in chemotaxis since dispersal may place bacteria near good food sources. From a broad perspective, elimination and dispersal is part of the population-level motile behavior.

Elimination and dispersal of bacteria should be thought of as a component of their overall motility.

18.2.5 Evolution of Bacteria

Mutations in *E. coli* occur at a rate of about 10^{-7} per gene, per generation. In addition to mutations that affect its physiological aspects (e.g., reproductive efficiency at different temperatures), *E. coli* bacteria occasionally engage in a type of "sex" called "conjugation," where small gene sequences are unidirectionally transferred from one bacterium to another. It seems that these gene sequences apparently carry good fitness characteristics in terms of reproductive capability, so conjugation is sometimes thought of as a transmittal of "fertility." To achieve conjugation, a pilus extends to make contact with another bacterium, and the gene sequence transfers through the pilus.

It is important to note that there are some very basic differences in evolution for higher organisms and bacteria. While conjugation apparently spreads "good" gene sequences, the "homogenizing effect" on gene frequency from conjugation is relatively small compared to how sex works in other organisms. This is partly since conjugation is relatively rare, and partly since the rate of reproduction is relatively high, on the order of hours depending on environmental conditions. Due to these characteristics, population genetics for *E. coli* may be dominated by selection sweeps triggered by the acquisition, via sex, of an adaptive allele.

18.2.6 Taxes in Other Swimming Bacteria

There are a wide range of foraging behaviors in bacteria, all of which can be modeled as optimization processes.

While most bacteria are motile and many types have analogous taxes capabilities to *E. coli* bacteria, the specific sensing, actuation, and decision-making mechanisms are different [384, 24]. For instance, while the proton-driven motor on *E. coli* rotates at a few hundred revolutions per second, Na^+-driven motors on some bacteria rotate at speeds up to 1000 revolutions per second, and on some species, the motor can turn in either direction or stop. Different types of bacteria can sense different phenomena and have different underlying decision-making, so they may search for and try to avoid different phenomena. Some bacteria can sense their own metabolic state and only respond to compounds currently required for growth and their pattern of responses may change based on their environment. Studies of the mechanisms for decision and control in various bacteria do, however, indicate that they have common features and hence, some have suggested that there was a single early evolutionary event that resulted in the swimming capability of bacteria. Swimming generally moves a bacterium to a more favorable environment for growth, or it maintains it in its current position, and hence, it gives the bacteria a survival advantage. Some scientists have suggested that the shapes of motile bacteria developed to allow efficient swimming. Some bacteria even change their shape to reduce the adverse effects of moving through more viscous media. Even though there can be significant differences between species, all swimming bacteria seem to have similar swimming patterns, where there is an alternation between smooth swimming and a change in direction (i.e., a type of saltatory search, a concept that is explained in Section 18.1.2). Next, several examples of other types of sensing and taxes in swimming bacteria are provided.

Some bacteria can search for oxygen, and hence their motility behavior is based on "aerotaxis," while others search for desirable temperatures resulting in "thermotaxis." Actually, the *E. coli* is capable of thermotaxis in that it seeks warmer environments with a temperature range of 20 deg. to 37 deg. C. Other bacteria, such as *Thiospirillum jenense*, search for or avoid light of certain wavelengths and this is called "phototaxis" Actually, the *E. coli* tries to avoid intense blue light, so it is also capable of phototaxis. Some bacteria swim along magnetic lines of force that enter the earth, so that when in the northern hemisphere, they swim towards the north magnetic pole, and in the southern hemisphere, they swim towards the south magnetic pole. (This is due to the presence of a small amount of magnetic material *in the cell* that essentially acts as a compass to passively reorient the cell.)

There are square-shaped bacteria that are propelled either forward or backward via flagella, and when multiple such bacteria naturally collide, their flagella can become "clumped," and this seems to be responsible for their tumbling. Hence, their motility behavior is characterized by forward movement, followed by either forward or backward movement, and an intermittent change in direction via tumbling [7]. *Vibrio alginolyticus* move differently when free-living versus living on a surface. Free-living *Vibrio alginolyticus* swims using a Na^+-driven motor on its flagella but when it is on the surface of a liquid, it senses the increased viscosity via the flagellar motor and then synthesizes many proton-driven flagella, which then allow the cell to move over surfaces [24]. The cells move as groups ("rafts"), since this is thought to help overcome viscous drag and surface tension. In other bacteria, flagella can be synthesized and discarded as they are needed.

18.2.7 Other Group Phenomena in Bacteria

A particularly interesting group behavior has been demonstrated for several motile species of bacteria, including *E. coli* and *S. typhimurium*, where intricate stable spatio-temporal patterns (swarms)[1] are formed in semi-solid nutrient media [84, 71, 83, 544, 24] (see Figure 18.8). When a group of *E. coli* cells is placed in the center of a semi-solid agar with a single nutrient chemo-effector (sensor), they move out from the center in a traveling ring of cells by moving up the nutrient gradient created by consumption of the nutrient by the group. Moreover, if high levels of the nutrient called succinate are used as the nutrient, then the cells release the attractant aspartate, so that they congregate into groups and hence, move as concentric patterns of groups with high bacterial density; see the concentric pattern of dots in Figure 18.8. (Note that many cells in those groups permanently lose motility.) The spatial order results from outward movement of the ring and the local releases of the attractant; the cells provide an attraction signal to each other so they swarm together. Pattern formation can be suppressed by a background of aspartate (since it seems that this will in essence scramble the chemical signal by eliminating its directionality). The pattern seems to form based on the dominance of two stimuli (cell-cell signaling and foraging).

Swarms arise due to communications, and can lead to more successful foraging (optimization).

The role of these patterns in natural environments is not understood; however, there is evidence that stress to the bacteria results in them releasing chemical signals that other bacteria are chemotactic towards. If enough stress is

[1] Actually, microbiologists reserve the term "swarming" for other characteristics of groups of bacteria. Here, we abuse the terminology and favor using the terminology that is used for higher forms of animals such as bees.

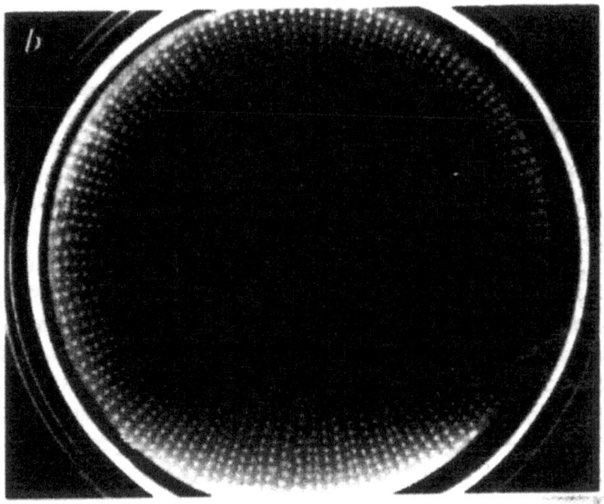

Figure 18.8: Swarm pattern of *E. coli* (figure taken from [84], © Macmillan Magazines, used with permission from Nature).

present, then a whole group can secrete the chemical signal strengthening the total signal, and hence, an aggregate of the bacteria forms. It seems that this aggregate forms to protect the group from the stress (e.g., by effectively hiding many cells in the middle of the group). It seems that the aggregates of the bacteria are not necessarily stationary; under certain conditions they can migrate, split, and fuse. This has led researchers to hypothesize that there may be other communication methods being employed that are not yet understood.

As another example, there are "biofilms" that can be composed of multiple types of bacteria (e.g., *E. coli*) that can coat various objects (e.g., roots of plants or medical implants). It seems that both motility and "quorum sensing" are involved in biofilm formation. A biofilm is a mechanism for keeping a bacterial species in a fixed location, avoiding overcrowding, and avoiding nutrient limitation and toxin production by packing them at a low density in a "polysaccharide matrix." Secreted chemicals provide a mechanism for the cells to sense population density, but motility seems to assist in the early stages of biofilm formation. It is also thought that chemotactic responses are used to drive cells to the outer edges of the biofilm, where nutrient concentrations may be higher.

In a variety of bacteria, including *E. coli*, complex patterns result primarily not from motility, but from reproduction [464]. In some bacteria, it seems that there is a type of signaling that occurs and results in the formation of regular patterns as the culture of bacteria grows. Formation of such patterns is sometimes thought of as a type of multicellular "morphogenesis." For example, the formation of the "fruiting bodies" by *Myxococcus xanthus* can be viewed as a type of morphogenesis, but one that seems to be primarily based on motility and cell deaths rather than reproduction [467].

Other types of bacteria exhibit group behaviors [334]. For instance, there are luminous bacteria that will emit no light until the population reaches a certain density. For instance, the bacteria *Vibrio fischeri* lives in the ocean at low concentrations and its secreted "autoinducer" chemical signal is quite dilute. However, the squid *Euprymna scolopes* selects these bacteria to grow in its light organ. When a sufficiently large population is cultivated in its light organ, the autoinducer chemical signals given off by each bacterium effectively add to result in a high concentration of this chemical and, when it reaches a certain threshold, each cell will switch on its luminescence property so that as a group they emit a visible light [334]. The squid, which is a nocturnal forager, benefits since the light camouflages it from predators below, since its light resembles moonlight and hence, effectively eliminates its shadow. The bacteria benefit by getting nourishment and shelter. The bacteria and squid are in a symbiont relationship (i.e., they live together to benefit each other).

Also, the soil-dwelling *streptomycete* colonies can grow a branching network of long fiber-like cells that can penetrate and degrade vegetation and then feed on the resulting decaying matter. (In terms of combinatorial optimization, you may think of finding optimal trees or graphs.) Under starvation conditions, they can cooperate to produce spores on a structure called an "aerial mycelium" that may be carried away.

As another example, in *Proteus mirabilis* the rod-shaped cells exist as "swimmers" that are driven by fewer than 10 flagella when they are in liquid media and they have chemotactic responses analogous to those of *E. coli*. If, however, these swimmers are placed on a solid surface, the swimmer cell "differentiates" (changes) into a "swarmer cell" that is an elongated rod (of roughly the same diameter) with more than 10, 000 flagella. On solid surfaces, the cells aggregate and exhibit swarm behavior in foraging via group chemotaxis. If they are then placed back in a liquid medium, there is a process of "consolidation" where swarmer cells split into swimmer cells. Moreover, when swarming they exhibit the "Dienes phenomenon," where swarms of the same type of bacteria try to avoid each other. (The mechanisms of this apparent territorial behavior are not well-understood.)

18.3 *E. coli* Bacterial Swarm Foraging for Optimization

Suppose that we want to find the minimum of $J(\theta)$, $\theta \in \Re^p$, where we do not have measurements, or an analytical description, of the gradient $\nabla J(\theta)$. Here, we use ideas from bacterial foraging to solve this "nongradient" optimization problem. First, suppose that θ is the position of a bacterium and $J(\theta)$ represents the combined effects of attractants and repellents from the environment, with, for example, $J(\theta) < 0$, $J(\theta) = 0$, and $J(\theta) > 0$ representing that the bacterium at location θ is in nutrient-rich, neutral, and noxious environments, respectively. Basically, chemotaxis is a foraging behavior that implements a

The bacterial foraging algorithm is a nongradient stochastic optimization method.

type of optimization where bacteria try to climb up the nutrient concentration (find lower and lower values of $J(\theta)$) and avoid noxious substances and search for ways out of neutral media (avoid being at positions θ where $J(\theta) \geq 0$).

18.3.1 An Optimization Model for *E. coli* Bacterial Foraging

To define our optimization model of *E. coli* bacterial foraging, we need to define a population (set) of bacteria, and then model how they execute chemotaxis, swarming, reproduction, and elimination/dispersal. After doing this, we will highlight the limitations (inaccuracies) in our model.

Population and Chemotaxis

Define a chemotactic step to be a tumble followed by a tumble or a tumble followed by a run. Let j be the index for the chemotactic step. Let k be the index for the reproduction step. Let ℓ be the index of the elimination-dispersal event. Let

$$P(j, k, \ell) = \{\theta^i(j, k, \ell) | i = 1, 2, \ldots, S\}$$

represent the positions of each member in the population of the S bacteria at the jth chemotactic step, k^{th} reproduction step, and ℓ^{th} elimination-dispersal event. Here, let $J(i, j, k, \ell)$ denote the cost at the location of the ith bacterium $\theta^i(j, k, \ell) \in \Re^p$ (sometimes we drop the indices and refer to the ith bacterium position as θ^i). Note that we will interchangeably refer to J as being a "cost" (using terminology from optimization theory) and as being a nutrient surface (in reference to the biological connections). For actual bacterial populations, S can be very large (e.g., $S = 10^9$), but $p = 3$. In our computer simulations, we will use much smaller population sizes and will keep the population size fixed. We will allow $p > 3$, so we can apply the method to higher dimensional optimization problems.

Let N_c be the length of the lifetime of the bacteria as measured by the number of chemotactic steps they take during their life. Let $C(i) > 0$, $i = 1, 2, \ldots, S$, denote a basic chemotactic step size that we will use to define the lengths of steps during runs. To represent a tumble, a unit length random direction, say $\phi(j)$, is generated; this will be used to define the direction of movement after a tumble. In particular, we let

$$\theta^i(j + 1, k, \ell) = \theta^i(j, k, \ell) + C(i)\phi(j)$$

so that $C(i)$ is the size of the step taken in the random direction specified by the tumble. If at $\theta^i(j + 1, k, \ell)$ the cost $J(i, j + 1, k, \ell)$ is better (lower) than at $\theta^i(j, k, \ell)$, then another step of size $C(i)$ in this same direction will be taken, and again, if that step resulted in a position with a better cost value than at the previous step, another step is taken. This swim is continued as long as it continues to reduce the cost, but only up to a maximum number of steps, N_s. This represents that the cell will tend to keep moving if it is headed in the direction of increasingly favorable environments.

Swarming Mechanisms

The above discussion was for the case where no cell-released attractants are used to signal other cells that they should swarm together. Here, we will also have cell-to-cell signaling via an attractant and will represent that with $J^i_{cc}(\theta, \theta^i(j,k,\ell))$, $i = 1, 2, \ldots, S$, for the ith bacterium. Let

Swarm optimization exploits a regional approximation to a gradient and helps it to climb over noise.

$$d_{attract} = 0.1$$

be the depth of the attractant released by the cell (a quantification of how much attractant is released) and

$$w_{attract} = 0.2$$

be a measure of the width of the attractant signal (a quantification of the diffusion rate of the chemical). The cell also repels a nearby cell in the sense that it consumes nearby nutrients and it is not physically possible to have two cells at the same location. To model this, we let

$$h_{repellent} = d_{attract}$$

be the height of the repellent effect (magnitude of its effect) and

$$w_{repellent} = 10$$

be a measure of the width of the repellent. The values for these parameters are simply chosen to illustrate general bacterial behaviors, not to represent a particular bacterial chemical signaling scheme. The particular values of the parameters were chosen with the nutrient profile in mind, which we will use later in Figure 18.10. For instance, the depth and width of the attractant is small relative to the nutrient concentrations represented in Figure 18.10. Let

$$\begin{aligned} J_{cc}(\theta, P(j,k,\ell)) &= \sum_{i=1}^{S} J^i_{cc}(\theta, \theta^i(j,k,\ell)) \\ &= \sum_{i=1}^{S}\left[-d_{attract}\exp\left(-w_{attract}\sum_{m=1}^{p}(\theta_m - \theta^i_m)^2\right)\right] \\ &+ \sum_{i=1}^{S}\left[h_{repellent}\exp\left(-w_{repellent}\sum_{m=1}^{p}(\theta_m - \theta^i_m)^2\right)\right] \end{aligned}$$

denote the combined cell-to-cell attraction and repelling effects, where $\theta = [\theta_1, \ldots, \theta_p]^\top$ is a point on the optimization domain and θ^i_m is the mth component of the ith bacterium position θ^i (for convenience, we omit some of the indices). An example for the case of $S = 2$ and the above parameter values is shown in Figure 18.9. Here, note that the two sharp peaks represent the cell locations, and as you move radially away from the cell, the function decreases and then increases (to model the fact that cells far away will tend not to be attracted, whereas cells close by will tend to try to climb down the cell-to-cell nutrient gradient towards each other and hence try to swarm). Note that as each cell moves, so does its $J_{cc}^{i}(\theta, \theta^{i}(j,k,\ell))$ function, and this represents that it will release chemicals as it moves. Due to the movements of all the cells, the $J_{cc}(\theta, P(j,k,\ell))$ function is *time-varying* in that, if many cells come close together, there will be a high amount of attractant and hence, an increasing likelihood that other cells will move towards the group. This produces the swarming effect. When we want to study swarming, the ith bacterium, $i = 1, 2, \ldots, S$, will hill-climb on

$$J(i,j,k,\ell) + J_{cc}(\theta, P)$$

(rather than the $J(i,j,k,\ell)$ defined above) so that the cells will try to find nutrients, avoid noxious substances, and at the same time try to move towards other cells, but not too close to them. The $J_{cc}(\theta, P)$ function dynamically deforms the search landscape as the cells move to represent the desire to swarm (i.e., we model mechanisms of swarming as a minimization process).

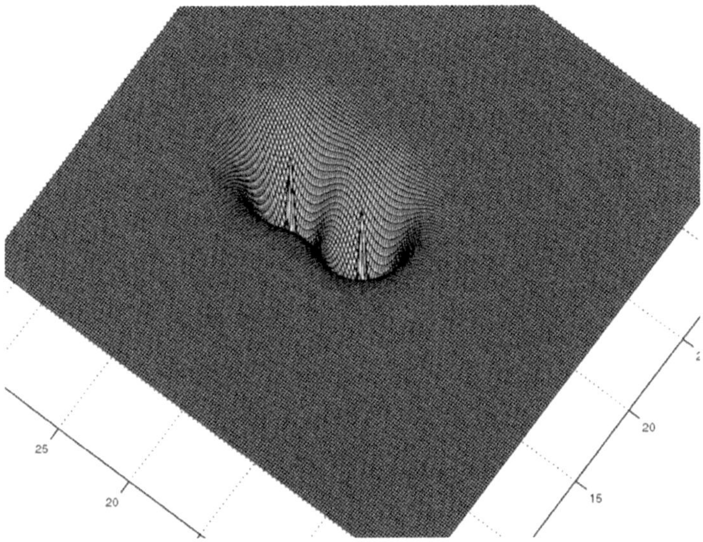

Figure 18.9: Cell-to-cell chemical attractant model, $S = 2$.

Reproduction and Elimination/Dispersal

After N_c chemotactic steps, a reproduction step is taken. Let N_{re} be the number of reproduction steps to be taken. For convenience, we assume that S is a positive even integer. Let

$$S_r = \frac{S}{2} \tag{18.1}$$

be the number of population members who have had sufficient nutrients so that they will reproduce (split in two) with no mutations. For reproduction, the population is sorted in order of ascending accumulated cost (higher accumulated cost represents that it did not get as many nutrients during its lifetime of foraging and hence, is not as "healthy" and thus unlikely to reproduce); then the S_r least healthy bacteria die and the other S_r healthiest bacteria each split into two bacteria, which are placed at the same location. Other fractions or approaches could be used in place of Equation (18.1); this method rewards bacteria that have encountered a lot of nutrients, and allows us to keep a constant population size, which is convenient in coding the algorithm.

Run length, reproduction, elimination, and dispersal all help to avoid local minima.

Let N_{ed} be the number of elimination-dispersal events, and for each such elimination-dispersal event, each bacterium in the population is subjected to elimination-dispersal with probability p_{ed}. We assume that the frequency of chemotactic steps is greater than the frequency of reproduction steps, which is in turn greater in frequency than elimination-dispersal events (e.g., a bacterium will take many chemotactic steps before reproduction, and several generations may take place before an elimination-dispersal event).

Foraging Model Limitations

Clearly, we are ignoring many characteristics of the actual biological optimization process in favor of simplicity and capturing the gross characteristics of chemotactic hill-climbing and swarming. For instance, we assume that consumption does not affect the nutrient surface (e.g., while a bacterium is in a nutrient-rich environment, we do not increase the value of J near where it has consumed nutrients) where clearly in nature, bacteria modify the nutrient concentrations via consumption. A tumble does not result in a perfectly random new direction for movement; however, here we assume that it does. Brownian effects buffet the cell, so that after moving a small distance, it is within a pie-shaped region of its start point at the tip of the piece of pie. Basically, we assume that swims are straight, whereas in nature they are not. Tumble and run lengths are exponentially distributed random variables, not constant, as we assume. Run-length decisions are actually based on the past 4 sec. of concentrations, whereas here we assume that at each tumble, older information about nutrient concentrations is lost. Although naturally asynchronous, we force synchronicity by requiring, for instance, chemotactic steps of different bacteria to occur at the same time, all bacteria to reproduce at the same time instant, and all bacteria that are subjected to elimination and dispersal to do so at the same time. We assume a constant population size, even if there are many nutrients and generations. We assume that the cells respond to nutrients in the environment in the same way that they respond to ones released by other cells for the purpose of signaling the desire to swarm. (A more biologically accurate model of the swarming behavior of certain bacteria is given in [544].) Clearly, other choices for the criterion of which bacteria should split could be used (e.g., based only on the concentration at the end of a cell's lifetime, or on the quantity of noxious substances that were encountered). We are also ignoring conjugation and other evolutionary characteristics. For instance, we assume that $C(i)$, N_s, and N_c remain the same for each generation. In nature it seems likely that these parameters could evolve for different environments to maximize population growth rates.

18.3.2 Bacterial Foraging Optimization Algorithm

For initialization, you must choose p, S, N_c, N_s, N_{re}, N_{ed}, p_{ed}, and the $C(i)$, $i = 1, 2, \ldots, S$. If you use swarming, you will also have to pick the parameters of the cell-to-cell attractant functions; here we will use the parameters given above. Also, initial values for the θ^i, $i = 1, 2, \ldots, S$, must be chosen. Choosing these to be in areas where an optimum value is likely to exist is a good choice. Alternatively, you may want to simply randomly distribute them across the domain of the optimization problem. The algorithm that models bacterial population chemotaxis, swarming, reproduction, elimination, and dispersal is given below (initially, $j = k = \ell = 0$). For the algorithm, note that updates to the θ^i automatically result in updates to P. Clearly, we could have added a more sophisticated termination test than simply specifying a maximum number of iterations.

1. Elimination-dispersal loop: $\ell = \ell + 1$
2. Reproduction loop: $k = k + 1$
3. Chemotaxis loop: $j = j + 1$
   (a) For $i = 1, 2, \ldots, S$, take a chemotactic step for bacterium i as follows.
   (b) Compute $J(i, j, k, \ell)$. Let
   $$J(i, j, k, \ell) = J(i, j, k, \ell) + J_{cc}(\theta^i(j, k, \ell), P(j, k, \ell))$$
   (i.e., add on the cell-to-cell attractant effect to the nutrient concentration).
   (c) Let $J_{last} = J(i, j, k, \ell)$ to save this value, since we may find a better cost via a run.
   (d) Tumble: generate a random vector $\Delta(i) \in \Re^p$ with each element $\Delta_m(i)$, $m = 1, 2, \ldots, p$, a random number on $[-1, 1]$.
   (e) Move: let
   $$\theta^i(j + 1, k, \ell) = \theta^i(j, k, \ell) + C(i)\frac{\Delta(i)}{\sqrt{\Delta^\top(i)\Delta(i)}}$$
   This results in a step of size $C(i)$ in the direction of the tumble for bacterium i.
   (f) Compute $J(i, j+1, k, \ell)$, and then let $J(i, j+1, k, \ell) = J(i, j+1, k, \ell) + J_{cc}(\theta^i(j + 1, k, \ell), P(j + 1, k, \ell))$.

(g) Swim (note that we use an approximation, since we decide swimming behavior of each cell as if the bacteria numbered $\{1, 2, \ldots, i\}$ have moved, and $\{i+1, i+2, \ldots, S\}$ have not; this is much simpler to simulate than simultaneous decisions about swimming and tumbling by all bacteria at the same time):

i. Let $m = 0$ (counter for swim length).

ii. While $m < N_s$ (if have not climbed down too long)

- Let $m = m + 1$.
- If $J(i, j+1, k, \ell) < J_{last}$ (if doing better), let $J_{last} = J(i, j+1, k, \ell)$ and let

$$\theta^i(j+1, k, \ell) = \theta^i(j+1, k, \ell) + C(i)\frac{\Delta(i)}{\sqrt{\Delta^\top(i)\Delta(i)}}$$

  and use this $\theta^i(j+1, k, \ell)$ to compute the *new* $J(i, j+1, k, \ell)$ as we did in (f) above.
- Else, let $m = N_s$. This is the end of the while statement.

(h) Go to next bacterium $(i+1)$ if $i \neq S$ (i.e., go to (b) above to process the next bacterium).

4. If $j < N_c$, go to step 3. In this case, continue chemotaxis, since the life of the bacteria is not over.

5. Reproduction:

(a) For the given k and ℓ, and for each $i = 1, 2, \ldots, S$, let

$$J^i_{health} = \sum_{j=1}^{N_c+1} J(i, j, k, \ell)$$

be the health of bacterium i (a measure of how many nutrients it got over its lifetime and how successful it was at avoiding noxious substances). Sort bacteria and chemotactic parameters $C(i)$ in order of ascending cost J_{health} (higher cost means lower health).

(b) The S_r bacteria with the highest J_{health} values die and the other S_r bacteria with the best values split (and the copies that are made are placed at the same location as their mother).

6. If $k < N_{re}$, go to step 2. In this case, we have not reached the number of specified reproduction steps, so we start the next generation in the chemotactic loop.

7. Elimination-dispersal: for $i = 1, 2, \ldots, S$, with probability p_{ed}, eliminate and disperse each bacterium (this keeps the number of bacteria in the population constant). To do this, if you eliminate a bacterium, simply disperse one to a random location on the optimization domain.

8. If $\ell < N_{ed}$, then go to step 1; otherwise end.

18.3.3 Guidelines for Algorithm Parameter Choices

The bacterial foraging optimization algorithm requires specification of a variety of parameters. First, you can pick the size of the population, S. Clearly, increasing the size of S can significantly increase the computational complexity of the algorithm. However, for larger values of S, if you choose to randomly distribute the initial population, it is more likely that you will start at least some bacterium near an optimum point, and over time, it is then more likely that many bacterium will be in that region, due to either chemotaxis or reproduction.

What should the values of the $C(i)$, $i = 1, 2, \ldots, S$, be? You can choose a biologically motivated value; however, such values may not be the best for an engineering application. If the $C(i)$ values are too large, then if the optimum value lies in a valley with steep edges, it will tend to jump out of the valley, or it may simply miss possible local minima by swimming through them without stopping. On the other hand, if the $C(i)$ values are too small, then convergence can be slow, but if it finds a local minimum, it will typically not deviate too far from it. You should think of the $C(i)$ as a type of "step size" for the optimization algorithm.

The size of the values of the parameters that define the cell-to-cell attractant functions J_{cc}^{i} will define the characteristics of swarming. If the attractant width is high and very deep, the cells will have a strong tendency to swarm (they may even avoid going after nutrients and favor swarming). On the other hand, if the attractant width is small, and the depth shallow, there will be little tendency to swarm and each cell will search on its own. Social versus independent foraging is then dictated by the balance between the strengths of the cell-to-cell attractant signals and nutrient concentrations.

Next, large values for N_c result in many chemotactic steps, and, hopefully, more optimization progress, but of course, more computational complexity. If the size of N_c is chosen to be too short, the algorithm will generally rely more on luck and reproduction, and in some cases, it could more easily get trapped in a local minimum ("premature convergence"). You should think of N_s as creating a bias in the random walk (which would not occur if $N_s = 0$), with large values tending to bias the walk more in the direction of climbing down the hill.

If N_c is large enough, the value of N_{re} affects how the algorithm ignores bad regions and focuses on good ones, since bacteria in relatively nutrient-poor regions die (this models, with a fixed population size, the characteristic where bacteria will tend to reproduce at higher rates in favorable environments). If N_{re} is too small, the algorithm may converge prematurely; however, larger values of N_{re} clearly increase computational complexity.

A low value for N_{ed} dictates that the algorithm will not rely on random elimination-dispersal events to try to find favorable regions. A high value increases computational complexity but allows the bacteria to look in more regions to find good nutrient concentrations. Clearly, if p_{ed} is large, the algorithm can degrade to random exhaustive search. If, however, it is chosen appropriately, it can help the algorithm jump out of local optima and into a global optimum.

18.3.4 Relations to Other Nongradient Optimization Methods

There are *algorithmic analogies* between the genetic algorithm and the above optimization model for foraging. There are analogies between the fitness function and the nutrient concentration function (both a type of "landscape"), selection and bacterial reproduction (bacteria in the most favorable environments gain a selective advantage for reproduction), crossover and bacterial splitting (the children are at the same concentration, whereas with crossover they generally end up in a region around their parents on the fitness landscape), and mutation and elimination and dispersal. However, the algorithms are not equivalent, and neither is a special case of the other. Each has its own distinguishing features. The fitness function and nutrient concentration functions are *not* the same (one represents likelihood of survival for given phenotypic characteristics, whereas the other represents nutrient/noxious substance concentrations, or for other foragers predator/prey characteristics). Crossover represents mating and resulting differences in offspring, something we ignore in the bacterial foraging algorithm (we could, however, have made less than perfect copies of the bacteria to represent their splitting). Moreover, mutation represents gene mutation and the resulting phenotypical changes, not physical dispersal in an environment.

From one perspective, note that all the typical features of genetic algorithms could augment the bacterial foraging algorithm by representing evolutionary characteristics of a forager in their environment. From another perspective, foraging algorithms can be integrated into evolutionary algorithms and thereby model some key survival activities that occur during the lifetime of the population that is evolving (i.e., foraging success can help define fitness, mating characteristics, etc.). For the bacteria studied here, foraging happens to entail hill-climbing via a type of biased random walk, and hence, the foraging algorithm can be viewed as a method to integrate a type of approximate stochastic gradient search (where only an approximation to the gradient is used, not analytical gradient information) into evolutionary algorithms. Of course, standard gradient methods, quasi-Newton methods, etc., depend on the use of an explicit analytical representation of the gradient, something that is not needed by a foraging or genetic algorithm. Lack of dependence on analytical gradient information can be viewed as an advantage (fewer assumptions), or a disadvantage (e.g., since, if gradient information is available, then the foraging or genetic algorithm may not exploit it properly).

You probably also recognize some similarities between certain features of the foraging algorithm and SPSA. What are they? What are the relationships to the nongradient methods of the last part? There are in fact many approaches to "global optimization" when there is no explicit gradient information available; however, it is beyond the scope of this book to evaluate the relative merits of foraging algorithms to the vast array of such methods that have been studied for many years. To start such a study, it makes sense to begin by considering the theoretical convergence guarantees for certain types of evolutionary algorithms, stochastic approximation methods, and pattern search methods (e.g., see [481] for work along these lines), and then proceed to consider foraging algorithms in this context. It also seems useful to consider how well the foraging algorithms will perform for time-varying nutrient landscapes, which occurs in the underlying biological problem and many engineering problems.

18.3.5 Example: Function Optimization via *E. coli* Foraging

As a simple illustrative example, we use the algorithm to try to find the minimum of the function in Figure 18.10 (note that the point $[15, 5]^\top$ is the global minimum point).

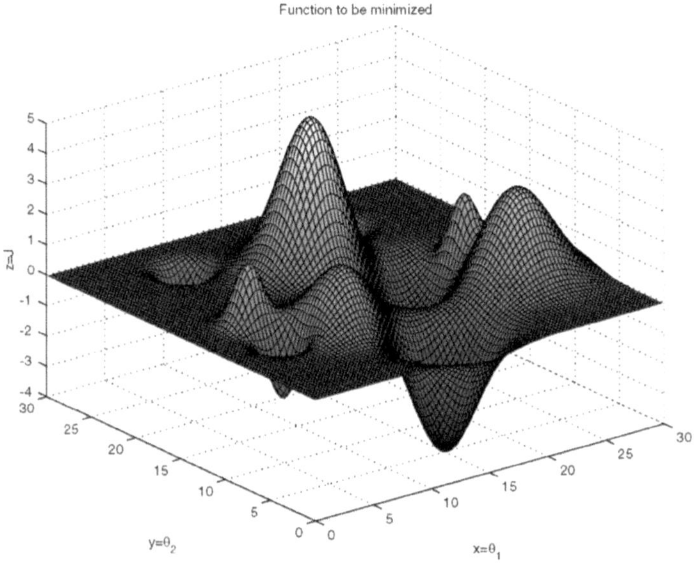

Figure 18.10: Nutrient landscape.

Nutrient Hill-Climbing: No Swarming

According to the above guidelines, choose $S = 50$, $N_c = 100$, $N_s = 4$ (a biologically motivated choice), $N_{re} = 4$, $N_{ed} = 2$, $p_{ed} = 0.25$, and the $C(i) = 0.1$, $i = 1, 2, \ldots, S$. The bacteria are initially spread randomly over the optimization domain. The results of the simulation are illustrated by motion trajectories of the bacteria on the contour plot of Figure 18.10, as shown in Figure 18.11. In the first generation, starting from their random initial positions, searching is occurring in many parts of the optimization domain, and you can see the chemotactic motions of the bacteria as the black trajectories where the peaks are avoided and the valleys are pursued. Reproduction picks the 25 healthiest bacteria and copies them, and then, as shown in Figure 18.11 in generation 2, all the chemotactic steps are in five local minima. This again happens in going to generations 3 and 4, but bacteria die in some of the local minima (due essentially to our requirement that the population size stay constant), so that in generation 3, there are four groups of bacteria in four local minima, whereas in generation 4, there are two groups in two local minima.

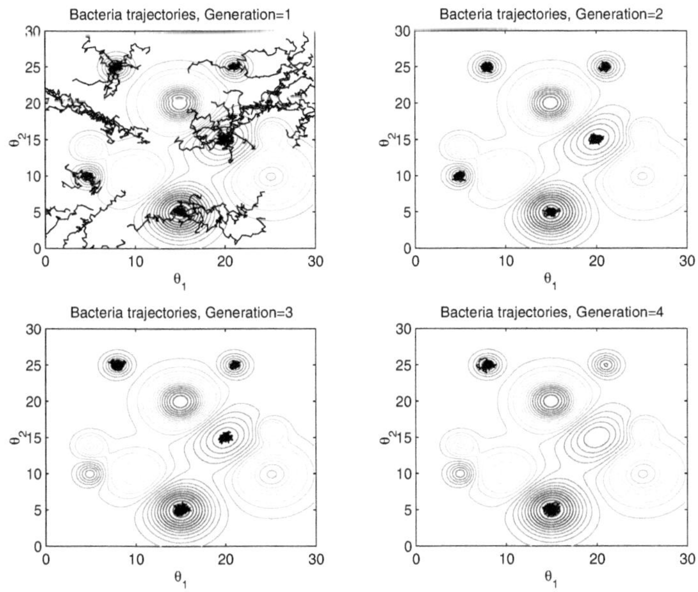

Figure 18.11: Bacterial motion trajectories, generations 1–4, on contour plots.

Next, with the above choice of parameters, there is an elimination-dispersal event, and we get the *next* four generations shown in Figure 18.12. Notice that elimination and dispersal shifts the locations of several of the bacteria and thereby the algorithm explores other regions of the optimization domain. However, qualitatively we find a similar pattern to the previous four generations where chemotaxis and reproduction work together to find the global minimum; this time, however, due to the large number of bacteria that were placed near the global minimum, after one reproduction step, all the bacteria are close to it (and remain this way). In this way, the bacterial population has found the global minimum.

Swarming Effects

Here we use the parameters defined earlier to define the cell-to-cell attraction function. Also, we choose $S = 50$, $N_c = 100$, $N_s = 4$, $N_{re} = 4$, $N_{ed} = 1$, $p_{ed} = 0.25$, and the $C(i) = 0.1$, $i = 1, 2, \ldots, S$. We will first consider swarming effects on the nutrient concentration function with contour map shown on Figure 18.13 which has a zero value at $[15, 15]^\top$ and decreases to successively more negative

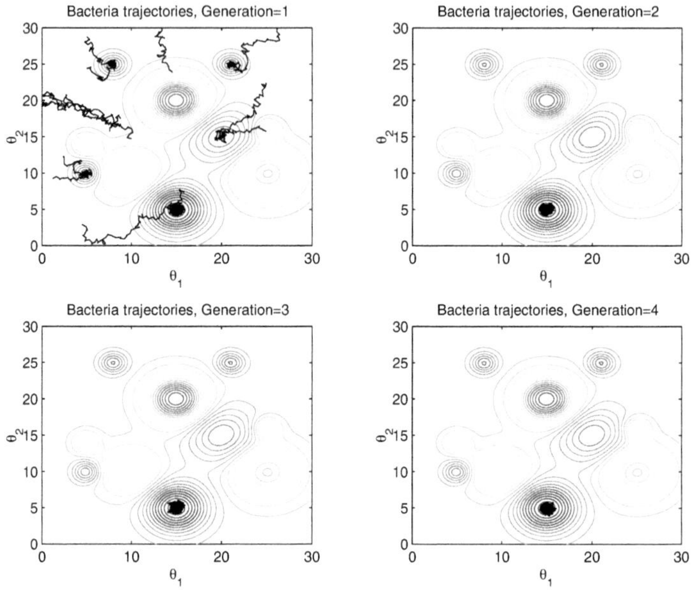

Figure 18.12: Bacterial motion trajectories, generations 1–4, on contour plots, after an elimination-dispersal event.

values as you move away from that point; hence, the cells should tend to swim away from the peak. We will initialize the bacterial positions by placing all the cells at the peak $[15, 15]^\top$. Using these conditions, we get the result in Figure 18.13. Notice that in the first generation, the cells swim radially outward, and then in the second and third generations, swarms are formed in a concentric pattern of groups. Notice that with our simple method of simulating health of the bacteria and reproduction, some of the swarms are destroyed by the fourth generation. We omit additional simulations that show the behavior of the swarm on the surface in Figure 18.10, since qualitatively the behavior is as one would expect from the above simulations. The interested reader can obtain the code mentioned above and further study the behavior of the algorithm.

18.4 Stable Social Foraging Swarms

In this section, we first overview some biology of swarms, with a focus on the honey bee in order to provide a concrete example. Then, we introduce a mathematical model for a generic swarm of agents. We conduct a mathematical analysis to prove stability (cohesiveness) of the swarm and perform simulations to provide insights into swarm dynamics.

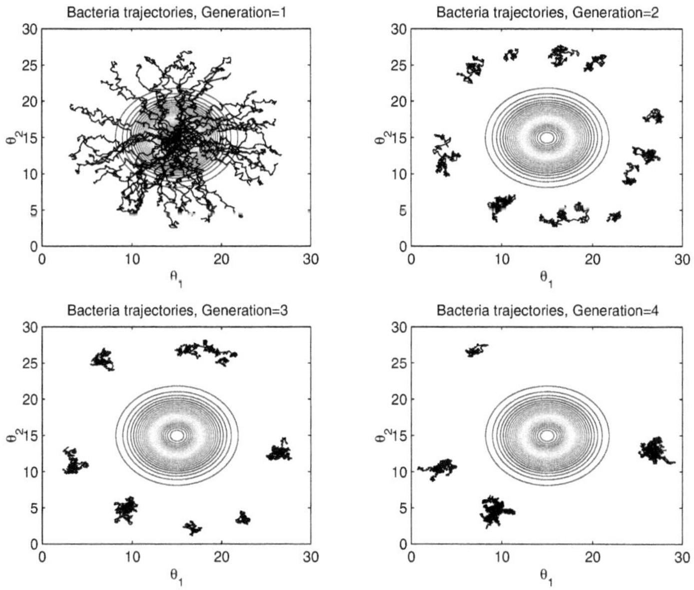

Figure 18.13: Swarm behavior of *E. coli* on a test function.

18.4.1 Example: The Biology of Honey Bee Swarms

There are many types of animals that swarm, including bacteria, insects, birds, fish, horses, and so on. (Groupings of different animals are typically given different names, but for convenience, here we will call them all swarms.) Such groups are composed of individuals with different physiological capabilities, and when operating as a group, they can achieve different types of emergent behaviors. Here, in order to be more concrete about how swarms operate, we will describe in more detail how and why one species, the honey bee, swarms.

First, when a hive splits, a group of bees will "cluster" around the queen on, for example, a nearby branch. This is a type of swarming where the group contracts together and forms a tightly packed group of relatively stationary bees. After forming such a cluster, the bees perform nest site selection. Upon liftoff, after the scouts generally reach agreement on a nest site, the honey bee swarm forms a spherical group that hovers near the place where the swarm had grouped while nest site selection occurred. Next, the group slowly starts to move in the direction of the new nest site, elongating laterally, and accelerating as a group to about 11 kilometers/hr. There is a type of "inertia" in getting the group moving due to the swarm dynamics. It seems that several factors contribute to in-transit swarm cohesiveness, two of which may involve pheromones. First, the queen releases a pheromone (but the queen is not in the lead so it does not seem likely that it can steer the bees that are ahead of her). Second, some evidence suggests that other bees in the swarm release Nasanov pheromones that help to maintain cohesiveness, but it seems that this is still not fully verified. How does the swarm know which *direction* to go? Again, there seem to be several factors. First, during the site selection process, it seems that those observing the final phase of the agreement process may know which direction to go. However, perhaps not all of the bees will know the direction, and this may explain the way that some bees in the swarm seem to wander somewhat within the swarm, apparently lacking knowledge of the direction to the nest. Alternatively, it seems possible that even if they knew the direction, they may not be able to easily navigate while in the swarm due to obstruction of navigation cues due to interference from many other bees. Another factor that certainly seems to influence the direction of movement of the swarm is the presence of the scouts that do know the way to the site, and which "pilot" the swarm by "streaking" through it in the general direction of the nest site. (You could think of this as an "aerial dance signal" to recruit bees in the proper direction.) Nevertheless, even with the possibility of several methods of group navigation, there are significant motions of bees within the swarm that are not directed towards the site. Aside from navigational obstructions, and possible bee-bee collisions, it also seems that it may be possible that there are conflicting indictions of the general direction to move, due to dynamic changes in the dominance of factors that guide the bee (e.g., pheromones, its own sense of direction, and the indications of scout piloting). Also, it should be clear that wind (perhaps even currents that are induced by the swarm) will disrupt a uniform flow of the swarm towards the nest site. At the same time, it should be clear that some type of "robust" group guidance and navigation is achieved since they successfully reach the site. Upon reaching the site the group stops (and it is not understood how that occurs), some scouts drop into the new nest entrance and release a pheromone, and this attracts the group to the new nest. A model of the clustering and in-transit motion of bees has been studied (see the "For Further Study" section at the end of the part for more details).

18.4.2 Swarm and Environment Models

Here, we describe the agent, communications, and the environment that the agents move in.

Agent Dynamics and Communications

Here, rather than focusing on the particular characteristics of one type of animal or autonomous vehicle, we consider a swarm to be composed of an interconnection of N "agents," each of which has point mass dynamics given by

$$\begin{aligned} \dot{x}^i &= v^i \\ \dot{v}^i &= \frac{1}{M_i} u^i \end{aligned} \tag{18.2}$$

where $x^i \in \Re^n$ is the position, $v^i \in \Re^n$ is the velocity, M_i is the mass, and $u^i \in \Re^n$ is the (force) control input for the ith agent. We use this simple linear model for an agent in order to illustrate the basic features of swarming and stability analysis of cohesion. Other nonlinear and stochastic models for agents in swarms have been considered in the literature (see the "For Further Study" section at the end of this part). Here, we will use $n = 3$ for swarms moving in a three-dimensional space. We will assume that each agent can sense information about the position and velocity of other agents, but only with some noise that we will define below.

The agents interact to form groups, and in some situations groups will split. Some think of having local interactions between agents, which "emerge" into a global behavior for the group. One way to represent which agents can interact with each other is via a directed graph (G, A) where $G = \{1, 2, \ldots, N\}$ is a set of nodes (the agents) and

$$A = \{(i, j) : i, j \in G, i \neq j\}$$

represents a "sensing/communication topology" (in general, then, each "link" (i, j) could be a dynamical system). For example, if $(i, j) \in A$, then it could be assumed that agent i can sense the position and velocity of agent j. In some vehicular systems it may be possible that A is fixed and independent of vehicle positions and velocities. Clearly, however, for biological systems it is often the case that A is not fixed, but changes dynamically based on the positions of the agents (e.g., so that only agents within a line-of-sight and close enough can be sensed). Here, for simplicity, we will assume that A does not change based on the positions and velocities of the agents, and that A is fully connected (e.g., that for all $i \in G$, $(i, j) \in A$). We also assume that there are no delays or noise in communicating, communications are not range constrained, and that there is infinite bandwidth. More general formulations may also include a "communications topology" that specifies which agents can send and receive messages to other agents. Such messages could represent a wide variety of communication capabilities of the agents (e.g., bee scout leadership in a swarm, sounds, and so on.).

Agent to Agent Attraction and Repulsion

Agent to agent interactions considered here are of the "attract-repel" type, where each agent seeks to be in a position that is "comfortable" relative to its neighbors (and for us, all other agents are its neighbors). Attraction indicates that each agent wants to be close to every other agent and provides the mechanism for achieving grouping and cohesion of the group of agents. Repulsion provides the mechanism where each agent does not want to be too close to every other agent (e.g., for animals to avoid collisions and excessive competition for resources). There are many ways to define attraction and repulsion, each of which can be represented by characteristics of how we define u^i for each agent, and we list a few of these below:

Agents want to be close to each other, but not too close.

- *Attraction:* There can be linear attraction and in this case we have terms in u^i of, for example, the form
$$-k_a\left(x^i - x^j\right)$$
where $k_a > 0$ is a scalar that represents the strength of attraction. If the agents are far apart, then there is a large attraction between them, and if they are close, there is a small attraction. In other cases, there can be nonlinear attraction terms that can be expressed in terms of nonlinear functions of $\left(x^i - x^j\right)$. Some attraction mechanisms are "local" (i.e., for range-constrained sensing, where the agent only tries to move to other agents that are close to it) and others which are "global" (i.e., where agents can be attracted to move near other agents no matter how far away they are). Attraction terms can be specified in terms of a variety of agent variables. For example, the above term is for positions, but we could have similar terms for velocity so that the agents will try to match the velocities of other agents. Moreover, the above term is "static" but we would have a local "dynamic" controller that tries to match agent variables.

- *Repulsion:* As with attraction, there are many types of repulsion terms, some local, global, static, or dynamic, each of which can be expressed in terms of a variety of agent variables. Sometimes repulsion is defined in terms that also quantify attraction, other times they quantify only a repulsion.

  - *Seek a "comfortable distance":* For example, a term in u^i may take the form
$$\left[-k\left(||x^i - x^j|| - d\right)\right]\left(x^i - x^j\right)$$
where $||x^i - x^j|| = \sqrt{\left(x^i - x^j\right)^\top \left(x^i - x^j\right)}$, $k > 0$ is the magnitude of the repulsion, and d can be thought of as a comfortable distance between the ith and j^{th} agents. Here, the quantity in the brackets sets the size of the repulsion. When $||x^i - x^j||$ is small (relative to d), the term in the bracket is positive so that agents i and j try to move away from each other (there is repulsion). When $||x^i - x^j||$ is big (relative to d), then the term in the brackets is negative so that the agents are attracted to each other. Balance between attraction and repulsion (a basic concept in swarm dynamics that is sometimes referred to as an "equilibrium," even though it may not be one in the stability-theoretic sense) may be achieved when $||x^i - x^j|| = d$ so that the term above is zero.

  - *Repel when close:* Another type of repulsion term in u^i, which may be used with, for example, a linear attraction term, may take the form
$$k_r \exp\left(\frac{-\frac{1}{2}||x^i - x^j||^2}{r_s^2}\right)\left(x^i - x^j\right) \tag{18.3}$$

where $k_r > 0$ is the magnitude of the repulsion, and $r_s > 0$ quantifies the region size around the agent from which it will repel its neighbors. When $\|x^i - x^j\|$ is big relative to r_s, the whole term approaches zero.

– *Hard repulsion for collision avoidance:* The above repulsion terms are "soft," in the sense that when the two agents are at the same location, the repulsion force is finite. In some cases, it is appropriate to use a repulsion term that becomes increasingly large as two agents approach each other. One such term, that does not have an attraction component, is in the form of

$$\left[\max\left\{\left(\frac{a}{b\|x^i - x^j\| - w} - \epsilon\right), 0\right\}\right](x^i - x^j) \tag{18.4}$$

Here, $w > 0$ affects the radius of repulsion of the agents, $a > 0$ is a gain on the magnitude of the repulsion, $b > 0$ can change the shape of the repulsion gain, and $\epsilon > 0$ can be used to define the term so that it only has a local influence. For instance, the radius of the repulsion is

$$R' = \frac{1}{b}\left(\frac{a}{\epsilon} + w\right)$$

For agent positions such that $\|x^i - x^j\| \geq R'$, the term in the brackets is zero. For given values of w, a, and b, you can choose ϵ to get any value of $R' > 0$. Note that a key feature of this repulsion term is that as $\|x^i - x^j\|$ goes from a large value to w/b, the value in the brackets goes to infinity to provide a "hard" repelling action and hence, avoid the possibility that two agents ever end up at the same position (i.e., to avoid collisions). Another way to define a hard repulsion is to consider agents to be solid "balls," where, if they collide, they do not deform.

For more ideas on how to define attraction and repulsion terms, see the "For Further Study" section at the end of this part.

Environment and Foraging

Next, we will define the environment that the agents move in. While there are many possibilities, here we will simply consider the case where they move over what we will call "resource profile" (e.g., nutrient profile) $J(x)$, where $x \in \Re^n$. We will however, think of this profile as being something where the agents want to be in certain regions of the profile and avoid other regions (e.g., where there are noxious substances). We will assume that $J(x)$ is continuous with finite slope at all points. Agents move in the direction of the negative gradient of $J(x)$

Agents follow resource profiles to meet their objectives.

$$-\nabla J(x) = -\frac{\partial J}{\partial x}$$

in order to move away from bad areas and into good areas of the environment (e.g., to avoid noxious substances and find nutrients). Hence, they will use a term in their u^i that holds the negative gradient of $J(x)$.

Clearly, there are many possible shapes for $J(x)$, including ones with many peaks and valleys. Here, we simply list two simple forms for $J(x)$ as follows:

- *Plane:* In this case, we have $J(x) = J_p(x)$ where

$$J_p(x) = R^\top x + r_o$$

where $R \in \Re^n$ and r_o is a scalar. Here, $\nabla J_p(x) = R$.

- *Quadratic:* In this case, we have $J(x) = J_q(x)$ where

$$J_q(x) = \frac{r_m}{2}\|x - R_c\|^2 + r_o$$

where r_m and r_o are scalars and $R_c \in \Re^n$. Here, $\nabla J_q(x) = r_m\,(x - R_c)$.

Below, we will assume that each agent can sense the gradient of the resource profile, but only with some noise.

It could be that the environment has many different types of agents in it, or the same types of agents with different objectives. In this case, there may be different resource profiles for each agent, or the agents may switch the profiles it follows, or strategies for following them. Different agents may have different capabilities to sense the profile (e.g., sensing only at a point, or sensing a range-constrained region) and move over it. If the agents are consuming food, this may change the shape of the profile, and of course, an agent may pollute its environment so it may affect the profile in that manner also. This would create a time-varying profile that is dependent on agent position, and possibly many other variables. See the "For Further Study" section at the end of this part for more discussion on nutrient profiles and, for example, their use in modeling bee swarming.

18.4.3 Stability Analysis of Swarm Cohesion Properties

Cohesion and swarm dynamics can be quantified and analyzed using stability analysis (e.g., via Lyapunov's method). You can pick agent dynamics, interactions, sensing capabilities, attraction/repulsion characteristics, and foraging environment characteristics, then quantify and analyze cohesion properties. Here, we will do this for a few simple cases in order to give a flavor of the type of analysis that is possible, and to provide insight into swarm properties and dynamics during social foraging.

Sensing, Noise, and Error Dynamics

First, let

$$\bar{x} = \frac{1}{N}\sum_{i=1}^{N} x^i$$

be the center of the swarm and

$$\bar{v} = \frac{1}{N}\sum_{i=1}^{N} v^i$$

be the average velocity (vector) which we view as the velocity of the group of agents. We assume that each agent can sense the distance from itself to $\bar{x}$, and the difference between its own velocity and $\bar{v}$. We assume that each agent knows its own velocity, but not its own position. Note that for some animals, its senses and sensory processing may naturally provide the distance to $\bar{x}$ and $\bar{v}$, but not all the individual positions and velocities of all agents that could be used to compute these. Also, we will consider below the case where there is noise in sensing these quantities.

The objective of each agent is to move so as to end up at or near $\bar{x}$ and have its velocity equal to $\bar{v}$; in this way, an emergent behavior of the group is produced where they aggregate dynamically and end up near each other and ultimately move in the same direction (i.e., they achieve cohesion). The problem is that since all the agents are moving at the same time, $\bar{x}$ and $\bar{v}$ are generally time-varying; hence, in order to study the stability of swarm cohesion, we study the dynamics of an error system with

$$e_p^i = x^i - \bar{x}$$

and

$$e_v^i = v^i - \bar{v}$$

Other choices for error systems are also possible and have been used in some studies of swarm stability. For instance, you could use $\tilde{e}_p^i = \sum_{j=1}^{N}\left(x^i - x^j\right)$. This corresponds to computing the errors to each other agent and then trying to get all those errors to go to zero. Note, however, that

$$\tilde{e}_p^i = N\left(x^i - \frac{1}{N}\sum_{j=1}^{N} x^j\right) = N\left(x^i - \bar{x}\right) = N e_p^i$$

The same relationship holds for velocity.

Given the above choices, the error dynamics are given by

$$\begin{aligned} \dot{e}_p^i &= e_v^i \\ \dot{e}_v^i &= \frac{1}{M_i}u^i - \frac{1}{N}\sum_{j=1}^{N}\frac{1}{M_j}u^j \end{aligned} \qquad (18.5)$$

The challenge is to specify the u^i so that we get good cohesion properties and successful social foraging.

Assume that each agent can sense its position and velocity relative to $\bar{x}$ and $\bar{v}$, but with some bounded errors. In particular, let $d_p^i(t) \in \Re^n$, $d_v^i(t) \in \Re^n$ be these errors for agent i, respectively. We assume that $d_p^i(t)$ and $d_v^i(t)$ are sufficiently smooth and are independent of the state of the system. Each agent will try to follow the resource profile J_p defined earlier. (We use the plane profile for the sake of illustration, as it will show how swarm dynamics operate over a simple but representative surface.) We assume that each agent senses the gradient of J_p, but with some sufficiently smooth error $d_f^i(t) \in \Re^n$. You may think of this as either a sensing error or as variations (e.g., high frequency ripples) on the resource profile. Below, we will refer to the signals $d_p^i(t)$, $d_v^i(t)$, and $d_f^i(t)$ as "noise" signals, but clearly there is no underlying probability space and all signals in this section are deterministic. You may think of these signals as being generated by, for example, a chaotic dynamic system.

We assume that all the sensing errors are bounded such that

$$\begin{aligned} \|d_p^i\| &\leq D_p \\ \|d_v^i\| &\leq D_v \\ \|d_f^i\| &\leq D_f \end{aligned}$$

where $D_p > 0$, $D_v > 0$, and $D_f > 0$ are known constants. Similar results to what we find below can be found for the more general case, where we have $\|d_p^i\| \leq D_{p_1} \left\|E^i\right\| + D_{p_2}$, $\|d_v^i\| \leq D_{v_1} \left\|E^i\right\| + D_{v_2}$, and $\|d_f^i\| \leq D_f$ where D_{p_1}, D_{p_2}, D_{v_1} and D_{v_2} are known positive constants and E^i is defined in Equation (18.10). See the "For Further Study" section at the end of this part.

Thus, each agent can sense noise-corrupted versions of e_p^i and e_v^i, as

$$\hat{e}_p^i = e_p^i - d_p^i$$
$$\hat{e}_v^i = e_v^i - d_v^i$$

Also, each agent can sense

$$\nabla J_p\left(x^i\right) - d_f^i$$

at the location x^i where the agent is located.

Suppose that in order to steer itself, each agent uses

Noise makes it more difficult to get cohesive behavior and degrades foraging effectiveness.

$$\begin{aligned} u^i &= -M_i k_a \hat{e}_p^i - M_i k_a \hat{e}_v^i - M_i k_v v^i \\ &+ M_i k_r \sum_{j=1, j\neq i}^{N} \exp\left(\frac{-\frac{1}{2}\|\hat{e}_p^i - \hat{e}_p^j\|^2}{r_s^2}\right)\left(\hat{e}_p^i - \hat{e}_p^j\right) \\ &- M_i k_f \left(\nabla J_p\left(x^i\right) - d_f^i\right) \end{aligned} \tag{18.6}$$

Here, we assume that each agent knows its own mass M_i and velocity v^i. The parameter $k_v > 0$ is the gain for a "velocity damping term." We think of the scalar $k_a > 0$ as the "attraction gain" that indicates how aggressive the agents are in aggregating. The gain k_r is a "repulsion gain," which sets how much the agents want to be away from each other. Note that use of the repulsion term assumes that the ith agent knows, within some errors, the relative distance of all other agents from the swarm center. Also, since

$$\hat{e}_p^i - \hat{e}_p^j = \left(\left(x^i - \bar{x}\right) - d_p^i\right) - \left(\left(x^j - \bar{x}\right) - d_p^j\right) = \left(x^i - x^j\right) - \left(d_p^i - d_p^j\right)$$

we are assuming that the ith agent knows its position (and velocity) relative to each other agent within some bounded errors. Note, however, that when x^i and x^j are far apart, the $\exp(\cdot)$ term is close to zero. (So in effect, if each agent did not use distant agents' values in the repulsion term, we would get approximately the same results.) Also note that if $D_p = D_v = 0$, there is no sensing error on attraction and repulsion, thus, $\hat{e}_p^i = e_p^i$, $\hat{e}_v^i = e_v^i$, and $e_p^i - e_p^j = x^i - x^j$, and we get a repulsion term of the form explained in Equation (18.3). The sensing errors create the possibility that agents will try to move away from each other when they may not really need to, and they may move towards each other when they should not. Clearly, this complicates the ability of the agents to avoid collisions with their neighbors. The last term in Equation (18.6) indicates that each agent wants to move along the negative gradient of the resource profile with the gain k_f proportional to the agent's desire to follow the profile.

Social Foraging in Noise: Groups Can Increase Foraging Effectiveness

Next, we will substitute this choice for u^i into the error dynamics described in Equation (18.5) and study their stability properties. First, however, we will study how the *group* can follow the resource profile in the presence of noise. To do this, consider $\dot{e}_v^i = \dot{v}^i - \dot{\bar{v}}$. First, note that

$$\begin{aligned}
\dot{\bar{v}} &= \frac{1}{N}\sum_{i=1}^{N}\frac{1}{M_i}u^i \\
&= -\frac{k_a}{N}\sum_{i=1}^{N}(\hat{e}_p^i + \hat{e}_v^i) - \frac{k_v}{N}\sum_{i=1}^{N}v^i \\
&\quad +\frac{k_r}{N}\sum_{i=1}^{N}\sum_{j=1,j\neq i}^{N}\exp\left(\frac{-\frac{1}{2}\|\hat{e}_p^i - \hat{e}_p^j\|^2}{r_s^2}\right)(\hat{e}_p^i - \hat{e}_p^j) \\
&\quad -\frac{k_f}{N}\sum_{i=1}^{N}(R - d_f^i)
\end{aligned} \tag{18.7}$$

Notice that

$$\frac{1}{N}\sum_{i=1}^{N}\hat{e}_p^i = \frac{1}{N}\sum_{i=1}^{N}((x^i - \bar{x}) - d_p^i) = \bar{x} - \frac{1}{N}N\bar{x} - \frac{1}{N}\sum_{i=1}^{N}d_p^i = -\frac{1}{N}\sum_{i=1}^{N}d_p^i$$

Also, the term due to repulsion in Equation (18.7) is zero as we show next. Note that

$$\sum_{i=1}^{N}\sum_{j=1,j\neq i}^{N}\exp\left(\frac{-\frac{1}{2}\|\hat{e}_p^i - \hat{e}_p^j\|^2}{r_s^2}\right)(\hat{e}_p^i - \hat{e}_p^j) =$$

$$\left[\sum_{i=1}^{N}\hat{e}_p^i\sum_{j=1,j\neq i}^{N}\exp\left(\frac{-\frac{1}{2}\|\hat{e}_p^i - \hat{e}_p^j\|^2}{r_s^2}\right)\right]$$

$$- \left[\sum_{i=1}^{N} \sum_{j=1, j\neq i}^{N} \exp\left(\frac{-\frac{1}{2}\|\hat{e}_p^i - \hat{e}_p^j\|^2}{r_s^2} \right) \hat{e}_p^j \right] \tag{18.8}$$

The last term in Equation (18.8)

$$\sum_{i=1}^{N} \sum_{j=1, j\neq i}^{N} \exp\left(\frac{-\frac{1}{2}\|\hat{e}_p^i - \hat{e}_p^j\|^2}{r_s^2} \right) \hat{e}_p^j = \sum_{j=1}^{N} \hat{e}_p^j \sum_{i=1, i\neq j}^{N} \exp\left(\frac{-\frac{1}{2}\|\hat{e}_p^j - \hat{e}_p^i\|^2}{r_s^2} \right)$$

and since

$$\exp\left(\frac{-\frac{1}{2}\|\hat{e}_p^i - \hat{e}_p^j\|^2}{r_s^2} \right) = \exp\left(\frac{-\frac{1}{2}\|\hat{e}_p^j - \hat{e}_p^i\|^2}{r_s^2} \right)$$

we have this the same as

$$\sum_{j=1}^{N} \hat{e}_p^j \sum_{i=1, i\neq j}^{N} \exp\left(\frac{-\frac{1}{2}\|\hat{e}_p^i - \hat{e}_p^j\|^2}{r_s^2} \right) = \sum_{i=1}^{N} \hat{e}_p^i \sum_{j=1, j\neq i}^{N} \exp\left(\frac{-\frac{1}{2}\|\hat{e}_p^i - \hat{e}_p^j\|^2}{r_s^2} \right)$$

but this last value is the same as the first term on the right-hand side of Equation (18.8). So overall its value is zero. This gives us

$$\dot{\bar{v}} = \frac{k_a}{N} \sum_{i=1}^{N} d_p^i + \frac{k_a}{N} \sum_{i=1}^{N} d_v^i + \frac{k_f}{N} \sum_{i=1}^{N} d_f^i - k_v \bar{v} - k_f R$$

Letting $\bar{d_p}(t) = \frac{1}{N} \sum_{i=1}^{N} d_p^i(t)$ and similarly for $\bar{d_v}(t)$ and $\bar{d_f}(t)$, we get

$$\dot{\bar{v}} = -k_v \bar{v} + \underbrace{k_a \bar{d_p} + k_a \bar{d_v} + k_f \bar{d_f} - k_f R}_{z(t)} \tag{18.9}$$

This is an exponentially stable system with a time-varying but bounded input $z(t)$ so we know that $\bar{v}(t)$ is bounded. To see this, choose a Lyapunov function

$$V_{\bar{v}} = \frac{1}{2} \bar{v}^\top \bar{v}$$

defined on $D = \{\bar{v} \in \Re^n \mid \|\bar{v}\| < r_v\}$ for some $r_v > 0$, and we have

$$\dot{V}_{\bar{v}} = \bar{v}^\top \dot{\bar{v}} = -k_v \bar{v}^\top \bar{v} + z(t)^\top \bar{v}$$

with

$$\left\| \frac{\partial V_{\bar{v}}}{\partial \bar{v}} \right\| = \|\bar{v}\|$$

Note that $\|z(t)\| \leq \left\|k_a \bar{d_p}\right\| + \left\|k_a \bar{d_v}\right\| + \left\|k_f \bar{d_f}\right\| + \|k_f R\| \leq \delta$, where $\delta = k_a D_p + k_a D_v + k_f D_f + k_f \|R\|$. If $\delta < k_v \theta r_v$ for all $t \geq 0$ for some positive constant $\theta < 1$, and all $\bar{v} \in D$, then it can be proven that for all $\|\bar{v}(0)\| < r_v$, and some finite T, we have

$$\|\bar{v}(t)\| \leq \exp\left[-(1-\theta) k_v t\right] \|\bar{v}(0)\|, \ \forall \ 0 \leq t < T$$

and

$$\|\bar{v}(t)\| \leq \frac{\delta}{k_v\theta}, \ \forall \, t \geq T$$

Since this holds globally, we can take $r_v \to \infty$ so these inequalities hold for all $\bar{v}(0)$. If δ and θ are fixed, with increasing k_v we get that $\|\bar{v}(t)\|$ decreases faster for $0 \leq t < T$ and smaller bound on $\|\bar{v}(t)\|$ for $t \geq T$. If δ gets larger with k_v and θ fixed, $\|\bar{v}(t)\|$ has larger bound for $t \geq T$; hence, if the magnitude of the noise increases, this increases δ and hence, there can be larger magnitude changes in the ultimate average velocity of the swarm (e.g., the average velocity could oscillate). Note that if in Equation (18.9) $z(t) \approx 0$ (e.g., due to noise that destroys the directionality of the resource profile R), then the above bound may be reduced, but the swarm could be going in the wrong direction.

Regardless of the size of the bound, it is interesting to note that while the noise can destroy the ability of an individual agent to follow a gradient accurately, the average sensing errors of the group is what changes the direction of the group's movement relative to the direction of the gradient of $J_p(x)$. In some cases when the swarm is large (N big), it can be that $\bar{d_p} \approx \bar{d_v} \approx \bar{d_f} \approx 0$ to give a zero average sensing error and the group will perfectly follow the proper direction for foraging. (This may be a reason why, for some organisms, large group size is favorable.) In the case when $N = 1$ (i.e., single agent), there is no opportunity for a cancellation of the sensor errors; hence, an individual may not be able to climb a noisy gradient as easily as a group, and in some cases, a group may be able to follow a profile where an individual cannot. This characteristic has been found in biological swarms [230]. From an optimization perspective, you should think of an individual trying to execute a gradient optimization method, which we know can result in it getting stuck in local minima. The group is producing a type of approximation to the gradient by a larger spatial sampling and attraction/repulsion terms. Intuitively, it filters out the noise and moves in the proper direction. Of course, the group itself can get stuck in a local minimum if the basin of attraction of that minimum is large.

Social groups can climb noisy gradients better than individuals.

It is also important to note that there is an intimate relationship between sensor noise and observations of biological swarms (e.g., in bee swarms [458]) in that there is a type of "inertia" of a swarm. Note that for large swarms (N big), there can be regions where the average sensor noise is small, so that agents in that region move in the right direction. In other regions there may be alignments of the errors and hence, the agents may not be all moving in the right direction so they may get close to each other and impede each other's motion, having the effect of slowing down the whole group. With no noise, the group inertia effect is not found, since each agent is moving in the right direction. The presence of sensor noise generally can make it more difficult to get the group moving in the right direction (e.g., for foraging, migration, or movement to a nest site). Large swarms can help move the group in the right direction, but at the expense of possibly slowing their movement initially in a transient period.

Cohesive Social Foraging in Noise

Next, we return to the problem of finding the error dynamics and then stability analysis by considering the $\dot{v}^i$ term of $\dot{e}_v^i = \dot{v}^i - \dot{\bar{v}}$ in the error dynamics of Equation (18.5). Note that

$$\begin{aligned}
\dot{v}^i = \frac{1}{M_i}u^i &= -k_a\hat{e}_p^i - k_a\hat{e}_v^i - k_v v^i \\
&+ k_r \sum_{j=1, j\neq i}^{N} \exp\left(\frac{-\frac{1}{2}\|\hat{e}_p^i - \hat{e}_p^j\|^2}{r_s^2}\right)\left(\hat{e}_p^i - \hat{e}_p^j\right) - k_f\left(\nabla J_p\left(x^i\right) - d_f^i\right) \\
&= -k_a e_p^i + k_a d_p^i - k_a e_v^i + k_a d_v^i - k_v v^i \\
&+ k_r \sum_{j=1, j\neq i}^{N} \exp\left(\frac{-\frac{1}{2}\|\hat{e}_p^i - \hat{e}_p^j\|^2}{r_s^2}\right)\left(\hat{e}_p^i - \hat{e}_p^j\right) - k_f\left(R - d_f^i\right)
\end{aligned}$$

Hence, we have

$$\begin{aligned}
\dot{e}_v^i = \dot{v}^i - \dot{\bar{v}} &= -k_a e_p^i - k_a e_v^i - k_v e_v^i + k_a\left(d_p^i - \bar{d}_p\right) + k_a\left(d_v^i - \bar{d}_v\right) \\
&+ k_r \sum_{j=1, j\neq i}^{N} \exp\left(\frac{-\frac{1}{2}\|\left(x^i - x^j\right) - \left(d_p^i - d_p^j\right)\|^2}{r_s^2}\right)\left(\left(x^i - x^j\right)\right. \\
&- \left.\left(d_p^i - d_p^j\right)\right) + k_f\left(d_f^i - \bar{d}_f\right)
\end{aligned}$$

To study the stability of the error dynamics, and hence, swarm cohesiveness, define

$$E^i = \left[{e_p^i}^\top, {e_v^i}^\top\right]^\top \tag{18.10}$$

and $E = \left[{E^1}^\top, {E^2}^\top, \ldots, {E^N}^\top\right]^\top$, and choose a Lyapunov function

$$V(E) = \sum_{i=1}^{N} V_i\left(E^i\right)$$

where

$$V_i\left(E^i\right) = {E^i}^\top P E^i$$

with $P = P^\top$ and $P > 0$ (a positive definite matrix). We know that

$$\lambda_{min}(P){E^i}^\top E^i \leq {E^i}^\top P E^i \leq \lambda_{max}(P){E^i}^\top E^i$$

Notice that with I an $n \times n$ identity matrix, we have

$$\dot{E}^i = \underbrace{\begin{bmatrix} 0 & I \\ -k_a I & -\left(k_a + k_v\right) I \end{bmatrix}}_{A} E^i + \underbrace{\begin{bmatrix} 0 \\ I \end{bmatrix}}_{B} g^i(E)$$

where

$$\begin{aligned} g^i(E) &= k_a\left(d_p^i - \bar{d_p}\right) + k_a\left(d_v^i - \bar{d_v}\right) \\ &+ k_r \sum_{j=1, j\neq i}^{N} \exp\left(\frac{-\frac{1}{2}\|\hat{e}_p^i - \hat{e}_p^j\|^2}{r_s^2}\right)\left(\hat{e}_p^i - \hat{e}_p^j\right) \\ &+ k_f\left(d_f^i - \bar{d_f}\right) \end{aligned} \tag{18.11}$$

Note that any matrix

$$\begin{bmatrix} 0 & I \\ -k_1 I & -k_2 I \end{bmatrix}$$

with $k_1 > 0$ and $k_2 > 0$ has eigenvalues given by the roots of $(s^2 + k_2 s + k_1)^n$, which are in the strict left half plane. Since $k_a > 0$ and $k_v > 0$, the matrix A above is Hurwitz (i.e., has eigenvalues all in the strict left half plane).

We have

$$\dot{V}_i = {E^i}^\top P \dot{E}^i + {\dot{E}^i}^\top P E^i = {E^i}^\top \left(PA + A^\top P\right) E^i + 2{E^i}^\top P B g^i(E) \tag{18.12}$$

Note that if $-Q = \left(PA + A^\top P\right)$, then Q is such that $Q = Q^\top$ and $Q > 0$, and the unique solution P of $PA + A^\top P = -Q$ has $P = P^\top$ and $P > 0$ as needed. Also, since $\|B\| = 1$, ${E^i}^\top Q E^i \geq \lambda_{min}(Q){E^i}^\top E^i$, and $\|P\| = \lambda_{max}(P)$ with $P = P^\top > 0$, we have

$$\begin{aligned} \dot{V}_i &\leq -\lambda_{min}(Q)\left\|E^i\right\|^2 + 2\left\|E^i\right\|\lambda_{max}(P)\|g^i(E)\| \\ &= -\lambda_{min}(Q)\left(\left\|E^i\right\| - \frac{2\lambda_{max}(P)}{\lambda_{min}(Q)}\|g^i(E)\|\right)\left\|E^i\right\| \end{aligned} \tag{18.13}$$

Suppose for a moment that for each $i = 1, 2, \ldots, N$, $\|g^i(E)\| < \beta$ for some known β. Then, if

$$\left\|E^i\right\| > \frac{2\lambda_{max}(P)}{\lambda_{min}(Q)}\|g^i(E)\| \tag{18.14}$$

we have that $\dot{V}_i < 0$. Hence, the set

$$\Omega_b = \left\{E : \left\|E^i\right\| \leq 2\frac{\lambda_{max}(P)}{\lambda_{min}(Q)}\|g^i(E)\|,\ i = 1, 2, \ldots, N\right\} \tag{18.15}$$

is attractive and compact. Also we know that within a finite amount of time, $E^i \to \Omega_b$. This means that we can guarantee that if the swarm is not cohesive, it will seek to be cohesive, but this can only be guaranteed if it is a certain distance from cohesiveness, as indicated by Equation (18.14).

The positions and velocities of all agents can oscillate, yet overall swarm cohesiveness can be maintained.

It remains to show that for each i, $\|g^i(E)\| < \beta$ for some β. Note that

$$\|g^i(E)\| \leq k_a\|d_p^i - \bar{d_p}\| + k_a\|d_v^i - \bar{d_v}\| + k_f\|d_f^i - \bar{d_f}\| + k_r \sum_{j=1, j\neq i}^{N} \exp\left(\frac{-\frac{1}{2}\|\psi\|^2}{r_s^2}\right)\|\psi\| \tag{18.16}$$

where $\psi = \hat{e}_p^i - \hat{e}_p^j = \left(x^i - x^j\right) - \left(d_p^i - d_p^j\right)$. Notice that $\frac{1}{N}\sum_{j=1}^{N}\|d_p^j\| \leq D_p$ since $\|d_p^j\| \leq D_p$. Also

$$d_p^i - \frac{1}{N}\sum_{j=1}^{N} d_p^j \leq \|d_p^i\| + \frac{1}{N}\|\sum_{j=1}^{N} d_p^j\| \leq \|d_p^i\| + \frac{1}{N}\sum_{j=1}^{N}\|d_p^j\|$$

$\|d_p^i - \bar{d_p}\| \leq 2D_p$, $\|d_v^i - \bar{d_v}\| \leq 2D_v$, and $\|d_f^i - \bar{d_f}\| \leq 2D_f$.

For the last term in Equation (18.16), note that as $\|x^i - x^j\|$ becomes large for all i and j, the agents are all far from each other and the repulsion term goes to zero. Also, the term due to the repulsion is bounded with a unique maximum point. To find this point, note that

$$\frac{\partial}{\partial\|\psi\|}\left(\|\psi\|\exp\left(\frac{-\frac{1}{2}\|\psi\|^2}{r_s^2}\right)\right) = \exp\left(\frac{-\frac{1}{2}\|\psi\|^2}{r_s^2}\right) - \frac{\|\psi\|^2}{r_s^2}\exp\left(\frac{-\frac{1}{2}\|\psi\|^2}{r_s^2}\right)$$

The maximum point occurs at a point such that

$$1 - \frac{\|\psi\|^2}{r_s^2} = 0$$

or when $\|\psi\| = r_s$. Hence, we have

$$\begin{aligned}\|g^i(E)\| &\leq 2k_a\left(D_p + D_v\right) + 2k_f D_f + k_r \sum_{j=1,j\neq i}^{N} \exp\left(-\frac{1}{2}\right) r_s \\ &= 2k_a\left(D_p + D_v\right) + 2k_f D_f + k_r r_s (N-1)\exp\left(-\frac{1}{2}\right) = \beta\end{aligned}$$

If you substitute this value for β into Equation (18.15), you get the set Ω_b that ultimately all the trajectories will end up in.

Cohesive Social Foraging with No Noise: Optimization Perspective

When there is no noise, tighter bounds and stronger results can be obtained. First, we can eliminate the effect of P via $\lambda_{max}(P)$ on the bound for the no-noise case. Assume there is no sensor noise so $D_p = D_v = D_f = 0$. Choose

$$\begin{aligned}u^i &= -M_i k_a e_p^i - M_i k_a e_v^i - M_i k_v v^i \\ &\quad + M_i k_r \left(B^\top P^{-1} B\right) \sum_{j=1,j\neq i}^{N} \exp\left(\frac{-\frac{1}{2}\|e_p^i - e_p^j\|^2}{r_s^2}\right)\left(e_p^i - e_p^j\right) \\ &\quad - M_i k_f R \end{aligned} \tag{18.17}$$

where $P = P^\top$, $P > 0$ was defined earlier, so P^{-1} exists. Also

$$\dot{V}_i \leq -\lambda_{min}(Q)\left\|E^i\right\|^2$$

$$
\begin{aligned}
&+ \quad 2E^{i^\top} PB \left(k_r B^\top P^{-1} B \sum_{j=1, j\neq i}^{N} \exp\left(\frac{-\frac{1}{2}\|e_p^i - e_p^j\|^2}{r_s^2} \right) (e_p^i - e_p^j) \right) \\
&= \quad -\lambda_{min}(Q) \left\|E^i\right\|^2 + 2k_r E^{i^\top} B \sum_{j=1, j\neq i}^{N} exp\left(\frac{-\frac{1}{2}\|e_p^i - e_p^j\|^2}{r_s^2} \right) (e_p^i - e_p^j) \\
&\leq \quad -\lambda_{min}(Q) \left\|E^i\right\|^2 + 2k_r \left\|E^i\right\| (N-1) \exp\left(-\frac{1}{2}\right) r_s
\end{aligned}
$$

So $\dot{V}_i < 0$ if $\left\|E^i\right\| > \frac{2k_r(N-1)r_s}{\lambda_{min}(Q)} \exp\left(-\frac{1}{2}\right)$. Let

$$
\Omega_b' = \left\{ E : \ \left\|E^i\right\| \leq \frac{2k_r r_s (N-1)}{\lambda_{min}(Q)} \exp\left(-\frac{1}{2}\right), \ i = 1, 2, \ldots, N \right\}
$$

Next, note that in the set Ω_b, we have bounded e_p^i and e_v^i but we are not guaranteed that $e_v^i \to 0$ for any i. Achieving $e_v^i \to 0$ for all i would be a desirable property, since this represents that $v^i = \bar{v}$ for all i so that the group will all move cohesively in the same direction. To study this, consider Ω_b', and consider a Lyapunov function $V^o(E) = \sum_{i=1}^N V_i^o\left(E^i\right)$ with

$$
V_i^o\left(E^i\right) = \frac{1}{2} k_a e_p^{i^\top} e_p^i + \frac{1}{2} e_v^{i^\top} e_v^i + k_r r_s^2 \sum_{j=1, j\neq i}^{N} \exp\left(\frac{-\frac{1}{2}\|e_p^i - e_p^j\|^2}{r_s^2} \right)
$$

Note that this Lyapunov function satisfies $V_i^o\left(E^i\right) \geq 0$. You should view the objective of the agents as being that of *minimizing* this Lyapunov function; they try to minimize the distance to the center of the swarm, match the average velocity of the group, and minimize the repulsion effect (to do that, the agents move away from each other). We have

$$
\begin{aligned}
\nabla_{e_p^i} V_i^o &= k_a e_p^i - k_r \sum_{j=1, j\neq i}^{N} \exp\left(\frac{-\frac{1}{2}\|e_p^i - e_p^j\|^2}{r_s^2} \right) (e_p^i - e_p^j) \\
\nabla_{e_v^i} V_i^o &= e_v^i
\end{aligned}
$$

so

$$
\begin{aligned}
&\dot{V}_i^o = \left(\nabla V_i^o\left(E^i\right)\right)^\top \dot{E}_i \\
&= \quad k_a e_p^{i^\top} e_v^i - k_r \sum_{j=1, j\neq i}^{N} \exp\left(\frac{-\frac{1}{2}\|e_p^i - e_p^j\|^2}{r_s^2} \right) (e_p^i - e_p^j)^\top e_v^i \\
&+ \quad e_v^{i^\top} \left(-k_a e_p^i - k_a e_v^i - k_v e_v^i + k_r \sum_{j=1, j\neq i}^{N} \exp\left(\frac{-\frac{1}{2}\|e_p^i - e_p^j\|^2}{r_s^2} \right) (e_p^i - e_p^j) \right) \\
&= \quad -(1 + k_v)\, e_v^{i^\top} e_v^i
\end{aligned}
$$

Hence,

$$\dot{V}^o = -(1+k_v)\sum_{i=1}^{N} \|e_v^i\|^2 \leq 0$$

on $E \in \Omega$ for a compact set Ω. Choose Ω so it is positively invariant, which is clearly possible, and so $\Omega_e \in \Omega$ where

When noise is not present, ultimately the velocity of all agents becomes the same and the swarm moves directly down the resource profile.

$$\Omega_e = \{E: \dot{V}^o(E) = 0\} = \{E: e_v^i = 0, \ i = 1, 2, \ldots, N\}$$

From LaSalle's Invariance Principle, we know that if $E(0) \in \Omega$ then $E(t)$ will converge to the largest invariant subset of Ω_e. Hence,

$$e_v^i(t) \to 0$$

as $t \to \infty$. When $R = 0$ (no resource profile effect), $\bar{v}(t) \to 0$ and hence $v^i(t) \to 0$ as $t \to \infty$ for all i (i.e., ultimately no oscillations in the average velocity). If $R \neq 0$, then $\dot{\bar{v}} = -k_v\bar{v} - k_f R$ and $\bar{v}(t) \to -\frac{k_f}{k_v}R$ as $t \to \infty$, and thus, $v^i(t) \to -\frac{k_f}{k_v}R$ for all i as $t \to \infty$, i.e, the group follows the profile. These results help to highlight the effects of the noise. The noise makes it so that the swarm may not follow the profile as well (but makes following it possible when it may not be possible for a single individual), and it destroys tight cohesion characterized by getting $e_v^i(t) \to 0$. Next, we will study additional characteristics of swarms by analyzing the results of this and the previous sections in more detail.

18.4.4 Cohesion Characteristics and Swarm Dynamics

Here, we will study the effects of various parameters on cohesion characteristics and then provide a simulation to provide insight into swarm dynamics, especially transient behavior. Suppose that u^i is given by Equation (18.6).

Effects of Parameters on Swarm Size

The size of Ω_b in Equation (18.15), which we denote by $|\Omega_b|$, is directly a function of several known parameters. Consider the following cases:

Attraction gains should be set high to get tight swarm cohesion, but not too high or the attraction will amplify the noise and swarm compactness can degrade.

- *No sensing errors:* If there are no sensing errors, i.e., $D_p = D_v = D_f = 0$, and if $Q = k_a I$, we obtain

  $$\Omega_b = \left\{E: \left\|E^i\right\| \leq \frac{2k_r r_s(N-1)}{k_a}\lambda_{max}(P)\exp\left(-\frac{1}{2}\right), \ i = 1, 2, \ldots, N\right\}$$

  If N, k_r, and r_s are fixed, then if k_a increases from zero, we get $\frac{\lambda_{max}(P)}{k_a} \to 1$ from above and we get a decrease in $|\Omega_b|$, but only up to a certain point.

- *Sensing errors:* There are several characteristics of interest:
  - *Noise cancellations:* In the special situation when $d_p^i = d_p^j$, $d_v^i = d_v^j$, and $d_f^i = d_f^j$ for all i and j, then $d_p^i - \bar{d}_p = d_v^i - \bar{d}_v = d_f^i - \bar{d}_f$ for all i and it is as if there is no error and $|\Omega_b|$ is smaller.

- *Repel effects:* For fixed values of N, k_a, and k_r if we increase r_s, each agent has a larger region from which it will repel its neighbors so $|\Omega_b|$ is larger. For fixed k_r, k_a, and r_s, if we let $N \to \infty$, then $|\Omega_b| \to \infty$ as we expect due to the repulsion. (The bound is conservative since it depends on the special case of all agents being aligned on a line so there are $N-1$ inter-agent distances that sum to make the bound large.)
- *Attraction can amplify noise:* Let $D_s = D_p + D_v$ and J quantify the size of the set Ω_b. Next, we study the special case of choosing $Q = k_a I$. Fix all values of the parameters except k_a and D_s. A plot of J versus k_a and D_s is shown in Figure 18.14, where the locus of points are those values of k_a that minimize J for each given value of D_s. This plot shows that if there is a set magnitude of the noise,

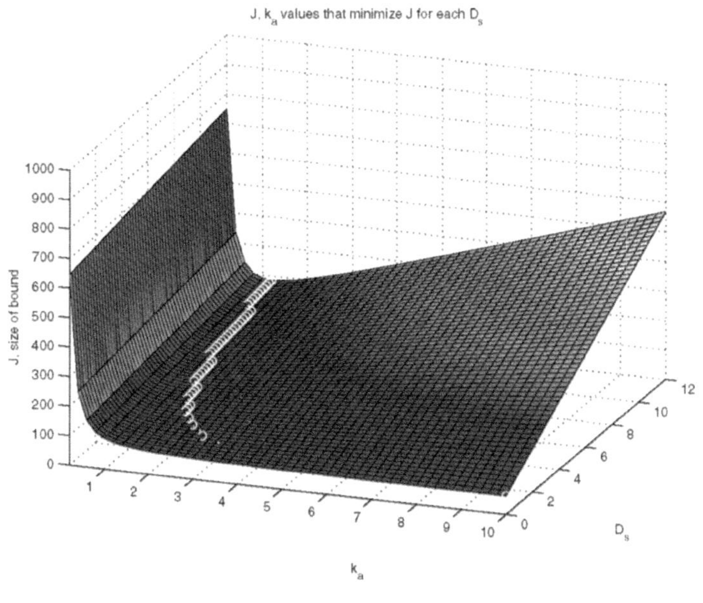

Figure 18.14: Values of k_a that minimize J for given values of noise magnitude D_s.

then to get the best cohesiveness (smallest Ω_b), k_a should not be too small (or it would not hold the group together), but also not too large since then the noise is also amplified by the attraction gain and poor cohesion results. Could you interpret the plot as a type of fitness function, and then come to conclusions about the evolution of the agent parameters?

- *Swarm size N:* In some situations, when N is very large, $\bar{d_p} \approx \bar{d_v} \approx \bar{d_f} \approx 0$ and there is no biasing of sensing errors so that on average they are zero and this reduces the above bound on $\|g^i(E)\|$.

For additional analysis of swarm properties, see the "For Further Study" section at the end of this part.

Swarm Dynamics: Individual and Group

Here, we will simply simulate a swarm for the no noise case in order to provide some insights into the dynamics, especially the transient behavior. We will in particular seek to study the individual motions and how they collectively move as a group to achieve cohesion, and the dynamics of the motion of the group. We use linear attraction, velocity damping, the Gaussian form for the repulsion term, and the resource profile with the shape of a plane. The parameters for the simulation are $N = 50$, $k_a = 1$, $k_r = 10$, $r_s^2 = 0.1$, $k_v = k_f = 0.1$, $R = [1, 2, 3]^\top$, and $r_o = 0$. Simulating for 10 sec. we get the agent trajectories in Figure 18.15.

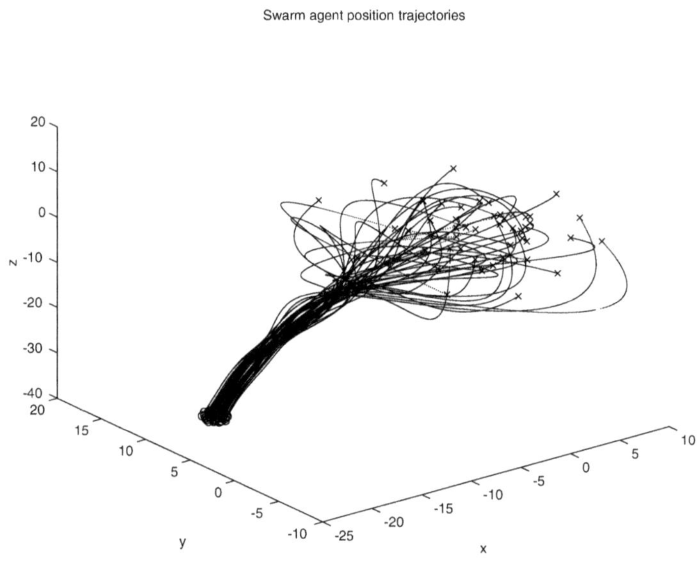

Figure 18.15: Agent trajectories in a swarm.

Notice that initially the agents move to achieve cohesion. By the end of the simulation, the agents are moving at the same velocity and as a group, with some constant inter-agent spacing between each pair of agents. Moreover, due to the choice of initial conditions (actually random), the group achieves a certain level of aggregation relatively quickly, then the *group* moves to follow the foraging profile. Of course, some orientation towards following the profile is achieved during the initial aggregation period, but in this simulation, significant reorientation towards following the resource profile occurs *after* there is tight aggregation.

18.5 Design Example: Robot Swarms

This problem is an extension of the one in Section 6.2, where we sought to develop a guidance algorithm for moving a robot from an initial position to a goal position in a factory while avoiding collisions with obstacles. Here, we consider one approach to guiding multiple robots through the same maze as was studied there, using the swarm approach discussed in the last section.

18.5.1 Robot Swarm Formulation

A robot swarm is simply a group of robots that move in some cohesive fashion in order to perform some task. Here, the task is simply to get the group to a certain location in a factory, so that they can perform some activity together there. In moving them from one location to another, the key challenge is to have them all avoid certain obstacles that appear in their paths. Here, we use the same maze as described in Section 6.2 with the same final goal position. We use the same "obstacle function" defined there to represent the positions of the obstacles and the same "goal function" to represent where we want the group to go. To solve this problem, we view the combined obstacle/goal functions as a cost function just as we did in the chapter on planning. Here, however, we take the view that the group of robots is engaged in social foraging over the cost function, which we think of as a nutrient profile. The obstacles are thought of as regions with a noxious substance, and increasing amounts of food are obtained by moving towards the goal position. We use a different model for the robot from the one considered in Section 6.2. Here, we use the model of a swarm agent from the last section with all the masses of the robots the same, at $M_i = 1$.

Cooperative guidance strategies based on swarms are useful for groups of robots.

We assume that each vehicle knows its own velocity. Here, we will first consider the case where each robot perfectly knows the swarm center and swarm average velocity. For some types of robotic systems, this would require each robot to know the positions and velocities of all the other robots so it can compute the swarm center and average velocity, but for others there may be a sensor which could directly sense these. We will also consider the case where there is noise in sensing, of the type described in Design Problem 18.7, so that each robot does not perfectly know the swarm center and average velocity.

18.5.2 Performance in Obstacle Avoidance and Noise Effects

Suppose that there are 30 robots in the swarm. Let $k_p = 1$, $k_v = 0.1$, $k_f = 0.1$, $k_r = 10$, and $r_s^2 = 1$. We pick some random initial locations and velocities, but within some fixed ranges. We pick $w_1 = 120$ and $w_2 = 0.1$ after some tuning. Also, to keep the simulation simple, we use an Euler approximation and simulate the swarm as a discrete-time system. For this, we use a sampling period of 0.01 and simulate for 80 sec. Clearly, due to the use of a different form of a "resource profile" and the discrete-time approach, the stability results of the previous section do not apply directly; however, the simulations will illustrate that the basic ideas still do apply.

The noise-free case is shown in Figure 18.16. This shows the position trajectories of the 30 robots in moving from an initial position to the point where they are around the goal position. As time goes on, the robots slow down and stop when they are near the goal position. If you study the trajectories you will see that they successfully avoid the obstacles and still maintain a cohesive group. The ultimate size of the group is set by the attraction and repulsion parameters, and the shape of the goal (and obstacle) functions (e.g., if they resulted in a very steep slope near the goal position, then the group of robots would be packed tighter in their final position).

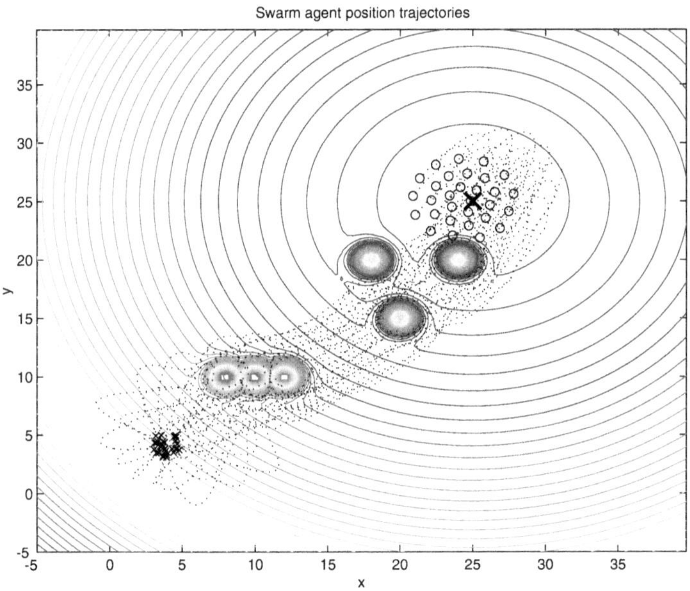

Figure 18.16: Robot swarm, robot position trajectories.

Figure 18.17 shows what happens if the robots sense with noise of the type discussed in Design Problem 18.7. Here, due to the sensing errors, there are actually collisions with some of the obstacles by some of the robots (actually due to noise in use of the obstacle and goal functions in the simulation). Moreover, the positions and velocities of the robots oscillate near the goal position. Clearly, noise can adversely affect cohesion and orderly operation of a group of robots.

18.5.3 Additional Robot Swarm Design Challenges

As outlined in Section 6.2.4, there are many additional challenges that can arise in autonomous vehicle guidance, and these are compounded in the case where we want to guide multiple autonomous robots. The challenges there included

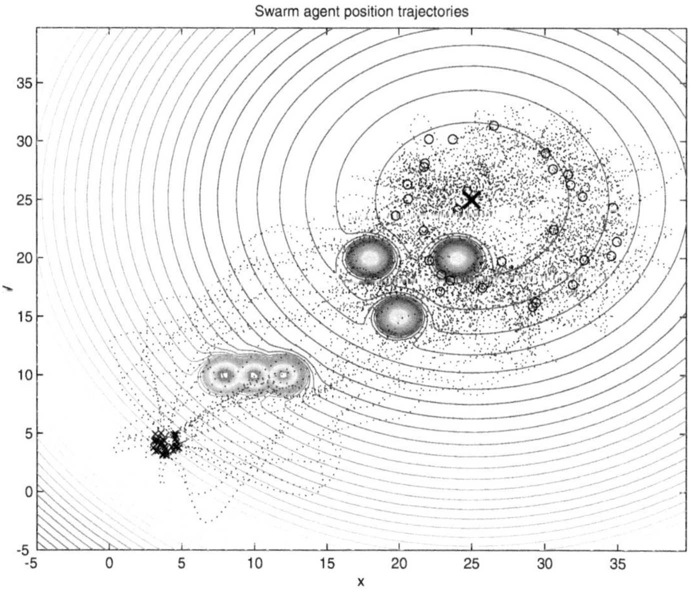

Figure 18.17: Robot swarm, robot position trajectories, noise case.

complex mazes that led to dead ends and circular loops, mobile obstacles, and uncertainty. Dead ends and circular loops that one, or some subset, of the robots get stuck in may be solved in the swarm case by having some robots that are moving in the right direction "pull" them out of the problem (via appropriate desire to stay grouped). However, it can also be that the ones that get stuck in a dead end or circular loop, hold back the others that are headed in the right direction. The problems with mobile obstacles are multiplied for the swarm case, since now the whole group must avoid such obstacles. It may be easier to avoid mobile obstacles if the robots cooperate by telling each other when they sense a mobile obstacle, but the very existence of a swarm creates more problems with mobile obstacles, since each robot in the group can be thought of as a mobile obstacle to avoid (in the last section, we considered point-sized vehicles and hence, avoided the whole issue of inter-robot collisions).

Uncertainty makes each of these issues more challenging, and the swarm can create more problems with uncertainty. For instance, suppose that the group of robots is connected via a communication network, with its "topology" specifying which robots are connected. If this communication network is less than perfect (e.g., via bandwidth constraints, delays, or topology changes), then clearly the maintenance of cohesive robot swarm behavior will be even more challenging. Some of these issues will be discussed in more detail in the "Challenge Problem" in the next chapter.

18.6 Exercises and Design Problems

Exercise 18.1 (*E. coli* Swarm Foraging):

(a) Illustrate for the simulations in the chapter the effects of changing N_c and the parameters of the cell-to-cell attractant function. Explain how to make poor choices for these parameters (e.g,. ones that result in swarming, but little optimization progress on the nutrient concentration profile) and good choices (e.g., ones where cells pull each other toward minima), and illustrate in each case the resulting behavior of the algorithm.

(b) Suppose that you use Rosenbrock's function in Equation (15.9) as a cost function. Make appropriate algorithm parameter choices and illustrate the performance of the algorithm for this function.

Exercise 18.2 (Foraging of Square-Shaped Bacteria):

(a) Based on the brief description the foraging behavior of square bacteria, develop an optimization algorithm that models its foraging behavior. To do this, specify how each bacterium moves forward and backward and changes directions. Ignore reproduction, elimination, and dispersal. Write a simulation of the algorithm.

(b) Use the algorithm to perform optimization over the multiple-extremum function in Figure 18.10. Show plots to illustrate the characteristics of the algorithm.

Exercise 18.3 (ODE Model of Dynamic Labor Force Allocation for Bees): In [89] the authors develop a nonlinear differential equation model of the key functional aspects of dynamic labor force allocation of honey bees and validate the model against T. Seeley and his colleagues' earlier experimental work (see [456]).

(a) Simulate this model, compare to Seeley's results, and explain the key aspects of the model and its properties.

(b) In what ways is the model inaccurate? What does it not represent?

(c) Can you modify the ODEs with some nonlinearities so that the simulations will fit Seeley's data better?

Exercise 18.4 (Effects of Parameters on Swarm Dynamics): In this problem, you will simulate the swarm in Section 18.4.4 for appropriate parameter choices to illustrate different behaviors (this is a noise-free case). Start with the parameter values given there and tune only the parameter indicated in each part.

(a) Pick a value for k_a that will result in a tighter packed ultimate swarm (i.e., a smaller ultimate swarm size).

(b) Pick a value for k_r that will result in a more loosely packed ultimate swarm (i.e., a larger ultimate swarm size).

(c) Repeat (b), but for the r_s parameter.

(d) Repeat (a)–(c) but for the case where there is noise in sensing. Hint: Simulate the noise by generating it as the output of a chaotic system (e.g., via Duffing's equation) so that it satisfies the smoothness requirement on the sensing errors.

Exercise 18.5 (Hard Repulsion, Collision-Avoidance, and Swarm Dynamics): Replace the Gaussian-type repulsion term in Section 18.4 with a repulsion term that will provide a "hard" constraint to avoid collisions between agents. Add velocity damping, a foraging plane, and an appropriate attraction term. Simulate the swarm and demonstrate via plots that there are no collisions both during the transient and in the steady-state. Investigate the effects of all parameters of the controller.

Design Problem 18.1 (Extensions to *E. coli* Swarm Foraging): Study how to find the minimum of the function in Figure 18.10, but model, code, and illustrate the performance and algorithmic characteristics of the algorithm for the following cases.

(a) Change the nutrient concentration to represent consumption of nutrients.

(b) Model Brownian effects so that the bacteria cannot swim straight.

(c) Make the tumble and run lengths exponentially distributed random variables consistent with what has been found in nature.

(d) Make run-length decisions based on the past 4 sec. of concentrations.

(e) Develop a fully asynchronous model.

(f) Allow a time-varying population size.

(g) A more biologically accurate model of the swarming behavior of certain bacteria is given in [544]. Simulate the partial differential equation model given there.

(h) Develop other criteria by which bacteria split. Add effects of conjugation and other evolutionary characteristics (e.g., evolve $C(i)$, N_s, and N_c).

Design Problem 18.2 (*M. xanthus* Swarm Foraging)*: Review the current literature and develop a cellular automaton model of the foraging of *M. xanthus*. Include the modeling of fruiting bodies and the full lifecycle of the bacteria.

Design Problem 18.3 (Robot Swarms): In this problem, you will investigate aspects of the robot swarm problem studied in Section 18.5.

(a) Explain via simulations, the effects of k_p, k_v, k_r, and r_s on the transient and ultimate (as $t \to \infty$) swarm behavior (i.e., show in simulation the effects of changing each of these parameters and, for example, explain whether the swarm size (diameter) increases or decreases).

(b) Design a new control that is based on a "hard repel" and demonstrate via simulations that you can design it so that there will be no inter-robot collisions and no collisions with obstacles. This may require retuning the parameters of the simulation.

(c) Repeat (a) and (b) except simulate the system as a continuous-time system. Hint: This will require using the gradient of the cost function that holds the goal and obstacle functions.

(d) Repeat (a), (b), and (c) but for the case where noise of the type described in Design Problem 18.7 is used.

(e) Define a communication topology and an "intelligent" strategy (e.g., one that uses planning, learning, and/or attention) for each of the robots to coordinate their actions to solve the problem. Simulate to evaluate performance.

Design Problem 18.4 (Social Foraging Strategies for Indirect/Direct Adaptive Control): In this problem, you will study the development of indirect and direct adaptive controllers for the process control problem studied in Sections 12.4 and 12.6, but where foraging algorithms are used for the optimization method. First, read the part in the chapter on the genetic adaptive control strategies in Section 16.5. Basically, you will simply replace the online genetic algorithm with a foraging algorithm. This changes how we view the operation of the algorithm. From a foraging perspective, we view θ^i as the location of the ith forager in its environment. In a foraging method, we will move the position of the forager θ^i so as to minimize $J(\theta^i)$. The particular manner used to adjust the $\theta^i(k-1)$ to find $\theta^i(k)$ will depend on the choice of the foraging algorithm steps (e.g., will it involve swarming, or other communications between individuals that tell each other when they are doing well?). In an indirect adaptive control strategy, we view foraging as searching for good model information. If a foraging strategy is used, we view $\theta(k)$ in Equation (16.2) as the forager who has found the best model information. With a foraging strategy, we could view the fixed-position members that were discussed in the chapter as "information centers," if the foragers were endowed with communication capabilities. If such centers have good model information, they will tend to attract foragers. Clearly, you can specify a direct adaptive control strategy based on social foraging in a similar way to how we did for direct genetic adaptive control.

Next, we show how to develop an indirect adaptive controller based on foraging for the tank problem. The problem given after that will focus on various extensions of the method.

Our forager's position in one dimension is given by θ_α and in the other dimension by θ_β so our i^{th} forager's position is $\theta^i = [\theta^i_\alpha, \theta^i_\beta]^\top$, $i = 1, 2, \ldots, S$. We choose $S = 10$ as the population size. The foraging strategy we employ is the one from Section 18.3, which is based on *E. coli* chemotaxis, but without swarming, elimination-dispersal, and reproduction. Hence, we only use the chemotactic hill-climbing strategy to adjust the parameters. At each time step, we take one foraging step, which for our foraging strategy means that we use either one tumble-tumble step or part of a "run." In this way, the foraging occurs while the control system operates with foraging (searching) for parameters occurring at each time step. For instance, if over one time step the cost did not decrease for an individual, then there is a tumble, and to do this, we generate a random direction and update the parameters (location of the forager) in that direction. If, however, the cost improved from the last step, then another step in the same direction taken last time is made (provided not too many steps in that direction have already been made). In this case, the forager is on a "run" in a good direction, down the cost function.

The step size $C(i, k)$ is set to be 0.05 for all bacteria for all times. The maximum number of steps along a good direction is $N_s = 4$, and $\theta^i_\alpha(0) = 2$, $\theta^i_\beta(0) = 0.5$, $i = 1, 2, \ldots, S$. We use the cost evaluation procedure with $J_s(\theta^i(k-1), N)$ in Equation (16.3) with $N = 100$. To keep things simple, we will simply use $\theta_\alpha(k) = \theta_\alpha(0)$ and $\theta_\beta(k) = \theta_\beta(0)$ for $k = 1, 2, \ldots, N+1$ (i.e., we wait till we have computed all the values needed for the cost evaluation before we start the adaptation procedure). Clearly, other initialization choices are possible. For instance, in the beginning you could use an expanding length window of data to evaluate the quality of the estimators.

The performance of the closed-loop system is illustrated in Figure 18.18, where we see that, after an initial transient period that results in part due to the poor initialization of estimators and the controller start-up method, we get reasonably good tracking of the reference input. Next, Figure 18.19 shows that the estimate of the tank liquid level is quite good, even though at times the individual estimates of the nonlinearities are not.

To further illustrate some properties of the adaptive controller, see Figure 18.20 where we plot the cost of the best individual in the population (the one that leads to the specification of the controller) and the index i of the best individual in the population for every time step. First, note that early in the simulation, cost is zero due to how we start up the controller. Then, when we start the controller at $t = 10$ sec., the cost jumps to a relatively high value; this represents that we have a poor initialization for the population. After some time, however, the foraging strategy is somewhat successful at adjusting the population members so that the estimation error decreases and hence, the best cost decreases. Note, however, that the cost does not *always* decrease over time. It can also increase and one cause of this can be the change in the reference input. Next, note that in the

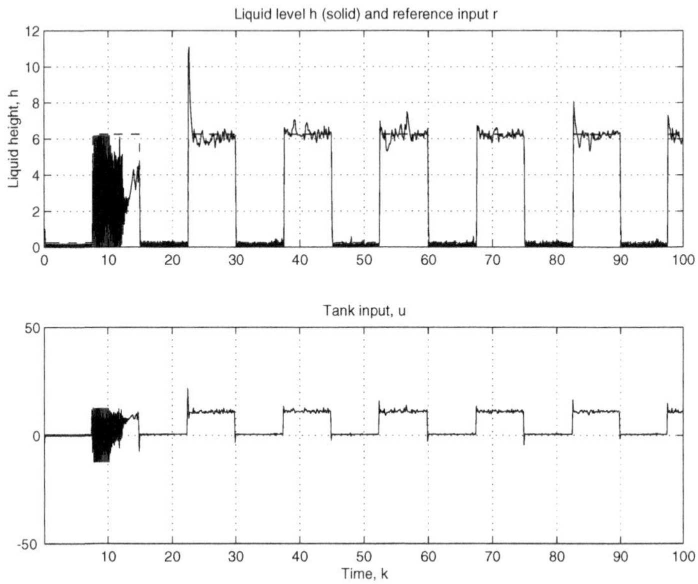

Figure 18.18: Indirect adaptive controller based on foraging.

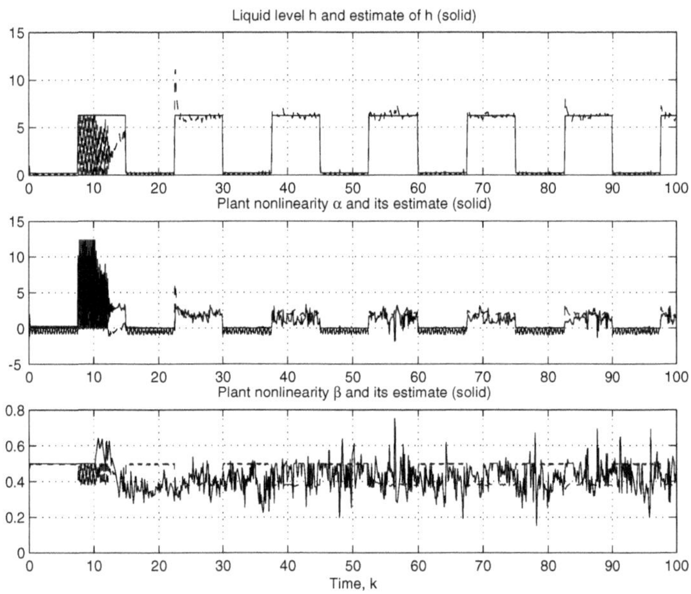

Figure 18.19: Indirect adaptive controller based on foraging, estimates of liquid level, and nonlinearities.

bottom plot we show the index of the best individual in the population at each step. Notice that there are some short stretches of time where the best individual does not change; however, there is a significant amount of switching between different members of the population that provide better estimates at different times.

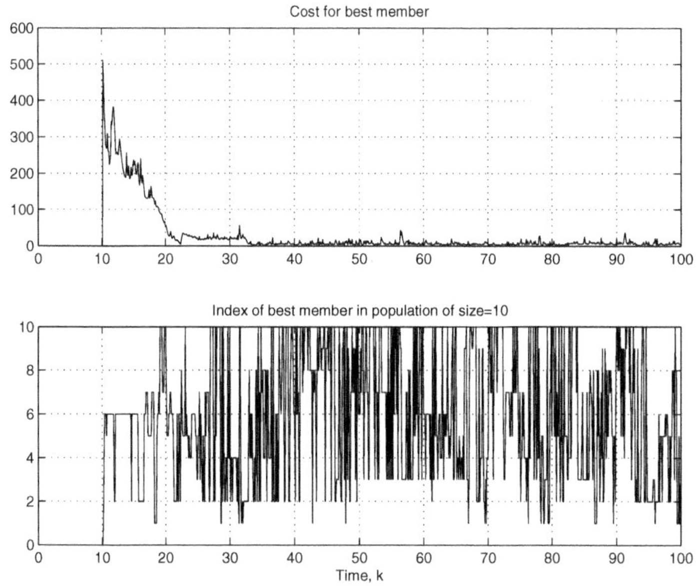

Figure 18.20: Indirect adaptive controller based on foraging, best cost, and index of best individual in the population.

(a) Develop an indirect adaptive controller for the process control problem using a foraging strategy for parameter adjustments. You may build your approach above where an *E. coli* chemotactic foraging strategy was simulated, but you must add some additional feature to the foraging algorithm that represents a situation in nature, where foragers can communicate to each other how well they are doing and subsequently, use that information to improve their foraging success. Regardless of which approach you use, verify the operation of your controller in the same manner as was done above. Study the effect of the choice of the reference input on the ability of the approximator mappings to match the underlying unknown nonlinearities. Provide plots to illustrate the quality of the matching as was done in the chapter.

(b) Develop a direct adaptive controller for the process control problem that is based on a foraging strategy that includes some type of communication of information between foragers, and subsequent use of that information to improve foraging success. Regardless of which foraging approach you use, verify the operation of your controller in the same manner as was done above. Study the effect of the choice of the reference input on the ability of the approximator mapping to match the "ideal" controller nonlinearity discussed in the chapter. Provide plots to illustrate the quality of the matching as was done in the chapter.

(c) Compare the performance of the indirect and direct methods and discuss. Evaluate the value of using fixed portions of the population of models or controllers. Study the effects of good population initialization.

Design Problem 18.5 (Foraging in Colombia)⋆: This problem continues with Design Problem 14.2. Suppose that you have a group of foragers that find more food at higher elevations. Design a foraging algorithm that successfully finds the highest point on the topographical map. Compare its performance to that of the genetic algorithm. Next, suppose that you have a forager who is searching for coffee beans in Colombia. Formulate a foraging landscape and develop a simulation that illustrates the success of foragers across that landscape.

Design Problem 18.6 (Foraging for Approximator Tuning)⋆: We can think of learning as a process of foraging for information. Can a foraging algorithm be used to tune an approximator structure? Is this a good idea? Why or why not? It is generally not possible to compute a gradient with respect to a structure change in an approximator (e.g., the change in the output with respect to the change in the number of neurons in a hidden layer). Can a foraging algorithm be used to tune the structure of an approximator? How? Develop an example and method, and test its performance relative to standard ones.

Design Problem 18.7 (Stable Social Foraging Swarms)⋆: Use the same model as in the chapter for a swarm of agents. Suppose that, however, $\|d_p^i\| \leq D_{p_1} \left\|E^i\right\| + D_{p2}$, $\|d_v^i\| \leq D_{v_1} \left\|E^i\right\| + D_{v_2}$, and $\|d_f^i\| \leq D_f$ where D_{p_1}, D_{p2}, D_{v_1} and D_{v_2} are known positive constants.

(a) Find conditions under which you are guaranteed to have the trajectories of the swarm uniformly ultimately bounded.

(b) Find bounds on the size of the trajectories of the error dynamics in terms of the parameters of the problem.

(c) Simulate the noise as the output of a chaotic system (e.g., via Duffing's equation), but one which still satisfies the above bounds. Study in simulation the effects of the parameters of the problem (e.g., k_a, k_r, and the noise bounds). Show simulations of the agent trajectories to illustrate the transient behavior of the swarm. To see how to do this for a *discrete-time* approximation to the continuous-time system, see Section 18.5.2.

Design Problem 18.8 (Modeling and Simulation of Honey Bee Clusters and In-Transit Swarms)*: Research the literature on how honey bees form clusters when a colony splits and then swarm together to a new nest site. Using the modeling ideas in this chapter, produce an ordinary differential equation model of both the clustering (aggregation process) and the in-transit swarm. Try to validate your model against the experimental results found in the literature.

Design Problem 18.9 (Predation and Aggregation)*: Read the paper entitled "Geometry for the Selfish Herd" by W.D. Hamilton in [240] that studies why an organism that is trying to protect itself will at times aggregate (group closely) with conspecifics (it also discusses other reasons for aggregation).

(a) Develop and simulate "jumping rules" (e.g., methods to generate Figure 1 in [240]). Characterize and analyze emergent aggregation behaviors by showing differences in aggregation behavior for different choices of rules. You could consider the case where all the "frogs" have the same rules, and the case where there are nonhomogeneous rules.

(b) Consider Hamilton's statement on p. 296 of [240]: "I know of no rule of jumping that can prevent them from aggregating." Why does Hamilton say this? Can you find a "rule" like the one he is talking about that can result in them not aggregating?

(c) Model and simulate a two-dimensional case, using some of the ideas in Section 3 of [240].

Design Problem 18.10 (Stable Dynamic Sphere Packing)*: There is a problem found in physics and biology of trying to pack objects on the surface of a sphere.

(a) Suppose that you seek to pack spherical swarm agents on a sphere. Suppose the agents are modeled as in the chapter via a double integrator but that a type of "hard-repel" term is used that makes them into spherical agents. Define the environment to be a sphere with the objective of each agent to be on the surface of the sphere by touching it at one point. Define appropriate attract-repel terms between agents so that only local interactions are allowed (i.e., so that every agent cannot be influenced by every other one, but only by its "neighbors"). Simulate the dynamic formation of packing on the sphere.

(b) Next, perhaps with the help of the literature in mathematics and physics, characterize the equilibria that represent the sphere being packed and show in simulation that these equilibria can be achieved. This is not a trivial problem. Are there equilibria that correspond to the agents moving about on the sphere?

(c) Provide conditions for Lyapunov stability or asymptotic stability of the equilibria specified in (b).

(d) Repeat (a)–(c) but for packing different shapes (e.g., an ellipsoid or a cube).

Design Problem 18.11 (Distributed Synchronization—From Fireflies to Simulations and Circuits)*: In this problem, you will investigate a biological phenomenon, simulate it, and then implement a circuit realization of it.

(a) Some types of fireflies exhibit a property, where they can synchronize their flashing. Investigate the literature on this and write a description of this phenomenon. Start with [496].

(b) Build a network of electronic fireflies according to the design in [194]. For this, build a line topology of at least four coupled oscillators.

(c) There are mathematical models of coupled nonlinear oscillators that can be used to represent the distributed sychronization. Study the one in [495], which has

$$\dot{\theta}_i = \omega_i + \frac{K}{N}\sum_{j=1}^{N} \sin(\theta_j - \theta_i)$$

for $j = 1, 2, \ldots, N$, where θ_i is the phase of oscillator i, ω_i is its natural frequency, and $K \geq 0$ is a coupling strength. Let $N = 10$, the $\omega_i = 1$, and $K = 1$. Develop a simulation of the N coupled oscillators and simulate for 20 sec. Plot the phase angles.

(d) Use the ideas in [495] to pick parameters and study "phase transitions" by illustrating these via a series of simulations.

(e) Explain the relationship between distributed synchronization and swarming (e.g., what is the "attraction/repulsion" function in the above distributed synchronization model?). Can you find conditions under which the distributed synchronization system is stable? What type of stability property does it possess?

(f) Model the system that you implemented in (b). Characterize and analyze the stability properties of the model. Compare to what is found in actual experiments.

Chapter 19

Competitive and Intelligent Foraging

Chapter Contents

Foraging sometimes involves simply going out and finding nutrients, but often a key aspect of foraging is competing for resources with other foragers. This can involve trying to find resources before another forager. In other cases, it could involve issues in fighting other animals for a resource, or perhaps one forager is the nutrient source for another forager and then there may be issues of pursuit and evasion. Clearly, as for the case of noncompetitive foraging, many aspects of the environment affect competitive foraging behavior and as always, evolution plays a fundamental role in foraging strategy design.

In this chapter, characteristics of competitive and cooperative foraging in nature are modeled using a game-theoretic perspective. The basic definitions and rules in game theory are introduced by showing how to set up and solve static finite two-player games (matrix games) for security and saddle point strategies. Then, it is shown how to define and solve bimatrix and infinite competitive (adversarial) games for Nash equilibria, minimax solutions, and Stackelberg solutions. Pareto-optimal solutions for cooperative games are discussed to help clarify the relationships between competitive and cooperative games. We apply the methods to the study of adversarial foraging games that involve resource competition, and then resource allocation when there is cooperation among players and hence, social foraging. Next, we define a model for a dynamic game, and discuss aspects of strategies and the information space.

This chapter complements the last one. We treat cooperative foraging here, but only to show another way to view social foraging, and with the intent of contrasting it with competitive foraging. This chapter is also important in that each forager views all the other foragers as part of its environment (plant) and tries to take actions (generate control inputs) in order to succeed. The other foragers are not simply treated as uncertainty (a disturbance). Certain aspects of the other foragers are modeled (e.g., the assumption that they are rational and hence, will try to maximize their own returns).

In the last section of this chapter, we introduce "intelligent" foraging which is one example of how other aspects of intelligence affect foraging (e.g., planning, attention, and learning). The last section is meant to challenge you to think about how to integrate all the methods of this book, and indeed, its main objective is to provide a "challenge problem" that you can help define and then solve (e.g., as a final project in a class).

19.1 Competition and Fighting in Nature

Competition for resources that are critical for survival affects many aspects of animal behavior and evolution. Some animals use poisoning for self-defense. Some evolve armor or grow spines. Some fight other animals for food resources or mates. Others simply seek to eliminate their competitors. Some recruit predators of their predators to obtain a defense. Some guard a resource-rich territory. Complex behaviors of groups of animals have evolved for self-defense. For instance, when a predator approaches a prey group, the normal response of the group of prey is to tighten and this can have an effect of maximizing mes-

Competition encountered in an environment affects organism evolution.

sage transmission speed by increasing an ability to detect low strength signals. Essentially, the size of the group of organisms can change based on actions of predators, so that a group-level distributed yet coordinated defensive action can be taken. There is a vast diversity of behaviors that have evolved to ensure that animals get the resources they need. Here, we will model and analyze some of these in a game-theoretic framework and hence, will refer to them as "foraging games."

19.1.1 Foraging Games

Limitations in availability of resources naturally leads to competition in foraging.

Here, we will model certain aspects of both competitive and cooperative behavior via a game theoretic approach. We are not concerned here with the dynamics and small time-scale strategies of one-on-one "battles" between two animals (e.g., how a lion may capture, handle, kill, and eat an antelope). Our focus will be on foraging for resources via the study of "foraging games." What is a foraging game? It is a game where players (animals, humans) seek and consume (capture, occupy) "resources" (e.g., food, prey, territory, shelter, mates). In a foraging game, the "players" from classical game theory are referred to as "foragers." The foragers in such games are typically mobile and decide "where to go and what to do" (e.g., go to a certain location and consume a certain resource type). There are many types of foraging games, which can arise in nature and we outline some characteristics of these as follows:

- *Environment and resource characteristics:* Different types of games arise depending on characteristics of the environment, resources, and foragers. Some resources may be higher priority than others and these priorities may depend on the forager and its current needs (e.g., its diet). Some resources are static in the sense that they do not move, while others may be able to move quickly. Other resources are only available for a limited amount of time, and then they naturally dissipate. The foraging environment (where the forager forages) affects resource availability and likelihood of the location of resources.

- *Forager characteristics:* Forager capabilities significantly affect the characteristics of the game. For example, how fast a forager can move, how efficient it can move in different environments, its consumption capabilities, fighting capabilities, and cognitive capabilities (e.g., memory, learning, and reasoning) all affect its ability to forage successfully. Moreover, foraging games are sometimes dominated by whether there is competition for resources between foragers, or cooperation between multiple foragers to obtain resources ("social foraging"). Such cooperation requires communication or use of shared information in some way and hence, demands that a forager has certain capabilities.

Foragers may compete for food, and one forager may be food (prey) for the other.

- *Competitive foraging:* A competitive foraging game is one in which the foragers compete with each other in an adversarial relationship in order to obtain resources (e.g., if one forager gets the resource, it is no longer available for the other forager). In the case where one forager is the predator and the other is its prey, we have a special foraging game that may be a "pursuit evasion game." We can then think of one forager as being a resource for another and one tries to capture or consume the other. It can also be the case that each forager is a resource for the other. Then, each tries to achieve a competitive advantage to capture and consume the other.

- *Cooperative foraging is social foraging:* A cooperative foraging game is one where foragers may share information to try to optimize resource acquisition for the group (or variance reduction in resource acquisition rates); the idea is that by working together, every forager does better. Cooperative foraging is the essence of "social foraging." A key concept in such cooperation is the "allocation" (distribution of a pattern) of activities that results in an allocation of resources to each group member (forager) to ensure that the *group* does as good as possible. This may involve, at times, sacrifices by one group member to increase the group's performance; however, over the long run it should be better, perhaps in an evolutionary sense, for the group to cooperate rather than compete. A group of cooperative foragers may compete with another group of cooperative foragers so that both elements of cooperative and competitive foraging are present. Indeed, there are often some elements of competition even in groups of cooperative foragers.

Cooperative and competitive foraging coexist in some foraging scenarios.

19.1.2 Intelligent Foragers

Attention, learning, and planning may help an individual forager increase energy intake per unit time spent foraging. Paying attention to the proper aspects of an environment can help the forager find nutrients and avoid predators. Learning helps you not to go back to places where you cannot find food (since it may not be likely that some food moved there). It helps the organism to remember where it has not yet gone for food (related to attention), where a past food source is, or what types of environmental "signs" typically indicate the presence of a good food source. Such learned information can be used by a planning system to further improve foraging performance. Planning involves reasoning over learned information, setting priorities, and optimizing choices locally (e.g., it allows the organism to actually perform an optimization of E/T over a short period of time, using less-than-certain learned information). Via learning and planning together, the forager can directly try to make locally optimal decisions, at least for the learned information.

"Intelligent" foraging involves the use of higher-level cognitive functions such as planning, attention, and learning.

Another relevant topic from planning theory, is the concept of "retrospective" and "prospective" coding, where it is thought that we recall where we have been and know where we have not been and use this to plan our activities. There is some evidence that when we begin planning, we use a retrospective encoding and then at some point, switch to a prospective encoding since then, we do not have to hold as much in short-term memory. Basically, early in the process, we may be taking actions that help us to learn about the environment, while later, after we have learned a significant amount about the environment, it may be best to use the learned information to decide what actions to take, and to take actions that may not be directed towards learning more, but towards other objectives.

Some learning theorists have thought of learning as foraging for information that is then stored in "cognitive maps." For example, they may think of operant conditioning as foraging for reinforcement. Particularly relevant is the concept of "sign tracking" that has been used by learning theorists to explain foraging, and this approach motivates the use of smooth cost functions that represent where food is, and where risks are. The theory of "behavioral regulation" proposes that there are homeostatic mechanisms for behavior, where an organism tries to choose an optimal distribution of activities for survival. It is thought that if the balance of activities is upset, behavior is assumed to change to correct the deviation from the homeostatic level (this is the "behavioral bliss point approach").

19.1.3 Evolution and Foraging

Foraging strategies are fine-tuned by evolution since more successful foragers tend to have more offspring that possess aspects of their successful foraging strategies. In a sense, evolution seeks to perform a robust optimization of a forager's strategy in the face of the following constraints:

Foraging strategies evolve, and there is a complex dynamic interaction between environmental and forager changes.

- Environment and resource characteristics (e.g., a typical environment where the forager lives, along with a typical spread of the resources).
- Forager capabilities (e.g., motor and cognitive).

The environment and resources can change. Indeed, in a predator-prey situation, both the predator and the prey are evolving and hence, "coevolution" can occur, which in some cases can be thought of as a type of "arms race."

Anyone familiar with evolutionary optimization will quickly see how it can be used for robust foraging strategy design. Hence, we will not concern ourselves with that here. Another area of relevant study from theoretical biology is "evolutionary game theory." See the "For Further Study" section at the end of this part.

19.2 Introduction to Game Theory

In this section, we introduce basic definitions for concepts that form the foundation for the game-theoretic view of competition and cooperation, and show how to compute several types of player strategies.

19.2.1 Strategies and Information for Decisions

We start by defining a simple "matrix game" and explaining how it is played. This leads us to define strategies and a discussion on what information is used in making decisions.

Players, Rules, and Payoffs

Suppose that we have two "players" ("decision-makers") that we denote by P_1 and P_2. Let θ^1 and θ^2 denote the "decision variables" of the two players, respectively. Let J_1 and J_2 denote the cost functions of the players (i.e., what they gain or lose for a given set of decisions). The cost J_1 could represent, for example, a payment in cash from P_1 to P_2. In this case P_1 (P_2) wants to minimize (maximize) J_1. The problem will be, however, that P_1 (P_2) cannot unilaterally minimize (maximize) the cost. Each player's losses and gains are also influenced by the actions of the other player; that is the essence of a competitive game.

If

$$J_1(\theta^1, \theta^2) + J_2(\theta^1, \theta^2) = 0$$

for all θ^1 and θ^2, then we have a "zero sum game" so that gains of one player are losses of the other and they are in an adversarial relationship. If $J_1(\theta^1, \theta^2) + J_2(\theta^1, \theta^2) = c$ for some known constant c, then we can simply redefine the cost functions to incorporate the value of c to obtain a zero sum game.

To keep things simple, initially suppose that

$$\theta^1 \in \{1, 2, \ldots, D_1\}$$

and

$$\theta^2 \in \{1, 2, \ldots, D_2\}$$

so that there are only a finite number of D_1 decisions for P_1 and D_2 decisions for P_2. For simplicity, we will refer to the different decisions (which sometimes we will call "strategies") as $\theta^1 = i$ and $\theta^2 = j$ for P_1 and P_2, respectively. We then denote the costs by $J_1(i, j)$ and $J_2(i, j)$.

Here, we will often think of $J_1(i, j)$ as being a cash payoff (clearly, many other interpretations for cost are possible) of P_1 to P_2 given the decisions i and j by P_1 and P_2, respectively. In the zero sum case $J_1(i, j) = -J_2(i, j)$ (player 2 gets all the payoff of player 1 and vice versa) and if $J_1(i, j) \leq 0$, this represents that player 2 pays player 1 if P_1 uses a decision i and P_2 uses a decision j. In the two-player case, it is sometimes convenient to represent the payoff functions $J_1(i, j)$ and $J_2(i, j)$ as $D_1 \times D_2$ matrices, J_1^{ij} and J_2^{ij}, respectively (then $J_1(i, j) = J_1^{ij}$, an element of the matrix, for all i and j). This will be especially useful below when we consider "matrix games" where J_1^{ij} and J_2^{ij} will be called "payoff matrices."

Each player tries to maximize its own gains in a competitive game, possibly at the expense of the other player.

Suppose that the game is only played once (i.e., P_1 and P_2 only make decisions once). The players make decisions with full information about payoffs, they make their decisions simultaneously, and are "rational." Assuming that P_1 and P_2 are "rational" players, means that P_1 tries to minimize $J_1(i, j)$ and

P_2 tries to maximize it. Particular values of θ^1 and θ^2 are called "actions." A "strategy" is an established way of acting that depends on the possible actions of another player (e.g., if the player's decisions depended on gathered information about the decision process).

Generally, players are concerned with what strategies to use in playing a game. A pair of decision strategies for a two-player finite game is denoted by (i, j). The "outcome" of the game is J_1^{ij}. An "optimal" strategy, or simply one that we choose from a fixed set of possibilities, will be something that will be denoted by (i^*, j^*).

Security Strategies, Saddle Point Strategies, and Information

Let $D_1 = 5$ and $D_2 = 3$. Suppose that

$$J_1^{ij} = \begin{bmatrix} -3 & 4 & 4 \\ 0 & -5 & 2 \\ -2 & 1 & -4 \\ 2 & 3 & -4 \\ 2 & -2 & -5 \end{bmatrix} \tag{19.1}$$

Note that rows of this matrix correspond to P_1 decisions and columns correspond to P_2 decisions.

For a security strategy, a player chooses so as to minimize its maximum loss.

In a "security strategy," a player makes decisions to secure losses against whatever the other player might do (i.e., it minimizes its maximum possible loss). Hence, P_1 picks row i^* such that any value in any column of this row is no bigger than the largest value of any other row $i \neq i^*$ (i.e., pick the row that minimizes the maximum size column value). For low values of D_1 and D_2, it is possible to specify the solution by inspection of the matrix in Equation (19.1) above. For larger values of m or n, you may want to write a computer program to solve for the security strategy.

For Equation (19.1), the list of maximum values for each row is

$$\begin{matrix} 4 \\ 2 \\ 1 \\ 3 \\ 2 \end{matrix}$$

and so the security strategy is for P_1 to pick $i^* = 3$ to minimize what it has to pay to P_2. The "loss ceiling" (i.e., the most it can lose) is 1, which is less than the other possible losses, and this is called the "security level" of P_1. Similarly, P_2 can adopt a security strategy by choosing the column j^* whose row values are smaller than the smallest value found for another column $j \neq j^*$. In this case, the list of minimum values is

$$-3 \quad -5 \quad -5$$

so that the security strategy for P_2 is $j^* = 1$ and P_2 secures gains at the "gain floor" (its security level) of -3 (i.e., he pays no more than 3). Clearly the security level of P_1 is never below the security level of P_2. The "outcome" of the game, which in this case is

$$J_1^{i^* j^*} = J_1^{31} = -2$$

will lie between the two security levels.

In the special case where the security levels of the two players are the same, the security strategies of the two players are in "equilibrium" with each other since they are "optimal" with respect to each other; in this case, they are called "saddle point strategies." A pair of strategies is said to be in a "saddle point equilibrium" if unilateral deviations by one player from its strategy will not benefit that player. In general, for a $D_1 \times D_2$ matrix game if (i^*, j^*) is the pair of chosen strategies, and if

$$J_1^{i^* j} \leq J_1^{i^* j^*} \leq J_1^{i j^*}$$

for all $i \in \{1, 2, \ldots, D_1\}$ and $j \in \{1, 2, \ldots, D_2\}$, then (i^*, j^*) constitutes a saddle point equilibrium. The value of $J_1^{i^* j^*}$ is called the "saddle point value." Is there a saddle point equilibrium for the above example matrix game?

Next, we discuss the issue of the quantity and type of information that is used by a player for its strategy, but only via our above simple example for security strategies. Note that if in the example above we changed the game so that P_1 plays first, then P_2, with P_2 *knowing* what P_1 had chosen, then the outcome will be different. We can actually deduce the outcome of the game by simple inspection of the matrix in Equation (19.1). First, note that it makes sense for P_1 to use a security strategy, since we assume that it does not know anything about the decision tendencies of P_2. This gives $i^* = 3$ just like above. If P_2 is informed of this choice, then it would pick $j^* = 2$ to get a gain of 1, which is more than what it got with a security strategy. This simple example shows that the best strategy to use depends on what information is available to the player, and when it is available. This is a fundamental principle that drives the design of strategies. Games of the type where players can use information from the process of the evolving game are called "dynamic" games, whereas the above example is called a "static game," since only a priori information is used to play it.

Strategies depend critically on what information is available to a player.

The development of strategies above assumes that each player is rational and that each player knows this about the other player. If we assume that a probability distribution is known by one player P_1 about the other player's decisions (i.e., that it knows the probability that it will make each decision), then we can design a so-called "mixed strategy" (as opposed to the cases above, which are sometimes called "pure strategies"), where the decisions each player makes are based on the outcome of random events (i.e., a derived probability distribution on P_1 decisions, given the information on P_2). For example, the design of the strategy of P_1 could involve choosing a probability distribution on its decisions so that it would be most likely to minimize what it pays to P_2.

The security strategy concept extends to having more than two players and then results in a multidimensional matrix game. For instance, we could imagine adding a third player to the above game (which we could call "nature") that makes random decisions, and then develop security strategies for P_1 and P_2. The above example, and several others in this section, are "noncooperative games," where there is a clear adversarial relationship between the players (e.g., one's losses are the other's gains). There are also games where there is not a diametrically opposed relationship between the two players or where they are truly "cooperative" and try to help each other achieve their goals (so long as they also do well). Some such games are called "cooperative games" and typically, they involve sharing information so that each gains as much as possible. These will be discussed below when we discuss the Pareto-optimal solution.

Extensive Forms and Decision Trees

The matrix form of a game is called its "normal form." The normal form does not depict a complete representation of a game, for example, if there are repeated steps of play where players use different information from past player decisions. An "extensive form" of a game involves creating a type of labeled "decision tree" to represent the game. We discuss the extensive form via a simple example.

The game we consider is depicted in Figure 19.1. Figure 19.1(a) shows the extensive form representation for the matrix game in Equation (19.1), so consider it first. The game starts at the "top" of the (upside down) tree and evolves to the tip of one of its branches by a sequence of decisions starting with P_1 in "level 1" of decision-making and then P_2 at level 2. In level 1, P_1 chooses among its five alternatives, each represented with a separate labeled branch, and in level 2, P_2 chooses among its three alternatives, represented by the three labeled branches. The outcomes for the various decisions are shown at the tips of the branches. For example, in the security strategy case, we had $(i^*, j^*) = (3, 1)$ so we get an outcome of -2 as shown in Figure 19.1(a). The dashed line encirclement of the "decision nodes" at level 2 depicts the "information set" of P_2. It represents that in Figure 19.1(a), P_2 does not know which branch (decision) P_1 has chosen. This is then equivalent to the two players *simultaneously* making their decisions (how we interpreted the matrix game).

Figure 19.1(b) shows the case discussed in the last subsection where P_1 makes the decision first, then P_2 makes its decision, knowing the choice of P_1. The information sets in this case are again shown with the dashed encirclements of the decision nodes at level 2. As we noted above, due to the change in available information, there is a change in strategy and hence, outcome. The dashed encirclements clearly represent the inherent difference in the two games in Figure 19.1, whereas the normal form representation does not. This is one more reason why the extensive form is a convenient and intuitive representation for games, at least finite games without too many decision alternatives so that the trees are not too "bushy."

Finally, note that in general, decision trees may have many levels (multiple decisions), for example, with P_1 and P_2 taking turns so that at odd levels, P_1 would act and at even levels, P_2 would act. Moreover, it should be clear how to make an extensive form for the case where there are more than two players.

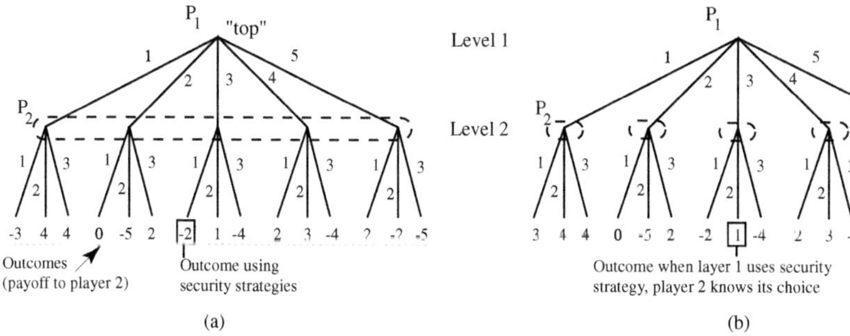

Figure 19.1: Example of an extensive representation, in this case in (a) for the matrix game in Equation (19.1) and in (b) for the same payoff matrix, but when P_1 chooses first and then P_2 knows its decision before choosing.

Decisions Versus Strategies

Figure 19.1 allows us to define the concept of a "strategy" relative to a "decision" more clearly. Suppose that we let $G^i(\cdot)$ denote the strategy for player i, where "·" is the information that is available for decision-making. Note that the security strategies for P_1 and P_2 in Figure 19.1(a) are simply

$$G^1 = 3, \; G^2 = 1$$

(Note that there is no argument for the G^i functions, since there is no information from the decision-making *process*, in addition to available a priori information that is used to make the choices.)

The player strategies for the game in Figure 19.1(b) are $G^1 = 3$ (the security strategy), and the strategy of P_2 is what we might call "pick the best payoff given the decision of P_1." The representation for this strategy could be

$$G^2(i) = \begin{cases} 2 & \text{if } i = 1 \\ 3 & \text{if } i = 2 \\ 2 & \text{if } i = 3 \\ 2 & \text{if } i = 4 \\ 1 & \text{if } i = 5 \end{cases}$$

(Of course, we could use $G^2(1) = 3$ also, since it results in the same payoff to P_2.) Note that since there is information used from the decision-making process, particularly the decision of P_1 at level 1, the strategy is defined in terms of what decision i that P_1 might use.

The definition of a strategy is in general quite different from the meaning of a "decision" (action) as long as there is more than one "information set" (i.e., a dashed encirclement in Figure 19.1). The strategy is a mapping that specifies what action to take (decision to make) depending on what is known, as specified via an information set. If a player has an ability to distinguish something about a decision process as it evolves, then to keep the strategy of the player independent of what the other player does, it must have different ways of reacting, depending on what the other player did. For example, this is perhaps clarified if we considered a strategy for P_1 that was simply based on a random choice for its options. In this case, P_2 would have to be ready to react no matter what P_1 did, and the above strategy $G^2(i)$ would define that.

Finally, note that if there are multiple levels, with many players, there are many options for how to define information sets, and hence, strategies for games. Finding strategies for the case where there are such dynamic games is, in general, a challenging problem.

19.2.2 Nash, Minimax, and Stackelberg Strategies

In this section, we introduce the Nash equilibrium strategy and provide an example of how to compute it. Moreover, we introduce infinite games (i.e., ones where there are an infinite number of strategies by at least one player), the concept of a reaction curve, and use an example to illustrate the basic ideas. This is followed by an introduction to minimax and Stackelberg strategy design.

Nash Equilibrium Strategies

Up till now, we had considered the zero sum case, and now we move beyond that to consider the nonzero sum case, that is, when it can be that

$$J_1(i,j) + J_2(i,j) \neq 0$$

Now, we have two payoff matrices J_1^{ij} and J_2^{ij} denoting losses of P_1 and P_2, respectively. Assume that both players are rational, so they try to minimize their losses. Unless otherwise stated, we assume that there is no cooperation and decisions are made independently.

The basic problem for each player is that the outcome resulting from their decision also depends on what the other player decides. So, what strategies should the players use? Recall that a pair of strategies is said to be in a "saddle point equilibrium" if unilateral deviations by one player from its strategy will not benefit that player. There are actually a variety of "equilibrium" solutions for a pair of strategies of a game.

If players use the Nash equilibrium solution, then they have no reason after playing the game to regret their decisions.

A strategy pair (i^*, j^*) is a noncooperative (Nash) equilibrium solution to a "bimatrix" game (J_1^{ij}, J_2^{ij}) if the inequalities

$$J_1^{i^*j^*} \leq J_1^{ij^*} \tag{19.2}$$

and

$$J_2^{i^*j^*} \leq J_2^{i^*j} \tag{19.3}$$

are both satisfied for all $i \in \{1, 2, \ldots, D_1\}$ and all $j \in \{1, 2, \ldots, D_2\}$. The pair $(J_1^{i^*j^*}, J_2^{i^*j^*})$ is the noncooperative (Nash) equilibrium outcome of the game.

For a given bimatrix game, there can be no Nash solutions, one Nash solution, or many Nash solutions. If $J_1^{ij} = -J_2^{ij}$ for all i and j, then we have a zero sum game, and a Nash solution is a saddle point equilibrium for the game.

As an example, let $D_1 = 5$ and $D_2 = 3$. Suppose that

$$J_1^{ij} = \begin{bmatrix} -1 & 5 & -3 \\ -2 & 5 & 1 \\ 4 & 3 & -2 \\ -5 & -1 & 5 \\ 3 & 0 & 2 \end{bmatrix}, \quad J_2^{ij} = \begin{bmatrix} -1 & 2 & -3 \\ 2 & -3 & 1 \\ -2 & 3 & 1 \\ -1 & 1 & -1 \\ 4 & -4 & 1 \end{bmatrix} \tag{19.4}$$

To solve for the Nash equilibria, consider candidate (i^*, j^*) pairs in turn and test if they satisfy both the inequalities in Equations (19.2) and (19.3). To do this, consider $(1,1)$ and see if J_1^{11} is less than all other row elements of column one to test Equation (19.2). Since it is not, $(1,1)$ cannot be a Nash solution. If you test $(1,2)$, you will also find it is not a Nash solution. However, if you test $(1,3)$, you will see that $J_1^{13} \leq J_1^{i3}$ for all i so it is a candidate, so test the inequality in Equation (19.3) and you will find that $J_2^{13} \leq J_2^{1j}$ for all j; hence, $(i^*, j^*) = (1,3)$ is indeed a Nash equilibrium. The Nash equilibrium outcome is $(-3,-3)$ so that both players *gain* 3. Show that $(4,1)$ is also a Nash solution, with an outcome of $(-5,-1)$, but that all others are not.

Note that a Nash equilibrium solution is special since, if the players adopt it, then they have no reason after playing the game to regret their decisions. Note, however, that there can be more than one Nash equilibrium so the question of which one to use arises. However, it is not possible to totally order the Nash strategies according to the values of their outcomes, because they are defined by *pairs* of numbers. We can, however, say that "one Nash strategy is better" than another if *both* outcomes are better than the other. Then, we will call a Nash strategy "admissible" if there is no better Nash strategy. For the above example, there were two Nash solutions and each one is admissible since one is not better than the other. Note that in this case, if P_1 picks $(1,3)$ and P_2 picks the other Nash solution (one thinks that the other is picking the other strategy to play by, since there is no reason to think that they would definitely pick the same Nash strategy to play without some type of cooperation), then the strategy pair that is employed is $(1,1)$, which as the above example showed, is not a Nash solution. In fact, $(1,1)$ results in an outcome of $(-1,-1)$, which is *worse* for both players. This creates a problem with implementing a Nash solution for a noncooperative game when the Nash solution is not unique.

If there is only one (admissible) Nash solution, this problem will not arise. However, if the two payoff matrices are the same, this problem can still arise. Why? There are then many cases where the situation can arise where there are multiple Nash equilibria, so that the players cannot use the solutions effectively without some type of cooperation. Essentially, this problem with the Nash solutions arises, since bimatrix games may not be "antagonistic," so that a noncooperative solution concept can be inappropriate (i.e., elements of cooperation can be reflected in the payoff matrices). We will revisit this issue when we discuss Pareto-optimal solutions below.

Infinite Games and Reaction Curves

An infinite game is one in which there are an infinite number of strategy choices by one or both of the players. The concept of Nash equilibria is also valid for this case, and we illustrate this via a simple example here.

First, suppose that the actions of P_1 can be

$$\theta^1 \in [-4, 4]$$

and for P_2 they can be

$$\theta^2 \in [-5, 5]$$

Suppose that cost functions for the two players are

$$J_1(\theta^1, \theta^2) = -\exp\left(-\frac{(\theta^1 - 2)^2}{8} - \frac{(\theta^2 - 4)^2}{2}\right)$$

and

$$J_2(\theta^1, \theta^2) = -\exp\left(-\frac{(\theta^1 - 1)^2}{1} - \frac{(\theta^2 + 1)^2}{6}\right)$$

each of which has one global minimum, and also for which, if you fix θ^1 (θ^2), there is a unique minimum point in the other value. (This will simplify the discussion.)

Define the "reaction curve" of P_1 to be

$$R_1(\theta^2) = \arg\min_{\theta_1} J_1(\theta^1, \theta^2)$$

where we are using the assumption of uniqueness of the minimum point, so that there is only one point at which the minimum is achieved. The reaction curve $R_1(\theta^2)$ defines how P_1 should react for every possible action of P_2 in order to minimize its losses. Similarly, the reaction curve of P_2 is

$$R_2(\theta^1) = \arg\min_{\theta_2} J_2(\theta^1, \theta^2)$$

and it defines how P_2 should react for every possible action of P_1 in order to minimize its losses. Contour plots of J_1 and J_2 are shown in Figure 19.2(a) for the above loss functions, along with the reaction curves $R_1(\theta^2)$ and $R_2(\theta^1)$.

It is interesting to note that any intersection point of the two curves in Figure 19.2(a) is a Nash equilibrium, so $(2, -1)$ is the unique Nash equilibrium. For other cost functions, it is possible that the curves take on different shapes and have multiple intersection points, with each intersection point corresponding to a Nash equilibrium. It is also possible that the reaction curves do not intersect, indicating that there are no Nash equilibria. In the case where there are multiple minimum points for each fixed value of θ^1 (or θ^2), we may not have connected "curves," but more generally reaction "sets" and in this case, it should be clear that there can be multiple or no intersection points. Figure 19.2(b) shows a different set of cost functions and reaction curves, but with one Nash equilibrium

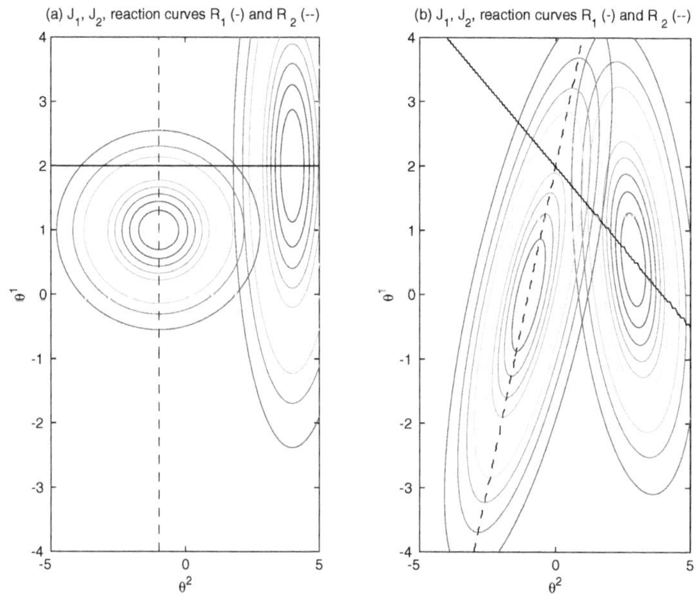

Figure 19.2: (a) Contour plots of J_1 and J_2 and reaction curves R_1 (solid) and R_2 (dashed); (b) same but for different cost functions J_1 and J_2.

corresponding to the intersection point $(2, 0)$. Note that the reaction curve for P_1 is the locus of points that is tangent to lines corresponding to fixed values of θ^2 across a range of such fixed values. Clearly, adjustment of the shape of these functions can easily result in an intersection point that lies outside the allowed ranges of decisions for the players and this would correspond to the case where there are no Nash equilibria.

Finally, note that since the plots in Figure 19.2 were generated on a digital computer, we actually discretized each of the axes and computed a finite number of cost values and points on the reaction curves (as is usual, discrete points are connected by lines in the plots to give a visual effect of continuity when it does not actually exist). Hence, we are actually discussing *finite* bimatrix games, but ones with many possible actions for each player.

Stable Nash Equilibria

It is possible to further refine the characterization of Nash equilibrium solutions and we do this in this section via a simple example. Suppose that for the game pictured in Figure 19.2(b), we have P_1 play first, then P_2, followed by P_1, and so on; hence the players alternate moves with P_1 going first. Suppose that we number the moves with an index k so that at $k = 1$, P_1 moves, at $k = 2$, P_2 moves, and so on. Suppose that each player knows the other's last move, and takes an action that minimizes its losses given this past move. For an arbitrarily chosen first decision by P_1, the "trajectory" in decision-space is shown in Figure 19.3. Since the reaction curves were already computed for this case, it is easy to construct the trajectory since, once P_1 makes a decision $\theta^1(k)$ at iteration k (k odd), we have

$$\theta^2(k+1) = R_2(\theta^1(k))$$

and once P_2 makes a decision $\theta^2(k)$ at iteration k (k even), we have

$$\theta^1(k+1) = R_1(\theta^2(k))$$

It is this iterative sequence that results in the trajectory. Clearly, if we have nonuniqueness of minimum points and hence, reaction sets rather than simple curves, then appropriate adjustments to this discussion would be needed.

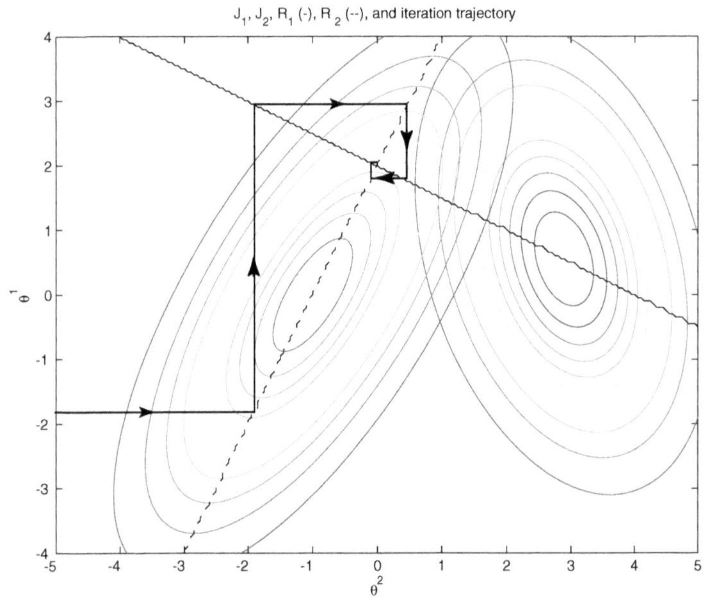

Figure 19.3: Contour plots of J_1 and J_2, reaction curves R_1 (solid) and R_2 (dashed), and iteration trajectory (arrows indicate direction of time).

If for every initial choice by P_1 (i.e., $\theta^1(1)$), the trajectory in the decision space moves to the point $(2, 0)$, then the point $(2, 0)$ is said to be a "stable Nash equilibrium." If only small variations in $\theta^1(1)$ from 2 (its ultimate value for the Nash equilibrium) are possible for it to still converge to 2, then $(2, 0)$ is said to be a "locally stable Nash equilibrium." If a Nash equilibrium point is not stable, it is said to be an "unstable Nash equilibrium." Different cost functions generally result in different reaction curves (sets), which in turn result in different trajectories and hence, different types of stability.

Note that stability is a qualitative property of the trajectories, whose generation depends on *the given iteration method.* Hence, we generally speak of stability relative to the given iteration method. Clearly, the ideas extend to the multiplayer case and then the iteration method depends on that, too. For example, the iteration may involve taking turns in some fixed or dynamically changing order so that the choice by one player at a current time may depend on the movements of some players at the last time instant, and the decisions others made at perhaps random points in the past.

Minimax Strategies

Next, we develop security strategies for each player of a bimatrix game which we will call "minimax strategies" (solutions). These solutions may or may not be the same as a Nash solution.

To find the minimax strategy, you simply find the security strategy for P_1 based on J_1^{ij} and the security strategy for P_2 based on J_2^{ij} and then, taken together, these directly provide a strategy pair that is the "minimax" strategy for the bimatrix game. (Of course, here, since we view the payoff matrices J_1^{ij} and J_2^{ij} as "loss" matrices, there is a slight difference in the mechanics of finding a security strategy for P_2, since now it also tries to minimize its loss under all possible actions of P_1, where before it tried to maximize its gain.) It is interesting to note that P_1 (P_2) does not need knowledge of J_2^{ij} (J_1^{ij}) to compute its strategy. Moreover, the minimax concept essentially ignores whether the opponent is rational or not. These facts can be important in practical applications.

The security levels of the players are called the minimax values of the bimatrix game. These minimax values are not better than those of a Nash equilibrium outcome, even if the Nash and minimax strategies are the same. Why? Also, note that if a minimax strategy is not a Nash strategy (i.e., not an equilibrium), it can still be quite useful in games where there are multiple Nash equilibria, when a player is not completely certain about values of the cost matrix, or whether the other player will be rational. On the other hand, these features can lead to very conservative strategies in some applications. Clearly, minimax strategies can be extended to a multiplayer game in an obvious way.

As an example, show that the minimax strategy for the cost matrices in Equation (19.4), where we had two Nash equilibria, is $(5, 3)$ with an outcome $(2, 1)$. Recall that for this case, the Nash solutions were $(1, 3)$ and $(4, 1)$ with outcomes $(-3, -3)$ and $(-5, -1)$. Note that the minimax strategy is not as good as either Nash solution. Note, however, that for the cost matrices

$$J_1^{ij} = \begin{bmatrix} 4 & 1 & -4 \\ -2 & 5 & 3 \\ -3 & 2 & -1 \\ 4 & 4 & 4 \\ -3 & -5 & 2 \end{bmatrix}, \quad J_2^{ij} = \begin{bmatrix} 2 & -1 & 0 \\ -2 & 4 & 0 \\ -3 & 0 & -1 \\ -3 & 3 & 4 \\ -3 & 0 & -5 \end{bmatrix}$$

there is one Nash solution and it is also a minimax solution. Show this.

As another example, consider the game in Figure 19.2(a). Show that the Nash equilibrium is a minimax solution. However, show that for the game of Figure 19.2(b), a minimax solution is $(4, -1)$. In this case, is the Nash solution a minimax solution? Is the minimax solution unique?

Stackelberg Strategies

Until now, for our solution concepts we had a type of symmetry where no one player dominated the decision process. What if one player can enforce her or his strategy on the other players? With this approach we get a "hierarchical equilibrium solution" concept introduced by Stackelberg. For such games, the "policy enforcer" will be called the "leader" and the other player(s) will be called follower(s). We assume that the players act rationally. We could have multiple levels of leaders and followers, but here we just consider the case where we have one leader and one follower.

The basic idea is that P_1 tries to pick i^* to minimize its loss, assuming that with i^* enforced on P_2, P_2 will pick its strategy in a rational way (by minimizing its losses), and then this choice defines the outcome, and hence, allows P_1 to rank its alternatives. In summary, P_1 evaluates its m alternatives by considering what P_2 will pick in response to each one. There is, however, an added complication. There may be more than one P_2 strategy that minimizes its losses for a given P_1 strategy. This creates the possibility that P_1 has more than one different possible loss for each enforced strategy, since P_2 can react in different (rational) ways. Here, we will adopt the convention that we will use a security strategy approach to resolve the ambiguity (i.e., P_1, in the face of several possible answers from P_2, will pick the alternative that will minimize its maximum possible losses).

To compute a strategy pair (i^*, j^*) that is a Stackelberg solution, we execute the following two steps:

1. *Compute Follower Reactions:* For each possible strategy choice i by P_1, compute the set of rational reactions by P_2 as

$$R(i) = \left\{ j^*(i) : j^*(i) = \arg\min_j J_2^{ij} \right\}$$

   This is done by having P_2 execute an optimization over its responses for each given i. Note that $|R(i)| \geq 1$ and if, for example, $|R(i)| = 2$ for some i, then there are two possible rational strategies for P_2 which give P_2 the same loss (i.e., two optima).

2. *Leader Finds Best Strategy:* The leader, taking into account the follower reactions, chooses its strategy via

$$i^* = \arg\min_i \max_{j \in R(i)} J_1^{ij}$$

   which is the P_1 strategy that achieves the lowest loss considering rational P_2 reactions, and which is secure against ambiguities in the response of

P_2 represented in $R(i)$. Note that in finding the "max" and "min" in the computations for i^*, it is possible that there is more than one maximizer and minimizer. For each case, however, it does not matter which you choose, since the "Stackelberg cost" (defined next) is the same. Hence, a normal approach is to resolve ties with an arbitrary choice.

The "Stackelberg strategy" of P_1 is i^*. The Stackelberg solution strategy pair is $(i^*, j^*(i^*))$ and the "Stackelberg cost" for the leader is $J_1^{i^* j^*(i^*)}$. Finally, it is interesting to note that the leader P_1 never does worse in a Stackelberg game relative to if it had played a Nash game. Why?

As an example, to find the Stackelberg solution for the cost matrices in Equation (19.4), with P_1 as the leader and P_2 as the follower, we find that: if $i = 1$, $R(i) = \{3\}$; if $i = 2$, $R(i) = \{2\}$; if $i = 3$, $R(i) = \{1\}$; if $i = 4$, $R(i) = \{1, 3\}$; and if $i = 5$, $R(i) = \{2\}$. The loss to P_1 in each case is then (this takes into account the security component for the $i = 4$ case): if $i = 1$, P_1 loses -3; if $i = 2$, P_1 loses 5; if $i = 3$, P_1 loses 4; if $i = 4$, P_1 loses $5 = \max\{-5, 5\}$; and if $i = 5$, P_1 loses 0. Hence, the Stackelberg solution is $(i^*, j^*) = (1, 3)$ and the Stackelberg cost for P_1 is -3. Note that if we had, for the computation of the P_1 losses, found equal values for more than one i, then there would have been more than one valid Stackelberg solution (but this does not create the problems that we found when we had multiple Nash solutions, since the Stackelberg costs would be the same).

As another example, consider the game in Figure 19.2(b). Clearly the Stackelberg solution will lie on the curve $R_2(\theta^1)$. Why? Show in this case that the Stackelberg solution is $(3.45, 0.75)$, which can be roughly approximated from the plots in Figure 19.4. How? To see how, note that the Stackelberg solution is the pair (θ^1, θ^2) such that θ_1 minimizes $J_1(\theta^1, R_2(\theta^1))$ and this is simply the point where the $R_2(\theta^1)$ curve is tangent to the contour plot of J_1.

19.2.3 Cooperation and Pareto-Optimal Strategies

In the cases above, we generally considered the games to be noncooperative so we assumed that there was a type of adversarial relationship between the two players. In the Nash game, each player is trying to do as well as possible, taking into consideration the other player's actions. In a minimax game, each player assumes the worst possible reactions of the other player (i.e., a highly adversarial, perhaps irrational, opponent) to pick the strategy. There are, however, games where the two players may be able to share information to try to do better; that is, they may cooperate. In a certain sense, the leader-follower game with Stackelberg solutions can be considered a type of cooperative game, if you view the leader-enforced strategy as information that is used by the follower. Whether a Stackelberg game is truly cooperative depends, however, on the application domain and particularly on whether the cost functions of the players have relationships such that the leading actions by P_1 and the subsequent following actions by P_2 are achieving (or trying to achieve) the same type of goal.

In a cooperative game, a player may give up some gains so that the group it is collaborating with gains more.

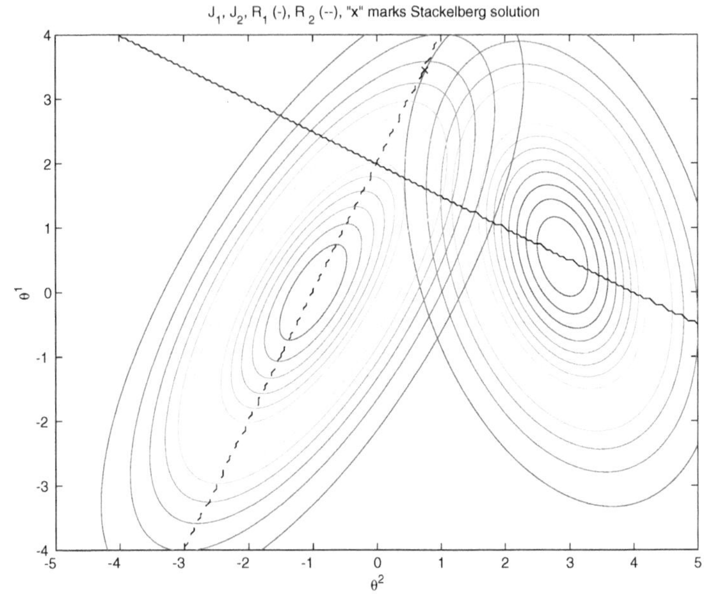

Figure 19.4: Contour plots of J_1 and J_2, reaction curves R_1 (solid) and R_2 (dashed), and "×" marks the Stackelberg solution.

At the other end of the spectrum, there are games that can be considered to be "cooperative" in the sense that each player is willing to share all information and help every other player do as well as possible, so long as it does not degrade its gains too much. A key concept for such games is the idea of a Pareto-optimal (equilibrium) solution, where no other such joint decision exists that can improve the outcome for P_1 or P_2 without degrading the outcome of the other.

Multiobjective Optimization and Pareto Optimality

Here, we begin by defining Pareto optimality in a general setting and then discuss its use in games via some examples. A *multiobjective optimization problem* is in the form of

$$\begin{aligned} \text{minimize: } & \{J_1(\theta), \ldots, J_N(\theta)\} \\ \text{subject to: } & \theta = [(\theta^1)^\top, \ldots, (\theta^N)^\top]^\top \in \Theta \end{aligned}$$

Here, we want to *simultaneously* minimize a *set* of cost functions (called a "cost function vector") by changing the same parameter vector θ. You can think of this as a type of general optimization problem, where you want to minimize not one cost function, but many, each representing the desire to achieve a different objective. Here, we assume that $\theta^i = [\theta^i_1, \ldots, \theta^i_n]^\top$, $i = 1, 2, \ldots, N$, so that decisions are $n \times 1$ vectors, rather than just scalars, and there are N costs to minimize (e.g., in a two-player game, we will have $N = 2$). Also, Θ is used to represent constraints on the decisions (e.g., constraints on the size of the values of the elements of θ^i, often characterized via functions and inequalities).

A decision vector $\theta^* \in \Theta$ is *Pareto optimal* if there does not exist any other $\theta \in \Theta$ such that

$$J_i(\theta) \leq J_i(\theta^*)$$

for all $i = 1, 2, \ldots, N$, and at the same time

$$J_j(\theta) < J_j(\theta^*)$$

for at least one index j. A cost function vector is called Pareto optimal, if the corresponding decision vector is Pareto optimal. Hence, intuitively, a cost function vector is Pareto optimal, if you cannot improve one cost value without degrading others.

It should be clear that there can be many Pareto optimal solutions. In multiobjective optimization, there is a need to specify *preferences* to be able to pick which Pareto optimal solution specifies an acceptable solution (e.g., one that balances the wins and losses of two players). In many approaches, it is hypothesized that there is a "decision maker" who will specify these preferences in some manner. Sometimes the decision maker will specify a "value function" that can be viewed as a specification of what the decision maker wants in terms of the minimizations of each of the cost functions (i.e., it specifies the trade-offs between players), and other times the decision maker has to be repeatedly asked for its preferences. There are many ways to define such value functions for the decision maker, and the examples below will discuss one way.

Before turning to games, to illustrate the idea of Pareto-optimal solutions, consider the case where θ^1 and θ^2 are scalars, so that θ is a 2×1 vector. Assume that we have quadratic costs so they are convex. Suppose in particular that we have

$$J_1(\theta) = J_1(\theta^1, \theta^2) = \left(\theta^1 - 2\right)^2 + \left(\theta^2 - 3\right)^2$$

$$J_2(\theta) = J_2(\theta^1, \theta^2) = \left(\theta^1 + 2\right)^2 + \left(\theta^2 + 2\right)^2$$

Contour plots of these two functions are shown in Figure 19.5. In this case, it is possible to determine Pareto-optimal solutions by inspection. To do this, ignore the plotted Pareto points in Figure 19.5 and note that the (unique global) minimum points on the two cost functions are at the centers of the two sets of concentric circles. Next, note that if a line on this contour plot is tangent to a contour of *both* costs, then the point of tangency for both costs is a Pareto-optimal solution. (Of course, there are only a finite number of contour lines drawn on the plot so you must imagine where the other ones are.) Where are these "tangency points"? Simple inspection shows that they are at the "$\times$" marks on the plot. Why are these Pareto-optimal solutions? Note that the gradient at such a point θ^* is $\frac{\partial J_1}{\partial \theta}\big|_{\theta=\theta^*}$ and suppose this is pointing in the same direction as $-\frac{\partial J_2}{\partial \theta}\big|_{\theta=\theta^*}$. Imagine that you are at some Pareto-optimal solution θ^* in Figure 19.5. The direction of the negative gradient is the direction to move from θ^* in order to get a steepest amount of decrease in the value of one cost function (the basic idea behind steepest descent gradient optimization). The key observation is that if you perturb θ^* along the gradient in direction of the minimum point for J_1 (J_2), the cost for J_1 (J_2) goes down, but the cost for J_2 (J_1) goes up. So, we cannot reduce one cost without increasing the other, which is the very definition of Pareto optimality. Note that there are actually an infinite number of points that lie on a line that is tangent to the contour of the two costs, and hence, an infinite number of Pareto-optimal solutions in this case, all of which lie on the line between the minimum points of the two cost functions. The set of all Pareto-optimal solutions is sometimes called the "family" of Pareto-optimal solutions.

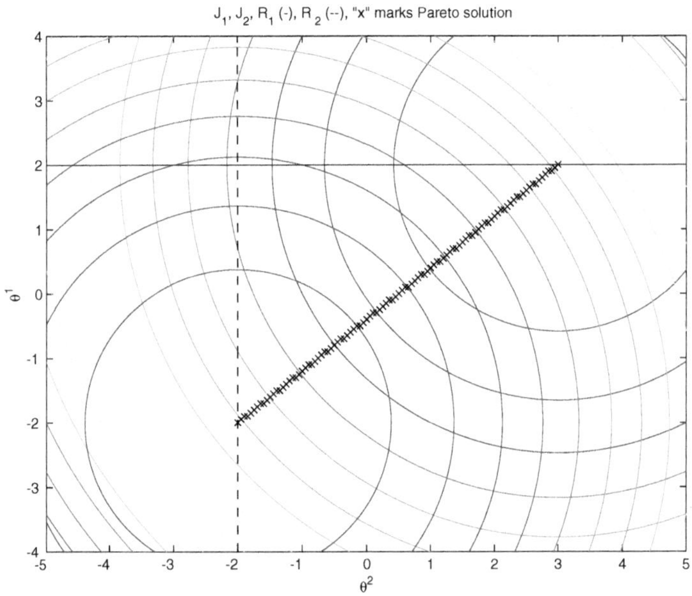

Figure 19.5: Example family of Pareto-optimal points for two quadratic cost functions ("×" marks Pareto solutions).

The scalarization approach is only one way to define the Pareto cost.

It is interesting to note that, for this example, if we define a "Pareto cost" to be

$$J_p(\theta) = pJ_1(\theta) + (1-p)J_2(\theta)$$

for $p \in [0, 1]$ (this is called the "scalarization" approach to constructing the Pareto cost), then the family of Pareto points is the set of (unique) global minima for $J_p(\theta)$ as p varies from zero to one, which is just the equation for the line between the two minimum points in Figure 19.5. Hence, we can view this Pareto cost as a "value function" for the underlying multiobjective optimization problem; it is, however, a special one, since it shows how, for this special quadratic case, it is possible to find *all* Pareto-optimal solutions *via standard optimization of one cost function.* Intuitively, note that the value of p scales the depth of one minimum and $(1-p)$, the depth of the other. We can think of the above weighting scheme via p as *interpolating* between the two minimum points on the two cost functions, with a specific value of p providing more value to minimizing one cost function over the other.

Pareto-Optimal Solutions for Games

Motivated by the above example, a standard way to define Pareto optimality for two-player bimatrix games is to define a new loss matrix (sometimes called the Pareto cost) that represents a combination of the losses for the two players,

$$J_p^{ij} = pJ_1^{ij} + (1-p)J_2^{ij}$$

for $p \in [0,1]$. This Pareto cost can be thought of as a "value function" for the decision maker in a multiobjective minimization problem. *Any* minimum point in the matrix J_p^{ij} is called a Pareto-optimal solution for the bimatrix game (and note that there may be several minimum points for any one value of p). If $p=1$ $(p=0)$, all the emphasis is placed on the two players collaborating to minimize the losses of P_1 (P_2). With appropriate values for the cost functions, the value of $p=\frac{1}{2}$ may represent equal emphasis on minimizing the losses of the two players.

As an example, suppose that we use the bimatrix game with payoff matrices in Equation (19.4). Note that these cost values are not specified via a convex function, so there are additional complexities that arise here with uniqueness of the Pareto solutions for fixed values of p. Also, in general, finding a minimum of the Pareto cost J_p^{ij} for all values of p may not provide all possible Pareto-optimal solutions. Why? Returning to our example, note that the minimum element in J_1^{41} is -5, which is also the minimum of J_p for the case where $p=1$. The minimum element $J_2^{52}=-4$, which corresponds to the minimum of J_p for $p=0$. As p varies from zero to one, the minimum point (points?) of J_p will move. Assuming J_p always has a unique minimum point for every value of p, as p varies continuously from zero to one, the minimum point will move continuously from J_2^{52} to J_1^{41}. (If there are multiple minimum points for different values of p, then clearly the situation can be significantly more complex and the family of Pareto points may not all lie on a trajectory.) This is depicted in Figure 19.6, where the indices and outcomes for a set of Pareto-optimal points are shown for this case (note, however, that here we are ignoring the possibility of multiple optima and just finding one for each value of p). Also shown are the losses that result for each player due to the choice of a Pareto-optimal strategy; due to the way that the p parameter results in an interpolation between J_1^{ij} and J_2^{ij} elements, the Pareto value will lie between the two costs as shown for this example.

It should be clear that it is not possible for the outcome to go below the minimum of the two smallest values in the two payoff matrices; hence, the very fact that cooperation is taking place means that one of the two players is

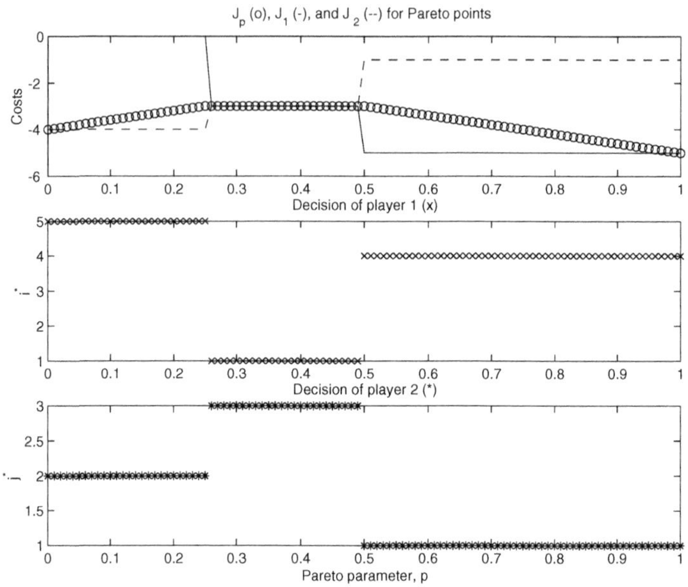

Figure 19.6: Example family of Pareto-optimal points for a bimatrix game (bottom two plots provide the indices of the points), and the resulting outcome (top plot).

sacrificing by losing more than if they were able to minimize their own loss. In applications, however, we are most often interested in the manner in which the Pareto-optimal solution "balances" the optimization to achieve a compromise that is the true goal (cooperation is for achieving a higher goal than individual selfish ones).

Finally, it is interesting to note that $(1, 3)$ and $(4, 1)$ are strategy pairs that we found earlier to be Nash equilibrium solutions for a *noncooperative* game. Recall that in our analysis of nonunique Nash equilibria, the essential problem was that bimatrix games could exhibit elements of "cooperation." Note that here for a cooperative version of the same bimatrix game, as seen in Figure 19.6, we find that for some values of p, the Pareto-optimal solution corresponds to each of the two Nash solutions. This shows in another way that the Nash solutions *required* cooperation of a certain type. For some ways of balancing objectives (values of p), cooperation results in one Nash solution, and for other values of p, it characterizes a different type of cooperation and hence, a different Nash solution.

Defining the Pareto Cost and Finding Pareto Solutions

In this section, we will outline a few problems that you encounter in trying to define Pareto costs and compute Pareto solutions. We will do this via the game in Figure 19.2(b) and, analogous to the above example, let

$$J_p(\theta^1, \theta^2) = pJ_1(\theta^1, \theta^2) + (1-p)J_2(\theta^1, \theta^2)$$

for $p \in [0, 1]$. To give insight into the shape of the cost surface J_p, see Figure 19.7, which is the case for $p = 0.5$. Clearly there are multiple local minima so finding the global one can be challenging. Also, note that the p parameter will in this case scale the "depth" of the two minima.

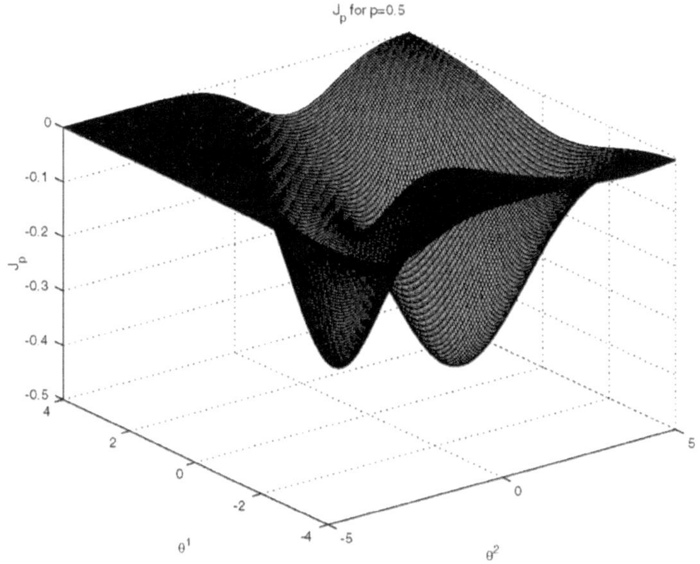

Figure 19.7: Pareto cost $J_p(\theta^1, \theta^2)$ for $p = 0.5$.

It can be difficult to pick an appropriate Pareto cost for an application.

Problems with Forming Pareto Costs via Scalarization: Figure 19.8 shows two possible Pareto solutions for this case, which are $(0, -1)$ (roughly for $p \in [0, 0.455]$) and $(0.55, 2.99)$ (roughly for $p \in (0.455, 1]$). Neither of these Pareto solutions corresponds to a Nash, minimax, or Stackelberg solution for this game. Intuitively, as p varies from zero to one, there is a point at which the deepest "well" in Figure 19.7, switches from one well to another. The "family" of Pareto solutions lies only on two isolated points. Hence, in this special case, there is the curious property that the Pareto points lie on the reaction curve of one or the other player. Hence, for some values of p, even with cooperation, P_1 (P_2) ends up gaining significantly and P_2 (P_1) loses significantly (i.e., cooperation here entails sacrifice by one player for the other). For some applications this may be satisfactory; however, for others, you may need to choose a different form for $J_p(\theta^1, \theta^2)$ that allows for a *smooth balancing* between the selfish objectives of P_1 and P_2.

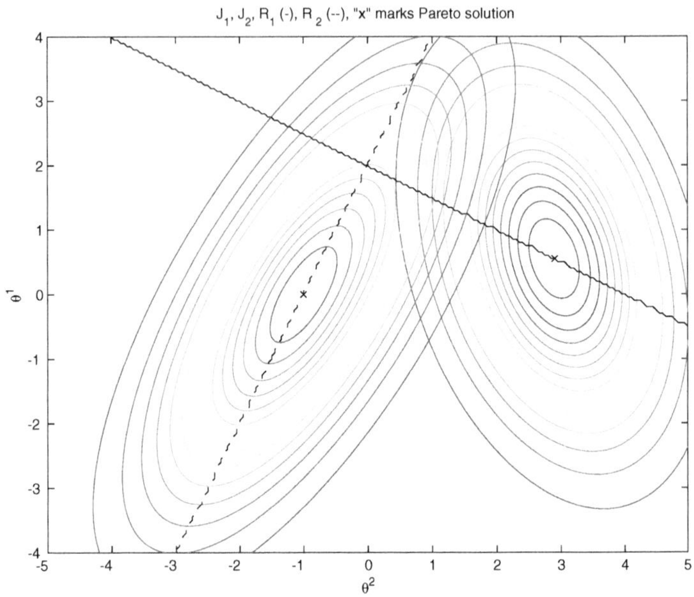

Figure 19.8: Contour plots of J_1 and J_2, reaction curves R_1 (solid) and R_2 (dashed), and "×" marks the two Pareto solutions.

Choice of Pareto cost affects how cooperation between players is achieved.

The Family of Pareto Points—Other Ways to Balance Cooperation: By inspection, it should also be clear that there are *many* other Pareto-optimal solutions. Can you sketch additional Pareto-optimal points on Figure 19.8? Would a point such that the gradients of both cost functions point in opposite directions be a Pareto point? If it were, then, where are such points on Figure 19.8? How would you *compute* all Pareto-optimal solutions for this case? A computationally intensive approach to approximating the set of all Pareto points is to simply directly apply the definition of Pareto optimality (given m and n, how many comparisons are needed to compute all Pareto solutions for a bimatrix game?). When we do this, we get all the Pareto points shown in Figure 19.9 (notice the rough edges on the contour plot due to the coarse discretization). Of course, these points include the ones that result from the scalarization approach of the above example. The others represent other ways to balance the two performance objectives. Notice that the family of Pareto points is not a point or a curve, but a set. In general, it is difficult to characterize or compute the *entire* set of Pareto points, except in certain special cases (e.g., like the one we examined earlier).

Essentially, the definition of our value function J_p via scalarization results in missing all these other Pareto-optimal points. Is this good or bad? If you consider the value function found via scalarization to specify your true preferences, then missing other Pareto-optimal solutions is not a problem. If, however, other

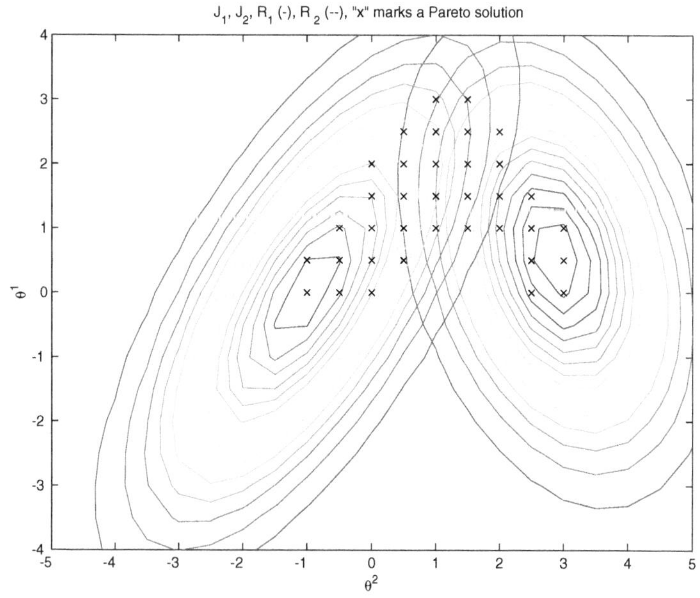

Figure 19.9: Family of Pareto points.

types of preferences are needed, it may be difficult to know how to parameterize and define the J_p function, so that conventional optimization can be used to find what you consider to be good Pareto points. One other way to parameterize the above function would be to have p and $(1-p)$ scale the minimum points of the functions; however, in practical examples you may not know the minimum points a priori. In short, this example shows some of the problems one can encounter by using a simple weighting scheme to turn a multiobjective optimization problem into a conventional single-objective optimization problem to find a Pareto-optimal solution. Other ways of forming the Pareto cost can result in a better balancing and hence, a more "fair" cooperation may be obtained.

19.3 Design Example: Static Foraging Games

In this section, we study the basics of competition and cooperation in foraging games by considering a static, full-information, one-stage (i.e., one decision) game to illustrate the Nash, minimax, Stackelberg, and Pareto solution concepts of Section 19.2.

19.3.1 Static Foraging Game Model

We start by formulating a very simple model of a foraging game; after its development it will become clear how to extend this model to more general cases.

We consider a two-forager ($N = 2$), static, discrete, full-information "foraging game on a line." This is the one-dimensional case, where we consider multiple resources to be distributed over the real line and the foragers move on that line to get the resources.

Foraging games have foragers as players and nutrients as payoffs.

Suppose that the resources are distributed in "cells" (bins) along the real line. Suppose that there are M different types of resources in Q cells and denote the initial distribution of resources of type m to be $r^m(q)$, $q = 1, 2, \ldots, Q$, $m = 1, 2, \ldots, M$. Here, we assume that $r^m(q) \geq 0$, $q = 1, 2, \ldots, Q$, but the model is easily extended to the negative resource case (where one could think of moving to avoid regions where resources are lost). Let D_1 (D_2) be the number of decisions that forager 1 (2) can make and $\theta^1 \in \{1, 2, \ldots, D_1\}$ ($\theta^2 \in \{1, 2, \ldots, D_2\}$) be those decisions, which correspond to forager 1 (2) moving to a cell q if $\theta^1 = q$ ($\theta^2 = q$), $q = 1, 2, \ldots, Q$. We assume that $D_1 = D_2 = Q$, so that each forager can move to any available cell.

Effort and Resource Consumption

Let z_1 (z_2) denote the effort allocated by forager 1 (2) to consume resources. For simplicity, we assume that the same amount of effort is expended for consumption of each resource type $m = 1, 2, \ldots, M$ when a forager goes to a cell. Let $P(q)$ be the set of foragers that decides to go to the same position q to consume resources there; hence,

Effort expenditure can be counted against nutrient returns in computing payoff.

$$P(q) = \left\{ i : \theta^i = q \right\}$$

Notice that $0 \leq |P(q)| \leq N$ for all q and $\sum_{i \in P(q)} z_i$, the total consumption effort at q, is zero if $|P(q)| = 0$.

Assume that α^m, $m = 1, 2, \ldots, M$, is used to model the depletion rate of resource m in the presence of consumption effort. We model the amount of resource of type m remaining at cell q after one play (one unit of expenditure of effort) as

$$r^m(q) e^{-\alpha^m \sum_{i \in P(q)} z_i}$$

This type of model, which is used in foraging theory, represents that initial expenditures of effort in a cell yield more resources than later ones. Hence, as resources diminish in a cell, there is a need for increasing amounts of effort to get the same return. With the exponential model, effort expenditure always provides a return on the investment; other models could represent complete depletion of a resource after a finite amount of effort.

Next, we define the amount of *consumption* given that a strategy pair (θ^1, θ^2) is played by foragers 1 and 2. To do this, note that if both foragers are in the same cell expending effort to consume the same resource, then they have to split the resource, since there is a type of competition for it. Here, we simply assume that if two foragers are at the same cell, then they split the resources evenly. Let the amount of consumption of resource m for decision pair (θ^1, θ^2) for foragers 1 and 2 be defined as follows:

1. *Foragers at different locations:* If $\theta^1 \neq \theta^2$, then for $i = 1, 2$,

$$C_i^m(\theta^1, \theta^2) = r^m(\theta^i)\left(1 - e^{-\alpha^m z_i}\right)$$

2. *Foragers at the same location:* If $\theta^1 = \theta^2 = \bar{\theta}$, then for $i = 1, 2$,

$$\begin{aligned} C_i^m(\theta^1, \theta^2) &= \frac{1}{|P(\bar{\theta})|} r^m(\bar{\theta})\left(1 - e^{-\alpha^m \sum_{i \in P(\bar{\theta})} z_i}\right) \\ &= \frac{1}{2} r^m(\bar{\theta})\left(1 - e^{-\alpha^m(z_1+z_2)}\right) \end{aligned} \tag{19.5}$$

Foragers going to the same location results in a type of competition.

So, in cases where forager 1 (2) goes to a cell that forager 2 (1) does not go to, $r^m(\theta^1)$ ($r^m(\theta^2)$) is the initial amount of resource of type m and $r^m(\theta^1)e^{-\alpha^m z_1}$ ($r^m(\theta^2)e^{-\alpha^m z_2}$) is the amount remaining after consumption. When both foragers go to the same cell, then they both expend effort, but they have to split the returns in half. This results in a resource conservation property of: "all that is consumed plus what is remaining is equal to what was initially there."

Forager Payoffs: Consumption, Energy, and Danger Avoidance

We assume that each forager has certain priorities to consume different resources. We denote these by p_1^m (p_2^m) for forager 1 (2), $m = 1, 2, \ldots, M$. You can think of these priorities as representing preferences or "tastes" for resources. One aspect of the cost to forager 1 (2) that it wants to minimize is given by the negative total consumption weighted by the priorities

$$\begin{aligned} J_{1c}^{ij} &= J_{1c}(\theta^1, \theta^2) = -\sum_{m=1}^{M} p_1^m C_1^m(\theta^1, \theta^2) \\ J_{2c}^{ij} &= J_{2c}(\theta^1, \theta^2) = -\sum_{m=1}^{M} p_2^m C_2^m(\theta^1, \theta^2) \end{aligned}$$

where $\theta^1 = i$, $\theta^2 = j$, and J_{1c}^{ij} and J_{2c}^{ij} constitute a matrix representation of the game. So the problem for forager 1 (2) is how to pick θ^1 (θ^2). The adversarial nature of the foraging game will dictate what to choose (e.g., in a competitive game, each forager may get less than if they cooperate).

Foraging often requires energy consumption to go to a cell from some initial location (e.g., for locomotion). Here, we will think of the foragers as being located at position "0" (i.e., on one edge outside the foraging area) initially. Then, we model the cost to move along the line to go to position i (j) for forager 1 (2) as

$$J_{1e}^i = J_{1e}(\theta^1) = w_{e1} i \quad \left(J_{2e}^j = J_{2e}(\theta^2) = w_{e2} j\right)$$

where $\theta^1 = i$, $\theta^2 = j$, $w_{e1} \geq 0$ and $w_{e2} \geq 0$ represent the unit amount of energy expenditure to move one unit (e.g., from cell 1 to cell 2), and we assume that energy expenditure is not affected by the actions of the other forager. When the energy to conduct foraging is taken into account, then it may be possible that even though a resource is plentiful, a forager may choose a closer one that would provide less return on its effort investment, so that it tries to maximize its amount of resource return for a certain investment in foraging energy. Clearly, there are many ways to model this basic aspect of foraging.

For many foragers, there are areas of the foraging environment that are more "dangerous" than others. This may be due to a predator who is trying to consume the forager, or other environmental characteristics (e.g., presence of a noxious chemical). Clearly, one could model the situation where one forager "consumes" the other. Here, we consider a simple model of location-dependent danger for forager 1 (2) with

$$J_{1d}^{i} \geq 0 \ \left(J_{2d}^{j} \geq 0\right)$$

where bigger values of the costs represent worse areas to be in and actions of the other forager do not affect the danger to a forager.

A forager generally wants to get as many high priority resources for a given energy investment, while avoiding as many dangers as possible. Hence, we can think of forager 1 (2) trying to minimize

$$J_1^{ij} = J_{1c}^{ij} + J_{1e}^{i} + J_{1d}^{i} \ \left(J_2^{ij} = J_{2c}^{ij} + J_{2e}^{j} + J_{2d}^{j}\right)$$

so that it maximizes the amount of resources it gets and minimizes the energy expenditure and exposure to dangers to get them. With this, if $J_{1c}^{ij} = J_{2c}^{ij} = 0$, $J_{1e}^{i} > 0$, and $J_{2e}^{j} > 0$ for all i and j, and $J_{1d}^{i} > 0$ and $J_{2d}^{j} > 0$ for all i and j, then the foragers would not even want to move to a location, since there would be no return of resources for an energy expenditure and exposure to danger (however, for our model we force them to move, so they are not able to choose the option of not playing the game). Generally, the foragers will move farther for resources that are more important to them, but that assumes there is not too much danger.

Notice that we have set this up as a static bimatrix game. Hence, each forager knows everything about the game (e.g., the payoffs, costs of movement, dangers, the other forager's objectives, etc.). Next, for the sake of illustration, we will provide a numeric example.

19.3.2 Competition and Cooperation for a Resource

Choose $D_1 = D_2 = Q = 21$, $M = 1$, $z_1 = z_2 = 1$, $\alpha^1 = 1$, $p_1^1 = p_2^1 = 1$, and $w_{e1} = w_{e2} = 0$ (no energy required for foraging). Assume that $J_{1d}^{i} = J_{2d}^{j} = 0$ for all i and j. The initial resource distribution is shown in Figure 19.10.

The cost functions J_1^{ij} and J_2^{ij} are plotted in Figures 19.11 and 19.12. Notice in Figure 19.11, that if you hold j constant, then forager 1 generally gets more consumption and hence, more payoff if it moves to where there are more resources; however, if both foragers move to the same location, they get less,

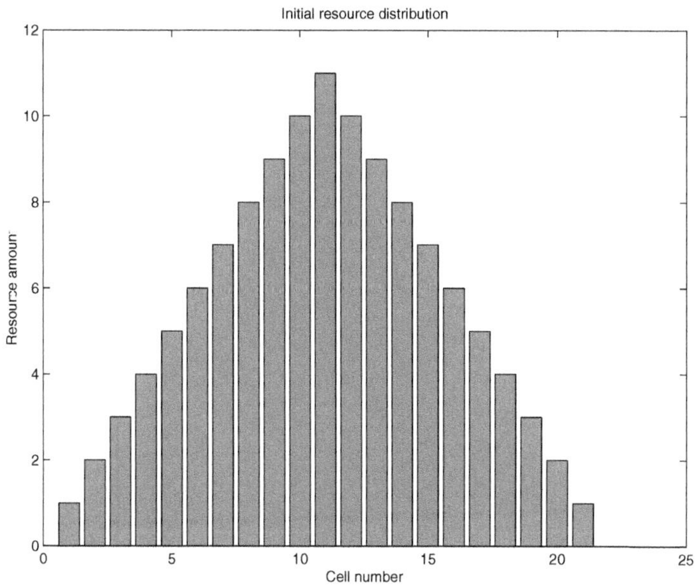

Figure 19.10: Example initial resource distribution.

since they will then compete for resources at that cell. This competition is represented by the ridges of increased cost (a competition cost) that cut diagonally through Figures 19.11 and 19.12.

In competitive foraging, forgers may gain less than if they foraged cooperatively.

First, suppose that we have an adversarial (noncooperative) game, so that the foragers do not coordinate where to go to forage. There are four Nash solutions

$$(10, 11), (11, 10), (11, 12), (12, 11)$$

Does this make sense? From Figure 19.10, the cell with the most resources is cell 11. In the presence of competition, one forager gets the most resources and the other gets the second highest amount possible and these are the four strategy pairs that represent this. Note, however, that the problem of nonunique Nash solutions arises. Forager 1 may pick 11, and with no communication and coordination, forager 2 may also pick that point and there will then be *less* payoff than if the above solutions were chosen. In fact, the minimax solution in this case is $(11, 11)$, since if each forager tries to minimize its maximum possible losses, then it will go to the location with a maximum number of possible resources, since if it goes to a cell with fewer resources, then the other forager can go there also and both would get even fewer resources. A Stackelberg solution with forager 1 as the leader is $(11, 12)$. Why?

Next, consider a cooperative foraging game. First, suppose that the two foragers cooperate by using a Pareto cost found via scalarization as $J_p^{ij} = pJ_1^{ij} + (1-p)J_2^{ij}$ with p as the Pareto parameter that balances the cooperation. Pareto

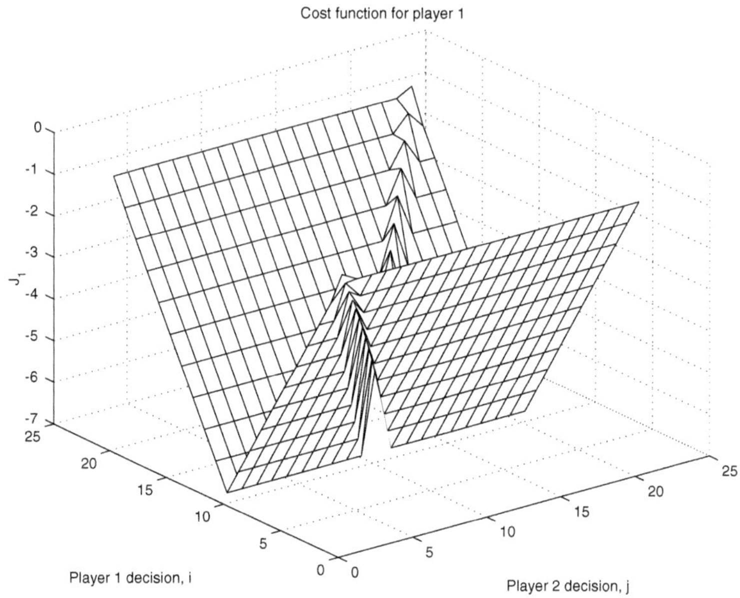

Figure 19.11: Cost for forager 1, J_1^{ij}.

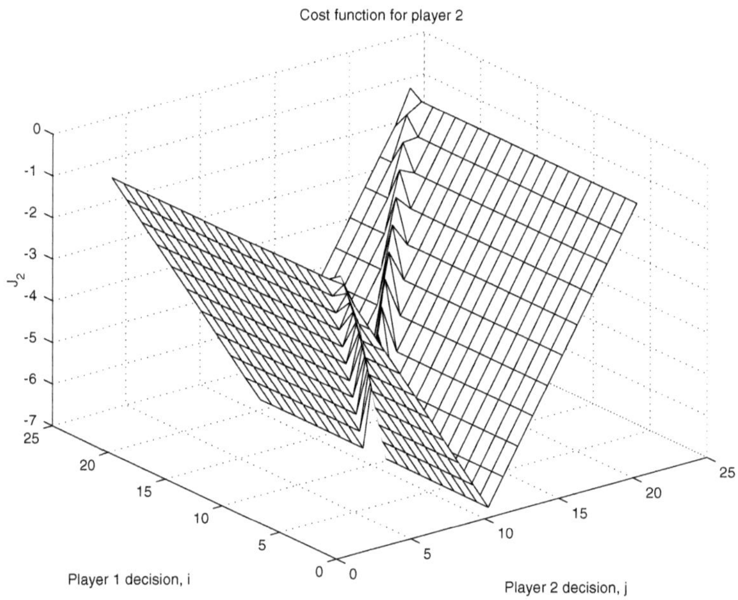

Figure 19.12: Cost for forager 2, J_2^{ij}.

points found in this case are shown in Figure 19.13. We get Pareto points (which are also Nash solutions) (10, 11) or (11, 10), depending on the Pareto parameter p so long as $p \in (0, 1)$. The two foragers would communicate to decide who goes to which location, which, as opposed to the Nash game, is possible since the two foragers are cooperating. The one that goes to position 11 will get the most resources. When p is close to zero, it favors forager 2, so forager 2 goes to position 11, and when p is close to one, it favors forager 1, so forager 1 goes to position 11. The p parameter can be used to balance the cooperation to favor one forager or the other. What happens in the case where $p = 0$? Then, the J_1^{ij} cost does not enter into J_p^{ij} since it is multiplied by $p = 0$. This means that forager 2 makes its best decision and goes to position 11, and forager 1 can go *anywhere* else. In Figure 19.13, it goes to position 1, simply due to how the code was written. The case for $p = 1$ is explained similarly.

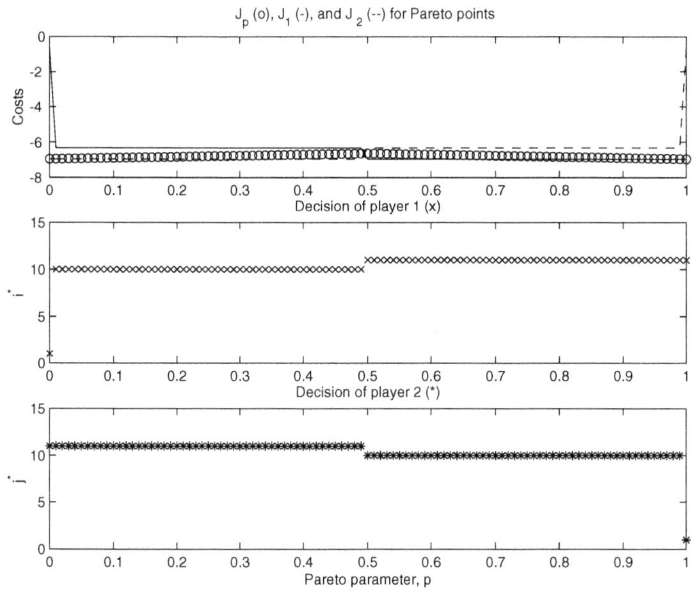

Figure 19.13: Set of all Pareto points for a cooperative foraging game, scalarized Pareto cost.

The scalarization approach is, however, only one way to form the Pareto cost. The set of all Pareto points for the game is shown in Figure 19.14, and you can see that the ones that arise from the above scalarization approach are a subset of all possible Pareto points. These other Pareto points represent different ways to balance the payoffs to each of the two foragers. First, notice that all the Nash solutions are a subset of the Pareto points. Why do the other Pareto points make sense (e.g., the one at (11, 20))? What type of Pareto cost might be used for those cases?

It is interesting to note that you can view a cooperative foraging game as one

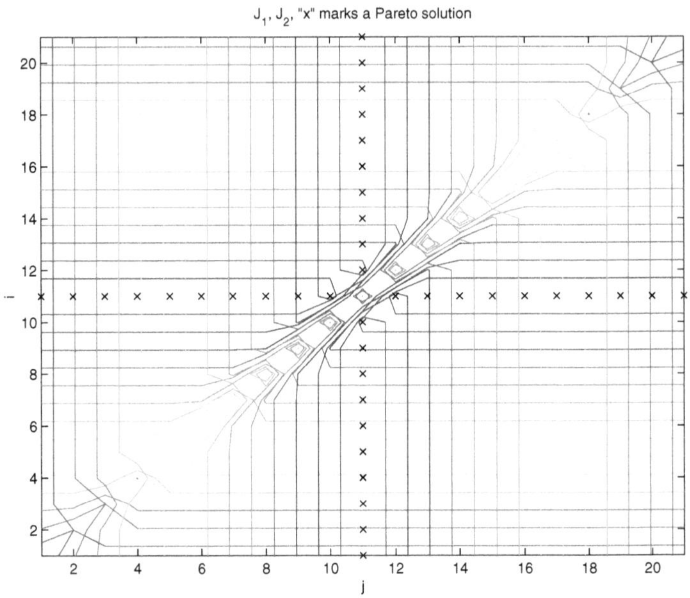

Figure 19.14: Set of all Pareto points for a cooperative foraging game.

where you try to *allocate* resources to all the foragers so that everyone "wins," with the relative payoffs given by which Pareto points you choose. In this case, scalarization provides a nice way to balance the allocation.

19.3.3 Energy Constraints and Multiple Resources

Energy expenditure considerations drastically influence foraging decisions.

Choose $M = 2$, $p_1^1 = p_1^2 = 1$, $p_2^1 = 1$, and $p_2^2 = 2$, so that forager 2 places a high priority on getting resource 2. We let $w_{e1} = w_{e2} = 0.1$, so that moving to cell locations with higher values of q costs more energy. As before, we have $D_1 = D_2 = Q = 21$, $z_1 = z_2 = 1$, $\alpha^1 = \alpha^2 = 1$, and $J_{1d}^i = J_{2d}^j = 0$ for all i and j. The initial resource distribution for the two resources is shown in Figure 19.15.

The cost functions J_1^{ij} and J_2^{ij} are plotted in Figures 19.16 and 19.17. The "ridge" arises as in the last section and it represents the case where the two foragers choose the same cell and hence, compete. Focus on Figure 19.16, and notice that, even though it sets an equal priority for both resources, the costs generally increase as i increases (ignoring the ridge) due to the presence of the J_{1e}^i term that represents the energy needed to forage at each position. This raises the cost of the second resource. Notice that in Figure 19.17, we have the presence of this same effect, *and* the effect of the higher priority of resource 2 for forager 2 so that for forager 2, even though it has to travel farther to get resource 2, since it is higher priority, it may be willing to do that.

Consider the competitive case first. The unique Nash solution is $(5, 14)$.

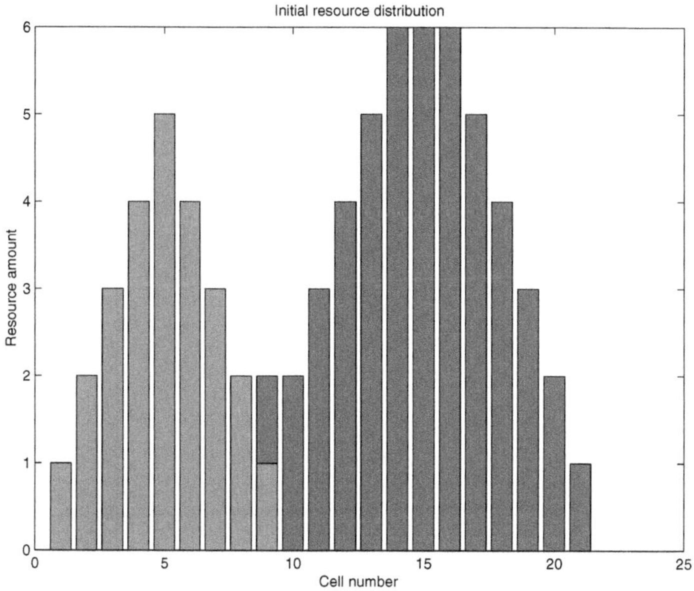

Figure 19.15: Initial resource distribution (darker shaded bars on the right are the second resource) with an "overlap" of resources in the middle designated by "stacking" the plots).

Essentially, with the above choices, forager 1 chooses resource 1 since it is close, but forager 2 picks resource 2, since its level of priority is high, so it is willing to expend the energy to get it. Notice that the maximum for the second resource is achieved at three contiguous positions, but forager 2 picks the smallest of these to minimize energy. The minimax and Stackelberg strategies are both $(5, 14)$. Why?

If the foragers enter into a cooperative game, with a scalarized Pareto cost $J_p^{ij} = pJ_1^{ij} + (1-p)J_2^{ij}$, then we get the Pareto solutions all at $(5, 14)$ for all p. In this case, the two foragers' objectives are so different that there is nothing to be gained by cooperation (and nothing to be lost by competition) and hence, there is really no need for communication.

Cooperative foraging strategies can be viewed as a type of allocation of foragers to resources to maximize payoff to the group.

19.4 Dynamic Games

Dynamic games are ones where players use information about how the game has evolved in order to make decisions. This notion is perhaps closer to the common notion of a game, where there are repeated observations, decisions, and actions by each player, and a resulting dynamic interaction between players in some "arena."

Dynamic games consider repeated decisions, actions, and observations by the players.

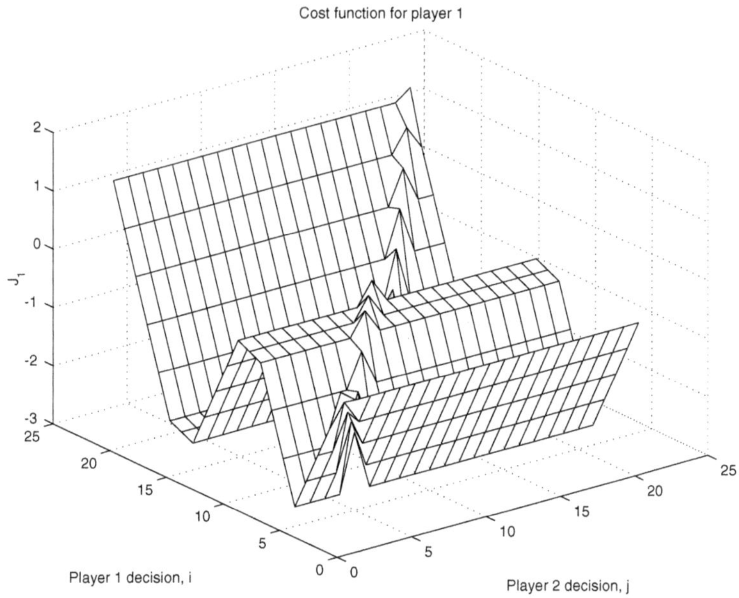

Figure 19.16: Cost for forager 1, J_1^{ij}.

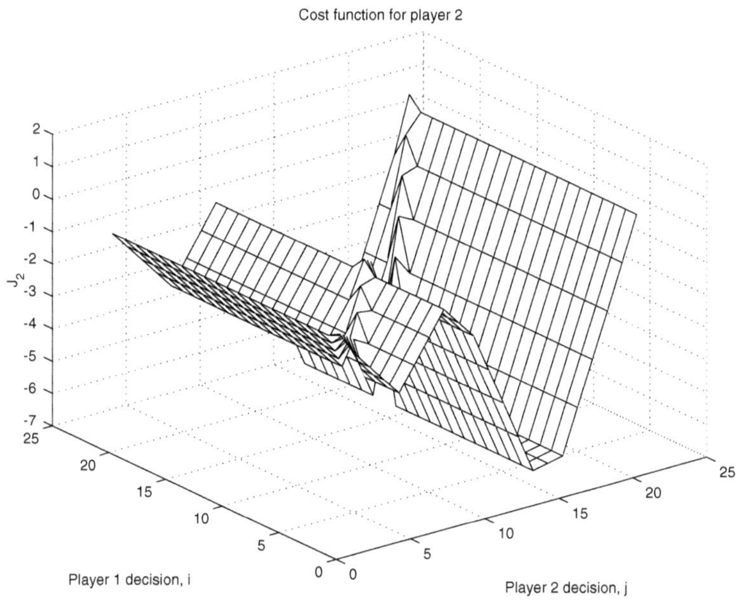

Figure 19.17: Cost for forager 2, J_2^{ij}.

19.4.1 Modeling the Game Arena and Observations

We assume that there are N players and use a discrete-time formulation. To define the dynamic game, we will produce a model of the game, including all the players, rules, and payoffs. First, let

$$x(k) \in X \subset \Re^{n_x}$$

denote the state of the game at time k, $k \geq 0$. The admissible controls (actions) by player i are for $k \geq 0$

$$u^i(k) \in U^i(k) \subset \Re^{n_u}$$

The outputs (measurements of what is happening in the game) are, for $k \geq 0$,

$$y^i(k) \in Y^i(k) \subset \Re^{n_y}$$

Let

$$u(k) = \left[(u^1(k))^\top, (u^2(k))^\top, \ldots, (u^N(k))^\top\right]^\top$$

and

$$y(k) = \left[(y^1(k))^\top, (y^2(k))^\top, \ldots, (y^N(k))^\top\right]^\top$$

Define the "arena" in which the game is played as f where

$$x(k+1) = f(x(k), u(k), k) \tag{19.6}$$

and suppose that the initial state of the game is $x(0) \in X$. This is a deterministic game model, but it can be time-varying. A stochastic game could be represented with $x(k+1) = f(x(k), u(k), w(k), k)$ where $w(k)$ is used to model stochastic effects.

The observations that player i can make about the arena of the game are specified by the function $y^i(k) = h^i(x(k), k)$ for $k \geq 0$, and if we let $h(x(k), k) = \left[(h^1)^\top, (h^2)^\top, \ldots, (h^N)^\top\right]^\top$, then

$$y(k) = h(x(k), k) \tag{19.7}$$

You could view h as part of the representation of the arena of the game as it models what can be observed by each player while the game is played. The dynamic game evolves by players iteratively making a sequence of decisions and taking a sequence of actions. Observations lead to decisions, which lead to actions, which lead to observations, and so on.

Let $J_i(x(k), u(k))$ denote the loss (cost) function of the ith player at the kth stage of play. When there are multiple stages of play (e.g., N_s stages), one typical choice for the loss of each player is the additive one,

$$J_i^{N_s} = \sum_{k=0}^{N_s-1} J_i(x(k), u(k)) \tag{19.8}$$

Hence, each player tries to choose a sequence of $u^i(k)$ that will minimize its own loss $J_i^{N_s}$ after N_s actions, within the constraints of the game listed above. One other typical choice for the loss function is

$$J_i^{N_s} = \sum_{k=0}^{N_s-1} J_i(x(k+1), x(k), u(k)) \tag{19.9}$$

so that losses are assigned based on the type of change in the state and the player actions, for N_s actions. Note that in general, player i does not know its own cost function $J_i(x(k), u(k))$ since it may not know $x(k)$ and $u(k)$. Moreover, use of other players' cost functions $J_j(x(k), u(k))$, $j \neq i$, in the strategy of a player i requires special assumptions.

19.4.2 Information Space and Strategies

How are player's strategies defined? This is a bit more complicated than in the static game case. Why? Because, now each player may make decisions based on "what they know and when they know it" and hence, it is not assumed that each player knows everything at one time and only one action is taken by each player at that time. To make this more precise, it should be clear that if a player has memory, it can store and recall past observations. Then, for any player i, at the kth stage of play, it may base its decisions to choose $u^i(k)$ on a subset of

$$\begin{aligned}&\{y^1(0), \ldots, y^1(k); \cdots; y^N(0), \ldots, y^N(k);\\ &u^1(0), \ldots, u^1(k-1); \cdots; u^N(0), \ldots, u^N(k-1)\}\end{aligned}$$

(clearly the elements of the subset are defined by what information is available to each player i and when it is available). Note that $u^i(k)$, $i = 1, 2, \ldots, N$, is not allowed in the above set, since these are the decisions that the players are trying to reach based on the available information at time k. Also, if $k = 0$, then there are no u^i elements available. Each such subset is called an "information structure" and the information structure of the game is the set of all such subsets that are used. Let

$$\begin{aligned} I^i(k) \quad \subset \quad & (Y^1(0) \times \cdots \times Y^1(k)) \times \cdots \times (Y^N(0) \times \cdots \times Y^N(k)) \times \\ & (U^1(0) \times \cdots \times U^1(k-1)) \times \cdots \times (U^N(0) \times \cdots \times U^N(k-1)) \end{aligned}$$

denote the "information space" of player i at time $k \geq 0$ (again, for $k = 0$, the information space along the u^i dimensions collapses, since there is no past decision information).

Note that the information space $I^i(k)$ is implemented via an appropriately defined communication network between the players and memory within each player to store past values. In an adversarial game, there may be no communication links between the players, but there may need to be memory to hold past information that was encountered. However, in a cooperative game, the sharing of information that is necessary for real-time cooperation comes via communication. A communication network has a "topology" of communication links between players, each of which may provide for uni- or bi-directional transfer of information between players. (It is sometimes useful to think of the communication network as a graph with nodes as players, arcs as communication links, and the topology as being defined via the interconnection pattern of the nodes.) There can be bandwidth constraints and random but bounded communication delays on any link. Moreover, the topology may be "dynamic," in that it may change based on aspects of the environment or player actions.

In some cases, the constraints of the game may specify $I^i(k)$, but other times, the designer may be able to choose it. For example, if all players only make decisions based on their own current observations of the arena of play and the previous actions of all other players and itself, then

$$I^i(k) \subset (Y^i(k)) \times (U^1(k-1) \times \cdots \times U^N(k-1))$$

A strategy for a player i at the kth stage of play is G_k^i, $k \geq 0$,

$$G_k^i : I^i(k) \to U^i(k)$$

The design of strategies involves designing *both* $I^i(k)$ and G_k^i. For example, if each player i can observe at stage k, only $y^i(k)$ (i.e., its only observation) and all actions $u^i(k-1)$, $i = 1, 2, \ldots, N$, the strategies of the players are defined via a G_k^i mapping for each player that specifies its actions,

$$u^i(k) = G_k^i(y^i(k), u^1(k-1), \ldots, u^N(k-1))$$

(and at $k = 0$, there are no elements in the u^i slots). The "full state feedback" case is when $y^i(k) = x(k)$ for all i and k and is the case where each player has "perfect information" about the arena of play. Note that for each fixed initial state $x(0)$ and fixed player strategies, unique sequences of $u^i(k)$, $x(k)$, and $y^i(k)$ and loss values are generated.

The standard concepts of saddle point, Nash, and Stackelberg equilibrium solutions can be extended to dynamic games. For instance, if the initial state is known, the strategies are fixed, and there are a fixed number of stages of play, then the above provides a normal form description. For this case, we get unique loss values for all the players for a given strategy. Note, however, that there are additional solution concepts, depending on the information space that is assumed. For instance, depending on whether you know the initial state, the sequence of states up to the current one, or only the current state, you get different solution concepts (e.g., the "feedback Nash solution"). Here, we will not consider these additional solution concepts and methods for solving for strategies for those cases. Our focus will be on using only the basic game concepts of the last section, but within the context of dynamic games. This choice is driven by practical issues, such as a desire not to consider only the well-understood "linear system with quadratic cost" case (it is covered in detail elsewhere), yet the need to avoid very general dynamic game formulations that cannot be solved analytically, or that sometimes demand computationally intractable solutions.

19.4.3 Decision and Action Timing

It is useful to clarify some issues related to timing of when decisions (actions) are taken in a dynamic game. First, we think of the game as proceeding over a finite number of N_s stages (time steps), or in some cases it may be appropriate to consider $k \to \infty$. Second, while we use a discrete-time model, the actions need not be *synchronous*, in the sense that you think of the index k as being associated with real time $t = kT$ and $t' = kT + T$ for a fixed sampling period T for all k. The actions of the players may occur *asynchronously*, in the sense that the real time duration between indices k and $k+1$ may be nondeterministic. Third, note that for the above model, all players act at each stage k. We can often, however, for specific applications, define a "null play" choice that corresponds to an action that is equivalent to doing nothing. Fourth, in some games, "simultaneous play" by two or more players may be possible.

Combining these four points, we see that we can represent a dynamic game where players can independently act at random points in time and do not have to be in "lockstep" with each other (e.g., in a two-player game with the two players taking turns). There is a new index $k+1$ whenever *any* player acts. If at that time no other player acts, we represent this via using their null plays. These issues of representation can be important in practical games, where the timing of play is a critical aspect of the arena of play.

19.5 Example: Dynamic Foraging Games

We begin by using the model of a dynamic game to represent a foraging game. Next, we use biomimicry of foraging in nature to specify some candidate foraging strategies.

19.5.1 Dynamic Foraging Game Model

To define the model of a dynamic foraging game, we explain each part of the dynamic game model in the last section.

State and Inputs

We have $N \geq 2$ foragers. The state $x \in \Re^{n_x}$ is composed of aspects of the foraging environment and the positions of the foragers in that environment. Assume that you have a two-dimensional foraging environment (a "foraging plane"). Extension to the three-dimensional case is straightforward. The position of the ith forager is given by

The state holds information on the environment and foragers.

$$x^i(k) = \left[x_1^i(k), x_2^i(k)\right]^\top \in \{1, 2, \ldots, Q_1\} \times \{1, 2, \ldots, Q_2\} = F$$

with $x_1^i(k)$ its horizontal and $x_2^i(k)$ its vertical position on a discrete grid. Here, Q_1 (Q_2) sets the upper boundary for horizontal (vertical) movements. The variable F is used here to denote the set of all points in the foraging environment.

The decisions by forager i are commands to move itself to each of the cells that are adjacent to the current position, and which resource type to consume there. That is, at time k, so long as the movement is in the valid foraging region so that $x^i(k) \in F$, we have that $u_p^i(k)$ is in the set

$$\left\{\left[x_1^i(k), x_2^i(k)\right]^\top, \left[x_1^i(k)+1, x_2^i(k)\right]^\top, \ldots, \left[x_1^i(k)+1, x_2^i(k)+1\right]^\top\right\} \bigcap F$$

which we will denote by $U_p^i(k)$. Here $U_p^i(k)$ is then the set of feasible moves at time k by forager i. The first element in the above set indicates that the forager should stay at the same location, the second indicates that it should move to the right horizontally, and not vertically, and so on (to all positions around the current one). Clearly, in this case, there are nine possible locations that any forager can move to at each step, *provided that* the forager is well within the region F. Disallowing movements outside the foraging region F is represented by the intersection with F. In particular, if a forager is at the upper horizontal boundary and tries to move to the right, we will say that it stays at the same place, thereby employing the basic idea of "projection" used in optimization theory. For convenience, we let $u_p(k) = \left[(u_p^1(k))^\top, (u_p^2(k))^\top, \ldots, (u_p^N(k))^\top\right]^\top$.

Assume that there are M resources that are indexed with the variable m. We represent the choice of resource by player i at time k as $u_r^i(k)$, $i = 1, 2, \ldots, N$, where

$$u_r^i(k) \in U_r^i(k) \subset \{1, 2, \ldots, M\}$$

represents the resource type m that forager i chooses to consume at time k, and $U_r^i(k)$ can be used to model the set of resources that it can choose from (extension to the case where each player can consume more than one resource at a time is straightforward). Define $u_r(k) = \left[u_r^1(k), u_r^2(k), \ldots, u_r^N(k)\right]^\top$. The decision $u^i(k)$ of forager i at time k is composed of a position choice and resource choice. In particular, we let

The input holds the decision of the foragers, where to go and what to do.

$$u^i(k) = \left[(u_p^i(k))^\top, u_r^i(k)\right]^\top \in U_p^i(k) \times U_r^i(k)$$

and $u(k) = \left[(u^1(k))^\top, (u^2(k))^\top, \ldots, (u^N(k))^\top\right]^\top$.

The distribution of resources is also part of the state. Extending the development of the static case, let

$$q = [q_1, q_2]^\top \in F$$

denote a cell in the foraging plane. Let z_i^m denote the effort allocation to consume resource m by forager i. Let

$$P^m(q) = \left\{i : u_p^i = q, u_r^i = m\right\}$$

be the set of foragers that decide to go to position q to consume resource m at time k. Notice that $0 \leq |P^m(q)| \leq N$, but below, we will only use $P^m(q)$ for $q = u_p^i$ for some $i = 1, 2, \ldots, N$, so $|P^m(q)| > 0$.

We use the depletion rate α^m, $m = 1, 2, \ldots, M$, for the mth resource. The amount of resource at time k of type m at cell q is $r^m(q, k)$ with $r^m(q, 0)$ the initial distribution. The resources change over time due to growth (e.g. plants), weather, disease, farming, and foraging. For foraging, resources may diminish due to consumption, and in some cases, such consumption may result in the increase of other resources (e.g., since the resources may be living, so foraging influences their competitive balance). In other cases, foraging for one type of resource at one time may make it possible to forage for other resources later (e.g., if one forager eats one type of resource and this gives rise to other resources due to, for example, a forager leaving behind remains). Suppose that we consider the effects due to foraging where resources diminish according to

$$r^m(q, k+1) = r^m(q, k) e^{-\alpha^m \sum_{i \in P^m(q)} z_i^m} \tag{19.10}$$

for all $q \in F$. For this equation, notice that $P^m(q)$ is a function of u. Let

Forager actions affect the environment and hence, subsequent decisions.

$$x_p(k) = \left[(x^1(k))^\top, (x^2(k))^\top, \ldots, (x^N(k))^\top \right]^\top$$

denote the vector of places where the foragers are located. Let

$$x_r(k) = \begin{bmatrix} r^1([1,1]^\top, k) \\ \vdots \\ r^1([Q_1, Q_2]^\top, k) \\ \vdots \\ r^M([1,1]^\top, k) \\ \vdots \\ r^M([Q_1, Q_2]^\top, k) \end{bmatrix}$$

be a vector that holds a vectorized representation of the resource distribution (maps). The state of the game is

$$x(k) = \begin{bmatrix} x_p(k) \\ x_r(k) \end{bmatrix}$$

Foragers naturally operate somewhat independently of each other, at least with respect to some of their decisions and timing of actions.

Finally, we need to define how to generate the next state (to define f in the game model in Equation (19.6)). First, note that $x_p(k+1) = u_p(k)$, so that at the next time instant, each forager will have moved to the position that it was commanded to move to at the current time step. This represents that we are assuming no dynamics and kinematics for our forager (e.g., constraints on how fast it can move, turn, etc.), or at least, that the time scale is sufficiently slow relative to such physical phenomena. Second, note that $x_r(k+1)$ is defined via Equation (19.10). This completes the definition of how to generate the next state given the current state and the current input to the game; however, we still need to clarify issues related to the timing of decisions by the players.

First, we assume that the real time between k and $k+1$ is fixed so that the real time is $t = kT$, where T is a sampling period. Then, the real time at the next sampling instant is $t' = kT + T$. So, we require that time proceeds according to a clock with a certain tick-length. This is necessary, due to how we model depletion of resources. Why? Because, it makes the effort allocation taken at each step for each resource by each player a constant as we had specified. (If we had random time lengths between decision times, then the fact that one forager makes a decision would affect the consumption rate of other foragers.) We have, however, still created a type of *asynchronous* game model, in the sense that if a forager does not make a move at time k, then it "chooses" its next position to be the same as its current one (the "null play" discussed in Section 19.4.2), so that it continues to forage at the same position. Additionally, the model allows for multiple (up to N) players to simultaneously take actions at each time step and the above formulas define how the state evolves with such simultaneous actions. So, our decisions occur asynchronously, but only at times given by the tick of some clock. If the clock period T is very small, then we can approximate fully asynchronous behavior.

Sensing and Outputs

The observations that forager i can make about the foraging environment at time k are denoted by $y^i(k)$. Clearly, the physiology of the animal constrains what sensing is possible. For instance, some animals can only sense via sampling chemicals in their immediate surrounding environment (e.g., certain bacteria), while others can sense light or sound and hence "see" for long distances. In terms of the mathematical representation, some possibilities for representing the feasible observations are the following:

1. *Full observations:* If for each forager, $i = 1, 2, \ldots, N$, and time k,

$$y^i(k) = x(k) \tag{19.11}$$

   then each forager can sense the distribution of all the resources over the entire foraging environment and the positions of all the other foragers at each time step k.

Sensing is never perfect, so foragers always make decisions under uncertainty.

2. *Resource observations and own position:* If

$$y^i(k) = h^i(x(k), k) = \begin{bmatrix} x^i(k) \\ x_r(k) \end{bmatrix}$$

   then each forager knows its own position and where all the resources are, but does not know the positions of the other foragers.

3. *Range-constrained sensing:* Let $S(q)$ denote the set of cell locations that a forager can sense resources in, or other forager positions, when it is located at cell q. This set can be used to specify characteristics of the sensing capabilities of the foragers. For example, suppose that foragers have constraints on how far they can sense resources that are independent of time and resource type (you could also make sensing range depend on

resource type and time). Then, as the forager moves, the set of cells that it can sense resources in changes. Suppose that this set of cells is defined via a circular region with radius R_s about the current location of the forager, provided that this sensing region is within the foraging region F. In this case, we could define

$$S(q) = \left\{\bar{q} : \sqrt{(q-\bar{q})^\top (q-\bar{q})} \leq R_s\right\} \bigcap F$$

First, form a vector of the forager locations, for foragers that can be sensed, from elements of x_p, as $x_p^{s_i}$ with elements $x^j(k)$, where $x^j(k) \in S(x^i(k))$ for all $j = 1, 2, \ldots, N$. Second, form a new vector of the currently sensed cells from elements of x_r, as $x_r^{s_i}$ with elements $r^m(q,k)$, where $q \in S(x^i(k))$ for all $m = 1, 2, \ldots, M$. If

$$y^i(k) = h^i(x(k), k) = \left[\begin{array}{c} x_p^{s_i} \\ x_r^{s_i}(k) \end{array}\right]$$

then the forager can sense resources in a region around its current location and it knows its own position, and the positions of the other foragers within its sensing range. If R_s is large enough, so that the forager can sense the whole environment no matter where it is in the environment, then this reduces to case 1 above.

Clearly, there are many other possible sensor models. For instance, sensing quality could depend on the resource, forager type, position in the environment, or time (e.g., to represent aging effects). The size of the sensing range may change over time. Different foragers may have different sensing capabilities.

Consumption, Energy, and Payoff to Foragers

Next, we must define the payoff for each forager. To do this, first define the amount of consumption of resource m by forager i, $i = 1, 2, \ldots, N$, at time k for a set of forager decisions u^1, u^2, $\ldots$, u^N, as

$$\begin{aligned} & C_i^m(u^1(k), u^2(k), \ldots, u^N(k)) \\ = & \frac{1}{|P^m(u_p^i(k))|}\left(r^m(u_p^i(k), k) - r^m(u_p^i(k), k+1)\right) \\ = & \frac{1}{|P^m(u_p^i(k))|} r^m(u_p^i(k), k)\left(1 - e^{-\alpha^m \sum_{i \in P^m(u_p^i(k))} z_i^m}\right) \end{aligned}$$

Notice that $|P^m(u_p^i(k))| > 0$. The factor $\frac{1}{|P^m(u_p^i(k))|}$ is used to represent that if there are $|P^m(u_p^i(k))|$ foragers at location $u_p^i(k)$ foraging for resource m at time k, then the returns from foraging are split evenly among those foragers. (Other definitions of splitting the resource returns could represent more capable foragers winning more returns when they forage next to some less capable forager.)

The cost due only to consumption for one move is, given that forager i has priority p_i^m for resource m,

$$J_{ic}(x(k+1), x(k), u(k)) = -\sum_{m=1}^{M} p_i^m C_i^m(u^1(k), u^2(k), \ldots, u^N(k))$$

Each forager must expend energy to forage, and we define this via

$$J_{ie}(x(k+1), x(k)) = w_{ie}\left(x^i(k+1) - x^i(k)\right)^\top \left(x^i(k+1) - x^i(k)\right)$$

where $w_{ie} \geq 0$ sets the amount of energy needed to move a certain distance. We assume that energy is independent of resource type being sought and consumed. The "danger" aspect could be modeled as in the last section, but we ignore this possibility here. Our total payoff to forager i at time k is

$$J_i(x(k+1), x(k), u(k)) = J_{ic}(x(k+1), x(k), u(k)) + J_{ie}(x(k+1), x(k))$$

If there are N_s steps in the game, we have a payoff $J_i^{N_s}$ for playing the entire multistage game as given in Equation (19.9). Each forager wants to minimize $J_i^{N_s}$ and thereby maximize consumption with minimal energy expenditure. This can require considerable finesse, as it may be a good strategy to give up payoffs at some points in time in order to realize more benefits at some other later time.

Information Space and Strategy Design Challenges

Designing a strategy involves picking the information space $I^i(k)$ and strategy G_k^i, and of course, there are many possibilities. Here, for the remainder of this section, we pick one simple approach and invite you to consider others. Suppose that Equation (19.11) holds and each forager only uses $y^i(k)$ so

$$I^i(k) = Y^i(k)$$

and we need to choose G_k^i where $u^i(k) = G_k^i(x(k))$. This corresponds to allowing each forager to observe all forager positions and resources in all cells at each time k. The forager does not, however, have memory so it cannot store, for example, sequences that characterize the pattern of resource depletion or motion of the other players (which could be useful in estimating the intent of other players).

Who knows what and when they know it significantly affects distributed decision-making.

Recall that the information space is also defined via specification of a communication network between the players. In an adversarial game, there may be no communications or communications may be present, but the adversaries may try to mislead each other so that they can gain more themselves. However, in many cooperative games, there may be significant sharing of information over the network. Recall that the network may be defined via a communication topology that says who can communicate with whom. (Think of the topology as a directed graph with nodes as the foragers and arrows pointing from any forager to a forager that can receive or sense information about it.) There can be bandwidth constraints for the communication links, or random but bounded delays in transmitting/sensing information. The topology and communication network characteristics may depend on the locations of the foragers (e.g., if two foragers move too far away from each other, then their communication link may get "broken" and, if two other foragers get close enough to each other, they may be able to "create" a communication link). Moreover, activities of the group of foragers may dictate how the topology should be configured and dynamically reconfigured. Here, we will assume that the communication topology enables the sharing of sensed information for $I^i(k)$ above. This will allow each forager to compute all the decisions of all the other foragers, so that they do not need to share information on $u(k)$.

19.5.2 Biomimicry for Foraging Strategies

Biomimicry may provide a way to define practical (computationally feasible and scalable) distributed and cooperative controllers for autonomous vehicles.

Here, we introduce the idea of using biomimicry of foraging strategies found in nature.

Rules, Planning, Learning

How do we define G_k^i? There are many approaches. One, which may correspond to how simply organisms find food in some environments, would be to use simple "rules" to search for and find food. For instance, such rules may use environmental cues to tell them where to move to be likely to find food.

Another more sophisticated approach is to observe the environment and use past information about how a typical environment holds food, and then to use this model in a planning strategy. The traditional approach in game theory in engineering is in fact to assume some ordering for player decisions and to then use dynamic programming to find optimal paths for N_s steps. Clearly, this can be computationally prohibitive, especially for high values of N, Q_i, L, and N_s. So typically, one approach is to use a "receding horizon" controller (i.e., a planning system), where you use a (perhaps simplified) model of the plant and simulate ahead in time, find the best input sequence, implement the first decision in each sequence, and then repeat. Clearly, this can also have computational problems, except perhaps for the case where you look ahead only one or two steps, or where the model used to simulate ahead in time has appropriate simplifications that reduce complexity, yet still lead to good decisions. (It is generally quite difficult to balance these objectives to get a good simplified model.) This is, moreover, the common approach that has been investigated in many contexts for many problems in the past (e.g., for linear systems with quadratic costs, "model predictive control" (MPC), and more general decision-making systems).

Some animals actually learn the strategy of their opponent, then take appropriate actions in order to optimize their gains. For this, they may have a model of a typical opponent and then observe their actions to guess what type of strategy they are currently using. Then, if they assume that the opponent will not switch strategies soon, they may be able to work more effectively against the opponent.

A Generic Saltatory Strategy

In order to be more concrete about biomimicry of competitive and cooperative foraging, suppose that we want to model "saltatory search" and foraging, where an animal alternates moving and thinking about where to move next. This takes into account physiological constraints for many animals, where they alternate between moving and stopping to sense and decide where to move next.

The set of steps we use to model a "saltatory strategy" is the following:

1. Play a static matrix game of the type studied in the last section at $k = 0$ to determine where each forager should seek to go. Call the resulting goal position $x_*^i(\bar{k})$, where $\bar{k}$ is the index of the times when the forager stops to sense and decide where to go next. Hence, the game played at $k = 0$ results in $x_*^i(0)$ for all $i = 1, 2, \ldots, N$ and these specify the first set of goal positions that the foragers try to move to. (Issues in path choice from the current forager position to the goal position are discussed below.)

2. If player i gets to $x_*^i(\bar{k})$ at time k', it plays a matrix game with all other players even if they did not get to their goal positions that time. This gives us $x_*^i(\bar{k}+1)$, the next set of goal positions. This can result in some foragers switching from one goal position to another if they evaluate that to be profitable.

3. Go back to step 2 if the termination condition (e.g., N_s time steps) is not achieved.

This is a type of "asynchronous" saltatory strategy, where the time it takes between decisions in the m index depends on how far it decides to move each time. How do the foragers move from $x_*^i(\bar{k})$ to $x_*^i(\bar{k}+1)$ (i.e., what path do they take)? This depends on many factors. If the foragers can sense and make decisions during movement, then they may try to move towards the new goal position along a path that will maximize their consumption. Depending on the forager's goals, it may be willing to make significant deviations from a straight-line path between $x_*^i(\bar{k})$ and $x_*^i(\bar{k}+1)$, where the goal is simply to minimize energy consumption to get to the goal position. For instance, the forager may compute a type of optimal path, one that minimizes energy consumption while maximizing resource consumption, between $x_*^i(\bar{k})$ and $x_*^i(\bar{k}+1)$, and thereby obtain more resources (i.e., it tries to do some consumption along the way to its goal cell). It should be clear that overall this strategy could be viewed as a type of planning (or receding horizon) strategy, since it does involve looking ahead in time; however, it is only using very simplified information to decide successive goal positions. Moreover, extension to the case where it looks ahead more than one step across the $\bar{k}$ index should be clear.

The type of game that is played at the times $\bar{k}, \bar{k}+1, \ldots$ depends on what information is used to play the game. For instance, a forager may inherently know that another forager will go to some region and then spend significant time there, since it will minimize its energy expenditures by staying in that region for some time. A forager may then try to "keep its distance" from another forager and pick a "foraging region" rather than a point (of course, this depends on the physical dimensions that correspond to our cell sizes). How can we represent this? One approach would be to define an "abstraction" of the resource distribution part of the state, where elements of the abstraction correspond to, for example, sums of resources of a certain type in sets of contiguous cells. Then decisions about where to go can depend on "super-cells" created by the abstraction. This will result in a computationally simpler game, since there are fewer super-cells to consider moving to. Moreover, it is possible to define a "nested" strategy where once a region is chosen, a game is played between all players that decide to go to that region (and multiple levels of abstraction, and hence, nested games can be played).

It is also the case that some types of foragers know what type of separation to keep with other foragers, especially in the case of cooperative (social) foraging. They do not want to be too close, so they crowd each other and each forager does not get enough resources to survive, but they do not want to be so far from each other that they cannot benefit from communicating with the other foragers so they can be led in the best directions towards the most profitable sites.

Coping with Complexity: Space and Time Abstractions

So, what are the basic concepts employed in foraging in nature that allow animals to overcome computational complexity in deciding how to forage? First, there is the prevailing fact that while foragers do not want to die, the overall species has extraordinary reproductive capability so if they do die, they can be replaced. This is a basic fact of life, but it may not have too much relevance in the case, where we use biomimicry to design automated systems in engineering (e.g., for cooperative robotics). Second, evolution essentially generates practical and robust foraging strategies by optimizing them in the face of complexity constraints (e.g., forager physiology that dictates how much memory it can have). Again, however, at the present time in engineering, it is often impractical to use such a fact. At other times, such as in the area of "evolutionary cooperative robotics," researchers try to evolve good behavior over time. Certainly, there may be the possibility that such evolution could take place a priori and in simulation rather than in actual hardware. Regardless, the problem is that there may be little guidance on how successful such an evolutionary foraging strategy design approach will be.

We cope with complexity via hierarchies and space or time abstractions.

Here, we simply assume that we will observe *existing* foraging strategies and use biomimicry to capture the essential principles of their operation that help the forager cope with complexity. While it is clear that there are many principles used to cope with complexity, and that it may be difficult to observe the basic principles used to cope with complexity for any given species, here we will focus on the principles that seem to arise when one studies how saltatory strategies operate as we discussed. There seem to be at least the two following general principles:

- *Spatial abstraction:* Decision-making often involves having animals "group" aspects of the foraging environment in order to make decisions. For instance, an animal may look in a few directions and pick the general direction that seems to have the most resources. In this way, it avoids being too greedy by favoring resource "peaks" that may have few resources around them, and hence, does a type of prediction since it knows that it will continue to forage and try to minimize energy consumption by not moving out of a region.

- *Time abstraction:* Next, there is a type of "asynchronous time decimation," where the animal does not decide what it will do at each small time step, but only where it should be (or what it should be doing) at certain future times. As they move toward their current goals, they solve the problem of what to do along the way.

These seem to be fundamental principles driving the design of practical foraging strategies, and how to cope with complexity in decision making.

Finally, note that complexity presents significant challenges in distributed decision making. Even for the simple strategy defined in the past section, complexity can be significant, especially for large N. For instance, for a full information cooperative game, there are three decision variables (horizontal and vertical position to move to and nutrient to seek there) for each of the N players and so computing the cost involves finding and computing an optimal value for a very large matrix game. (How large is the matrix for the case described above?)

19.6 Challenge Problems: Intelligent Social Foraging

Bacteria forage according to relatively simple rules that dictate how they climb up nutrient surfaces, or aggregate for the purpose of survival. Their biochemical "brain" is very simple, essentially a bag of molecules where chemical reactions implement foraging "decisions." Recall that we assume that there is a "cognitive spectrum" of intelligence in making foraging decisions. For instance, some higher animals have central nervous systems and via these they can achieve planning, attention, and learning. We can think of such animals as "intelligent foragers." Intelligent foragers typically also have an ability to communicate so that they can achieve "social foraging." For instance, they may work together as a group to improve their foraging success and survival chances. Some animals are of relatively low intelligence, but enhance their foraging success with communications (e.g., bacteria and ants) to gain an "emergent" intelligence for the group. Other animals have significant intelligence but may not exploit their communication capabilities to a great extent, since a "loner" approach in foraging in their environment is more successful.

Multiple vehicle guidance and decentralized decision-making problems provide nice classes of applications to integrate the methods of this book.

Here, we first focus on intelligent individual (i.e., nonsocial) foraging by explaining how to use planning, attention, and learning methods for foraging. To do this, and in order to be concrete, we discuss yet another nongradient optimization method that is based on the use of "surrogate models" (in a foraging problem, a model of the foraging landscape that includes information about where predators and prey are). We challenge the reader to solve a particular type of intelligent foraging problem with a surrogate model method, by presenting it as a "design challenge problem" for students, where they can integrate earlier methods from this book. (For this problem, there may or may not be multiple foragers and competition.) Next, we discuss the social foraging problem by introducing it as a second design challenge problem, which also requires the use of coordination (and hence communications) among multiple agents. The main challenge is to implement distributed rule-based, planning, attention, or learning strategies in order to coordinate the behavior of a group of foraging agents. Varying levels of intelligence, distribution, and communications are discussed. Aspects of competition are discussed, and in general, there are multiple teams of foragers that compete for various resources in the environment.

Finally, note that other problems not related to foraging could be used in place of the problems defined in this section to provide a challenge problem for the student (e.g., process-wide control/automation for a factory).

19.6.1 Intelligent Foraging

In this section, we explain how to develop a nongradient optimization method that relies on planning, attention, and learning and hence, provides an algorithmic approach to intelligent (nonsocial) foraging. This is only one of many possible methods that can be envisioned for the solution of this problem. We discuss intelligent foraging in the context of this method simply to be concrete about the ideas, the connections to optimization, and how a simulation might be constructed.

Surrogate Model Method Representation of Intelligent Foraging

A surrogate model method simultaneously learns an approximation to the cost function and uses it to guide where to search the cost function.

For some optimization problems, it is not only the case that it is impossible to compute or know the analytical gradient, but it can also be the case that it is *very expensive* to compute a value of the cost function for each point in the optimization space (e.g., if the computation requires extensive simulations using a very complex model, use of experimental apparatus, or physical rearrangement of physical elements in an environment to determine the value). One approach to such problems is to use a "surrogate model" to represent the cost function. The idea is that each time you compute a value of the cost function, you use the pairing between the test point in the optimization domain and the cost function value that is computed to form a training data pair, and then construct an approximator for the cost function (perfectly analogous to how we did this in Part III, but here the focus is on learning the mapping implemented by the cost function). How could an approximation of a cost function be useful? The key is to note that it can be difficult to compute points on the actual cost function, but it can be very easy to compute test points on the approximation to the actual cost function. (If the approximation is good, the values will be close.) Moreover, if the approximation is reasonably good, it can suggest points that are good candidates for testing on the actual cost function. Remember the response surface methodology of Section 15.2, where we approximated a cost function with an approximator and used it to pick an optimal design point. The difference here is that our surrogate model method will operate in real time via sequential acquisition of information (not in the "batch-mode" that RSM typically uses).

The surrogate model method proceeds as follows:

1. Pick a test point (or set of test points) for J and compute J at this point (these points). Note that the method can be "set-based" so that it computes in parallel the cost function at several test points (e.g., via parallel processing).

2. Store the pairing(s) between the test point(s) and value(s) of J in a training data set G for an approximator f for J.

3. Construct an approximator (interpolator) for the data in G (perhaps removing some points as others are added). This approximator retuning can be achieved via repeated application of recursive least squares over a linear in the parameters approximator, or via application of a Levenberg-Marquardt method to training a nonlinear in the parameter approximator.

4. Perform an optimization *over the approximator surface* (not the cost function) to find a minimum point on that surface. (You may use gradient methods or pattern search methods to perform this optimization.) Call this a new test point, compute J at this point (and for a set-based method, perhaps at a pattern of points around it), and add this (these) to the training data set. Go back to step 3.

You can think of this as constructing an approximation of the cost function to guide you to make choices about where to explore the cost function to find the minimum (the "search" proceeds via the optimization over the learned surface with periodic updates via sampling the actual cost function and updating the approximation surface). You can think of the optimization process over the approximator surface as providing a strategy for picking points to include in the training data set G. Clearly, when applied to specific problems, the strategy for picking points to include in G may be constrained by the problem at hand (e.g., if applied to a foraging problem, the surrogate model may be a representation of some aspect of the environment and you may be constrained in choosing candidate points on the surrogate model by how fast the vehicle can move and what sensing resources it has—e.g., whether it can sense at a distance).

If you use a pattern search method, it may make sense to think of points on the pattern as predictions about J, and the selection of points as a selection of a plan, and the approximation process as learning. In this way, you can think of the surrogate model method as defining a class of learning-planning methods, where it is possible to choose a whole variety of pattern search methods as the basis for planning and gradient optimization methods as a basis for learning (of

A surrogate model method can be thought of as a method for integrated learning and planning.

course, as pointed out in the last section, it is possible to use the pattern search methods as a basis for learning).

Example: Intelligent Foraging Over Nutrient Surfaces

Here, we provide a brief explanation of one way to define aspects of the environment as a cost function that can then be used in a surrogate model method to emulate intelligent foraging.

- Define a nutrient surface, analogous to how we did it for bacteria. This is the surface that we want to learn about and plan over.
- Add an attentional map defined over that same domain that simply says where we have searched and where we have not (so if we have visited one region, then change the map to indicate that, and make it so that climbing down the attentional map surface corresponds to looking in unexplored areas; in the theory of surrogate optimization, this is sometimes called a "merit function").
- Add the nutrient surface to the attentional map and call that the cost function. Use planning over the currently learned map so that there is a type of look-ahead for the forager to decide where to move; however, it can only make its decisions on the learned map, not the actual one (but it does know the entire attentional map). This way, in seeking to minimize the cost, it will try to achieve competing objectives: (i) try to find the lowest point which corresponds to good food, and (ii) try to search the entire surface (as dictated by the attentional map). It balances a desire to search far and wide, possibly finding a better food source, with the desire to eat now. The attentional map helps it to avoid getting stuck in a local minimum.

What are the key elements to coding this? First, it seems logical to use a set-based method, some pattern of sensed points of the nutrient cost function, placed around the current position of the forager. Second, vehicle dynamics should be kept simple but must be present in some form (e.g., it cannot be that you can move the vehicle in one time step an arbitrary distance across the optimization domain—this is a key difference from standard nongradient optimization methods). One way to model this is to use a "momentum-term" (see gradient methods) in the optimization algorithm update formula. Third, it seems logical to use a linear in the parameter approximator for the nutrient function, perhaps with RLS to compute the approximator update at each step (so run RLS for each point in the pattern, at every time step). Fourth, an RBF could be used for the attentional map, with a simple strategy to update it based on where the forager actually visits (e.g., it could simply change the map to represent that the pattern of sensed points was there). Fifth, we see then that planning corresponds simply to the optimization over the surrogate model (which here would be the approximation of the nutrient cost function, plus the attentional map), and since you want the planning method to consider many directions from the current forager position, it seems logical to try a pattern search method (e.g., simple coordinate search or multidirectional search over the combined approximator/attentional map). It should be possible to show a movie of learning a nutrient map, the updates to the attentional map, and it should be that these maps would give good insights into design of the strategy. How computationally complex will the method be? Well, this depends on the resolution you choose for your approximator for the nutrient map, and the attentional map. Also, it depends on the optimization method that you use for the approximator surface, how big the set is in the set-based method, and how long your planning horizon is.

19.6.2 Intelligent Social Foraging

In this section, you are asked to consider the wide range of possibilities for how to design the control and guidance algorithms for automating a group of intelligent social foraging *vehicles*. Hence, this section serves to specify a "capstone" design problem for this book that challenges the student to integrate the various methods to solve a particularly challenging problem. The design challenge problem of the previous subsection is relatively simple compared to the one discussed here (and is included as only a part of the more general problem here).

The intelligent social vehicular foraging problem also provides a glimpse into potential topics for further investigation. For instance, there has been little discussion on the relevance of language and communications in groups of intelligent systems, let alone aspects related to learning language. There has been little discussion of distributed learning by groups, distributed planning, distributed attention, etc. Moreover, there has been little discussion on learning and evolution of the *structure* of controllers and estimators. This challenge problem provides a framework to study such issues (of course, you may want to start by more thoroughly studying each of these topics in isolation, before confronting the more challenging social foraging problem).

The main problem for the student will be how to even attack this problem, provide a solution that shows you understand the basic methods of this book, and yet integrate the methods to solve a meaningful problem. This is a design problem where you help design the problem! Perhaps your instructor will help you, but this still makes the problem more challenging, since your objectives will not be explicitly listed. You will have to invent them, and the very design of the objectives is something you will be graded on. You should be careful to adopt a scientific approach to solving this problem. You should not simply construct some ad hoc combination of earlier methods, so that you can get lucky in simulation to show that it works. You must evaluate the methods fairly, provide biological motivations if appropriate (and if you like doing that), and it would be especially nice if you can augment your study with mathematical analysis that verifies the operation of the system (e.g., in the spirit of the stability analysis that we have studied for neural/fuzzy control, attentional systems, adaptive control, or swarm cohesion). If you have the opportunity to implement your methods on a group of vehicles, this can provide another way to study the validity of your approach. Keep in mind that standard engineering/scientific principles apply here, as they were discussed in Part I.

The problem statement, which is very simple, is given in Design Problem 19.5. So, what is the first step to solve this problem? Read this section as its basic focus is to give you ideas on how to integrate methods and concepts from this book to specify decision-making strategies for foraging. It may also give you ideas for how to extend some of the methods in the book, and you may be particularly attracted to doing so for a method that you were particularly intrigued with. Next, study the current literature. See the "For Further Study" section at the end of this part for some ideas on where to start; however, you should search the library or Internet for other current literature. Next, see the Web site for this book where some relevant literature and ideas are posted. Finally, work hard and have fun!

Vehicles, Environment, and Objectives

Here, we define the challenge problem.

Groups of Vehicles—Dynamics, Communications, and Control Structure: The first challenge is to define the type of vehicle that you will use. In particular, you need to define the following:

- *Vehicle dynamics:* For instance, if you use an automobile as your vehicle, what are the differential equations that you will use to simulate its motion? You can choose the vehicle type. It could be any type of autonomous land (e.g., automobile, truck, cross-country), water (surface or underwater), air (e.g., helicopter or airplane), or space vehicle (e.g., mobile satellite, explorer vehicle, etc.). You probably want to pay attention to the complexity of the dynamics. If you use extremely complex dynamics, and later try to define a sophisticated control strategy with many vehicles, your simulation may be too computationally complex. Hence, it is likely that you will want to use simple point-mass dynamics for your vehicle, and perhaps saturations on turn rates and velocity.

- *Vehicle sensors/actuators:* You will need to define which sensors and actuators you need for "inner-loop" control (e.g., in order to force the vehicle to track a trajectory that it chooses to follow; for example, doing heading regulation for the tanker ship). Moreover, you need to define what it can sense about its environment (e.g., types of prey, elevation of the earth, motion of predators, etc.), and how it can change its environment (e.g., by killing a prey, or building a bridge to be able to cross a river).

- *Vehicle communications:* You need to define the characteristics of the communications that each vehicle is capable of. Do they all have the same capabilities? Are the communications noisy or bandlimited? Are there random but bounded communication delays? Are the communication capabilities range-limited, so that if one vehicle moves too far away, it will not be able to communicate with some other vehicles (e.g., simply due to the distance, or possibly due to being behind some obstacle)?

- *Hierarchy and distribution in the group of vehicles:* Building on the last point, you need to define the allowable communication channels between vehicles, whether the structure of these channels (i.e., the topology of the interconnections between all vehicles) can change over time, and whether there is a type of hierarchy where some vehicles command others to perform tasks. For example, there may be no leaders in the group and all the vehicles may be able to communicate. Perhaps there is a single leader who is not endowed with any special communication capabilities, but who may behave differently. Perhaps there is a leader with special communication capabilities, multiple leaders with different objectives, or a hierarchy of leaders and followers for command and control. In some problems, you may be able to design the hierarchy or change it during operation of the system, and in others you may be given the hierarchy and have to work with it with no changes.

What types of vehicles are actually used in groups to achieve some objective? While there are clearly military applications, there are also commercial ones. For example, an "automated highway system" can be viewed as a group of autonomous vehicles that operate within a type of command and control structure (see Figure 1.13 on page 40). Or, groups of autonomous vehicles could be used in pollution clean-up, farming, exploration, or inventory control in manufacturing systems.

Environment Model and Goals: Your vehicles need to operate in some environment and hence, your simulation will need to represent its characteristics. For instance, you may want to consider the inclusion of the following:

- *Media:* What media do your vehicles move through? Air, water, outerspace? Are there disturbances that affect the motion of the vehicle? Solar pressure? Wind or water currents? For land vehicles, are there hills and roads?

- *Predator/prey (or noxious substance/nutrient) characteristics:* If your vehicles seek to consume prey while avoiding predators, what are the characteristics of your predators and prey? How fast can they move? What types of evasive or pursuit strategies do they use? What is their energy value if they are consumed? What is the probability that you will encounter them? What are the predator/prey densities? You could set this up as a *game*, where two students design teams of predators to operate in a single environment and compete for food sources. Alternatively, each predator could be prey for the other. Regardless, other predator and prey strategies help to define the environment for a set of vehicles.

- *Environmental changes:* Do the characteristics of the environment change over time? Do the characteristics of the media change? Does predator/prey density change? Do predator and prey strategies change?

What is the overall goal of the group of vehicles? The goal may involve characteristics, such as the following:

- *Energy consumption:* Find and ingest as much energy as possible (in the form of prey or nutrients of some type), while avoiding getting killed and eaten by some predator.

- *Achieving goal positions:* In some problems, the goal is simply for the group of vehicles to navigate their environment and achieve some goal positions.

- *Gathering information:* The group of vehicles may simply want to create as accurate a picture as possible about the environment that they operate in. For example, such an objective may be useful in space exploration via a set of vehicles.

- *Changing the environment:* There may be a collective goal to modify the environment that the group of vehicles operate in. For instance, in cooperative robotics, the problem of how to use a group of autonomous vehicles to move an object has been studied. For other problems, there may be a desire to eliminate targets, cultivate land, build a home, or gather food into a certain location.

It should be clear that there are certain principles that govern trade-offs in achieving goals. For example, typically the desire to achieve a wide-area search (e.g., to find prey) competes with a desire to focus activities in a single local region (e.g., in consuming prey). You should focus on uncovering such fundamental principles/trade-offs and illustrating them via simulations.

As an example, in the IVHS application the media is air, friction with a road that may turn, wind can influence dynamics, and there are hills and valleys. Temperature, snow, and rain may also affect vehicle dynamics. Destinations can be thought of as goal positions.

Elements of Distributed Decision-Making

While each of the foraging vehicles could have neural networks that implement various control functions, and hence, there would be a distributed neural network for instinctual control, here, we discuss the cases where groups of vehicles share information and implement distributed rule-based, planning, or attentional schemes for cooperative control that are not necessarily implemented by neural networks. In a sense, this section indicates how methods of Part II can be extended to the social foraging problem. Hence, the focus is not on use of the methods there for the development of controllers for a single control loop, but how to coordinate the use of information to meet the objectives of the group of vehicles (for "outer-loop" control, or guidance) that possibly has an opposing team (or teams) of vehicles.

A key component of the problem of foraging for many organisms is the search for food, and hence, this can be a key component for the case where groups search for food. Hence, in all the elements of decision-making, there is an element of distributed search. There are, however, several other key components including cooperative identification of prey, cooperative avoidance of predators, and cooperative attack of prey. The approaches outlined below show different ways to look at these basic elements of the group foraging problem.

Distributed Rule-Based Foraging: Some organisms (e.g., ants) actually use simple rules to specify how to forage for food, and when taken together, a group of such organisms seems to have an "emergent intelligence." For example, such rule-based behavior can indicate when to move in certain directions, and when all the organisms follow such simple rules, the group appears to move with purposeful behavior, acting as if they were a single organism rather than many individual ones. The key difference, compared with the rule-based systems in Part II, is that communications with other organisms are possible. Rules basically have two parts: antecedents and consequents. Each of these can be different for rule-based cooperative control in the following manner:

- *Using neighbor's information in rule antecedents:* The type of information that arrives for use by each organism depends on what type of communications the organism is capable of achieving. How should the information from other group members be used? Sometimes it simply would be used as an additional term in the antecedent of the rules for behavior of an organism. For instance, without communications, an organism may simply move greedily about looking for food so its rules' antecedents simply depend on direct sensing of environmental variables. If there are communications from neighbors, this may modify the behavior. How? It could be there are rules that indicate that the organism is supposed to look for food, but also follow its neighbors. If such a desire is followed by many organisms in a group, swarm behavior may emerge (individual rules can lead to interesting higher-level emergent patterns of behavior). Notice that in this case, the organism is using rules that depend not only on the environment, but also on communications from neighbors (e.g., a rule might say to move towards food, if you do not crowd a neighbor).

- *Rules for sending information to neighbors:* Rule consequents may contain not only information about which direction to move to get food, but also a specification of what to communicate to your neighbors about your experiences, actions, or goals. For instance, if a forager decides to pursue some prey, it may send a signal to some of its neighbors to come help it (e.g., in the case of some fish when they try to kill and ingest a much larger animal). Alternatively, an organism may communicate its intent to search some region for food and rules in other organisms may then trigger to indicate that they should not also look there (e.g., based on expected prey densities).

This provides a simple introduction as to how rules could be used in cooperative control (e.g., we did not discuss the fact that inference strategies could be communicated and shared between organisms). It is important to emphasize that even a set of simple rules implemented on each organism (identical or different rules) can lead to seemingly intelligent emergent behavior. That is why, by working together in simple ways, great things can be achieved. In this case, it can be that they simply enjoy greater foraging success. But, it is important to recognize that even what are thought of as higher-level cognitive capabilities can be achieved as the result of many simple communicating organisms (e.g., think of trail-laying by ants as implementing a type of learning to improve the success of the colony of ants). Moreover, note that rule-based strategies may be employed in hierarchies, and in conjunction with the planning and attention methods discussed below.

Distributed Planning for Foraging: You should think of distributed planning as an advanced form of distributed rule-based cooperative control, where models are used for prediction, and optimization is used for plan selection. To fully exploit the capabilities of distributed planning you may need higher bandwidth communication channels, since you may want to communicate models, plans, or plan selection strategies between organisms. The following show some ideas for how to achieve distributed planning:

- *Sharing models:* While operating independent of others, an organism that employs planning would use its own model to decide the best way to forage. In a cooperative strategy, an organism could get model information from other organisms in the group. For example, the model may indicate where other organisms have searched or what they have found. Sharing of model information between organisms essentially results in a type of adaptive planning, since models can be updated online while the planning by a single organism that uses that model is taking place. (Again, we see that learning emerges via the cooperation.)

- *Sharing plans or sets of plans:* It is also possible that an organism may share its current plan, or set of possible plans, with other organisms in the group. These may indicate where it intends to move, or the set of possible places that it intends to move. Or, it may indicate its strategies for attacking a prey or avoiding a predator.

- *Sharing plan selection strategies:* The cost function that is minimized in order to select plans could be communicated to other organisms to indicate aspects of the planning strategy that an organism is using (e.g., to indicate its intent to try to minimize the use of a certain resource).

Clearly, distributed planning can become very complex and complex behavior can emerge from even simple planning strategies. It should be clear that hierarchical planning strategies may be useful, with similar sharing of information between planning strategies, and the possibility that subordinates execute steps of plans specified by higher-level organisms.

Distributed Attention for Foraging: If one organism is given the task of attending to a set of mobile objects, it allocates its cognitive resources by dynamically refocusing its attentional focus (and perhaps the field of view of its sensors). Suppose that we want a group of social foraging vehicles to attend to a group of mobile predators/prey. Consider the following approaches to this problem:

- *Distributed agreement on focus regions:* One approach is to try to develop a strategy to divide the region into subregions and have each organism focus only in that subregion. This is, however, a difficult problem since the foraging group can move and there may be less than perfect communications between the organisms. This can result in one organism being responsible for attending to a subregion with too many mobile predators/prey to cope with (i.e., the capacity condition for that organism may not be satisfied so that there is no way it can succeed in its task). Or, it could be that there are too many organisms focusing on one subregion so that their cognitive resources are essentially wasted.

- *Leaders and hierarchical strategies:* For the above case, our intent was to consider a predator/prey focused on if *any* organism focused on it. This did not require global communications. What can we achieve if we have global communications? First, this enhances the possibilities to allocate *organisms* to groups of predators and prey. (We think of allocating whole organisms that in turn allocate their cognitive resources over time to attending to certain predators/prey; notice that the global communications enables the implementation of a type of hierarchy in the group's attentional strategy.) Second, this enables the group of organisms to construct a composite "snapshot" of its predator/prey environment for use in making foraging decisions.

Clearly, it may be possible to enhance the effectiveness of the attentional strategies with rule-based, planning, and learning capabilities (e.g., via distributed adaptive model predictive control as an attentional strategy). We will discuss distributed learning in more detail in the next section.

Distributed Learning

This section indicates how methods of Part III could be used to augment the strategies in Part II for use in the social foraging problem. The focus here is again on the distributed implementation of the methods, where special problems arise due to, for instance, the lack of global information.

Distributed Learning in Groups of Foragers: Learning can affect all aspects of each of the strategies discussed in the last section, from the functionalities that focus on the search for prey to ones concerned with cooperative avoidance of predators. Learning should be thought of as gathering and storing information to change future behavior. Hence, rather than try to list all the ways that learning can be used in social foraging, we will focus on what types of information can be learned, and why it might be useful in enhancing foraging success. Some ideas include the following:

- *Learning characteristics of the environment:* Individuals or the group may try to learn as much about the environment as possible. For instance, they may remember locations and quality of food sources, evasive maneuvers of prey, attack patterns of predators, etc. Learning of these characteristics may be facilitated by certain instincts in each of these cases (e.g., built-in expectations about food distribution and density or understanding typical rates of movement of predators and prey).

- *Learning foraging strategies from other group members:* It may be possible for one forager to cooperatively forage with other foragers and at the same time learn how its neighbors (or predators) succeed at foraging to improve its own performance, and at the same time improve the performance of the group (i.e., it may obtain gains that are not at the expense of others in the group).

- *Learning how to communicate:* It may be possible that a forager may learn how to communicate with the group to enhance foraging success.

To make these ideas more concrete, we briefly discuss a specific example of how learning can be integrated with planning and attention in foraging.

Distributed and Integrated Planning, Attention, and Learning: You can imagine that there are many ways to combine learning, planning, and attention to achieve effective group foraging strategies. Here, only one is outlined, essentially expanding a bit on the concepts in Section 19.6.1. You could certainly generate many more; in the next section, we will discuss the issue of ranking the quality of such proposed group foraging strategies via an evolutionary perspective. Here, we ignore the quality of the resulting strategy and simply outline a *way* that what you might call "intelligent" social foraging could be achieved.

Suppose that we have a group of foragers searching on an (x, y) plane for nutrients. Consider an individual forager. Suppose that this forager has a dynamical "attentional map" that indicates, for instance, where the forager has not looked for food (i.e., where it needs to pay attention in case there is food present), and where it has (and hence, where it may have found food and need to have a mechanism to track it). Corresponding concepts work for predators. The attentional map amounts to a type of memory, and hence, its dynamic updating corresponds to a type of learning. The forager can use this map to try to focus its attention in the proper way to make sure that it has as accurate a view as possible of the predator/prey environment, where this process is basically driven by the inherent goal of survival and reproduction.

Next, suppose that the forager maintains a "cognitive map" of its environment that it learns while moving throughout the environment. It stores information about physical characteristics of its environment (e.g., locations of rivers or forests), and perhaps also the location and characteristics of predators and prey. Next, suppose that we endow our forager with an ability to plan using both its cognitive map of the environment, and its attentional map. That is, think of the maps as constituting a type of model of the foraging environment that is learned during the foraging activity, or during the lifetime of the forager. Now, with a planning capability, the forager can use its current best information about the environment in order to project into the future and pick the best way to forage (e.g., it may predict how a prey will react to its movements or how a predator may behave). Clearly, this predictive capability depends critically on the quality of the learned information (and hence on the *process* of learning), and its own abilities to simultaneously consider a large number of possible predications and responses by the environment and predators/prey (e.g., the amount of memory and computational throughput of the forager directly constrain the plan generation and selection process).

Now, suppose that each forager has the attention, learning, and planning capabilities, and it has a certain type of ability to communicate with its neighbors. For example, suppose that each forager can communicate its own attentional and cognitive map to any other forager that is within a fixed distance from its current position. Can the group achieve more effective foraging? This seems quite plausible, since foragers will generally have more accurate attentional and cognitive maps without expending more energy (i.e., if the communications are cheaper than gathering information independently). This improved information should directly affect the quality of the attentional strategies and predictions made in planning, and hence, the quality of the foraging decisions that are made. It should be clear that if you increase the allowable range of communications, there should be corresponding improvement in the quality of information in each forager, and hence, an overall improvement in the quality of foraging by the group (assuming of course, that intergroup competition does not dominate). Clearly, it is also the case that if a forager has a poor sensor or is not honest, then if this information is passed to other team members, the overall foraging success could degrade. Indeed, when there are two teams, a key strategy may be how to deceive the opponent so that they make bad decisions.

Evolution of Foragers

In this final section, we make a few remarks about the relevance of evolution as discussed in Part IV to the intelligent social foraging problem (Chapter 18 serves as a concrete introduction to this rather philosophical section, since it integrates foraging concepts and evolution for *E. coli*).

Evolving Foraging Strategies: A key idea here is that via natural selection, the environment both influences an organism's physiology, and helps to define foraging success. Hence, the environment dictates what is an optimal group foraging strategy via the process of evolution. It should be the case then that optimal social vehicular foraging can only be achieved by careful consideration of vehicular "physiology" (e.g., amount of computing resources and communication capability) and the vehicle's environment.

It should be immediately clear that it is possible to simulate the effects of an evolutionary process on the design (redesign, "tweaking") of a social foraging strategy, and even the communications infrastructure that is used by the group of foragers. Clearly, foraging success would be a part of the fitness function, and neural, rule-based, planning, attentive, and learning strategies could all be encoded and evolved. Just because this is possible, does not mean that it is easy to do. Moreover, the focus in such studies should be on uncovering principles, rather than on the ad hoc construction of complex simulations that loosely emulate biological processes, but which may perhaps get lucky and provide an occasional good solution.

What principles? We must recognize that the design of social foraging strategies can be very difficult. Hence, it would be nice if we had some design principles for social foraging vehicles. Consider the following ways to uncover intuitions about design principles:

- *Designing parameters of the decision-making elements:* The complexity of the social foraging problem can make it quite difficult to pick some design parameters of the decision-making strategies, ones that may be relatively easy to pick in a single-forager problem. For instance, there should be an evolved optimal prediction horizon for planning strategies, and optimal resolution needed for the attentional and cognitive maps discussed in the last section.

- *Achieving balance between decision-making functionalities:* It is very difficult to know how much sophistication is needed for each type of decision-making mechanism (e.g., planning, learning, attention), and whether it is possible to use a complex learning strategy, but a simple planning strategy.

- *Evolving simple designs:* There should be a way to optimize the "cognitive complexity" (onboard computational resources) of the vehicle, to provide an effective yet simple engineering design. The question is whether we can use evolutionary design principles to realize the "keep it simple" principle in engineering design.

- *Studying trade-offs between computational and communication resources:* Could we study questions about whether it is possible to evolve a balance between how many communications are needed (e.g., bandwidth, communication range, level of locality) and how many learning, planning, and attentional capabilities are needed? Is it better to use lots of communications with simple decision-makers, or limited communications with sophisticated decision-making? Clearly, some aspects of such a study may be constrained by the particular vehicular problem being studied.

- *Coevolution:* What we call the "environment" includes intelligent adversaries (predators and prey) and for some organisms, a type of "arms race" occurs where one predator evolves some capability, then another evolves a way to counteract it, and so on. This is called coevolution. Suppose that you work on a student team, where each student designs an artificial organism that is both a predator and prey for ones designed by other students. Could we evolve an "evolutionary stable strategy" (see the "For Further Study" section at the end of this part) for these organisms? Could you do this for two *groups* of vehicles that compete?

- *Darwinian design of the software:* Could an analogous approach to the one described in Section 15.7.2, be used to synthesize/tune the software for cooperative foraging strategies (i.e., where the software is constructed via implementation in a test bed and evolution)?

Evolving Vehicular Hardware: Suppose that the actual robot or vehicular *hardware* can evolve. What are the implications? Here, we simply ask the following questions:

- How do you make the hardware replicate itself with fecundity and variation, and instill inheritance into the process? How do you implement (un)natural selection via environmental influences? A central problem is one of available resources to support the fecundity and the waste that results from selection. Biological evolution is based on selection of a few of many to be the ones that reproduce; the rest are in a sense "wasted." This resource/waste problem may be a key limitation that may drive any truly evolutionary strategy to exist on the molecular scale. But, considering nanotechnology advances, molecular vehicles or robots may not be out of the question.

- Could you make this emulate evolution of biological organisms, however simple they might be?

- Would this be useful for understanding biological evolution?

- Could you argue that your hardware is alive?

- What is the engineering utility of performing hardware evolution for populations? Could it be a way to make a group of vehicles or robots more adaptable to changes in its environment?

- If you can achieve some of these objectives, then will biological evolution have spawned another type of evolution? Or is this just a natural progression that is expected from evolution that can be thought of as being subsumed in what we now think of as evolution? Evolution did spawn

> many other types of optimization processes as we have discussed in this book, so why not another? It would seem that if it could, it would constitute an important event in evolutionary time.

Via combined hardware-software evolution, learning, planning, attention, rule-based, and neural systems approaches, could you implement a truly "intelligent" system?

19.7 Exercises and Design Problems

Exercise 19.1 (Properties of Static Games): This problem provides exercises to support the tutorial introduction to game theory in this chapter. You may use the code available at the Web site for the book to solve the problems.

(a) Using the examples given in the chapter, define a bimatrix game that has no Nash solution.

(b) Find a bimatrix game that has two Nash solutions.

(c) Find a game that has a Nash solution that is not stable.

(d) Find a bimatrix game where a Nash solution is the same as the minimax solution.

(e) Find a two-player game that has only one Pareto solution.

Exercise 19.2 (Static and Iterative Foraging Games): For the static foraging game studied in the chapter:

(a) Investigate the effect of changing the amount of energy it takes to travel to get a resource. (To do this, assume it is zero and show the results, then slowly increase the cost of travel to get a resource and each time, study the choices made in foraging.)

(b) Add a third resource type and develop a simulation to illustrate properties of the Nash, minimax, and Pareto solutions.

(c) Iterate the static foraging game and show how the resource profiles decrease. Study the effects of parameters of the costs on the rate of decrease (i.e., the rate that resources are eliminated from the environment).

Design Problem 19.1 (Static Foraging Games—Extensions):

(a) Extend the model developed in the chapter to a two-dimensional foraging plane. Define all details of the model.

(b) Introduce a way for each animal to find its path in the plane (e.g., via a shortest path method over the cells, with a cost from the energy to travel). Simulate to show that it finds the shortest path.

(c) Define foraging strategies, both cooperative and competitive, and evaluate their performance in simulation. Study the "iterated" case where multiple steps are taken.

Design Problem 19.2 (Evolution of Cooperation and the Iterated Prisoner's Dilemma)*: Read [35] and the chapter on evolution of strategies in [36].

(a) Produce a mathematical model of a two-person iterated prisoner's dilemma. Explain clearly what the allowable costs are for the game to represent an iterated prisoner's dilemma and why it is representative of some of the typical problems found in cooperation (and give specific examples of where iterated prisoner's dilemmas arise, not just in prisons).

(b) Using the "tit-for-tat," "tit-for-two-tats," and random strategies [35], plus two others of your choice, simulate 200 iterations of the strategies playing against each other. Compare. Discuss.

(c) Develop, using the ideas from [35, 36], an evolutionary algorithm for the strategies (i.e., one that in some way eliminates, after a certain number of iterations, strategies that are not performing well and generates extra copies of ones that perform well). Run the algorithm in a Monte Carlo simulation and explain the results. Which strategy wins? What does it mean for a strategy to be an "evolutionary stable strategy"?

(d) Explain the relevance of the theory of cooperation from [35, 36] on cooperation in multivehicle applications. Develop a specific simulation to illustrate your ideas.

Design Problem 19.3 (Evolutionary Stable Strategies)*: Read the parts of [245, 534, 213] relevant to evolutionary stable strategies and evolutionary dynamics. Write a brief introduction to evolutionary stable strategies using a matrix games approach to introduce the main ideas. Develop a simple example and simulation to illustrate the key ideas. Repeat, but for evolutionary dynamics. Explain the relevance of these ideas to engineering applications in general, and vehicular applications in particular.

Design Problem 19.4 (Challenge Problem: Foraging Games)*: Develop a dynamic foraging game, in two or more dimensions (i.e., it cannot be a foraging game on a line, like the one we studied in the chapter). Write out the full details of the mathematical model and strategies that you choose. The foraging strategy could involve the use of rules, planning, attention, or learning. Justify the choice for your strategy and quantify its computational complexity. Evaluate the performance of the strategy in simulation. Depending on your strategy (e.g., how it is parameterized), you may also want to study its evolution via a genetic algorithm.

Design Problem 19.5 (Challenge Problems: Design of Intelligent Social Foraging Strategies)*: The problems below are designed to integrate a number of the methods studied in this book.

(a) Develop a surrogate model method for intelligent foraging and evaluate its performance in simulation. Clearly, a key aspect of this challenge problem is how to formulate a problem that you can solve, and one that is amenable to at least simulation-based analysis of performance. You may need to consider nontraditional performance measures.

(b) Define a foraging problem (nonsocial) different from (a), and design a decision-making strategy for each forager. Include aspects of competition by having more than one noncooperative forager. Evaluate the design in simulation.

(c) Define a social vehicular foraging problem and design a decision-making strategy for each vehicle that leads to foraging success for the group. Include two opposing (adversarial) teams. Evaluate the design in simulation. Several approaches to this problem are discussed in Section 19.6.2.

Chapter 20

For Further Study

Introductions to solitary and social foraging theory are given in [490] and [214], respectively. Biology foundations for intelligent social multiagent applications can be found in the emerging field of cognitive ecology [157]. Game theory is introduced in a variety of books, but one that has a treatment more consistent with a control-theoretic viewpoint is [47]. The literature on cooperative control for multivehicle systems is rapidly growing, so the reader should search it for other relevant and more recent studies.

Foraging, Search, and Optimization: Foraging theory is described in [490, 214] and the section on search strategies of foraging animals is based on [390]. A good reference on dynamic programming is [55], but there are several other more recent books. Animal behavior, including foraging theory and its ecological validity, is discussed in [12, 490, 214] and the behavioral ecology of finding resources is also discussed in [54].

A seminal treatment of the biology of social organisms, "sociobiology," is given in [540]. Sensing, guidance, and navigation by organisms is discussed in [160, 161]; such issues significantly affect foraging. General issues in time and energy management for animal behaviors are discussed in [112]. Aspects of intelligent foraging are discussed in [188, 152, 453]. Neural substrates of planning and learning are considered in [453], and some of the discussions there are relevant to foraging concepts. An overview of the biology and behavioral ecology of swarms is given in [403]. The view that social foraging evolved for improving climbing of noisy gradients of nutrient resources is introduced in [230].

Simulation of group behavior of organisms is discussed in the areas of swarm intelligence and artificial life [73, 6, 434, 313]. In these areas, distributed optimization and learning-evolution synergy has been studied in a variety of contexts. "Ant colony optimization" is an optimization method that was developed by modeling foraging in ant colonies, and it is discussed in [73] (the discussion on the Argentine ants was taken from [25]). There, the focus is on biomimicry for solution of combinatorial optimization algorithms (e.g., shortest path algorithms). In Section 18.2.7, we discuss some other biomimicry concepts for combinatorial optimization methods (e.g., the *streptomycete*). The reader should be aware that there is a very large literature on combinatorial optimization methods, including many good books [400, 218]. A relevant book on self-organization in biological systems is [88], and this book also discusses foraging of several types of organisms, synchronization of firefly flashing, and other types of self-organization properties of groups of organisms (e.g., construction of structures such as honeycombs). The book [276] models swarming as an optimization process (see also [107]).

The problem of how to search for an object has been studied for many years, and there is a field called "search theory" that grew out of applications in WWII. The two key seminal references for search theory and its methods are [494, 283] where the authors discuss theoretical principles of search in a stochastic environment when sensors are not perfect. There is significant focus on how to allocate search effort across the environment. The methods that are developed have been used profitably in applications, and have even been related to search strategies of animals and used to explain animal search strategies [160]. The reader who wishes to gain a deeper understanding of search by autonomous agents is encouraged to see [494, 283] or perhaps the survey article [58]; however, there has been a recent growing interest of how to apply search theory to multiple communicating autonomous vehicles so the reader should be aware that articles continue to appear in this topical area (see, e.g., [138]).

Bacterial Foraging: The description of the biological details of the *E. coli* bacteria and their motile behavior were taken from [342, 384, 8, 62, 61, 459, 60]. The chemotactic pathways are simulated in [369] and elsewhere. More details on the flagellar motor are given in [141, 369, 61]. Computer simulations of *growth* of *E. coli* populations are explained in [287]. Pattern formation in *E. coli* and *S. typhimurium* is discussed in [84, 71, 83, 544, 24, 464] (and a mathematical swarm model and simulations are provided in [544]). Integral feedback control mechanism models of internal bacterial decision-making processes are studied in [553].

An overview of tactic responses that were used to explain motile behavior of bacteria other than *E. coli* is given in [24]. Motility behavior of square bacteria is discussed in [7]. The view of bacteria as multicellular organisms is given in [463]. Motile behavior of bacterial swarms of *M. xanthus* and some related bacteria are described in [334, 24, 467, 432]. A swarm motility model for *M. xanthus* that is based on a high-frequency gene mutation, called the "Pied piper model," is given in [281, 282] (computer simulations of some aspects of this model are given in [282]). The discussion on *P. mirabilis* is based on [52]. Simulations of *myxobacteria* based on a stochastic cellular automata approach are described in [491, 492, 323].

While we may often think of studying learning or other higher functions for organisms that have a neural network, some have studied behavioral plasticity properties of *Paramecium aurelia* as classical learning characteristics [208].

Finally, note that the conjugation of bacteria has been modeled in genetic algorithms and used for optimization [381]. There, taxes and foraging were not considered, just the particular mechanism of "sex" for gene transfer and how this affects the optimization process of the genetic algorithm.

Social Foraging of Insects: See [88, 73] and the references therein for discussion on several types of social insects. Some very good references on social insects include [539, 246, 540]. Evolution/optimization of social insect colony composition is discussed in [539]. In [246], the idea of viewing each individual in a social insect colony as an energy unit that is expended to get food for the group is discussed at length. The book [145] discusses information processing in social insects. General references on bees, several of which focus on foraging, include [456, 454, 455].

Swarms: Swarming of honey bees is studied in [86, 14, 458, 455, 542, 541, 457]. For more on factors that contribute to in-transit swarm cohesiveness, see [14, 458, 455, 542].

General references relevant to swarming in the biology literature include [403, 163, 374]; swarm modeling and analysis is studied in [231, 393, 533, 159, 232, 233, 229, 364, 330]. An article that has a model like the one we used here is given in [393], where the author also provides examples of modeling actual biological swarms. Physicists take an approach to modeling swarms where they model each individual as a particle, which they usually call a *self-driven* or *self-propelled particle*, and they study the collective behavior due to their interactions [430, 510, 511, 110, 524, 114, 115, 113, 525, 116, 358, 468, 310].

Swarms have also been studied in the context of engineering applications, particularly in collective robotics, where there are teams of robots working together by communicating over a communication network [41, 500]. Particularly relevant to the treatment in this part, is the work in [433], which considers a distributed control approach based on artificial force laws between individual robots and robot groups that they call "social potential functions." For an overview of autonomous search strategies by robots and animals, see [209]. Special cases of swarms have also been studied in "intelligent vehicle highway systems" [56, 502, 501, 123]. Also quite relevant is the study of stable formation control for robots [148], aircraft, and in cooperative control for uninhabited autonomous (air) vehicles [142, 143, 164, 391, 308, 215, 270, 503, 394]. Early work on swarm stability is in [264, 57]. Later work, where there is asynchronism and time delays, appears in [326, 325, 327, 329, 199, 328]. Particularly relevant is the study in [308], where the authors use virtual leaders and artificial potentials and the work in [392, 38], where climbing noisy gradients is studied. See also the work in [204, 202, 200], where the authors provide a class of attraction/repulsion functions and provide conditions for swarm stability (size of ultimate swarm size and ultimate behavior). The work in [203, 201, 205, 324] represents progress in the direction of combining the study of aggregating swarms and how, during this process, decisions about foraging or threat avoidance can affect the collective/individual motion of the swarm/swarm members (i.e., typical characteristics influencing social foraging). There, results are provided on swarm stability (cohesiveness) in the presence of an attractant/repellent profile. Finally, note that the literature in the area of swarms is rapidly expanding; the reader should consult the current literature to find recent developments.

Evolution, Cooperation, and Competition: An introduction to the "iterated prisoner's dilemma" and cooperation is given in [35]. That work is expanded on in [36], where the author also explains a number of topics relevant to the work in this part (e.g., his landscape theory of aggregation where general political and economic alignments in large groups are modeled). A biologist's evolutionary perspective on cooperation is given in [437]. Cooperation and competition in foraging are also discussed in [490, 214].

Game theory is introduced in a variety of books, but the one that most affected the development here was [47]. The classic reference for dynamic programming is [55], but there are several more recent books. For more information on combinatorial optimization and complexity theory, see [400, 218]. Predatory impacts on swarms are discussed in [403]. Competition in insects is discussed in [246]. An area relevant to the topics of this part is evolutionary game theory [472, 473] and the analysis of the stochastic dynamics of population evolution (e.g., stability analysis of "replicator dynamics") [245, 534, 213]. For this work, authors study the dynamics of the evolutionary process, sometimes using concepts based on ideas used in this book (e.g., Lyapunov stability).

Intelligent Social Foraging: Cooperative Robotics Applications: Use of some of the ideas of the static and dynamic games formulation in the last chapter for vehicular applications, is studied in [191, 190]. Cooperative search with multiple vehicles is studied in [50]. The use of surrogate optimization for cooperative control of multiple autonomous vehicles is studied in [554]. A task load balancing approach to cooperative control for multiple vehicles is studied in [178]. A distributed task scheduling approach to cooperative control is studied in [212]. The work in [212] uses ideas from [211] and Chapter 7. The literature in the area of intelligent networked autonomous vehicular teams is, however, relatively immature and currently expanding; hence, as part of the "challenge" problem in Chapter 19.6, it is recommended that you familiarize yourself with the current literature. The reader interested in the application to groups of mobile robots may find the book [148] useful for developing models and nonlinear controls for the individual mobile robots. A natural choice for a vehicle model for some applications is the "Dubins car;" for instance, see how it is used in [178, 212] for on application.

Bibliography

[1] S. Abe and M.-S. Lan. Fuzzy rules extraction directly from numerical data for function approximation. *IEEE Trans. on Systems, Man, and Cybernetics*, 25(1):119–129, January 1995.

[2] D.Y. Abramovitch. Fuzzy control as a disruptive technology. *IEEE Control Systems Magazine*, 19:100–102, June 1999.

[3] D.H. Ackley. *A Connectionist Machine for Genetic Hillclimbing*. Kluwer Academic Press, Boston, MA, 1987.

[4] D.H. Ackley. A case for lamarckian evolution. In C. Langton, editor, *Artificial Life III, SFI Studies in the Sciences of Complexity, Proc. Vol. XVII*. Addison-Wesley, CA, 1994.

[5] D.H. Ackley and M. Littman. Interactions between learning and evolution. In C. Langton, C. Taylor, J.D. Farmer, and S. Rasmussen, editors, *Artificial Life II, SFI Studies in the Sciences of Complexity, Proc. Vol. X*. Addison-Wesley, CA, 1991.

[6] C. Adami. *Introduction to Artificial Life*. Springer-Verlag, NY, 1998.

[7] M. Alam, M. Claviez, D. Oesterhelt, and M. Kessel. Flagella and motility behavior of square bacteria. *EMBO Journal*, 13(12):2899–2903, 1984.

[8] B. Alberts, D. Bray, J. Lewis, M. Raff, K. Roberts, and J.D. Watson. *Molecular Biology of the Cell*. Garland Publishing, NY, 2nd edition, 1989.

[9] J.S. Albus. A new approach to manipulator control: Cerebellar model articulation controller (CMAC). *Journal of Dynamic Systems, Measurement, and Control*, 97:220–227, Sept. 1975.

[10] J.S. Albus. Outline for a theory of intelligence. *IEEE Trans. on Systems, Man, and Cybernetics*, 21(3):473–509, May/Jun. 1991.

[11] J.S. Albus and A. Meystel. *Engineering of Mind: An Intelligent Systems Perspective*. John Wiley and Sons, NY, 2000.

[12] J. Alcock. *Animal Behavior: An Evolutionary Approach*. Sinauer Associates, Inc., Sunderland, MA, 1998.

[13] L.M. Aldeman. Computing with DNA. *Scientific American*, 279(2):54–61, Aug. 1998.

[14] J.T. Ambrose. Swarms in transit. *Bee World*, 57:101–109, 1976.

[15] B.D.O. Anderson and J.B. Moore. *Optimal Control: Linear Quadratic Methods*. Prentice-Hall, Englewood Cliffs, NJ, 1990.

[16] J.R. Anderson and L.J. Schooler. Reflections of the environment in memory. *Psychological Science*, 2:396–408, 1985.

[17] A. Angsana and K.M. Passino. Distributed fuzzy control of flexible manufacturing systems. *IEEE Trans. on Control Systems Technology*, 2(4):423–435, December 1994.

[18] P. Antsaklis, editor. *Special Issue on Hybrid Systems: Theory and Applications*. Proceedings of the IEEE, Vol. 88, No. 7 July, 2000.

[19] P.J. Antsaklis, W. Kohn, A. Nerode, and S. Sastry, editors. *Hybrid Systems II*. Lecture Notes in Computer Science, LNCS 999, Springer-Verlag, NY, 1995.

[20] P.J. Antsaklis and K.M. Passino. Towards intelligent autonomous control systems: Architecture and fundamental issues. *Journal of Intelligent and Robotic Systems*, 1:315–342, 1989.

[21] P.J. Antsaklis and K.M. Passino, editors. *An Introduction to Intelligent and Autonomous Control*. Kluwer Academic Publishers, Norwell, MA, 1993. This book available for free download at author Web site.

[22] P.J. Antsaklis, K.M. Passino, and S.J. Wang. An introduction to autonomous control systems. *IEEE Control Systems Magazine*, 11(4):5–13, June 1991. also appears in M.M. Gupta and D.H. Rao, ed., Neuro-Control Systems: Theory and Applications, IEEE Press, NJ, 1994.

[23] R.C. Arkin. *Behavior-Based Robotics*. MIT Press, Cambridge, MA, 1998.

[24] J.P. Armitage. Bacterial tactic responses. *Advances in Microbial Physiology*, 41:229–290, 1999.

[25] S. Aron, J.L. Deneubourg, S. Goss, and J.M. Pasteels. Functional self-organization illustrated by inter-nest traffic in ants: The case of the Argentine ant. In W. Alt and G. Hoffmann, editors, *Biological Motion, Lecture Notes in Biomathematics, Vol. 89*, pages 533–547. Springer-Verlag, Berlin, 1990.

[26] W.R. Ashby. *Design for a Brain: The Origin of Adaptive Behavior*. Wiley, NY, 2nd edition, 1960.

[27] K.J. Åström, J.J. Anton, and K.E. Arzen. Expert control. *Automatica*, 22(3):277–286, March 1986.

[28] K.J. Åström and T. Hägglund, editors. *PID Control: Theory, Design, and Tuning*. Instrument Society of America Press, Research Triangle Park, NC, 2nd edition, 1995.

[29] K.J. Åström and B. Wittenmark. *Computer Controlled Systems: Theory and Design*. Prentice-Hall, Englewood Cliffs, NJ, 1984.

[30] K.J. Åström and B. Wittenmark. *Adaptive Control*. Addison-Wesley, Reading, MA, 1995.

[31] M. Athans, D. Castanon, K. Dunn, C.S. Greene, W.H. Lee, N.R. Sandell, and A.S. Willsky. The stochastic control of the F-8C aircraft using a multiple model adaptive control (MMAC) method - Part 1: equilibrium flight. *IEEE Transactions on Automatic Control*, 22(5):768–780, 1977.

[32] M. Athans and P.L. Falb. *Optimal Control*. McGraw-Hill, NY, 1966.

[33] D.P. Atherton. *Nonlinear Control Engineering: Describing Function Analysis and Design*. Van Nostrand Reinhold, Berkshire, England, 1982.

[34] T. Audesirk and G. Audesirk. *Biology: Life on Earth*. Prentice Hall, Englewood Cliffs, NJ, 5th edition, 1999.

[35] R. Axelrod. *The Evolution of Cooperation*. Basic Books, Perseus Books Group, USA, 1980.

[36] R. Axelrod. *The Complexity of Cooperation: Agent-Based Models of Competition and Collaboration*. Princeton University Press, NJ, 1997.

[37] R. Babuška. *Fuzzy Modeling for Control*. Kluwer Academic Publishers, Boston, 1998.

[38] R. Bachmayer and N.E. Leonard. Vehicle networks for gradient descent in a sampled environment. In *Proc. of Conf. Decision Control*, pages 113–117, Las Vegas, Nevada, December 2002.

[39] T. Back. *Evolutionary Algorithms in Theory and Practice*. Oxford University Press, NY, 1996.

[40] K. Baker. *Introduction to Sequencing and Scheduling*. Wiley, NY, 1974.

[41] T. Balch and R.C. Arkin. Behavior-based formation control for multirobot teams. *IEEE Trans. on Robotics and Automation*, 14(6):926–939, December 1998.

[42] W. Banzhaf, P. Nordin, R.F. Keller, and F.D. Francone. *Genetic Programming: An Introduction*. Morgan Kaufmann Pub., San Francisco, CA, 1998.

[43] A. Baraldi and P. Blonda. A survey of fuzzy clustering algorithms for pattern recognition-parts i & ii. *IEEE Transactions on Systems, Man, and Cybernetics-Part B*, 29(6):778–801, 1999.

[44] J. Barraquand and J.-C. Latombe. Robot motion planning: A distributed representation approach. *Int. Journal of Robotics Research*, 10(6):51–72, Dec. 1991.

[45] A.R. Barron. Universal approximation bounds for superpositions of a sigmoid function. *IEEE Trans. on Information Theory*, 39(3):930–945, May 1993.

[46] G. Bartolini, G. Casalino, F. Davoli, M. Mastretta, R. Minciardi, and E. Morten. Development of performance adaptive fuzzy controllers with applications to continuous casting plants. In M. Sugeno, editor, *Industrial Applications of Fuzzy Control*,

pages 73–86. Elsevier, Amsterdam, The Netherlands, 1985.

[47] T. Basar and G.J. Olsder. *Dynamic Noncooperative Game Theory.* SIAM, Philadelphia, 1998.

[48] C. Batur, A. Srinivasan, and C.-C. Chan. Automated rule-based model generation for uncertain complex dynamic systems. *Engineering Applications in Artificial Intelligence*, 4(5):359–366, 1991.

[49] C. Batur, A. Srinivasan, and C.C. Chan. Fuzzy model based fuzzy predictive control. In *Proc. of the 1st Int. Conf. on Fuzzy Theory and Technology*, pages 176–180, 1992.

[50] M. Baum and K.M. Passino. A search-theoretic approach to cooperative control for uninhabited air vehicles. In *AIAA GNC Conference*, Monterey, CA, 2001.

[51] M. Bech and L. Smitt. Analogue simulation of ship maneuvers. Technical report, Hydro-og Aerodynmisk Laboratorium, Lyngby, Denmark, 1969.

[52] R. Belas. Proteus mirabilis and other swarming bacteria. In J.A. Shapiro and M. Dworkin, editors, *Bacteria as Multicellular Organisms*, pages 183–219. Oxford University Press, NY, 1997.

[53] R.K. Belew. Evolution, learning, and culture: Computational metaphors for adaptive algorithms. *Complex Systems*, 4:11–49, 1990.

[54] W.J. Bell. *Searching Behavior: The Behavioral Ecology of Finding Resources.* Chapman and Hall, London, 1991.

[55] R. Bellman. *Dynamic Programming.* Princeton Univ. Press, NJ, 1957.

[56] J. Bender and R. Fenton. On the flow capacity of automated highways. *Transport. Sci.*, 4:52–63, February 1970.

[57] G. Beni and P. Liang. Pattern reconfiguration in swarms—convergence of a distributed asynchronous and bounded iterative algorithm. *IEEE Trans. on Robotics and Automation*, 12(3):485–490, June 1996.

[58] S.J. Benkoski, M.G. Monticino, and J.R. Weisinger. A survey of the search theory literature. *Naval Research Logistics*, 38:469–494, 1991.

[59] J.M. Benyus. *Biomimicry: Innovation Inspired by Nature.* Quill William Morrow, NY, 1997.

[60] H.C. Berg. *Random Walks in Biology, New Expanded Edition.* Princeton Univ. Press, Princeton, NJ, 1993.

[61] H.C. Berg. Motile behavior of bacteria. *Physics Today*, pages 24–29, Jan. 2000.

[62] H.C. Berg and D.A. Brown. Chemotaxis in *escherichia coli* analysed by three-dimensional tracking. *Nature*, 239:500–504, Oct. 1972.

[63] H. Bersini. Immune network and adptive control. In *Proceedings of the First European Conference on Artificial Life*, pages 465–470, 1991.

[64] H. Bersini. Reinforcement and recruitment learning for adaptive process control. In *Proceedings of the 1992 IFAC/IFIP/IMACS on Artificial Intelligence in Real Time Control*, pages 331–337, 1992.

[65] H. Bersini. The endogenous double plasticity of the immune network and the inspiration to be drawn for engineering artifacts. In *Artificial Immune Systems and Their Applications*, pages 22–44. Springer-Verlag, Germany, 1999.

[66] H. Bersini and F. Varela. Hints for adaptive problem solving gleaned from immune networks. In *Proceedings of the First Conference on Parallel Problem Solving from Nature*, pages 343–354, 1990.

[67] H. Bersini and F. Varela. The immune learning mechanisms: reinforcement, recruitment and their applications. In *Computing with Biological Metaphors*, pages 166–192. Chapman and Hall, London, 1994.

[68] D.P. Bertsekas. *Nonlinear Programming.* Athena Scientific Press, Belmont, MA, 1995.

[69] D.P. Bertsekas and J.N. Tsitsiklis. *Neuro-Dynamic Programming.* Athena Scientific Press, Belmont, MA, 1996.

[70] S.A. Billings and Q.M. Zhu. Model validation tests for multivariable nonlinear models including neural networks. *Int. Journal of Control*, 62(4):749–766, 1995.

[71] Y. Blat and M. Eisenbach. Tar-dependent and independent pattern

formation by *salmonella typhimurium*. *J. Bacteriology*, 177:1683–1691, April 1995.

[72] B.W. Boehm. A spiral model of software development and enhancement. *IEEE Computer*, 21:61–72, May 1988.

[73] E. Bonabeau, M. Dorigo, and G. Theraulaz. *Swarm Intelligence: From Natural to Artificial Systems*. Oxford Univ. Press, NY, 1999.

[74] G.E.P. Box. Evolutionary operation: A method for increasing industrial productivity. *Appl. Statist.*, 6:81–101, 1957.

[75] M. S. Branicky. Multiple lyapunov functions and other analysis tools for switched and hybrid systems. *IEEE Trans. on Automatic Control*, 43(4):475–482, April 1998.

[76] R.A. Brooks. Planning collision-free motions for pick-and-place operations. *Int. Journal of Robotics Research*, 2(4):19–44, 1983.

[77] R.A. Brooks. A robust layered control system for a mobile robot. *IEEE Trans. on Robotics and Automation*, 2(1), March 1986.

[78] R.A. Brooks, editor. *Cambrian Intelligence: The Early History of the New AI*. MIT Press, Cambridge, MA, 1999.

[79] K. Brown. Seeds of concern. *Scientific American*, 284(4):52–57, April 2001.

[80] M. Brown and C. Harris. *Neurofuzzy Adaptive Modeling and Control*. Prentice-Hall, Englewood Cliffs, NJ, 1994.

[81] S.C. Brown and K.M. Passino. Intelligent control for an acrobot. *Journal of Intelligent and Robotic Systems: Theory and Applications*, 18:209–248, 1997.

[82] L. Brownston, R. Farrell, E. Kant, and N. Martin. *Programming Expert Systems in OPS5*. Addison Wesley, Reading, MA, 1986.

[83] E.O. Budrene and H.C. Berg. Complex patterns formed by motile cells of *escherichia coli*. *Nature*, 349:630–633, Feb. 1991.

[84] E.O. Budrene and H.C. Berg. Dynamics of formation of symmetrical patterns by chemotactic bacteria. *Nature*, 376:49–53, 1995.

[85] C. Bundesen. A theory of visual attention. *Psychological Review*, 97:523–547, 1990.

[86] C.G. Butler, R.K. Callow, and J.R. Chapman. 9-hydroxydec-2-enoic acid, a pheromone stabilizing honeybee swarms. *Nature*, 201:733, Feb. 1964.

[87] A.G. Cairns-Smith. *Evolving the Mind: On the Nature of Matter and the Origin of Consciousness*. Cambridge University Press, Cambridge, 1996.

[88] S. Camazine and et al. *Self-Organization in Biological Systems*. Princeton Univ. Press, NJ, 2001.

[89] S. Camazine and J. Sneyd. A model of collective nectar source selection by honey bees: self-organization through simple rules. *Journal of Theoretical Biology*, 149:547–571, 1991.

[90] S. Cameron. Obstacle avoidance and path planning. *Industrial Robot*, 21(5):9–14, 1994.

[91] N.A. Campbell. *Biology*. Benjamin Cummings Pub. Co., Menlo Park, CA, 1996.

[92] R. Carelli, E.F. Camacho, and D. Patiño. A Neural Network Based Feedforward Adaptive Controller for Robots. *IEEE Trans. on Systems, Man, and Cybernetics*, 25(9):1281–1288, September 1995.

[93] J. Carlson and J. Doyle. Highly optimized tolerance: A mechanism for power laws in designed systems. *Phys. Rev. E*, 60:1412–1427, 1999.

[94] J. Carlson and J. Doyle. Highly optimized tolerance: Robustness and design in complex systems. *Phys. Rev. Lett.*, 84:2529–2532, 2000.

[95] C.G. Cassandras. *Discrete Event Systems*. R.D. Irwin, Inc. and Aksen Associates, Inc., Homewood, IL, 1993.

[96] E. Charniak. Bayesian networks without tears. *AI Magazine*, 12(4):50–63, 1991.

[97] E. Charniak and D. McDermott. *Introduction to Artificial Intelligence*. Addison Wesley, Reading, MA, 1985.

[98] B.-S. Chen and Y.-M. Cheng. A structure-specified h^∞ optimal control design for practical applications: A genetic approach. *IEEE Trans. on Control Systems Technology*, 6(6):707–718, Nov. 1998.

[99] B.-S. Chen, C.-H. Lee, and Y.-C. Chang. H^∞ Tracking Design of Uncertain Nonlinear SISO Systems: Adaptive Fuzzy Approach. *IEEE Trans. on Fuzzy Systems*, 4(1):32–43, February 1996.

[100] C.T. Chen. *Linear System Theory and Design.* Saunders College Publishing, Fort Worth, TX, 1984.

[101] F.-C. Chen and C.-C. Liu. Adaptively Controlling Nonlinear Continuous-Time Systems Using Multilayer Neural Networks. *IEEE Trans. on Automatic Control*, 39(6):1306–1310, June 1994.

[102] S. Chen, S.A. Billings, and P.M. Grant. Recursive hybrid algorithm for non-linear system identification using radial basis function networks. *Int. Journal of Control*, 55(5):1051–1070, 1992.

[103] S. Chen, S.A. Billings, and W. Luo. Orthogonal least squares methods and their application to non-linear system identification. *Int. Journal of Control*, 50(5):1873–1896, 1989.

[104] S.L. Chiu. Selecting input variables for fuzzy models. *Journal of Intelligent and Fuzzy Systems*, 4:243–256, 1996.

[105] J.Y. Choi and J.A. Farrell. Nonlinear adaptive control using networks of piecewise linear approximators. In *IEEE Conf. on Decision and Control*, pages 1671–1676, Phoenix, AZ, December 1999.

[106] D. Christiansen. Pigeon-guided missiles that bombed out. *IEEE Spectrum*, page 46, August 1987.

[107] M. Clerc and J. Kennedy. The particle swarm—explosion, stability, and convergence in a multidimensional complex space. *IEEE Trans. on Evolutionary Computation*, 6(1):58–73, February 2002.

[108] A.R. Conn, K. Scheinberg, and Ph.L. Toint. Recent progress in unconstrained nonlinear optimization without derivatives. *Mathematical Programming*, 79:397–414, 1997.

[109] R.W. Conway, W.L. Maxwell, and L.W. Miller. *Theory of Scheduling.* Addison-Wesley, Reading, Massachussetts, 1967.

[110] Z. Csahok and T. Vicsek. Lattice-gas model for collective bilogical motion. *Physical Review E*, 52(5):5297–5303, November 1995.

[111] R. F. Curtain and H. J. Zwart. *An Introduction to Infinite Dimensional Linear System Theory.* Springer-Verlag, NY, 1995.

[112] I.C. Cuthill and A.I. Houston. Managing time and energy. In J.R. Krebs and N.B. Davies, editors, *Behavioural Ecology*, pages 97–120. Blackwell Science, London, 4th edition, 1997.

[113] A. Czirok, A.-L. Barabasi, and T. Vicsek. Collective motion of self-propelled particles: Kinetic phase transition in one dimension. *Physical Review Letters*, 82(1):209–212, January 1999.

[114] A. Czirok, E. Ben-Jacob, I. Cohen, and T. Vicsek. Formation of complex bacterial colonies via self-generated vortices. *Physical Review E*, 54(2):1791–1801, August 1996.

[115] A. Czirok, H.E. Stanley, and T. Vicsek. Spontaneously ordered motion of self-propelled partciles. *Journal of Physics A: Mathematical, Nuclear and General*, 30:1375–1385, 1997.

[116] A. Czirok and T. Vicsek. Collective behavior of interacting self-propelled particles. *Physica A*, 281:17–29, 2000.

[117] E. Czogała and W. Pedrycz. On identification in fuzzy systems and its applications in control problems. *Fuzzy Sets and Systems*, 6:73–83, 1981.

[118] E. Czogała and W. Pedrycz. Control problems in fuzzy systems. *Fuzzy Sets and Systems*, 7:257–273, 1982.

[119] S. Daley and K.F. Gill. A design study of a self-organizing fuzzy logic controller. *Proc. Institute of Mechanical Engineers*, 200(C1):59–69, 1986.

[120] S. Daley and K.F. Gill. Atitude control of a spacecraft using an extended self-organizing fuzzy logic controller. *Proc. Institute of Mechanical Engineers*, 201(C2):97–106, 1987.

[121] S. Daley and K.F. Gill. Comparison of a fuzzy logic controller with a P+D control law. *Trans. ASME, Journal of Dynamical System, Measurement, and Control*, 111:128–137, June 1989.

[122] A.R. Damasio. How the brain creates the mind. *Scientific American*, 281(6):112–117, Dec. 1999.

[123] S. Darbha and K.R. Rajagopal. Intelligent cruise control systems and traffic

flow stability. *Transportation Research Part C*, 7:329–352, 1999.

[124] D. Dasgupta, editor. *Artificial Immune Systems and Their Applications.* Springer-Verlag, Germany, 1999.

[125] D. Dasgupta. Special issue on artificial immune systems. *IEEE Trans. on Evolutionary Computation*, 6(3):225–313, 2002.

[126] R. Dawkins. *The Selfish Gene.* Oxford University Press, NY, 2nd edition, 1989.

[127] R. Dawkins. *The Bling Watchmaker: Why the Evidence of Evolution Reveals a Universe without Design.* W.W. Norton and Comp., NY, 1996.

[128] R. Dawkins. *Climbing Mount Improbable.* W.W. Norton and Comp., NY, 1996.

[129] R. Dawkins. *The Extended Phenotype: The Long Reach of the Gene.* Oxford University Press, NY, 1999.

[130] P. Dayan and L.F. Abbott. *Theoretical Neuroscience: Computational and Mathematical Modeling of Neural Systems.* MIT Press, Cambridge, MA, 2001.

[131] J.H. D'Azzo and C.H. Houpis. *Linear Control System Analysis and Design: Conventional and Modern.* McGraw-Hill, NY, 1995.

[132] C.W. de Silva. An analytical framework for knowledge-based tuning of servo controllers. *Engineering Applications in Artificial Intelligence*, 4(3):177–189, 1991.

[133] C.W. de Silva. Considerations of hierarchical fuzzy control. In H.T. Nguyen, M. Sugeno, R. Tong, and R.R. Yager, editors, *Theoretical Aspects of Fuzzy Control*, pages 183–234. Wiley, NY, 1995.

[134] C.W. de Silva. *Intelligent Control: Fuzzy Logic Applications.* CRC Press, Boca Raton, FL, 1995.

[135] C. Canudas de Wit, B. Siciliano, and G. Bastin, editors. *Theory of Robot Control.* Springer-Verlag London Limited, 1996.

[136] T. Dean, J. Allen, and Y. Aloimonos. *Artificial Intelligence: Theory and Practice.* Addison Wesley Pub., Menlo Park, CA, 1995.

[137] T. Dean and M.P. Wellman. *Planning and Control.* Morgan Kaufman, San Mateo, CA, 1991.

[138] R.F. Dell, J.N. Eagle, G.H.A. Martins, and A.G. Santos. Using multiple searchers in constrained-path moving-target search problems. *Naval Research Logistics*, 43:463–480, 1996.

[139] D.C. Dennett. *Darwin's Dangerous Idea: Evolution and the Meanings of Life.* Touchstone Press, NY, 1995.

[140] J.E. Dennis and V. Torczon. Direct search methods on parallel machines. *SIAM Journal Optimization*, 1(4):448–474, Nov. 1991.

[141] D.J. DeRosier. The turn of the screw: The bacterial flagellar motor. *Cell*, 93:17–20, 1998.

[142] J.P. Desai, J. Ostrowski, and V. Kumar. Controlling formations of multiple mobile robots. In *Proc. of IEEE International IEEE Conference on Robotics and Automation*, pages 2864–2869, Leuven, Belgium, May 1998.

[143] J.P. Desai, J. Ostrowski, and V. Kumar. Modeling and control of formations of nonholonomic mobile robots. *IEEE Trans. on Robotics and Automation*, 17(6):905–908, December 2001.

[144] A.A. Desrochers. *Intelligent Robotic Systems for Space Exploration.* Kluwer Academic Pub., Boston, MA, 1992.

[145] C. Detrain, J.L. Deneubourg, and J.M. Pasteels, editors. *Information Processing in Social Insects.* Birkhauser Verlag, Boston, 1999.

[146] Y. Diao and K.M. Passino. Immunity-based hybrid learning methods for approximator structure and parameter adjustment. *Engineering Applications of Artificial Intelligence*, 15(6):587–600, Dec. 2002.

[147] J.L. Diez, J.L. Navarro, and A. Sala. Identification for local-model control with fuzzy clustering. In *Proceedings of the IFAC Workshop on Advanced Fuzzy-Neural Control*, pages 99–104, Valencia, Spain, Oct. 15–16 2001.

[148] W. Dixon, D. Dawson, E. Zergeroglu, and A. Behal. *Nonlinear Control of Wheeled Mobile Robots.* Springer-Verlag, London, 2001.

[149] M. Dogruel and Ü. Özgüner. Controllability, reachability, stabilizability and state reduction in automata. In *IEEE International Symposium on Intelligent Control*, Glasgow, Scotland, 1992.

[150] M. Dogruel and Ü. Özgüner. Stability of hybrid systems. In *IEEE International Symposium on Intelligent Control*, Columbus, OH, 1994.

[151] M. Dogruel and Ü. Özgüner. Modeling and stability issues in hybrid systems. In P. Antsaklis, W. Kohn, A. Nerode, and S. Sastry, editors, *Hybrid Systems II*, pages 148–165. Lecture Notes in Computer Science, LNCS 999, Springer-Verlag, NY, 1995.

[152] M. Domjan. *The Principles of Learning and Behavior.* Brooks/Cole Pub., NY, 4th edition, 1998.

[153] R.C. Dorf and R.H. Bishop. *Modern Control Systems.* Addison-Wesley, Reading, MA, 1995.

[154] J.C. Doyle, B.A. Francis, and A.R. Tannenbaum. *Feedback Control Theory.* Macmillan, NY, 1992.

[155] D. Driankov, H. Hellendoorn, and M. Reinfrank. *An Introduction to Fuzzy Control.* Springer-Verlag, NY, 1993.

[156] G. Dudek and et al. A taxonomy for swarm robots. In *IEEE/RSJ Int. Conf. on Intelligent Robots and Systems*, Yokohama, Japan, July 1993.

[157] R. Dukas, editor. *Cognitive Ecology: The Evolutionary Ecology of Information Processing and Decision Making.* The Univ. of Chicago Press, Chicago, IL, 1998.

[158] H.F. Durrant-Whyte. Consistent integration and propagation of disparate sensor observations. *Int. J. of Robotics Research*, 6(3):3–24, 1987.

[159] R. Durrett and S. Levin. The importance of being discrete (and spatial). *Theoretical Population Biology*, 46:363–394, 1994.

[160] D.B. Dusenbery. *Sensory Ecology: How Organisms Acquire and Respond to Information.* W.H. Freeman and Company, NY, 1992.

[161] D.B. Dusenbery. *Life at Small Scale.* Scientific American Library, NY, 1996.

[162] J.C. Eccles. *Evolution of the Brain: Creation of Self.* Routledge, NY, 1989.

[163] L. Edelstein-Keshet. *Mathematical Models in Biology.* Brikhäuser Mathematics Series. Random House, NY, 1989.

[164] M. Egerstedt and X. Hu. Formation constrained multi-agent control. *IEEE Trans. on Robotics and Automation*, 17(6):947–951, December 2001.

[165] J. M. Elzebda, A. H. Nayfeh, and D. T. Mook. Development of an analytical model of wing rock for slender delta wings. *AIAA Journal of Aircraft*, 26:737–743, August 1989.

[166] F. Hayes-Roth, et al. *Building Expert Systems.* Addison Wesley, Reading, MA, 1985.

[167] S. Fabri and V. Kadirkamanathan. Dynamic structure neural networks for stable adaptive control of nonlinear systems. *IEEE Trans. on Neural Networks*, 7(5):1151–1167, September 1996.

[168] J. D. Farmer. A rosetta stone for connectionism. *Physica D*, 42:153–187, 1990.

[169] J.D. Farmer, N.H. Packard, and A.S. Perelson. The immune system, adaptation, and machine learning. *Physica*, 22D:187–204, 1986.

[170] J.A. Farrell. Motivations for local approximators in passive learning control. *Journal of Intelligent Systems and Control*, 1(2):195–210, 1996.

[171] J.A. Farrell. Neural control. In W. Levine, editor, *The Control Handbook*, pages 1017–1030. CRC Press, Boca Raton, FL, 1996.

[172] J.A. Farrell. Persistence of excitation conditions in passive learning control. In *IFAC*, pages 313–318, 1996.

[173] J.A. Farrell. Persistence of excitation conditions in passive learning control. *Automatica*, 33(4):699–703, 1997.

[174] J.A. Farrell. Stability and approximator convergence in nonparametric nonlinear adaptive control. *IEEE Trans. on Neural Networks*, 9(5):1008–1020, 1998.

[175] J.A. Farrell. Wavelet-based system identification for nonlinear control. *IEEE Trans. on Automatic Control*, 44(2):412–417, 1999.

[176] J.A. Farrell and W. Baker. Learning control systems. In P.J. Antsaklis and K.M. Passino, editors, *An Introduction to Intelligent and Autonomous Control Systems*, pages 237–262. Kluwer Academic Publishers, Norwell, MA, 1993.

[177] R.E. Fenton and R.J. Mayhan. Automated highway studies at The Ohio State University: An overview. *IEEE Trans. on Vehicular Technology*, 40(1):100–113, February 1991.

[178] J. Finke, K.M. Passino, and A. Sparks. Cooperative control via task load balancing for networked uninhabited autonomous vehicles. In *Proc. of Conf. Decision and Control*, Maui, HI, December 2003.

[179] D. Fogel. *Evolutionary Computation: Toward a New Philosophy of Machine Intelligence.* IEEE Press, NY, 1995.

[180] D.B. Fogel. *System Identification Through Simulated Evolution: A Machine Learning Approach to Modeling.* Ginn Press, Needham Heights, MA, 1991.

[181] C. Foias, H. Özbay, and A. Tannenbaum. *Robust Control of Infinite Dimensional Systems: Frequency Domain Methods.* Lecture Notes in Control and Information Sciences, No. 209, Springer-Verlag, London, 1996.

[182] S. Forrest and A. S. Perelson. Genetic algorithms and the immune system. In *Parallel Problem Solving from Nature.* Springer-Verlag, Berlin, 1991.

[183] G.F. Franklin, J.D. Powell, and A. Emami-Naeini. *Feedback Control of Dynamic Systems.* Addison-Wesley, Reading, MA, 1994.

[184] G.F. Franklin, J.D. Powell, and M.L. Workman. *Digital Control of Dynamic Systems.* Addison-Wesley, Reading, MA, 1990.

[185] S. Freeman and J.C. Herron. *Evolutionary Analysis.* Prentice Hall, Englewood Cliffs, NJ, 1998.

[186] S. French. *Sequencing and Scheduling.* Wiley, 1982.

[187] B. Friedland. *Control System Design.* McGraw-Hill, NY, 1986.

[188] S.L. Friedman and E.K. Scholnick. An evolving "blueprint" for planning: Psychological requirements, task characteristics, and social-cultural influences. In S.L. Friedman and E.K. Scholnick, editors, *The Developmental Psychology of Planning: Why, How, and When Do We Plan?*, pages 3–22. Lawrence Erlbaum Assoc., Pub., Mahwah, NJ, 1997.

[189] M. Frixione, G. Vercelli, and R. Zaccaria. Diagrammatic reasoning for planning and intelligent control. *IEEE Control Systems Magazine*, 21(2):34–53, April 2001.

[190] S. Ganapathy and K.M. Passino. Distributed agreement strategies for cooperative control: Modeling and scalability analysis. In *Conference on Cooperative Control and Optimization*, Gainesville, Florida, Dec. 2002.

[191] S. Ganapathy and K.M. Passino. Agreement strategies for cooperative control of uninhabited autonomous vehicles. In *Proceedings of the American Control Conference*, Denver, Colorado, June 2003.

[192] C.E. Garcia, D.M. Prett, and M. Morari. Model predictive control: Theory and practice—a survey. *Automatica*, 25(3):335–348, 1989.

[193] E. Garcia-Benitez, S. Yurkovich, and K.M. Passino. Rule-based supervisory control of a two-link flexible manipulator. *Journal of Intelligent and Robotic Systems*, 7(2):195–213, 1993.

[194] W. Garver and F. Moss. Electronic fireflies. *Scientific American*, pages 128–130, Dec. 1993.

[195] M.H. Garzon and R.J. Deaton. Biomolecular computing and programming. *IEEE Trans. on Evolutionary Computation*, 3(3):236–250, Sept. 1999.

[196] V. Gazi. Control and stabilization of discrete event systems with limited look-ahead policies. Master's thesis, The Ohio State University, 1998.

[197] V. Gazi, M.L. Moore, K.M. Passino, W. Shackleford, F. Proctor, and J.S. Albus. *The RCS Handbook: Tools for Real Time Control Systems Software Development.* John Wiley and Sons, Pub., NY, 2000.

[198] V. Gazi and K.M. Passino. Direct adaptive control using dynamic structure fuzzy systems. In *Proceedings of the American Control Conf.*, pages 1954–1958, Chicago, IL, June 2000.

[199] V. Gazi and K.M. Passino. Stability of a one-dimensional discrete-time asynchronous swarm. In *Proc. of the joint IEEE Int. Symp. on Intelligent Control/IEEE Conf. on Control Applications*, pages 19–24, Mexico City, Mexico, September 2001.

[200] V. Gazi and K.M. Passino. A class of attraction/repulsion functions for stable swarm aggregations. In *Proc. of Conf. Decision Control*, pages 2842–2847, Las Vegas, Nevada, December 2002.

[201] V. Gazi and K.M. Passino. Stability analysis of social foraging swarms: Combined effects of attractant/repellent profiles. In *Proc. of Conf. Decision Control*, pages 2848–2853, Las Vegas, Nevada, December 2002.

[202] V. Gazi and K.M. Passino. Stability analysis of swarms. In *Proc. American Control Conf.*, pages 1813–1818, Anchorage, Alaska, May 2002.

[203] V. Gazi and K.M. Passino. Stability analysis of swarms in an environment with an attractant/repellent profile. In *Proc. American Control Conf.*, pages 1819–1824, Anchorage, Alaska, May 2002.

[204] V. Gazi and K.M. Passino. Stability analysis of swarms. *IEEE Trans. on Automatic Control*, 48(4):692–697, April 2003.

[205] V. Gazi and K.M. Passino. Stability analysis of social foraging swarms. *IEEE Trans. Systems, Man, and Cybernetics-Part B: Cybernetics*, 34(1):539–557, 2004.

[206] M.S. Gazzaniga, R.B. Ivry, and G.R. Mangun. *Cognitive Neuroscience: The Biology of the Mind.* W. W. Norton and Co., NY, 1998.

[207] S.S. Ge, T.H. Lee, and C.J. Harris. *Adaptive Neural Network Control of Robotic Manipulators.* World Scientific, Singapore, 1998.

[208] B. Gelber. Retention in paramecium aurelia. *Journal Comp. Physiol. Psychology*, 51:110–115, 1958.

[209] E. Gelenbe, N. Schmajuk, J. Staddon, and J. Reif. Autonomous search by robots and animals: A survey. *Robotics and Autonomous Systems*, 22:23–34, 1997.

[210] S.B. Gershwin. *Manufacturing System Engineering.* PRT Prentice Hall, Englewood Cliffs, NJ, 1994.

[211] A. Gil and K.M. Passino. Stability analysis of network-based cooperative resource allocation strategies. In *Proc. of Conf. Decision and Control*, Maui, HI, December 2003.

[212] A. Gil, K.M. Passino, and A. Sparks. Cooperative scheduling of tasks for networked uninhabited autonomous vehicles. In *Proc. of Conf. Decision and Control*, Maui, HI, December 2003.

[213] H. Gintis. *Game Theory Evolving: A Problem-Centered Introduction to Modeling Strategic Interaction.* Princeton Univ. Press, Princeton, NJ, 2000.

[214] L. Giraldeau and T. Caraco. *Social Foraging Theory.* Princeton Univ. Press, Princeton, NJ, 2000.

[215] F. Giulietti, L. Pollini, and M. Innocenti. Autonomous formation flight. *IEEE Control Systems Magazine*, 20(6):34–44, December 2000.

[216] D. Goldberg. *Genetic Algorithms in Search, Optimization and Machine Learning.* Addison-Wesley, Reading, MA, 1989.

[217] G.H. Golub and J.H. Ortega. *Scientific Computing and Differential Equations: An Introduction to Numerical Methods.* Academic Press, Boston, 1992.

[218] M. Gondran and M. Minoux. *Graphs and Algorithms.* John Wiley and Sons, NY, 1984.

[219] G.C. Goodwin and K.S. Sin. *Adaptive Filtering Prediction and Control.* Prentice-Hall, Englewood Cliffs, NJ, 1984.

[220] S.J. Gould. *Full House: The Spread of Excellence from Plato to Darwin.* Three Rivers Press, NY, 1996.

[221] P. Graham and R. Newell. Fuzzy identification and control of a liquid level rig. *Fuzzy Sets and Systems*, 26:255–273, 1988.

[222] P. Graham and R. Newell. Fuzzy adaptive control of a first-order process. *Fuzzy Sets and Systems*, 31:47–65, 1989.

[223] P. Gray. *Psychology.* Worth Pub., NY, 3rd edition, 1999.

[224] J.R. Gremling and K.M. Passino. Genetic adaptive failure estimation. In *Proceedings of the American Control Conference*, pages 908–912, Albuquerque, NM, June 1997.

[225] J.R. Gremling and K.M. Passino. Genetic adaptive state estimation for a jet engine compressor. In *Proceedings of the IEEE International Symposium on Intelligent Control*, pages 131–136, Istanbul, Turkey, September 1997.

[226] J.R. Gremling and K.M. Passino. Genetic adaptive parameter estimation. *Int. Journal of Intelligent Control and Systems*, 3(4):465–503, 1999.

[227] J.R. Gremling and K.M. Passino. Genetic adaptive state estimation. *Engineering Applications of Artificial Intelligence*, 13(6):611–623, 2000.

[228] M.J. Grimble and M.A. Johnson. *Optimal Control and Stochastic Estimation*. Wiley, NY, 1988.

[229] P. Grindrod. Models of individual aggregation or clustering in single and multi-species communities. *Journal of Mathematical Biology*, 26:651–660, 1988.

[230] D. Grunbaum. Schooling as a strategy for taxis in a noisy environment. *Evolutionary Ecology*, 12:503–522, 1998.

[231] D. Grünbaum and A. Okubo. Modeling social animal aggregations. In *Frontiers in Theoretical Biology*, volume 100 of *Lecture Notes in Biomathematics*, pages 296–325. Springer-Verlag, NY, 1994.

[232] S. Gueron and S.A. Levin. The dynamics of group formation. *Mathematical Biosciences*, 128:243–264, 1995.

[233] S. Gueron, S.A. Levin, and D.I. Rubenstein. The dynamics of herds: From individuals to aggregations. *J. of Theoretical Biology*, 182:85–98, 1996.

[234] M. Gupta and N. Sinha, editors. *Intelligent Control: Theory and Practice*. IEEE Press, Piscataway, NJ, 1995.

[235] M. Guven and K.M. Passino. Avoiding exponential parameter growth in fuzzy systems. *IEEE Trans. on Fuzzy Systems*, 9(1):194–199, Feb. 2001.

[236] M.L. Hadjili and V. Wertz. Generalized predictive control using takagi-sugeno fuzzy models. In *Proceedings of the IEEE Int. Symp. on Intelligent Control*, pages 405–410, Cambridge, MA, Sept. 1999.

[237] M.L. Hadjili, V. Wertz, and G. Scorletti. Fuzzy model based predictive control. In *Proceedings of the IEEE Conf. on Decision and Control*, pages 2927–2929, Tampa, FL, Dec. 1998.

[238] M. Hagan, H. Demuth, and M. Beale, editors. *Neural Network Design*. PWS Publishing, Boston, MA, 1996.

[239] T.R. Halliday and P.J.B. Slater, editors. *Animal Behavior, Volume 1: Causes and Effects*. W.H. Freeman and Company, NY, 1983.

[240] W.D. Hamilton. Geometry for the selfish herd. *Journal of Theoretical Biology*, 31:295–311, 1971.

[241] D.O. Hebb. *The Organization of Behavior*. Wiley, NY, 1949.

[242] H. Hellendoorn and D. Driankov. *Fuzzy Model Identification: Selected Approaches*. Springer-Verlag, NY, 1997.

[243] G.E. Hinton and S.J. Nowlan. How learning can guide evolution. *Complex Systems*, 1:495–502, 1987.

[244] L. Ho, editor. *Discrete Event Dynamic Systems: Analyzing Complexity and Performance in the Modern World*. IEEE Press, NY, 1992.

[245] J. Hofbauer and K. Sigmund. *Evolutionary Games and Population Dynamics*. Cambridge University Press, Cambridge, England, 1998.

[246] B. Holldobler and E.O. Wilson. *The Ants*. Belknap Press of Harvard Univ. Press, Cambridge, MA, 1990.

[247] C. H. Hsu and E. Lan. Theory of wing rock. *AIAA Journal of Aircraft*, 22:920–924, 1985.

[248] F.-Y. Hsu and L.-C. Fu. A new design of adaptive robust controller for nonlinear systems. In *Proceedings of the American Control Conference*, June 1995.

[249] J. E. Hunt and D. E. Cooke. An adaptive distributed learning system based on the immune system. In *Proceedings of the IEEE International Conference on Systems, Man, and Cybernetics*, pages 2494–2499, 1995.

[250] J. E. Hunt and D. E. Cooke. Learning using an artificial immune system.

Journal of Network and Computer Applications: Special Issue on Intelligent Systems: Design and Application, 19:189–212, 1996.

[251] K.J. Hunt. Induction of decision trees for rule based modelling and control. In *Proc. of the IEEE Int. Symp. on Intelligent Control*, pages 306–311, Glasgow, Scotland, August 1992.

[252] K.J. Hunt, D. Sbarbaro, R. Zbikowski, and P.J. Gawthrop. Neural networks for control systems: A survey. In M.M. Gupta and D.H. Rao, editors, *Neuro-Control Systems: Theory and Applications*, pages 171–200. IEEE Press, Piscataway, NJ, 1994.

[253] H. Ichihashi. Efficient algorithms for acquiring fuzzy rules from examples. In H.T. Nguyen, M. Sugeno, R. Tong, and R.R. Yager, editors, *Theoretical Aspects of Fuzzy Control*, pages 261–280. Wiley, NY, 1995.

[254] P.A. Ioannou and J. Sun. *Robust Adaptive Control*. Prentice-Hall, Englewood Cliffs, NJ, 1996.

[255] S. Isaka, A. Sebald, A. Karimi, N. Smith, and M. Quinn. On the design and performance evaluation of adaptive fuzzy controllers. In *Proc. of the IEEE Conf. on Decision and Control*, pages 1068–1069, Austin, TX, December 1988.

[256] H. Ishibuchi, K. Nozaki, and N. Yamamoto. Selecting fuzzy rules by genetic algorithm for classification problems. In *Proc. of the 2nd IEEE Int. Conf. on Fuzzy Systems*, pages 1119–1124, San Francisco, CA, March 1993.

[257] A. Isidori. *Nonlinear Control Systems*. Springer-Verlag, NY, 1995.

[258] W. Jacek. *Intelligent Robotic Systems: Design, Planning, and Control*. Kluwer Academic / Plenum Pub., NY, 1999.

[259] J.-S.R. Jang. ANFIS: Adaptive-Network-Based Fuzzy Inference System. *IEEE Trans. on Systems, Man, and Cybernetics*, 23(3):665–685, May/Jun. 1993.

[260] J.-S.R. Jang. Input selection for ANFIS learning. In *Proceedings of IEEE International Conference on Fuzzy Systems*, New Orleans, September 1996.

[261] J.-S.R. Jang and C.-T. Sun. Neuro-fuzzy modeling and control. *Proc. of the IEEE, Special Issue on Fuzzy Logic in Engineering Applications*, 83(3):378–406, March 1995.

[262] J.-S.R. Jang, C.-T. Sun, and E. Mizutani. *Neuro-Fuzzy and Soft Computing: A Computational Approach to Learning and Machine Intelligence*. Prentice Hall, Englewood Cliffs, NJ, 1997.

[263] D.L. Jenkins and K.M. Passino. An introduction to nonlinear analysis of fuzzy control systems. *Journal of Intelligent and Fuzzy Systems*, 7(1):75–103, 1999.

[264] K. Jin, P. Liang, and G. Beni. Stability of synchronized distributed control of discrete swarm structures. In *Proc. of IEEE International IEEE Conference on Robotics and Automation*, pages 1033–1038, San Diego, California, May 1994.

[265] A. Juditsky, H. Hjalmarsson, A. Benveniste, B. Deylon, L. Ljung, J. Sjoberg, and Q. Zhang. Nonlinear black-box modeling in system identification: Mathematical foundations. *Automatica*, 31(12):1691–1724, December 1995.

[266] A. Kandel and G. Langholz, editors. *Fuzzy Control Systems*. CRC Press, Boca Raton, FL, 1993.

[267] E.R. Kandel. Cellular mechanisms of learning and memory. In E.R. Kandel, J.H. Schwartz, and T.M. Jessell, editors, *Essentials of neural science and behavior*. Appleton and Lange, Norwalk, CT, 1995.

[268] E.R. Kandel, J.H. Schwartz, and T.M. Jessell, editors. *Essentials of neural science and behavior*. Appleton and Lange, Norwalk, CT, 1995.

[269] E.R. Kandel, J.H. Schwartz, and T.M. Jessell, editors. *Principles of neural science*. McGraw-Hill, NY, 2000.

[270] W. Kang, N. Xi, and A. Sparks. Formation control of autonomous agents in 3*d* space. In *IEEECRA*, pages 1755–1760, San Francisco, CA, April 2000.

[271] C. Karr and E. Gentry. Fuzzy control of pH using genetic algorithms. *IEEE Trans. on Fuzzy Systems*, 1(1):46–53, 1993.

[272] S. Kauffman. *At Home in the Universe: The Search for the Laws of Self-Organization and Complexity.* Oxford University Press, NY, 1995.

[273] S. Kaufman. *The Origins of Order: Self-Organization and Selection in Evolution.* Oxford University Press, NY, 1993.

[274] M. Kawato and D. Wolpert. Internal models for motor control. In G.R. Bock and J.A. Goode, editors, *Sensory Guidance of Movement*, pages 291–307. John Wiley and Sons, Pub., NY, 1998.

[275] C.T. Kelley. *Iterative Methods for Optimization.* SIAM, Philadelphia, PA, 1999.

[276] J. Kennedy, R.C. Eberhart, and Y. Shi. *Swarm Intelligence.* Morgan Kaufmann Pub., San Francisco, CA, 2001.

[277] H.K. Khalil. *Nonlinear Systems.* Prentice-Hall, Englewood Cliffs, NJ, 3rd edition, 1999.

[278] O. Khatib. Real-time obstacle avoidance for manipulators and mobile robots. *Int. J. of Robotics Research*, 5(1):90–98, 1986.

[279] Y.H. Kim and F.L. Lewis. *High-Level Feedback Control with Neural Networks.* World Scientific, Singapore, 1998.

[280] G.J. Klir and B. Yuan. *Fuzzy Sets and Fuzzy Logic: Theory and Applications.* Prentice-Hall, Englewood Cliffs, NJ, 1995.

[281] A. Koch and D. White. The social lifestyle of myxobacteria. *BioEssays*, 20:1030–1038, 1998.

[282] A.L. Koch. The strategy of *myxococcus xanthus* for group cooperative behavior. *Antonie van Leeuwenhoek*, 73:299–313, 1998.

[283] B.O. Koopman. *Search and Screening: General Principles with Historical Applications.* Pergamon Press, NY, 1980.

[284] J.R. Koza. *Genetic Programming.* MIT Press, Cambridge, MA, 1992.

[285] J.R. Koza. *Genetic Programming II.* MIT Press, Cambridge, MA, 1994.

[286] L.G. Kraft and D.P. Campagna. A comparison between CMAC neural network control and two traditional adaptive control systems. *IEEE Control Systems Magazine*, 10(3):36–43, April 1990.

[287] J.-U. Kreft, G. Booth, and J.W.T. Wimpenny. BacSim, a simulator for individual-based modelling of bacterial colony growth. *Microbiology*, 144:3275–3287, 1998.

[288] K. Kristinsson and G. Dumont. System identification and control using genetic algorithms. *IEEE Transactions on Systems, Man, and Cybernetics*, 22(5):1033–1046, 1992.

[289] M. Krstic, I. Kanellakopoulos, and P. Kokotovic. *Nonlinear and Adpative Control Design.* Wiley, NY, 1995.

[290] P.R. Kumar and T.I. Seidman. Dynamic instabilities and stabilization methods in distributed real-time scheduling of manufacturing systems. *IEEE Trans. Automatic Control*, 35(3):289–298, March 1990.

[291] P.R. Kumar and P.P. Varaiya. *Stochastic Systems: Estimation, Identification, and Adaptive Control.* Prentice-Hall, Englewood Cliffs, NJ, 1986.

[292] B.C. Kuo. *Automatic Control Systems.* Prentice-Hall, Englewood Cliffs, NJ, 1991.

[293] H.J. Kushner and G.G. Yin. *Stochastic Approximation Algorithms and Applications.* Springer-Verlag, NY, 1997.

[294] A. Kusiak. *Intelligent Manufacturing Systems.* Prentice Hall, Englewood Cliffs, NJ, 1990.

[295] T.-Y. Kwok and D.-Y. Yeung. Constructive algorithms for structure learning in feedforward neural networks for regression problems. *IEEE Trans. on Neural Networks*, 3(8):630–645, May 1997.

[296] W.A. Kwong and K.M. Passino. Dynamically focused fuzzy learning control. *IEEE Trans. on Systems, Man, and Cybernetics*, 26(1):53–74, February 1996.

[297] W.A. Kwong, K.M. Passino, E.G. Laukonen, and S. Yurkovich. Expert supervision of fuzzy learning systems for fault tolerant aircraft control. *Proc. of the IEEE, Special Issue on Fuzzy Logic in Engineering Applications*, 83(3):466–483, March 1995.

[298] J.C. Lagarias, J.A. Reeds, M.H. Wright, and P.E. Wright. Convergence properties of the nelder-mead simplex method in low dimensions. *SIAM Journal Optimization*, 9(1):112–147, 1998.

[299] J. Layne. Fuzzy model reference learning control. Master's thesis, Department of Electrical Engineering, The Ohio State University, 1992.

[300] J.R. Layne and K.M. Passino. Fuzzy model reference learning control. In *IEEE Conf. on Control Applications*, pages 686–691, Dayton, OH, September 1992.

[301] J.R. Layne and K.M. Passino. Fuzzy model reference learning control for cargo ship steering. *IEEE Control Systems Magazine*, 13(6):23–34, December 1993.

[302] J.R. Layne and K.M. Passino. Fuzzy model reference learning control. *Journal of Intelligent and Fuzzy Systems*, 4(1):33–47, 1996.

[303] J.R. Layne, K.M. Passino, and S. Yurkovich. Fuzzy learning control for anti-skid braking systems. *IEEE Trans. on Control Systems Technology*, 1(2):122–129, June 1993.

[304] C.-H. Lee and S.-D. Wang. A self-organizing adaptive fuzzy controller. *Fuzzy Sets and Systems*, pages 295–313, 1996.

[305] M.A. Lee and H. Takagi. Integrating design stages of fuzzy systems using genetic algorithms. In *Proc. of the 2nd IEEE Int. Conf. on Fuzzy Systems*, pages 612–617, San Francisco, CA, March 1993.

[306] Y.-C. Lee, C. Hwang, and Y.-P. Shih. A combined approach to fuzzy model identification. *IEEE Trans. on Systems, Man, and Cybernetics*, 24(5):736–744, May 1994.

[307] W.K. Lennon and K.M. Passino. Genetic adaptive identification and control. *Engineering Applications of Artificial Intelligence*, 12(2):185–200, April 1999.

[308] N.E. Leonard and E. Fiorelli. Virtual leaders, artificial potentials and coordinated control of groups. In *Proc. of Conf. Decision Control*, pages 2968–2973, Orlando, FL, December 2001.

[309] D. Levin and J. Katz. Dynamic load measurements with delta wings undergoing self-induced roll oscillations. *AIAA Journal of Aircraft*, 21:30–36, January 1984.

[310] H. Levine and W.-J. Rappel. Self-organization in systems of self-propelled particles. *Physical Review E*, 63(1):017101–1–017101–4, January 2001.

[311] W. Levine, editor. *The Control Handbook.* CRC Press, Boca Raton, FL, 1996.

[312] I.B. Levitan and L.K. Kaczmarek. *The Neuron: Cell and Molecular Biology.* Oxford University Press, Oxford, 2nd edition, 1997.

[313] S. Levy. *Artificial Life: A Report from the Fontier where Computers Meet Biology.* Vintage Books, NY, 1992.

[314] F.L. Lewis. *Optimal Control.* Wiley, NY, 1986.

[315] F.L. Lewis. *Optimal Estimation.* Wiley, NY, 1986.

[316] F.L. Lewis, A. Yeşildirek, and K. Liu. Multilayer neural-net robot controller with guaranteed tracking performance. *IEEE Trans. on Neural Networks*, 7(2):1–12, March 1996.

[317] R.M. Lewis and V. Torczon. Pattern search algorithms for bound constrained minimization. *SIAM Journal Optimization*, 9(4):1082–1099, 1999.

[318] M.D. Lezak. *Neuropsychological Assessment.* Oxford University Press, NY, 1995.

[319] Y. Li and C. Lau. Development of fuzzy algorithms for servo systems. *IEEE Control Systems Magazine*, 9(2):65–72, April 1989.

[320] C.T. Lin and C.S.G. Lee. Neural network-based fuzzy logic control and decision system. *IEEE Trans. on Computers*, 40(12):1320–1336, December 1991.

[321] C.-C. Liu and F.-C. Chen. Adaptive control of non-linear continuous-time systems using neural networks—general relative degree and mimo cases. *International Journal of Control*, 58(2):317–335, 1993.

[322] G.P. Liu, V. Kadirkamanathan, and S.A. Billings. Variable neural networks for adaptive control of nonlinear systems. *IEEE Trans. on Systems, Man, and Cybernetics-Part C: Applications and Reviews*, 29(1):34–43, 1999.

[323] Y. Liu and K.M. Passino. Biomimicry of social foraging behavior for distributed optimization: Models, principles, and emergent behaviors. *Journal of Optimization Theory and Applications*, 115(3):603–628, December 2002.

[324] Y. Liu and K.M. Passino. Stable social foraging swarms in a noisy environment. *IEEE Trans. on Automatic Control*, 49(1):30–44, 2004.

[325] Y. Liu, K.M. Passino, and M.M. Polycarpou. Stability analysis of one-dimensional asynchronous mobile swarms. In *Proc. of Conf. Decision Control*, pages 1077–1082, Orlando, FL, December 2001.

[326] Y. Liu, K.M. Passino, and M.M. Polycarpou. Stability analysis of one-dimensional asynchronous swarms. In *Proc. American Control Conf.*, pages 716–721, Arlington, VA, June 2001.

[327] Y. Liu, K.M. Passino, and M.M. Polycarpou. Stability analysis of m-dimensional asynchronous swarms with a fixed communication topology. In *Proc. American Control Conf.*, pages 1278–1283, Anchorage, Alaska, May 2002.

[328] Y. Liu, K.M. Passino, and M.M. Polycarpou. Stability analysis of m-dimensional asynchronous swarms with a fixed communication topology. *IEEE Transactions on Automatic Control*, 48(1):76–95, 2003.

[329] Y. Liu, K.M. Passino, and M.M. Polycarpou. Stability analysis of one-dimensional asynchronous swarms. *IEEE Trans. Automatic Control*, 48(10):1848–1854, Oct. 2003.

[330] M. Lizana and V. Padron. A specially discrete model for aggregating populations. *Journal of Mathematical Biology*, 38:79–102, 1999.

[331] L. Ljung. *System Identification: Theory for the User.* Prentice-Hall, Englewood Cliffs, NJ, 2nd edition, 1999.

[332] C.F. Van Loan. *Introduction to Scientific Computing: A Matrix-Vector Approach Using MATLAB.* Prentice-Hall, Englewood Cliffs, NJ, 1999.

[333] K. A. Loparo, M. R. Buchner, and K. S. Vasudeva. Leak detection in an experimental heat exchanger process: A multiple model approach. *IEEE Transactions on Automatic Control*, 36(2):167–177, 1991.

[334] R. Losick and D. Kaiser. Why and how bacteria communicate. *Scientific American*, 276(2):68–73, 1997.

[335] G. Lowe, M. Meister, and H. Berg. Rapid rotation of flagellar bundles in swimming bacteria. *Nature*, 325:637–640, Oct. 1987.

[336] S. Lucidi and M. Sciandrone. On the global convergence of derivative free methods for unconstrained optimization. *Technical Report, Univ. di Roma "La Sapienza"*, Dec. 1997.

[337] D.G. Luenberger. *Linear and Nonlinear Programming.* Addison-Wesley, Reading, MA, 1984.

[338] A.D. Lunardhi and K.M. Passino. Verification of qualitative properties of rule-based expert systems. *Int. Journal of Applied Artificial Intelligence*, 9(6):587–621, Nov./Dec. 1995.

[339] R.C. Luo and M.G. Kay. Multisensor integration and fusion in intelligent systems. *IEEE Trans. on Systems, Man, and Cybernetics*, 19(5):901–931, 1989.

[340] Ü. Özgüner M. Dogruel and S. Drakunov. Sliding mode control in discrete state and hybrid systems. *IEEE Transactions on Automatic Control*, 41(3):414–419, 1996.

[341] J.M. Maciejowski. *Multivariable Feedback Design.* Addison-Wesley, Reading, MA, 1989.

[342] M.T. Madigan, J.M. Martinko, and J. Parker. *Biology of Microorganisms.* Prentice Hall, NJ, 8th edition, 1997.

[343] M. Maggiore, R. Ordóñez, K.M. Passino, and S. Adibhatla. Estimator design in jet engine applications. *Engineering Applications of Artificial Intelligence*, 16:579–593, 2003.

[344] K.F. Man, K.S. Tang, and S. Kwong. *Genetic Algorithms.* Springer-Verlag, NY, 1999.

[345] J. Marczyk. *Principles of Simulation-Based Computer-Aided Engineering.* Artes Graficas Torres, 1999.

[346] E.B. Mason. *Human Physiology.* Benjamin/Cummings Pub., Menlo Park, CA, 1983.

[347] P.S. Maybeck and R.D. Stevens. Reconfigurable flight control via multiple model adaptive control method. *IEEE Transactions on Aerospace and Electronic Systems*, 27(3):470–479, May 1991.

[348] E. Mayr. *The Growth of Biological Thought: Diversity, Evolution, and Inheritance.* Harvard University Press, Cambridge, MA, 1982.

[349] K.I.M. McKinnon. Convergence of the nelder-mead simplex method to a nonstationary point. *SIAM Journal Optimization*, 9(1):148–158, 1998.

[350] J.M. Mendel. Fuzzy logic systems for engineering: A tutorial. *Proc. of the IEEE, Special Issue on Fuzzy Logic in Engineering Applications*, 83(3):345–377, March 1995.

[351] R.H. Meyers and D.C. Montgomery, editors. *Response Surface Methodology: Process and Product in Optimization Using Designed Experiments.* John Wiley and Sons, Pub., NY, NY, 1995.

[352] Z. Michalewicz. *Genetic Algorithms + Data Structure = Evolution Programs.* Springer-Verlag, Berlin, 1992.

[353] Z. Michalewicz. *Genetic Algorithms + Data Structure = Evolution Programs.* Springer-Verlag, Berlin, 3rd edition, 1996.

[354] Z. Michalewicz, K. Deb, M. Schmidt, and T. Stidsen. Test-case generator for nonlinear continuous parameter optimization techniques. *IEEE Trans. on Evolutionary Computation*, 4(3):197–215, 2000.

[355] Z. Michalewicz and D. Fogel. *How to Solve It: Modern Heuristics.* Springer-Verlag, Berlin, 2000.

[356] A.N. Michel and J.A. Farrell. Associative memories via artificial neural networks. *IEEE Control Systems Magazine*, 10(3):6–17, April 1990.

[357] A.N. Michel and R.K. Miller. *Qualitative Analysis of Large Scale Dynamical Systems.* Academic Press, NY, 1977.

[358] A.S. Mikhailov and D.H. Zanette. Noise-induced breakdown of coherent collective motion in swarms. *Physical Review E*, 60(4):4571–4575, October 1999.

[359] K.R. Miller. *Finding Darwin's God: A Scientist's Search for Common Ground Between God and Evolution.* Cliff Street Books, NY, 1999.

[360] R.K. Miller and A.N. Michel. *Ordinary Differential Equations.* Academic Press, NY, 1982.

[361] W.T. Miller, R.S. Sutton, and P.J. Werbos, editors. *Neural Networks for Control.* The MIT Press, Cambridge, MA, 1991.

[362] W.T. Miller III, R.P. Hewes, F.H. Glanz, and L.G. Kraft III. Real-time dynamic control of an industrial manipulator using a neural-network-based learning controller. *IEEE Trans. on Robotics and Automation*, 6(1):1–9, Feb. 1990.

[363] M. Mitchell. *An Introduction to Genetic Algorithms.* MIT Press, Cambridge, MA, 1996.

[364] A. Mogilner and L. Edelstein-Keshet. A non-local model for a swarm. *Journal of Mathematical Biology*, 38:534–570, 1999.

[365] M.L. Moore, V. Gazi, K.M. Passino, W. Shackleford, and F. Proctor. Design and implementation of complex control systems using the nist-rcs software library. *IEEE Control Systems Magazine*, 19(3):12–28, Dec. 1999.

[366] M.L. Moore, J. Musachio, and K.M. Passino. Genetic adaptive control for an inverted wedge. In *Proc. of the American Control Conf.*, pages 400–404, San Diego, CA, June 1999.

[367] M.L. Moore, J. Musachio, and K.M. Passino. Genetic adaptive control for an inverted wedge: Experiments and comparative analysis. *Engineering Applications in Artificial Intelligence*, 14(1):1–14, 2001.

[368] H. Moravec. Rise of the robots. *Scientific American*, 281(6):124–135, Dec. 1999.

[369] C.J. Morton-Firth. *Stochastic simulation of cell signalling pathways.* PhD thesis, Univ. of Cambridge, Cambridge England, 1998.

[370] V.G. Moudgal, W.A. Kwong, K.M. Passino, and S. Yurkovich. Fuzzy learning control for a flexible-link robot. *IEEE Trans. on Fuzzy Systems*, 3(2):199–210, May 1995.

[371] V.G. Moudgal, K.M. Passino, and S. Yurkovich. Rule-based control for a flexible-link robot. *IEEE Trans. on Control Systems Technology*, 2(4):392–405, December 1994.

[372] M.C. Mozer and M. Sitton. Computational modeling of spatial attention. In H. Pashler, editor, *Attention*, pages 341–393. Psychology Press, East Sussex, UK, 1998.

[373] T.F. Murphy, S. Yurkovich, and S.-C. Chen. Intelligent control for paper machine moisture control. In *Proc. of the IEEE Conf. on Control Applications*, pages 826–833, Dearborn, MI, September 1996.

[374] J.D. Murray. *Mathematical Biology*. Springer-Verlag, New York, 1989.

[375] R. Murray-Smith and T.A. Johansen, editors. *Multiple Model Approaches to Nonlinear Modeling and Control*. Taylor and Francis, UK, 1996.

[376] K.S. Narendra and A.M. Annaswamy. *Stable Adaptive Systems*. Prentice-Hall, Englewood Cliffs, NJ, 1989.

[377] K.S. Narendra and J. Balakrishnan. Improving transient response of adaptive control systems using multiple models and switching. *IEEE Transactions on Automatic Control*, 39(9):1861–1866, 1994.

[378] K.S. Narendra and K. Parthasarathy. Identification and control of dynamical systems using neural networks. *IEEE Trans. on Neural Networks*, 1(1):4–27, 1990.

[379] K.S. Narendra and K. Parthasarathy. Identification and Control of Dynamical Systems Using Neural Networks. *IEEE Trans. on Neural Networks*, 1(1):4–27, 1990.

[380] K.S. Narendra and M.A. Thathachar. *Learning Automata: An Introduction*. Prentice-Hall, Englewood Cliffs, NJ, 1989.

[381] N.E. Nawa and T. Furuhashi. Fuzzy system parameters discovery by bacterial evolutionary algorithm. *IEEE Trans. on Fuzzy Systems*, 5(7):608–616, October 1999.

[382] A. H. Nayfeh, J. M. Elzebda, and D. T. Mook. Analytical study of the subsonic wing-rock phenomenon for slender delta wings. *AIAA Journal of Aircraft*, 26(9):805–809, 1989.

[383] R.E. Neapolitan. *Learning Bayesian Networks*. Pearson Prentice-Hall, Englewood Cliffs, NJ, 2004.

[384] F.C. Neidhardt, J.L. Ingraham, and M. Schaechter. *Physiology of the Bacterial Cell: A Molecular Approach*. Sinauer Associates, Inc., Pub., Massachusetts, 1990.

[385] J.A. Nelder and R. Mead. A simplex method for function minimization. *Computer Journal*, 7:308–313, 1965.

[386] S.C. Ng, S.H. Leung, C.Y. Chung, A. Luk, and W.H. Lau. The genetic search approach: A new learning algorithm for adaptive iir filtering. *IEEE Signal Processing Magazine*, 13(6), Nov. 1996.

[387] N.J. Nilsson. *Artificial Intelligence: A New Synthesis*. Morgan Kaufmann Pub., San Francisco, CA, 1998.

[388] S. Nolfi and D. Floreano. *Evolutionary Robotics: The Biology, Intelligence, and Technology of Self-Organizing Machines*. MIT Press, Cambridge, MA, 2000.

[389] H.N. Nounou and K.M. Passino. Fuzzy model predictive control: Stability issues and applications. In *Proceedings of the IEEE Int. Symp. on Intelligent Control*, pages 423–428, Cambridge, MA, Sept. 1999.

[390] W.J. O'Brien, H.I. Browman, and B.I. Evans. Search strategies of foraging animals. *American Scientist*, 78:152–160, Mar./Apr. 1990.

[391] P. Ögren, M. Egerstedt, and X. Hu. A control Lyapunov function approach to multi-agent coordination. In *Proc. of Conf. Decision Control*, pages 1150–1155, Orlando, FL, December 2001.

[392] P. Ogren, E. Fiorelli, and N.E. Leonard. Formations with a mission: Stable coordination of vehicle group maneuvers. *Proc. Symposium on Mathematical Theory of Networks and Systems*, August 2002.

[393] A. Okubo. Dynamical aspects of animal grouping: swarms, schools, flocks, and herds. *Advances in Biophysics*, 22:1–94, 1986.

[394] R. Olfati-Saber and R.M. Murray. Distributed cooperative control of multiple vehicle formations using structural potential functions. In *Proc. IFAC World Congress*, Barcelona, Spain, June 2002.

[395] A. Ollero and A.J. Garcia-Cerezo. Direct digital control, auto-tuning and supervision using fuzzy logic. *Fuzzy Sets and Systems*, 12:135–153, 1989.

[396] R. Ordóñez and K.M. Passino. Stable multiple-input multiple-output adaptive fuzzy/neural control. *IEEE Trans. on Fuzzy Systems*, 7(3):345–353, 1999.

[397] R. Ordóñez, J. Zumberge, J.T. Spooner, and K.M. Passino. Adaptive fuzzy control: Experiments and comparative analyses. *IEEE Trans. on Fuzzy Systems*, 5(2):167–188, 1997.

[398] H. Özbay. *Introduction to Feedback Control Theory*. CRC Press, Boca Raton, FL, 2000.

[399] R. Palm, D. Driankov, and H. Hellendoorn. *Model Based Fuzzy Control*. Springer-Verlag, NY, 1997.

[400] C.H. Papadimitriou and K. Steiglitz. *Combinatorial Optimization: Algorithms and Complexity*. Prentice-Hall, Englewood Cliffs, NJ, 1982.

[401] R. Parasuraman. The attentive brain: Issues and prospects. In R. Parasuraman, editor, *The Attentive Brain*, pages 3–15. MIT Press, Cambridge, MA, 1998.

[402] D. Park, A. Kandel, and G. Langholz. Genetic-based new fuzzy reasoning models with application to fuzzy control. *IEEE Trans. on Systems, Man and Cybernetics*, 24(1):39–47, 1994.

[403] J.K. Parrish and W.M. Hamner, editors. *Animal Groups in Three Dimensions*. Cambridge Univ. Press, Cambridge, England, 1997.

[404] H. Pashler, editor. *Attention*. Psychology Press, East Sussex, UK, 1998.

[405] K. M. Passino and Ü. Özgüner. Modeling and analysis of hybrid systems: Examples. In *IEEE International Symposium on Intelligent Control*, Arlington, VA, 1991.

[406] K.M. Passino. Intelligent control for autonomous systems. *IEEE Spectrum*, 32(6):55–62, June 1995.

[407] K.M. Passino. Biomimicry, mathematics, and physics for control and automation: Conflict or harmony? In *Proceedings of the IFAC Workshop on Advanced Fuzzy-Neural Control*, pages 211–216, Valencia, Spain, Oct. 15–16 2001.

[408] K.M. Passino and P.J. Antsaklis. A system and control theoretic perspective on artificial intelligence planning systems. *Int. Journal of Applied Artificial Intelligence*, 3:1–32, 1989.

[409] K.M. Passino and K.L. Burgess. *Stability Analysis of Discrete Event Systems*. John Wiley and Sons Pub., NY, 1998.

[410] K.M. Passino and A.D. Lunardhi. Qualitative analysis of expert control systems. In M. Gupta and N. Sinha, editors, *Intelligent Control: Theory and Practice*, pages 404–442. IEEE Press, Piscataway, NJ, 1996.

[411] K.M. Passino, A.N. Michel, and P.J. Antsaklis. Lyapunov stability of a class of discrete event systems. *IEEE Transactions on Automatic Control*, 37:269–279, February 1994.

[412] K.M. Passino and S. Yurkovich. *Fuzzy Control*. Addison Wesley Longman, Menlo Park, CA, 1998. This book available for free download at author Web site.

[413] J. Pearl. *Heuristics: Intelligent Search Strategies for Computer Problem Solving*. Addison-Wesley, Reading, MA, 1984.

[414] J. Pearl. *Probabilistic Reasoning in Intelligent Systems*. Morgan Kaufmann, San Francisco, 1988.

[415] P. Peleties and R. DeCarlo. Modeling of interacting continuous time and discrete event systems: An example. In *Proc. of the 26th Allerton Conf. on Communication, Control, and Computing*, pages 1150–1159, Oct. 1988.

[416] P. Peleties and R. DeCarlo. A modeling strategy with event structures for hybrid systems. In *Proc. of the 28th Conf. on Decision and Control, Tampa, FL*, pages 1308–1313, Dec. 1989.

[417] A.S. Perelson. Immune network theory. *Immunological Review*, 110:5–36, 1989.

[418] J.R. Perkins and P.R. Kumar. Stable, distributed, real-time scheduling of flexible manufacturing/assembly/disassembly systems. *IEEE Transaction on Automatic Control*, 34:139–148, February 1989.

[419] M. Pinedo. *Scheduling: Theory, Algorithms, and Systems*. Prentice Hall, Upper Saddle River, NJ, 2002.

[420] S. Pinker. *The Language Instinct: How the Mind Creates Language*. W. Morrow and Comp. Inc., NY, 1994.

[421] S. Pinker. *How the Mind Works*. W.W. Norton and Comp., NY, 1997.

[422] T. Poggio and F. Girosi. Networks for approximation and learning. *Proc. of the IEEE*, 78(9):1481–1497, 1990.

[423] M.M. Polycarpou. Stable adaptive neural control scheme for nonlinear systems. *IEEE Trans. on Automatic Control*, 41(3):447–451, March 1996.

[424] M.M. Polycarpou and P.A. Ioannou. Identification and control of nonlinear systems using neural network models: Design and stability analysis. Electrical Engineering–Systems Report 91-09-01, University of Southern California, September 1991.

[425] M.M. Polycarpou and M.J. Mears. Stable Adaptive Tracking of Uncertain Systems Using Nonlinearly Parameterized On-Line Approximators. *Int. Journal of Control*, 1997. Special Issue on Neural Network Control.

[426] L.L. Porter and K.M. Passino. Genetic model reference adaptive control. In *Proc. of the IEEE Int. Symp. on Intelligent Control*, pages 219–224, Chicago, August 1994.

[427] L.L. Porter and K.M. Passino. Genetic adaptive observers. *Engineering Applications of Artificial Intelligence*, 8(3):261–269, 1995.

[428] L.L. Porter and K.M. Passino. Genetic adaptive and supervisory control. *Int. Journal of Intelligent Control and Systems*, 12(1):1–41, 1998.

[429] T. Procyk and E. Mamdani. A linguistic self-organizing process controller. *Automatica*, 15(1):15–30, January 1979.

[430] E.M. Rauch, M.M. Millonas, and D.R. Chialvo. Pattern formation and functionality in swarm models. *Physics Letters A*, 207:185–193, October 1995.

[431] R. Reed. Pruning algorithms—a survey. *IEEE Trans. on Neural Networks*, 4(5):740–747, Sept. 1992.

[432] H. Reichenbach. Biology of the myxobacteria: Ecology and taxonomy. In M. Dworkin and D. Kaiser, editors, *Myxobacteria II*, pages 13–62. American Society for Microbiology, Washington, DC, 1993.

[433] J.H. Reif and H. Wang. Social potential fields: A distributed behavioral control for autonomous robots. *Robotics and Autonomous Systems*, 27:171–194, 1999.

[434] M. Resnick. *Turtles, Termites, and Traffic Jams: Explorations in Massively Parallel Microworlds*. MIT Press, Cambridge, MA, 1994.

[435] F. Van Der Rhee, H. Van Nauta Lemke, and J. Dijkman. Knowledge based fuzzy control of systems. *IEEE Trans. on Automatic Control*, 35(2):148–155, February 1990.

[436] M. Ridley. *Evolution*. Blackwell Science Inc., Cambridge, MA, 2nd edition, 1982.

[437] M. Ridley. *The Origins of Virtue: Human Instincts and the Evolution of Cooperation*. Penguin Books, NY, NY, 1996.

[438] M. Ridley. *The Cooperative Gene: How Mendel's Demon Explains the Evolution of Complex Beings*. The Free Press, NY, 2001.

[439] H. Robbins and S. Monro. A stochastic approximation method. *Ann. Math. Stat.*, 29:400–407, 1951.

[440] T. Ross. *Fuzzy Logic in Engineering Applications*. McGraw-Hill, NY, 1995.

[441] G.A. Rovithakis and M.A. Christodoulou. Adaptive control of unknown plants using dynamical neural networks. *IEEE Trans. on Systems, Man, and Cybernetics*, 24(3):400–412, March 1994.

[442] G.A. Rovithakis and M.A. Christodoulou. Direct Adaptive Regulation of Unknown Nonlinear Dynamical Systems via Dynamic Neural Networks. *IEEE Trans. on Systems, Man, and Cybernetics*, 25(12):1578–1594, December 1995.

[443] W. Rugh. Analytical framework for gain scheduling. *IEEE Control Systems Magazine*, 11(1):79–84, January 1991.

[444] S. Russell and P. Norvig. *Artificial Intelligence: A Modern Approach*. Prentice Hall, Englewood Cliffs, NJ, 1995.

[445] N. Sadegh. A Perceptron Network for Functional Identification and Control of Nonlinear Systems. *IEEE Trans. on Neural Networks*, 4(6):982–988, November 1993.

[446] T. Samad and G. Balas, editors. *Software-Enabled Control: Information Technology for Dynamical Systems*. IEEE Press, Piscataway, NJ, 2003.

[447] R.M. Sanner and J. J.E. Slotine. Gaussian Networks for Direct Adaptive Control. *IEEE Trans. on Neural Networks*, 3(6):837–863, November 1992.

[448] S. Sastry and M. Bodson. *Adaptive Control: Stability, Convergence, and Robustness*. Prentice-Hall, Englewood Cliffs, NJ, 1989.

[449] S.R. Schach. *Software Engineering*. Aksen Assoc. Inc. Pub., Irwin, Homewood, IL, 1993.

[450] D.L. Schacter. The seven sins of memory: Insights from psychology and cognitive neuroscience. *American Psychologist*, 54(3):182–203, March 1999.

[451] E. Scharf and N. Mandic. The application of a fuzzy controller to the control of a multi-degree-of-freedom robot arm. In M. Sugeno, editor, *Industrial Applications of Fuzzy Control*, pages 41–62. Elsevier, Amsterdam, The Netherlands, 1985.

[452] W.M. Schubert and R.F. Stengel. Parallel synthesis of robust control systems. *IEEE Trans. on Control Systems Technology*, 6(6):701–706, Nov. 1998.

[453] W. Schultz, P. Dylan, and P.R. Montague. A neural substrate of prediction and reward. *Science*, 275:1593–1598, 1997.

[454] T.D. Seeley. *Honeybee ecology*. Princeton University Press, Princeton, NJ, 1985.

[455] T.D. Seeley. The honeybee as a superorganism. *American Scientist*, 77:546–553, 1989.

[456] T.D. Seeley. *The Wisdom of the Hive: The Social Physiology of Honey Bee Colonies*. Harvard University Press, Cambridge, Mass, 1995.

[457] T.D. Seeley and R.A. Morse. Dispersal behavior of honey bee swarms. *Psyche*, 84:199–209, 1977.

[458] T.D. Seeley, R.A. Morse, and P.K. Visscher. The natural history of the flight of honey bee swarms. *Psyche*, 86:103–113, 1979.

[459] J.E. Segall, S.M. Block, and H.C. Berg. Temporal comparisons in bacterial chemotaxis. *Proc. of the National Academy of Sciences*, 83:8987–8991, Dec. 1986.

[460] C.Y. Seong and B. Widrow. Neural dynamic optimization for control systems: Part i: Background, part ii: Theory, part iii: Applications. *IEEE Trans. on Systems, Man, and Cybernetics, Part B: Cybernetics*, 31(4):482–513, 2001.

[461] J.S. Shamma and M. Athans. Analysis of nonlinear gain-scheduled control systems. *IEEE Trans. on Automatic Control*, 35(8):898–907, August 1990.

[462] J.S. Shamma and M. Athans. Gain scheduling: Potential hazards and possible remedies. *IEEE Control Systems Magazine*, 12(2):101–107, April 1992.

[463] J.A. Shapiro. Bacteria as multicellular organisms. *Scientific American*, 258:62–69, 1988.

[464] J.A. Shapiro. Multicellularity: The rule, not the exception. In J.A. Shapiro and M. Dworkin, editors, *Bacteria as Multicellular Organisms*, pages 14–49. Oxford University Press, NY, 1997.

[465] M. Shaw and D. Garlan. *Software Architecture*. Prentice Hall, Englewood Cliffs, NJ, 1996.

[466] C.Y. Shieh and S.S. Nair. A new self tuning fuzzy controller design and experiements. In *Proc. of the 2nd IEEE Int. Conf. on Fuzzy Systems*, pages 309–314, San Francisco, CA, March 1993.

[467] L.J. Shimkets and M. Dworkin. Myxobacterial multicellularity. In J.A. Shapiro and M. Dworkin, editors, *Bacteria as Multicellular Organisms*, pages 220–244. Oxford University Press, NY, 1997.

[468] N. Shimoyama, K. Sugawa, T. Mizuguchi, Y. Hayakawa, and M. Sano. Collective motion in a system of motile elements. *Physical Review Letters*, 76(20):3870–3873, May 1996.

[469] J. Sjoberg, Q. Zhang, L. Ljung, A. Benveniste, B. Deylon, P. Glorennec,

H. Hjalmarsson, and A. Juditsky. Nonlinear black-box modeling in system identification: A unified overview. *Automatica*, 31(12):1725–1750, December 1995.

[470] J.E. Slotine and W. Li. *Applied Nonlinear Control.* Prentice-Hall, Englewood Cliffs, NJ, 1991.

[471] D.J. Smith, S. Forrest, and A.S. Perelson. Immunological memory is associative. In *Artificial Immune Systems and Their Applications*, pages 105–114. Springer-Verlag, Germany, 1999.

[472] J. Maynard Smith. *Evolution and the Theory of Games.* Cambridge University Press, Cambridge, 1982.

[473] J. Maynard Smith. *Did Darwin Get it Right? Essays on Games, Sex, and Evolution.* Chapman and Hall, NY, 1988.

[474] R. Solomon. Evolutionary algorithms and gradient search: Similarities and differences. *IEEE Trans. on Evolutionary Computation*, 2(2):45–55, 1998.

[475] I. Sommerville. *Software Engineering.* Addison-Wesley Pub. Co., Reading, MA, 1992.

[476] J.C. Spall. Multivariate stochastic approximation using a simultaneous perturbation gradient approximation. *IEEE Trans. on Automatic Control*, 37:332–341, 1992.

[477] J.C. Spall. Implementation of the simultaneous perturbation algorithm for stochastic optimization. *IEEE Trans. on Aerospace and Electronic Systems*, 34(3):817–823, 1998.

[478] J.C. Spall. An overview of the simultaneous perturbation method for efficient optimization. *Johns Hopkins APL Technical Digest*, 19(4):482–492, 1998.

[479] J.C. Spall. Stochastic optimization, stochastic approximation, and simulated annealing. In J.G. Webster, editor, *Wiley Encyclopedia of Electrical and Electronics Engineering*, pages 529–542. Wiley, NY, 1999.

[480] J.C. Spall. *Introduction to Stochastic Search and Optimization: Estimation, SImulation, and Control.* John Wiley and Sons, NY, 2002.

[481] J.C. Spall, S.D. Hill, and D.R. Stark. Some theoretical comparisons of stochastic optimization approaches. In *Proc. the American Control Conf.*, Chicago, IL, June 2000.

[482] G. Sperling and E. Weichselgartner. Episodic theory of the dynamics of spatial attention. *Psychological Review*, 102:503–532, 1995.

[483] M. Spong and M. Vidyasagar. *Robot Dynamics and Control.* Wiley, NY, 1989.

[484] J.T. Spooner, M. Maggiore, R. Ordóñez, and K.M. Passino. *Stable Adaptive Control and Estimation for Nonlinear Systems: Neural and Fuzzy Approximator Techniques.* John Wiley and Sons Pub., NY, 2002.

[485] J.T. Spooner and K.M. Passino. Stable adaptive fuzzy control for an automated highway system. In *Proc. of the IEEE Int. Symp. on Intelligent Control*, pages 531–536, Monterey, CA, August 1995.

[486] J.T. Spooner and K.M. Passino. Stable adaptive control using fuzzy systems and neural networks. *IEEE Trans. on Fuzzy Systems*, 4(3):339–359, August 1996.

[487] J.T. Spooner and K.M. Passino. Decentralized adaptive control of nonlinear systems using radial basis neural networks. *IEEE Trans. on Automatic Control*, 44(11):2050–2057, 1999.

[488] M. Srinivas and L.M. Patnaik. Genetic algorithms: A survey. *IEEE Computer Magazine*, pages 17–26, June 1994.

[489] R.F. Stengel. Toward intelligent flight control. *IEEE Trans. on Systems, Man, and Cybernetics*, 23(6):1699–1717, Nov./Dec. 1993.

[490] D.W. Stephens and J.R. Krebs. *Foraging Theory.* Princeton Univ. Press, Princeton, NJ, 1986.

[491] A. Stevens. Simulations of the gliding behavior and aggregation of myxobacteria. In W. Alt and G. Hoffmann, editors, *Biological Motion, Lecture Notes in Biomathematics, Vol. 89*, pages 548–555. Springer-Verlag, Berlin, 1990.

[492] A. Stevens. A stochastic cellular automaton, modeling gliding and aggregation of myxobacteria. *SIAM J. Applied Mathematics*, 61(1):172–182, 2000.

[493] B.S. Stewart, C.F. Liaw, and C.C. White. A bibliography of heuristic search through 1992. *IEEE Transactions on Systems, Man, and Cybernetics*, 24(2):268–293, 1994.

[494] L.D. Stone. *Theory of Optimal Search*. Academic Press, NY, 1975.

[495] S.H. Strogatz, R.E. Mirollo, and P.C. Matthews. Coupled nonlinear oscillators below the synchronization threshold: Relaxation by generalized landau damping. *Physical Review Letters*, 68(18):2730–2733, May 4 1992.

[496] S.H. Strogatz and I. Stewart. Coupled oscillators and biological synchronization. *Scientific American*, pages 102–109, Dec. 1993.

[497] C.-Y. Su and Y. Stepanenko. Adaptive Control of a Class of Nonlinear Systems with Fuzzy Logic. *IEEE Trans. on Fuzzy Systems*, 2(4):285–294, November 1994.

[498] M. Sugeno and T. Yasukawa. A fuzzy-logic-based approach to qualitative modeling. *IEEE Trans. on Fuzzy Systems*, 1(1):7–31, February 1993.

[499] R.S. Sutton and A.G. Barto, editors. *Reinforcement Learning: An Introduction*. MIT Press, Cambridge, MA, 1998.

[500] I. Suzuki and M. Yamashita. Distributed anonymous mobile robots: Formation of geometric patterns. *SIAM Journal on Computing*, 28(4):1347–1363, 1999.

[501] D. Swaroop. *String Stability of Interconnected systems: An Application to Platooning in Automated Highway Systems*. PhD thesis, Departnent of Mechanical Engineering, University of California, Berkeley 1995.

[502] D. Swaroop, J.K. Hedrick, C.C. Chien, and P. Ioannou. A comparison of spacing and headway control laws for automatically controlled vehicles. *Vehicle System Dynamics*, 23:597–625, 1994.

[503] P. Tabuada, G.J. Pappas, and P. Lima. Feasable formations of multi-agent systems. In *Proc. American Control Conf.*, pages 56–61, Arlington, VA, June 2001.

[504] T. Takagi and M. Sugeno. Fuzzy identification of systems and its applications to modeling and control. *IEEE Trans. on Systems, Man, and Cybernetics*, 15(1):116–132, January 1985.

[505] H. Takahashi. Automatic speed control device using self-tuning fuzzy logic. In *Proc. of the IEEE Workshop on Automotive Applications of Electronics*, pages 65–71, Dearborn, MI, October 1988.

[506] K.S. Tang, K.F. Man, S. Wong, and Q. He. Genetic algorithms and their applications. *IEEE Signal Processing Magazine*, 13(6), Nov. 1996.

[507] R. Tanscheit and E. Scharf. Experiments with the use of a rule-based self-organising controller for robotics applications. *Fuzzy Sets and Systems*, 26:195–214, 1988.

[508] S.A. Teukolsky, W.T. Vetterling, and B.P. Flannery. *Numerical Recipes in C*. Cambridge University Press, Cambridge, England, 1992.

[509] A. Thompson, P. Layzell, and R.S. Zebulum. Explorations in design space: Unconventional electronics design through artificial evolution. *IEEE Trans. on Evolutionary Computation*, 3(3):167–196, Sept. 1999.

[510] J. Toner and Y. Tu. Long-range order in a two-dimensional dynamical *xy* model: How birds fly together. *Physical Review Letters*, 75(23):4326–4329, December 1995.

[511] J. Toner and Y. Tu. Flocks, herds, and schools: A quantitative theory of flocking. *Physical Review E*, 58(4):4828–4858, October 1998.

[512] V. Torczon. On the convergence of the multidirectional search algorithm. *SIAM Journal Optimization*, 1(1):123–145, Feb. 1991.

[513] V. Torczon. On the convergence of pattern search algorithms. *SIAM Journal Optimization*, 7(1):1–25, 1997.

[514] V. Torczon and M. Trosset. Using approximations to accelerate engineering design optimization. In *Proceedings of the 7th AIAA/USAF/NASA/ISSMO Symposium on Multidisciplinary Analysis and Optimization*, pages 738–748, St. Louis, Missouri, 1998.

[515] J.Z. Tsien. Building a brainier mouse. *Scientific American*, 282(4):62–68, April 2000.

[516] S. Tzafestas and N. Papanikolopoulos. Incremental fuzzy expert PID control. *IEEE Trans. on Industrial Electronics*, 37(5):365–371, October 1990.

[517] S.G. Tzafestas, editor. *Methods and Applications of Intelligent Control.* Kluwer Academic Pub., Norwell, MA, 1997.

[518] V.I. Utkin. *Sliding Modes in Control Optimization.* Springer-Verlag, Berlin, 1992.

[519] V.I. Utkin, J. Guldner, and J. Shi. *Sliding Mode Control in Electromechanical Systems.* Taylor and Francis, London, 1999.

[520] K. Valavanis and G. Saridis. *Intelligent Robotic Systems: Theory, Design, and Applications.* Kluwer Academic Publishers, Norwell, MA, 1992.

[521] H.R. van Nauta Lemke and D.-Z. Wang. Fuzzy PID supervisor. In *Proc. of the IEEE Conf. on Decision and Control*, pages 602–608, Fort Lauderdale, FL, December 1985.

[522] A.H. van Zomeren and W.H. Brouwer. *Clinical Neuropsychology of Attention.* Oxford University Press, NY, 1994.

[523] A. Varšek, T. Ubančič, and B. Filipič. Genetic algorithms in controller design and tuning. *IEEE Trans. on Systems, Man and Cybernetics*, 23(5):1330–1339, Sept./Oct. 1993.

[524] T. Vicsek, A. Czirok, E. Ben-Jacob, I. Cohen, and O. Shochet. Novel type of phase transition in a system of self-driven particles. *Physical Review Letters*, 75(6):1226–1229, August 1995.

[525] T. Vicsek, A. Czirok, I.J. Farkas, and D. Helbing. Application of statistical mechanics to collective motion in biology. *Physica A*, 274:182–189, 1999.

[526] M. Vidyasagar. *Nonlinear Systems Analysis.* Prentice-Hall, Englewood Cliffs, NJ, 1993.

[527] M.D. Vose. *The Simple Genetic Algorithm.* MIT Press, Cambridge, MA, 1999.

[528] B.H. Wang and G. Vachtsevanos. Learning fuzzy logic control: An indirect control approach. In *Proc. 1st IEEE Int. Conf. on Fuzzy Systems*, pages 297–304, San Diego, CA, March 1992.

[529] L.-X. Wang. A Supervisory Controller for Fuzzy Control Systems that Guarantees Stability. *IEEE Trans. on Automatic Control*, 39(9), September 1994.

[530] L.-X. Wang. *Adaptive Fuzzy Systems and Control: Design and Stability Analysis.* Prentice-Hall, Englewood Cliffs, NJ, 1994.

[531] L.-X. Wang. *A Course in Fuzzy Systems and Control.* Prentice-Hall, Englewood Cliffs, NJ, 1997.

[532] L.-X. Wang and J.M. Mendel. Fuzzy basis functions, universal approximation and orthogonal least-squares learning. *IEEE Trans. on Neural Networks*, 3(5):1–8, September 1992.

[533] K. Warburton and J. Lazarus. Tendency-distance models of social cohesion in animal groups. *Journal of Theoretical Biology*, 150:473–488, 1991.

[534] J. Weibull. *Evolutionary Game Theory.* The MIT Press, Cambridge, MA, 1995.

[535] G. Weiss, editor. *Multiagent Systems: A Modern Approach to Distributed Artificial Intelligence.* MIT Press, Cambridge, MA, 1999.

[536] D. White and D. Sofge, editors. *Handbook of Intelligent Control: Neural, Fuzzy and Adaptive Approaches.* Van Nostrand Reinhold, NY, 1992.

[537] P. Whitfield. *From So Simple a Beginning.* Macmillan Pub. Co. NY, 1993.

[538] C.D. Wickens and J.G. Hollands. *Engineering Psychology and Human Performance.* Prentice-Hall, NJ, 3rd edition, 2000.

[539] E.O. Wilson. *The Insect Societies.* Belknap Press of Harvard Univ. Press, Cambridge, MA, 1971.

[540] E.O. Wilson. *Sociobiology.* Belknap Press of Harvard Univ. Press, Cambridge, MA, 1975.

[541] A.T. Winfree. *The Timing of Biological Clocks.* Scientific American Library, NY, 1987.

[542] P.C. Witherell. A review of the scientific literature relating to honey bee bait hives and swarm attractants. *American Bee Journal*, 125:823–829, 1985.

[543] D. Wolpert and W.G. Macready. No free lunch theorems for optimization. *IEEE Trans. on Evolutionary Computation*, 1(1):67–82, 1997.

[544] D.E. Woodward, R. Tyson, M.R. Myerscough, J.D. Murray, E.O. Budrene, and H.C. Berg. Spatio-temporal

patterns generated by salmonella typhimurium. *Biophysi. J.*, 68:2181–2189, 1995.

[545] Q.M. Wu and C.W. de Silva. Model identification for fuzzy dynamic systems. In *Proc. of the American Control Conf.*, pages 2246–2247, San Francisco, CA, June 1993.

[546] T. Yabuta and T. Yamada. Neural network controller characteristics with regard to adaptive control. *IEEE Trans. on Systems, Man, and Cybernetics*, 22(1):170–177, January/February 1992.

[547] T. Yamazaki. *An Improved Algorithm for a Self-Organizing Controller and Its Experimental Analysis*. PhD thesis, London University, 1982.

[548] L. Yao and W.A. Sethares. Nonlinear parameter estimation via the genetic algorithm. *IEEE Transactions on Signal Processing*, 12(1):927–935, April 1994.

[549] X. Yao. Evolving artificial neural networks. *Proc. of the IEEE*, 87(9):1423–1447, Sept. 1999.

[550] C. Yap. Algorithmic motion planning. In J. Schwartz and C. Yap, editors, *Advances in Robotics, Vol. 1*. Lawrence Erlbaum Associates, NJ, 1995.

[551] H. Ye, A. N. Michel, and L. Hou. Stability theory for hybrid dynamical systems. *IEEE Trans. on Automatic Control*, 43(4):461–474, April 1998.

[552] A. Yeşildirek and F.L. Lewis. Feedback linearization using neural networks. *Automatica*, 31(11):1659–1664, 1995.

[553] T.-M. Yi, Y. Huang, M.I. Simon, and J.C. Doyle. Robust perfect adaptation in bacterial chemotaxis through integral feedback control. *PNAS*, 97(9):4649–4653, April 25 2000.

[554] C. Zhang and R. Ordóñez. Decentralized adaptive coordination of UAVs using surrogate optimization. In *Proceedings of the American Control Conference*, Denver, CO, June 2003.

[555] L. Zhen and L. Xu. Fuzzy learning enhanced speed control of an indirect field-oriented induction machine drive. *IEEE Trans. on Control Systems Technology*, 8(2):270–278, March 2000.

[556] Y.F. Zheng. Integration of multiple sensors into a robotic system and its performance evaluation. *IEEE Trans. on Robotics and Automation*, 5(5):658–669, 1989.

[557] K. Zhou, J.C. Doyle, and K. Glover. *Robust and Optimal Control*. Prentice-Hall, Englewood Cliffs, NJ, 1996.

[558] M. Zigmond, F. Bloom, S. Landis, J. Roberts, and L. Squire, editors. *Fundamental Neuroscience*. Academic Press, NY, 1999.

[559] H.J. Zimmerman. *Fuzzy Set Theory–and Its Applications*. Kluwer Academic Press, Boston, 2nd edition, 1991.

[560] J. Zumberge and K.M. Passino. A case study in intelligent control for a process control experiment. In *Proc. IEEE Int.. Symp. on Intelligent Control*, pages 37–42, Dearborn, MI, September 1996.

Index

GPSR Compliance

The European Union's (EU) General Product Safety Regulation (GPSR) is a set of rules that requires consumer products to be safe and our obligations to ensure this.

If you have any concerns about our products, you can contact us on ProductSafety@springernature.com

In case Publisher is established outside the EU, the EU authorized representative is:

Springer Nature Customer Service Center GmbH
Europaplatz 3
69115 Heidelberg, Germany

Batch number: 10683412

Printed by Printforce, the Netherlands